F THE ELEMENTS

				III	IV	V	VI	VII	0
									2 He 4.00
				5 B 10.81	6 C 12.01	7 N 14.01	8 O 16.00	9 F 19.00	10 Ne 20.18
				13 Al 26.98	14 Si 28.09	15 P 30.97	16 S 32.06	17 Cl 35.45	18 Ar 39.95
27 Co 58.93	28 Ni 58.69	29 Cu 63.55	30 Zn 65.39	31 Ga 69.72	32 Ge 72.59	33 As 74.92	34 Se 78.96	35 Br 79.90	36 Kr 83.80
45 Rh 102.91	46 Pd 106.42	47 Ag 107.87	48 Cd 112.41	49 In 114.82	50 Sn 118.71	51 Sb 121.75	52 Te 127.60	53 I 126.90	54 Xe 131.29
77 Ir 192.22	78 Pt 195.08	79 Au 196.97	80 Hg 200.59	81 Tl 204.38	82 Pb 207.2	83 Bi 208.98	84 Po (209)	85 At (210)	86 Rn (222)

Atomic weights are based on carbon-12;
values in parentheses are for the most stable or the most familiar isotope.
† Symbol is unofficial
Elements shown in color are known to be essential to life

62 Sm 150.36	63 Eu 151.96	64 Gd 157.25	65 Tb 158.93	66 Dy 162.50	67 Ho 164.93	68 Er 167.26	69 Tm 168.93	70 Yb 173.04	71 Lu 174.97

94 Pu (244)	95 Am (243)	96 Cm (247)	97 Bk (247)	98 Cf (251)	99 Es (252)	100 Fm (257)	101 Md (258)	102 No (259)	103 Lr (260)

CHEMISTRY AND THE LIVING ORGANISM

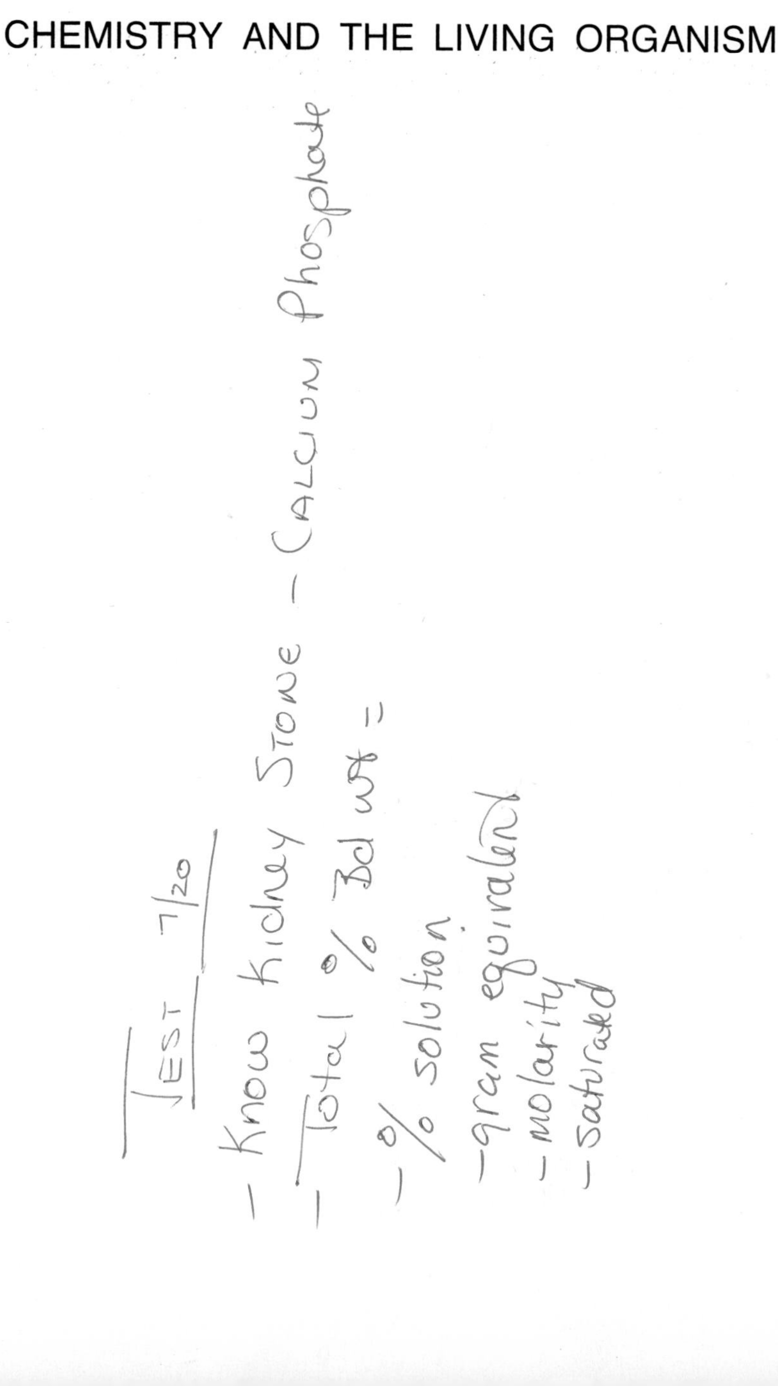

TO THE STUDENT: *A Study Guide for the textbook is available through your college bookstore under the title* Study Guide to Accompany Chemistry and the Living Organism *by Molly M. Bloomfield. The Study Guide can help you with course material by acting as a tutorial, review, and study aid. If the Study Guide is not in stock, ask the bookstore manager to order a copy for you.*

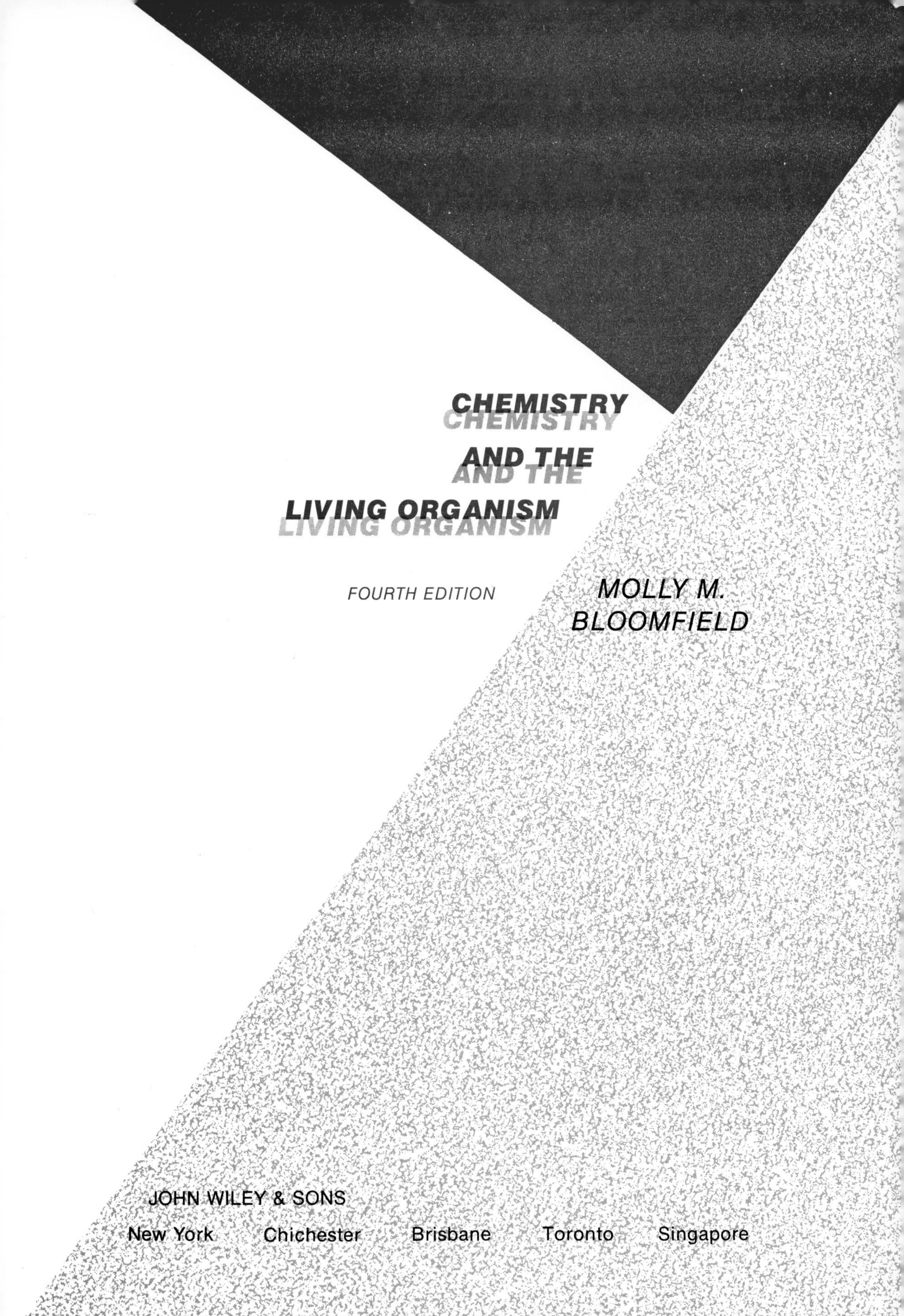

CHEMISTRY AND THE LIVING ORGANISM

FOURTH EDITION

MOLLY M. BLOOMFIELD

JOHN WILEY & SONS

New York Chichester Brisbane Toronto Singapore

Library of Congress Cataloging in Publication Data:

Bloomfield, Molly M., 1944-
Chemistry and the living organism.

Includes index.
1. Chemistry. 2. Biological chemistry.
I. Title.
QD33.B672 1987 540 86-22410
ISBN 0-471-84673-2

Printed in the United States of America

10 9 8 7 6 5 4 3 2 1

To Stefan, Rebecca, and Jonathan

PREFACE TO THE FOURTH EDITION

This fourth edition of *Chemistry and the Living Organism* follows very closely the third edition in providing a highly motivating and student-oriented approach to the study of chemistry. As in previous editions, my goal is to present an introduction to the basic principles of general, organic, and biological chemistry in a style that is easy to understand and enjoyable to read. This new edition updates the discussions in the third edition and provides additional motivational and skill-building material for the student.

The book is written for a one- or two-term survey course for students in the allied health sciences and related fields. One difficulty in writing such a textbook is the lack of agreement on the topics that should be included in this type of survey course. In recognition of this problem, the book includes more topics than can be covered in one term, thus providing more flexibility in choosing those topics best meeting the students' needs. The text is written in a very conversational style, and fundamental chemical concepts are illustrated with examples relevant to the student's own life. Each chapter begins with a set of learning objectives, helping the student identify the important concepts within the chapter. The presentation of those concepts involving mathematical operations is always accompanied by several worked-out examples. These examples are set off from the main text by a shaded background, allowing them to be easily referenced when the student is reviewing for an exam. Each set of examples, in turn, is followed by several exercises (with answers in the back of the book) that students may use to check their understanding of the concepts. As a further study aid, each chapter contains a summary that reviews all the key equations used in the chapter.

All chapters conclude with exercises and problems that drill the student on specific concepts and mathematical operations, and test the student's basic understanding through applied problems in a practical context. In this edition, the exercises and problems have been expanded to include more drills. Additionally, several new sections of integrated problems now require the student to use appropriate principles from previous chapters to solve realistic problems based on professional applications. The comprehensive glossary at the end of the text has also been expanded to include definitions of all boldfaced and medical terms used in the text.

Through the use of dramatic chapter-opening stories, pertinent examples throughout each chapter, and end-of-chapter problems with clear practical applications, the book constantly emphasizes the relevance of otherwise abstract chemical principles to the student's personal and professional life. To keep students abreast of the most current developments in scientific and medical research, many of the chapter-opening stories and in-chapter examples have been rewritten to reflect these important advances. For example, reference is now made to the events in Chernobyl, the importance of LDL receptors in cholesterol metabolism, new treatments for hypothermia and cancer, the continuing acid-rain controversy, and new research in such fields as genetic engineering and trace-element nutrition.

The organization of the text remains the same as in the third edition. In particular, in recognition of the extreme public interest in matters relating to nuclear energy and radioactivity, the two chapters covering radioactivity and its medical implications continue to appear early in the text. Also, the growing public awareness (and occasional misinformation) regarding trace-element nutrition is reflected in this text by a chapter devoted entirely to a comprehensive treatment of this important subject. However, if time constraints prevent these chapters from being covered in the term, they may be skipped without loss of continuity.

To meet the changing needs of our students and society at large, the revision of this text must be an ongoing process. I look forward to hearing comments, criticisms, and suggestions from the students, professors, and other concerned individuals who may use this text.

Supplemental Materials

The following supplemental materials are available for use with the fourth edition:

Student Study Guide The student study guide contains a brief summary of each section of the text and a list of important terms appearing in that section. The study guide features many worked-out examples and self-test questions (with answers) for each chapter.

Laboratory Manual The revised fourth edition of the laboratory manual, written by Joseph Bauer of William Rainey Harper College, contains 26 experiments carefully coordinated with the text.

Teacher's Manual The teacher's manual contains the answers to all end-of-chapter problems in the textbook, answers to the laboratory exercises, and a list of chemicals and equipment for the laboratory experiments.

Acknowledgments

The continual improvement of this text reflects the comments and suggestions of many people. I would specifically like to thank
Joseph Bauer, William Rainey Harper College;
Robert Becker, Ralph Quatrano, Jean Peters, and Daniel Selivonchick of Oregon State University;
Lawrence Stephens, Elmira College;
Robert Shine, Rampano College of New Jersey;
David Schroeder, Emporia State University;
John Albright, Texas Christian University;
Patricia McCoy, Jefferson Community College;
Sarah Margaret Willoughby, the University of Texas at Arlington;
Barbara Gage, Prince George's Community College;
Florence Haimes, California State University, San Francisco;
Carl Bonhorst, Portland State University;
James Stewart, Cypress College;
Miriam Smith, Pasadena City College;
Mahesh Sharma, Columbus College;
Carol Swezey, Purdue University;
C. R. Winkel, Ricks College;
Leslie Loew, State University of New York at Binghamton;
Robert Hawthorne, Jr., Purdue University—North Central Campus;
Henry Benz, Normandale Community College;
David Shaw, Madison Area Technical College;
W. J. Wasserman and Margaret Goodrich of Seattle Central College;
William Leoschke, Valparaiso University;
Thomas Rowland, University of Puget Sound;
and Kenneth Wright, North Idaho College.

This edition contains many new chapter-opening case histories, requiring substantial professional expertise. I am grateful for the generous help I have received from Allan Lefohn, ALS Associates; Dr. Anita Jansen and Dr. James Riley, Corvallis Clinic; Dr. William Lloyd and Dr. Michael Huntington, Good Samaritan Hospital, Corvallis, Oregon; Dwight Fullerton and Brian Dodd, Oregon State University; Judy Ladd, Benton County Health Department; and Charles Vaughan and James Pex, Oregon State Police Department Crime Laboratory.

At John Wiley, I would especially like to thank my chemistry editor, Dennis Sawicki; production manager, Ruth Greif; production supervisor, Ellen Brown; picture editor, Stella Kupferberg; designer, Janice Noto; and chief illustrator, John Balbalis, for the time and effort that they put into this edition.

None of this would be possible without the continued support and help of my family. I want to thank my husband Stefan for his editing and computer skills and his willingness to give up his free time, and my children, Rebecca and Jonathan, for seeing me through yet another edition.

Corvallis, Oregon

Molly M. Bloomfield

CONTENTS

I. INTRODUCTION

II. A CHEMICAL BACKGROUND

CHEMISTRY AND THE LIVING ORGANISM

SECTION ONE
INTRODUCTION

CHAPTER ONE
PKU— A CASE FOR UNDERSTANDING CHEMISTRY

1.1 Why Study Chemistry?

Chemistry is the study of the composition and interaction of substances.

Now, this may sound like a pretty general definition for such a specialized field of study. However, the broad scope of this definition is one way of indicating just how thoroughly chemistry is involved in each of our lives. For example, you drink water from your tap at home without a second's hesitation, for someone has added chemicals to the water to insure its safety. You seldom need to use an iron, thanks to the development of chemicals that give your clothes a permanent press. Just picture your daily routine: You wake up in the morning under sheets made of synthetic fibers that were chemically produced in a factory, or sheets made of cotton fibers—which were created through chemical reactions in the blossom of the cotton plant. You put on clothes made largely of synthetic materials, brush your teeth with a toothpaste containing fluoride, and eat a breakfast fortified with minerals and synthetic vitamins. You may drive to school in a car powered by the energy released through chemical reactions in the engine, or perhaps you pedal a bicycle—powered by the energy released through chemical reactions in your muscles. And now you're reading this textbook, whose paper was created through a chemical process and whose ink is a blend of chemicals. Literally every part of your life is closely related to the field of chemistry, whether it be the synthetic chemistry of the test tube and the modern laboratory, or the natural chemistry that makes up all of nature.

Chemistry also affects each of our lives in very personal ways. For example, your physical appearance is governed by chemicals. Chemical substances called hormones help determine your height, your weight, your build, and your sexual characteristics. Your good health depends upon chemicals that preserve the food you eat, chemicals that protect you from disease, and chemicals (in the form of food) that supply your body with the nutrients it needs to function properly. Chemicals influence your behavior and your emotional feelings. Much of your memory may be chemical; your thoughts and experiences may be stored in your brain in the form of chemical compounds. What we are trying to say is that your entire life is chemical, so a basic knowledge of chemistry will allow you to be more aware of your total self and the way in which you interact with your environment.

1.2 A Case for Understanding Chemistry

To illustrate how a knowledge of chemistry might be helpful in understanding the events surrounding us, let's turn the clock back to the 1950s and see what happened to a certain family from a small town in Ohio. Don't concern yourself too much right now with the chemistry described in this story. We will come back and discuss this case later in the book, after we have developed the basic vocabulary and background necessary to understand the chemical processes we will now be describing.

In 1958, Billy was brought home from the hospital as a happy, healthy baby. People who came to visit the family commented on how fair his skin and hair looked compared with the rest of the family. As Billy entered his fourth month, his mother started noticing that he no longer watched his mobile as it turned above the crib, and he rarely returned her smiles. Billy was slow in learning to sit up by himself. But then again, his brother had also been late in doing such things and he had turned out to be a very active young child. As Billy grew older, however, his parents became increasingly concerned about his development and behavior. He had become irritable and would have temper tantrums for no reason at all. Although his parents worked very hard to teach Billy to talk, he was able to learn only a few words. Furthermore, his mother noticed a strange musty odor about him when she changed his diapers, and his skin was often inflamed and flakey. As Billy neared the age of three, his parents began to admit that he was retarded. He still didn't walk or talk, and he was becoming uncontrollable.

Finally, a friend convinced them that the best thing for Billy and for themselves would be to take him to a clinic for diagnosis. At the clinic Billy was given a set of tests, which indicated that he had a disorder called phenylketonuria—PKU for short. Billy's parents were quite upset, and asked if there was any hope for Billy. The doctor replied that nothing could be done to reverse the retardation. When untreated, PKU causes irreversible brain damage, and Billy's IQ was found to be 40. (The average person's IQ is 100, and individuals with an IQ below 70 are considered retarded.) However, the doctor told them that Billy could be placed on a special diet which would improve his behavior and his skin condition. She further explained that PKU was a recessive inherited disease, which meant that each parent must be a carrier of the defective gene for that disease. Therefore, there was a 1 out of 4 chance that any other child that they might have would inherit the disease (Figure 1.1). Nevertheless,

Figure 1.1 This family illustrates the characteristic genetic distribution of phenylketonuria. The parents are both carriers of the trait. The laws of probability state that each child has a 25% chance of being normal, a 50% chance of being a carrier of the trait, and a 25% chance of actually having PKU. The son at the right is normal, the two daughters are carriers, and the son in the wheelchair has PKU. (Courtesy Willard R. Centerwall, M.D. From *Phenylketonuria,* Frank L. Lyman, Ed., Charles C Thomas, 1963.)

the doctor did not discourage Billy's parents from having more children. She explained that the disorder was now understood and could be treated, and that if PKU were diagnosed soon after birth, children having the disease could lead normal lives. Most states now had laws requiring the testing of newborn infants for PKU, so new cases of PKU were being diagnosed soon after birth.

Billy was taken home, and his parents placed him on a special diet. Within a few weeks, his parents were pleased to find that he no longer had the musty odor, that his skin and hair color darkened, that his behavior improved, and that he even began to smile. To her relief, Billy's mother found that she was now able to take care of him.

Billy's sister Susan was born eight years later. Before she was brought home from the hospital, a sample of blood was taken from her heel and placed on a piece of filter paper for testing. Three weeks later Susan was brought back to the clinic to have the test repeated. The doctor then told Susan's parents that the tests had revealed high levels of a substance called phenylalanine in Susan's blood. This meant that Susan, like Billy, had PKU. However, unlike Billy, the outlook for Susan was very hopeful. In order to reduce the abnormally high level of phenylalanine in her blood, Susan was immediately put on a low phenylalanine diet.

To her parents' delight, Susan is growing into a healthy, active child with an above average IQ. As she grows into maturity and her brain completes its development, Susan's diet can become more varied. The contrast between Billy and Susan is remarkable, and it resulted entirely from high levels of one chemical compound in the blood (Figure 1.2).

The physical symptoms of PKU result from a chemical imbalance within the human body. You are probably aware of instances in which the chemical balance that exists within your own body has been disrupted. Hangovers or muscle cramps are the unpleasant results of some common minor disruptions in the body's chemistry.

The cause of the chemical disruption occurring in PKU is an error in the body's process for converting phenylalanine into another substance called tyrosine. Both of these chemicals, in proper amounts, are required for the normal functioning of the body—especially for the proper formation of nerve cells in the rapidly growing brain of a young child. It is the abnormally high level of phenylalanine and the accompanying low level of tyrosine that cause the various symptoms observed in Billy. If PKU is diagnosed soon after birth, a child can be placed on a special diet that will allow his brain to develop under normal chemical conditions. Otherwise the brain will be growing in an abnormal chemical environment, leading to severe and irreversible mental damage and retardation. Even then, however, some of the other clinical symptoms such as skin disease, musty odor, abnormally light skin and hair color, seizures, and destructive behavior are reversible, and will improve once the child is placed on the proper diet. This is possible because the body systems responsible for these abnormalities can begin to function normally when they are placed in the correct chemical environment.

Figure 1.2 Both of these children suffer from phenylketonuria. The eleven-year-old boy, whose condition was not treated when he was an infant, is severely retarded; his three-year-old sister, whose phenylketonuria was treated from birth, is normal. (Photo by Willard R. Centerwall, M.D. From *Phenylketonuria,* Frank L. Lyman, Ed., Charles C Thomas, 1963.)

There are other examples of similar diseases which will be discussed in later chapters, but the point of this story is to emphasize the extreme importance of the proper chemical balance in the body, from conception throughout life. As researchers become more knowledgeable about the chemistry of living organisms, they will be able to control many more of the diseases that result from chemical irregularities.

1.3 This Textbook

If you had a PKU child in your family, you would certainly want to learn as much as you could about the disease, its cause, symptoms, and treatment. To understand the material written about PKU you would first need to learn the vocabulary used in such discussions, and perhaps you would want to read some general books on chemistry, biology, and anatomy to gain a good background in this subject.

Actually, this is an approach with which you should be quite familiar. You might want to learn how to change the spark plugs on a car, adjust the gears on a bicycle, or cook a Chinese dinner. In each case you would have to be familiar with the vocabulary used in the instruction manual, and would need to have at least some general knowledge about cars, bicycles, or kitchens. In the same way, before we can completely understand the many ways in which chemistry affects our lives, we must first become familiar with the vocabulary used in chemical discussions, and with some of the basic principles and laws that govern the chemical reactions in living organisms. In Section II of this text we will introduce many of the vocabulary terms and basic concepts that you will need to know. In Sections III and IV we will use this new vocabulary to discuss the many chemical substances and processes that are essential to life.

SECTION TWO
A CHEMICAL BACKGROUND

CHAPTER TWO

MATTER

Phil could hardly believe the results of his investigation. Only a week ago a young woman had told a terrifying tale of being stopped by an unknown bearded man as she drove through the woods. The man made her get out of the car and hand over her keys. When she instead pretended to throw her keys into the bushes, he angrily shot her in the arm. Still in a rage, he then poked the gun through the driver's window, shot her three children, and ran into the woods. The mother speeded to the hospital with two seriously wounded children and one dead daughter.

Phil was the forensic scientist on call from the police lab that evening. He spent the whole night photographing and collecting evidence from the woman's car. The mysteries quickly began building up. By vacuuming different sections of the car he found traces of gunpowder everywhere except the place it should have been—in the area near the driver's window. Phil also found blood splatters on the door frame and rock panel on the passenger's side of the car, which meant that the door had to have been open when one of the shots was fired. By carefully measuring the length and width of various blood droplets, he was able to calculate the angle from which the blood had fallen and to determine that one of the children was actually outside the car (and probably lying on the ground) when shot.

Other startling evidence came from the gunpowder patterns he later measured around the wounds. When a gun is fired, unexploded gunpowder is shot out of the barrel along with the bullet. If the gun is fired at close range, this gunpowder will imbed itself in clothing and skin; the more closely packed the pattern of such imbedded gunpowder, the closer the gun was to the victim. By carefully measuring the gunpowder patterns, Phil could tell that the children were shot from much closer range than would have been possible through the car window and that the mother's arm had been shot from extremely close range.

Phil's growing suspicions were confirmed when bullet casings later found in the mother's apartment were compared to those found in the car. Each gun leaves its own different set of marks on the cartridge casings that are worked through its mechanism and ejected from the gun. Microscopic examination showed that the casings from the woman's apartment had markings identical to those of the casings found in the car. It was Phil's investigative report that resulted in the mother being tried for the murder and attempted murder of her children.

Careful and accurate measurements are critical to the work of a forensic scientist such as Phil. But in fact most scientists are really detectives of a sort. Measurements are the key to their work. In this chapter we will introduce the vocabulary needed to carefully describe and measure the substances around us.

LEARNING OBJECTIVES

By the time you have finished this chapter, you should be able to:

1. Define *mass* and identify which of two given objects has the greater mass.
2. Define the following terms: *element, compound, atom,* and *molecule.*
3. State the difference between a compound and a mixture.
4. State the difference between homogeneous and heterogeneous mixtures and give three examples of each.
5. State the difference between a chemical change and a physical change.
6. State the difference between precise measurements and accurate measurements.
7. State the units of length, mass, volume, and temperature in the metric and SI systems.
8. Perform calculations with experimental data, maintaining the correct number of significant figures.
9. Convert measurements between the metric and English systems.
10. Define *density* and *specific gravity* and calculate these quantities when told the mass of a given volume of a substance.

MATTER

2.1 What Is Matter?

The physical world in which we live is made up of matter. **Matter** is scientifically defined as anything that has mass and occupies space. Of course, that definition doesn't do you much good unless you know what mass is. You probably have a general feeling about the concept of mass and would certainly be able to tell which has greater mass—a brick or a feather. The **mass** of an object is a measure of how hard it is to start the object moving, or how hard it is to change its speed or direction once it is moving. For example, a bowling ball is harder to push than a balloon because the bowling ball has a greater mass. The mass of an object is constant no matter where in the universe it is found (Figure 2.1). Imagine a bowling ball and a balloon both floating around weightlessly in the cabin of an orbiting spacecraft. The balloon would bounce harmlessly off the instrument panel, but the bowling ball could badly damage the delicate equipment.

The term **weight** is probably more familiar to you than mass, but you might not be certain of the difference between these terms. Weight is a

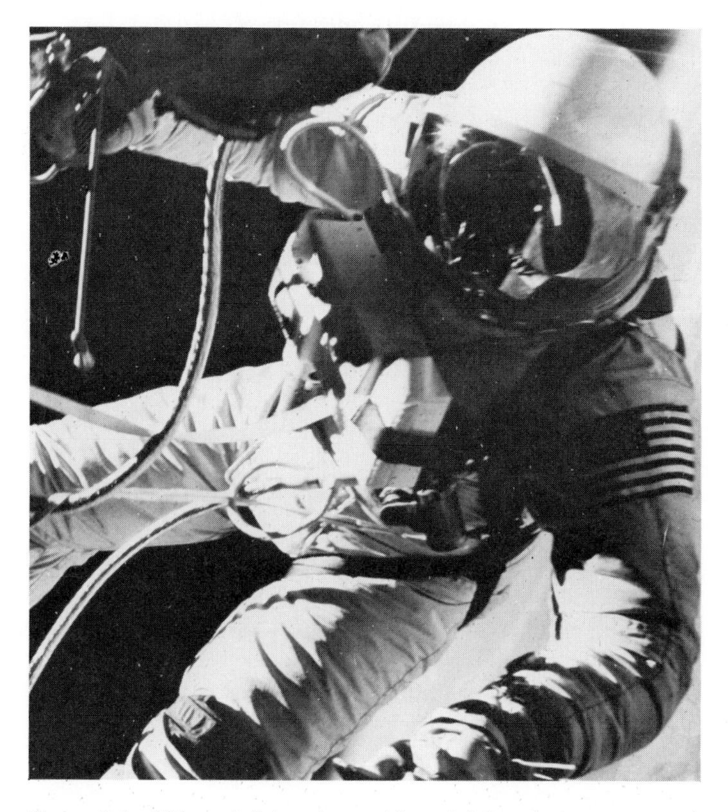

Figure 2.1 Although this astronaut is weightless in outer space, he has the same mass as on earth. (Courtesy NASA.)

measure of the force or attraction of gravity on an object; the mass of an object, however, does not depend on gravitational attraction, so it never changes. For example, because the moon's gravitational pull is roughly one-sixth that of the earth, an astronaut who weighs 180 pounds on the earth would weigh only 30 pounds on the moon. But the astronaut would have the same mass in either location. It is common and accepted practice, however, to use the terms *mass* and *weight* interchangeably and we will do so in this book.

2.2 Composition of Matter

Matter is composed of extremely small particles called **atoms.** The diameter of an atom is about eight-billionths of an inch (0.00000002 cm, or 2×10^{-8} cm—see Appendix 1 if you are confused by this notation). It is very difficult to imagine anything so small. To give you an example, a single page of this textbook is about 500,000 atoms thick.

The last 100 years have seen a tremendous body of knowledge develop about the atom, its structure, and the principles governing its behavior. Although photographs of individual atoms have been taken with specially designed electron microscopes (Figure 2.2), scientists have been able to determine the structure of atoms only from careful observations of their behavior.

As early as 400 B.C., the Greeks pictured the atom as being indivisible, but the work of many scientists over the last 80 years has shown that the atom is made up of smaller particles. Dozens of subatomic particles have

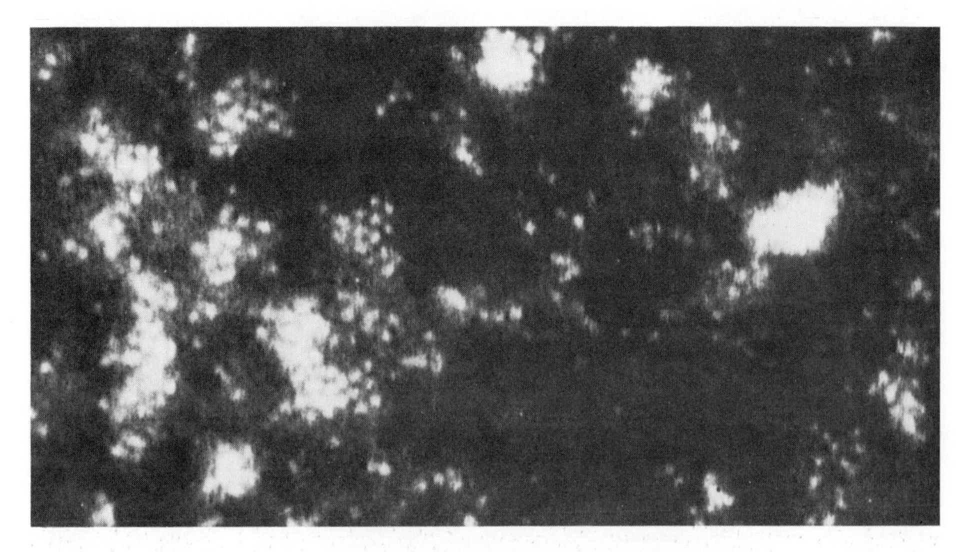

Figure 2.2 Single atoms of uranium and uranium microcrystals are seen as white spots on this micrograph taken with the University of Chicago Scanning Transmission Electron Microscope. (Courtesy Albert V. Crewe.)

now been identified, but only three are important for our discussions: the **proton,** the **neutron,** and the **electron.** It is the number of these particles and the way in which they are arranged that give each atom its particular chemical properties. We will discuss these important subatomic particles in greater detail in Chapter 5.

2.3 Classes of Matter: Elements, Compounds, and Mixtures

Any living or nonliving matter can be classified as either a pure element, a pure compound, or a mixture of these two (Figure 2.3). An **element** is a pure substance that cannot be broken down into simpler substances by ordinary chemical processes. There are currently 108 known elements, the heaviest of which are synthetically produced.

An **atom** is the smallest unit of an element having the properties of that element. A **molecule** is a chemical unit containing two or more atoms joined together. The atoms making up the molecule can be of the same element, or can be different elements. For example, atmospheric oxygen is found in the form of a molecule containing two atoms of oxygen; a molecule of water consists of two atoms of hydrogen and one atom of oxygen (Figure 2.4). When atoms of two or more different elements combine chemically, a **compound** is formed. The properties of a compound are totally different from those of the elements that make it up. For example, we have just mentioned that two atoms of the element hydrogen, which is a gas that can burn, will react with one atom of the element oxygen, which is a gas we breathe, to form the compound called water. But water, as you know, won't burn, and no living organism can stay alive by breathing water. (You might think that fish breathe water, but they only filter water through their gills to remove the oxygen gas that is dissolved in the water.)

Compounds can be broken down to simpler substances by chemical means. But no matter how a compound is formed or broken down, one

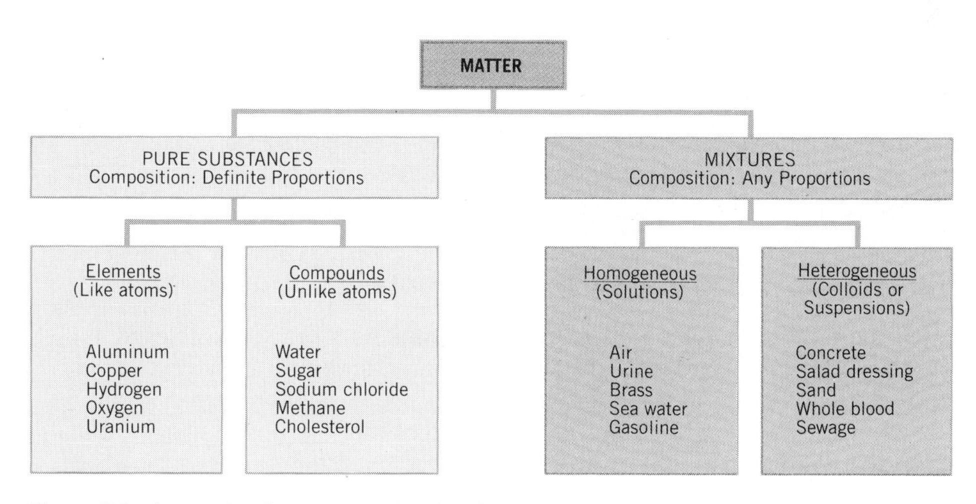

Figure 2.3 A sample of matter can be classified as either a pure substance or a mixture.

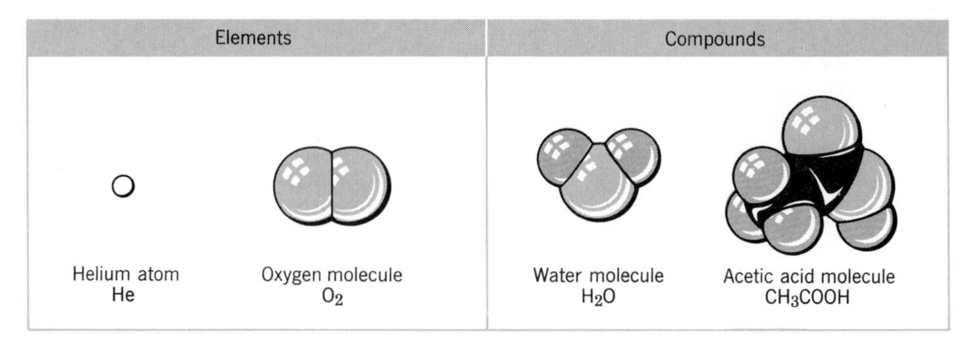

Figure 2.4 Many elements (such as helium) exist as single atoms, but other elements (such as oxygen) exist in molecular form. Compounds (such as water and acetic acid) exist as molecules containing atoms of more than one element.

fact will always be true: *A compound is composed of specific elements in a definite proportion by weight.* This is what makes a compound different from a mixture and is known as the **Law of Definite Proportions.**

A **mixture** can be made up of two or more substances (elements or compounds) mixed together in any proportion. When two substances chemically combine to form a compound, they form a new substance having different properties. But in a mixture all the substances keep their own individual properties.

Sugared coffee, for example, is a mixture of sugar and coffee; the proportion of sugar to coffee in your cup depends upon how sweet you like your coffee. The air that we breathe, the water that we drink, the ground on which we walk, and the gasoline that we put into our cars are each mixtures of various substances. Mixtures may be **homogeneous,** meaning they are so uniform that you can't tell one part from another, or **heterogeneous,** meaning that one part may be different from another (Figure 2.3). For example, a well-mixed cup of coffee with sugar in it is a homogeneous mixture. It will have the sugar molecules evenly distributed throughout the coffee, and each sip will taste the same. On the other hand, coffee with sugar in it that has not been totally dissolved and stirred is an example of a heterogeneous mixture. You would certainly be able to tell the difference between the first few sips of coffee and the last.

2.4 Names and Symbols for the Elements

The table on the inside back cover of this book lists the known elements together with their symbols. The **symbol** of an element is a shorthand way to indicate one atom of that element. Ordinarily, the symbol given to an element would be the first letter in the name of that element; for example, H for hydrogen, C for carbon, and O for oxygen. It often happens, however, that more than one element has a name starting with the same letter. In such cases, the first two letters of the name are used for the symbol, with the first letter capitalized and the second lowercase; for example, Co for cobalt and Ca for calcium. Note that because chlorine and chromium share the same first two letters, these two elements have the symbols Cl and Cr, respectively. Many of the names of the elements come from Latin or Greek and, in a few cases, the symbol of an element corresponds to an original word that is different from the common English name. For example, the symbol Fe for iron comes from the Latin word *ferrum,* the symbol Ag for silver comes from the Latin word *argentum,* and the symbol Pb for lead comes from the Latin word *plumbum.*

2.5 Physical and Chemical Changes

In this book we will be studying two types of changes that matter can undergo: physical change and chemical change (Figure 2.5). Common examples of substances undergoing physical change include water boiling to form steam, sugar dissolving in a cup of coffee, hamburger grease solidifying on a plate, and wood being split by an axe. A **physical change** is one in which a substance changes form, but keeps its chemical identity. For example, water still keeps all of its chemical properties whether it is in the form of ice, water, or steam. When water boils and forms steam, the water has simply undergone a physical change and can be returned to its previous state by collecting and cooling the steam. Similarly, wax dripping down a candle undergoes a physical change, turning from a liquid to a solid as it cools. However, the wax keeps the same chemical identity whether it is a liquid or a solid.

Chemical changes are occurring when you fry an egg, allow your bicycle to rust in the rain, burn wood in a campfire, accelerate your car from a stoplight, or digest your dinner. In a **chemical change,** the starting materials (reactants) are "used up," and different substances (products) are formed in their place. Obviously a fried egg has a different appearance and taste from a raw one. When it cools on your plate, the egg does not return to the raw state. Similarly, if you were to collect smoke from a campfire, the substances in the smoke would have none of the properties of the wood you put on the fire. The basic difference to remember, therefore, is that a *physical change involves only a change of form, whereas a chemical change involves a change in the basic chemical composition of the substances involved.*

MEASURING MATTER

2.6 Scientific Method

In Chapter 1 we stated that chemistry is the study of the composition and interaction of matter. In order to answer specific questions about the nature of matter, chemists observe the behavior of matter. Such observations are most often made under carefully controlled laboratory conditions, allowing the experiment to be repeated to see if the results are reproducible—that

(a) (b)

(c) (d)

Figure 2.5 Physical changes [(a) and (b)] involve a change in form, whereas chemical changes [(c) and (d)] involve a change in the basic nature of the substance. [(a) Stefan Bloomfield; (b,d) Phyllis Lefohn; (c) Mimi Forsyth/Monkmeyer.]

is, to see if the same results occur each time the experiment is run. The observations that chemists make during experiments are carefully recorded as data. If patterns are noticed in the data that are collected, this may allow the scientist to propose a hypothesis, or suggested explanation, of why matter behaves as it does. This hypothesis must then be tested by further experiments. If data from these further experiments show the hypothesis to be incorrect, then it must be discarded and another one proposed. But if the data continue to support the hypothesis, it then becomes a theory or explanation of the behavior of matter. Such theories may be accepted by scientists for years, even centuries, only to be shown to be incorrect as new data finally become available.

This careful manner of studying nature—through observation, development of an hypothesis, and testing of the hypothesis—is called the **scientific method.** It is a way of dealing with our surroundings that we all use, even if we don't go about it in as precise a way as a scientist might.

Doctors use the scientific method in their diagnosis and treatment of a sick patient. They collect data about the disease by means of blood tests, measurements of temperature and blood pressure, EKGs, etc. They then make a diagnosis (their hypothesis) and go on to test their hypothesis by prescribing a treatment. If the patient does not respond to the treatment, they must discard their hypothesis and collect more data—by further testing or perhaps exploratory surgery—and then propose a new or revised diagnosis and treatment.

2.7 Accuracy and Precision

The testing of an hypothesis often requires the careful measurement of the behavior of matter. The usefulness of experimental data collected in the laboratory depends upon its accuracy and precision. An **accurate** measurement is one that is correct; the closer a measurement comes to the real value, the more accurate it is. You can see, therefore, that an accurate measurement depends upon the measuring device; it must be carefully calibrated and in good working order. A speedometer that registers 55 mph when you are going 65 mph is not very accurate, and it may result in an incorrect hypothesis that you are driving at the speed limit.

A **precise** measurement is one that is reproducible, so that repeating the measurement produces values for the data that are very close to each other. For example, repeated readings of 135.5, 136.0, and 135.0 pounds on a bathroom scale are fairly precise measurements and may convince a woman to begin a diet or to skip desserts. But precise measurements are not always accurate. This woman, for example, might find that she weighs 127.3 pounds on her doctor's scale. The measurements on her bathroom scale were fairly precise, but not very accurate (Figure 2.6).

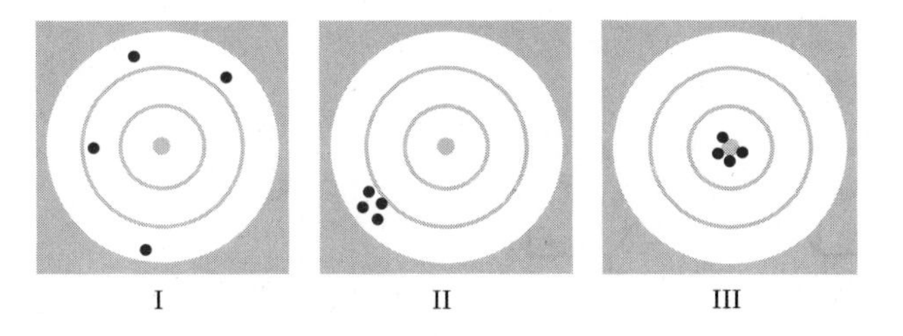

Figure 2.6 These three targets illustrate the relationship between precision and accuracy. The marksman shooting at target I was neither precise nor accurate. The marksman for target II was precise but not too accurate. The marksman for target III was both precise and accurate.

2.8 Significant Figures

The form in which experimental data is written down indicates the precision with which the measurement is made. When making a measurement, a scientist records all the digits that are certain, plus one that is uncertain. The digits in such a measurement are called **significant figures** or **significant digits.** For example, suppose we want to measure the length of a block of wood with the following two rulers:

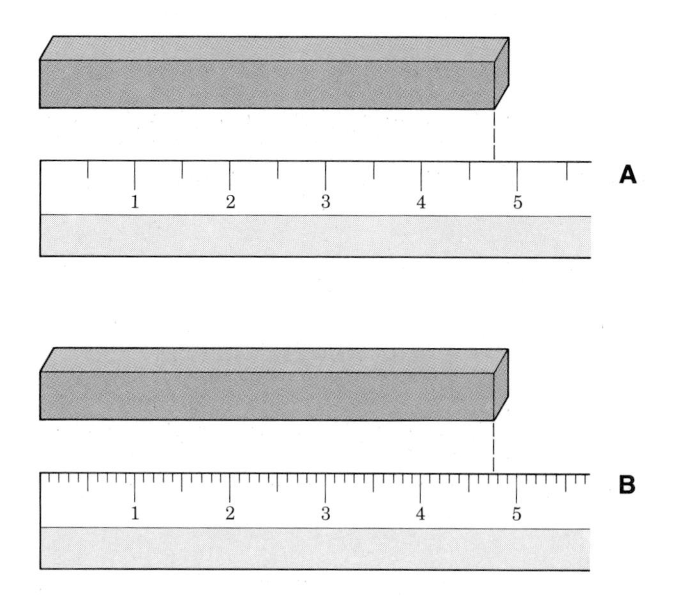

Using ruler A, we know that the block is over 4 centimeters long, and we can estimate the total length to the nearest tenth of a centimeter: 4.8 centimeters. Other persons using ruler A to measure the block would also have to estimate the length over 4 centimeters and might record a value of 4.7 or 4.9 centimeters. Therefore, our recorded value of 4.8 centimeters is uncertain by 0.1 centimeter. The number of significant figures in this measurement is two (one digit that we know for certain and one that is uncertain).

Using ruler B, we know that the block is at least 4.7 centimeters long and can estimate the length to the nearest hundredth of a centimeter: 4.74 centimeters. This measurement is uncertain by about 0.01 centimeter, so the measurement of 4.74 centimeters has three significant figures.

Keeping track of significant figures is very important in chemical calculations, especially if you use a calculator. Appendix 2 explains how to keep track of significant figures in mathematical calculations.

UNITS OF MEASURE

2.9 Unit Factor Method

In performing the many calculations needed in any area of science, it is very important to keep careful track of the units of measure represented by each of the numbers being used. For example, rather than just writing down the number 15, you should write 15 grams, or 15 feet, or 15 gallons, or whatever units of measure the number 15 happens to represent. Doing this will not only prevent confusion in the middle of a calculation, but it can remind you of the steps you must perform to finish solving the problem. For example, suppose you must do a calculation to determine the speed of some object. If you know that speed is commonly measured in such units as miles per hour (mi/hr), feet per second (ft/sec), or meters per second (m/sec), this tells you that you will eventually need to divide some measurement of distance (in the appropriate units) by a measurement of time (in the appropriate units). For this reason, you will find the calculations required in this textbook much easier to perform if you make a point of keeping your numbers properly labeled with their unit of measure.

To perform chemical calculations, we make use of two properties of the number "one." The first property is that any number, when multiplied by one, remains the same. More generally, any kind of quantity remains the same when multiplied by the number one. For example,

$$36 \times 1 = 36$$

$$7 \text{ apples} \times 1 = 7 \text{ apples}$$

$$92 \text{ miles/hour} \times 1 = 92 \text{ miles/hour}$$

The second property of the number one is that this number can be written as 2/2, or 156/156, or as the quotient of any number divided by itself. More generally, the number one can be represented by any quantity divided by itself. For example, the number one can be represented by 5 apples/5 apples, or 156 camels/156 camels. Going one step further, we can take any equation and divide one side by the other side to obtain a ratio equal to the number one. To see why this is true, consider the following familiar equation:

$$12 \text{ inches} = 1 \text{ foot}$$

If we divide both sides of the equation by the right-hand side (1 ft), we will have

$$\frac{12 \text{ in.}}{1 \text{ ft}} = \frac{1 \text{ ft}}{1 \text{ ft}} \quad \text{or} \quad \frac{12 \text{ in.}}{1 \text{ ft}} = 1$$

In exactly the same way we can represent the number one by (12 eggs/1 doz eggs) or (60 min/1 hr). Such ratios that are equivalent to the number one are called **unit factors** or **conversion factors,** and they are the key to most of the calculations in this textbook. We will use conversion factors as a means of changing the units that are given to us in a problem to units that we want in the answer.

To illustrate how unit factors or conversion factors are used in a calculation, let's look at a problem you could easily solve: How many inches are there in 2 feet? You would immediately say 24 inches, which you would have calculated by multiplying $2 \times 12 = 24$. But what did you do when you started with a distance that you called "2", and said that it is the same as a distance that you called "24"? What you actually did was to use the conversion factor that we derived above.

$$2 \text{ ft} \times \frac{12 \text{ in.}}{1 \text{ ft}} = \frac{2 \times 12}{1} \frac{(\cancel{\text{ft}})(\text{in.})}{(\cancel{\text{ft}})} = 24 \text{ in.}$$

You feel confident that the distance you started out with (2 ft) is the same distance that you ended up with (24 in.) because all you really did was to multiply your initial distance by a unit factor—that is, by the number one. When you make a point of keeping close track of the units of measure for each of the numbers in the problem (as we just did), you see that similar units in the numerator and denominator "cancel," leaving you with an answer in the desired unit of measure.

2.10 The SI System of Units

Scientists throughout the world have long used the metric system for the measurement of matter. However, in 1960 at an international meeting of scientists in Paris, a revised set of units was proposed. This revised set, based largely on the metric system, is called the **SI** (Systeme Internationale) **system of units** and has been accepted by scientists as a means of easily exchanging data throughout the world. Table 2.1 lists the basic units of the SI system. In this book we will use the SI units as well as the more common metric units in many of our discussions.

The United States is one of only four countries in the world which do not use the metric system, but that situation is slowly changing. The Metric Conversion Act of 1975 calls for the voluntary conversion of units to the metric system throughout the United States. We are beginning to see more

Table 2.1 Base Units in the SI System

Quantity	Name	Symbol
Length	meter	m
Mass	kilogram	kg
Time	second	s
Temperature	kelvin	K
Amount of substance	mole	mol
Electric current	ampere	A
Luminous intensity	candela	cd

Figure 2.7 Increasingly, we are encountering metric system units in our everyday lives. (Top left, Jane Scherr/Jeroboam; bottom left, Mark Antman/The Image Works; right, K. Bendo.)

and more examples of the metric system being used in our daily lives (Figure 2.7).

Using the familiar English system of measurements, we all have learned that 4 quarts make a gallon, 12 inches make a foot, and 16 ounces make a pound. This is a complicated system of measurements to use, because there is no consistent relationship between the number of smaller units needed to make up a larger unit of measure. By contrast, the great advantage of both the metric and SI systems is that all units of measure are related to their subunits by multiples of 10, and standard prefixes are used to indicate the number of multiplications or divisions by 10 that are required. The idea of such a numerical prefix is not new to you; we all know that the prefix *tri-*, as in the words tricycle, tripod, or trio, indicates that the object being named has three of something. In the same way, the prefix *kilo-* used in the metric and SI systems means 1000. One kilometer, then, is 1000 meters, and one kilogram is 1000 grams. Table 2.2 lists some of the more common prefixes used in the metric and SI systems. Table 2.3 shows some relationships between units of measure in the metric and English systems. The back overleaf of the textbook contains a list of other common conversion factors. The use of decimals in the metric system makes calculations much easier to perform than in the English system. In Sections 2.11 to 2.14 we will look at several examples of conversions between units in the metric system and the English system.

2.11 Length

The unit measure of length in the SI system is the **meter.** (*Metre* has been chosen as the preferred international spelling for this unit, but *meter* is the spelling that has been used in the United States and is the spelling we will use in this book.) This unit was defined in 1790 as one ten-millionth of the distance from the north pole to the equator. Although this standard has

Table 2.2 Some Prefixes Used in the Metric and SI Systems

Prefix	Symbol	Multiple	Example[a]
kilo-	k	$1000 = 10^3$	kilometer, km = 1000 meters
hecto-	h	$100 = 10^2$	hectometer, hm = 100 meters
deka-	da	$10 = 10^1$	dekameter, dam = 10 meters
			meter, m (basic unit)
deci-	d	$0.1 = 10^{-1}$	decimeter, dm = 0.1 meter
centi-	c	$0.01 = 10^{-2}$	centimeter, cm = 0.01 meter
milli-	m	$0.001 = 10^{-3}$	millimeter, mm = 0.001 meter
micro-	μ	$0.000001 = 10^{-6}$	micrometer, μm = 0.000001 meter
nano-	n	$0.000000001 = 10^{-9}$	nanometer, nm = 0.000000001 meter

[a] Conversions between numbers in the SI and metric systems are exact numbers and may be written with as many significant figures as necessary.

Table 2.3 Some Relationships between Units in the Metric System and the English System of Measure[a]

	Metric Units	Conversion Factor	English Units
Length	Meter	1.0 m = 3.3 ft	Feet
Mass	Kilogram	1.0 kg = 2.2 lb	Pound
Volume	Liter	1.00 liter = 1.06 qt	Quart

[a] For other commonly used conversion factors, see the back overleaf of this book.

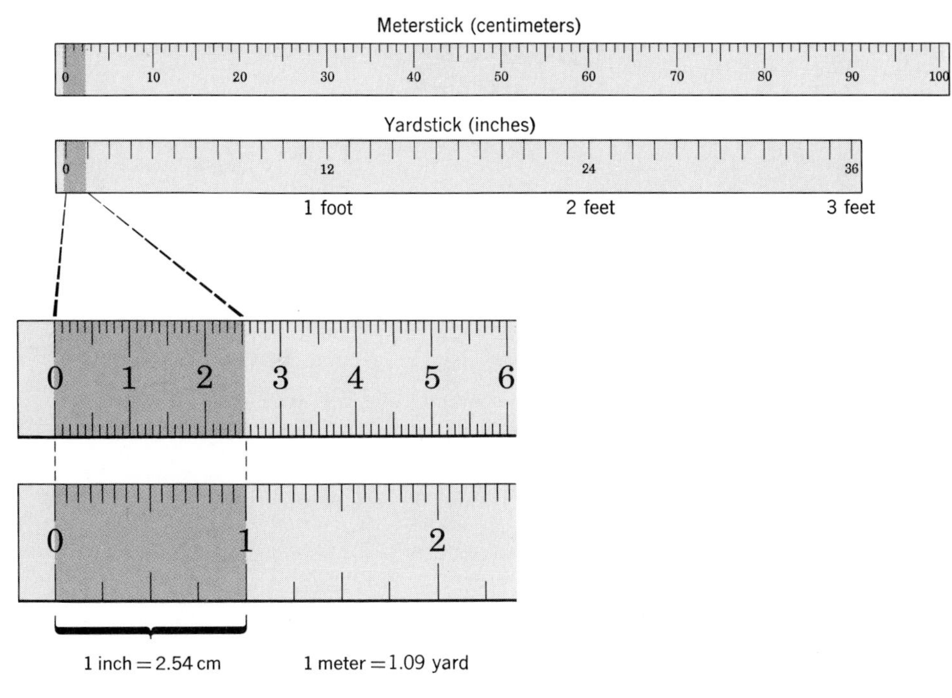

Figure 2.8 A meter is just slightly longer than a yard, and an inch is 2.54 centimeters.

been redefined three times since then, its length has remained pretty much unchanged. One meter is equal to 39.37 inches and is, therefore, just slightly longer than a yard (Figure 2.8).

A kilometer is the unit that is used to measure large distances, such as the distance between cities (1 kilometer = 0.621 mile). There are 1000 meters in a kilometer. Units commonly used that are smaller than a meter are the millimeter and the centimeter.

$$1 \text{ centimeter (cm)} = 10 \text{ millimeters (mm)}$$
$$1 \text{ meter (m)} = 100 \text{ cm}$$
$$1 \text{ m} = 1000 \text{ mm}$$

EXAMPLE 2-1

1. How many centimeters are there in 4 meters?

 We might rewrite the question as follows:

$$4 \text{ meters} = (?) \text{ centimeters}$$

 Because the problem requires us to change meters into centimeters, we must look for a unit factor that shows the relationship between these two units of measure. Using the equality 1 meter = 100 centimeters, we can write two unit factors that might be useful:

$$\frac{1\ \text{m}}{100\ \text{cm}} \quad \text{and} \quad \frac{100\ \text{cm}}{1\ \text{m}}$$

The second unit factor is the correct one to choose because it will allow units of measure to cancel, thereby giving us an answer in centimeters.

$$4\ \cancel{\text{m}} \times \frac{100\ \text{cm}}{1\ \cancel{\text{m}}} = \frac{4 \times 100}{1}\ \text{cm} = 400\ \text{cm}$$

2. A newborn baby measures 18.2 inches in length, but this measurement must be written in centimeters on the hospital chart. How many centimeters long is the baby?

$$18.2\ \text{inches} = (?)\ \text{centimeters}$$

The relationship between inches and centimeters is 1.00 in. = 2.54 cm. This gives us the unit factor that we need:

$$18.2\ \cancel{\text{in.}} \times \frac{2.54\ \text{cm}}{1.00\ \cancel{\text{in.}}} = 18.2 \times 2.54\ \text{cm} = 46.2\ \text{cm}$$

Exercise 2-1

1. Make the following conversions (use the back overleaf of the textbook for help with the conversion factors):

(a) 48 mm = (?) cm
(b) 3.2 m = (?) mm
(c) 0.03 km = (?) m
(d) 635 cm = (?) in.
(e) 7.5 mi = (?) km
(f) 27.3 m = (?) yd

2. The route for the Boston marathon measures 26.2 miles. Alberto Salazar won the race in 1982 with a time of 2 hours, 8 minutes, and 51 seconds.

(a) How many kilometers long is this race?
(b) On the average, how many minutes did it take Salazar to run one mile?
(c) On the average, how many minutes did it take Salazar to run one kilometer?

2.12 Mass

The unit of measure of mass in the metric system is the **gram.** The SI standard for this mass is a block of platinum metal weighing exactly one

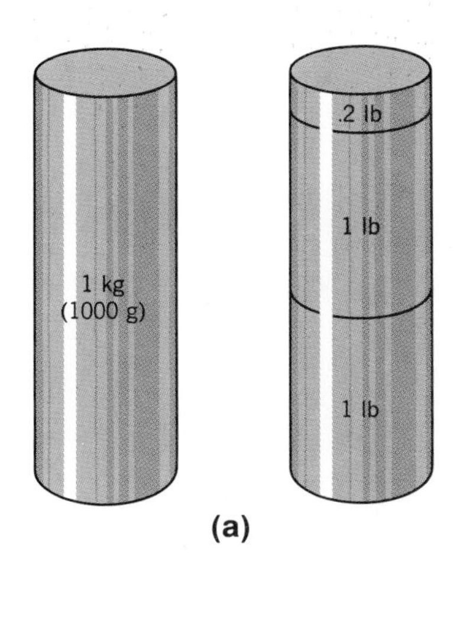

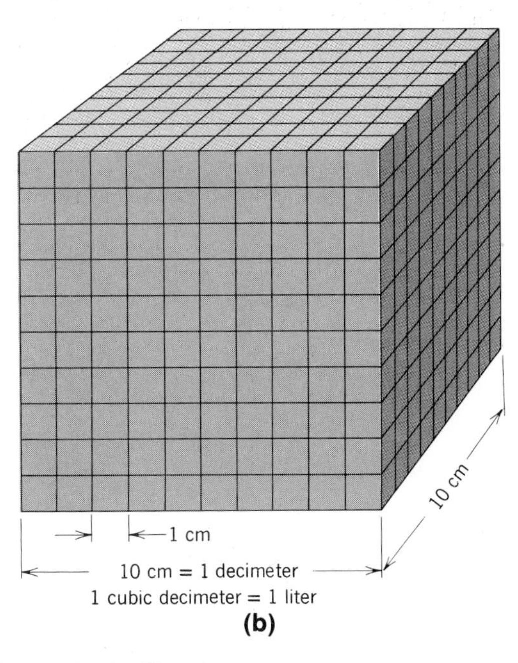

Figure 2.9 (a) A cylinder that weighs 1 kilogram in the SI system will weigh 2.2 pounds in the English system. (b) This box has a volume of one liter. (Volume = 10 cm × 10 cm × 10 cm = 1000 cm^3 = 1000 ml = 1 liter.)

kilogram (1000 grams) that is kept in a vault in France by the International Bureau of Weights and Measures. Units of mass commonly used are the kilogram, gram, and milligram (Figure 2.9a).

$$1 \text{ kg} = 1000 \text{ g}$$
$$1 \text{ g} = 1000 \text{ mg}$$
$$1.0 \text{ kg} = 2.2 \text{ lb}$$

The mass of a sample is determined by comparing the weight of the sample to the weight of standard masses on a balance, such as shown in Figure 2.10.

EXAMPLE 2-2

1. How many milligrams are there in 0.024 gram?

We know that 1 gram = 1000 milligrams, so we can write two unit factors that show the relationship between grams and milligrams:

$$\frac{1 \text{ g}}{1000 \text{ mg}} \quad \text{or} \quad \frac{1000 \text{ mg}}{1 \text{ g}}$$

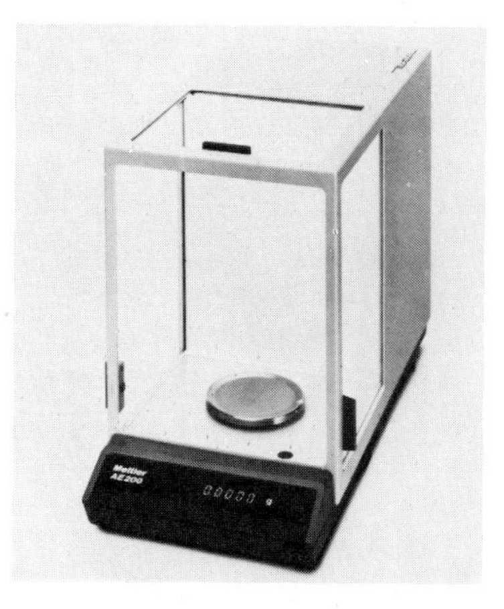

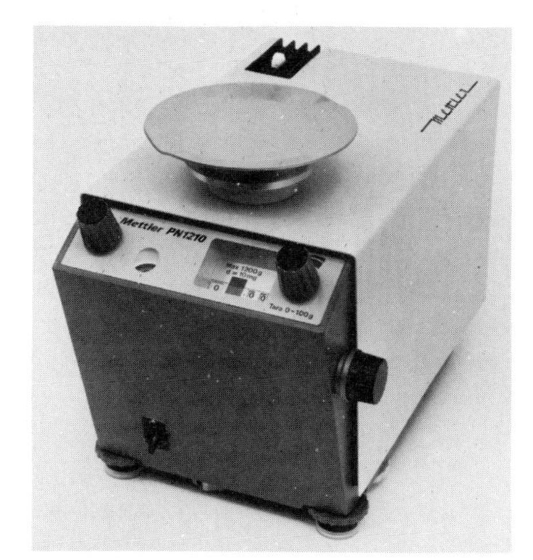

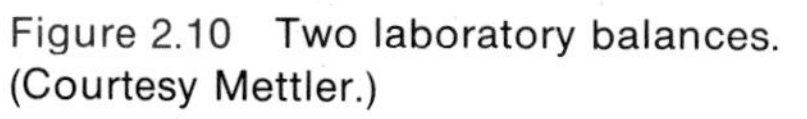

Figure 2.10 Two laboratory balances. (Courtesy Mettler.)

Our problem states that 0.024 g = (?) mg, so the second unit factor is the one to use.

$$0.024\ \cancel{g} \times \frac{1000\ \text{mg}}{1\ \cancel{g}} = 0.024 \times 1000\ \text{mg} = 24\ \text{mg}$$

2. Labels on drugs given to infants often list the dosage per kilogram of body weight. What is the weight in kilograms of a 16.5-lb baby?

$$16.5\ \text{lb} = (?)\ \text{kg}$$

From Table 2.3, the relationship between pounds and kilograms is

$$1.0\ \text{kg} = 2.2\ \text{lb}$$

Using the appropriate conversion factor, we obtain

$$16.5\ \cancel{\text{lb}} \times \frac{1.0\ \text{kg}}{2.2\ \cancel{\text{lb}}} = \frac{16.5}{2.2}\ \text{kg} = 7.5\ \text{kg}$$

3. A doctor prescribes a dose of 0.1 g of medication. How many 25-mg tablets are required to fill the prescription?

This problem must be solved in two steps. First,

$$0.1\ \text{g} = (?)\ \text{mg}$$

The relationship between grams and milligrams is

$$1\ \text{g} = 1000\ \text{mg}$$

So, using the appropriate conversion factor yields

$$0.1\ \cancel{g} \times \frac{1000\ \text{mg}}{1\ \cancel{g}} = 0.1 \times 1000\ \text{mg} = 100\ \text{mg}$$

Second, we must answer the question,

$$100 \text{ mg} = (?) \text{ tablets}$$

We are told that 1 tablet = 25 mg, so we can write

$$100 \cancel{\text{mg}} \times \frac{1 \text{ tablet}}{25 \cancel{\text{mg}}} = \frac{100}{25} \text{ tablets} = 4 \text{ tablets}$$

Exercise 2-2

1. Do the following conversions:

(a) $253\ \mu g = (?)$ mg
(b) 3.2 kg = (?) g
(c) 0.005 kg = (?) mg
(d) 0.34 kg = (?) oz
(e) 681 g = (?) lb
(f) 24.0 oz = (?) g

2. A premature baby born after 26 weeks of gestation weighed 832 g. How many pounds did the baby weigh?

2.13 Volume

You will often find the volume of an object stated in terms of some unit of length. For example, to calculate the volume of a box we might multiply the length × width × height. The SI unit of volume is the cubic meter, abbreviated m^3. This is a fairly large unit of measure—a box with a volume of 1 m^3 would hold about 1060 quarts of milk. For most practical purposes, therefore, we will use the metric unit of volume—the **liter**—which is only slightly larger than a quart (Figure 2.9b). (The international spelling is *litre,* but in this book we will use the American spelling.)

$$1 \text{ m}^3 = 1000 \text{ liters}$$
$$1 \text{ liter} = 1000 \text{ milliliters (ml)}$$
$$1 \text{ ml} = 1 \text{ cubic centimeter (cm}^3 \text{ or cc)}$$

You will often find laboratory measuring devices such as syringes or pipets labeled in either cubic centimeters or milliliters (Figure 2.11).

EXAMPLE 2-3

1. A patient excretes 2.65 quarts of urine. How many liters of urine is this?

$$2.65 \text{ qt} = (?) \text{ liters}$$

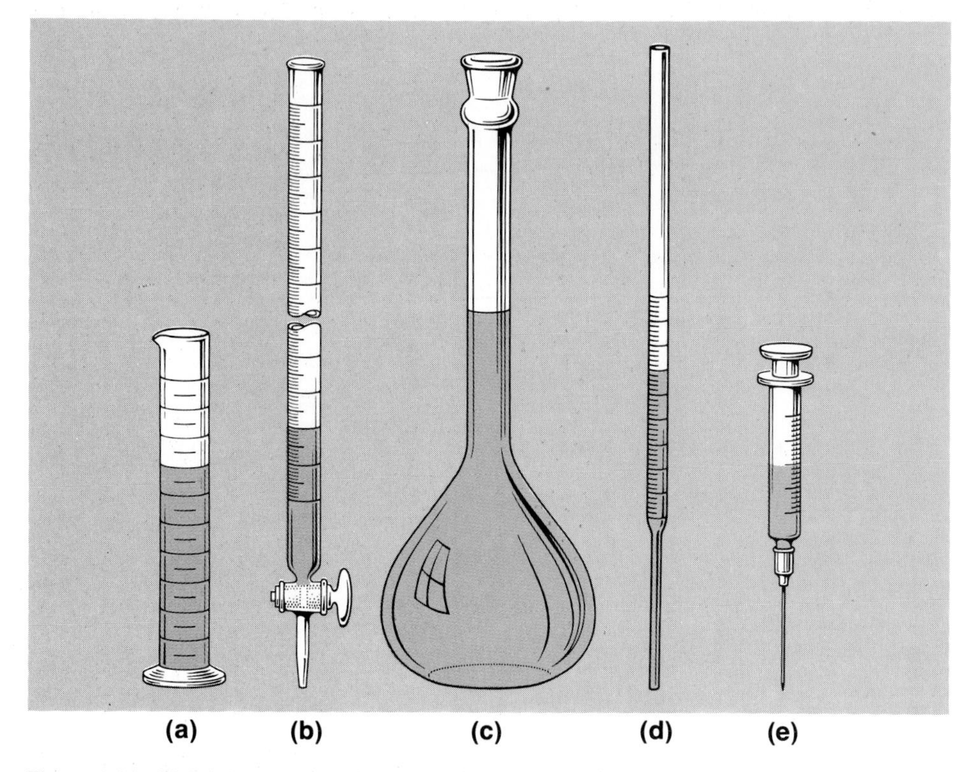

Figure 2.11 This laboratory equipment is commonly used to measure volume: (a) graduated cylinder, (b) buret, (c) volumetric flask, (d) pipet, and (e) syringe.

The relationship between quarts and liters is

$$1.06 \text{ qt} = 1.00 \text{ liter}$$

Therefore, we can write

$$2.65 \cancel{\text{qt}} \times \frac{1.00 \text{ liter}}{1.06 \cancel{\text{qt}}} = \frac{2.65 \times 1.00}{1.06} \text{ liter} = 2.50 \text{ liters}$$

2. A bottle of antibiotic contains 0.075 liter. How many 5.0-ml doses does it contain?

This problem must be solved in two steps. First,

$$0.075 \text{ liter} = (?) \text{ milliliters}$$

The relationship between liters and milliliters is

$$1 \text{ liter} = 1000 \text{ ml}$$

Using the appropriate conversion factor, we obtain

$$0.075 \text{ liter} \times \frac{1000 \text{ ml}}{1 \text{ liter}} = 0.075 \times 1000 \text{ ml} = 75 \text{ ml}$$

Second, we must calculate

$$75 \text{ ml} = (?) \text{ doses}$$

We are told that 1 dose = 5.0 ml, so we can write

$$75 \text{ ml} \times \frac{1 \text{ dose}}{5.0 \text{ ml}} = \frac{75}{5.0} \text{ doses} = 15 \text{ doses}$$

Exercise 2-3

1. Do the following conversions:

   (a) 2.5 liters = (?) ml
   (b) 345 ml = (?) liters
   (c) 25 ml = (?) cc
   (d) 343 gal = (?) liters
   (e) 5.3 liters = (?) qt
   (f) 94 ml = (?) pint

2. Which is a better buy: three 12-oz bottles of root beer for $1, or one 1-liter bottle of root beer for $1? (*Hint:* How many liters are there in 12 oz?)

2.14 Temperature

Many instruments have been developed to measure the temperature of a substance. The instrument with which you are probably most familiar is the mercury thermometer, a thin graduated glass tube with a mercury-containing bulb on the end. As the bulb is heated, the mercury expands and rises in the tube. The numbers marked on the glass tube depend upon the particular temperature scale chosen for the thermometer. We are accustomed to seeing temperatures expressed in the **Fahrenheit (°F)** scale, which is part of the English system of measurement. On this scale, the freezing point of water is 32°F and the boiling point of water is 212°F (a difference of 180 degrees). In the metric system, the **Celsius (°C)** temperature scale is used. On this scale, the freezing point of water is 0°C and the boiling point of water is 100°C (a difference of 100 degrees). This means that a degree on the Celsius scale is almost twice as big (180/100, or 1.8 times as big) as a degree on the Fahrenheit scale (Figure 2.12 and Table 2.4).

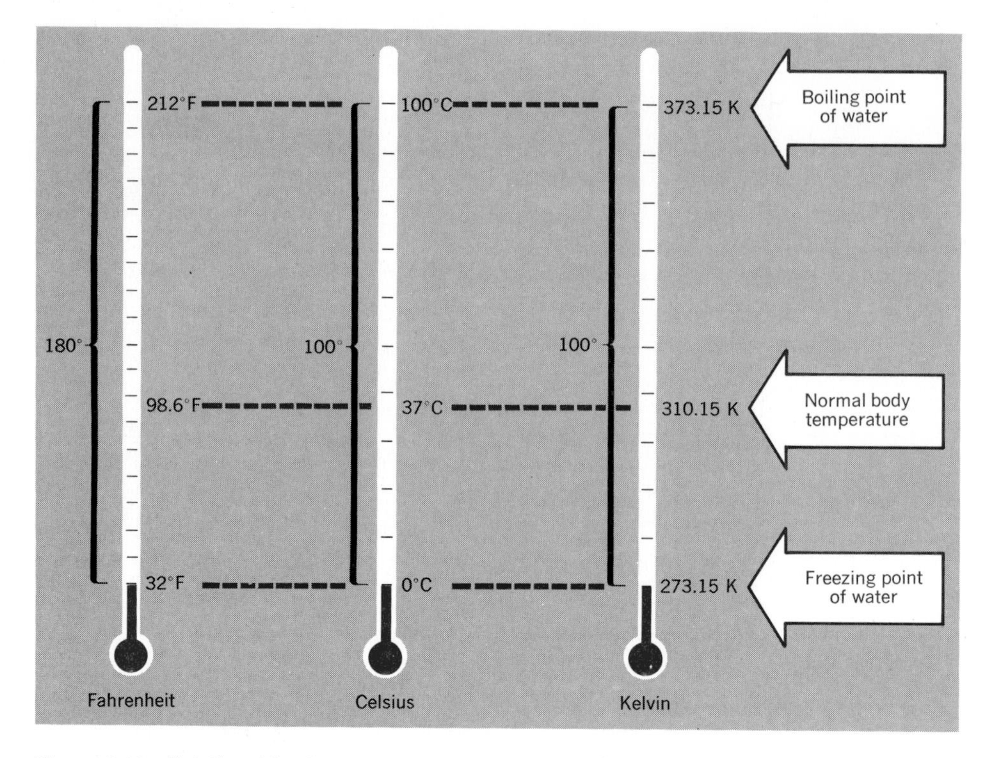

Figure 2.12 **Relationships between the Fahrenheit, Celsius, and Kelvin temperature scales.**

To convert temperatures between the Celsius and Fahrenheit temperature scales, we use one of the following equations*:

$$°C = \frac{100°C}{180°F}(°F - 32°F) \quad \text{or} \quad °C = \frac{5°C}{9°F}(°F - 32°F) \qquad (1)$$

$$°F = \frac{180°F}{100°C}(°C) + 32°F \quad \text{or} \quad °F = \frac{1.8°F}{1°C}(°C) + 32°F \qquad (2)$$

A third temperature scale is the **Kelvin (K)** temperature scale, which belongs to the SI system. (Note that Kelvin temperatures are written as K, not °K.) The freezing point of water on the Kelvin scale is 273.15 K, and the boiling point is 373.15 K. As with the Celsius scale, this is a difference of 100 degrees. You may wonder how a number such as 273.15 was chosen for the freezing point of water. This number was experimentally determined to make zero degrees on the Kelvin scale the lowest

---

* *For those of you who have trouble remembering the conversion between Celsius and Fahrenheit, the 40/40 method might be helpful:*
1. Add 40 to the starting temperature.
*2. Multiply by 5/9 going from °F to °C. (Remember, **f**ive-ninths **f**rom **F**ahrenheit.) Likewise, multiply by 9/5 going from °C to °F.*
3. Subtract 40 to get the answer.

Table 2.4 A Comparison of Temperatures on the Fahrenheit and Celsius Scales

	Fahrenheit	Celsius
A cold winter day	−5°F	−21°C
Freezing point of water	32°F	0°C
Room temperature	68°F	20°C
Normal body temperature	98.6°F	37°C
A hot summer day	100°F	38°C
Boiling point of water	212°F	100°C

temperature theoretically possible to reach. For this reason, zero degrees Kelvin is known as **absolute zero.**

To convert temperatures between the Kelvin and Celsius temperature scales, we use the following equation:

$$K = {}^\circ C + 273.15 \tag{3}$$

EXAMPLE 2-4

1. 50°F = (?)°C

Substituting in eq. (1) shown above, we get

$$^\circ C = \frac{5^\circ C}{9^\circ F}(50^\circ F - 32^\circ F)$$

$$^\circ C = \frac{5^\circ C}{9^\circ F}(18^\circ F) = 10^\circ C$$

2. 45°C = (?)°F

Substituting in eq. (2) shown above, we obtain

$$^\circ F = \frac{1.8^\circ F}{1^\circ C}(45^\circ C) + 32^\circ F$$

$$^\circ F = 81^\circ F + 32^\circ F = 113^\circ F$$

3. 20°C = (?)K

Substituting in eq. (3) shown above (and rounding off our conversion factor to the nearest whole number), we get

$$K = 20 + 273 = 293$$

4. 345 K = (?)°C

Again substituting in eq. (3), we obtain

$$345 = °C + 273$$
$$345 - 273 = °C$$
$$72° = °C$$

Exercise 2-4

1. Complete the following conversions:

(a) $-85°F = (?)\ °C$
(b) $40.5°C = (?)\ °F$
(c) $132°F = (?)\ °C$
(d) $32.0°C = (?)\ °F$
(e) $142\ K = (?)\ °C$
(f) $37°C = (?)\ K$

2. Immediately after he won the 1982 Boston Marathon, Alberto Salazar collapsed. His body temperature at that moment was measured to be 31°C. What was his body temperature in °F?

3. Lead melts at 621.3°F. What is its melting point in degrees Celsius? In degrees Kelvin?

2.15 Density and Specific Gravity

The **density** of a substance is defined as the mass of that substance per unit of volume (most often expressed as g/cc).

$$\text{Density} = \frac{\text{mass}}{\text{volume}}$$

The density of copper is 8.92 g/cc, iron 7.86 g/cc, calcium 1.54 g/cc, water 1.00 g/cc, and ethyl alcohol 0.789 g/cc.

EXAMPLE 2-5

1. Calculate the density of nickel if a 5.00-cc sample weighs 44.5 g.

$$\text{Density of nickel} = \frac{44.5\ \text{g}}{5.00\ \text{cc}} = 8.90\ \text{g/cc}$$

2. What volume would a 42.5-g sample of a liquid occupy if its density is 1.70 g/cc?

$$42.5 \text{ g} = (?) \text{ cc}$$

The value for the density can be used as a unit factor to solve this problem.

$$42.5 \not{\text{g}} \times \frac{1.00 \text{ cc}}{1.70 \not{\text{g}}} = 25.0 \text{ cc, or } 25.0 \text{ ml of liquid}$$

Exercise 2-5

1. What is the density of lead if a 6.10-cc sample weighs 68.9 g?
2. What volume will a 3.6-g sample of a liquid occupy if its density is 1.2 g/cc?
3. What would a bar of lead weigh if its dimensions were 3.00 cm by 5.50 cm by 2.00 cm? (Use the density value for lead from question 1.)

Specific gravity is the ratio of the mass of a substance to the mass of an equal volume of water (measured at the same temperature). More simply, specific gravity compares the density of a substance with the density of water at that temperature, and is most commonly used to compare the density of other liquids to that of water.

$$\text{Specific gravity} = \frac{\text{density of sample}}{\text{density of water (1.00 g/cc at 4°C)}}$$

For example, carbon tetrachloride has a specific gravity of 1.59 and, therefore, is more dense than water. Ethyl alcohol, with a specific gravity of 0.789, is less dense than water.

The specific gravity of a liquid can be measured with a **hydrometer,** a weighted bulb-shaped instrument that has a scale calibrated in specific gravities at a given temperature (Figure 2.13). The specific gravity of urine is measured with a **urinometer,** which is a hydrometer calibrated for the specific gravities of urine. The specific gravity of urine increases with the amount of solid waste present in the urine, and normal readings range from 1.005 to 1.030. Low urine specific gravity may indicate kidney damage or diabetes insipidus, whereas high values may be caused by diabetes mellitus or dehydration.

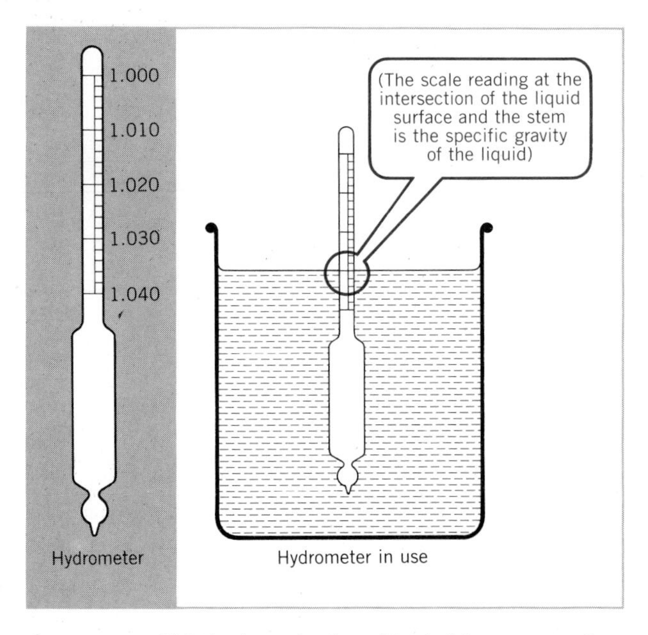

Figure 2.13 This hydrometer is calibrated to measure the specific gravity of liquids that are more dense than water.

EXAMPLE 2-6

A 50.0-ml sample of urine weighs 50.5 grams. (a) What is the density of the urine? (b) What is its specific gravity? (c) Is the urine more or less dense than water?

(a) $\text{Density} = \dfrac{\text{mass}}{\text{volume}}$

$= \dfrac{50.5\ \text{g}}{50.0\ \text{cc}}$ (*Note:* 1 ml = 1 cc.)

$= 1.01\ \text{g/cc}$

(b) $\text{Specific gravity} = \dfrac{\text{Density of sample}}{\text{Density of water}}$

$= \dfrac{1.01\ \text{g/cc}}{1.00\ \text{g/cc}} = 1.01$

(c) The urine is slightly more dense than water.

Exercise 2-6

A 200-ml sample of ethylene glycol weighs 222 g.

(a) What is the density of ethylene glycol?
(b) What is its specific gravity?
(c) Which is more dense: water or ethylene glycol?

CHAPTER SUMMARY

CHAPTER SUMMARY

Matter is anything that has mass and occupies space. The mass of an object is a measure of how hard it is to change its speed or direction of movement. The weight of an object is a measure of the gravitational attraction on the object. The weight of an object can change, but the mass is always constant. Pure substances are either elements, which can't be broken into simpler substances by ordinary chemical means, or compounds, which are composed of two or more elements chemically combined in definite proportions by weight. An atom is the smallest unit of an element having the properties of that element. A molecule contains two or more atoms chemically joined together. Mixtures are made up of two or more substances combined in any proportion. Mixtures can be homogeneous, meaning uniform throughout, or heterogeneous, meaning nonuniform. Matter can undergo physical changes, which are changes in form, or chemical changes, which result in changes in the chemical composition of the substances involved.

Scientists develop theories about the nature and behavior of matter by using the scientific method. Data collected to support these theories should be both accurate and precise, with the precision of a measurement indicated by how many significant figures appear in the number being recorded. Scientists measure matter using units in the SI system, a modified metric system. All units in the SI system are related to their subunits by multiples of 10, and prefixes are used to indicate the power of 10 required. The density of a substance is the mass of that substance per unit of volume. The specific gravity of a substance compares the density of the substance to the density of water.

Important Equations

$$°C = \frac{5°C}{9°F}(°F - 32°F) \qquad \text{Density} = \frac{\text{mass}}{\text{volume}}$$

$$°F = \frac{1.8°F}{1°C}(°C) + 32°F \qquad \text{Specific gravity A} = \frac{\text{density of A}}{\text{density of } H_2O}$$

$$K = °C + 273.15$$

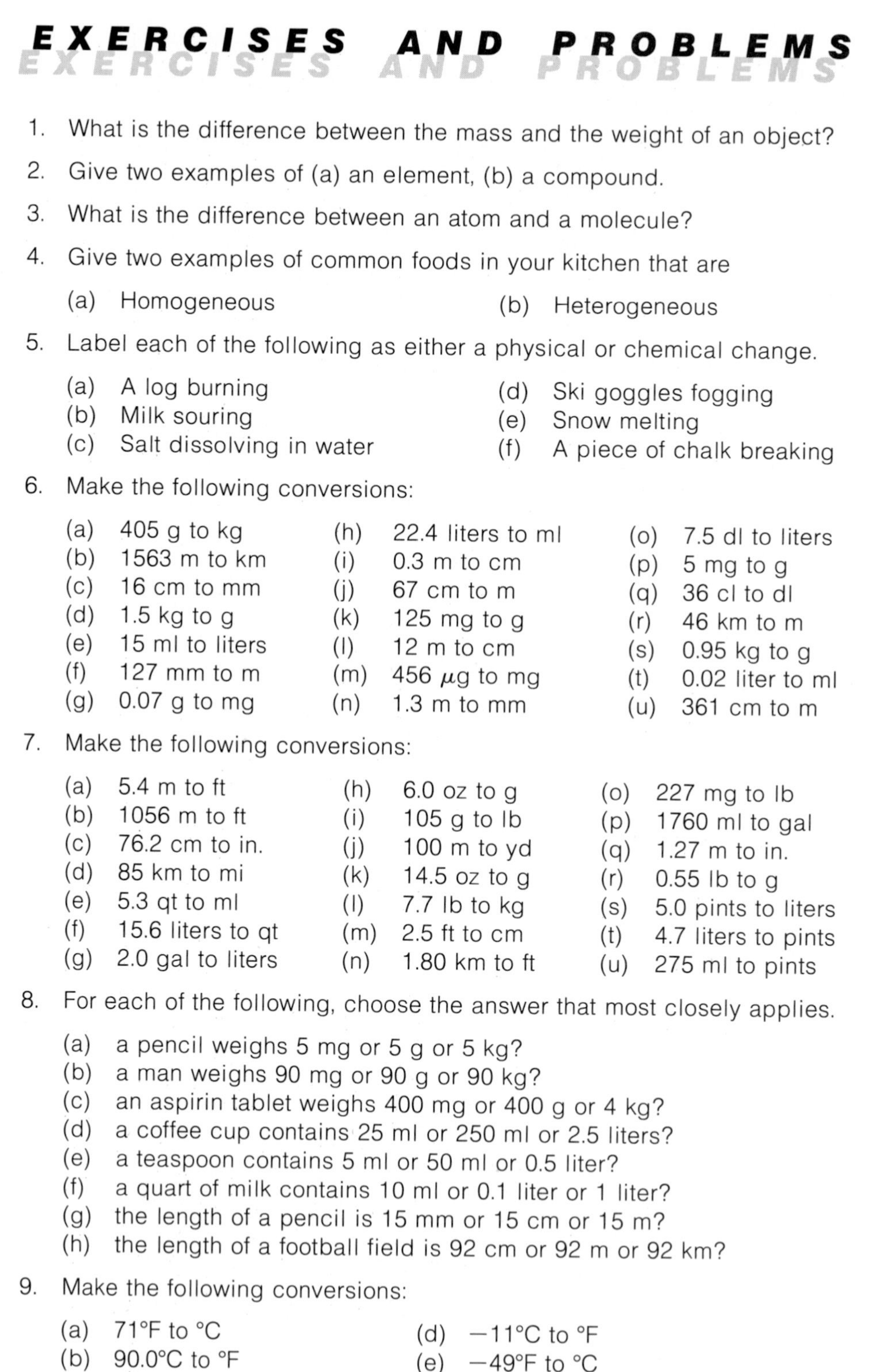

EXERCISES AND PROBLEMS

1. What is the difference between the mass and the weight of an object?
2. Give two examples of (a) an element, (b) a compound.
3. What is the difference between an atom and a molecule?
4. Give two examples of common foods in your kitchen that are

   (a) Homogeneous
   (b) Heterogeneous

5. Label each of the following as either a physical or chemical change.

   (a) A log burning
   (b) Milk souring
   (c) Salt dissolving in water
   (d) Ski goggles fogging
   (e) Snow melting
   (f) A piece of chalk breaking

6. Make the following conversions:

   (a) 405 g to kg
   (b) 1563 m to km
   (c) 16 cm to mm
   (d) 1.5 kg to g
   (e) 15 ml to liters
   (f) 127 mm to m
   (g) 0.07 g to mg
   (h) 22.4 liters to ml
   (i) 0.3 m to cm
   (j) 67 cm to m
   (k) 125 mg to g
   (l) 12 m to cm
   (m) 456 μg to mg
   (n) 1.3 m to mm
   (o) 7.5 dl to liters
   (p) 5 mg to g
   (q) 36 cl to dl
   (r) 46 km to m
   (s) 0.95 kg to g
   (t) 0.02 liter to ml
   (u) 361 cm to m

7. Make the following conversions:

   (a) 5.4 m to ft
   (b) 1056 m to ft
   (c) 76.2 cm to in.
   (d) 85 km to mi
   (e) 5.3 qt to ml
   (f) 15.6 liters to qt
   (g) 2.0 gal to liters
   (h) 6.0 oz to g
   (i) 105 g to lb
   (j) 100 m to yd
   (k) 14.5 oz to g
   (l) 7.7 lb to kg
   (m) 2.5 ft to cm
   (n) 1.80 km to ft
   (o) 227 mg to lb
   (p) 1760 ml to gal
   (q) 1.27 m to in.
   (r) 0.55 lb to g
   (s) 5.0 pints to liters
   (t) 4.7 liters to pints
   (u) 275 ml to pints

8. For each of the following, choose the answer that most closely applies.

   (a) a pencil weighs 5 mg or 5 g or 5 kg?
   (b) a man weighs 90 mg or 90 g or 90 kg?
   (c) an aspirin tablet weighs 400 mg or 400 g or 4 kg?
   (d) a coffee cup contains 25 ml or 250 ml or 2.5 liters?
   (e) a teaspoon contains 5 ml or 50 ml or 0.5 liter?
   (f) a quart of milk contains 10 ml or 0.1 liter or 1 liter?
   (g) the length of a pencil is 15 mm or 15 cm or 15 m?
   (h) the length of a football field is 92 cm or 92 m or 92 km?

9. Make the following conversions:

   (a) 71°F to °C
   (b) 90.0°C to °F
   (c) 151 K to °C
   (d) −11°C to °F
   (e) −49°F to °C
   (f) 302°F to K

10. The normal boiling point is shown for each of the following compounds. What is each boiling point in Kelvin? in degrees Fahrenheit?

(a)	Silicon dioxide	1610°C
(b)	Ethanol	78.5°C
(c)	Hydrogen sulfide	−60.7°C

11. For each of the following three liquids calculate (1) the density, (2) the specific gravity, (3) the weight of 250 ml of the liquid, (4) the volume in ml of 1.0 g of the liquid.

(a)	liquid A:	100 ml weighs 79 g
(b)	liquid B:	1 liter weighs 1.1 kg
(c)	liquid C:	500 ml weighs 490 g

12. State the number of significant figures in each of the following measurements:

(a)	0.0125 g	(d)	10.7 m	(g)	3000 mg
(b)	150 ml	(e)	4005.00 cm^3	(h)	0.00250 mm
(c)	12.060 km	(f)	3.20×10^{-2} mg	(i)	1604.732 kg

13. Your laboratory partner was to measure the length of a bar of iron (approximately 10 in.) as accurately as possible using a meter stick marked in millimeters. Which of the following measurements would you accept: 25.65 cm, 25.6 cm, 25 cm, 256.5 mm, 256 mm?

14. A prescribed injection of a drug is 1.5 ml. If the syringe is graduated in cubic centimeters, to what mark on the syringe would you draw the drug?

15. Whiskey is sold in bottles marked 4/5 of a quart. How many liters of whiskey are there in 4/5 of a quart? How many ml?

16. A sign on the freeway outside San Francisco reads "Los Angeles, 412 miles." How many kilometers would that be?

17. The daily dosage for ampicillin to treat an ear infection is 100 mg per kilogram of body weight. What is the daily dose for a 22-lb baby with an ear infection?

18. A circus fat lady, "Dolly Dimples," was suffering from a life-threatening heart condition and had to lose weight. She went from 553 pounds to 152 pounds in a 14-month period. What was her average weight loss in kilograms per month?

19. The sperm whale has the largest brain (about 9 kg) of any animal that has ever lived. This whale can dive to depths of more than 1.5 kilometers while holding its breath for over one and one-half hours.

    (a) What is the approximate weight of a sperm whale's brain in grams? in pounds?
    (b) What is the depth to which the whale can dive in meters? in miles?

20. In an experimental treatment of ingrown toenails, the toe is sprayed with liquid nitrogen (N_2) at 77 K. The frozen skin dies, scales, and sloughs off, leaving the toenail unharmed and eliminating the need for surgery and the removal of the toenail. What is the temperature of the liquid N_2 (a) in °C (b) in °F?

21. A weather forecaster on the television news states that it is 22°C in Paris and 32°C in Rome. What are the temperatures in Paris and Rome in °F?

22. Diamond has the highest melting point of any element, 3550°C. What is its melting point in °F? K?

23. A California couple were found dead in their hot tub, victims of hyperthermia or heatstroke. Doctors and hot tub manufacturers recommend heating a tub to 39°C to 40°C; the water in the couple's tub was found to be 46°C. What is the recommended tub temperature and the couple's tub temperature in °F?

24. (a) What is the density of mercury if 15.0 ml weighs 204 g?
    (b) What is the weight in kilograms of 250 ml of mercury?

25. Would you expect gasoline to have a specific gravity greater than or less than water? Explain your answer.

26. A sample of water has a mass of 455 g at 25°C. The density of water at 25°C is 0.997 g/ml. What is the volume of this sample of water in milliliters? in liters?

CHAPTER THREE
ENERGY

Hannah Segal, an energetic 81-year-old, lived alone in her Cleveland home. She volunteered time at her local library and never missed her Tuesday bridge game or Friday night bingo. One day in early December, however, things were quite different; she came into the library that morning confused, drowsy, and forgetful. A librarian, noticing how Hannah was stumbling as she walked among the shelves, became alarmed at this sudden change in her behavior and decided to take her to the hospital. Hannah's condition continued to deteriorate even as they drove to the hospital, and she fell into a coma soon after arrival. The emergency room physician diagnosed a stroke and told Hannah's family that her chances of survival were slim. Even if she were to recover, he continued, the lasting effects of the stroke would probably require that she live in a nursing home. An alert nurse, however, noticed that Hannah's body was unusually cold to the touch. Using a special thermometer, she found that Hannah's temperature was 89°F (or 32°C). Hannah had not suffered a stroke but rather was suffering from hypothermia, a lowering of the body's inner temperature. This condition can be fatal without prompt warming of the patient and administration of intravenous fluids to prevent dehydration and restore the body's chemical balance. After 12 hours of such treatment, Hannah's body temperature was back to normal; after several days, she was able to return to her home, anxious to resume her regular activities.

Hypothermia is a condition that can be fatal when the body drops as few as six degrees below its normal temperature of 98.6°F (a drop of three degrees Celsius from the normal 37°C). Surprisingly, a person can survive a temperature drop of 40 to 50°F in the hands and feet, but only a small drop in the temperature of the body core can cause death. Hypothermia begins as soon as the body starts to lose energy faster than it can be produced. The hands and feet are affected first; a drop of only 3°F in body temperature will reduce manual dexterity to the point that a person is unable to perform the basic tasks necessary for survival. In addition, as the temperature of the body is lowered, the brain becomes numb and the central nervous system is depressed, resulting in progressive confusion, slurred speech, stumbling, seizures and, ultimately, coma and death.

Hypothermia has been diagnosed for centuries as the cause of death for such people as mountain climbers, skiers, and boaters who were inadequately protected against the cold. This type of hypothermia is called exposure hypothermia. Because heat flows from a warm region to a colder region, hypothermia can occur under ordinary conditions even when the air temperature is relatively mild. Being wet increases the flow of heat from the body, because water will conduct heat about 240 times better than still air. The speed with which hypothermia develops varies considerably with the energy reserves of the individual and the nature of the situation. However, anything that one can do to prevent heat loss is extremely important. The proper choices of clothing, insulation, or shelter, and the minimization of muscular activity will all help prevent the onset of hypothermia.

Only recently, however, has hypothermia been recognized as a common condition that affects elderly people who remain indoors, but who are inadequately clothed to protect themselves against house temperatures of 65°F or below. This form of hypothermia is called accidental hypothermia and can also be caused by lack of activity of the thyroid gland, or from side effects of phenothiazines (drugs widely prescribed in the treatment of psychiatric conditions and for nausea and vomiting). Young people are able to counteract heat loss by regulating the amount of blood flow through the skin and by shivering. For unknown reasons, shivering, which increases heat production by about five times, is often absent in elderly people. Also, older people seem to lose their perception of cold as hypothermia sets in. Commonly, when elderly people have been found dead in their homes, their deaths have been attributed to natural causes or falls; now it is suspected that hypothermia has often been the actual cause of such deaths.

The need for a person to maintain a stable body temperature in a cold environment is only one example of the complex energy interactions between living systems and their environments. An understanding of these interactions requires an examination of the different types of energy that are involved and the energy changes that can occur.

LEARNING OBJECTIVES

By the time you have finished this chapter, you should be able to:

1. Define *energy*.
2. Describe the difference between kinetic energy and potential energy and give several examples of each.
3. State which molecules have greater kinetic energy: those in a sample of water at 20°C, or in the same sample of water at 100°C?
4. List four different forms of energy.
5. Define *calorie* and use experimental data to calculate the number of calories in a food sample.
6. Describe five types of electromagnetic energy and list them in order of increasing wavelength and increasing energy.

ENERGY

3.1 What Is Energy?

Energy is a topic that seems to come up all the time. Sometimes it's in terms of the world's energy supply—referring to the petroleum and coal taken from the earth. Sometimes it's in discussions of nuclear energy, either peaceful energy from nuclear reactors or destructive energy from nuclear weapons. Sometimes it's in more personal terms, as in "I just don't seem to have any energy today!" Even though you may use the word "energy" daily, you may have only a vague notion of precisely what energy is. Some days you may wake up feeling energetic, ready to get things accomplished—and the tasks you have completed by the end of the day will have resulted from your expenditure of energy. It is precisely the ability to accomplish something—the capacity to do work—that defines **energy.** A rushing stream, a rock poised at the top of a hill, the gasoline in a car, the muscles in an arm, each has the capacity to cause change, to do work on an object. Therefore, each is said to possess energy.

Although scientists define energy as the capacity to do work, work can mean different things to different people. Painting a fence, shoveling snow, or chopping logs may be work to some people, but fun and relaxation to others. For this reason, scientists have given the term *work* a mathematical meaning that does not depend on an individual's opinion. **Mechanical work** is defined to be the application of a force through a distance. Force is a push or pull on an object that causes the object to start moving, or to change speed or direction once it is moving. If you push a car and set it into motion, you have done work on the car. However, if you push a car but are unable to get it to move, you have not done any mechanical work.

From this definition, we see that we can't measure the energy contained in an object until it has been transferred from that object to another one. If we use the above example, the energy expended by your muscles can be measured only when this energy is transferred from your muscles to the car that you are pushing.

3.2 Kinetic Energy

Energy can take many forms, but it is often convenient to classify it as either energy of motion or energy of position. The name applied to energy of motion is **kinetic energy.** A car traveling at 20 mph, a rock hurtling through the air, or steam from a water boiler all have energy by virtue of motion. That is, each possesses kinetic energy. The moving car could do a great deal of damage if it hit a parked car, the rock could break a window, and the steam could push the pistons of a steam engine. Although it is obvious that both the car and the rock have motion and, therefore, possess kinetic energy, you may wonder about the steam. The rock and the car are large objects whose motion we can see, but if we had a "super" microscope that could let us see at the level of atoms and molecules, we would realize that all matter is in constant motion. The molecules of water in steam are moving extremely rapidly and, for this reason, possess a great deal of kinetic energy (Figure 3.1).

Figure 3.1 The kinetic energy possessed by molecules of steam is used to power this steam engine. (Courtesy Canadian Pacific Railroad.)

Kinetic energy can be measured, and its value will depend upon the mass of the particle and the particle's speed or velocity. Algebraically,

$$\text{Kinetic energy} = \frac{1}{2} \times \text{mass} \times (\text{velocity})^2$$

or

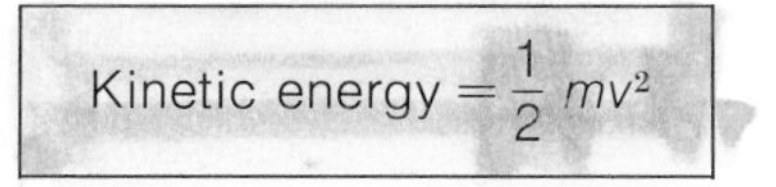

$$\text{Kinetic energy} = \frac{1}{2} mv^2$$

Although you may not have thought of it in this way before, you already have a good general feeling for the kinetic energy possessed by an object. For example, you would probably rather have an eight-year-old throw a baseball for you to catch than have a major league pitcher try it. In each case, the baseball would have the same mass, but would be traveling at a different speed and, therefore, would have different kinetic energy. The ball thrown by the child would be traveling at a much slower speed; it would have less kinetic energy and would do less damage to your hand (Figure 3.2). As a second example, would you rather have your parked car hit by a bicycle going 5 mph or a dump truck going 5 mph? In this example the speed of the two objects is the same, but the much larger mass of the dump truck gives it a

Figure 3.2 Which ball would you rather catch? (Left, Stefan Bloomfield; right, Focus on Sports.)

much larger kinetic energy and a much greater capacity to do work—that is, to crumple your fender.

A baseball, a car, and a dump truck are all objects that we can weigh, whose speed we can determine, and whose kinetic energy we can then calculate. But how can we determine the kinetic energy of particles that we can't see, such as the molecules of water in steam. The temperature of a substance allows us to measure the kinetic energy of its particles. Scientifically, **temperature** is a measure of average kinetic energy. The term *average kinetic energy* is used because all molecules, as with all people, are not alike. At any temperature, some molecules will be moving very rapidly, some very slowly, and the majority somewhere in between. But just as we can talk about the average behavior or performance of a group of students, so can we refer to the average behavior of a group of atoms or molecules (Figure 3.3). As the average kinetic energy of a group of molecules increases, so does the temperature of that substance. From experience, you know that the hotter a pan of water, the more severely the water can burn your hand, or the faster it can cook vegetables. You now understand that the hotter water has a higher average kinetic energy and, therefore, has a greater ability to damage tissues or cook vegetables.

3.3 Potential Energy

A rock hurtling through the air has kinetic energy; when it strikes a window, it can break it. But a rock perched on the top of a cliff also possesses energy; it has the capacity to do work. If it fell off the cliff onto a passing car, it could do severe damage. The rock on the top of the cliff is said to possess **potential energy,** or energy of position. Such energy is not in use, but it is stored and has the capacity to do work when it is converted to other forms of energy. For example, the water stored behind a dam

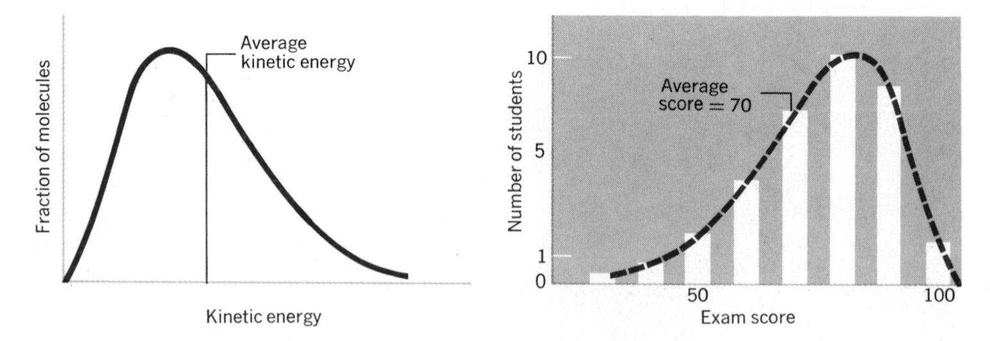

Figure 3.3 At any given temperature each gas molecule will have a specific kinetic energy, just as each student will have a specific score on an exam. However, just as there is an average score for the exam, there is an average kinetic energy for the gas molecules. Temperature is a measure of this average kinetic energy.

possesses potential energy. When released in a controlled fashion, it can drive turbines to produce electrical energy or, if released in an uncontrolled fashion by the collapse of the dam, the water could display its energy by destroying everything in its path (Figure 3.4).

Figure 3.4 The rocks in the top photo possess potential energy. These rocks could do significant damage if they slid off the hillside unto a passing truck (bottom). (Top, Grant Heilman; bottom, Milan Chuckovich)

The food that we eat is another example of a substance possessing potential energy. The energy stored in food is called **chemical energy**. As the food molecules are broken down or metabolized in our bodies to simpler molecules with lower chemical energy, we use the released energy to contract muscles or maintain body heat.

3.4 Heat Energy

Energy can be transferred from one place to another in many forms, among which are electricity, sound, light, and heat. Heat is energy that is transferred from one place to another because of a difference in temperature. Scientists originally measured heat in units called calories. A **calorie (cal)** is the amount of energy that must be transferred to one gram of water to raise its temperature from 14.5°C to 15.5°C. The calorie is now defined in terms of the SI unit of heat energy, the **Joule (J).**

$$1 \text{ cal} = 4.1840 \text{ J}$$

The definition of a calorie comes from a property of water called specific heat. The **specific heat** of a substance is defined to be the amount of heat energy necessary to raise the temperature of one gram of the substance by one degree Celsius.

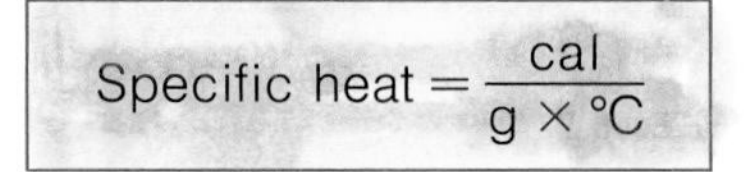

$$\text{Specific heat} = \frac{\text{cal}}{\text{g} \times {}^\circ\text{C}}$$

Liquid water has one of the highest specific heats known: 1 cal/g°C (rounded to one digit). This means that it takes a lot of heat energy to change the temperature of water. For example, one calorie of heat energy would have about twice the effect on the temperature of a sample of ethyl alcohol (specific heat = 0.586 cal/g°C) than on the same size sample of water.

To calculate the number of calories required to raise the temperature of a cup of water (250 grams) from 25°C to 100°C to make a cup of coffee, we use the following equation:

$$\text{calories} = \text{grams} \times \text{temperature change } (\Delta t) \times \text{specific heat}$$

$$\text{cal} = \text{g} \times \Delta t \times \frac{1 \text{ cal}}{\text{g}^\circ\text{C}}$$

So, in this case,

$$\text{cal} = 250 \text{ g} \times (100^\circ\text{C} - 25^\circ\text{C}) \times \frac{1 \text{ cal}}{\text{g}^\circ\text{C}}$$

$$\text{cal} = 250 \times 75 \text{ cal} = 18{,}750 \text{ cal}$$

A calorie is actually a very small unit of energy, so we often find it more convenient to talk in terms of 1000 calories, called a **kilocalorie (kcal).** The food Calorie (note the capital C) that you count when you are on a diet is actually a kilocalorie. That 1-ounce bag of potato chips that you had for a snack contains 160 food Calories or 160 kilocalories. To avoid confusion, in this book we will always refer to food energy content in terms of kilocalories.

The instrument that is used to determine the calorie content of potato chips or other substances is called a **calorimeter,** and is shown in Figure 3.5. The sample to be tested is placed in the inner chamber and then is burned completely. The energy released from this burning warms the water in the surrounding container, allowing the number of calories transferred to be calculated from the rise in water temperature.

The minimum amount of energy required daily to maintain the basic continuous processes of life in the body at rest is called the **basal metabolism rate (BMR).** This amount of energy varies for each individual; a woman weighing 121 lb (55 kg) will require about 1400 kcal, and a man weighing 143 lb (65 kg) about 1600 kcal. Calories that are consumed in excess of that amount either will be used to maintain body heat and to supply the energy needed for the work an individual does, or will be deposited as fat. About 3500 excess kilocalories will produce one pound of body fat. Table 3.1 gives examples of the energy used in various activities.

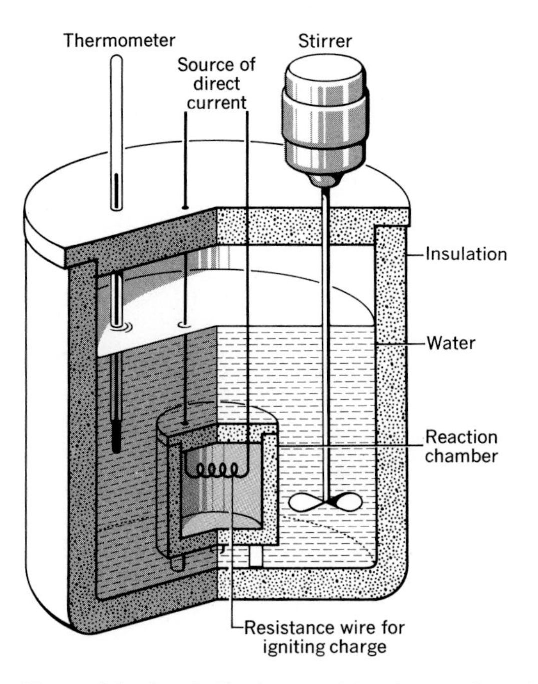

Figure 3.5 A calorimeter consists of a reaction chamber that is surrounded by water held in an insulated shell. As the sample is burned in the reaction chamber, the surrounding water is stirred and its temperature change measured.

Table 3.1 Energy Expenditures for Everyday Activities

Activity	Kcal per Hour per Pound of Body Weight	Activity	Kcal per Hour per Pound of Body Weight
Watching TV	0.6	Dishwashing	1.0
Eating a meal	0.7	Walking (3 mph)	1.5
Standing	0.8	Playing ping-pong	2.7
Driving a car	1.0	Running	4.0
Typing rapidly	1.0	Swimming (2 mph)	4.5

EXAMPLE 3-1

1. What is the average calorie content of a peanut, if the temperature of 1000 grams of water in a calorimeter increases by 50°C when 10 peanuts are burned?

$$\text{calories} = \text{grams} \times \Delta t \times \text{specific heat}$$

$$= 1000\text{ g} \times 50°\text{C} \times \frac{1\text{ cal}}{\text{g}°\text{C}}$$

$$= 50{,}000\text{ cal produced by 10 peanuts}$$

$$= 5000\text{ cal produced by one peanut}$$

2. How many calories are required to raise the temperature of a 150-ml sample of water (density of water = 1.0 g/ml at 15°C) from 15 to 20°C?

We first must calculate the number of grams in our sample of water.

$$150\text{ ml} \times \frac{1.0\text{ g}}{1\text{ ml}} = 150\text{ g}$$

Then,

$$\text{calories} = 150\text{ g} \times 5°\text{C} \times \frac{1\text{ cal}}{\text{g}°\text{C}}$$

$$= 750\text{ cal}$$

Exercise 3-1

1. How many kilocalories of energy are contained in 1.2 kilograms of body fat?

2. A glass of cold water is a refreshing drink on a hot day. Once in your body, of course, the water is quickly warmed to body temperature. How many kilocalories of energy must be transferred within your body to warm 500 g of water from 4° to 37°C?

3.5 Electromagnetic Energy

Light represents another form of energy that always surrounds us. The light that we see is only a tiny part of an entire range of electromagnetic energy, from low-energy radio waves to very high-energy X rays and gamma rays. What we see as white light is actually made up of the whole spectrum of colors that appears in a rainbow. The location of this visible spectrum in the **electromagnetic spectrum** is shown in Figure 3.6.

Light waves may be compared with ocean waves, with their crests and troughs. The **wavelength** of light (represented by the Greek letter lambda, λ) is the distance from one crest to the next crest or from trough to trough, or from middle to middle for that matter (Figure 3.7). Each wavelength of light corresponds to a specific level of energy. The longer

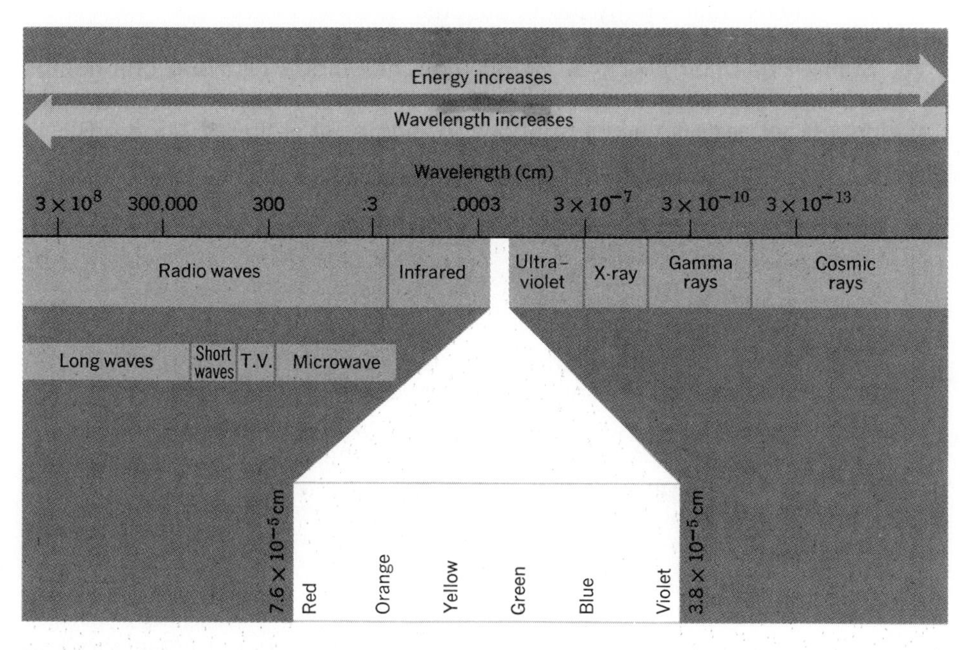

Figure 3.6 The electromagnetic spectrum. Visible light makes up only a very small part of this spectrum.

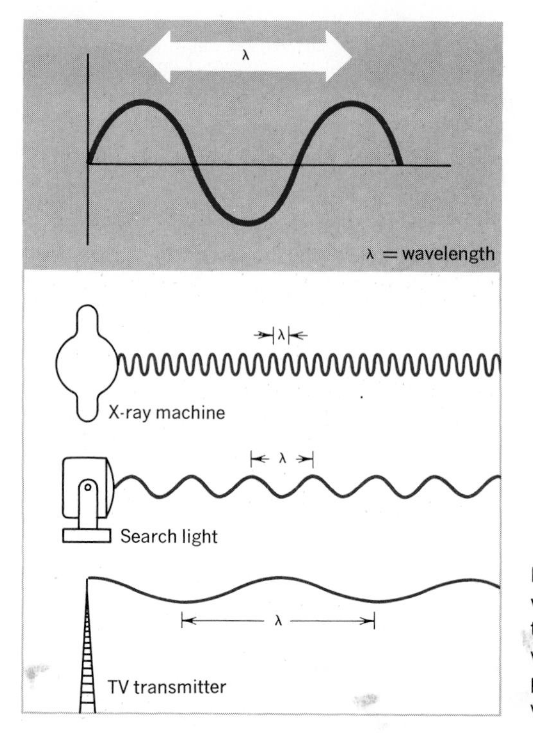

Figure 3.7 Light travels in waves. The wavelength of light can be defined as the distance from crest to crest. Each wavelength is associated with a particular energy: the longer the wavelength, the lower the energy.

the wavelength, the lower the energy of the light. This relationship can be expressed mathematically by the equation

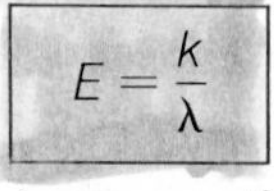

$$E = \frac{k}{\lambda}$$

where E is the energy of the electromagnetic radiation, and k is a constant related to the speed of light. Look at the visible spectrum shown in Figure 3.6. Visible light with the highest energy and the shortest wavelength is found in the purple region, whereas red light has the lowest energy and the longest wavelength.

Energy in the blue region of the visible spectrum is used to treat infants suffering from jaundice or hyperbilirubinemia. The livers of these infants cannot excrete bilirubin, a compound formed from the breakdown of red blood cells. When untreated, high levels of bilirubin can lead to nerve damage and death. The excess bilirubin in these infants collects under their skin, giving the skin a yellow color. When the skin is exposed to the energy of blue light, however, the bilirubin molecule is converted to a form that can then be excreted by the infant's liver.

Microwave Radiation

You're certainly familiar with the use of long-wavelength, low-energy radiation to send radio and television signals. Slightly shorter wavelength, or **microwave,** radiation is becoming increasingly popular as a means of rapid cooking. Microwaves interact with food by setting the molecules within the food into motion, causing the food to heat uniformly. This is quite different from conventional cooking, in which the molecules on the outside of the food are heated first, and the heat is then transferred gradually to the molecules in the center of the food. Because microwave ovens uniformly heat all of the food at the same time, cooking time is greatly reduced. You will often find warning signs posted in areas where microwave ovens are being used, because these cooking devices can interfere with the operation of electronic cardiac pacemakers.

Infrared Radiation

Radiation in the **infrared (IR)** range cannot be seen, but can be felt as heat and can be detected by a thermometer. Infrared radiation from the sun warms us, and light fixtures that give off this radiation are used to keep food warm in restaurants. Although infrared radiation doesn't affect ordinary photographic film, special film can detect the infrared radiation given off by warm objects. Such film has been used to locate thermal pollution from power plants and to map the density of vegetation in certain regions (Figure 3.8). Two species of snakes have infrared sensors that can detect low-density infrared radiation given off by their prey. Using these sensors, the snakes can pinpoint and capture their prey in the dark, and can locate hiding places having comfortable living temperatures.

Ultraviolet Radiation

Ultraviolet (UV) light has shorter wavelengths and, therefore, higher energy than visible light. It is penetrating radiation and can harm living tissue by causing changes in certain biological systems or by causing irreparable damage to cells. Fortunately, most of the ultraviolet radiation coming to the earth from the sun is absorbed by the ozone layer in the upper atmosphere and never reaches us. Because some wavelengths of ultraviolet light are very effective in killing bacteria, many sterilizing units use this type of radiation. It is perhaps surprising, then, that UV light is also quite important

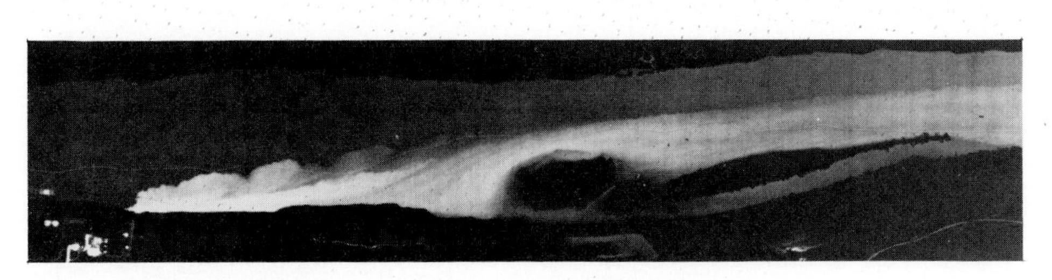

Figure 3.8 Infrared-sensitive film was used to record the heat discharge from a power plant on the Connecticut River. The power plant and an oil tanker with its hot engine room are at the lower left, and the glowing cloud running down the bank is the warm water being discharged from the cooling system of the plant. (Courtesy Environmental Analysis Department/HRB-Singer, Inc.)

to our bodies. It causes the production within the skin of vitamin D, a compound necessary for the prevention of rickets.

Unfortunately, too much ultraviolet exposure can produce the familiar burn we call sunburn. Prolonged exposure to the sun over the years can cause wrinkling of the skin, thick warty growths on the skin (a condition called keratosis) and, in some cases, skin cancer. The cells in the skin do have some protection against damage by ultraviolet light. First, the dead horny layer of cells on the surface of the skin absorbs some of the UV light. Second, exposure to the sun triggers cells in the skin to produce more of the skin pigment called melanin, which acts to further block the ultraviolet light. This increase in protective skin pigment, of course, is what we call a suntan. However, an effective tan takes three to five days to develop and, even then, reduces the penetration of UV light by only one-half. To gain additional protection, you need to apply sun screens containing chemicals that prevent the most damaging shorter-wavelength UV light from penetrating the skin.

X Rays and Gamma Rays

X rays and **gamma rays** have shorter and shorter wavelengths, higher energies, and greater penetrating power. X rays originate from specially constructed X-ray tubes, whereas gamma rays come from natural sources. The shortest X rays can pass through steel walls, and gamma rays can pass through a 10-inch thick lead plate. Because X rays will not pass as easily through bones and teeth as they will through tissue, they are a very useful diagnostic tool for the medical and dental professions. However, penetration of cells by high-energy radiation such as X rays or gamma rays can disrupt normal chemical processes, causing cells to grow abnormally and to die. Because effects of such radiation build up over time, dosages and repeated exposure to X rays must be carefully controlled. Cancer cells are more sensitive to radiation than normal cells, so X rays and gamma rays can be used to treat certain forms of cancer. Medical radiation equipment is designed to control the exact degree of penetration of the radiation and to concentrate the radiation on the cancerous area to minimize the damage to normal tissue.

CONSERVATION OF ENERGY*

3.6 Law of Conservation of Energy

Energy can be converted from one form to another and will ultimately end up as heat energy. For example, electrical energy is converted into heat energy in our furnaces, stoves, and toasters, and into light energy and heat energy in our lamps. Chemical energy is converted into heat energy when gas is burned in stoves or furnaces, or into mechanical energy and heat energy when gasoline is burned in car engines. Although energy can be

---

** This section is optional and may be skipped without loss of continuity.*

converted from one form to another, the total amount of energy in the universe is constant. This means that energy can neither be created nor destroyed, but only changed in form. In other words, the total amount of energy at the end of a reaction or process must equal the total amount of energy at the beginning. This is a statement of the **Law of Conservation of Energy,** or the **First Law of Thermodynamics.**

You know that gasoline is burned to provide energy for powering automobiles. However, you might be surprised to learn that only about 20% of this energy is used in actually moving the car. Where does the rest of the energy go? (The Law of Conservation of Energy says that it cannot just disappear.) In this particular case, the rest is lost as heat energy—*lost* in the sense that this energy does not do any useful work.

In the same way that gasoline provides energy for a car, the food that we eat provides energy for our cells. But in this case too, the conversion of the chemical energy in the food into energy that our cells can use is not 100% efficient. For example, only about 44% of the energy that can be transferred from a molecule of sugar is converted into energy that our cells can use to perform work. The rest ends up as heat energy that helps to maintain body temperature. (Actually, a transformation of energy in any type of system will result in the production of some waste heat.) Much to many people's unhappiness, our bodies must obey the Law of Conservation of Energy. If we take in more food energy than our bodies need, this excess energy doesn't just disappear, but is stored within our tissues, mainly in the form of unwanted fat. To get rid of this excess stored energy, we must take in less food energy than our bodies require so that our cells will start using the fatty tissue as a source of energy to maintain body activities.

3.7 Entropy

You might be wondering why we need a constant supply of new energy from the sun to maintain life on earth if energy is conserved in all natural processes. Although the First Law of Thermodynamics tells us that the total quantity of energy remains unchanged, it tells us nothing about the quality of that energy. Only concentrated forms of energy can be used to do work, and work is essential to the maintenance of living organisms. The energy from the sun reaches us in the very concentrated form of light energy, but this energy is quickly converted to heat, a less concentrated form of energy.

If you think about it for a moment, you will realize that all naturally occurring (or spontaneous) processes tend to go in one direction. Left alone, water always runs downhill, heat flows from a hot to a cold object, gases flow from regions of high pressure to regions of low pressure, and people grow old. We would be quite shaken if we were to see water flowing uphill on its own, but we still might ask what it is that prevents naturally occurring processes from reversing direction. The answer to this question is fairly obvious when the final state of the system is at a lower energy level than the initial state. Water flows downhill from higher to lower potential energy, and heat flows from a state of higher to lower kinetic energy. It is quite reasonable that there should be a general tendency for all substances to reach a state of lower energy.

But there are other spontaneous processes with which you are familiar that don't seem to depend upon the energy of the system. For example, if you were to put a drop of ink into a glass of water, the ink would immediately begin to spread throughout the liquid, eventually giving the water a uniform tint (Figure 3.9). This process certainly occurs in a single direction. You would be astonished if you ever saw a glass of tinted liquid suddenly change so that all of the color moved through the liquid to form a single concentrated spot of dye. Yet, it is impossible to point out any change of energy taking place in this process.

This example leads us to conclude that there is another factor besides energy levels that determines the direction of spontaneous processes. That factor is the amount of randomness or disorder of the system. The term used to describe the disorder of a system is **entropy.** The more random and disordered a situation is, the greater its entropy. All spontaneous reactions go toward a condition of greater randomness, greater disorder, and greater entropy. We can now see that the drop of ink, which was originally in a small compact drop, will spontaneously spread throughout the liquid into a random and disordered arrangement of ink molecules, which increases the entropy of the system. The driving force behind the spreading of the drop of ink is the tendency for the entropy of the system to increase.

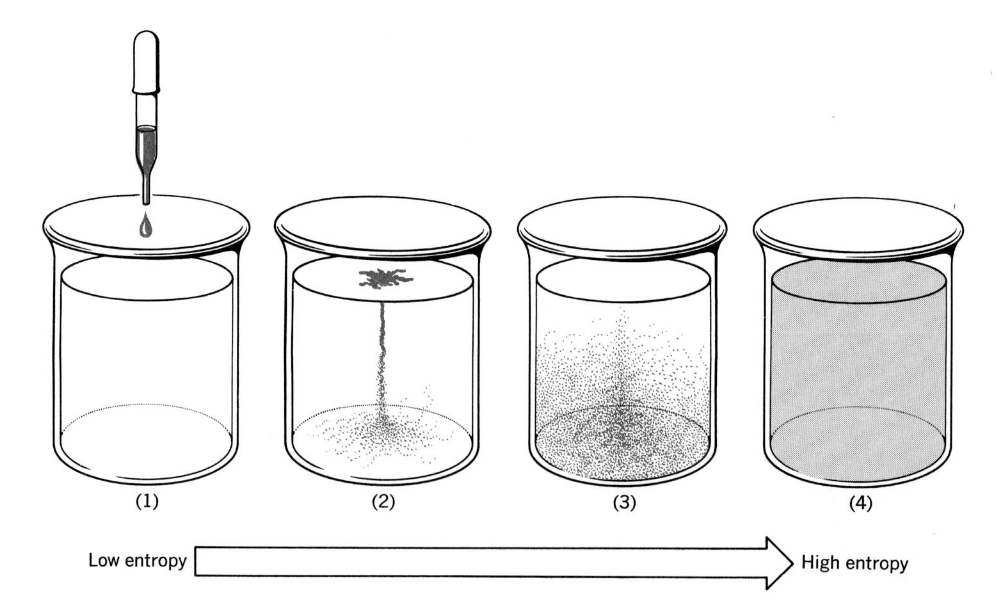

Figure 3.9 Ink dropped into a glass of water will quickly change from a state of low entropy (the droplet) to a state of high entropy (spread throughout the water).

The **Second Law of Thermodynamics** states that the entropy of the universe is increasing. This law is interwoven through all of science and into each of our lives. Even the writers of nursery rhymes had an intuitive feeling for the Second Law when they wrote, "All the king's horses and all the king's men couldn't put Humpty together again." You now see that the breaking of an egg is an irreversible process that leads to increased disorder of the egg parts and, therefore, greater entropy (Figure 3.10).

By now it may have occurred to you that there are some naturally occurring processes that can be reversed. Water can be pumped uphill to a storage reservoir, heat can be pumped from cold to hot areas in a refrigerator, and air can be pumped from low pressure to high pressure in a bicycle tire. The key word in each of these examples is *pump*. Each requires that energy be added; that is, work must be done. This means that each of these seemingly "reversed" processes must be coupled with, or related to, another reaction in which energy is produced. Electrical energy is needed to power the water pump or to run the refrigerator, and chemical energy is needed to give our muscles the strength to push a bicycle tire pump. But in the reactions that are required to produce this energy, entropy is increased. So, if we now look at both of the reactions required to pump the water uphill (that is, the process of actually pumping thewater uphill and the process of generating the energy necessary to do this pumping), we would find that the total entropy of the combined processes has increased.

Figure 3.10 The writer of this nursery rhyme had an intuitive understanding of the concept of entropy.

An awareness of entropy is very important in understanding the functioning of living organisms. Such organisms are composed of cells, which are highly complex structures. Therefore, there is a natural tendency for these cells to break down and increase the entropy of the system (Figure 3.11). In order to maintain the structure and functioning of the living system, energy must be added to counteract the natural drive toward increased entropy. We will see that this energy, which comes from the food that organisms eat, actually originates on the sun. If we want to consider the coupled reactions of this system, we see that the life processes that build up and maintain the complicated structures in living organisms depend upon energy-producing reactions on the sun. It is rather amazing to realize that each of our bodies maintains its highly complex structures and functions only at the expense of huge entropy increases on the far-off sun.

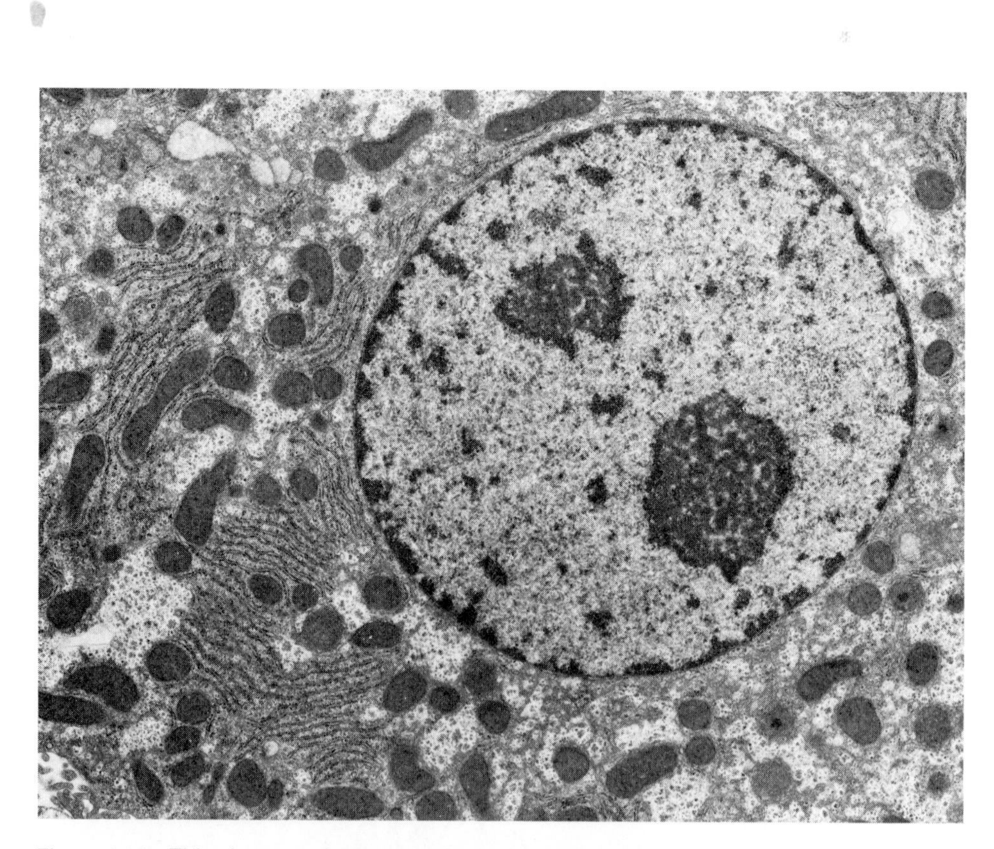

Figure 3.11 This electron micrograph of a liver cell illustrates the highly organized, complex structures found in living organisms. To function normally, the cell must constantly expend energy to counteract the natural tendency toward increased entropy and the breakdown of these complex structures. (Dr. Robert L. Wood.)

CHAPTER SUMMARY

Energy is the capacity to cause a change or to do work on an object. Kinetic energy is energy of motion, and temperature is a measure of the average kinetic energy of the particles of a substance. Potential energy is stored energy or energy of position. There are many forms of energy. Heat is energy that is transferred from one place to another because of a difference in temperature and can be measured in calories. Electromagnetic radiation is energy that travels in waves. The shorter the wavelength, the higher the energy of the radiation. Radiation of all wavelengths has many uses and can have both beneficial and harmful effects on living organisms.

Important Equations

Kinetic Energy $= \frac{1}{2} mv^2$

Specific Heat $= \frac{\text{cal}}{\text{g} \times {}^\circ\text{C}}$

Energy $= \frac{k}{\lambda}$

calories $=$ g $\times \Delta t \times$ specific heat

EXERCISES AND PROBLEMS

1. Which one in each of the following pairs has the greater kinetic energy?

   (a) 100 g of water or 100 g of steam
   (b) A car moving at 25 mph or at 50 mph
   (c) A football player or a sprinter, both running 100 yards in 12 seconds.

2. Which one in each of the following pairs has the greater potential energy?

   (a) Snow in the mountains or ocean water
   (b) A pendulum at the top or the bottom of its swing
   (c) A resting bow and arrow or a drawn bow and arrow

3. How many calories are there in:

   (a) 3.5 kilocalories
   (b) 125 food Calories
   (c) 0.01 kcal
   (d) 0.25 kcal
   (e) 146 J
   (f) 135 KJ

4. Calculate the number of calories and kilocalories transferred in each of the following sets of conditions:

	Grams of water	*Initial temperature*	*Final temperature*
(a)	25.0 g	20.0°C	27.0°C
(b)	50.0 g	24.0°C	32.0°C
(c)	1000 g	25.0°C	76.0°C
(d)	500 g	45.0°C	23.0°C

5. Which light contains more energy: orange light or yellow light?
6. List the following in order of (a) increasing energy and (b) increasing wavelength:

   X rays, short waves, ultraviolet light, yellow light, microwaves
7. Discuss the potential and kinetic energy of a skier as she:

   (a) stands in the lift line
   (b) travels to the top of the lift
   (c) stands at the top of the mountain
   (d) skis down the hill to the bottom
8. How much more kinetic energy would a car possess traveling at 40 mi/hr than it would traveling at 10 mi/hr?
9. Which has more kinetic energy: a 13.5-g rifle bullet traveling at 1000 km/hr or a 200-g baseball traveling at 93.2 mi/hr? What is the difference in their kinetic energy?
10. When four cashews are burned in a calorimeter, the temperature of the 1000 g of water changes from 25° to 69°C. How many kilocalories of energy are transferred from one cashew? How many food Calories?
11. How many kilocalories of energy are transferred from one gram of butter if burning 10 grams of butter raises the temperature of 1000 grams of water in a calorimeter from 20° to 90°C?
12. A man is completely immersed in a tub containing 60 liters of water. In one hour he raises the temperature of the water from 32.0°C to 33.5°C. How many kilocalories of energy did the man give off in one hour?
13. In which place would you probably get a more severe sunburn on a clear day: summer skiing in the Rocky Mountains or swimming at a Delaware beach? Explain your answer.
14. Ultraviolet light can be classed into two groups: UV-A with longer wavelengths and UV-B with shorter wavelengths. Which type of ultraviolet light do you think is more likely to cause skin cancer? Give a reason for your answer.

*15. In a nuclear power plant, only 40% of the energy released in the nuclear reaction is converted to electrical energy. Why doesn't this violate the First Law of Thermodynamics?

*16. Describe the changes in entropy of the molecules of a sugar cube as it dissolves in water.

---

* *Problems marked with an asterisk are from optional sections in the chapter.*

CHAPTER FOUR
THE THREE STATES OF MATTER

Cliff Brown was nearing retirement at the age of 65, but was starting to wonder if he'd ever be able to enjoy the fishing and hunting he'd been looking forward to. Over the past five years he had been increasingly bothered by a shortness of breath and continual coughing. He knew that his heavy cigarette smoking was to blame for many of these symptoms, but had never been able to cut down to less than two packs a day. But now his condition seemed to get worse—he was trying to recover from a bad cold, but felt extremely tired all the time and was finding it more and more difficult to breathe.

After seeing his doctor, Cliff was sent to the hospital to have a series of tests done on his pulmonary system (the parts of his body involved in lung function). Chest X rays showed that his lungs were enlarged and his diaphragm flattened. Other tests revealed that his ability to exhale air was well below normal, and that the movement of oxygen (O_2) into his blood and carbon dioxide (CO_2) out of the blood was not occurring at the proper rate. Blood tests verified this imbalance of oxygen and carbon dioxide in his blood. Oxygen in his blood was measured at an abnormally low tension, or pressure, of 40 mm Hg (normal is 80 to 100 mm Hg), and carbon dioxide was measured at a very high tension of 70 mm Hg (normal is 40 mm Hg). The diagnosis was clear: Cliff was suffering from emphysema, one of the chronic obstructive pulmonary diseases (COPD). Emphysema involves the deterioration of the tiny, bubblelike air sacs, called alveoli, in the lungs. These sacs make possible the transport of oxygen into the blood and carbon dioxide out of the blood. In emphysema the alveoli lose their ability to expand and contract. Eventually they break open, forming larger, inflexible sacs that cannot as easily exchange the oxygen and carbon dioxide (Figure 4.1).

Cliff was admitted to the hospital for treatment that would help ease the imbalance of these two gases in his blood and would reduce the amount of work required for him to breathe. This therapy provided Cliff with air to breathe that had a concentration of oxygen slightly higher than that found in normal room air. This extra oxygen was administered to Cliff at a very slow rate of 2 liters per minute because a much higher rate could actually have been quite dangerous for him. To understand this, you have to know that the brain keeps track of the concentration of carbon dioxide in the blood. High blood levels of carbon dioxide cause the brain to increase the rate of breathing. However, in an emphysema patient who has chronically high blood levels of carbon dioxide, the brain undergoes a change to become sensitive to blood O_2 levels instead of blood CO_2 levels. Low concentrations of oxygen in the blood of these persons will trigger more rapid respiration. Therefore, if oxygen is administered to an emphysema patient at a high rate, the blood oxygen level will rapidly increase and the brain will not receive a signal for the need to breathe. This means that the patient may actually stop breathing and could easily die.

Cliff remained in the hospital for a week. His upper respiratory infection was cleared up with antibiotics, and the oxygen tension in his blood was increased to an adequate level. At the end of the week Cliff was taken off the oxygen treatment and his blood gases were monitored for 48 hours. This check revealed that Cliff's disease had progressed so far that he was unable to maintain an adequate oxygen tension in his blood by his own breathing efforts. To allow Cliff to return to his home, he was supplied with a home oxygen unit containing liquid oxygen. This machine had long tubes that enabled Cliff to breathe oxygen as he moved around the house; he also had a portable unit that he could fill with oxygen and carry over his shoulder when he wanted to leave the house for several hours.

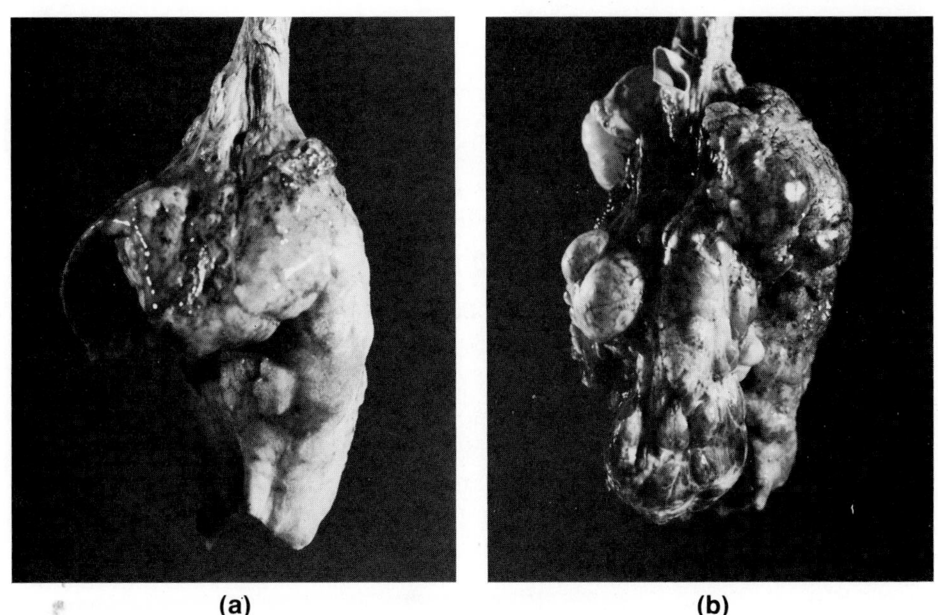

(a) (b)

Figure 4.1 (a) A normal lung. (b) The lung of a patient who suffered from emphysema. (Cuerden Advertising Design, Denver.)

You might wonder how the oxygen that Cliff needed to breathe could be stored as a liquid for his use. Liquid and gas are two of the three possible states, or forms, in which matter can exist. The other state, of course, is solid. In this chapter we will closely examine these different states of matter to see how they differ and how they can be converted from one to another. We will then be able to discuss some of the physical laws that describe the behavior of matter when it is in the gaseous state.

LEARNING OBJECTIVES

By the time you have finished this chapter, you should be able to:

1. State the difference between the three states of matter.
2. Define the *melting point* and the *boiling point* of a substance.
3. Describe the changes that occur at the molecular level as an ice cube is warmed from −5° to 110°C.
4. Define *heat of fusion* and *heat of vaporization.*
5. Describe the five properties of ideal gases stated by the kinetic–molecular theory.
6. List five units used to measure pressure.

7. State, in your own words, five laws of gas behavior and give everyday examples of each.
8. Perform calculations using Boyle's law, Charles' law, the general gas law, and Dalton's law.

STATES OF MATTER

4.1 Solids

As we have just stated, matter can be found in the form of a gas, a liquid, or a solid (Figure 4.2). These three states differ in the distance between, the attraction between, and the amount of movement of the particles making up the substance. In a solid, the attractive forces between the particles are relatively strong, and the particles are packed close together in a rigid structural arrangement, giving the solid a definite shape and volume. There are two types of solids, differing in the arrangement of particles that make them up. The particles of **amorphous solids,** such as ordinary glass and most plastics, are "frozen" in a disordered state or randomly arranged pattern—much like the arrangement of the particles in the liquid state. But the particles of **crystalline solids** form a three-dimensional structure in which particles are arranged in a regular, repeating pattern. Examples of crystalline solids are quartz, table salt, diamond, and snowflakes (Figure 4.3). Particles in a crystalline solid have specific positions and fixed neighboring particles, giving the solid a rigid structure. It is important to realize, however, that these particles are not totally motionless. They move back and forth and up and down, or vibrate around a fixed point (Figure 4.4).

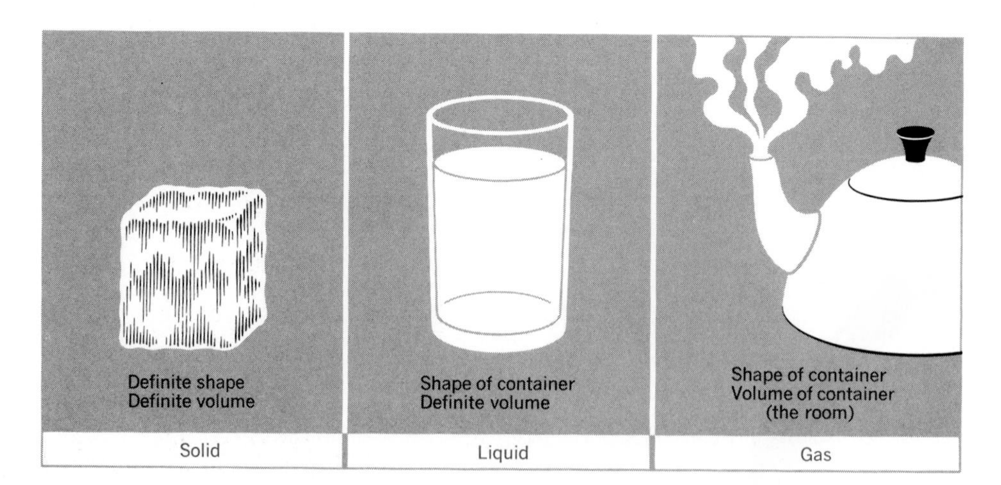

Figure 4.2 The three states of matter: water can exist as a solid, liquid, or gas.

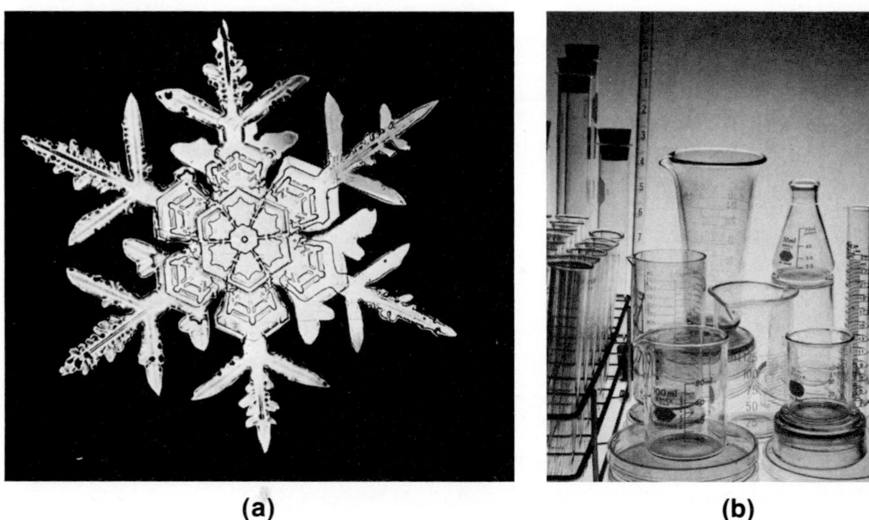

Figure 4.3 (a) The regular, intricate shapes of snowflakes result from the orderly arrangement of water molecules in ice crystals. (b) Glass can be cooled to form any shape because it is an amorphous solid whose particles are randomly arranged. (a, courtesy of The American Museum of Natural History; b, Runk/Schoenberger—Grant Heilman.)

When the temperature is low, these vibrations are not enough to overcome the attraction between the particles of the solid. But if energy is added in the form of heat, the kinetic energy of the particles increases. Their vibrations become more and more violent until the solid finally begins to break apart or melt. The specific temperature at which this occurs is called the **melting point** of the solid. The greater the attraction between the particles of a solid, the higher the melting point. When a crystalline solid has reached its melting point, additional heat energy does not go into increasing the average kinetic energy of the particles—that is,

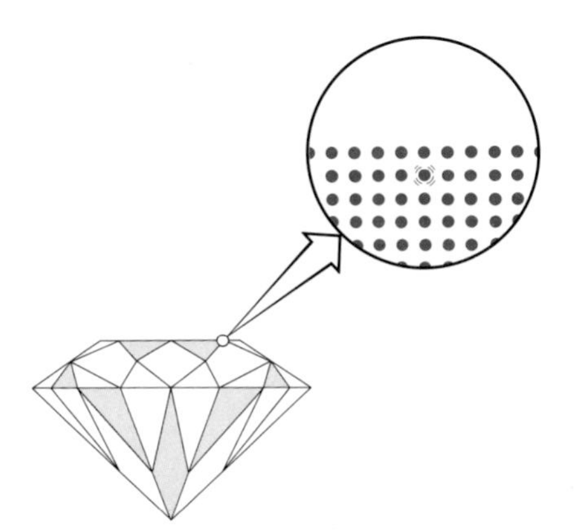

Figure 4.4 The particles in a solid are very closely packed together, but they are still able to vibrate around a fixed point.

into increasing the temperature—but rather goes entirely into melting the solid. The amount of energy that must be added to change a solid to a liquid at the melting point is called the **heat of fusion** (expressed in cal/g). The heat of fusion of water, for example, is 80 cal/g. (Note that this same amount of energy will be given off when the liquid again becomes a solid.) Only after the solid is completely melted will the temperature of the substance again begin to rise as additional heat energy is added (Figure 4.5).

4.2 Liquids

The particles in a liquid are not held together as tightly or as rigidly as they are in a solid. Although the particles are fairly close together, they can move from place to place by slipping past one another (Figure 4.6). This, for example, allows you to walk through a liquid but not a solid. A liquid can flow from one place to another and take on the shape of its container, even though the volume of the liquid does not change.

Although a liquid will flow, its particles are still strongly attracted to one another. The strength of this attraction determines two properties of a

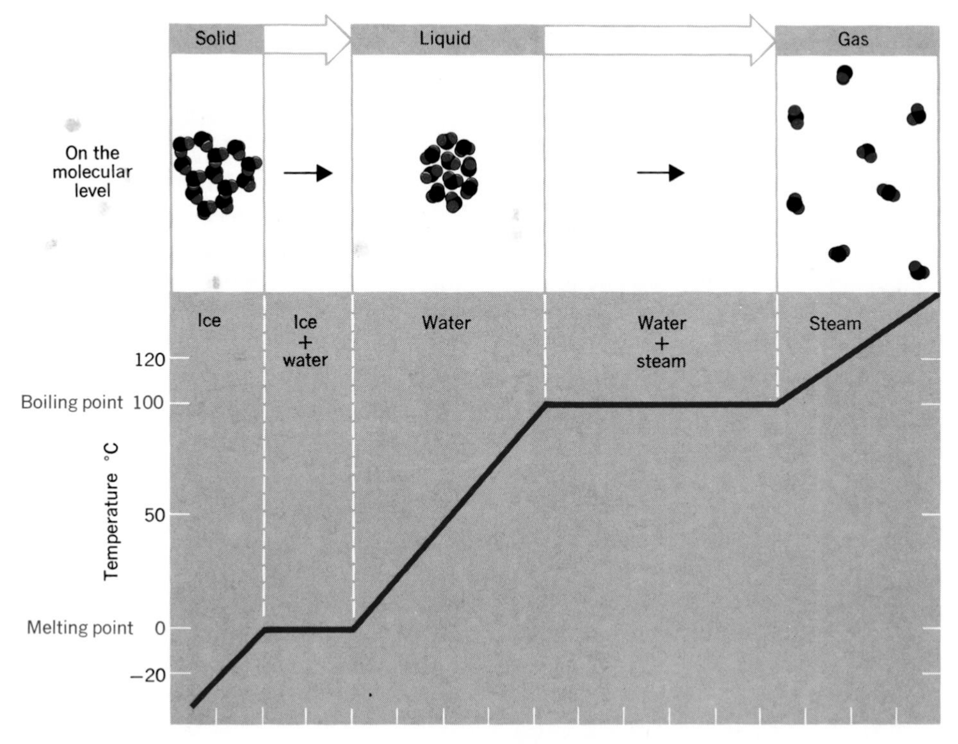

Figure 4.5 This heating curve shows the phase changes of water from solid to liquid to gas as heat is added at a constant rate.

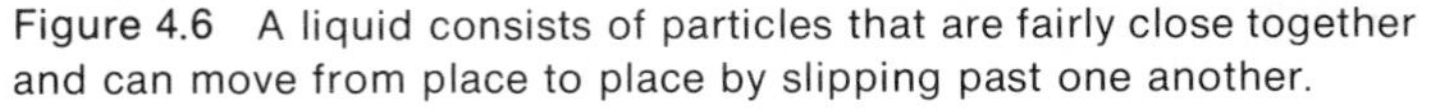

Figure 4.6 A liquid consists of particles that are fairly close together and can move from place to place by slipping past one another.

liquid: its viscosity and its surface tension. **Viscosity** is a measure of how easily the liquid flows. The higher the attraction between the particles of the liquid, the greater the viscosity or syrupy nature of the liquid. Molasses is more viscous than water, which in turn is more viscous than gasoline. **Surface tension** is the resistance of particles on the surface of the liquid to the expansion of the liquid (Figure 4.7). Water has a high surface tension, whereas gasoline has a low surface tension. Both the viscosity and surface tension of a liquid will decrease with an increase in temperature (that is, as the kinetic energy of the particles increases).

Figure 4.7 The high surface tension of water prevents the water molecules in this glass from spilling over the rim. (Photo by Stefan Bloomfield.)

Very energetic particles near the surface of a liquid can break away from the surface. If these particles don't collide with air molecules and return to the surface of the liquid, they will escape from the liquid completely. This is the process called **evaporation.** Because the very energetic particles are lost through evaporation, the average kinetic energy of the particles left behind will be lower—that is, the liquid will be cooler. This principle allows your body to rid itself of heat through the evaporation of sweat (Figure 4.8). Similarly, when alcohol is applied to the skin, the resulting cool feeling is caused by the rapid evaporation of alcohol molecules, leaving behind a cooler liquid on the skin. Note that as more heat energy is added to a liquid, the kinetic energy of the particles will increase, which will in turn increase the rate of evaporation.

Condensation is the opposite of evaporation. As gas particles are cooled, their kinetic energy decreases to a point at which the particles condense (or return to the liquid state). Good examples of condensation are the formation of dew on the grass, or water on the windshield of a car, on a chilly morning.

Adding more heat energy to a liquid will eventually increase the motion of the particles to the point that violent collisions will occur within the liquid, and bubbles of vapor (gas) will be formed. This is the process called **boiling.** The particular temperature at which boiling occurs under conditions of normal atmospheric pressure is called the **normal boiling point** of a substance. At this temperature, additional energy added to the substance goes into pulling the particles apart—that is, changing the liquid to a gas (see again Figure 4.5).

The amount of energy required to change a substance from a liquid to a gas at its boiling point is called the **heat of vaporization** (expressed in cal/g). The heat of vaporization of water, for example, is 539 cal/g. (Note again that this same amount of energy will be given off when the gas

Figure 4.8 The body is able to cool itself through the evaporation of sweat. (Phyllis Greenberg/Photo Researchers.)

condenses to become a liquid.) Only when all of the liquid has become a gas will the temperature of the substance again increase as more heat energy is added.

4.3 Gases and the Kinetic-Molecular Theory

Particles in the gaseous state are moving very rapidly (almost 1000 miles per hour at room temperature) in a completely random fashion (Figure 4.9). The physical properties of gases can be explained by the **kinetic-molecular theory of gases.** First, this theory states that gases are made up of very small particles with a great deal of space between them. This fact explains why it is easier to run through air than through water. Second, the particles of gas travel at high speeds in straight lines until they collide with one another or with the sides of the container. This explains why gas molecules will move throughout the container, taking on the shape of the container. Third, the collisions that occur are completely elastic; that is, a gas particle doesn't lose any energy in a collision and bounces off at its original speed. Fourth, there is no attraction or repulsion between the particles of a gas. And fifth, the kinetic energy of the gas particles changes with temperature. When heated, the particles move faster and faster. When cooled, the particles move slower and slower until, theoretically, a temperature is reached when all motion stops. This temperature, called **absolute zero,** is equal to −273.15°C or 0 K.

The kinetic-molecular theory describes the behavior of an *ideal* gas. Actually, real gases may not always behave as described, especially under conditions of very high pressure or very low temperature. But under less extreme conditions, real gases seem to behave like ideal gases.

4.4 Changes in State

A **change in state** occurs when matter goes from one state to another, such as when a solid melts or a liquid evaporates. Another change of

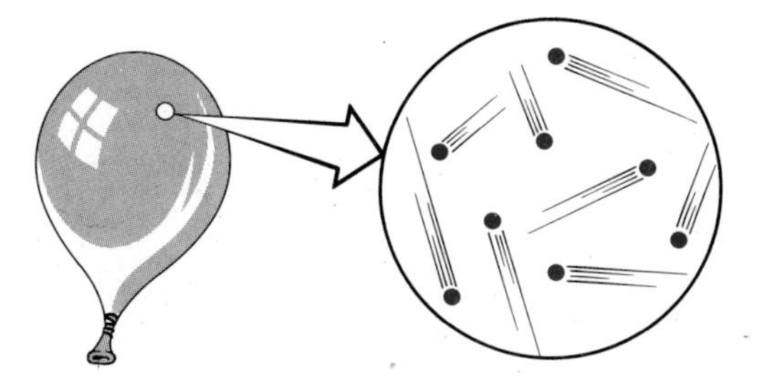

Figure 4.9 A gas consists of widely separated particles that are moving very rapidly in a random, chaotic fashion.

state that can occur is the process of **sublimation:** the change of a solid directly into a gas. If the particles of a solid are held together very weakly, the most energetic particles at the surface of the solid can evaporate, and such solids tend to sublime—to "vanish into thin air." Examples of substances that sublime are moth balls (naphthalene) and dry ice (solid carbon dioxide).

Changes in state are physical changes for which energy must either be added to make the process occur (as in melting or boiling), or will be given off as the process occurs (as in condensation or freezing). A change for which energy must be added if the process is to occur is said to be **endothermic.** A change in which energy is given off as the process occurs is called **exothermic.** Melting, therefore, is an endothermic process, and freezing an exothermic one.

LAWS OF GAS BEHAVIOR

A great deal of study has led scientists to formulate laws describing the behavior of gases. Although you may not know the precise mathematical statement of these laws, you certainly are familiar with their general form from everyday experience. For example, an aerosol can will explode when heated, and a tire may blow out when driven at high speeds on a very hot day (because increasing the temperature of a gas increases its pressure). When you push down on a bicycle pump, the plunger goes down in the air cylinder (because increasing the pressure on a gas decreases its volume). When you accidentally burn the cookies you were baking, the whole house soon smells like burned cookies (because a gas will diffuse to all regions of a container). When you open a bottle of carbonated beverage, you hear a hissing sound (because the solubility of a gas in a liquid depends upon the pressure).

4.5 Units of Pressure

Before we can discuss the exact mathematical forms of the laws mentioned above, we need to take a look at how pressure is measured. The pressure exerted by a gas is nothing more than the collisions of the billions of gas particles against the sides of the container. **Pressure** is defined as a force that is exerted per unit of area. The following are some of the units that are used to measure pressure.

Millimeters of Mercury

One of the oldest units of pressure is **millimeters of mercury (mm Hg).** This unit comes from the use of a column of mercury to measure atmospheric pressure (Figure 4.10). One millimeter of mercury (1 mm Hg) is the atmospheric pressure sufficient to support a column of mercury 1 millimeter in height.

Atmosphere

The height of the column of mercury that can be supported by atmospheric

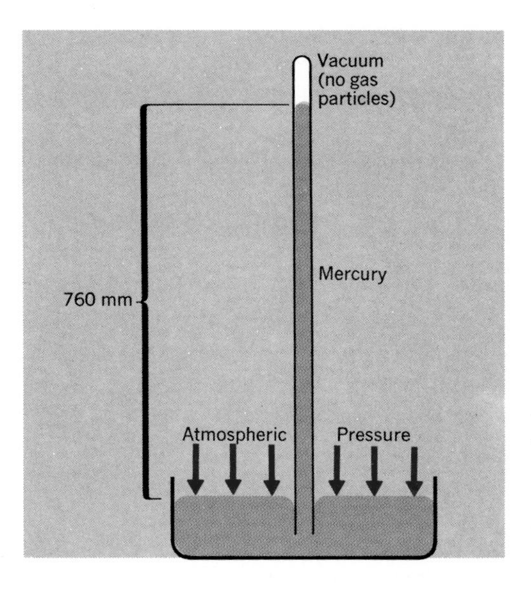

Figure 4.10 The mercury barometer is an instrument used to measure atmospheric pressure. The height of the column will vary from place to place and from time to time as the atmospheric pressure changes. At sea level the atmosphere will support a column of mercury about 760 mm high, but at an altitude of three and one-half miles the column would be only about 380 mm high.

pressure depends upon the temperature and the geographic location where the measurement is being made. **Standard atmospheric pressure** is the average pressure exerted by the earth's atmosphere at sea level and 0°C. One standard **atmosphere (atm)** will support a column of mercury 760 mm high at 0°C.

$$1 \text{ atm} = 760 \text{ mm Hg}$$

The conditions called "standard temperature and pressure (STP)" are 0°C and 1 atmosphere. In the English system, pressure is measured in pounds per square inch (psi). One atmosphere is equivalent to 14.7 pounds per square inch.

$$1 \text{ atm} = 14.7 \text{ psi}$$

Torr

The **torr (torr)** is a small unit of pressure, named in honor of the seventeenth century Italian physicist, Evangelista Torricelli, the inventor of the mercury barometer. One torr equals 1/760 atm. Therefore,

$$1 \text{ atm} = 760 \text{ torr} \quad \text{and} \quad 1 \text{ torr} = 1 \text{ mm Hg}$$

Pascal

The SI unit of pressure is the **pascal (Pa),** named in honor of the seventeenth century French scientist, Blaise Pascal. The relationship between pascal and torr is

$$1 \text{ torr} = 133.3 \text{ Pa}$$

Chemists find it convenient to use measurements in units of torr and atmospheres in their calculations, so these are the units we will use most often in this book.

EXAMPLE 4-1

1. How many torr equal 2.75 atmospheres?

   To solve this problem, we use the conversion factor 1 atm = 760 torr. Therefore,

$$2.75\ \cancel{\text{atm}} \times \frac{760\ \text{torr}}{1\ \cancel{\text{atm}}} = 2090\ \text{torr}$$

2. Assume that the barometer reading in your chemistry laboratory is 740 mm Hg. What is this pressure in torr and atmospheres?

   To solve these two problems, we use the following equalities:

   (a) 1 torr = 1 mm Hg

   Therefore,

$$740\ \cancel{\text{mm}}\ \text{Hg} \times \frac{1\ \text{torr}}{1\ \cancel{\text{mm}}\ \text{Hg}} = 740\ \text{torr}$$

   (b) 1 atm = 760 mm Hg

   Therefore,

$$740\ \cancel{\text{mm}}\ \text{Hg} \times \frac{1\ \text{atm}}{760\ \cancel{\text{mm}}\ \text{Hg}} = 0.974\ \text{atm}$$

Exercise 4-1

A Canadian weather broadcast reported the air pressure inside a cyclone to be 660 mm Hg. What is this pressure in atmospheres and torr?

4.6 Boyle's Law (the Relationship Between Pressure and Volume)

In the seventeenth century, the British chemist Robert Boyle discovered that the volume of a gas varies inversely with its pressure if the temperature is kept constant (Figure 4.11). In other words, if you increase the pressure, the volume decreases. Decrease the pressure and the volume increases. We can use the kinetic-molecular theory to explain Boyle's law. When the volume of a gas is decreased, there is less space for the particles to move around in, so they will collide with the sides of the container much more often. This will appear as an increase in pressure.

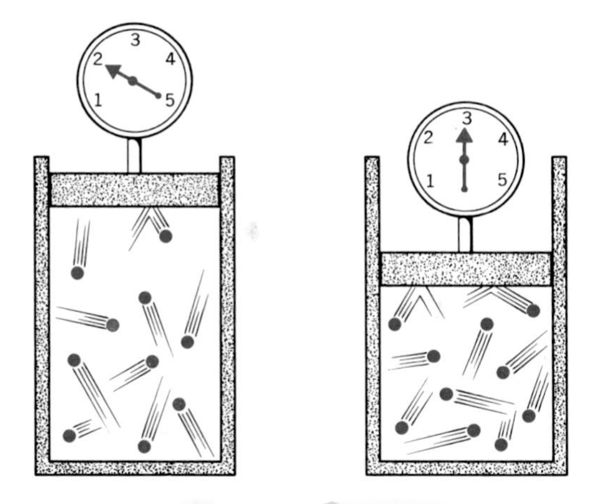

Figure 4.11 Boyle's law states that when the volume of a gas decreases, the pressure will increase (if the temperature of the gas remains constant).

An illustration of Boyle's law at work in our bodies is provided by examining how we breathe (Figure 4.12). Your lungs are located in the thoracic cavity, surrounded by the ribs and a muscular membrane called the diaphragm. To inhale, the diaphragm contracts and flattens out, increasing the volume of the thoracic cavity. This lowers the air pressure in the cavity below that of the atmosphere, causing air to flow into the lungs. To exhale, the diaphragm relaxes and pushes up into the thoracic

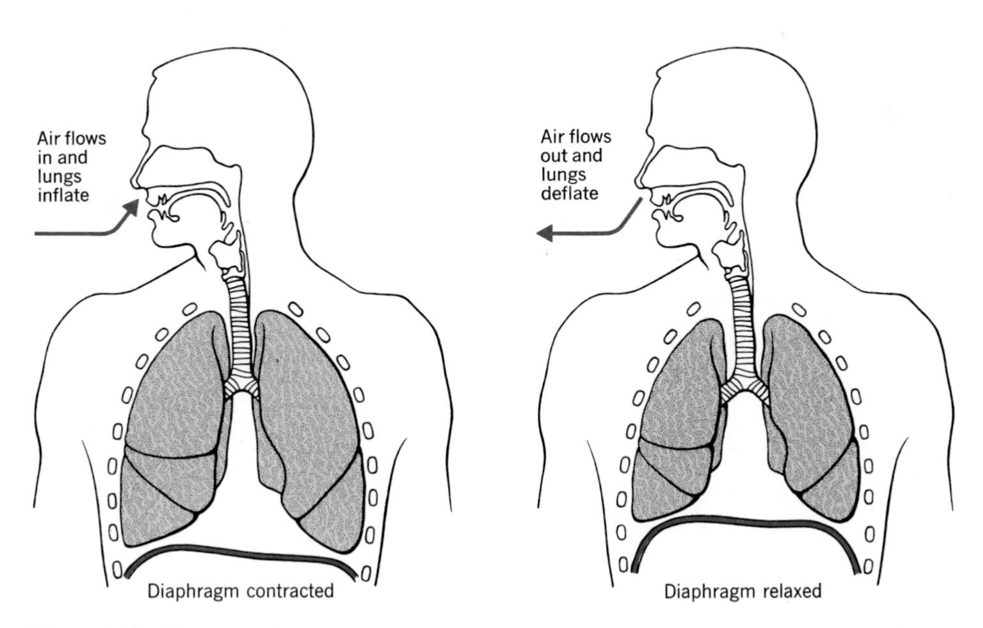

Figure 4.12 Breathing is an example of Boyle's law in action.

cavity, increasing the pressure above that of the atmosphere and causing air to flow out of the lungs.

We have just described Boyle's law in words, but this inverse relationship can also be stated mathematically as follows:

$$PV = k \qquad (k \text{ is a constant})$$

And, when given two conditions of pressure and volume,

$$\boxed{P_1V_1 = P_2V_2}$$

where P_1 and V_1 are the starting conditions of pressure and volume, and P_2 and V_2 are the final conditions. The following examples show ways to think through various types of problems, as well as how to use the equation for Boyle's law.

EXAMPLE 4-2

1. Imagine that a gas is inside a cylinder with a movable piston. If the volume of the gas is 3 liters when the pressure is 760 torr, what would the volume be if the pressure were increased to 1140 torr while the temperature was held constant?

   (a) The first step is to identify the quantities that are stated in the problem.

$$P_1 = 760 \text{ torr} \qquad V_1 = 3 \text{ liters}$$
$$P_2 = 1140 \text{ torr} \qquad V_2 = ?$$

   (b) Now let's think for a moment about the problem. Will the starting volume of 3 liters increase or decrease if the pressure is increased from 760 to 1140 torr? Boyle's law states that the volume of a gas will decrease if the pressure is increased. So we must multiply the starting volume by some fraction or ratio of the pressures that will decrease the volume.

$$3 \text{ liters} \times \frac{760 \cancel{\text{torr}}}{1140 \cancel{\text{torr}}} = 2 \text{ liters}$$

   (c) Now let's check our answer using the equation for Boyle's law:

$$P_1V_1 = P_2V_2$$
$$760 \text{ torr} \times 3 \text{ liters} = 1140 \text{ torr} \times V_2$$
$$\frac{760 \cancel{\text{torr}} \times 3 \text{ liters}}{1140 \cancel{\text{torr}}} = V_2$$
$$2 \text{ liters} = V_2$$

2. Suppose that a 2.5-liter tank contains oxygen at a pressure of 44 atm. What pressure would this amount of oxygen exert in a 55-liter tank at the same temperature?

(a) Identify the given quantities:

$$V_1 = 2.5 \text{ liters} \qquad P_1 = 44 \text{ atm}$$
$$V_2 = 55 \text{ liters} \qquad P_2 = ?$$

(b) Reason the problem through. Will the starting pressure increase or decrease if the volume is increased from 2.5 to 55 liters? We know from Boyle's law that if the volume is increased, the pressure will decrease. So we must multiply the starting pressure by a ratio of volumes that will decrease the pressure.

$$44 \text{ atm} \times \frac{2.5 \text{ liters}}{55 \text{ liters}} = 2.0 \text{ atm}$$

(c) Or, using the Boyle's law equation, we have

$$44 \text{ atm} \times 2.5 \text{ liters} = P_2 \times 55 \text{ liters}$$

$$\frac{44 \text{ atm} \times 2.5 \text{ liters}}{55 \text{ liters}} = P_2$$

$$2.0 \text{ atm} = P_2$$

Exercise 4-2

1. Suppose that the volume of a balloon is 3.50 liters in New York City when the atmospheric pressure measures 760 torr. What is the atmospheric pressure in Mexico City (which is about 7000 feet above sea level) if the same balloon has a volume of 4.43 liters there? (Assume that the temperature is the same in both cities.)
2. A sample of oxygen has a volume of 840 ml at a pressure of 800 torr. What would be the volume of this sample if the pressure were reduced to standard atmospheric pressure while the temperature was held constant?

4.7 Charles' Law (the Relationship Between Volume and Temperature)

In the early nineteenth century, the French physicist Jacques Charles discovered that the volume of a gas varies directly with Kelvin temperature when the pressure is constant. That is, when the temperature of a gas increases, the volume increases (Figure 4.13). On the molecular level, increasing the temperature means increasing the kinetic energy of the gas particles. Because these particles will now be moving faster, they will collide with the sides of the container much more often. The only way that the pressure can remain constant under these circumstances is to increase the volume of the container. Similarly, when gas particles are cooled, they slow down. In order for the pressure to remain constant, the volume must decrease so that the gas particles can hit the sides of the container just as often as before.

Charles' law can be expressed mathematically by the following direct relationship:

$$V = kT \quad (k \text{ is a constant})$$

And, when given two conditions of volume and temperature,

$$\boxed{\frac{V_1}{T_1} = \frac{V_2}{T_2}}$$

where V_1 and T_1 are the starting conditions of volume and temperature (in degrees Kelvin), and V_2 and T_2 are the final conditions.

EXAMPLE 4-3

1. Imagine that a gas occupies a volume of 2.0 liters at a temperature of 27°C. To what temperature in degrees Celsius must the gas be cooled to reduce its volume to 1.5 liters if the pressure is held constant?

   (a) Identify the known quantities:

   $V_1 = 2.0$ liters $T_1 = 300$ K (remember that the temperature *must* be in degrees Kelvin, K = °C + 273)

   $V_2 = 1.5$ liters $T_2 = ?$

   (b) Reason the problem through. From Charles' law we know that to decrease the volume of a gas the temperature must also decrease. So we must multiply the starting temperature by a ratio of the volumes that will decrease the temperature.

   $$300 \text{ K} \times \frac{1.5 \text{ liters}}{2.0 \text{ liters}} = 225 \text{ K}$$

   The answer in degrees Celsius would be

   $$225 = {}^\circ\text{C} + 273$$
   $${}^\circ\text{C} = 225 - 273 = -48^\circ\text{C}$$

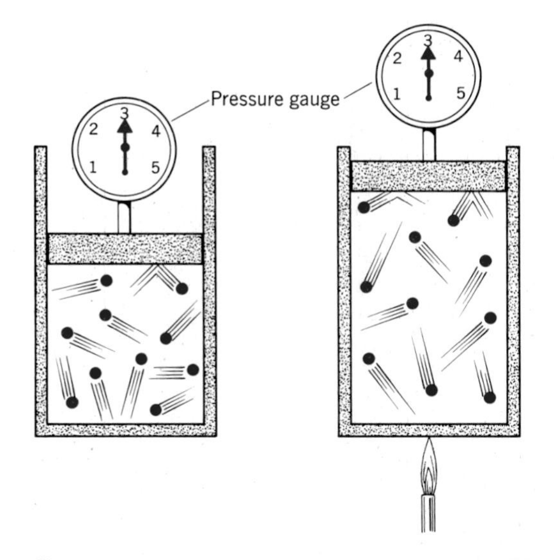

Figure 4.13 Charles' law states that when the temperature of a gas is increased, the volume of the gas will increase (if the pressure remains constant).

(c) Or use the Charles' law equation

$$\frac{2.0 \text{ liters}}{300 \text{ K}} = \frac{1.5 \text{ liters}}{T_2}$$

$$2.0 \text{ liters} \times T_2 = 1.5 \text{ liters} \times 300 \text{ K}$$

$$T_2 = \frac{1.5 \text{ liters} \times 300 \text{ K}}{2.0 \text{ liters}} = 225 \text{ K} = -48°\text{C}$$

2. Imagine that you are skiing on a beautiful, clear day with a temperature outside of −3°C. The cold air that you breathe is warmed to a body temperature of 37°C as it travels to your lungs. If you inhaled 400 ml of air at −3°C, what volume would it occupy in your lungs? (Assume that the pressure is constant.)

(a) Identify the known quantities:

$T_1 = -3 + 273 = 270 \text{ K}$ $\quad V_1 = 400 \text{ ml}$
$T_2 = 37 + 273 = 310\text{K}$ $\quad V_2 = ?$

(b) Reason the problem through. The temperature of the gas is increasing as it enters your lungs, so from Charles' law we know that the volume must also increase. Therefore, we must multiply the starting volume by a ratio of temperatures that will increase the volume.

$$400 \text{ ml} \times \frac{310 \not{K}}{270 \not{K}} = 459 \text{ ml}$$

(c) Or, using the Charles' law equation, we obtain

$$\frac{400 \text{ ml}}{270 \text{ K}} = \frac{V_2}{310 \text{ K}}$$

$$\frac{400 \text{ ml} \times 310 \not{K}}{270 \not{K}} = V_2$$

$$459 \text{ ml} = V_2$$

Exercise 4-3

1. Suppose that a balloon having a volume of 2.0 liters at room temperature (20°C) is taken outside on a warm summer day. What is the temperature (in °C) outside if the volume of the balloon outside becomes 2.1 liters? (Assume that the pressure remains constant.)
2. An anesthesiologist administers a gas at 20°C to a patient whose body temperature is 37°C. What is the change in volume in milliliters of a 1.20-liter sample of gas as it goes from room temperature to body temperature? (Assume that the pressure remains constant.)

4.8 General Gas Law

We can combine the laws stated by Charles and Boyle to write a general mathematical equation that will allow us to predict gas behavior under various different conditions of pressure, temperature, and volume:

$$\boxed{\frac{P_1V_1}{T_1} = \frac{P_2V_2}{T_2}}$$

EXAMPLE 4-4

A cylinder contains 250 ml of carbon dioxide at 27°C and 810 torr. What volume would the carbon dioxide occupy at standard temperature and pressure (1 atm, or 760 torr, and 0°C)?

(a) Identify all the known quantities:

$P_1 = 810$ torr $\quad T_1 = 27 + 273 = 300$ K $\quad V_1 = 250$ ml
$P_2 = 760$ torr $\quad T_2 = 0 + 273 = 273$ K $\quad V_2 = ?$

(b) We can solve this problem by reasoning in two steps. First, notice that the pressure is decreasing. From Boyle's law we know that this should make the volume increase. So we multiply the volume by a ratio of pressures that would increase the volume. Second, notice that the temperature is decreasing. From Charles' law we know that a decrease in temperature causes the volume to decrease. So the ratio of temperatures to use is the one that would decrease the volume.

$$250 \text{ ml} \times \frac{810 \text{ torr}}{760 \text{ torr}} \times \frac{273 \text{ K}}{300 \text{ K}} = 242 \text{ ml}$$

(c) Or, using the combined gas law equation, we have

$$\frac{810 \text{ torr} \times 250 \text{ ml}}{300 \text{ K}} = \frac{760 \text{ torr} \times V_2}{273 \text{ K}}$$

$$\frac{810 \text{ torr} \times 250 \text{ ml} \times 273 \text{ K}}{300 \text{ K} \times 760 \text{ torr}} = V_2$$

$$242 \text{ ml} = V_2$$

Exercise 4-4

Suppose that a 40.0-g sample of gas occupies a volume of 8.00 liters at standard temperature and pressure.

(a) What volume would this gas occupy at 20.0°C and 860 torr?
(b) At what temperature would the volume decrease by one-half if the pressure were 1.50 atm?
(c) What would be the pressure of the gas in torr if the temperature decreased to −15.0°C as the volume decreased to 6.50 liters?

Another way to combine the various relationships we've been discussing is given by the equation of state of an ideal gas, also known as the ideal gas law:

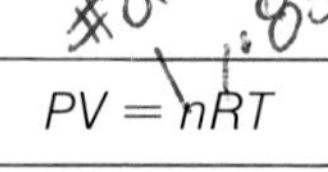

$$PV = nRT$$

where R is a number called the universal gas constant, and n is the number of gas particles expressed in a unit called the mole (to be discussed in Chapter 9).

4.9 Graham's Law of Effusion

In the nineteenth century the Scottish chemist Thomas Graham noticed that when the temperature and pressure are held constant, different gases effuse at different rates. **Effusion** is the movement of a gas through a tiny hole to a region of lower pressure (Figure 4.14). (Don't confuse effusion with diffusion: diffusion of a gas is its ability to mix spontaneously with other gases and to spread throughout a container.) Graham observed that lighter (less dense) gases effuse more rapidly than heavier (more dense) gases. Mathematically, **Graham's law** states that a gas effuses at a rate inversely proportional to the square root of its density when the pressure and temperature of the gas are held constant.

$$\text{Effusion rate} = \frac{k}{\sqrt{\text{density}}}$$

where k is a constant of proportionality. To compare the rate of effusion of gas A with gas B, we use the following relationship:

$$\frac{\text{Effusion rate (gas } A)}{\text{Effusion rate (gas } B)} = \sqrt{\frac{\text{density } (B)}{\text{density } (A)}}$$

For example, if the density of one gas was four times the density of another gas, the less dense gas would effuse twice as fast. Mathematically,

$$\text{Let the density of gas } A = 1$$
$$\text{Let the density of gas } B = 4$$

Then,

$$\frac{\text{Effusion rate } (A)}{\text{Effusion rate } (B)} = \sqrt{\frac{4}{1}} = 2$$

4.10 Henry's Law

When you open a bottle of carbonated beverage and listen to the hiss of the escaping carbon dioxide, you are observing Henry's law in action. At the beginning of the nineteenth century the English chemist William Henry

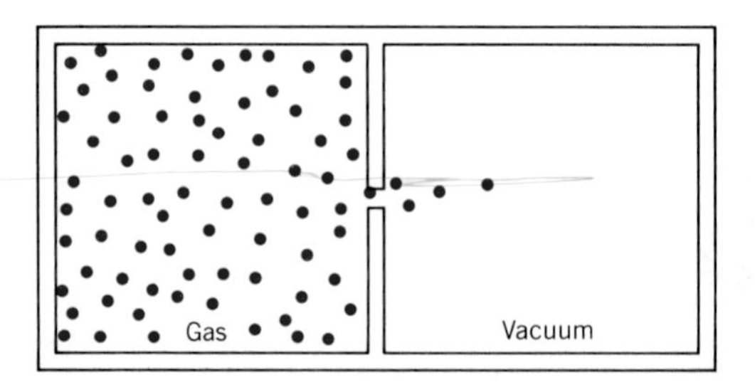

Figure 4.14 Effusion is the movement of a gas through a small opening to a region of lower pressure.

discovered that the solubility of a gas in a liquid at a given temperature is directly proportional to the pressure of that gas on the liquid. That is, the higher the pressure, the more gas that will dissolve in a liquid when the temperature does not change. Carbonated beverages are bottled under high pressure, and when you open the cap, you reduce the pressure in the bottle. This, then, lowers the solubility of the carbon dioxide and permits the gas to escape from the beverage.

At the beginning of this chapter, we described a person suffering from emphysema. Such patients can benefit from an application of Henry's law. The amount of oxygen dissolved in their blood increases when these patients breathe air containing higher than normal amounts of oxygen and, therefore, having a higher oxygen pressure.

Another example of a change in solubility which is caused by a change in pressure is the bends, a disorder that occurs when deep-sea divers are brought to the surface too quickly. Deep-sea divers breathe air (mainly nitrogen and oxygen) under high pressure, which increases the solubility of the nitrogen in the blood. If the diver is brought to the surface too quickly, the sudden drop in pressure causes the dissolved nitrogen to leave the blood and form tiny bubbles (air emboli), which cause severe pain in the arms, legs, and joints. The only cure for the bends is slow decompression, which allows the dissolved nitrogen to escape slowly without forming bubbles (Figure 4.15). Divers who must work for long periods at great pressure breathe a mixture of helium and oxygen instead of air. Helium, like nitrogen, has no dangerous effects on the body and has the advantage of being only about 40% as soluble in the blood as nitrogen. Also, because helium is a much lighter gas than nitrogen, helium will diffuse from the bloodstream into the lungs at a higher rate than will

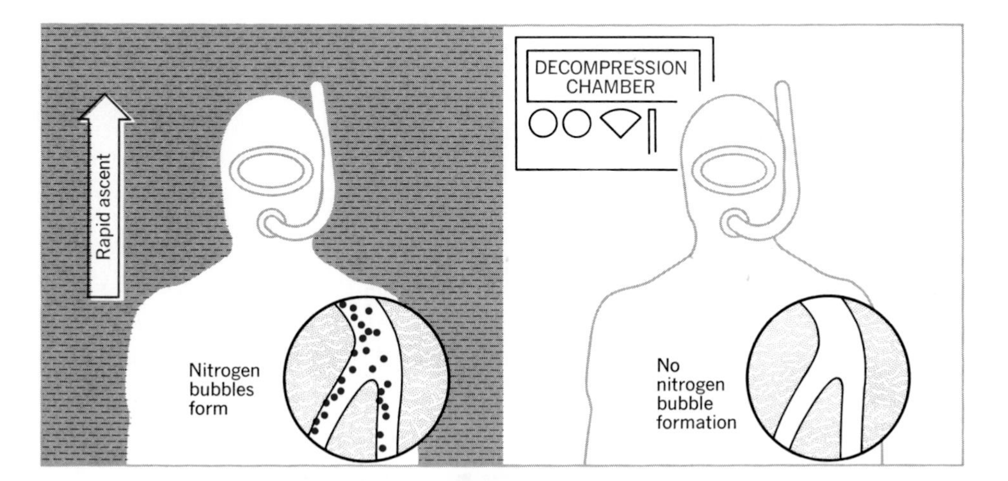

Figure 4.15 The bends is a disorder resulting from a change in the solubility of nitrogen in the blood caused by a rapid change in pressure.

nitrogen. As a result, the use of helium rather than nitrogen greatly reduces the risk of the bends.

4.11 Dalton's Law of Partial Pressures

We have seen that the pressure of a gas comes from the collisions of the gas particles against the walls of the container. There are two ways we can increase the frequency of collisions and, therefore, increase the pressure. One way is to increase the temperature of the gas. This will increase the kinetic energy of the gas particles and the number of collisions that occur. The second way is to increase the number of particles of gas in the container.

The pressure exerted by a gas at constant temperature does not depend on the type of gas particles, but instead depends directly on the number of particles of gas that are present. For example, we can double the pressure in a container by adding an equal number of particles of the same gas or of a different gas (Figure 4.16). In a mixture of gases, the pressure exerted by each gas is called the **partial pressure, P,** of that gas and depends only upon the number of particles of that gas present. **Dalton's law** states that the total pressure of a mixture of gases is equal to the sum of the partial pressures of each of the gases in the mixture. For a mixture of four gases, *A, B, C,* and *D,*

$$P_{\text{total}} = P_A + P_B + P_C + P_D \tag{1}$$

The earth's atmosphere is a mixture of nitrogen, oxygen, argon, and other gases found in small amounts. By Dalton's law then,

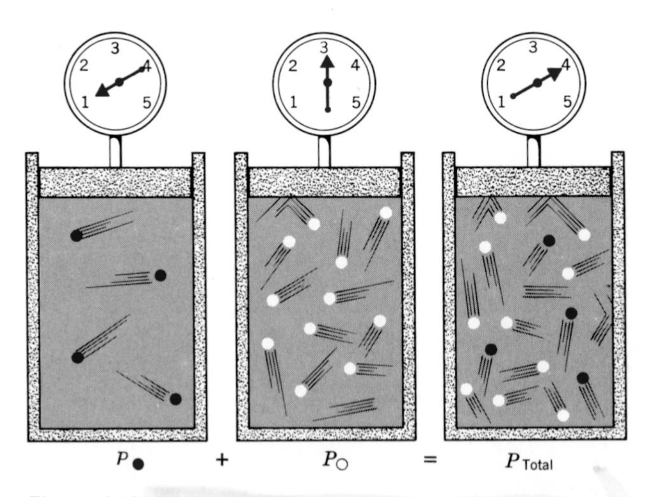

Figure 4.16 Dalton's law states that the total pressure in a container will equal the sum of the partial pressures of the gases in that container.

$$P_{\text{atmosphere}} = P_{N_2} + P_{O_2} + P_{Ar} + P_{\text{other}} \qquad (2)$$

When gases are prepared in a laboratory, they are often collected over water in an apparatus such as the one shown in Figure 4.17. The bottle will contain both the gas being collected and water vapor, which are together called a wet gas. Water vapor will exert a pressure just like any other gas, and the partial pressure of the water vapor in the bottle is called the **vapor pressure.** The amount of water vapor that is present and, therefore, the vapor pressure depend only upon the temperature of the water. As you can see from Table 4.1, the vapor pressure of water increases as the temperature increases. The total pressure of the wet gas in the jar (which equals atmospheric pressure if the water levels are equal) is the sum of the partial pressure of the gas collected and the vapor pressure of water.

$$P_{\text{wet gas}} = P_{\text{dry gas}} + P_{\text{water}}$$

EXAMPLE 4-5

1. What is the partial pressure of oxygen in the air if $P_{N_2} = 593.0$ torr, $P_{Ar} = 7.0$ torr, and $P_{\text{other}} = 0.2$ torr, when the atmospheric pressure is measured at 760.0 torr?

   Substituting into eq. (2), we get

   $$760.0 \text{ torr} = 593.0 \text{ torr} + P_{O_2} + 7.0 \text{ torr} + 0.2 \text{ torr}$$

   So that,

   $$P_{O_2} = 760.0 \text{ torr} - 600.2 \text{ torr}$$
   $$P_{O_2} = 159.8 \text{ torr}$$

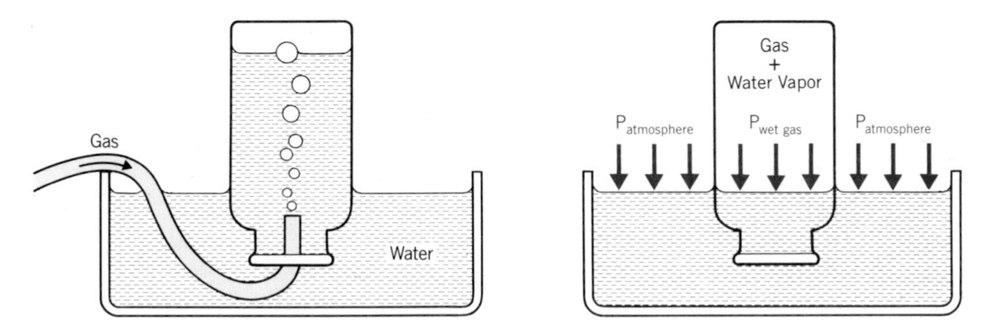

Figure 4.17 When a gas is collected over water, the bottle will contain both gas and water vapor. The total pressure of the wet gas in the bottle is the sum of the pressure of the dry gas and the pressure of the water vapor.

Table 4.1 Vapor Pressure of Water

Temperature (°C)	Vapor Pressure (torr)	Temperature (°C)	Vapor Pressure (torr)
0	4.6	40	55.3
10	9.2	50	92.5
15	12.8	60	149.4
20	17.5	70	233.7
25	23.8	80	355.1
30	31.8	90	525.8
35	42.2	100	760.0

2. A cylinder contains a mixture of oxygen and nitrous oxide (N_2O) which is used as an anesthetic. The pressure gauge on the tank reads 1.20 atm. If the partial pressure of the oxygen is 137 torr, what is the partial pressure of the nitrous oxide?

The total pressure in torr can be determined using the relationship 1 atm = 760 torr. Thus,

$$1.20 \text{ atm} \times \frac{760 \text{ torr}}{1 \text{ atm}} = 912 \text{ torr}$$

Using Dalton's Law, we obtain

$$P_{N_2O} + P_{O_2} = P_{total}$$
$$P_{N_2O} + 137 \text{ torr} = 912 \text{ torr}$$
$$P_{N_2O} = 775 \text{ torr}$$

Exercise 4-5

A student collects oxygen gas at 25°C by displacement of water until the levels of the water inside and outside the container are equal (Figure 4.17). If we assume that the barometer reads 755 torr, what is the partial pressure of the oxygen gas in the container?

4.12 The Diffusion of Respiratory Gases

The diffusion of respiratory gases (oxygen and carbon dioxide) within our bodies is directly related to the partial pressures of these gases (Figure 4.18). These gases will diffuse from a region of higher partial pressure to a region of lower pressure. For example, venous blood entering the lungs from the tissues has been depleted of its oxygen supply and is carrying the waste product carbon dioxide from the cells. Oxygen will diffuse from the lungs ($P_{O_2} = 104$ torr) to the blood ($P_{O_2} = 40$ torr), and carbon dioxide will diffuse from the blood ($P_{CO_2} = 45$ torr) to the lungs ($P_{CO_2} = 40$ torr) to be exhaled. The oxygen, most of which is held by the carrier molecule hemoglobin, is then transported by the arterial blood to the tissues. Tissue cells are constantly using oxygen, so the partial pressure of oxygen (or the oxygen tension) in the cells is low. Because the arterial blood has a higher oxygen tension, the oxygen diffuses from the blood ($P_{O_2} = 95$ torr) to the tissues ($P_{O_2} = 35$ torr). Carbon dioxide, which is produced in the tissues by the cells, will diffuse from the cells ($P_{CO_2} = 50$ torr) to the bloodstream ($P_{CO_2} = 45$ torr) to be carried back to the lungs.

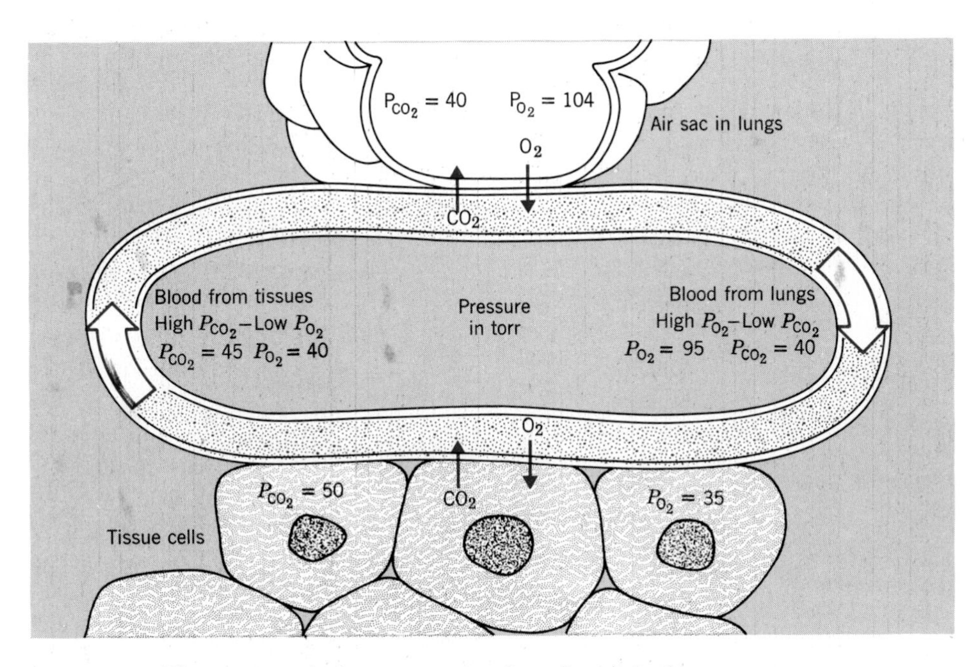

Figure 4.18 The movement of oxygen and carbon dioxide in the blood and tissues depends upon the partial pressure of each gas.

CHAPTER SUMMARY

Matter can exist in three states: solid, liquid, and gas. Solids have a rigid structure, with definite shape and volume. Liquids have a fixed volume, but take on the shape of their container. Gases have no fixed volume or shape. Adding energy to a solid increases the kinetic energy of the particles until the solid structure breaks down or melts. The temperature at which the solid melts to form a liquid is called the melting point, and the amount of energy that must be added to convert one gram of the substance from solid to liquid (at the melting point) is called the heat of fusion. Adding energy to a liquid will increase the kinetic energy of the liquid particles until the liquid boils. The temperature at which this occurs under conditions of normal atmospheric pressure is called the boiling point. The amount of energy needed to convert a gram of liquid to gas (at the boiling point) is called the heat of vaporization.

The kinetic-molecular theory of gases describes the physical properties of an ideal gas as follows:

1. Gases are made up of very small particles, with a great deal of space between them.
2. The particles travel at high speeds in straight lines.
3. The collisions of the particles with each other and with the container are completely elastic.
4. There is no attraction or repulsion between the particles.
5. The kinetic energy of the particles increases with temperature.

Real gases behave as ideal gases except under conditions of very high pressure or very low temperature. The pressure of a gas depends upon the number of particles of gas in the container, not the nature of the particles. The partial pressure of a gas is the pressure that the gas would exert if it were the only gas in the container.

Boyle's law states that if the pressure of a gas is increased, the volume will decrease when the temperature is constant. Charles' law states that the volume of a gas will increase when the temperature increases, if the pressure remains constant. Graham's law states that the lighter the gas, the faster the rate of effusion. Henry's law says that the higher the pressure of a gas over a liquid, the greater the solubility of the gas in the liquid. Dalton's law states that the total pressure of the gases in a container is equal to the sum of the partial pressures of each gas.

Important Equations

Boyle's Law: $P_1V_1 = P_2V_2$

Charles' Law: $\frac{V_1}{T_1} = \frac{V_2}{T_2}$

General Gas Law: $\frac{P_1V_1}{T_1} = \frac{P_2V_2}{T_2}$

Ideal Gas Law: $PV = nRT$

Graham's Law: $\frac{\text{Effusion }(A)}{\text{Effusion }(B)} = \sqrt{\frac{\text{density }(B)}{\text{density }(A)}}$

Dalton's Law: $P_{\text{Total}} = P_A + P_B + P_C + P_D$

EXERCISES AND PROBLEMS

EXERCISES AND PROBLEMS

1. Complete the following conversions:

   (a) 5.0 atm = (?) mm Hg
   (b) 190 mm Hg = (?) atm
   (c) 70 torr = (?) mm Hg
   (d) 2.7 atm = (?) torr
   (e) 63 torr = (?) Pa
   (f) 0.9 atm = (?) kPa
   (g) 737 torr = (?) psi
   (h) 890 torr = (?) atm

2. Ethanol, or ethyl alcohol, has a melting point of −117°C and a boiling point of 78.5°C. Draw a heating curve for ethanol (similar to Figure 4.5) and label the following:

   (a) Melting point
   (b) Boiling point
   (c) Region A—Ethanol is a gas
   (d) Region B—Ethanol is a liquid
   (e) Region C—Ethanol is a solid
   (f) Region D—Ethanol exists as solid and liquid
   (g) Region E—Ethanol exists as liquid and gas

3. Describe in your own words what happens on the molecular level when water is cooled from 110° to −5°C.

4. How much energy will be given off when 5 grams of water are frozen at 0°C?

5. At room temperature ammonia is a gas, water is a liquid, and sugar is a solid. Which substance has the strongest attraction between its particles? Which has the weakest?

6. Which substance in each of the following pairs is more viscous?

   (a) Motor oil or gasoline
   (b) Water or honey
   (c) Water at 10°C or water at 70°C

7. State in your own words the kinetic-molecular theory of gas behavior.

8. Propose a reason why real gases do not obey the kinetic–molecular theory under conditions of very low temperature.

9. State each of the following laws in your own words and then write

them in the form of an equation. Tell what conditions or variables are held constant for each law.

(a) Boyle's Law
(b) Charles' Law
(c) Graham's Law
(d) Henry's Law
(e) Dalton's Law

10. Everyone has observed that if the sun comes out immediately after a rain shower, the puddles dry up much faster than if it remains cloudy. Explain, on the molecular level, why this occurs.

11. Why does splashing your face with water cool you on a hot day?

12. Which would cause a more severe burn: water at 100°C or steam at 100°C? Explain your answer.

13. If there is a danger of a frost in late spring, grape growers will sprinkle their vineyards with water to protect their grapes from freezing. Suggest a reason why this is effective.

14. Rooms that must be kept sterile are often kept at a pressure slightly higher than atmospheric pressure. Explain how this procedure helps to keep out dust and microorganisms.

15. When the barometer reads 720 torr, a sample of oxygen occupies a volume of 250 ml. What volume will the sample occupy when the barometer reads 750 torr and the temperature remains constant?

16. A 1.0-liter sample of gas is collected under a pressure of 912 torr and 0°C. What will the volume of the gas be at standard temperature and pressure?

17. A diver collecting samples at a depth of 100 meters exhales a bubble having a volume of 100 ml. The pressure at this depth is 11 atm. What will the volume of the bubble be when it reaches the surface of the ocean (if we assume the water temperature is constant)?

18. If a sample of nitrogen occupies 505 ml at −23°C, what volume will it occupy at 23°C if the pressure is held constant?

19. If a sample of ammonia occupies a volume of 1.2 liters at 45°C, to what temperature in °C must the gas be lowered to reduce its volume to 1.0 liter when the pressure is held constant?

20. What volume will a gas occupy at 27°C and 760 mm Hg, if it occupies 2.8 liters at 30°C and 909 mm Hg?

21. When collected under the conditions of 27°C and 800 mm Hg, a gas occupies 400 ml. To what temperature must the gas be cooled to reduce its volume to 320 ml when the pressure falls to 720 mm Hg?

22. A 415-ml sample of gas is collected at STP.
    (a) Calculate the volume the gas would occupy under the following new conditions:
        (1) 3.20 atm and 0.00°C
        (2) 760 torr and −23.0°C
        (3) 400 torr and 100°C

(b) Calculate the pressure of the gas under the following new conditions:
    (1) 1.25 liters and 0.00°C
    (2) 415 ml and 45.0°C
    (3) 138 ml and −100°C
(c) Calculate the temperature of the gas under the following new conditions:
    (1) 830 ml and 1 atm
    (2) 0.415 liter and 404 kPa
    (3) 2.49 liters and 507 torr

23. Helium has a density of 0.00016 g/cc, and air has a density of 0.0012 g/cc at 25°C and 1 atm. If two balloons, one filled with helium and the other with an equal amount of air, have a hole of the same size poked in each of them, which balloon will deflate faster? Give a reason for your answer.

24. Assume that you have two flasks containing equal amounts of water into which equal amounts of carbon dioxide have been introduced. The pressure in one container is 1 atm and in the other container it is 2.5 atm. Which container would have more carbon dioxide dissolved in the water? Give the reason for your answer.

25. At the Children's Hospital in Boston, doctors performed open-heart surgery on a one-day-old baby in a hyperbaric chamber, which simulates conditions equivalent to 66 ft below sea level. Performing the operation in the hyperbaric chamber gave the doctors two minutes to do a critical part of the surgery without risk of brain damage to the baby, whereas they would have had only one minute under normal atmospheric conditions. Explain why.

26. The pressure in a bottle containing a mixture of oxygen and nitrogen is 1.00 atm. The partial pressure of the oxygen is 76 mm Hg. What is the partial pressure of the nitrogen?

27. A cyclopropane-oxygen mixture can be used as an anesthetic. If the partial pressure of cyclopropane is 255 torr and the partial pressure of oxygen is 855 torr, (a) What is the total pressure in the tank? (b) What is the total pressure expressed in atmospheres? (c) Which gas has a greater number of molecules in the tank?

28. Exhaled breath is a mixture of nitrogen, oxygen, carbon dioxide, and water vapor. What is the vapor pressure of the water in exhaled breath at 37°C (body temperature) if the partial pressure of the oxygen is 116 torr, of the nitrogen 569 torr, and of the carbon dioxide 28 torr? (Assume that the atmospheric pressure is 1 atm.)

29. A chemistry student collected a 250-ml sample of methane over water at 20.0°C.

    (a) What is the partial pressure of the methane if the barometer reading in the classroom is 756.2 torr?
    (b) Calculate the volume that the sample of dry methane (the methane without the water vapor) would occupy at STP.

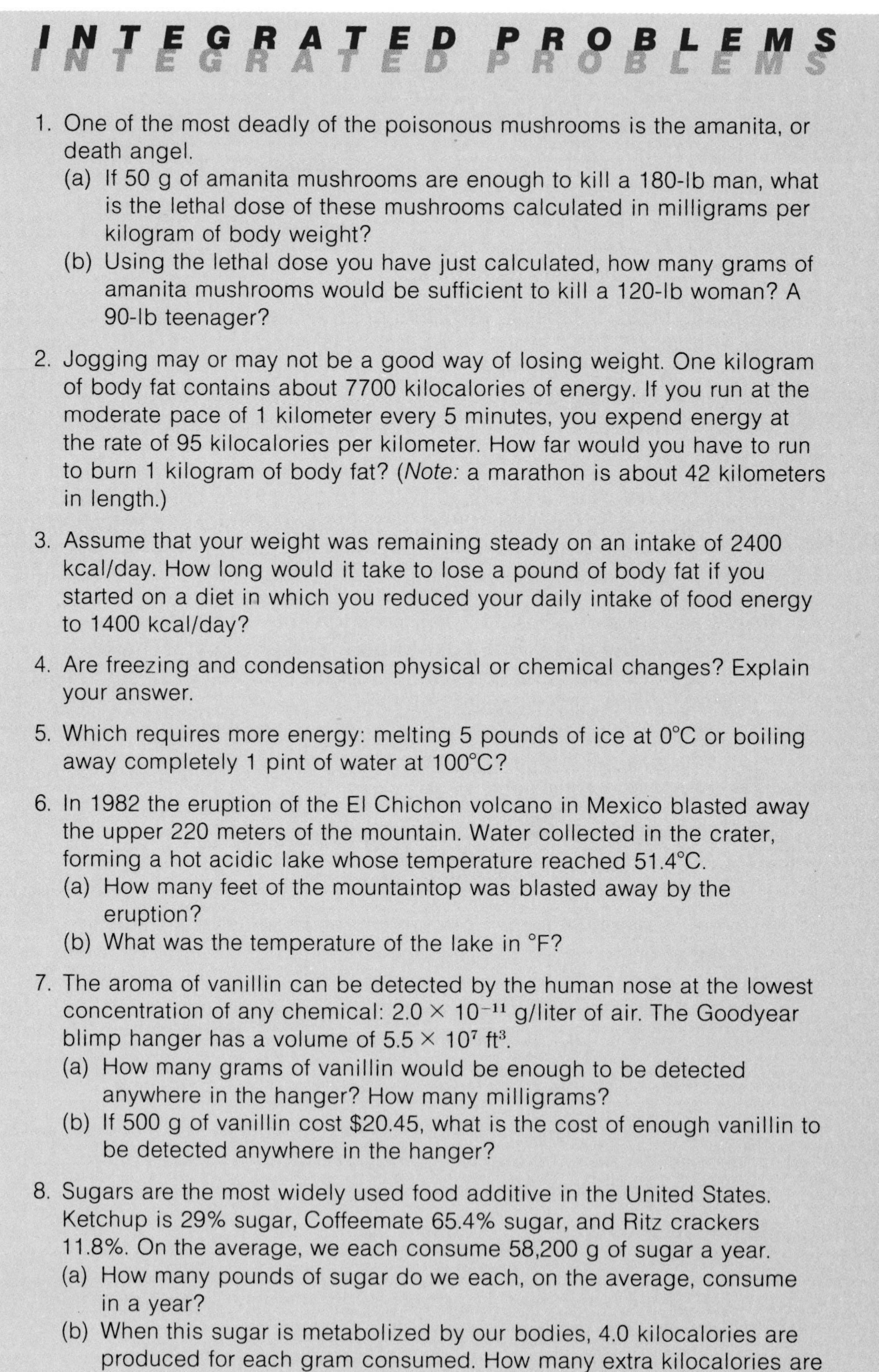

INTEGRATED PROBLEMS

INTEGRATED PROBLEMS

1. One of the most deadly of the poisonous mushrooms is the amanita, or death angel.
   (a) If 50 g of amanita mushrooms are enough to kill a 180-lb man, what is the lethal dose of these mushrooms calculated in milligrams per kilogram of body weight?
   (b) Using the lethal dose you have just calculated, how many grams of amanita mushrooms would be sufficient to kill a 120-lb woman? A 90-lb teenager?

2. Jogging may or may not be a good way of losing weight. One kilogram of body fat contains about 7700 kilocalories of energy. If you run at the moderate pace of 1 kilometer every 5 minutes, you expend energy at the rate of 95 kilocalories per kilometer. How far would you have to run to burn 1 kilogram of body fat? (*Note:* a marathon is about 42 kilometers in length.)

3. Assume that your weight was remaining steady on an intake of 2400 kcal/day. How long would it take to lose a pound of body fat if you started on a diet in which you reduced your daily intake of food energy to 1400 kcal/day?

4. Are freezing and condensation physical or chemical changes? Explain your answer.

5. Which requires more energy: melting 5 pounds of ice at 0°C or boiling away completely 1 pint of water at 100°C?

6. In 1982 the eruption of the El Chichon volcano in Mexico blasted away the upper 220 meters of the mountain. Water collected in the crater, forming a hot acidic lake whose temperature reached 51.4°C.
   (a) How many feet of the mountaintop was blasted away by the eruption?
   (b) What was the temperature of the lake in °F?

7. The aroma of vanillin can be detected by the human nose at the lowest concentration of any chemical: 2.0×10^{-11} g/liter of air. The Goodyear blimp hanger has a volume of 5.5×10^{7} ft^3.
   (a) How many grams of vanillin would be enough to be detected anywhere in the hanger? How many milligrams?
   (b) If 500 g of vanillin cost $20.45, what is the cost of enough vanillin to be detected anywhere in the hanger?

8. Sugars are the most widely used food additive in the United States. Ketchup is 29% sugar, Coffeemate 65.4% sugar, and Ritz crackers 11.8%. On the average, we each consume 58,200 g of sugar a year.
   (a) How many pounds of sugar do we each, on the average, consume in a year?
   (b) When this sugar is metabolized by our bodies, 4.0 kilocalories are produced for each gram consumed. How many extra kilocalories are we consuming each week from this added sugar?

CHAPTER FIVE
ATOMIC STRUCTURE

In 1918 Ruth Adams, then 16 years old, was excited to have finally found a job. She was hired by the Radium Luminous Materials Corporation of Orange, New Jersey, to apply luminous paint to watch dials and instrument dials. As you can imagine, this was painstaking work that required great precision. To help her get started, the other women showed Ruth how to keep a fine point on her brush by turning the bristles on her tongue and lips. The paint she used was invented by the president of the company and contained phosphorescent zinc sulfide with a small amount of radium and adhesive added.

At that time, neither Ruth nor anyone else realized the dangers that radioactive material posed to human tissues. However, it wasn't long before the possible harm of internal exposure to radium was very clearly brought to the public's attention. In 1925, the *New York Times* reported that five watch dial painters had died, and ten others had been stricken with "radium necrosis," a general breakdown of the bone tissues. But still, so little was known about the nature of radium and its effects on human tissues that the company, when brought to trial, defended itself with the claim that small quantities of radium were actually beneficial to health!

The watch dial painters who died in the 1920s suffered a variety of symptoms that included severe anemia, tumors of the sinuses, and inflammation of the bone marrow in the jaw and other bone structures of the body. Although many of the women who worked in this industry during that time are still living and in good health, many others died after developing bone cancer 20 to 30 years later. Ruth Adams worked in the watch dial factory for six years before leaving to get married. She showed no ill effects until some 25 years later, when she died of bone cancer which had spread throughout her body.

The disabilities or deaths of the women who were exposed to radium resulted from the destruction of their bone tissue, especially the tissue of the bone marrow, which produces the blood cells for the body. You might wonder what special property of radium causes it to collect in the bones of the body, as opposed to other tissues. We will see that radium (Ra) has chemical properties very similar to those of calcium (Ca), which is the main component of bones and teeth. Unfortunately, the body is unable to tell the difference between the toxic radium and the essential calcium. Therefore, both elements are deposited in the bones, where the radioactive radium does its damage. To understand why calcium and radium have similar chemical properties, we must study the structure of the atom.

By the time you have finished this chapter, you should be able to:

1. Describe the structure of an atom, listing three subatomic particles, their relative mass, their charge, and their location in the atom.
2. Define *atomic number* and *mass number,* and given the atomic number and mass number of any element, indicate the number of protons, neutrons, and electrons in an atom of that element.
3. Describe how isotopes of an element differ.
4. Define *atomic weight.*
5. Describe the Bohr model of the atom and indicate the maximum number of electrons possible in energy levels 1 through 7.
6. State the difference between an atom in the ground state and an excited atom and explain how a substance can be luminous.
7. Use the periodic table to state the symbol, atomic number, atomic weight, and electron configuration of an element.

8. Given one element in a period or group, name another member of that same period or group.
9. State the number of valence electrons in an atom of any representative element.
10. Identify on a periodic table the elements that are metals, nonmetals, metalloids, representative elements, transition elements, and inner transition elements.
11. Define *periodicity* and predict how atomic size will change across a period and down a group.
12. Define *ionization energy* and *electron affinity* and predict how each will change across a period or down a group.

ATOMIC STRUCTURE

5.1 The Parts of the Atom

In Chapter 2 we stated that atoms are composed of many types of particles, but that we would be focusing our attention on three such particles: **protons, neutrons,** and **electrons.** Protons and electrons are electrically charged particles, whereas a neutron is neutral (that is, it has no charge). A proton is assigned the smallest unit of positive charge (1+) that will just cancel the negative charge on an electron (1−). Interestingly, the terms *positive* and *negative,* which are used to describe the opposing effects of electrical charges, were first used by Ben Franklin more than 50 years before the discovery of the electron and proton. A proton will repel other protons (charges that are alike repel one another) and will attract electrons (unlike charges attract one another). See Table 5.1.

An atom consists of a small dense **nucleus** that contains protons and neutrons. Because we have just said that protons repel one another, you might wonder what holds the nucleus together. Although nobody completely understands nuclear forces, the neutrons seem to play an important role in binding together the positive protons. The electrons are found in the region surrounding the nucleus, but for the most part an atom

Table 5.1 Subatomic Particles

Name of Particle	Location in the Atom	Charge	Symbol	Relative Mass (amu)
Proton	Nucleus	1+	p, ${}^{1}_{1}p$, ${}^{1}_{1}\mathrm{H}$	1
Electron	Around the nucleus	1−	e, e^-, ${}^{0}_{-1}e$	$\frac{1}{1837}$
Neutron	Nucleus	0	n, ${}^{1}_{0}n$	1

is just empty space. To give you an idea of the relative positions of the subatomic particles in an atom, suppose you were in a large baseball stadium. If we were to let a flea on second base represent the nucleus of the atom, the nearest electron would be found somewhere in the top deck of the stands (Figure 5.1).

5.2 Atomic Number

The special characteristic that determines which element an atom represents is the number of protons in that atom. The **atomic number** of an element is the term used to describe the number of protons in the nucleus of any atom of that element. (The atomic numbers of the elements are listed on the inside back cover of this book.) Because the electrical charge on a proton just cancels the electrical charge on an electron, we can see that the atomic number of an element also tells us how many electrons there are in a neutral atom of that element. For example, the element sodium has an atomic number of 11. Therefore, each neutral atom of sodium will have 11 protons in the nucleus and 11 electrons surrounding the nucleus. But, no matter how many electrons (or neutrons) there may be, the identity of an element is always determined by the number of protons in its nucleus—that is, by its atomic number.

Figure 5.1 An atom is mostly empty space. If we imagine a flea on second base to be the nucleus of an atom, the nearest electron would be a speck of dust somewhere in the top deck of the stands. (Peter Menzel/Stock, Boston.)

5.3 Mass Number

Protons, electrons, and neutrons are extremely small particles. Protons and neutrons have a mass of 1.7×10^{-24} grams; electrons are even lighter, having a mass only $\frac{1}{1837}$ that of a proton. In fact, the mass of the electrons in an atom is so small in comparison with the mass of the protons and neutrons that it is ignored when calculating the mass of an atom. We define the **mass number** of an atom to equal the number of protons plus the number of neutrons in the nucleus of the atom.

$$\text{Mass number} = \text{Protons} + \text{Neutrons}$$

or

$$M = p + n$$

Chemists often use shorthand methods to express the mass number and the atomic number of an atom. For example, an atom of carbon (C) has six protons and six neutrons. The atomic number (Z) is 6, and the mass number (M) is 12. Chemists may refer to atoms of carbon having the mass number of 12 as carbon-12, ^{12}C or $^{12}_{6}C$, where the upper number is the mass number and the lower number is the atomic number. In general, then, the symbol $^{M}_{Z}X$ gives you the symbol of the element (X), the mass number (M), and the atomic number (Z).

EXAMPLE 5-1

1. State the number of protons, neutrons, and electrons in a neutral atom of each of the following:

(a) $^{14}_{6}C$ (b) Cobalt-60 (c) ^{235}U

(a) Using the table on the inside back cover of this book, we can identify this element as carbon. The symbol $^{14}_{6}C$ tells us that the mass number is 14 and the atomic number is 6. The atomic number gives both the number of protons and the number of electrons in a neutral atom.

$$p = 6 \quad \text{and} \quad e^- = 6$$

We know that $M = p + n$

therefore $14 = 6 + n$

$8 = n$

(b) Looking up cobalt, we find that the atomic number is 27. The mass number is 60. From the atomic number we then have that

$$p = 27 \quad \text{and} \quad e^- = 27$$

$$60 = 27 + n$$

$$33 = n$$

(c) U is the symbol for the element uranium, which has an atomic number of 92. This atom of uranium has a mass number of 235. From the atomic number,

$$p = 92 \qquad \text{and} \qquad e^- = 92$$

From the mass number,

$$235 = 92 + n$$
$$143 = n$$

2. State three ways of writing the symbol for an atom of iodine with 78 neutrons.

From the inside back cover we learn that the symbol for iodine is I and its atomic number is 53. Therefore, the mass number for such an atom is

$$M = 78 + 53 = 131$$

The three different notations would be

Iodine-131, ^{131}I, and $^{131}_{53}I$

Exercise 5-1

1. State the number of protons, neutrons, and electrons in a neutral atom of each of the following:

   (a) ^{222}Ra (b) Chromium-51 (c) $^{203}_{80}Hg$

2. What is the atomic number, mass number, and symbol for an atom having 15 protons and 16 neutrons?

5.4 Isotopes

In the early nineteenth century, John Dalton developed an atomic theory based on the idea that each atom of a given element is exactly alike. It was not until 100 years later that Fredrick Soddy proved part of this theory to be wrong by showing that the element neon consisted of not one, but two types of atoms. Some of these neon atoms had a mass number of 20, and some had a mass number of 22. (A third type of neon atom with the mass number of 21 also exists.) Neon has an atomic number of 10, so each atom of neon must have 10 protons in the nucleus. Therefore, these two different types of neon atoms must have different numbers of neutrons in their nuclei. Soddy invented the term **isotopes** to describe atoms of an element containing different numbers of neutrons in the nucleus (Figure 5.2). For neon, then, the isotopes are:

	Atomic Number	Mass Number	Number of p	Number of n
Neon-20	10	20	10	10
Neon-21	10	21	10	11
Neon-22	10	22	10	12

A few of the elements have only one type of atom, but most elements have two or more naturally occurring isotopes; in fact, tin (Sn) has 10!

5.5 Atomic Weight

Because atoms are so small and light, it is impractical to measure the mass of a small number of atoms. However, it is possible to measure the relative masses of atoms of different elements. A scale based on these relative measurements has been developed, with the most common isotope of carbon being arbitrarily assigned a value of exactly 12 **atomic mass units (amu).** It has been determined by indirect measurement that the mass of one atom of carbon-12 is 1.992×10^{-23} g (a *very* small mass!) Therefore, one atomic mass unit equals one-twelfth of this mass: 1 amu = 1.660×10^{-24} g.* If an isotope of another element were one-half as heavy as the carbon-12 isotope, then its mass in atomic mass units would be $1/2 \times 12$ or 6; if an isotope were 2.84 times as heavy as the carbon-12 isotope, its mass would be $2.84 \times 12 = 34.08$ amu.

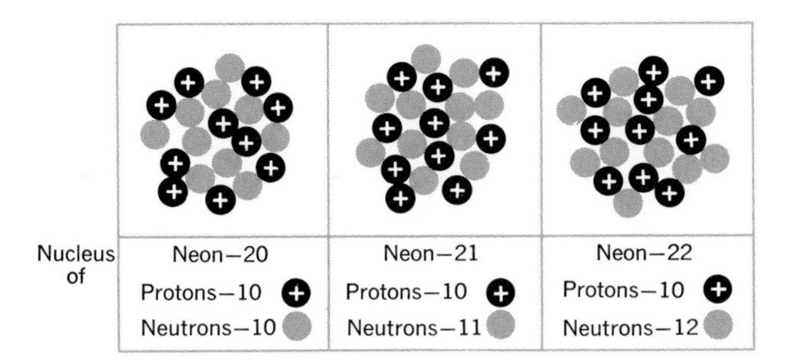

Figure 5.2 The element neon has three types of atoms, which differ in the number of neutrons in the nucleus. These atoms are isotopes of neon.

* *Some scientists use the term dalton (after John Dalton) for one atomic mass unit: 1 dalton = 1 amu.*

Most elements have at least two naturally occurring isotopes, so a sample of any of these elements will contain a mixture of the different isotopes. Table 5.2 lists the share of the sample, or percent abundance, of the naturally occurring isotopes of several elements. Suppose, for example, you were asked for the weight of a single atom of chlorine. According to Table 5.2, sometimes the mass of a single chlorine atom will be 35 amu and sometimes 37 amu, depending on the isotope that happens to be chosen. However, because any sample of chlorine will contain both of these kinds of chlorine atoms, we can approach this question by speaking instead of the *average* weight of a chlorine atom. If 75.53% of the atoms in the sample have a mass of 35.0 amu, and 24.47% have a mass of 37.0 amu, then the average mass of a chlorine atom taken from this sample will be

$$(0.7553 \times 35.0 \text{ amu}) + (0.2447 \times 37.0 \text{ amu}) = 35.5 \text{ amu}$$

For questions dealing with the total mass of this sample, therefore, we could act as if each chlorine atom had a mass of 35.5 amu. This number is called the atomic weight of chlorine. More generally, the **atomic weight** of an element is the weighted average of the masses of the naturally occurring isotopes of the element, expressed in atomic mass units. (Note that such a number really ought to be called the "atomic mass" of the element, but this more correct term is seldom used.)

The atomic weight of each of the elements is listed on the inside back cover of this book. The most important thing to know about atomic weights is that when we take samples of any two elements in such a way that the masses of the two samples have the same ratio (one to the other) as do their atomic weights, then the two samples will each contain the same number of atoms.

EXAMPLE 5-2

The calculation of a weighted average is not a very difficult procedure. In fact, teachers often use such a procedure in assigning final grades in a course. For example, suppose a teacher were to tell you that each of two mid-terms would determine 25% of your final grade, and the final exam would be worth 50% of the final grade. If you received 76 on your first mid-term, 64 on your second mid-term, and 90 on your final exam, your grade for the course would be 80. This final score is calculated by multiplying each grade by its percent weight in the final grade, and then adding up these products.

$$\begin{aligned} 76 \times 25\% &= 76 \times 0.25 = 19 \\ 64 \times 25\% &= 64 \times 0.25 = 16 \\ 90 \times 50\% &= 90 \times 0.50 = \underline{45} \\ & \qquad\qquad\qquad\quad 80 \end{aligned}$$

Table 5.2 The Relative Abundance of the Isotopes of Several Elements

Isotope	Percent Natural Abundance	Isotope	Percent Natural Abundance
Hydrogen-1	99.99%	Silicon-28	92.21%
Hydrogen-2	0.01%	Silicon-29	4.70%
		Silicon-30	3.09%
Carbon-12	98.89%		
Carbon-13	1.11%	Chlorine-35	75.53%
		Chlorine-37	24.47%
Nitrogen-14	99.63%		
Nitrogen-15	0.37%	Zinc-64	48.89%
		Zinc-66	27.81%
Oxygen-16	99.76%	Zinc-67	4.11%
Oxygen-17	0.04%	Zinc-68	18.57%
Oxygen-18	0.20%	Zinc-70	0.62%
Fluorine-19	100.00%	Bromine-79	50.54%
		Bromine-81	49.46%

A similar procedure is followed in calculating the atomic weight of an element. The mass of each isotope is multiplied by its percent abundance, and the products are then added up to give the atomic weight. For example, boron has two isotopes, boron-10 and boron-11, whose percentage abundances are 19.6% and 80.4%, respectively. The atomic weight of boron is, then,

$$\begin{aligned} 10.0 \text{ amu} \times 19.6\% &= 10.0 \times 0.196 = 1.96 \\ 11.0 \text{ amu} \times 80.4\% &= 11.0 \times 0.804 = \underline{8.844} \\ & \qquad 10.804 \text{ or } 10.8 \text{ amu} \end{aligned}$$

Exercise 5-2

Calculate the atomic weight of silicon using the isotopes listed in Table 5.2.

ELECTRON CONFIGURATION

In the beginning of this chapter we described an atom of any element as containing a small, dense, positive nucleus surrounded by negative electrons. You will remember that a neutral atom contains the same number of electrons as there are protons in the nucleus. Therefore, the number of electrons in a neutral atom of an element will be equal to the atomic number of that element. Because protons and electrons are too small to be seen with even the most powerful microscopes, we must depend on various theoretical models to describe the arrangement of the electrons around the nucleus.

5.6 The Bohr Model

One of the first models of the atom, called the planetary model, was developed in the early 1900s by Niels Bohr, who based his theory on the experimental data of Ernest Rutherford. Over the past 75 years this theory has been modified quite a bit, but it is helpful for us to understand Bohr's simple model. He pictured the atom as consisting of a small, dense nucleus surrounded by electrons traveling in orbits, similar to planets traveling around the sun (Figure 5.3). In his model the electrons could occupy only certain positions around the nucleus. Each of these positions corresponded to a certain energy value, so these energy positions were called **energy levels** or **energy shells.** Electrons closest to the nucleus are in the lowest possible energy state. As electrons are found farther from the nucleus, they have higher and higher energy states.

We might compare this idea of electron energy levels to the potential energy you possess as you climb the stairs to your seat in a football stadium. At the bottom of the stairs, you are at the lowest possible energy position. You must use a specific amount of energy to climb up one step. Once there you then possess greater potential energy relative to the ground. Also, there is no way that you can stand between two steps—you must use a specific amount of energy to get from one step to another step,

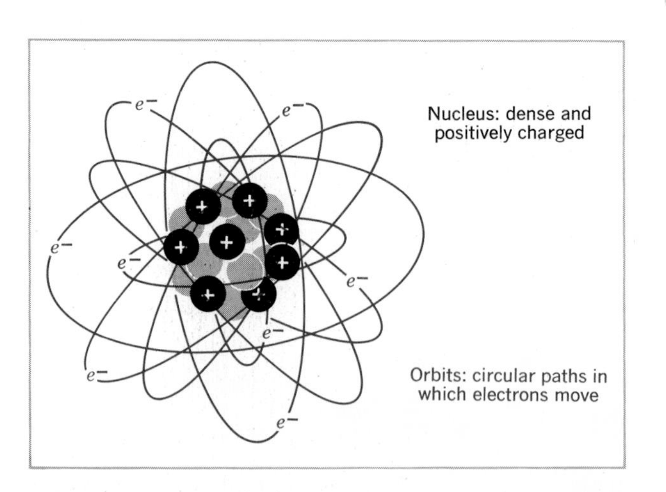

Figure 5.3 Bohr's planetary model of the atom has a dense, positively charged nucleus, with electrons moving in circular paths, called orbits, around the nucleus.

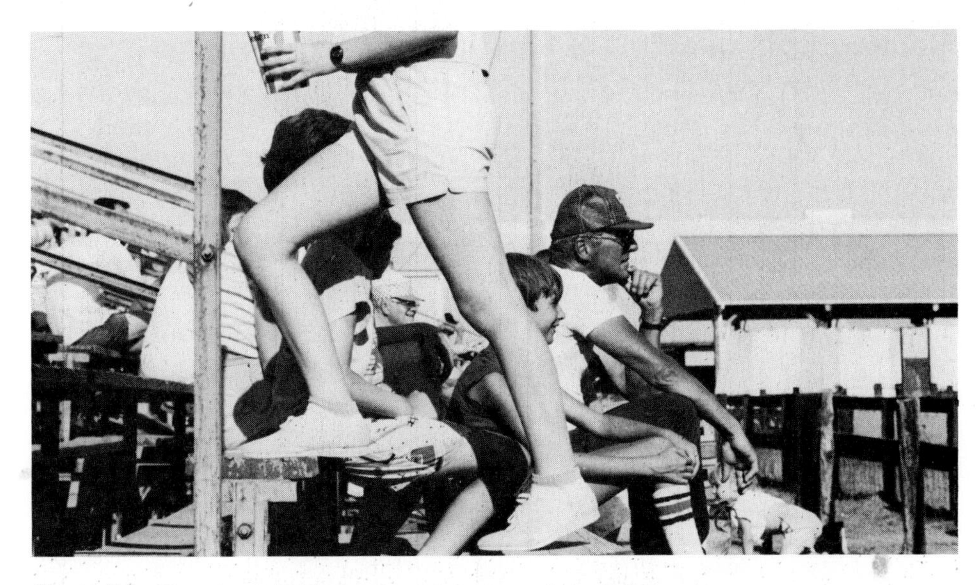

Figure 5.4 We can compare the energy possessed by electrons in an atom to the potential energy that is gained as you climb the stairs to your seats in a stadium. The higher you go, the more potential energy you possess. Also, you must exert a specific amount of energy to climb each stair, and there is no way to stand between steps. (Stefan Bloomfield)

and you are not able to stop part way in between (Figure 5.4). Similarly, an electron is not able to exist in between any of the specific energy levels of the atom. To move from one energy level to another, an electron must gain (or lose, depending on whether it is moving away from or closer to the nucleus) an amount of energy exactly equal to the energy difference between the two levels.

The energy levels within an atom are labeled with numbers, with the first energy level being the one closest to the nucleus. As the number of the energy level increases, it is found farther from the nucleus and contains electrons having greater and greater energy. Each energy level can contain only a certain maximum number of electrons, but in nature this maximum number is, in most cases, never realized (Table 5.3).

Table 5.3 Energy Levels in the Atom

Energy level number (n)	1	2	3	4	5	6	7
Maximum number of electrons allowed in theory ($2n^2$)	2	8	18	32	50	72	98
Maximum number of electrons actually found in nature	2	8	18	32	32	18	8

5.7 Electron Configuration of the First Twenty Elements

We have learned that each neutral atom of a given element has a specific number of electrons equal to the atomic number of that element. For the first twenty elements, these electrons will occupy positions in energy levels according to the following rules.

1. The number of electrons in a neutral atom of an element is equal to the atomic number.
2. Beginning with energy level 1, the electrons will fill in the lowest possible energy positions that are available. When any level has been filled, electrons go on to fill the next level. For example, sodium has atomic number 11 and, therefore, has 11 electrons. Two of these electrons will be in the first energy level, which fills the first level. The second level will be filled with the next 8 electrons. This leaves one remaining electron, which will be found in the third energy level.
3. For the first 18 elements, this pattern is closely followed. (See Table 5.4.) But for elements 19 and 20, the nineteenth and twentieth electrons do not go in the third level, even though this level has room for more than 8 electrons. Instead, these electrons are found in the fourth energy level.

In order to know how to assign electrons for elements having atomic numbers greater than 20, additional rules are required. Fortunately, for our study of living organisms it is not too important that we understand these rules, because the elements that make up the majority of living matter have atomic numbers less than 21.

5.8 Formation of Ions

In a normal atom, all of the electrons will be found in the lowest possible energy levels. Such an atom is said to be in the **ground state.** If the atom absorbs energy (such as heat or light) from an external source, some electrons might jump to a higher energy level. When this occurs, each such electron absorbs an amount of energy exactly equal to the energy difference between the two levels. An atom having one or more electrons in a higher than normal energy state is called an **excited atom.** Excited atoms are unstable, and in such atoms electrons will drop back to lower energy levels by giving off energy, usually in the form of radiant energy (UV, IR, or light waves). The light given off by an atom as an electron drops from one energy level to another has a specific energy or wavelength corresponding to the energy difference between the two levels.

If enough energy is supplied to an atom, one of its electrons may jump completely away from the atom, leaving the atom with one less electron than it has protons. This new particle is then no longer a neutral atom, but is rather a positively charged particle called a positive ion. An **ion** is a

Table 5.4 Electron Configuration of Elements with Atomic Numbers under 21

Element	Atomic Number	Energy Level Electron Configuration						
		1	2	3	4	5	6	7
Hydrogen, H	1	1						
Helium, He	2	2						
Lithium, Li	3	2	1					
Beryllium, Be	4	2	2					
Boron, B	5	2	3					
Carbon, C	6	2	4					
Nitrogen, N	7	2	5					
Oxygen, O	8	2	6					
Fluorine, F	9	2	7					
Neon, Ne	10	2	8					
Sodium, Na	11	2	8	1				
Magnesium, Mg	12	2	8	2				
Aluminum, Al	13	2	8	3				
Silicon, Si	14	2	8	4				
Phosphorus, P	15	2	8	5				
Sulfur, S	16	2	8	6				
Chlorine, Cl	17	2	8	7				
Argon, Ar	18	2	8	8				
Potassium, K	19	2	8	8	1			
Calcium, Ca	20	2	8	8	2			

positively or negatively charged particle that is formed when an atom loses or gains electrons.

5.9 Luminescence

When an element or compound becomes excited by absorbing energy from an external source, and then gives off that energy as visible light, it is said to be **luminescent.** There are two kinds of luminescence: fluorescence, in which the element or compound stops giving off light as soon as the energy being supplied to the compound is stopped; and phosphorescence, in which the element or compound continues to give off light for a short time after the incoming energy has stopped. There are many examples of luminescence with which you are familiar. Laundry whiteners and brighteners make use of fluorescent dyes which become attached to clothing in the laundering process. These compounds absorb energy from the ultraviolet light in sunlight and then give off light in the visible region. Through both reflection and fluorescence, then, the clothing gives off more visible light than falls on it—this makes the clothing appear "brighter." Red reflective tape, which is often attached to bicycles and cars to increase their visibility, contains compounds that absorb nonred light and give off the red-orange light we see (Figure 5.5). The luminous dial you may have on your watch contains tiny amounts of radioactive material mixed with a powdered phosphor. The radioactive material releases a small but steady amount of energy, which is absorbed by the atoms in the phosphor. When the electrons in these atoms return to the ground state, they release the visible light which you see as a faint glow on the watch dial.

Living organisms make use of luminescence for sexual attraction, protection, and hunting. The firefly uses bioluminescence to attract a mate. This insect uses chemical energy to excite electrons in a special compound that produces the firefly's blink as the electrons return to the ground state. Green plants use the reverse of this process in the production of food: energy from sunlight excites electrons in molecules of the plant's green

Figure 5.5 The reflective trim on these bicycle tires contains a fluorescent material, greatly increasing the visibility of these bicycles at night. (Courtesy 3M Company.)

pigment, called chlorophyll. As these electrons return to the ground state, the energy is not released in the form of light, but instead is finally trapped as chemical energy in a molecule of sugar. This process of using the sun's energy to produce a sugar molecule from carbon dioxide and water is called photosynthesis and will be studied in detail in Chapter 19.

5.10 The Quantum Mechanical Model of the Atom*

The simple planetary model of the atom developed by Bohr did not explain some of the experimental data that was later collected, and in the late 1920s Erwin Schrödinger, P. A. M. Dirac, Werner Heisenberg, and others developed a new model of the atom called the quantum mechanical model. The model is based on fairly complicated mathematical concepts, and we will look at only some of the ways this theory changed the idea of the atom developed by Bohr.

Although the Bohr atom described electrons orbiting around the nucleus, scientists found it impossible to accurately pinpoint the position and speed of an electron. Therefore, the electron's position could not be described as following a definite path. Instead, scientists could only specify regions in which there is a large probability of finding one or two electrons. These probability regions are known as atomic **orbitals.** Each energy level was found to have energy sublevels consisting of one or more atomic orbitals.

Atomic Orbitals

Each orbital is a particular region around the nucleus of the atom in which there is a high probability of finding an electron with a specific energy. An orbital can contain zero, one, or two electrons.

The first energy level contains one atomic orbital, labeled the 1*s* orbital. The probability region described by this orbital is a sphere with its center at the nucleus (Figure 5.6). The second energy level contains four atomic orbitals. One of the orbitals is spherical in shape, and is called the 2*s* orbital. The probability regions of the other three orbitals are shaped like dumbbells that cross each other at right angles (Figure 5.6). These orbitals, called the 2*p* orbitals, have slightly more energy than the 2*s* orbital.

Each higher energy level will contain one *s* and three *p* orbitals. In addition to these, there are two other types of orbitals: the *d* orbitals, which appear in groups of five starting from energy level 3, and the *f* orbitals, which appear in groups of seven starting from energy level 4. We will not discuss the shapes of the *d* and *f* orbitals in this textbook.

In addition to filling the lowest possible energy level, there are several other rules that describe how electrons are found in orbitals. First, an orbital will hold no more than two electrons and can hold two electrons only if they are spinning in opposite directions. (Electrons behave as though they are spinning around an axis, and scientists use arrows— $\uparrow$ for clockwise and

---

* *This section is optional and can be skipped without loss of continuity.*

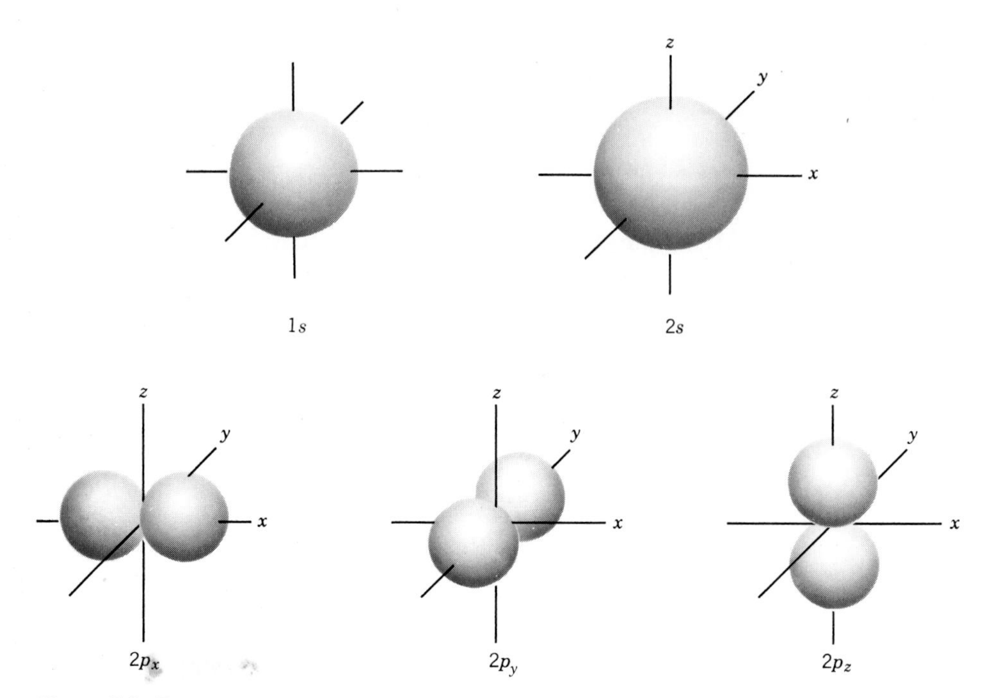

Figure 5.6 Representations of the probability regions (areas in which electrons are most likely to be found) for the 1*s*, 2*s*, and 2*p* orbitals.

↓ for counterclockwise—to indicate the direction of this spin.) Second, electrons will not pair up in an orbital if there is another orbital of the same energy available; that is, they pair up only when all orbitals of the same energy contain at least one electron. Figure 5.7 shows the relative energy levels of the orbitals in atoms having two or more electrons.

Electron Configurations

There are several ways to indicate the arrangement of electrons, or the **electron configuration,** in an atom of an element. Let's use boron (atomic number 5) as an example. The atomic number tells us that a neutral atom of boron has 5 electrons. Using Figure 5.7, we see that two of the electrons will fill in the 1*s* orbital, two more the 2*s* orbital, and the remaining electron will be found in one of the 2*p* orbitals. We can write this electron configuration as follows:

$$\text{B} \quad 1s^2 2s^2 2p^1$$

A second way to indicate the electron configuration of boron is to use an **orbital diagram,** in which each orbital is represented by a circle and each electron by an arrow.

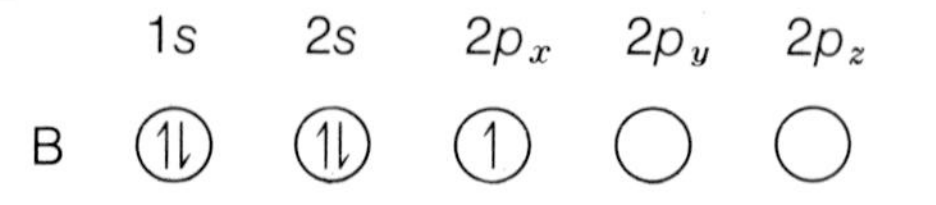

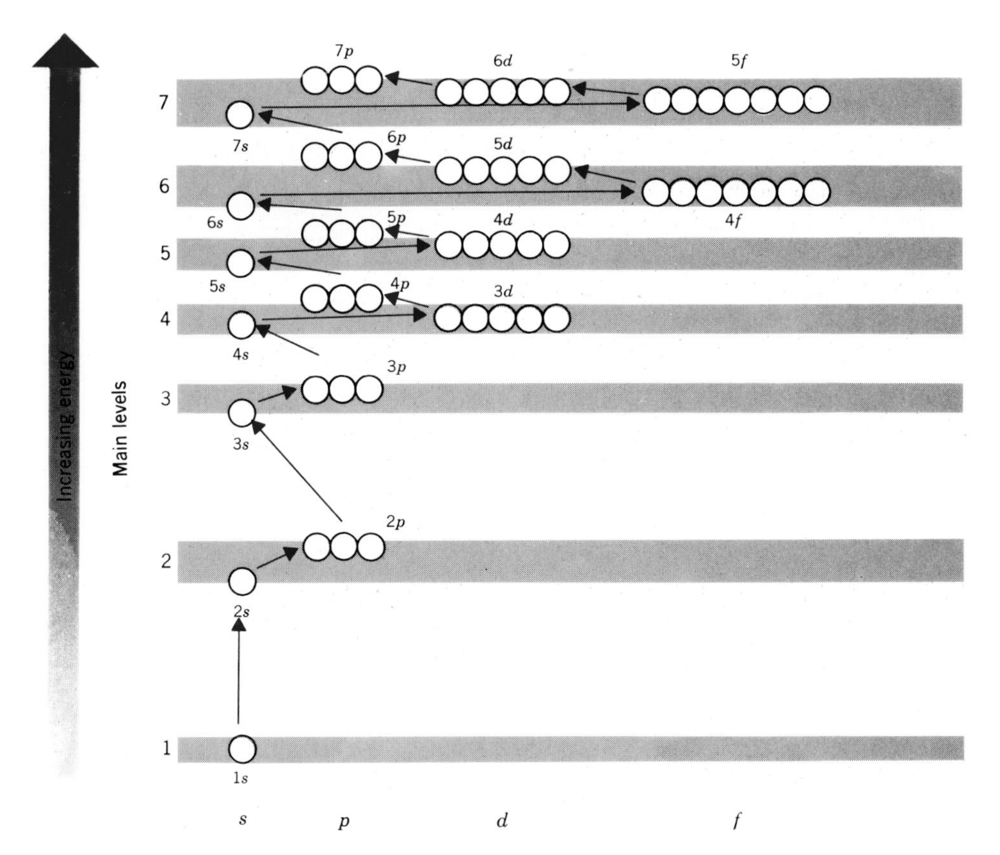

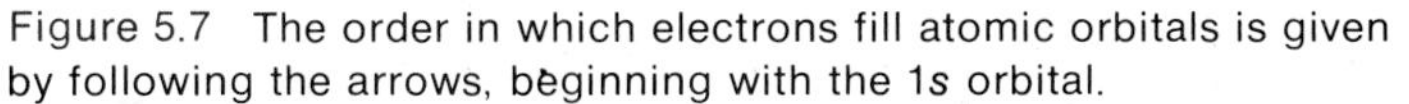

Figure 5.7 The order in which electrons fill atomic orbitals is given by following the arrows, beginning with the 1*s* orbital.

EXAMPLE 5-3

1. Write the orbital diagram and the electron configuration of carbon, C.

   Carbon's atomic number is 6, so a neutral atom of carbon contains six electrons. Using Figure 5.7 as a guide, we can fill in the orbitals. The 1*s* orbital is filled with two electrons, leaving four electrons still to be assigned. Next, the 2*s* orbital is filled with two electrons. These electrons will pair up in the 2*s* orbital before the 2*p* orbitals are filled, because the 2*p* orbitals have slightly more energy than the 2*s* orbital. The next electron is found in the $2p_x$ orbital. The last electron will not pair up in the $2p_x$, but will instead be found in a different 2*p* orbital having the same energy.

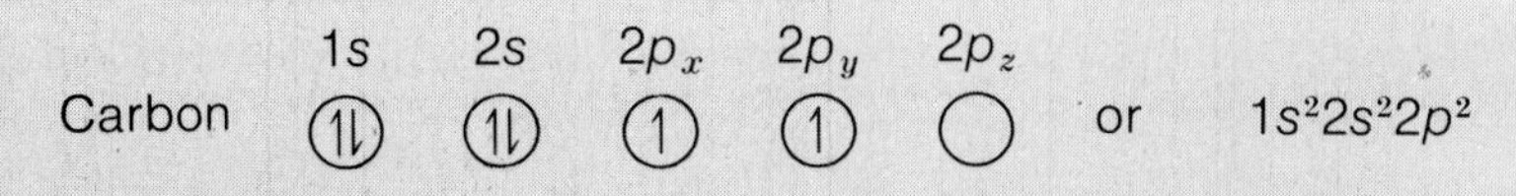

2. What is the orbital diagram and electron configuration of calcium, Ca?

Calcium has an atomic number of 20, so a neutral atom of calcium will have 20 electrons. Again using Figure 5.7 as a guide, we can fill in the orbitals. The 1*s* orbital is filled with two electrons; the 2*s* orbital is filled with two electrons and the three 2*p* orbitals are filled with six electrons, making a total of ten electrons that we have so far placed. Ten electrons remain. The 3*s* orbital is filled with two electrons, and the three 3*p* orbitals with six electrons. We now have only two electrons remaining. You will notice from Figure 5.7 that the 4*s* orbital has less energy than the 3*d* orbital. Therefore, the 4*s* orbital will be filled first with two electrons, completing the electron configuration for calcium. (Note that the 3*d* orbitals will be filled in elements having atomic numbers 21 through 30, and the 4*p* orbitals will then be filled in the elements having atomic numbers 31 through 36.)

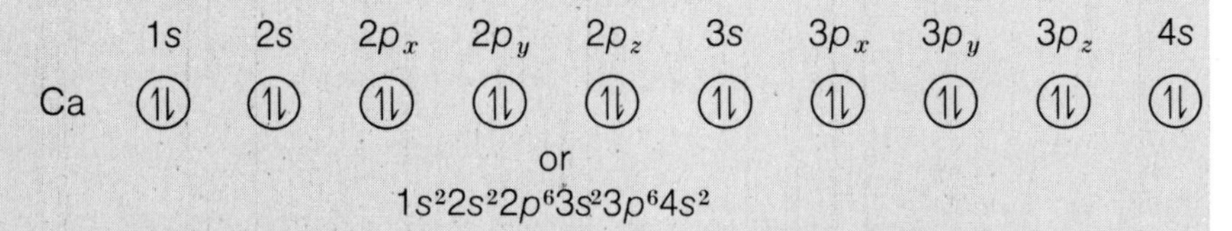

You might be deciding that electron configurations aren't too difficult to determine if you have Figure 5.7 to follow, but how will you ever be able to do it without this chart? Figure 5.8 gives you a way to remember the order of the subshells. Just list the subshells in order of increasing energy level and then follow the ordering shown by the diagonal arrows beginning at the bottom. Try using Figure 5.8 to do Exercise 5-3.

Exercise 5-3

Write the orbital diagram and the electron configuration for a neutral atom of the following elements:

(a) Sodium (b) Phosphorus (c) Chlorine

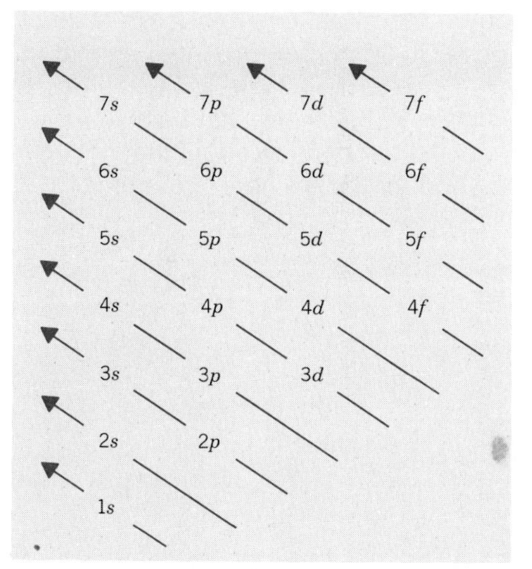

Figure 5.8 To remember the order in which electrons fill the orbitals, write the orbitals as shown and then follow the diagonal arrows beginning at the bottom right.

THE PERIODIC TABLE OF ELEMENTS

5.11 The Periodic Table

When discussing radium, we mentioned that the chemical properties of each element depend upon that element's electron configuration. It would certainly be quite a chore to memorize the electron configuration of all the chemical elements. But take heart; there is a way of writing down the chemical elements so that their similarities are easily seen. This method makes use of the **periodicity,** or repeating nature, of the chemical properties of the elements, which was first noticed by the Russian scientist Dmitri Mendeleev in 1869. He had devised a table showing each of the then-known elements arranged according to their atomic weight and noticed that the chemical properties of the elements seemed to show periodic relationships. That is, various similarities in chemical properties repeated themselves among the elements as he read down the table. However, Mendeleev also realized that there seemed to be gaps in this table which would have to be filled in if the periodic relationships were to hold exactly. He wisely left these spaces blank, suspecting that these gaps represented elements that had not yet been discovered. About 45 years later it was found that if Mendeleev's table were slightly rearranged so that the elements were listed by atomic number rather than by atomic weight, this would eliminate some of the irregularities left in the table. The resulting arrangement of elements is known as the **periodic table of the elements,** and modern-day periodic tables can be drawn so as to give a large amount of information about each element. A periodic table is shown on the inside front cover of this book.

Let's closely examine the periodic table. You will see that each element appears in a box containing specific information about that element. In this case the box lists the symbol of the element, the atomic number, and the atomic weight.

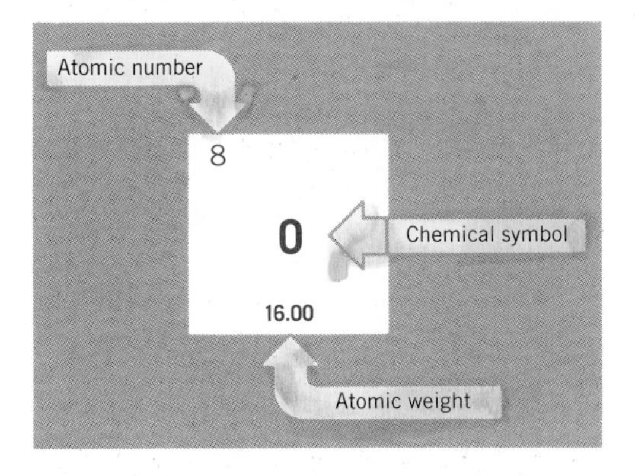

5.12 Periods and Groups

Each horizontal row of the periodic table is called a **period.** The periods are numbered from one to seven, corresponding to the seven energy levels of an atom that can contain electrons. This means that any element in, say, period four will have its outermost electrons assigned to the fourth energy level. For example, find sodium (symbol Na, atomic number 11) on the table. Sodium has only one electron in its outermost energy level and, because sodium appears on period three of the periodic table, it contains that one electron in the third energy level.

Each column of the periodic table is called a **group** or **chemical family.** Mendeleev noticed that the various members of a group show similar chemical behavior. We will see that it is the number of electrons in the outermost energy level of an atom that determines the chemical behavior of that element. These outermost energy level electrons are called **valence electrons;** each member of a chemical family has the same number of valence electrons (Table 5.5). Notice the columns of the periodic table labeled with the Roman numerals I through VII (Figure 5.9). These seven groups and group 0 are called the **representative elements** (also called the main group or the characteristic elements). The Roman numeral above these columns indicates the number of valence electrons for each member of the group. For example, the members of group II are beryllium (Be), magnesium (Mg), calcium (Ca), strontium (Sr), barium (Ba), and radium (Ra). Each neutral atom of these elements will contain two valence electrons. These two electrons will be located on the fourth energy level in an atom of calcium, and on the seventh energy level in an atom of radium.

The three rows of ten elements in the middle of the table are called

Table 5.5 Electron Configurations of Three Chemical Families

Group Number	Chemical Family	Element	Atomic Number	Electron Configuration 1	2	3	4	5	6	7
I	Alkali metals	Lithium	3	2	1					
		Sodium	11	2	8	1				
		Potassium	19	2	8	8	1			
		Rubidium	37	2	8	18	8	1		
		Cesium	55	2	8	18	18	8	1	
		Francium	87	2	8	18	32	18	8	1
II	Alkaline earth metals	Beryllium	4	2	2					
		Magnesium	12	2	8	2				
		Calcium	20	2	8	8	2			
		Strontium	38	2	8	18	8	2		
		Barium	56	2	8	18	18	8	2	
		Radium	88	2	8	18	32	18	8	2
VII	Halogens	Fluorine	9	2	7					
		Chlorine	17	2	8	7				
		Bromine	35	2	8	18	7			
		Iodine	53	2	8	18	18	7		
		Astatine	85	2	8	18	32	18	7	

transition elements or **transition metals;** they often show similar chemical properties not only within groups, but also along periods (Figure 5.9). This similarity in chemical behavior results from the fact that as we move along a period across the transition elements, electrons are not being added to the outermost energy level of the atoms, but rather are being added to the next-to-the-outermost level. (The 3*d* orbitals are

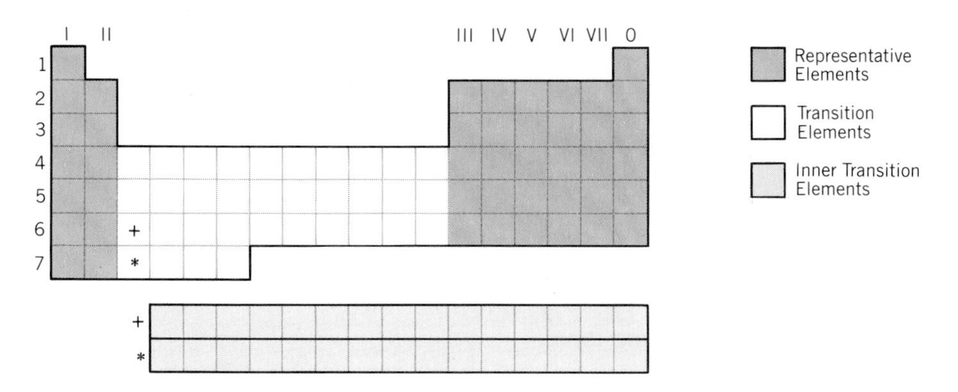

Figure 5.9 The locations of the representative elements, the transition elements, and the inner transition elements are shown on this periodic table.

being filled in the transition elements of period 4, the 4*d* orbital in the transition elements of period 5, etc.) The outermost level of each of the transition elements contains only one or two electrons. For example, iron (Fe), cobalt (Co), and nickel (Ni), located on the fourth row of the periodic table, all have two electrons on the same outermost energy level (the fourth). Although they contain different total numbers of electrons, they are very similar in chemical behavior and are often grouped together for this reason.

We have seen that the same number of electrons in the outermost energy level of the transition elements masks the difference in the number of electrons in the next-to-outermost level. This is doubly true for the **inner transition elements,** the two rows of 14 elements at the bottom of the table. Each of these rows is named after the element it follows in the main body of the periodic table (Figure 5.9). The differences in the electron arrangements of the inner transition elements are hidden beneath two identical outer shell electron configurations. (Electrons in these elements are filling the *f* orbitals—the 4*f* in the lanthanide series and the 5*f* in the actinide series.) As a result, these elements have very similar properties.

Some of the groups on the periodic table are known by common names. The elements in group I are called the **alkali metals,** those in group II the **alkaline earth metals,** and the elements in group VII the **halogens.** The elements of group 0 are called the **noble gases.** Some of these gases can be forced to form compounds with oxygen or with members of the halogen family, but otherwise they are extremely unreactive. Each noble gas, with the exception of helium (He), has an electron configuration of eight electrons in the outermost energy level, and the chemical stability of these elements comes from this stable arrangement of electrons. Although helium contains only two electrons, these two valence electrons completely fill the first energy level of this atom, giving helium its stability.

We have said that members of the same chemical family will show similar chemical behavior. For example, chlorine (Cl_2) and iodine (I_2), both members of the halogen family, are highly effective in killing bacteria and are used as disinfectants and antiseptics, respectively. Germanium (Ge) and silicon (Si), members of group IV, are both widely used as semiconductors in transistors. Hard water problems can be caused by either calcium (Ca^{2+}) or magnesium (Mg^{2+}) ions because both of these group II ions undergo similar chemical reactions with soap. Two radioactive wastes of special concern are strontium-90 and cesium-137. Strontium, like radium, is in group II, so it too will have chemical properties similar to calcium. Strontium-90 in the atmosphere falls onto the grass, is eaten by cows, and is consumed by humans when they drink milk. As with radium, the human body will mistake strontium-90 for calcium and will deposit it in bones and teeth. Cesium is in group I, and cesium-137 will mistakenly be used in the body in much the same way as sodium, an element that is essential for many body functions.

5.13 Metals, Nonmetals, and Metalloids

Various physical and chemical properties of elements allow us to classify the elements of the periodic table as **metals** or **nonmetals** (Figure 5.10).

You are probably familiar with many of the properties of metals. Metals are generally shiny, dense, malleable (able to be hammered or rolled into sheets), and ductile (able to be drawn out into wires). Metals have high melting points: all except mercury are solids at room temperature. Metals are also excellent conductors of electricity and heat.

Nonmetals, on the other hand, tend to be of low density and brittle when solid. They are poor conductors of heat and, except for the graphite form of carbon, are poor conductors of electricity. Most nonmetals have low melting points and are gases at room temperature. All the group 0 elements are gases; the other gaseous nonmetals are hydrogen, oxygen, nitrogen, fluorine, and chlorine.

The heavy, jagged line on the periodic table divides those elements that are metals from the nonmetals, with the metals lying on the left of the line and clearly making up the majority of elements. However, there is no sharp distinction between the elements having metallic properties and those having nonmetallic properties; the elements next to the jagged line may exhibit properties of both metals and nonmetals. Such elements are known as **metalloids** or **semimetals.** For example, arsenic (As, atomic number 33) is a brittle gray solid. When freshly cut, however, it has a bright metallic shine. Arsenic acts as a metal in forming compounds with oxygen and chlorine, but acts as a nonmetal in other chemical reactions. The most important physical property of metalloids is their ability to conduct electricity. Because they do not do this as well as metals, they are known as **semiconductors.** This behavior, especially in silicon and germanium, has led to the development of microcircuitry on silicon chips, and the rapid growth of solid state electronics and hand-held calculators and microcomputers.

You will notice that hydrogen at the top of group I does not fit into the division of metals and nonmetals. Hydrogen displays nonmetallic properties under normal conditions, although its outer shell electron configuration is the same as that for the group I metals. It has been found, however, that hydrogen will show metallic properties similar to the alkali metals under conditions of extremely high pressure.

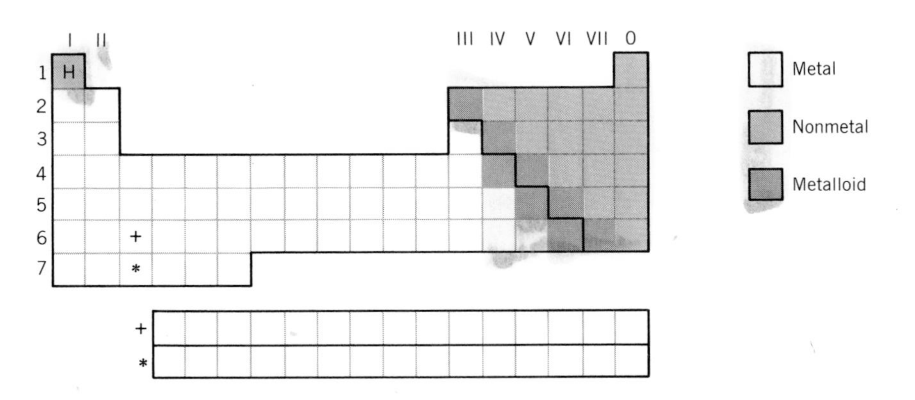

Figure 5.10 The locations of the metals, nonmetals, and metalloids are shown on this periodic table.

5.14 The Periodic Law

Various physical and chemical properties of atoms vary periodically with their atomic number. That is, a particular pattern of properties will be repeated as the atomic number increases. This fact is often referred to as the **periodic law.** An examination of three of these periodic properties will help us to better understand the chemical behavior of the elements, especially the different chemical behavior of metals and nonmetals.

Atomic Size

One property of the elements that shows a periodic relationship is the size of their atoms. From Figure 5.11 you can see that, with only a few exceptions, the atomic radius decreases as you move across a period of the periodic table (for example, from Li to F.) The group I, group II, and group VII elements are labeled on the graph, and this will help you see that as you move down a group or chemical family, the atomic radius increases as the atomic number increases. One result of this pattern is that the metallic elements on the left-hand side of the table will have large atomic radii, and the nonmetals on the upper right-hand part of the table will have small radii.

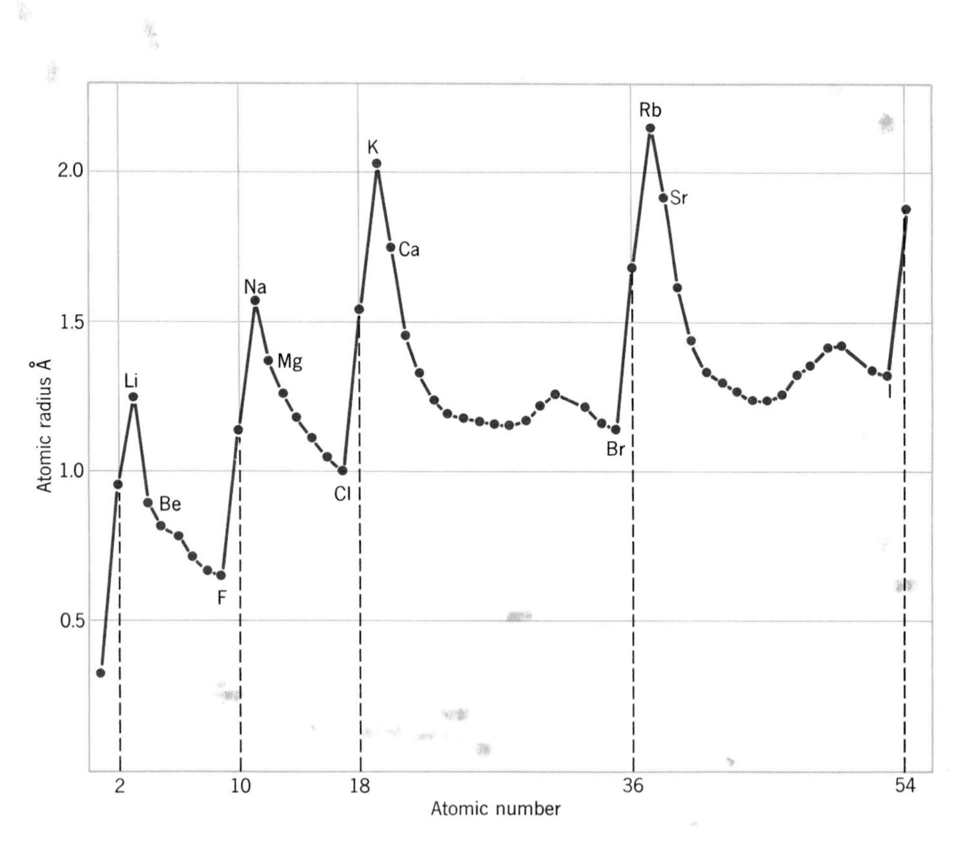

Figure 5.11 A graph of the periodic trends of atomic radii. With a few irregularities, atomic radii decrease across a period and increase as the atomic numbers increase down a chemical family.

Ionization Energy

A second important property of an element is its ionization energy. The **ionization energy** of an element is the amount of energy that must be added to an atom to remove one electron from its outermost energy level. Put another way, the ionization energy indicates how strongly the nucleus attracts the valence electrons. The stronger the attraction for these valence electrons, the greater is the ionization energy. It is easy to see that the size of an atom will affect how strongly the nucleus attracts the valence electrons. In small atoms the nucleus and the valence electrons are close together and, therefore, the electrical attraction is strong. In larger atoms the outermost energy level is farther from the nucleus and, therefore, the attraction for the valence electrons is weaker. So the trend is for the ionization energy to increase across a period as the radius decreases. On the other hand, as we move down any group, the atomic radii are increasing. Therefore, the attraction of the nucleus will become weaker and the ionization energy will decrease (Figure 5.12).

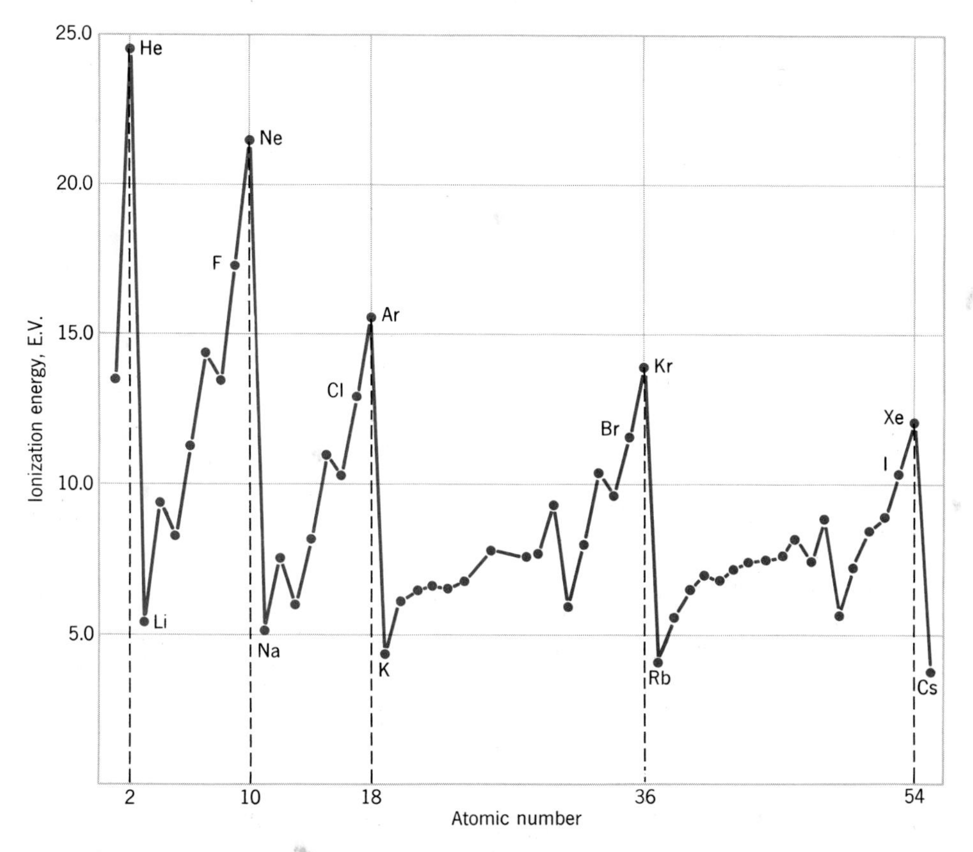

Figure 5.12 A graph of the periodic trends of ionization energies. Ionization energies tend to increase across a period and to decrease as the atomic numbers increase down a chemical family.

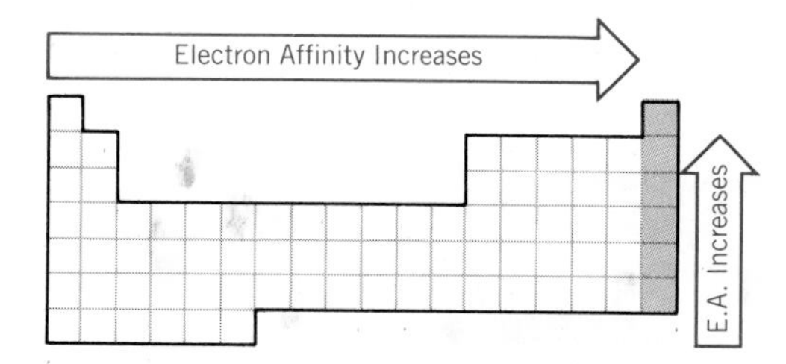

Figure 5.13 A periodic table showing the trends of electron affinities. Electron affinities tend to increase across a period and to decrease as the atomic numbers increase down a chemical family. The noble gases (shown in color) have low electron affinities.

Notice that the metals generally have lower ionization energies than do the nonmetals, so that it is easier to remove a valence electron from a metal than from a nonmetal. Also, Figure 5.12 shows that the noble gases, with their very stable configuration of electrons, all have very high ionization energies. This particular electron configuration somehow makes it extremely difficult to remove an electron from the outermost energy level of the atom.

Electron Affinity

A third property showing a periodic relationship among the elements is electron affinity. The **electron affinity** of an element is the amount of energy that is released when an electron is added to a neutral atom of that element. This property indicates the ability of an atom to attract additional electrons. The larger the electron affinity, the larger the attraction of the atom for additional electrons. Trends in electron affinities are similar to the trends in ionization energy. The electron affinity increases across a period and decreases down a group (Figure 5.13). Metals have low electron affinities and nonmetals have high electron affinities. As we might expect from their stable electron configurations, the noble gases have low electron affinities.

CHAPTER SUMMARY

The atom consists of a small, dense nucleus containing positive protons and neutral neutrons, and a region surrounding the nucleus containing negative electrons. The atomic number of an element indicates the number of protons in the nucleus of each atom of that element. The mass number is the sum of the number of protons and neutrons in the nucleus of an atom. Atoms of the same element that differ in the number of neutrons in their nuclei are called isotopes. The atomic weight of an element is a weighted average of the masses of the naturally occurring isotopes of that element, expressed in atomic mass units.

The Bohr model pictures the atom as consisting of a small, dense, positive nucleus surrounded by negative electrons. The electrons will be found only in certain allowed energy regions called energy levels. Each energy level can hold no more than a certain number of electrons, and specific rules are followed in determining the electron configuration of an atom. The energy level closest to the nucleus contains electrons with the least amount of energy. To move from one energy level to another, an electron must absorb or give off energy exactly equal to the energy difference between the two energy levels. If an atom absorbs energy from an outside source, an electron in that atom can move from a lower energy level to a higher energy level. Such an atom is said to be excited. If the atom then gives off energy in the form of visible light, it is said to be luminescent. If enough energy is absorbed by an atom, one of its electrons may jump completely away from the atom, forming a positively charged ion.

The elements show periodicity, or a repeating nature, in their chemical and physical properties. Three such properties are atomic size, ionization energy, and electron affinity. If the elements are arranged according to their atomic numbers on a periodic table, members of each vertical column, called a group or family, will display similar properties. Each member of a group will have the same number of valence, or outer energy level, electrons. Elements on the periodic table can be divided into metals and nonmetals and can be further classified as representative elements, transition elements, and inner transition elements.

Important Equations

Atomic number (Z) = protons (p)

Mass number (M) = protons (p) + neutrons (n)

EXERCISES AND PROBLEMS

EXERCISES AND PROBLEMS

1. Complete the following chart:

	Subatomic Particle	*Location*	*Charge*	*Relative Mass*
1.	protons	N	1+	1
2.	neutrons	N	0	1
3.	electrons	ATN	1−	$\frac{1}{1837}$

2. Complete the following chart (assume the atom is neutral):

	Atomic Number	Symbol	Mass Number	Number of p	e^-	n
(a)	3		7			
(b)		S				16
(c)				10		10
(d)	20					20
(e)	35		80			
(f)		Sr	88			
(g)		Sn				66
(h)		Hg	200			
(i)	88					138

3. What is the atomic number, mass number, and symbol for the following atoms?
   (a) An atom with 8 protons and 8 neutrons
   (b) An atom with 90 protons and 142 neutrons
   (c) An atom with 47 protons and 60 neutrons

4. Calculate the atomic number and the mass number and use two different shorthand methods to denote an atom of each of the following elements.

	Name	*No. of p*	*No. of n*
(a)	chlorine (Cl)	17	18
(b)	cobalt (Co)	27	33
(c)	hydrogen (H)	1	2

5. State the number of protons, neutrons, and electrons in a neutral atom of each of the five isotopes of zinc (Table 5.2).

6. Using the information with which you answered question 5, describe how isotopes differ.

7. How many electrons would all together weigh one gram?

8. Describe the Bohr model of the atom.

9. What is the maximum number of electrons possible in the fourth energy level? In the sixth?

10. Explain, on the atomic level, why a watch dial glows in the dark.

11. Fill in the following table.

	Symbol	Atomic Number	Atomic Weight	Electron Configuration
(a) Lithium	Li	3	6.94	$1s^2 2s^1$
(b) Nitrogen	N	7	14.01	$1s^2 2s^2 2p^3$
(c) Neon	Ne	10	20.18	$1s^2 2s^2 2p^6$
(d) Magnesium	Mg	12	24.31	$1s^2 2s^2 2p^6 3s^2$
(e) Aluminum	Al	13	26.98	$1s^2 2s^2 2p^6 3s^2 3p$
(f) Chlorine	Cl	17	35.45	$1s^2 2s^2 2p^6 3s^2 3p$

12. What is the electron configuration of a ground state atom containing 16 electrons?

13. Element number 116 has not yet been discovered or synthesized. To what chemical group would element 116 belong?

14. State the number of valence electrons in a neutral atom of each of the following elements.

    (a) Rubidium
    (b) Indium
    (c) Phosphorus
    (d) Krypton
    (e) Beryllium
    (f) Silicon
    (g) Bromine
    (h) Selenium

15. Identify the following as (1) either a metal, metalloid, or nonmetal and (2) either a representative element or transition element.

    (a) Strontium
    (b) Selenium
    (c) Iron
    (d) Germanium
    (e) Copper
    (f) Fluorine
    (g) Silicon
    (h) Potassium

16. For each of the following pairs predict which element has (1) the larger radius and (2) the larger ionization energy.

    (a) Na and P
    (b) C and O
    (c) Li and Rb
    (d) As and F
    (e) Ne and Xe
    (f) N and Sb
    (g) Sr and Si
    (h) Fe and Br

17. For each of the following pairs, predict which element has the higher electron affinity.

    (a) Na and Cl (b) Cl and I (c) Cl and Ar

18. Which has a higher ionization energy?

    (a) A sodium atom, or a sodium ion with a charge of 1+?
    (b) A magnesium ion with a charge of 1+, or a magnesium ion with a charge of 2+?

    Give a reason for your answers.

19. Which has the higher electron affinity, a chlorine atom or a chloride ion with a charge of 1−?
    Give a reason for your choice.

20. Arrange the following elements in order from the most metallic to the least metallic:

    (a) Sulfur, Chlorine, Silicon, Phosphorus
    (b) Tin, Rubidium, Silver, Palladium

21. Look at the following electron configuration for a magnesium atom, atomic number 12.

energy level, n	1	2	3	4	5
electron configuration	2	8	1	1	

Is this atom in the ground state? Why or why not?

22. Data for the nuclei of three neutral atoms are shown below. Which two atoms would have similar chemical properties?

    Atom A contains 4 protons and 5 neutrons
    Atom B contains 8 protons and 8 neutrons
    Atom C contains 12 protons and 12 neutrons

    Give a reason for your choice.

23. Using Table 5.2, calculate the atomic weight of each of the following to three significant figures.

    (a) Chlorine
    (b) Zinc

24. In nature, the element lithium exists as a mixture of two isotopes: 92.58% of the atoms have a mass number of 7 and 7.42% have a mass number of 6.

    (a) For each isotope of lithium, calculate
        (1) The atomic number
        (2) The mass number
        (3) The number of protons
        (4) The number of neutrons
        (5) The nuclear charge

    (b) Calculate the atomic weight of lithium

*25. Write the electron configuration for a neutral atom of each of the following elements (include the orbital diagram for parts a, d, and f).

    (a) Beryllium
    (b) Iodine
    (c) Silicon
    (d) Sulfur
    (e) Selenium
    (f) Copper
    (g) Molybdenum
    (h) Calcium
    (i) Radium
    (j) Potassium

** Problems marked with an asterisk are from optional sections in the chapter.*

CHAPTER SIX

THE ATOM AND RADIOACTIVITY

Jeff looked down at the Russian landscape far below his airplane window, thinking how strange this all was. His parents had insisted he return immediately to the United States from his exchange program in Kiev because of some great nuclear disaster that had supposedly occurred near him, but he certainly hadn't been aware of any such emergency. Everything had seemed normal in Kiev as the city prepared for its May Day national holiday. The Soviets had assured the American Embassy in Moscow that Kiev was safe for travel.

However, by the time he met his parents at the Seattle airport, Jeff had read several of the magazine and newspaper articles available on the plane. He now knew about the explosion at the Chernobyl nuclear reactor and the cloud of radioactive material that was still spreading over other nearby countries. As the week passed, Jeff heard a great deal more about the Chernobyl accident and the dangers posed to those living in the vicinity of the nuclear power plant. This contrasted sharply with the information that had been available to the residents of Kiev. He listened closely to discussions of the medical consequences of exposure to such nuclear fallout.

It wasn't long before the local newspapers had large headlines reading "Radiation Found in Milk," "Soviet Radiation Reaches Northwest," and "Radiation Levels Triple in Portland." But the articles went on to say that the radiation levels measured were extremely small and were no cause for concern at the present time. Clearly the public was concerned, however. Health-food stores in Oregon and Washington, where the fallout from Chernobyl's radioactive cloud was measurably increasing the background radiation, were completely sold out of potassium iodide and were having a hard time keeping up with the demand for bottled water. (When taken before or right after exposure to radioiodine, potassium iodide is thought to reduce the amount of this radionuclide taken up by the thyroid.) Having carefully read the newspaper articles, Jeff wondered why there would be this kind of public panic over exposure to such low levels of radiation.

The radioactive cloud that circled the world in 1986 after the explosion at Chernobyl contained mainly iodine-131 and cesium-137. But radionuclides can be detected every day in many of the common foods that we eat. For example, a kilogram of beef can contain 100 to 5000 picocuries (pCi) of cesium-137. The amount of iodine-131 (47 to 64 picocuries/liter) finally detected in some samples of Oregon milk after the Chernobyl accident was well below the established limit (15,000 pCi/l) above which regulatory action would be taken. In fact, if you were to drink 10 quarts of this milk, you would receive about the same added radiation dosage as you would get from living for one year at an elevation 100 feet higher than you currently live, or from moving for a month from a wooden house to a brick house, or from flying for about 1½ hours on a commercial jet airplane.

Unlike bullets or chemical poisons, radiation cannot be felt, smelled, or tasted. Yet, as we will discuss in the next two chapters, radiation exposure can be equally harmful to the body. The world has seen clearly that single, high doses of radiation can cause severe or life-threatening illness within minutes or days. The extent of ill effects from moderate levels of radiation depends on the amount of exposure, the type of radiation, and the susceptibility of those exposed. But what about the health effects of long-term exposure to low levels of radiation? Jeff did enough reading to realize this was still a subject of debate among scientists, although most seemed to agree that there is a possible increased risk of cancer from such long-term exposure. It was clear to him that a lot more study was needed on this important question.

Do you know what makes a substance radioactive? Why do radioactive substances occur naturally? What produces the tremendous energy released in a nuclear reaction? How is it that radioactivity can both cause and cure cancer? Radioactive materials are becoming more and more important in improving the standards of human life—from medical diagnosis and treatment to prevention of food spoilage to the production of energy. It is important, however, that we each know the answers to these types of questions because our local and national governments continually make policy decisions concerning radioactive materials and nuclear energy that will affect us and future generations.

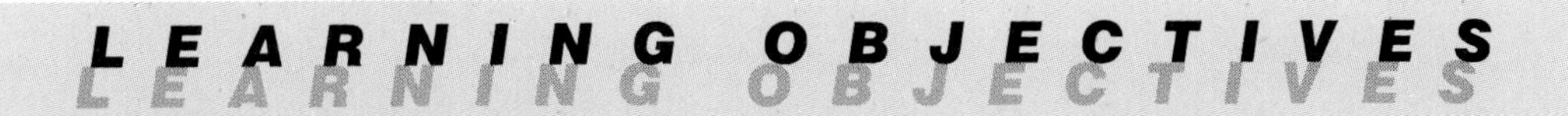

By the time you have finished this chapter, you should be able to:

1. Define *radioactivity, radionuclide,* and *decay series.*
2. Describe three types of radiation given off by radioactive material.
3. Write a balanced nuclear equation for a radioactive decay.
4. Define *half-life* and calculate the amount of radionuclide remaining after a given number of half-lives.
5. Define *nuclear transmutation* and write a balanced nuclear equation for a given transmutation.
6. Compare and contrast nuclear fission and nuclear fusion.
7. State the difference between a burner reactor and a breeder reactor and explain the advantages and disadvantages of using each to generate electricity.

RADIOACTIVITY

6.1 What is Radioactivity?

Some atoms have nuclei that are unstable. This is especially true of elements that have high atomic numbers and, therefore, a high number of positive protons in the nucleus (remember, charges that are alike repel one another). Such an unstable nucleus will decay; that is, it will "spit out" particles, leaving behind a new nucleus called a daughter nucleus. A daughter nucleus may or may not be stable. An unstable daughter nucleus will decay again, and this process will continue until a stable daughter nucleus is formed (Figure 6.1). Such a series of decays is called a **decay series** or **disintegration series** (Figure 6.2). The decay of a nucleus can

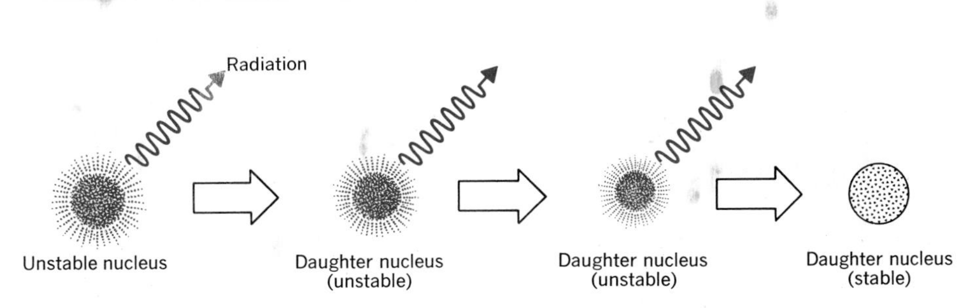

Figure 6.1 . Radioactivity is the term used to describe the emission of radiation from an unstable nucleus.

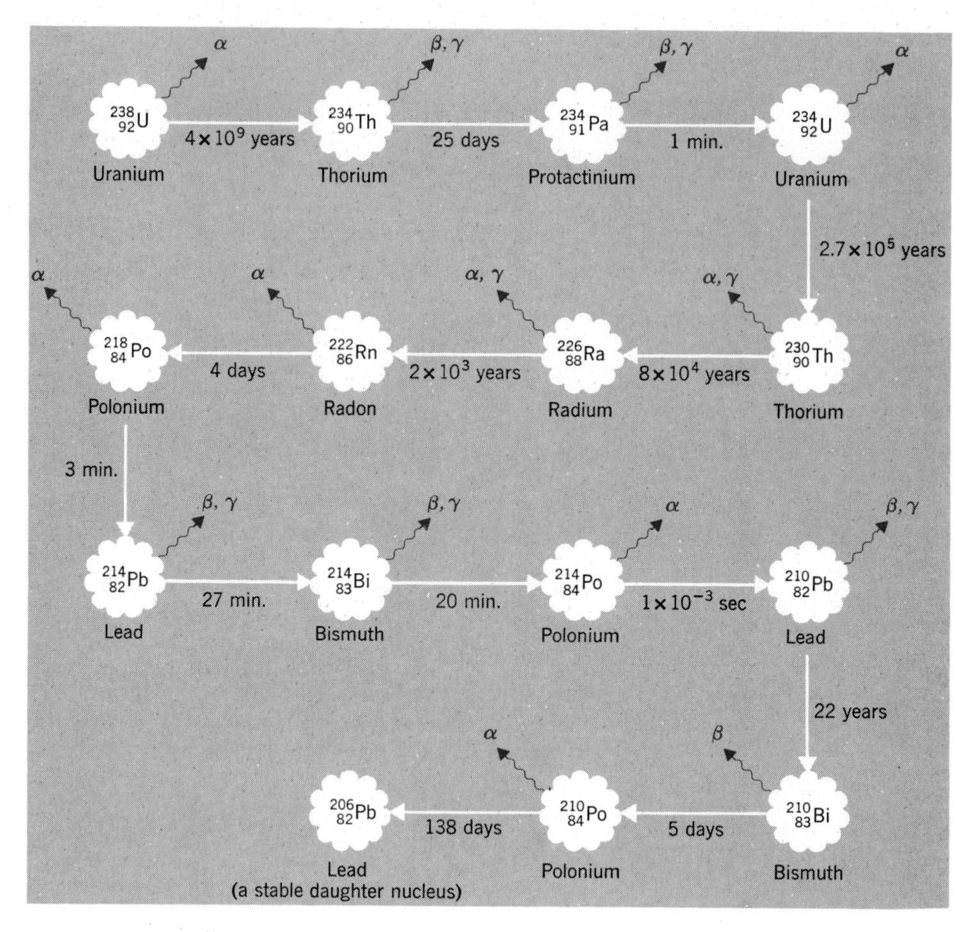

Figure 6.2 The decay series of uranium-238.

result in the emission of several different kinds of radiation (Table 6.1). **Radioactivity** is the term used to describe the giving off, or emission, of such radiation from isotopes of certain elements or their compounds, and these radioactive isotopes are called **radionuclides.**

6.2 Alpha Radiation (α Particles)

One way in which a nucleus can become more stable is by giving off alpha radiation. **Alpha radiation** consists of streams of alpha particles, each of which is made up of two protons and two neutrons (the nucleus of a helium atom, $^{4}_{2}$He). By giving off an alpha particle, the atomic number of the nucleus is reduced by two, and the mass number is reduced by four. A well-known source of alpha radiation is the most abundant isotope of uranium, uranium-238, which decays by giving off an alpha particle to form an atom of thorium-234. A shorthand way of representing this decay, called a nuclear equation, is as follows:

$$^{238}_{92}\text{U} \longrightarrow \,^{234}_{90}\text{Th} + \,^{4}_{2}\text{He}$$

Table 6.1 Particles and Radiation Emitted by Radionuclides

Particle/ Radiation	Type	Charge	Symbol
Alpha	Particle, helium nucleus	2+	α, ${}^{4}_{2}He$
Beta	Particle, electron	1−	β, ${}^{0}_{-1}e$
Gamma	Electromagnetic radiation	0	γ
Neutron	Particle	0	${}^{1}_{0}n$
Proton	Particle	1+	${}^{1}_{1}p$, ${}^{1}_{1}H$
Positron	Particle	1+	${}^{0}_{1}e$

The starting unstable nucleus is shown on the left-hand side of the arrow, and the products resulting from the radioactive decay of this nucleus are shown on the right-hand side. To be sure that this nuclear equation is correctly written, we must check that the number of protons and neutrons on one side of the arrow is equal to the number of protons and neutrons on the other side. In other words, the sum of the mass numbers on each side of the arrow must be equal, and the sum of the atomic numbers on each side of the arrow must also be equal. For the decay of uranium-238,

$$\text{Mass number:} \quad 238 = 234 + 4$$
$$\text{Atomic number:} \quad 92 = 90 + 2$$

By the way, the thorium atom produced by the decay of uranium-238 is itself unstable and will decay to form a new nucleus (Figure 6.2).

Alpha particles are the largest particles emitted by radioactive substances and have very little penetrating power. Even when traveling through air they lose energy very quickly through collisions with air molecules and will stop within a few inches. They can be stopped by a piece of paper and cannot penetrate even the dead layer of cells on the surface of your skin. An intense external dose of alpha radiation, however, would produce a burn on the skin. And alpha particles can do a great deal of damage if they are emitted inside the body, which might result from inhaling or swallowing an alpha emitter.

6.3 Beta Radiation (β Particles)

Beta radiation, like alpha radiation, consists of streams of particles. In this case, the particle is an electron (${}^{0}_{-1}e$) that is produced inside the nucleus. During beta decay, a neutron in the nucleus is changed into a proton and an electron. This electron (or β particle) is ejected from the nucleus, leaving behind a daughter nucleus having the same mass number as the original atom, but a different atomic number. For example, the thorium atom that is produced by the alpha decay of uranium-238 is a beta emitter.

$$^{234}_{90}\text{Th} \longrightarrow {}^{234}_{91}\text{Pa} + {}^{0}_{-1}e$$

To check if this nuclear equation is written correctly, we have

$$\text{Mass number:} \quad 234 = 234 + 0$$
$$\text{Atomic number:} \quad 90 = 91 + (-1) = 90$$

Beta particles are 7000 times lighter than alpha particles and have much more penetrating power. Beta particles can pass through a piece of paper, but will be stopped by a piece of wood. Beta radiation can penetrate the dead outer layer of your skin; it will be stopped within the skin layer, causing damage to the skin tissue and making it appear burned (Figure 6.3). As is the case with alpha particles, beta particles hitting the skin from the outside cannot penetrate to internal organs, but the effect on internal organs can be severe if a beta emitter is taken internally.

6.4 Gamma Radiation (γ Rays)

Gamma rays are not particles, but are high-energy radiation similar to X rays (you might look again at the energy spectrum shown in Figure 3.6). Quite often the daughter nucleus produced by an alpha or beta emitter will be in a high-energy, or excited, state. It can release this energy in the form of gamma radiation to become more stable. The release of gamma rays almost always occurs together with alpha or beta radiation. For

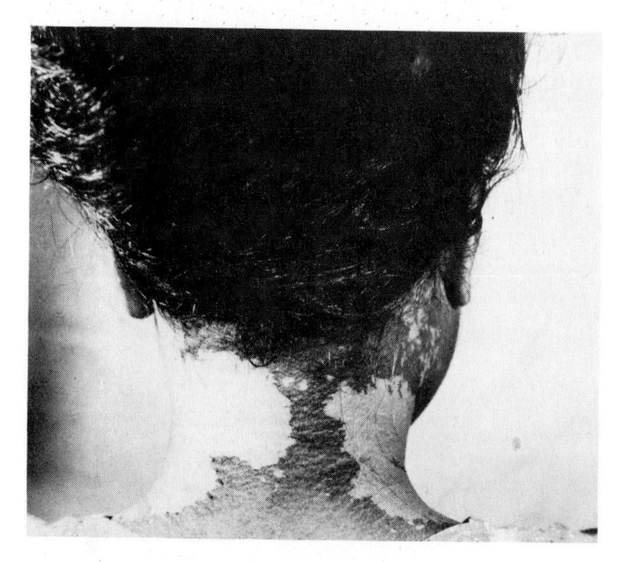

Figure 6.3 The burn on the neck of this Rongelap native was caused by accidental exposure to over 2000 rads of beta radiation from radioactive fallout. (Courtesy Brookhaven National Laboratory.)

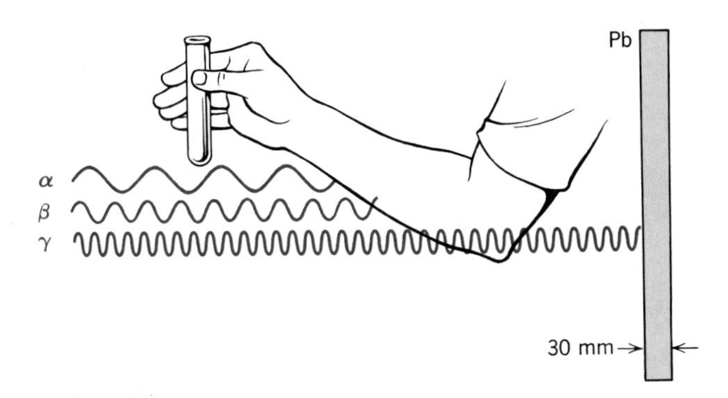

Figure 6.4 Each type of radiation (alpha, beta, or gamma) has a different penetrating power.

example, radium-226 has a radioactive nucleus that releases alpha and gamma radiation when it decays.

$$^{226}_{88}\text{Ra} \longrightarrow {}^{222}_{86}\text{Rn} + {}^{4}_{2}\text{He} + \gamma$$

Because they have such high energy, gamma rays easily pass through paper and wood, but can be stopped by lead blocks or thick concrete walls. Gamma rays will completely penetrate the human body, causing cellular damage as they pass through (Figure 6.4).

EXAMPLE 6-1

1. Determine the symbol of the element that belongs in the following decay.

$$^{3}_{1}\text{H} \longrightarrow (?) + {}^{0}_{-1}\text{e}$$

The sum of the mass numbers on both sides of the arrow must be equal, so the mass number of the unknown element is

$$3 = M + 0$$
$$M = 3$$

The sum of the atomic numbers on both sides of the arrow must be equal, so the atomic number of the unknown element is

$$1 = Z + (-1)$$
$$Z = 1 + 1 = 2$$

From the back inside cover of this book, we find that the element having an atomic number of 2 is helium, He.

The complete decay, therefore, is

$$^{3}_{1}H \longrightarrow {}^{3}_{2}He + {}^{0}_{-1}e$$

2. Polonium-214 is an alpha and gamma emitter that decays to form an isotope of lead. Write the shorthand representation of this decay.

   (a) Use the inside back cover and Table 6.1 to help write the necessary symbols

   Polonium-214 $= {}^{214}_{84}Po$
   Alpha particle $= {}^{4}_{2}He$
   Gamma ray $= \gamma$
   Lead $= {}^{M}_{82}Pb$ (we have to determine the mass number of the lead)

   (b) We can now write a representation of this decay, and use it to determine the mass number of the lead.

$$^{214}_{84}Po \longrightarrow {}^{M}_{82}Pb + {}^{4}_{2}He + \gamma$$

   The sum of the mass numbers on each side of the arrow must be equal, therefore,

$$214 = M + 4$$
$$M = 210$$

   (c) The correct representation is

$$^{214}_{84}Po \longrightarrow {}^{210}_{82}Pb + {}^{4}_{2}He + \gamma$$

Exercise 6-1

1. Write the symbol that belongs in the following nuclear equations:

   (a) (?) $\longrightarrow {}^{222}_{86}Rn + {}^{4}_{2}He$
   (b) $^{28}_{13}Al \longrightarrow$ (?) $+ {}^{0}_{-1}e$
   (c) $^{227}_{89}Ac \longrightarrow {}^{223}_{87}Fr +$ (?)

2. Write the nuclear equations for

   (a) The beta decay of $^{45}_{20}Ca$ (the decay of the $^{45}_{20}Ca$ isotope by emission of a beta particle).
   (b) The beta decay of $^{14}_{6}C$.
   (c) The alpha decay of samarium-149.

3. Cesium-137 is a radioactive waste produced by nuclear

power plants. It decays by giving off beta and gamma radiation. Write the nuclear equation for this decay.

6.5 Half-Life

Each radioactive substance has its own special rate of decay—that is, the number of emissions, or disintegrations, that occur each minute. This rate of decay is expressed by a number called the half-life of the isotope. The **half-life ($t_{1/2}$)** of a radionuclide is the length of time it takes for one-half of the atoms in a given sample to undergo radioactive decay. This means that after one half-life, one-half of a sample of the radioactive isotope will have decayed to form a new substance, and one-half will remain. After two half-lives, one-half of one-half (or one-fourth) will remain. After three half-lives, one-half of one-fourth (or one-eighth) will remain (Figure 6.5). For example, nitrogen-13 has a half-life of 10 minutes. If you have 1 gram of nitrogen-13, 10 minutes later 0.5 gram of nitrogen-13 will have decayed to form carbon-13, and 0.5 gram of nitrogen-13 will remain.

As another example, a radionuclide of technetium is widely used in medical diagnosis. This isotope has a half-life of 6 hours, which is quite favorable for purposes of diagnosis but which requires the laboratory to constantly renew its supply. If there were 1 gram of this isotope in the laboratory at 6:00 P.M. on Friday, only 0.25 gram would remain for use on Saturday morning (two half-lives later). Continuing in this way, only 1 milligram would still be found among the decay products when the lab opened on Monday morning (Figure 6.6).

Half-lives of the different radionuclides can be as short as fractions of a second, or as long as billions of years. These half-lives give an indication of how stable the isotope is. Most artificially produced radionuclides are highly unstable and have very short half-lives. Table 6.2 gives some examples of these isotopes and their half-lives.

The half-lives of radioactive elements can be very useful tools for discovering the age of archaeological objects. Carbon-14, which has a half-life of 5730 years, is used to date once-living material. This carbon isotope is created in the earth's upper atmosphere when nitrogen atoms are bombarded by cosmic rays (which are streams of particles pouring into

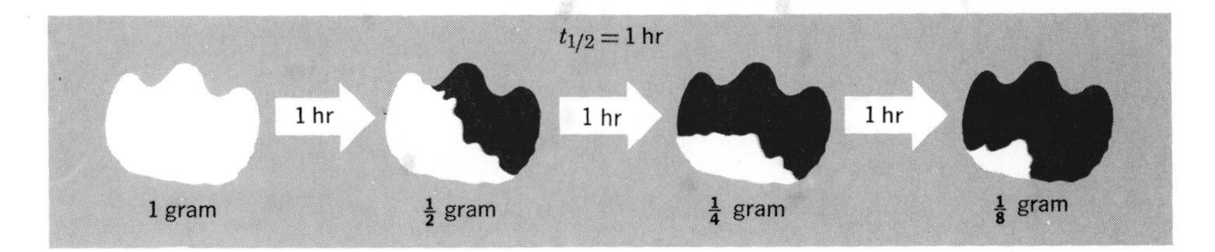

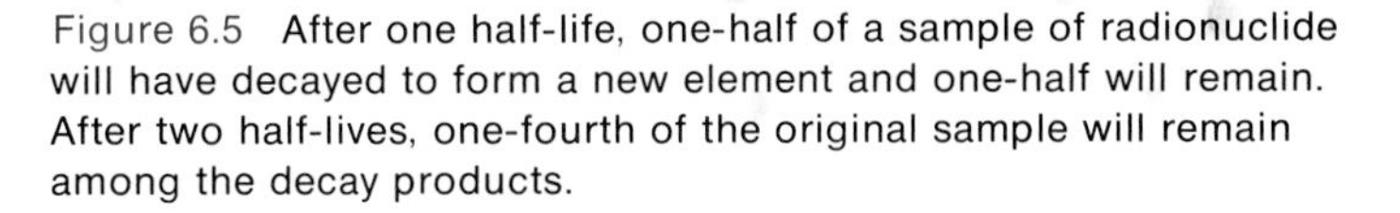

Figure 6.5 After one half-life, one-half of a sample of radionuclide will have decayed to form a new element and one-half will remain. After two half-lives, one-fourth of the original sample will remain among the decay products.

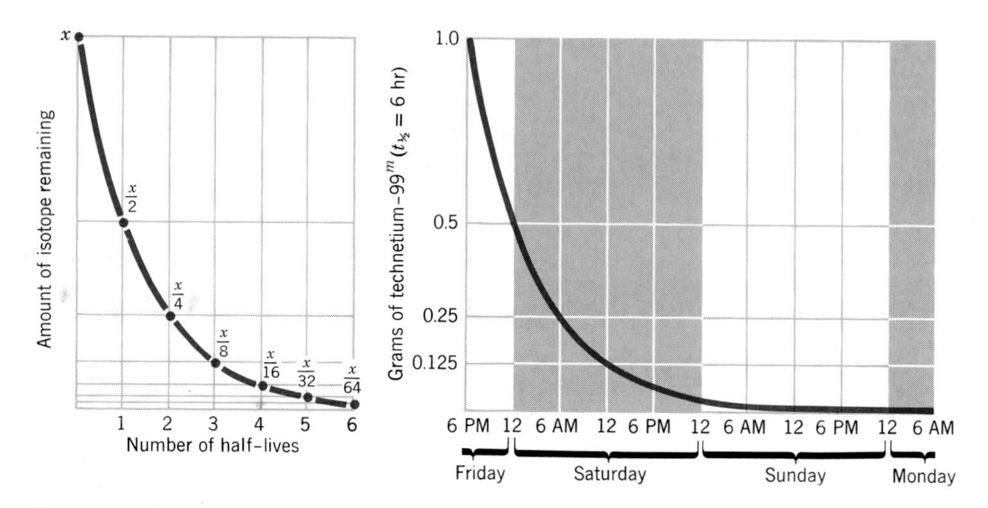

Figure 6.6 The half-life of a radionuclide is the amount of time it takes for one-half of the atoms in a given sample to undergo radioactive decay. The half-life of technetium-99^m is six hours. If you had a one-gram sample of technetium-99^m at 6 PM Friday you would have only one-half of a gram at 12 midnight, and by 6 AM Monday morning only one milligram would be left among the decay products.

Table 6.2 Some Radionuclides and Their Half-Lives

Element	Isotope	Half-Life	Radiations Given Off
Hydrogen	${}^{3}_{1}H$	12 years	Beta
Carbon	${}^{14}_{6}C$	5730 years	Beta
Phosphorus	${}^{32}_{15}P$	14 days	Beta
Potassium	${}^{40}_{19}K$	1.28×10^9 years	Beta and gamma
Cobalt	${}^{60}_{27}Co$	5 years	Beta and gamma
Strontium	${}^{90}_{38}Sr$	28 years	Beta
Technetium	${}^{99m}_{43}Tc$	6 hours	Gamma
Iodine	${}^{131}_{53}I$	8 days	Beta and gamma
Cesium	${}^{137}_{55}Cs$	30 years	Beta
Polonium	${}^{214}_{84}Po$	1.6×10^{-4} seconds	Alpha and gamma
Radium	${}^{226}_{88}Ra$	1600 years	Alpha and gamma
Uranium	${}^{235}_{92}U$	7.1×10^8 years	Alpha and gamma
	${}^{238}_{92}U$	4.5×10^9 years	Alpha
Plutonium	${}^{239}_{94}Pu$	24,400 years	Alpha and gamma

the atmosphere from the sun and outer space). The procedure of carbon-14 dating assumes that the ratio of carbon-14 to the stable carbon-12 isotope remains constant in living organisms. When such an organism dies, the total amount of carbon-12 it has accumulated during its life becomes fixed and will not change. However, half of the carbon-14 it has accumulated will be gone in 5730 years. Therefore, measuring the ratio of carbon-14 to carbon-12 in material that was once alive is a fairly accurate way to date objects that have died within the last 40,000 years.

Although carbon-14 is very useful for dating once-living objects, radionuclides with much longer half-lives must be used to date geologic periods in the earth's history. Uranium-238 ($t_{1/2} = 4.5$ billion years) decays through a decay series shown in Figure 6.2 to produce the stable lead-206 isotope. By measuring the ratio of uranium-238 to lead-206 in certain rock formations, scientists have estimated the age of the oldest rocks yet found on earth to be 4.55 billion years.

EXAMPLE 6-2

Phosphorus-32 is often used to study chemical reactions in biological research. The half-life of ^{32}P is 14 days. If a research laboratory received 500 mg of phosphorus-32, how many milligrams of ^{32}P would remain for use after 70 days?

(a) If 1 half-life equals 14 days, then 70 days would be

$$70\ \cancel{\text{days}} \times \frac{1\ \text{half-life}}{14\ \cancel{\text{days}}} = 5\ \text{half-lives}$$

(b) Using the graph in Figure 6.6, we see that after 5 half-lives the amount of ^{32}P remaining would be $\frac{X}{32}$, where X is the original amount of the isotope. In this case, $X = 500$ mg.

$$^{32}\text{P remaining} = \frac{500\ \text{mg}}{32} = 15.6\ \text{mg}$$

Exercise 6-2

Iodine-123 is used in medicine to diagnose problems of thyroid function; it has a half-life of 13.3 hours. How soon does a hospital have to reorder iodine-123 if it receives a shipment of 5.12 g and reorders when 10.0 mg remain?

6.6 Nuclear Transmutation

A **nuclear transmutation** is a reaction in which a particle collides with a nucleus to produce a different nucleus. Some transmutations occur in nature, and many are produced in laboratories. For example, carbon-14 is produced in the upper atmosphere, and can be produced in laboratories, by bombarding nitrogen-14 with neutrons.

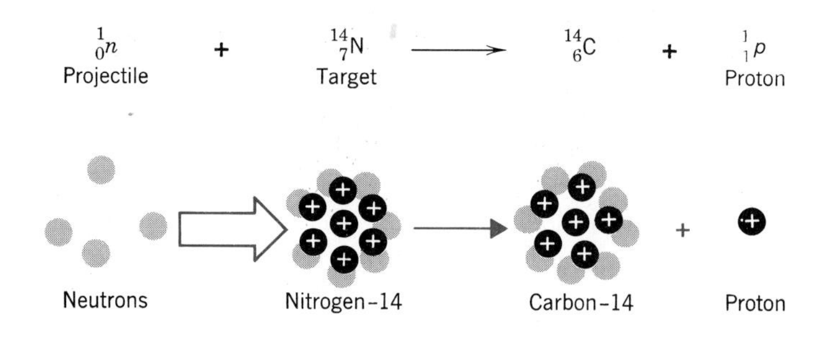

Carbon-14 is a beta emitter that can be substituted into biological compounds and traced as it travels through a living system.

$$^{14}_{6}C \longrightarrow ^{14}_{7}N + ^{0}_{-1}e$$

By using carbon dioxide "labeled" with carbon-14, Melvin Calvin, in the 1940s, was able to obtain a detailed picture of the chemical pathways of photosynthesis (the process that plants use to make sugar molecules from carbon dioxide and water).

A nuclear transmutation was produced in the laboratory for the first time in 1919 by Ernest Rutherford, who bombarded nitrogen gas with alpha particles.

$$^{14}_{7}N + ^{4}_{2}He \longrightarrow ^{18}_{9}F$$

The fluorine nucleus produced in this transmutation is very unstable and rapidly decays to form oxygen-17 and a proton.

$$^{18}_{9}F \longrightarrow ^{17}_{8}O + \underset{\text{Proton}}{^{1}_{1}p}$$

This experiment led to the discovery of the proton. The existence of the neutron also was shown by means of a nuclear transmutation. In 1932, James Chadwick bombarded beryllium-9 with alpha particles, causing the following transmutation:

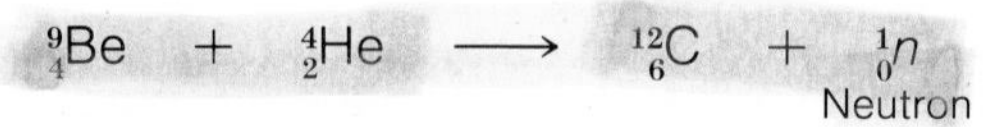

$$^{9}_{4}Be + ^{4}_{2}He \longrightarrow ^{12}_{6}C + \underset{\text{Neutron}}{^{1}_{0}n}$$

The neutrons that were produced in this transmutation caused additional nuclear reactions in nuclei with which they collided.

Before 1940, the uranium atom was the heaviest atom known. However, the invention of the cyclotron and other nuclear accelerators allowed scientists to create very high-energy projectiles, and in 1940, E. M.

McMillan and P. H. Abelson produced neptunium (^{93}Np) by bombarding uranium with a stream of high-energy deuterons (the nucleus of hydrogen-2, $^{2}_{1}H$).

$$^{238}_{92}U + ^{2}_{1}H \longrightarrow ^{239}_{92}U + ^{1}_{1}p$$

$$^{239}_{92}U \xrightarrow[t_{1/2}=23.5 \text{ min}]{} ^{239}_{93}Np + ^{0}_{-1}e$$

The neptunium produced in this way is a radioactive beta emitter and decays to form plutonium-239. Plutonium is an alpha emitter with a half-life of 24,400 years and is an important fuel used in atomic reactors and bombs.

$$^{239}_{93}Np \xrightarrow[t_{1/2}=2.33 \text{ days}]{} ^{239}_{94}Pu + ^{0}_{-1}e$$

EXAMPLE 6-3

Insert the bombarding particle for the following nuclear transmutation.

$$^{238}_{92}U + (?) \longrightarrow ^{246}_{98}Cf + 4\,^{1}_{0}n$$

The sum of the mass numbers on both sides of the arrow must be equal, so the mass number of the bombarding particle is

$$238 + M = 246 + 4(1) \quad [\text{4 neutrons are produced}]$$
$$M = 250 - 238 = 12$$

The sum of the atomic numbers on both sides of the arrow must be equal, so the atomic number of the bombarding particle is

$$92 + Z = 98 + 4\,(0)$$
$$Z = 98 - 92 = 6$$

The element with an atomic number of 6 is carbon. Therefore, the bombarding particle is $^{12}_{6}C$.

Exercise 6-3

Complete the following nuclear transmutations:

(a) $^{27}_{13}Al + ^{1}_{0}n \longrightarrow (?) + ^{4}_{2}He$
(b) $^{35}_{17}Cl + ^{1}_{0}n \longrightarrow (?) + ^{1}_{1}p$
(c) $^{7}_{3}Li + (?) \longrightarrow ^{7}_{4}Be + ^{1}_{0}n$
(d) $(?) + ^{1}_{0}n \longrightarrow ^{199}_{79}Au + ^{0}_{-1}e$

THE NUCLEUS AND ENERGY

Energy produced from atomic nuclei has deeply affected our modern world. From such peaceful uses as the generation of electricity in nuclear power reactors to such destructive uses as the explosive power in nuclear weapons, this energy source has imposed new responsibilities on the citizens of this planet. To participate in decisions regarding the future use of this nuclear technology, it is important that you understand how such vast amounts of energy can be produced from the nuclei of atoms and what the hazards accompanying such energy production are.

6.7 Nuclear Fission

When bombarded with neutrons, the nuclei of several isotopes (^{235}U, ^{233}U, and ^{239}Pu) are capable of breaking apart or undergoing **fission** to form smaller, more stable nuclei. For example,

$$^{1}_{0}n + ^{235}_{92}U \longrightarrow ^{236}_{92}U \longrightarrow ^{139}_{56}Ba + ^{94}_{36}Kr + 3^{1}_{0}n + \text{Energy}$$

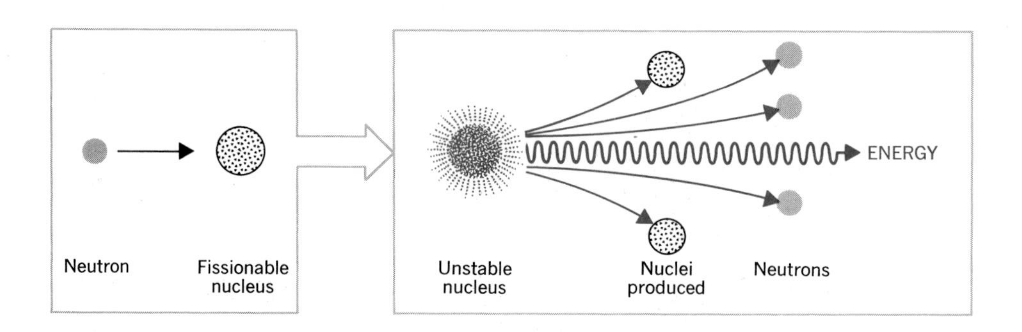

This fission process yields great amounts of energy. When 1 kilogram of ^{235}U or ^{239}Pu is fissioned, the amount of energy released is equivalent to 20,000 tons of TNT: Atomic bombs dropped on Japan at the end of World War II fissioned about 1 kilogram of material.

In nuclear fission, an atom of fissionable fuel (such as uranium-235) is struck by a neutron and breaks into two small fragments that fly apart at high speeds. The kinetic energy of these fast-moving particles is converted into heat as they collide with surrounding molecules. In addition to these two fragments, two or three neutrons are released, and these neutrons can react with other ^{235}U nuclei. If this process continues, a **chain reaction** occurs (Figure 6.7). If the rate of fission in a chain reaction is uncontrolled and extremely rapid, then all of the energy is released in a very short time—resulting in the explosion of an atomic bomb. But if the rate of fission in the chain reaction is held to a slower, constant rate, the result is the controlled energy release of a nuclear reactor. A chain reaction cannot occur, however, unless there are at least a minimum number of fissionable nuclei (called the **critical mass**) present in one place.

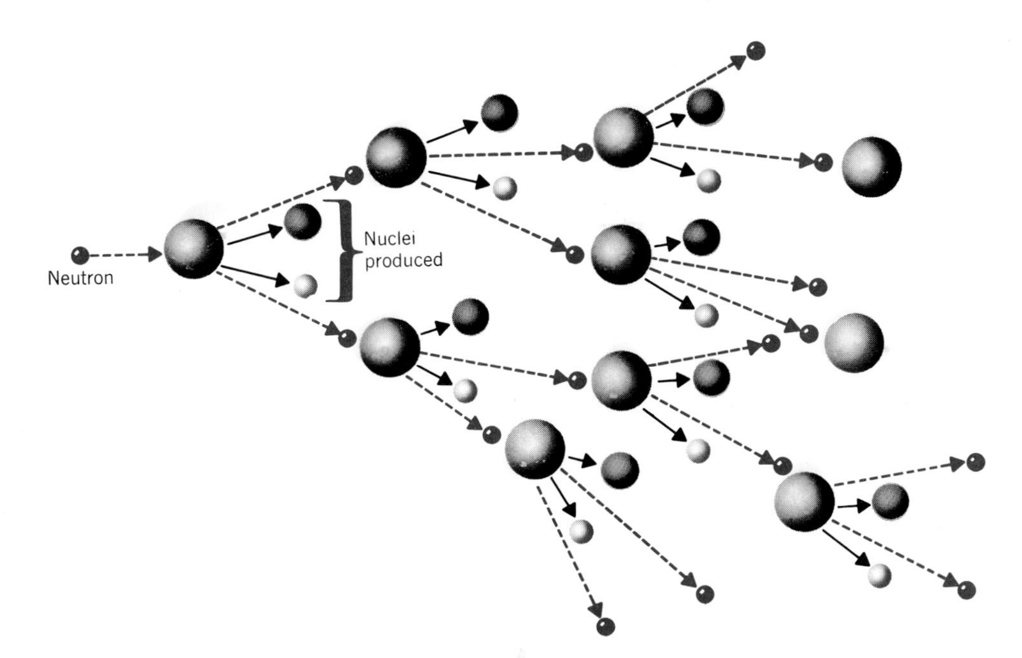

Figure 6.7 A nuclear chain reaction. When a nucleus undergoes fission, two or three neutrons are produced that are capable of reacting with other nuclei. If a critical mass of fissionable nuclei is present, a chain reaction will occur.

6.8 Nuclear Reactors

Burner Reactor

The first atomic reactor was built in 1942, in an abandoned squash court under the grandstands of the University of Chicago football field. This reactor was a **burner reactor,** meaning one that uses up its nuclear fuel as the reaction occurs. Today, several hundred burner reactors have been built around the world, in various sizes and designs depending upon their purpose. The basic components of all nuclear reactors are the same (Figure 6.8). These components include:

1. The fuel, which contains significant amounts of one of the fissile (easily fissioned) isotopes—^{235}U, ^{233}U, or ^{239}Pu—of which ^{235}U is the only naturally occurring isotope.
2. A moderator, such as graphite or water, which slows down the neutrons produced in the fission process.
3. Control rods made of cadmium or boron steel, which will absorb neutrons and which can be moved in and out of the reactor to control the rate of the reactions.
4. A heat transfer fluid such as water, liquid sodium, or pressurized

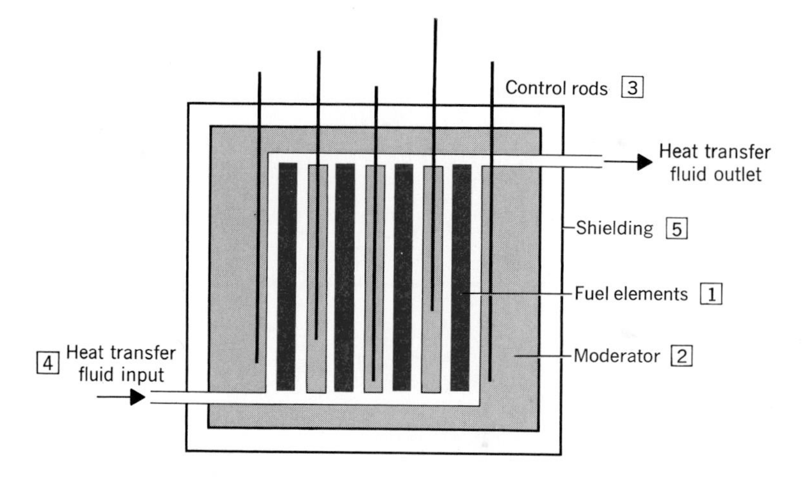

Figure 6.8 The core of a nuclear reactor.

gas, which removes the heat from the reactor core and transfers it to a steam generating system.

5. Shielding, which consists of an internal thermal shield to protect the walls of the reactor from radiation damage, and an external biological shield of high-density concrete to protect the workers from radiation.

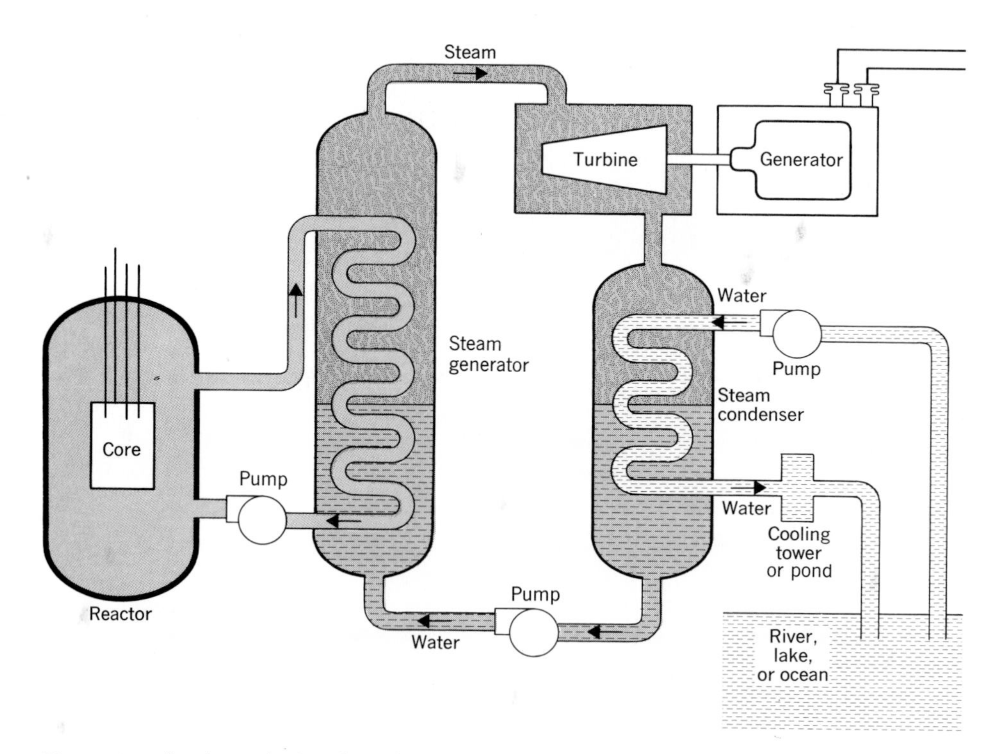

Figure 6.9 A schematic drawing of a nuclear power plant.

Figure 6.10 This nuclear plant generates electric power. The reactor is located in the building on the left; the tower on the right is used to cool the water before it is returned to the river. (Courtesy of Arkansas Power and Light.)

The exact reactor design will depend upon its use, which could be electric power generation, the production of plutonium-239, the propulsion of ships or rockets, or the production of heat for desalinization, drying, evaporation, or other industrial uses (Figures 6.9 and 6.10).

Breeder Reactors

In burner reactors, the ^{235}U fuel is used up as energy is produced. This uranium isotope is the only naturally occurring fissile fuel and is found only in extremely small amounts in uranium ore. Because the world's supply of this fuel may be relatively quickly exhausted, the true promise of nuclear power currently hinges on the development of **breeder reactors.** A breeder reactor is one in which more fissile fuel is produced than is used up in the heat generation process.

A breeder reactor uses uranium-238 and thorium-232, both of which are relatively abundant isotopes. By positioning them so that they are bombarded by some of the neutrons produced in the fission of uranium-235, new fissile isotopes are formed.

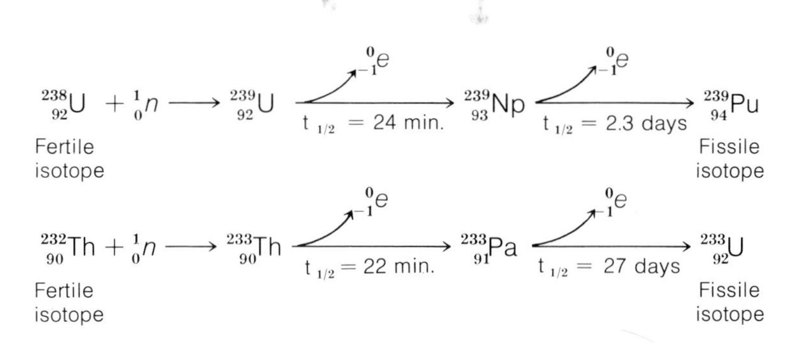

If the technology of breeder reactors can be sufficiently developed to produce large-scale, cost-efficient reactors, and if techniques can be perfected to safely transport and process large amounts of plutonium-239 and uranium-233, then the supplies of fertile uranium-238 and thorium-232 will allow the production of energy equal to hundreds of times the world's fossil fuel reserves.

Nuclear Wastes

One of the critical problems in the use of reactors for energy production is the question of safety. As we will see in the next chapter, exposure to large amounts of radioactive material can have damaging effects on living organisms, and there is still much debate in the scientific community about the long-term harmful effects of low levels of radiation on living organisms. Reactor safety systems must be able to protect against malfunctions that could result in accidents ranging from small leaks of radioactive gas, to a major release of radioactive materials that could contaminate an entire surrounding area.

Long-lived radioactive wastes are produced in the fission process, and must be safely stored for long periods of time—at least 20 half-lives. For example, strontium-90, a common and biologically dangerous waste (with a relatively short half-life of 28.1 years), should be kept in safe storage for about 600 years. Various alternatives for permanent waste disposal continue to be carefully studied. Technology is being developed to trap radioactive gases and to store them until they have decayed, to concentrate liquid wastes, and to dispose of highly radioactive wastes in solid form. But the nuclear emergency at Pennsylvania's Three Mile Island reactor in 1979 and the explosion at the Russian Chernobyl reactor in 1986 has led to governmental reviews of regulatory procedures and to increased pressure by citizens' groups for more stringent safety regulations. Such regulations may eventually make the power generated by nuclear power plants so expensive that nuclear fission will no longer be economically competitive with other forms of power generation.

6.9 Nuclear Fusion—A Captured Sun

Enormous amounts of energy are given off when small nuclei (such as those of hydrogen, helium, or lithium) combine to form heavier, more stable nuclei. This process is called **nuclear fusion.** Nuclear fusion reactions are the source of the energy released by the sun and by the hydrogen bomb. The fusion reaction occurring on the sun involves the combination of four hydrogen atoms and releases four times as much energy per gram of fuel as does a fission reaction. The following fusion reaction is the one that is most likely to be used here on earth because it occurs at the lowest temperature and because deuterium can be obtained from the sea at a very low cost.

The use of nuclear fusion is proving to be much more difficult than the development of nuclear fission because temperatures of 100 million

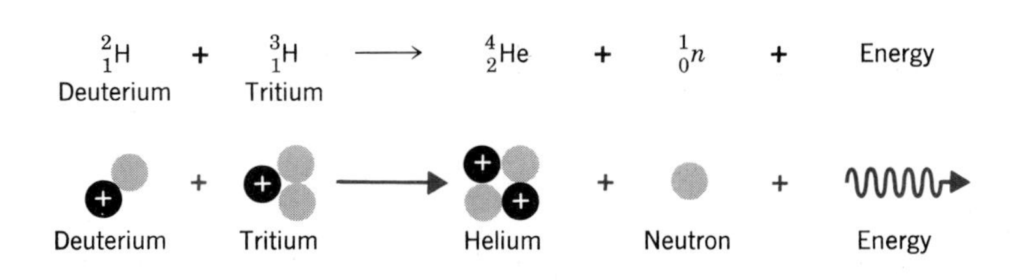

degrees Celsius or more are required for the fusion reaction to occur. If fusion reactors can be developed at a cost comparable to fission reactors, however, this process has great advantages over fission as a source of the world's energy supply. It would be more efficient, the fuel would be cheap and almost inexhaustible, there is no possibility of runaway accidents from a melting of the reactor core (as could occur in fission reactors), and there is very little radioactive waste. Tritium, which is one of the least toxic radioactive substances, may be produced, but it can be returned to the system as a fuel.

CHAPTER SUMMARY

The nuclei of atoms with high atomic numbers are often unstable and will emit particles to form a new daughter nucleus. This process is called radioactive decay. There are several types of natural radioactive radiation, but the three most common are alpha, beta, and gamma radiation. Each radionuclide has its own characteristic rate of decay, called the half-life of the isotope. Nuclear transmutation, the collision of a particle such as a neutron or alpha particle with a nucleus to produce a new nucleus, occurs in nature and has been used in the laboratory to produce elements with atomic numbers greater than 92.

When bombarded with neutrons, the nuclei of several elements are capable of undergoing fission to produce smaller, more stable nuclei and large amounts of energy. The energy generated in a nuclear fission reaction can be produced in a controlled fashion in a nuclear reactor. The fusion of several small nuclei into a heavier, more stable nucleus also generates large amounts of energy and is the source of energy released by the sun.

EXERCISES AND PROBLEMS

1. What causes a substance to be radioactive?
2. What is the difference between the three most common radiations given off by radioactive material?
3. What is the meaning of the term *nuclear transmutation?* Give an example.

4. What is the difference between atomic fission and atomic fusion? Which process is presently being used in power plants?

5. Why is the breeder reactor preferable to the burner reactor for meeting the world's long-range energy requirements?

6. Fill in the correct symbol, including the mass number and atomic number.

(a) $^{144}_{60}\text{Nd} \longrightarrow {}^{140}_{58}\text{Ce} + \quad (?)$

(b) $(?) \longrightarrow {}^{90}_{38}\text{Sr} + {}^{0}_{-1}\text{e} + \gamma$

(c) $(?) \longrightarrow {}^{206}_{82}\text{Pb} + {}^{4}_{2}\text{He}$

(d) $^{214}_{82}\text{Pb} \longrightarrow {}^{214}_{83}\text{Bi} + \quad (?) \quad + \gamma$

(e) $^{150}_{64}\text{Gd} \longrightarrow \quad (?) \quad + {}^{4}_{2}\text{He}$

(f) $^{234}_{91}\text{Pa} \longrightarrow \quad (?) \quad + {}^{0}_{-1}\text{e} + \gamma$

(g) $(?) \longrightarrow {}^{8}_{4}\text{Be} + {}^{0}_{-1}\text{e}$

(h) $^{245}_{96}\text{Cm} \longrightarrow \quad (?) \quad + {}^{4}_{2}\text{He}$

(i) $^{125}_{50}\text{Sn} \longrightarrow {}^{125}_{51}\text{Sb} + \quad (?)$

(j) $^{13}_{7}\text{N} \longrightarrow {}^{13}_{6}\text{C} + \quad (?)$

7. Write balanced equations for the following nuclear reactions:

(a) Alpha decay of beryllium-8
(b) Beta decay of iodine-135
(c) Beta decay of oxygen-20
(d) Alpha decay of berkelium-245
(e) Neutron emission by bromine-88
(f) Proton emission by potassium-37

8. Insert the correct bombarding particle for the following nuclear transmutations.

(a) $^{27}_{13}\text{Al} + \quad (?) \quad \longrightarrow {}^{30}_{15}\text{P} + {}^{1}_{0}n$

(b) $^{252}_{98}\text{Cf} + \quad (?) \quad \longrightarrow {}^{257}_{103}\text{Lw} + 5\ {}^{1}_{0}n$

(c) $^{46}_{20}\text{Ca} + \quad (?) \quad \longrightarrow {}^{47}_{20}\text{Ca}$

9. Complete the following nuclear transmutations:

(a) $(?) \quad + {}^{0}_{-1}\text{e} \longrightarrow {}^{54}_{24}\text{Cr}$

(b) $^{96}_{42}\text{Mo} + \quad (?) \longrightarrow {}^{1}_{0}n \quad + {}^{97}_{43}\text{Tc}$

(c) $^{54}_{26}Fe + ^{1}_{0}n \longrightarrow (?) + ^{1}_{1}p$

(d) $^{98}_{42}Mo + ^{1}_{0}n \longrightarrow (?) + ^{0}_{-1}e$

10. Strontium-90 is a beta emitter with a half-life of 28 years.
    (a) Write the shorthand representation for this decay.
    (b) Draw a half-life graph for the decay of 100 grams of strontium-90.
    Use your graph to answer the following:
    (c) How long would it take for $\frac{3}{4}$ of the sample to decay?
    (d) How many grams would be left after 4 half-lives?
    (e) How many grams would remain after 252 years?
    (f) How many half-lives would it take to have less than 1 gram remaining?

11. It has been suggested that radioactive wastes be stored for at least 20 half-lives. How long would that be for cesium-137, a radioactive waste produced in nuclear power plants?

12. Iodine-131 is used to destroy thyroid tissue in the treatment of an overactive thyroid. If a hospital received a shipment of 20.0 g of iodine-131, how much I-131 would remain in the laboratory after 20 days?

13. Sodium-24 is a beta and gamma emitter. Write the balanced nuclear equation for this decay. If 12.5 mg of ^{24}Na remains from a 200-mg sample after exactly 60 hours, what is the half-life of ^{24}Na?

14. If you had $10 to gamble in a slot machine in Las Vegas, but could bet only half of it the first hour, and half of the remainder in each succeeding hour, at the end of what hour would you have $1.25 left? (Neglect any earnings.) How is this problem similar to the half-life of a radioactive substance?

15. Radon-222 is an alpha particle emitter that decays to polonium-218. Write the balanced nuclear equation for this radioactive decay.

16. Bismuth-210 is a beta emitter that decays to form an isotope of polonium. Write the balanced nuclear equation for this radioactive decay.

17. When the thorium-230 isotope decays, it gives off an alpha particle and gamma radiation. Write the balanced nuclear equation for this radioactive decay.

CHAPTER SEVEN

RADIOACTIVITY AND THE LIVING ORGANISM

It was a beautiful spring afternoon, and 14-year-old Rebecca White was delighted by this opportunity to ride her horse, Nicky, out in the hillside pasture. As they were trotting near the woods at the end of the pasture, however, a startled deer dashed out in front of them. The suddenly frightened horse shied away and in doing so slipped on the wet ground and fell down to his knees. Rebecca had no time to recover; she tumbled off, landing hard on her left side. A sharp pain shot through her side, and for a moment she was unable to breathe. For a while she lay on the ground, almost in a daze, as Nicky nudged her as if to ask if she were all right. When finally she could move again, Rebecca slowly sat up and noticed that she had landed on the point of a rock sticking out of the ground. It was more than ten minutes before she was able to stand up, but even then the pain kept shooting through her side and she couldn't move her left arm very well. Nicky also was limping slightly on his left foreleg, so they both walked very slowly back to the barn.

When she arrived back, Rebecca assured everyone that she was all right. Her concern about Nicky led her to ignore her pain as she checked over his leg and made certain that he would be taken care of. After 20 minutes, however, the pain became so intense that Rebecca collapsed, crying, on a bale of hay. Realizing something was seriously wrong, Rebecca's father quickly drove her to the local hospital.

After carefully examing Rebecca, the emergency room physician ordered X rays to be taken of the ribs on Rebecca's left side. The X rays revealed that two ribs were badly broken. Unfortunately, these ribs were located directly above the spleen, and the doctor became concerned that the spleen might also have been damaged in the fall. The spleen is an organ, located behind the left side of the stomach, which stores blood, filters old red blood cells from the blood, and produces white blood cells. If the spleen were damaged, internal hemorrhaging could result—causing blood loss, shock and, if undetected and untreated, death.

Until the late 1960s, the only way to tell if a person's spleen had been damaged was to perform an abdominal tap. But even this procedure, which consists of drawing out abdominal fluid with a long needle to see if there is blood in it, would give a negative result if the spleen had ruptured within its elastic capsule of smooth muscle. But now, by administering a compound containing a radionuclide that collects in the liver and spleen, and using a special camera that detects gamma radiation given off by that compound, the doctor was able to determine quickly if Rebecca's spleen had been damaged. To prepare Rebecca for the test, 8 millicuries of technetium sulfur colloid (about the same radiation exposure as a series of three abdominal X rays) were injected into the vein in Rebecca's arm. After five minutes, Rebecca was placed on a table; the gamma camera was positioned over her abdomen, and the pattern of gamma radiation emissions from her body was viewed from several angles and precisely recorded on a sensitized photographic film (Figure 7.1). The resulting images of Rebecca's liver and spleen indicated to the doctor that the spleen had not been damaged. Luckily, Rebecca's recovery would involve only the healing of her broken ribs.

The use of radioactive materials is increasingly becoming a standard procedure in many areas of medical diagnosis. Such materials are also now playing important roles in the treatment of disease. It may seem just a bit contradictory, perhaps, to realize that these same materials can also critically damage and kill living organisms. In this chapter, we will discuss the different effects of radiation on living organisms and describe the use of radiation in the diagnosis and treatment of disease.

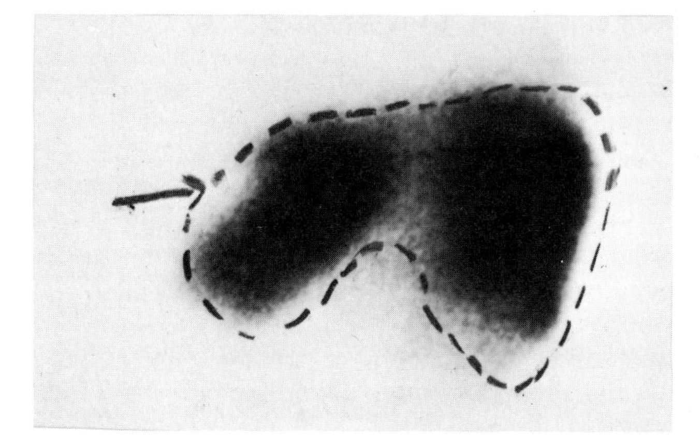

Figure 7.1 A gamma scan showing the posterior view of the liver and spleen. The liver is on the right and the spleen on the left. The area of decreased uptake of radioactive material at the arrow indicates a laceration or rupture of the spleen has occurred. (Courtesy Department of Nuclear Medicine, Albany General Hospital, Albany, Oregon)

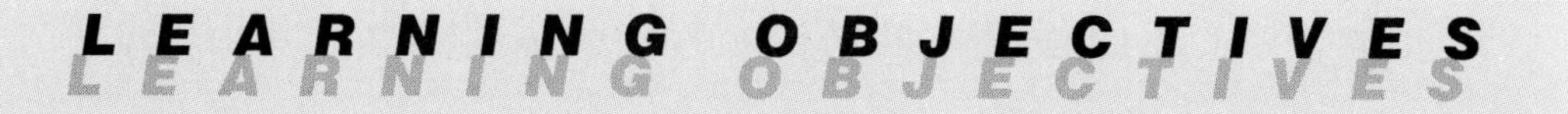

By the time you have finished this chapter, you should be able to:

1. Define *ionizing radiation.*
2. Describe two ways in which ionizing radiation can damage living tissue.
3. Define the following terms: *becquerel, curie, rad, gray, rem, sievert,* and *LET.*
4. Describe two methods for detecting ionizing radiation.
5. List three sources of background radiation.
6. Describe two clinical procedures that use ionizing radiation to aid in diagnosis.
7. Describe three different techniques for using radioactive materials in medical therapy.

THE EFFECTS OF RADIATION ON LIVING CELLS

7.1 Ionizing Radiation

Why are radioactive substances potentially so harmful? It is because alpha, beta, gamma, and cosmic radiation, when they hit living tissue, are each capable of producing unstable and highly reactive charged particles called ion pairs. For this reason, these radiations are called **ionizing radiation.**

In addition, such radiation can transfer so much energy to the molecules in living tissue that these molecules will vibrate apart, forming high-energy, uncharged fragments called free radicals (Figure 7.2). Free radicals are even more reactive than ions; they are capable of pulling other molecules apart and can completely disrupt living cells. Ions and free radicals may recombine with one another (resulting in little damage), or may combine with other molecules to form new substances foreign to the cell and, therefore, potentially dangerous. The new substances produced in this way are often highly reactive chemically, but are not themselves radioactive.

If we focus for a moment on an individual cell of the body, there are two ways in which ionizing radiation can cause great damage. First, the damage may be caused by direct action—a direct hit on a biologically important molecule that can cause the molecule to split into biologically useless fragments (Figure 7.3). The most vital molecule in a living cell is the DNA molecule, which carries all the "blueprints" necessary for the cell to divide and reproduce. If the DNA is destroyed, the cell cannot divide and will die. When such cells die without replacement, the entire irradiated tissue will eventually die. Furthermore, if this tissue is essential to the organism, the entire organism may die prematurely.

Even if the DNA molecule is only damaged and not destroyed completely, it may cause the cell to divide abnormally into new cells with altered DNA (Figure 7.4). Such cells are known as **mutant** cells. A mutant cell may have its DNA so altered that it is no longer under the body's control. The cell may then begin to grow and divide in an uncontrolled fashion, destroying the normal cells around it. Cells that behave in this manner are called **cancerous** or **malignant.**

Direct hits of ionizing radiation on biologically important molecules are not the only way that radiation can damage cells. Ionizing radiation can also cause damage through indirect action (see Figure 7.3). Animal cells are about 80% water, and ionizing radiation can cause water molecules in

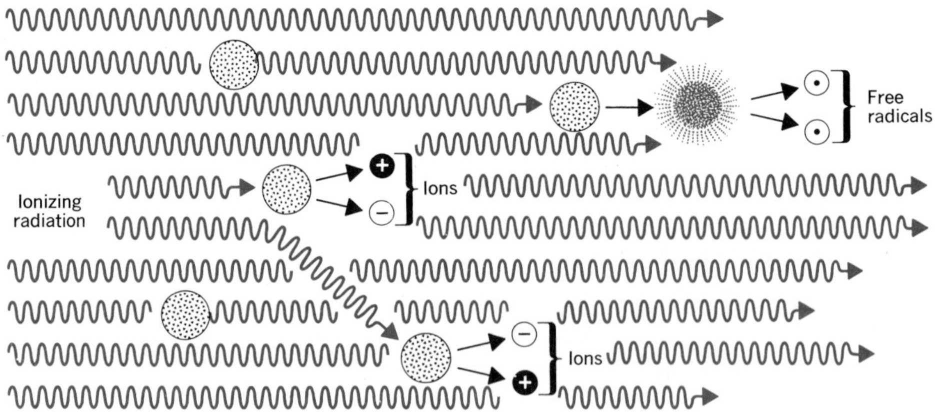

Figure 7.2 Ionizing radiation, when it hits living tissue, is capable of producing unstable and highly reactive charged particles called ions, and even more reactive uncharged particles called free radicals.

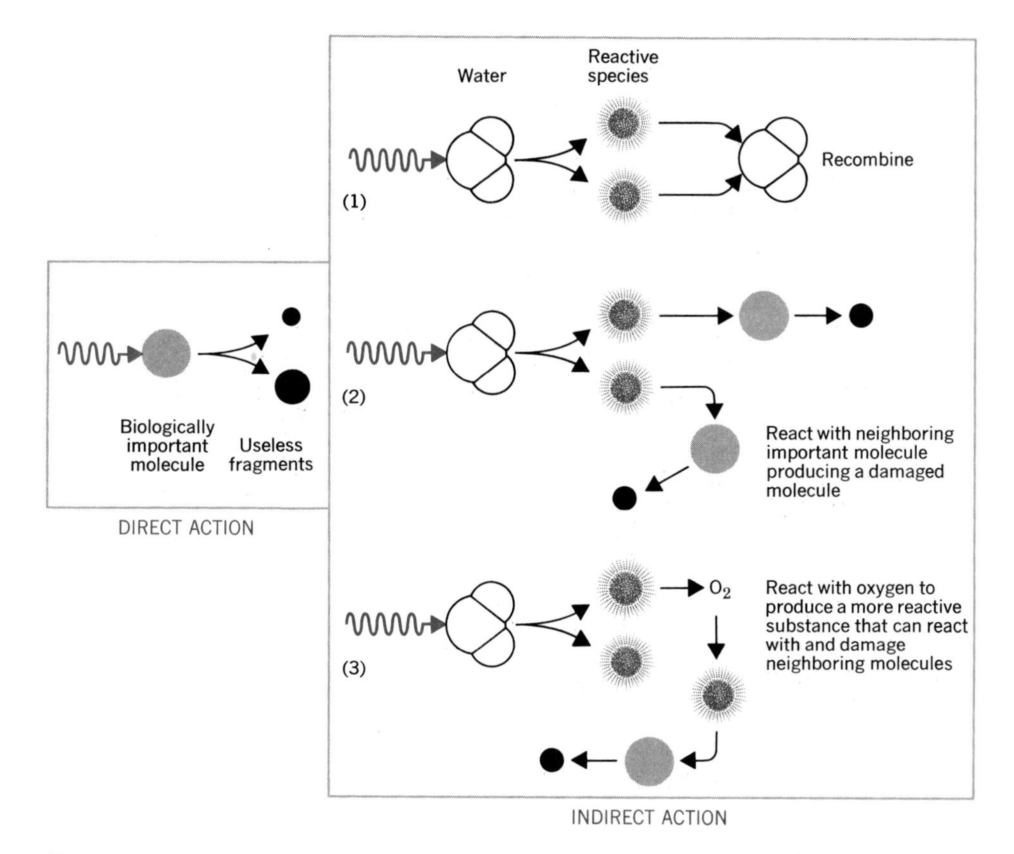

Figure 7.3 Ionizing radiation can damage living tissue by direct action or by indirect action.

living tissue to form positive and negative ions, or to form highly reactive free radicals. Such free radicals can recombine to form water (H_2O, which would be harmless), can combine to form hydrogen (H_2, which can be tolerated in small amounts by living cells), or can combine to form hydrogen peroxide (H_2O_2, which is a highly toxic substance). This may be the reason that radiation sickness resembles hydrogen peroxide poisoning in many respects. Free radical fragments can also react with oxygen in the cell to produce a free radical that is even more undesirable than hydrogen peroxide.

7.2 Radiation Dosage

There are many scientific units used to measure various aspects of ionizing radiation. These different units help answer specific questions about the nature of the radioactive source and its effects on living tissues. Some of these units are summarized in Table 7.1.

The first aspect of ionizing radiation that we might wish to describe is the activity level of the source of that radiation. That is, how often do disintegrations occur per unit of time? The SI unit used to describe the activity of a source is the **becquerel (Bq),** named after the French physicist

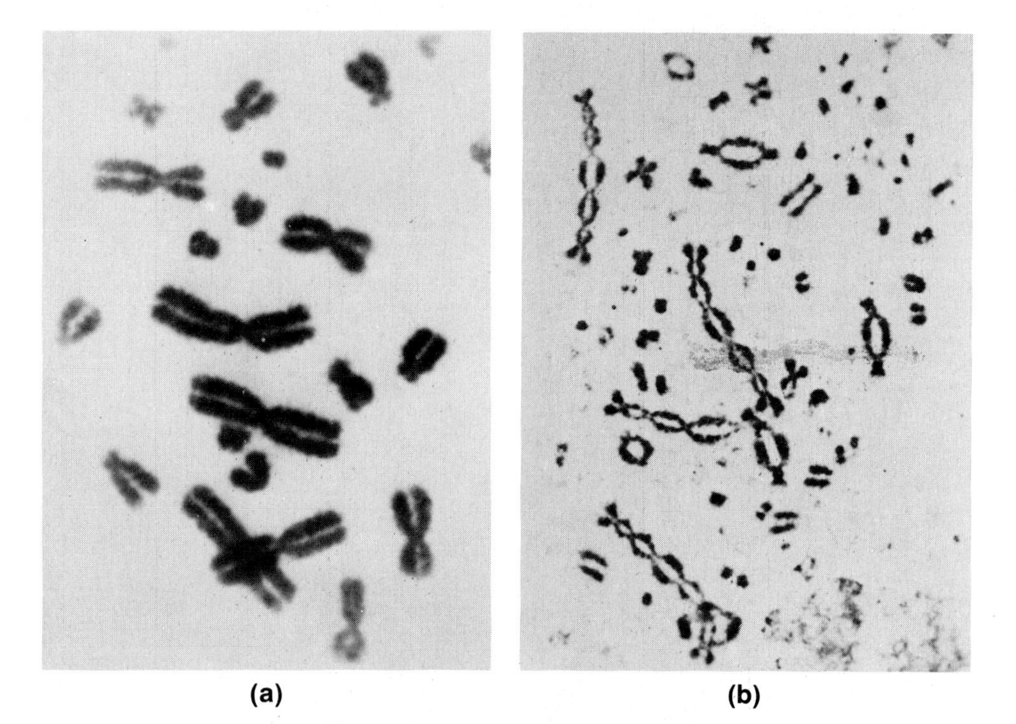

Figure 7.4 The effect of radiation on living cells. (a) A normal cell nucleus containing 23 pairs of chromosomes, composed of DNA molecules and protein. (b) Ionizing radiation has caused the chromosomes in this cell nucleus to double and triple abnormally. (Courtesy Argonne National Laboratory.)

Antoine Becquerel (1852–1908), who discovered radioactivity. A becquerel is equal to one disintegration or nuclear transformation per second. An older unit of activity is the **curie (Ci),** named after Marie Curie (1867–1934), the discoverer of radium. One curie is equal to 3.7×10^{10} disintegrations per second, which is the rate of disintegration of one gram of radium-226. It is important to realize that neither of these units tells us anything about the type of radiation produced, or the effect that this radiation will have on tissue or other matter. They describe only the rate at which radiation is produced. In radiation therapy, hospitals often use a sample of cobalt-60 as their source of gamma radiation. If a source is rated at 1.4 Ci, it will deliver $1.4 \times 3.7 \times 10^{10}$, or 5.2×10^{10}, disintegrations per second. A 2-Ci cobalt-60 source will deliver to a patient twice as many gamma rays per unit of time as a 1-Ci cobalt-60 source.

To study the effects of ionizing radiation on living tissue, we must try to measure the amount of energy that has been absorbed by the tissue. The dose actually absorbed by a tissue will be influenced by many factors, including the nature of the radioactive source, the energy of the radiation, the distance from the source to the tissue, the nature of the tissue itself, and the duration of exposure. The **rad** (**r**adiation **a**bsorbed **d**ose) is a unit used to measure the amount of energy that is absorbed by irradiated tissue. Regardless of the type of radiation, one rad is an absorbed dose of radiation that results in the transfer of 100 ergs of energy to a gram of

Table 7.1 Units Used to Measure Ionizing Radiation

Measurement	Unit	Definition
Rate of disintegration	curie (Ci)	1 Ci = 3.7×10^{10} disintegrations per second
	becquerel (Bq) (SI unit)	1 Bq = 1 disintegration per second
Absorbed dose of radiation	rad	1 rad = absorbed radiation that liberates 100 ergs per gram of tissue
	gray (Gy) (SI unit)	1 Gy = 100 rad
	rem	1 rem = absorbed dose that produces same biological effect as 1 rad of X rays
	sievert (Sv) (SI unit)	1 sievert = 100 rem
Energy transferred to tissue by absorbed dose	LET	LET = amount of energy transferred to tissue per unit of path length

irradiated tissue. The SI unit corresponding to the rad is the **gray (Gy),** where 1 Gy = 100 rad. Most mammals will not survive a dose of 1000 rads of radiation to the entire body (this means the delivery of 1000 × 100 ergs, or 100,000 ergs, of energy per gram of the mammal's tissue). Now, the erg is a very small unit of energy—100,000 ergs equals 0.0024 calorie. Why is such a small amount of energy all that is needed for a lethal dose of radiation? As we discussed earlier, it is the ability of ionizing radiation to change important biological molecules through direct or indirect action that results in the death of cells.

Scientists have discovered that the absorption of one rad of different types of ionizing radiation will produce different effects on identical biological systems. For example, it takes twice the dose of cobalt-60 gamma radiation as it does a specific level of X rays to produce the same disruption in a certain type of pollen grain. Therefore, in order to study the cumulative effects of different types of radiation on humans, a unit called the **rem** (**r**oentgen **e**quivalent **m**an) was devised. For each type of radiation, the rem is defined as the absorbed dose of radiation which will produce the same biological effect as one rad of therapy X rays. The SI unit of absorbed dose that corresponds to the rem is the **sievert (Sv),** with 1 sievert = 100 rem. When a dose is stated in rem, there is no need to specify the type of radiation, because the biological effects will be the same for a rem of each type. For example, 700 rem of any type of radiation over the whole body is a dose sufficient to kill 50% of a population of rats within 30 days (Table 7.2). One rad of beta radiation has a rem of 1, but one rad of alpha radiation has a rem of 10 to 20.

Why should one rad of different types of radiation have such different

Table 7.2 LD_{50}/30-Day Dose Values in Rems for Four Mammals

Animal	Rem
Dog	300
Man	500 (estimated)
Mouse	600
Rat	700

effects on living tissue? One way in which these radiations are different is the manner in which the energy of the radiation is deposited in the tissues. For example, alpha particles and neutrons (because they are relatively heavy particles) "crash" through tissues, producing dense clusters of ions along fairly short paths. X rays and gamma rays, on the other hand, "zip" through tissues, leaving isolated ion pairs at scattered points along a much longer path. This difference is described by the **linear energy transfer (LET),** or the average energy released per unit of path length, of a radiation. High LET radiation (neutrons and alpha particles) produces closely packed groups of ion pairs and does most of its damage by direct action. When a biologically important molecule is hit by high LET radiation, the damage is usually so great that it can't be repaired at all. By contrast, low LET radiation (beta particles, X rays, and gamma rays) leaves behind sets of ion pairs well spread out through the tissue and does most of its damage through indirect action. As a result, damage to biologically important molecules caused by sublethal doses of low LET radiation is much more likely to be repaired by the cell than is damage by high LET radiation.

7.3 Detecting Radiation: Radiation Dosimetry

You cannot hear, feel, taste, or smell low levels of radiation. (Do you feel anything when you have a dental X ray?) Because such radiation is potentially dangerous, various instruments and detection systems have been designed to measure radiation dose and exposure. Depending on the particular instrument, radiation dosage may be measured by recording the amount of ionization in a gas, the amount of chemical reaction produced, the heat produced when the radiation gives off energy, or the light produced in a luminescent material.

A **Geiger–Müller** counter is a widely used radiation detecting instrument. The detector consists of a gas-filled ionization chamber in the shape of a tube. Radiation that enters the tube will produce ions in the gas, allowing an electrical current to flow. This current will either generate a click in a speaker or will turn a numerical counter to indicate the number of radiations that have entered the tube (Figure 7.5). The Geiger–Müller counter is sensitive to beta radiation. Most gamma rays will pass through the gas without affecting it, and all but the most energetic alpha particles will be stopped before even entering the tube. Although the counter will indicate the number of particles entering the tube, it tells us nothing about the energy of those particles.

The **scintillation counter** is the most widely used instrument for

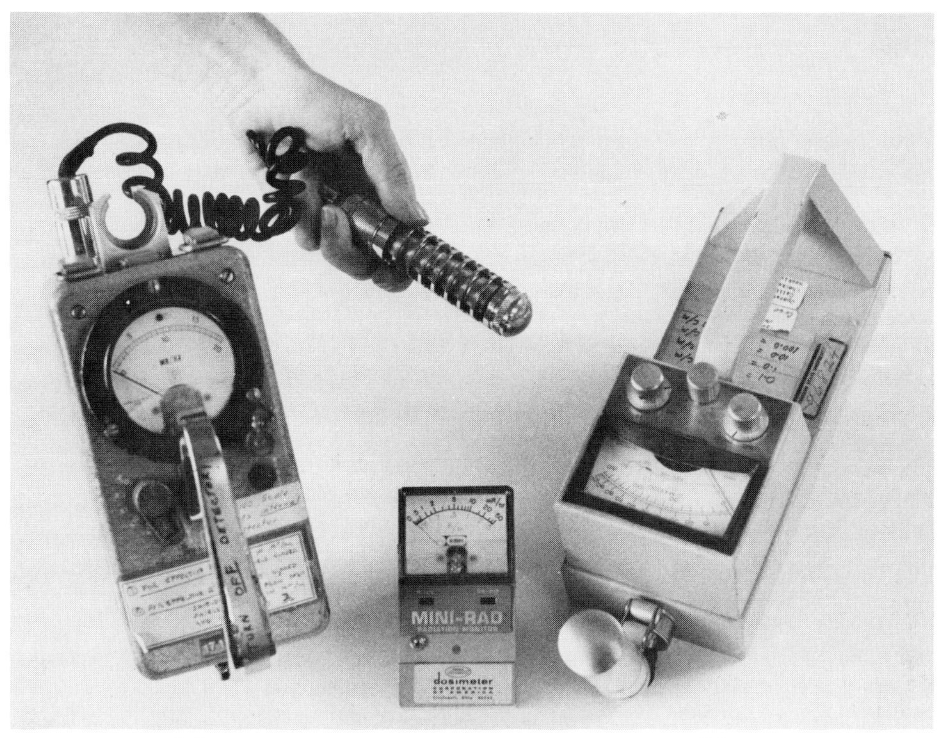

(a) (b)

Figure 7.5 Two portable Geiger–Müller counters [(a) and (b)] frequently used in laboratories. The small dosimeter in the middle can be worn on a person's belt.

determining both the number of radiations and the dose rate at the same time. Incoming radiations strike a surface that has been coated with special chemicals, producing tiny flashes of light. The number of these flashes and their brightness (which depends on the energy of the radiation) is electronically detected and converted to electric pulses that can be counted, measured, and recorded.

We have seen that it is important to carefully monitor the exposure of researchers, X-ray technicians, nurses, and others who work with ionizing radiation. To make this as easy as possible, small portable devices have been designed to accurately measure radiation dose. Two very commonly used devices are **pocket dosimeters** and **film badges** worn by the individual. The dosimeter has a built-in scale indicating the radiation exposure of the person wearing it. The film in the badge must be developed to obtain a reading. The degree of darkening in the resulting negative indicates the total amount of radiation to which the wearer has been exposed (Figure 7.6).

7.4 Protection against Radiation

In Chapter 6, we saw that various kinds of radiation differ in their penetrating power. Alpha and beta radiation have low penetrating power and are easily stopped; X rays and gamma rays have high penetrating

Figure 7.6 These badge dosimeters are each analyzed after specific lengths of time, and the exposure to millirem doses of radiation is carefully logged to prevent overexposure to radioactive materials or X rays. (Courtesy Landauer.)

power and require lead or thick concrete to be stopped. Because lead is such an effective shield, doctors and technicians often use lead aprons to protect parts of the patient's body from radiation during medical or dental X rays.

In addition to the use of shielding, another important safety factor is the distance from the source of the radiation. The farther you are from the source, the lower the intensity of the exposure. Because the rays spread out in a cone shape as they move away from the source, the number of radiations striking any object will decrease as the object moves away (Figure 7.7). In more mathematical language, the intensity of radiation on a given surface area decreases with the square of the distance from the source. Scientists call such a relationship an **inverse square law.** X-ray technicians minimize their exposure both by being as far away from the radiation source as possible, and by standing behind a wall when taking X rays.

Even with good shielding, the most sensible safety precaution is to minimize your exposure to ionizing radiation because some of the harmful effects of ionizing radiation are known to build up over time. Studies on survivors of the atomic bombs dropped on Hiroshima and Nagasaki (as well as other cases) have shown that a one-time exposure to high doses of ionizing radiation can cause cancer in humans. But the effects of long-term, low-level exposure are not totally known or understood. Living cells have an amazing ability to repair much of the damage from ionizing radiation: enzymes can repair DNA, and chemicals such as vitamins C, E, and A help prevent the damaging effects of free radicals. Low level

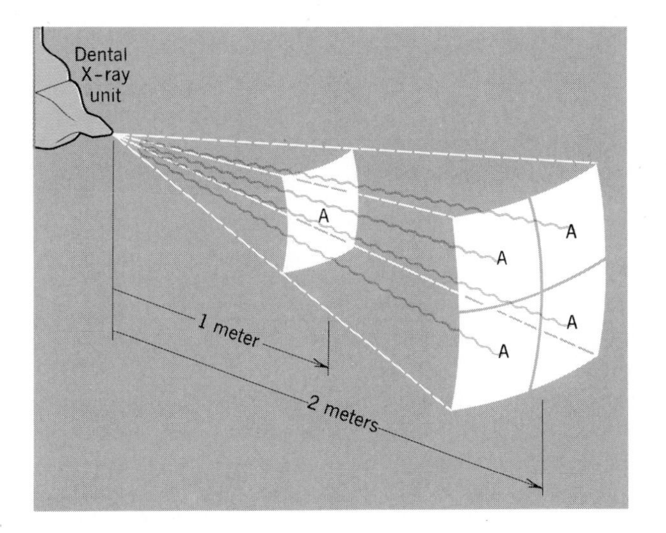

Figure 7.7 An illustration of the inverse square law. The same number of radiations hitting area *A* at one meter from the radioactive source will hit an area four times larger at two meters from the source. Therefore, the intensity of the radiation at two meters is one-fourth the intensity at one meter.

exposure may increase the rate of aging, infertility, and the development of cataracts and cancer, but much depends upon the age and genetic background of the individual.

7.5 Background Radiation

There is no way that we can totally escape exposure to ionizing radiation. In fact, the average human body undergoes several hundred thousand radioactive disintegrations per minute as a result of the natural radioisotopes (mostly potassium-40) found in the body. Seventy of the 350 naturally occuring isotopes are radioactive, and small amounts of radioactive materials are found in the soil we walk on, the food we eat, the water we drink, and the air we breathe.

On the average, we each receive a very small dose (0.184 rem, or about 200 millirem) of background radiation per year. Forty-five percent of this radiation is from outer space and natural materials in the soil, water, and air; 50% comes from medical radiation such as medical and dental X rays; 2% is from radioactive fallout and pollution from nuclear power plants; and 3% is from occupational and other miscellaneous sources (Table 7.3). We cannot control the natural background radiation, but we can control radiation exposure from other sources. Very few people would argue that we should stop all uses of radioactive materials, for these materials are playing a large role in improving living conditions around the world. But when considering decisions that might increase the level of background radiation, it becomes important to weigh the benefits to humanity against the risk. Such decisions are especially difficult because so little is known about the long-term effects on health of very low levels of radiation.

Table 7.3 Summary of the Annual per Capita Radiation Dose in the United States

Radiation Source	Average Dose Rate (millirem/year)	
Environmental		
Natural		
Cosmic rays	28	82
Terrestrial radiation		
External	26	
Internal	28	
Global fallout		4
All other (nuclear reactors, fuel processing, nuclear tests, etc.)		0.3
Medical		
X rays		
Medical diagnosis	77	92
Dental diagnosis	1.4	
Radiopharmaceutical	13.6	
Occupational		1
Miscellaneous (TV, consumer products, air travel, etc.)		5
	Total	184

Data from "Annual Dose Rates from Important Significant Sources of Radiation Exposure in the United States." National Research Council, 1980.

RADIONUCLIDES AND MEDICINE

One area in which the benefits of ionizing radiation appear to far outweigh the risk is the use of X rays and radionuclides in medical diagnosis, treatment, and research.

7.6 Medical Diagnosis

Major advances in medical diagnosis have been made possible by being able to trace the paths taken by specific chemical compounds in living systems. **Radioactive tracers** are chemicals that contain radioactive atoms and that have the same chemical nature and behavior as the compounds they are meant to trace. A living system treats the tracer just as it would the normal compound, but the radioactive atoms make it possible to follow the compound's path through the system.

In 1923 Georg von Hevesy was the first to use a radioactive tracer in his study of the uptake of lead-212 in plants. Artificially produced radionuclides were first used in 1936, when Joseph G. Hamilton and Robert S. Stone of the University of California at Berkeley used radioactive sodium produced in a cyclotron to study the uptake and excretion of sodium. Since then, the use of natural and artificially produced radionuclides has greatly increased.

In early tracer diagnostic procedures, the movement of the radionuclide was followed with a geiger counter. The 1950s saw the development of instruments called scanners. These devices have a sensitive measuring head that moves slowly back and forth across the patient's body to detect radioactivity, producing a picture that is built up line by line. In the 1970s computer programs were developed to analyze the scanning data in order to produce much clearer and sharper pictures. The gamma camera now in use does not have to move across the patient; it records all the radiation coming from a specified region of the patient, producing images in several seconds to a few minutes (Figures 7.8 and 7.1).

In 1979, Allan Cormack and Godfrey Hounsfield were awarded the Nobel prize in Medicine for developing a sophisticated X-ray machine called a computerized axial tomography (or CAT) scanner—now called a CT scanner—which is revolutionizing medical diagnosis. Ordinary X rays produce two-dimensional images whose details are often obscured by overlapping tissues, and in which certain types of tissue do not sufficiently stand out. For some diagnostic procedures, this requires that a series of X rays be taken from various angles. For a CT scan, the patient lies in a doughnut-shaped detector while the X-ray tube sweeps thousands of beams through the patient's body. (The patient is exposed, however, to no more radiation than with a standard series of X rays.) The rays strike the scanner's sensitive detectors, producing signals that are analyzed by a computer (Figure 7.9). With the computer manipulating the data, cross-sectional pictures of the patient's body can be made from any angle, complete with different shades or colors on the image to represent specific

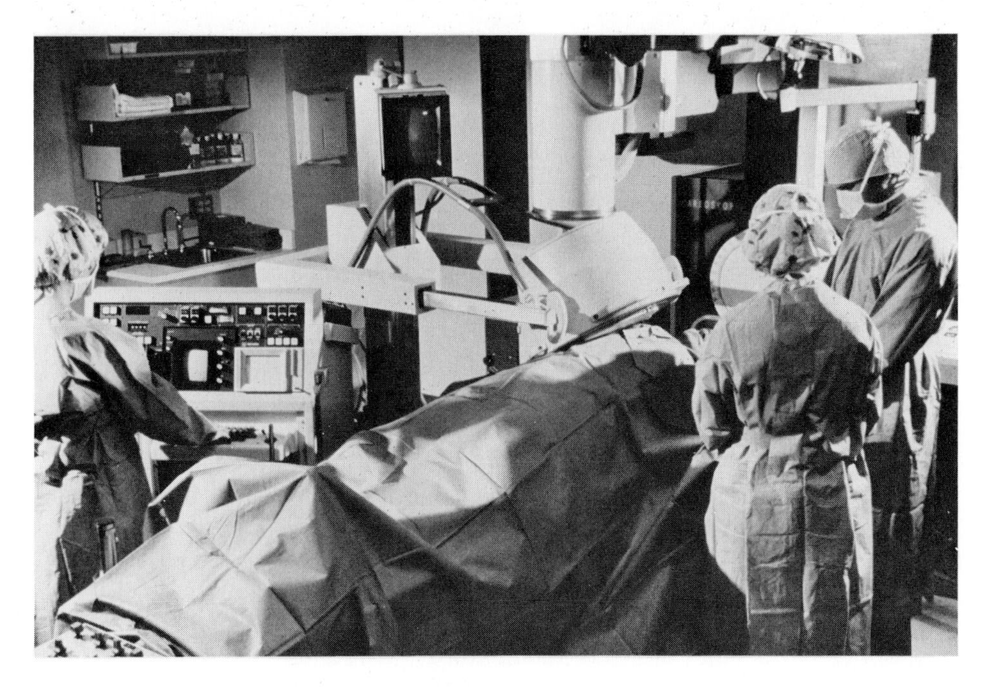

Figure 7.8 A gamma camera in use. (Spectrum One: Nuclear Cardiology System; courtesy of Ohio-Nuclear, Inc.)

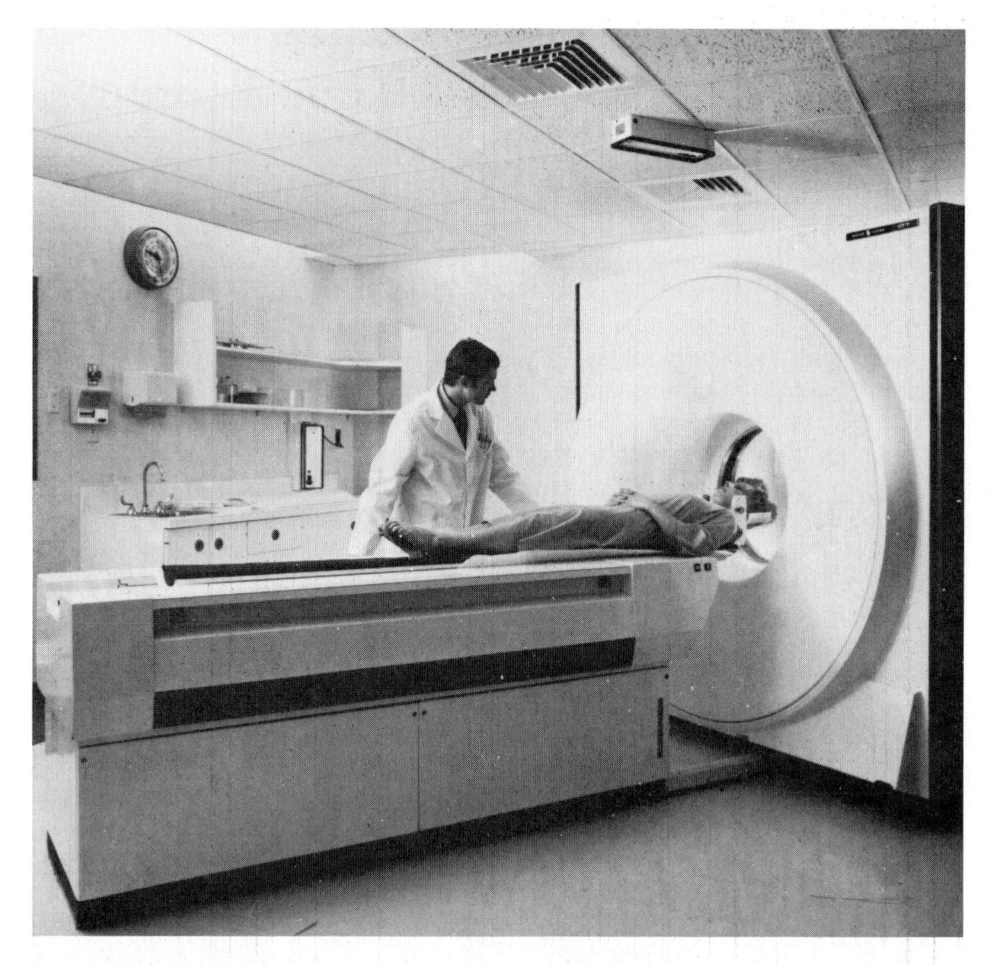

Figure 7.9 A CT scanner can complete an entire scan of the head or body in 5 seconds, producing a cross-sectional image showing all types of tissues. (Courtesy General Electric, Medical Systems Division.)

types of tissue or other body materials ranging in density from fluid to bone (Figure 7.10). Using a CT scan, doctors can identify and locate to within a fraction of a millimeter tumors, blood clots, herniated discs in the spinal column, and obstructions in the bile ducts. The CT scanner is eliminating the need for many invasive diagnostic tests and some exploratory surgery.

Several other very promising nonsurgical techniques for looking inside the body are beginning to have clinical applications. Magnetic resonance imaging (MRI) provides a means of analyzing the chemical composition of a tissue. MRI is proving very useful for looking at soft tissue, especially for finding cancerous tumors that don't show up on X rays because their density is only slightly different from that of the surrounding tissue. MRI is also used in the early diagnosis of muscle disease, such as muscular dystrophy, and several enzyme deficiencies. Positron-emission tomography (or PET scanning) differs from the other scanning techniques, which produce only static images. By tracing radioactive glucose, a PET scan produces an active image that shows the changes in metabolic activity in

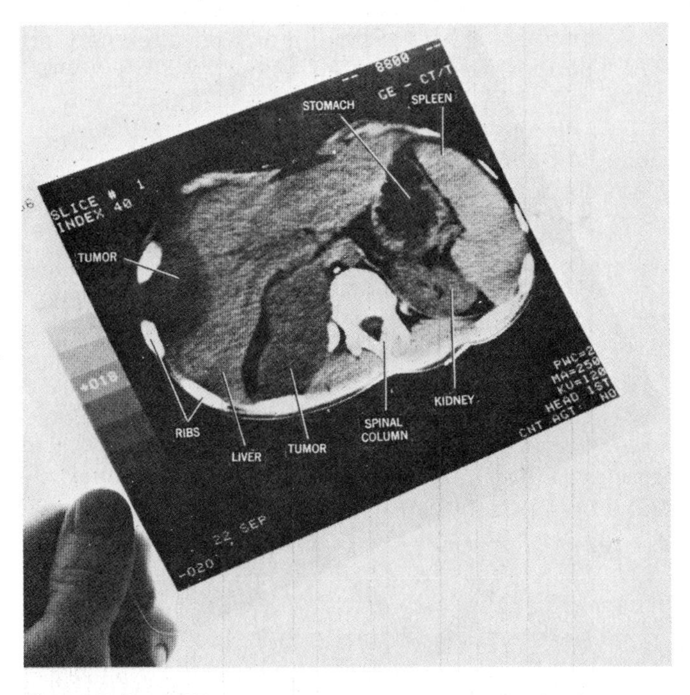

Figure 7.10 A CT scan showing a tumor on the left. (Courtesy General Electric, Medical Systems Division.)

any cross section of the body. Its main use is in medical research, whereas CT and MRI are becoming increasingly important clinical tools.

7.7 Radionuclides Used in Diagnosis

Technetium-99^m

Over the past 20 years many different radionuclides have been used to produce images of various body organs. However, technetium-99^m, because of its more suitable chemical and physical properties, is now replacing many of these other radionuclides in common diagnostic procedures. Technetium is one of four elements lighter than uranium that do not occur naturally and, therefore, must be artificially produced. It is obtained from the decay of molybdenum-99.

$$^{99}_{42}\text{Mo} \xrightarrow[t_{1/2} = 67 \text{ hr}]{} {}^{99m}_{43}\text{Tc} + {}^{0}_{-1}e + \gamma$$

The *m* after the mass number means that the isotope is **metastable,** or in an energy state higher than normal. When it decays, technetium-99^m releases this excess energy as gamma rays and forms the lower-energy isotope technetium-99.

$$^{99m}_{43}\text{Tc} \xrightarrow[t_{1/2} = 6.02 \text{ hr}]{} {}^{99}_{43}\text{Tc} + \gamma$$

The advantages of technetium-99^m over other isotopes are (1) the energy of the gamma rays are best for detection by the cameras now in use, (2) no alpha or beta radiations are given off (which would increase the absorbed dose to the patient), and (3) the 6-hour half-life allows enough time for the isotope to localize in the body after injection, and yet allows the isotope to be administered in amounts that limit the radiation exposure to that of a comparable X-ray procedure.

Various technetium compounds are used in diagnostic procedures. For example, a compound of technetium and sulfur is taken up by the normal cells of the liver, spleen, and bone marrow (and was used, you will recall, to look for spleen damage in the opening case history of this chapter). Cancerous cells, however, do not take up this compound, so areas of the liver or spleen containing a cancerous tumor will appear as areas of decreased activity on a scan (Figure 7.11a).

Technetium diphosphonate is used to detect whether cancer cells have spread, or **metastasized,** to the bones. Bone cells will absorb the diphosphonate and incorporate it into the structure of the bone in those places where bone tissue is being destroyed by cancerous cells. In this case, the highest radioactivity will be found in areas of the metastasized cells (Figure 7.11b).

Technetium-99^m, in several forms, is also allowing doctors to determine the location and extent of both old and new damage to the heart. Such procedures can help the doctor to decide what therapy the patient should or should not receive. Radioactive technetium (as with calcium) deposits in the mitochondria of heart cells that are irreversibly damaged. Thallium-201, which is produced in a cyclotron, is being used together with technetium to diagnose heart damage. Thallium, unlike technetium, acts like potassium

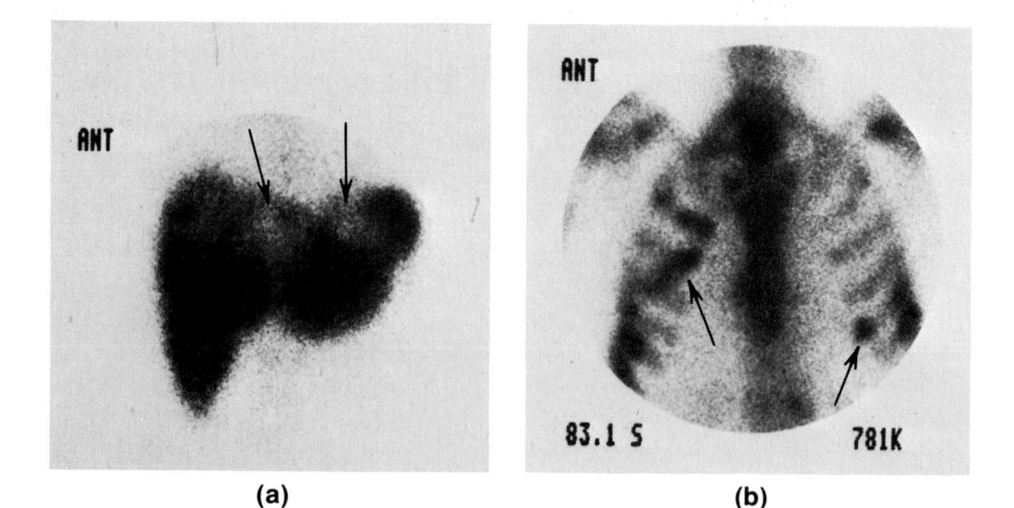

Figure 7.11 (a) This gamma scan taken after administration of Tc-99^m sulfur colloid shows light patches on the liver and spleen, indicating the presence of cancerous tumors. (b) The dark areas on the ribs in this gamma scan indicate the active uptake of Tc-99^m HDP by metastasized prostate cancer cells. (Courtesy William K. Lloyd, M.D., Special Imaging Department, Good Samaritan Hospital, Corvallis, OR.)

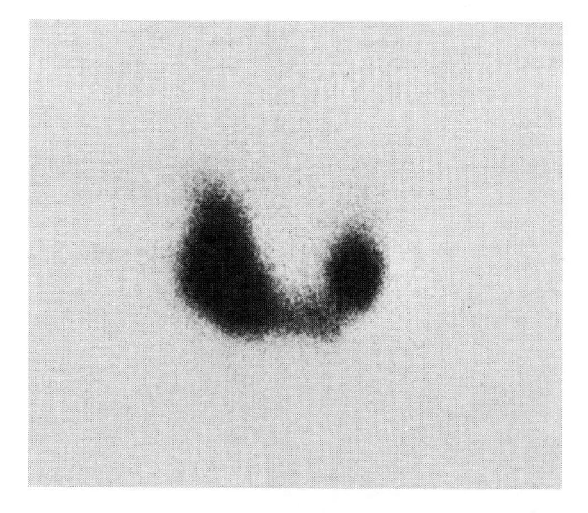

Figure 7.12 The lobe of the thyroid on the right of this scan has not taken up iodine-123, which indicates the presence of tumor cells. (Courtesy William K. Lloyd, M.D., Special Imaging Department, Good Samaritan Hospital, Corvallis, OR)

and concentrates in normal heart muscle. It can be used, for example, to determine whether a patient who enters the hospital with chest pains, but with few other symptoms of a heart attack, has a deficiency in the blood supply to the heart that might result in a future heart attack. Both of these radionuclides can be administered intravenously, and their use is safer, more comfortable and less expensive than the previous diagnostic technique of inserting a catheter through a vein into the heart.

Iodine-131

Early studies using iodine as a tracer made use of iodine-128, which has a half-life of 25 minutes. In 1938, searching for a tracer with a longer half-life, Hamilton was able to produce iodine-131 with the use of a cyclotron. This isotope is a beta and gamma emitter with a half-life of eight days and is a very versatile tracer.

One important use of radioiodine is in the measurement of thyroid function. Thyroxine, an iodine compound manufactured in the thyroid, is released in the blood to help control the use of nutrients by the body. To test the functioning of the thyroid gland, a patient can be given a small amount of iodine-131 in the form of the iodide ion. Because any iodine in the body will be concentrated in the thyroid gland, the doctor can monitor the amount of radioiodine in the patient's blood to determine the rate at which the thyroid takes up the iodine. The uptake of radioiodine is monitored over a 24-hour period, giving the doctor an indication of how well the thyroid gland is functioning. Too rapid an uptake means the patient is suffering from an overactive or hyperthyroid; too slow an uptake means the patient could be suffering from an underactive or hypothyroid.

Another isotope of iodine, iodine-123, can be used to determine whether a growth is cancerous or benign. The iodine-123 will not collect in cancerous cells, but will be taken up by other thyroid cells (Figure 7.12).

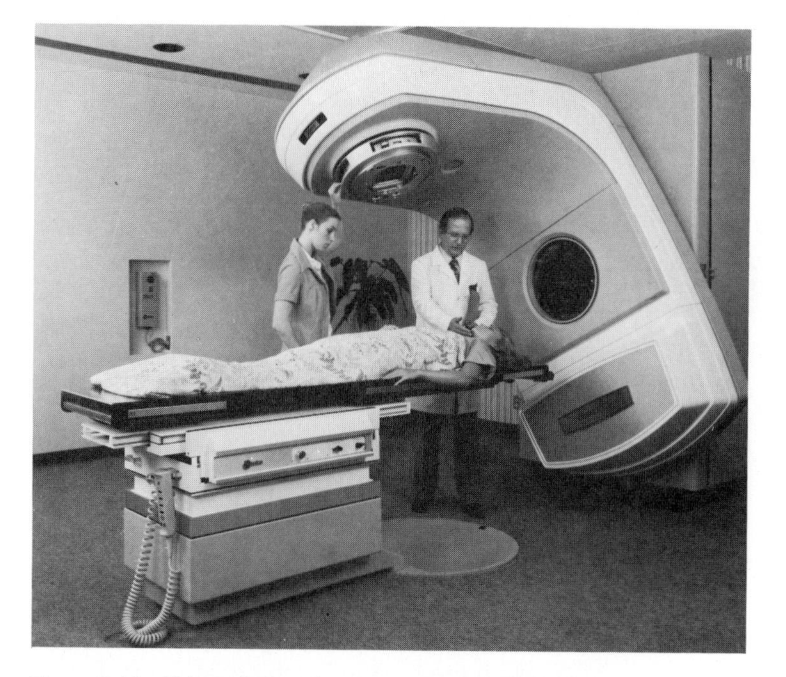

Figure 7.13 This radiation therapy unit is a 6-MEV linear accelerator. Lead blocks can be placed on a tray between the patient and the radiation source to give the treatment area a specified size and shape. (Courtesy Varian Associates, Palo Alto, CA.)

7.8 Radiation Therapy

Cancer cells are less able to repair ionization damage caused by ionizing radiation than are normal cells. This is the key fact that allows doctors to use radiation in killing cancerous tissue while leaving normal tissue relatively unharmed.

Teletherapy is the term used to denote the use of high-intensity radiation to destroy cancerous tissue that can't be removed by surgery (or which, because of its sensitivity to radiation or its location, is better treated with irradiation—perhaps in combination with surgery or chemotherapy). In such therapy, radiation from an X-ray machine, a cobalt-60 source, or a particle accelerator is focused into a beam that is directed at the cancerous tissue. In the treatment of internal cancers, the patient is carefully positioned on a treatment table; the source of radiation is then rotated around the patient (or directed at the patient from different angles) so that the radiation will produce minimal skin tissue damage, but will be constantly focused on the tumor cells (Figure 7.13).

Brachytherapy is the medical procedure of inserting a radionuclide (in the form of a seed) by means of a needle into the area to be treated. Seeds containing gold-198 or iridium-192 are sometimes implanted in the tumors of patients with advanced inoperable tumors in order to slow tumor growth. They are also implanted in tumors located near body surfaces, such as in the mouth, breast, prostate, and cervix, in order to destroy the cancer.

Radiopharmaceutical therapy involves administering radionuclides in chemical forms designed so that they will be concentrated in specific areas of the body. For example, because radioiodine is concentrated in the thyroid, it can be used to destroy thyroid cells in the treatment of an overactive or hyperthyroid. Phosphorus-32 is used to treat the disease polycythemia vera, which causes an abnormal increase in the number of red blood cells. There is no known cure for this disease, but the radiophosphorus destroys the cells that produce red blood cells. This will slow down the formation of red blood cells and will temporarily alleviate the symptoms of the disease.

CHAPTER SUMMARY

Radiations given off by radioactive materials are called ionizing radiation because they are capable of producing ions in material through which they pass. Ionizing radiation can cause damage to living tissue in two ways: through direct damage to biologically important molecules, or by reacting with water to form chemically reactive particles that can disrupt cellular functions. Ionizing radiation can be measured in different units such as the becquerel, curie, rad, gray, rem, sievert, and LET. These various units of measure give specific information about the nature of the radioactive source and its effect on living organisms. Various instruments such as the Geiger–Müller counter, scintillation counter, dosimeter, and film badge can be used to detect the presence of, or the amount of exposure to, ionizing radiation. To protect yourself from the dangerous effects of ionizing radiation, you should try to minimize your exposure to such radiation, stay as far as possible from the source, and use lead or other shielding when working with ionizing radiation. The amount of radiation occurring naturally in the environment is called background radiation.

Although ionizing radiation in large doses can be harmful to living tissue, it has proven extremely useful in the diagnosis and treatment of disease. A CT scan uses X rays to create very detailed images of the interior of the human body, thereby helping the physician locate tumors and blood clots. Radionuclides injected into the body are used to analyze liver, kidney, spleen, and thyroid functions and to locate metastasized cancers. The ionizing radiation given off by radioactive elements can also be used to destroy cancerous cells and tumors.

EXERCISES AND PROBLEMS

1. Explain how ionizing radiation can be both the cause of cancer and the means of stopping or slowing its growth.
2. Beta radiation is low LET radiation. Explain how beta radiation can damage living cells. What are the chances of a cell recovering from a dose that is sublethal to the organism?
3. Alpha radiation is high LET radiation. Explain how its effects on living

cells are different from beta radiation. Is a cell more likely to recover from sublethal doses of alpha or beta radiation?

4. You are working in a laboratory which you fear has been contaminated with a radioactive beta emitter. How might you try to detect this contamination?

5. A radiation therapy cobalt-60 source is rated at 1.7 Ci. How many disintegrations will it deliver in 15 seconds?

6. Both the rad and the rem are units used to describe the absorbed dose of radiation. What is the difference between these two units?

7. While running an experiment, you expose a group of 200 mice to 600 rem of radiation. How many mice would you expect to die within 30 days?

8. You are setting up a dental office in a building that is being remodeled. The contractor asks you for your preferences regarding the placement of the operating controls for the X-ray machine. As the dentist, what would be your order of preference? Give the reasons for your order.
   First possibility: On a wall 5 feet from the examining chair, with convenient access to the patient.
   Second: On a wall 5 feet from the patient but separated from the patient by a concrete dividing wall.
   Third: On a wall 11 feet from the patient.

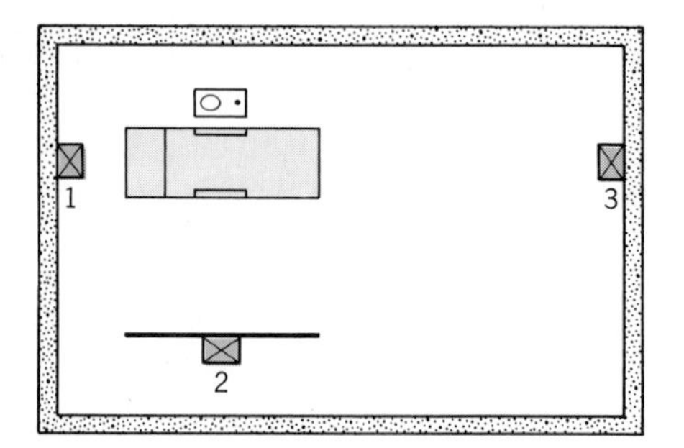

9. Why is the image produced by a CT scan superior to the image produced by ordinary X rays?

10. After giving her patient a complete examination, a doctor feels that the patient may be suffering from an abnormal thyroid. How might radioisotopes be used to confirm the doctor's diagnosis?

11. Strontium-90 decays as follows:

$$^{90}_{38}Sr \longrightarrow {}^{90}_{39}Y + {}^{0}_{-1}e$$

   Iridium-192 decays as follows:

$$^{192}_{77}Ir \longrightarrow {}^{192}_{78}Pt + {}^{0}_{-1}e + \gamma$$

   Assume that a patient is suffering from an inoperable cancer in his nose. Which of the two above radioisotopes is better for treating this tumor? Give a reason for your choice.

CHAPTER EIGHT
COMBINATIONS OF ATOMS

A light wind was blowing as the afternoon freight train rumbled through a small town in southeastern Louisiana. The train was just starting to gather speed about two miles outside of town when a freight car suddenly jumped the tracks, pulling 18 cars with it. One of those 18, a tank car containing 30 tons of liquid chlorine (Cl_2), lay on its side with a huge gash ripped open. The chlorine, which had immediately vaporized to a greenish-yellow gas, now poured out of the tank car and was carried by the breeze back toward the town.

The Harrison family lived in a nearby farmhouse, unaware of the approaching danger. Within minutes of the train accident the irritating odor of chlorine began to fill the house. Mrs. Harrison suddenly found she was having difficulty breathing, and her two small children began retching and vomiting. Her husband came running from the barn with his eyes streaming and quickly loaded his family into their car. As they rushed off,

Mrs. Harrison noticed that her 11-month-old son Randy was having a great deal of trouble breathing. Their desperate drive to the hospital was accompanied by the sound of the volunteer fire department's siren screaming out the emergency signal, and the sight of the sheriff's car helping to evacuate people from nearby farms.

The deadly cloud of chlorine gas forced nearly 1000 people to flee from their homes, offices, and schools. Hundreds of farm animals died from exposure to the gas, and no one could live in several square miles of the countryside for several days. Fifty people were treated at the hospital for severe irritation caused by the chlorine, and ten, including the Harrison family, were hospitalized with critical poisoning. Unfortunately, Randy died from the effects of the chlorine gas before his family could even reach the hospital.

Chlorine, at room temperature, is a greenish-yellow gas that has a characteristic irritating and suffocating odor. In low concentrations it will irritate mucous membranes and the respiratory system and in high concentration will cause difficulty breathing—leading in extreme cases to death from suffocation.

Another dangerous substance is the element sodium (Na). This element is an alkali metal which is so highly reactive that it is never found in a pure state in nature. (When isolated in the pure form, however, it is a soft, silvery metal that can easily be cut with a knife.) Great care must be taken in handling sodium to be sure it is kept away from water. When it does contact water, sodium reacts extremely vigorously, releasing hydrogen gas (H_2) that can be ignited by the heat from this reaction.

Chlorine and sodium, then, are both highly reactive elements that are potentially dangerous to living tissue. However, suppose you drop a piece of freshly cut sodium into a container of chlorine gas and warm the container. You will soon see a white powder start to form—a chemical reaction is taking place. The product of this reaction is sodium chloride (NaCl), the substance more commonly known as table salt. But table salt has none of the properties of the reactants, sodium and chlorine. In fact, sodium chloride is an essential part of our diets. It plays an important role in maintaining the proper amount of water in our cells and tissues and is needed for the contraction of muscles and the transport of nerve impulses.

Sodium chloride is a chemical compound, a homogeneous substance produced by the reaction of different elements. This chemical reaction, like many other spontaneous processes in nature, results in the formation of a more stable substance. In this case, the elements sodium and chlorine are both relatively unstable and will react spontaneously (the addition of heat will only speed up the reaction) to form sodium chloride, which is very stable.

By the time you have finished this chapter, you should be able to:

1. State the octet rule.

2. Describe an ionic bond and an ionic compound.
3. Describe a covalent bond and a covalent compound.
4. Describe a single, double, and triple covalent bond and give examples of each.
5. Draw the electron dot diagram for an ionic or a covalent compound formed from the representative elements.
6. Define *electronegativity* and describe how electronegativity changes on the periodic table.
7. Describe the difference between a polar and a nonpolar covalent bond.
8. Define *polyatomic ion.*
9. Define *oxidation number* and determine the oxidation numbers of the elements in a given compound or polyatomic ion.
10. Write the formula for an ionic or covalent compound when given its name and write the name of the compound when given its formula.
11. Given its shape, predict whether a simple molecule will be polar or nonpolar.
12. Describe the hydrogen bond.

IONIC BONDING

8.1 The Octet Rule

In the early 1900s, Richard Abegg, J. J. Thomson, C. N. Lewis, and Irving Langmuir developed a new model of atomic structure. They based this model on the observation that the elements in groups I through VII enter into chemical combinations that involve the loss, gain, or sharing of electrons in such a way as to end up with a total of eight valence electrons. (You might remember that this is the electron configuration of the noble gases, which are extremely stable.) The tendency of these elements to react so as to attain an outer octet—that is, eight valence electrons—is called the **octet rule.** Although there are many exceptions to this rule among the heavier elements, we will find it quite useful for predicting the composition of chemical compounds produced from reactions between lighter elements (atomic numbers 1 to 22).

8.2 Transfer of Electrons: The Ionic Bond

There are two ways in which chemical elements can attain a stable octet of electrons: by gaining or losing electrons, or by sharing electrons. The first of these ways, the transfer of electrons from one atom to another, causes electrically neutral atoms to become ions—and the force of attraction between these ions is called an **ionic bond.** Such a transfer occurs when an element such as chlorine, which has a very strong

Table 8.1 Electron Configurations of Atoms and Ions

	Atomic Number	Electron Configuration 1	2	3		Atomic Number	Electron Configuration 1	2	3
Sodium, Na	11	2	8	1	Chlorine, Cl	17	2	8	7
Sodium ion, Na^+	11	2	8		Chloride ion, Cl^-	17	2	8	8
Neon, Ne	10	2	8		Argon, Ar	18	2	8	8

attraction for additional electrons (that is, high electron affinity), reacts with an element such as sodium, which has a weak attraction for its valence electron (that is, low ionization energy). If an atom of sodium loses its single valence electron, it will form a positive sodium ion, Na^+. Notice that a sodium ion has 10 electrons (eight in the outer shell)—the same number as an atom of neon (Ne), the noble gas closest to sodium on the periodic table. Similarly, if an atom of chlorine gains an electron (such as might be lost from a sodium atom), it will form a negative chloride ion, Cl^-. Such a chloride ion will have 18 electrons (eight in the outer shell), the same number as the closest noble gas, argon (Ar) (Table 8.1 and Figure 8.1).

In general, we can expect that elements with few valence electrons (the metals in groups I, II, and III) will lose electrons when they react with elements that have almost eight valence electrons (the nonmetals in groups VI and VII). The ions formed by such a transfer are attracted to each other (remember that unlike charged particles attract), and it is this attraction between ions that forms the ionic bond. When, for example, we talked about dropping a chunk of freshly cut sodium into a container of chlorine gas, many billions of atoms were involved in this reaction (that is, in the transfer of electrons). The attraction formed between the positive and negative ions causes these ions to be grouped in an orderly three-dimensional pattern called a **crystal lattice.** This entire large grouping of ions is then called an **ionic compound** (Figure 8.2). In a crystal of sodium chloride each sodium ion is surrounded by six chloride ions, and each chloride ion is surrounded by six sodium ions. Because there is one

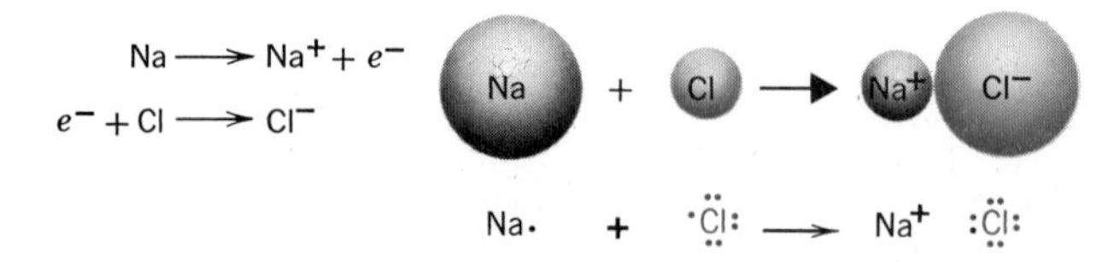

Figure 8.1 In the formation of sodium chloride, a sodium atom will lose one electron to a chlorine atom, producing the positive sodium ion, Na^+, and the negative chloride ion, Cl^-.

sodium ion, Na^+, for every chloride ion, Cl^-, a crystal of sodium chloride is electrically neutral. And we will find this always to be true—the ratio of ions in the crystal lattice of an ionic compound will always result in an electrically neutral compound.

An ionic compound contains no specific molecules that we can identify. No ion is attracted exclusively to another ion, but rather each ion is attracted to all the oppositely charged ions surrounding it. Again, you can see that the ionic bond is not a thing or a substance, but is simply the force of attraction between oppositely charged ions.

In general, ionic compounds are formed between metals and nonmetals. For a representative element, knowing its group on the periodic table allows us to predict the number of electrons its atoms must gain or

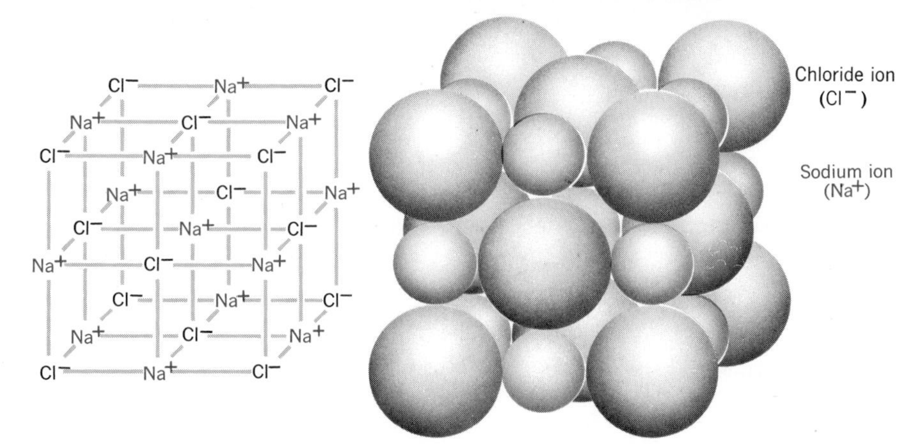

Figure 8.2 A sodium chloride crystal consists of sodium and chloride ions in a closely packed arrangement. (Photo courtesy American Museum of Natural History.)

lose and, therefore, the charge on the ions that will form (Table 8.2). The transition metals, however, are harder to predict, and they can often form more than one ion. For example, copper can form either Cu^{+} or Cu^{2+}. Table 8.3 lists some common ions formed by the transition metals and other metals that form more than one ion.

8.3 Lewis Electron Dot Diagrams

We have just stated that valence electrons are transferred from one atom to another to form an ionic bond. In Chapter 5 we learned how to write the complete electron configuration of an atom. In the rest of this book, however, we will be focusing our attention primarily on valence electrons. For this reason, we will find it convenient to use **Lewis electron dot diagrams** (named after G. N. Lewis) to represent an atom's configuration of valence electrons. As you may guess from the name, in these diagrams a dot is used to represent each valence electron, and these dots are placed around the symbol of the element. On each side of the element's symbol there is room for two dots. Beginning on any side, the dots are first placed singly, then paired for elements having more than four valence electrons. Table 8.4 shows the electron dot diagrams for the elements in period 2. Remember that the group number of a representative element tells you the number of valence electrons for each member of that group or family.

Table 8.2 Some Common Ions of the Representative Elements

Group I	Group II	Group III	Group V	Group VI	Group VII
Li^{+}	Be^{2+}	Al^{3+}	N^{3-}	O^{2-}	F^{-}
Na^{+}	Mg^{2+}		P^{3-}	S^{2-}	Cl^{-}
K^{+}	Ca^{2+}			Se^{2-}	Br^{-}
Rb^{+}	Sr^{2+}			Te^{2-}	I^{-}
Cs^{+}	Ba^{2+}				

Table 8.3 Some Common Ions of the Transition and Post-Transition Metals

Metal	Ion	Metal	Ion	Metal	Ion
Chromium	Cr^{2+}, Cr^{3+}	Copper	Cu^{+}, Cu^{2+}	Tin	Sn^{2+}, Sn^{4+}
Manganese	Mn^{2+}, Mn^{3+}	Zinc	Zn^{2+}	Lead	Pb^{2+}, Pb^{4+}
Iron	Fe^{2+}, Fe^{3+}	Silver	Ag^{+}		
Cobalt	Co^{2+}, Co^{3+}	Cadmium	Cd^{2+}		
Nickel	Ni^{2+}				

Table 8.4 Electron Dot Diagrams for the Elements of Period 2

Element	Symbol	Atomic Number	Group	Valence *e*	Electron Dot Diagram
Lithium	Li	3	I	1	Li·
Beryllium	Be	4	II	2	Be:
Boron	B	5	III	3	·B:
Carbon	C	6	IV	4	·Ċ:
Nitrogen	N	7	V	5	·Ṅ: (with a dot below)
Oxygen	O	8	VI	6	·Ȯ: (with two dots below)
Fluorine	F	9	VII	7	:Ḟ: (with two dots below)
Neon	Ne	10	0	8	:N̈e: (with two dots below)

EXAMPLE 8-1

1. Write the electron dot diagram for phosphorus.

   Phosphorus is in group V and, therefore, will have five valence electrons. Place the first four dots on each side of the symbol and then pair up the fifth dot with one of the first four. (*Note:* Any of the following dot diagrams are equally correct.)

   $\cdot\ddot{P}\cdot$ (one dot below) or $:\dot{P}\cdot$ (one dot below) or $\cdot\dot{P}\cdot$ (two dots below) or $\cdot\dot{P}:$ (one dot below)

2. Use Lewis diagrams to show the reaction between sodium and chlorine to form sodium chloride.

   The Lewis diagram for sodium with its one valence electron is ·Na and that for chlorine with its seven valence electrons is $\cdot\ddot{\underset{\cdot\cdot}{Cl}}:$.

   $$Na\cdot + \cdot\ddot{\underset{\cdot\cdot}{Cl}}: \longrightarrow Na^{+}:\ddot{\underset{\cdot\cdot}{Cl}}:^{-}$$

3. Write the electron dot diagram for the reaction between magnesium and chlorine.

   Magnesium is in group II and has two valence electrons. Chlorine, in group VII, has seven valence electrons. To attain a stable octet of electrons, magnesium must lose two electrons and chlorine must gain one electron. Therefore, two chlorines must react with each magnesium.

$$:\ddot{\underset{..}{Cl}}\cdot + \cdot Mg\cdot + \cdot\ddot{\underset{..}{Cl}}: \longrightarrow Mg^{2+} + 2\left[:\ddot{\underset{..}{Cl}}:^{-}\right]$$

(The 2 written before the brackets indicates that there are two chloride ions.)

Exercise 8-1

1. Write the Lewis electron dot diagrams for the following elements:

   (a) Rb (b) Si (c) I

2. Use Lewis electron dot diagrams to show the reaction between the following elements:

   (a) Potassium and iodine
   (b) Magnesium and iodine

8.4 Writing Chemical Formulas

Chemical formulas are a shorthand way of representing chemical compounds. The formula contains two pieces of information: it tells us the types of atoms or ions and the ratio of these atoms or ions in the compound. The type of atom is indicated by the symbol of the element, and the ratio of atoms is shown by subscripts following the symbol. For ionic compounds, the symbol of the positive ion is always placed first in the formula.

$CaCl_2$

Symbols show the type of atoms or ions present: calcium and chlorine

Subscript shows the ratio of atoms or ions: 2 chlorines to 1 calcium

As we mentioned before, all ionic compounds are electrically neutral. The ratio of ions used in the chemical formula is the lowest set of whole numbers that will give us such electrical neutrality. This means that the formula for calcium chloride would not be expressed as, say, $Ca_{1/2}Cl$ or Ca_6Cl_{12}. Note also that the charges on the ions are not written as part of the chemical formula.

EXAMPLE 8-2

Write the formula for the compound formed between magnesium and chlorine.

In Example 8-1, we wrote the electron dot diagram for the compound formed between these two elements. We found that we need one magnesium atom for every two chlorine atoms. Therefore, the formula is $MgCl_2$.

Exercise 8-2

Write the chemical formulas for the compounds formed in question (2) of Exercise 8-1.

COVALENT BONDING

8.5 Sharing Electrons: The Covalent Bond

Elements that are alike in their attraction for electrons cannot form ionic bonds. They must complete their octet and become stable by sharing electrons, thereby forming a **covalent bond.** A covalent bond results when two positive nuclei attract the same electrons, thus holding the two nuclei close together. When two or more atoms share electrons through covalent bonds, a single (electrically neutral) unit called a **molecule** is formed. **Covalent compounds** are composed of molecules, which in turn are composed of atoms held together by covalent bonds. Covalent bonds are normally formed between nonmetallic elements (remember that nonmetals are elements having high ionization energies and, therefore, strong attraction for their valence electrons).

Some nonmetallic elements exist in nature not as atoms, but as **diatomic molecules**—two atoms of the element covalently bonded together (Table 8.5). Chlorine, for example, exists as the diatomic molecule

Table 8.5 Elements That Are Found as Diatomic Molecules

Element	Formula	Element	Formula
Hydrogen	H_2	Fluorine	F_2
Nitrogen	N_2	Chlorine	Cl_2
Oxygen	O_2	Bromine	Br_2
		Iodine	I_2

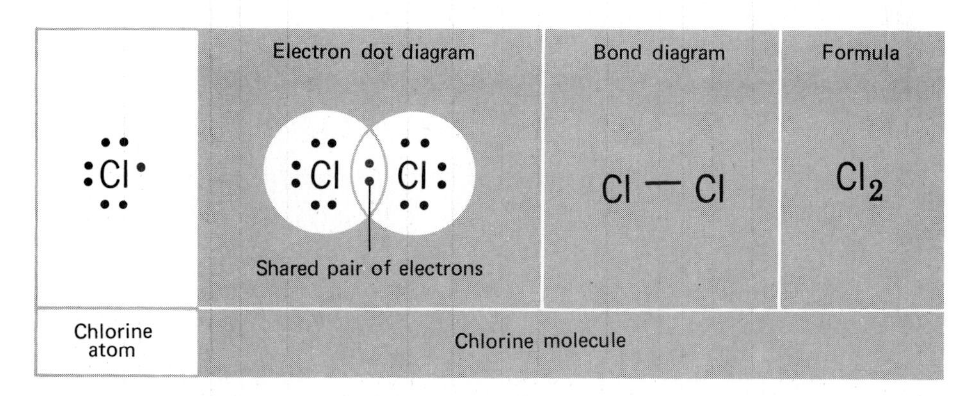

Figure 8.3 The diatomic chlorine molecule contains a single covalent bond.

Cl_2. Each chlorine atom has seven valence electrons and needs one more electron to reach the stable electron configuration of argon. Two chlorine atoms can share a pair of electrons (thus forming a covalent bond) to reach this stable octet. A shorthand way of representing a covalent bond is to draw a dash between the chemical symbols of the elements involved; such a representation is called a **bond diagram** (Figure 8.3).

Hydrogen is another element that is found as a diatomic molecule, H_2. Each hydrogen atom has one valence electron and needs one more electron to reach the stable electron configuration of the closest noble gas, helium. Two hydrogen atoms, then, will share a pair of electrons to become stable (Figure 8.4).

8.6 Multiple Bonds

Often a stable octet of electrons can be attained only if more than one pair of electrons is shared between two nuclei. A single shared pair of electrons results in a **single covalent bond,** which we have represented by a dash between the symbols of the elements. A **double bond** is formed when two pairs of electrons are shared between two nuclei, and is represented by two dashes, =. Carbon dioxide (CO_2) is an example of a compound containing double covalent bonds. The carbon atom must share two pairs of electrons with each oxygen atom to reach a stable octet of electrons (Figure 8.5).

Nitrogen is found in the atmosphere as the diatomic molecule N_2. Each nitrogen atom has five valence electrons and needs to share three electrons to reach a stable octet. This happens when each of two nitrogen atoms shares three electrons with the other atom. These three shared pairs of electrons form a **triple bond,** represented by three dashes (Figure 8.6). Nitrogen accounts for 80% of the gases in the atmosphere and is stable and relatively unreactive. Because of its unreactivity, nitrogen in the form of N_2 is useless to most forms of life. There is only one type of organism that

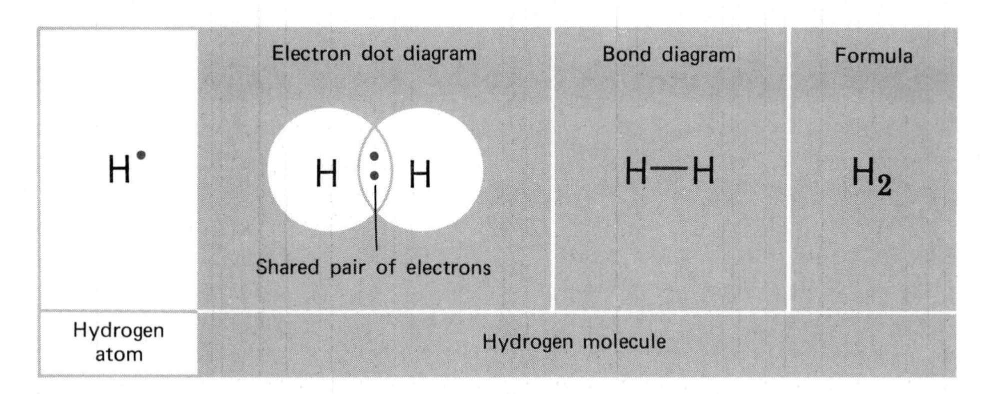

Figure 8.4 The diatomic hydrogen molecule contains a single covalent bond.

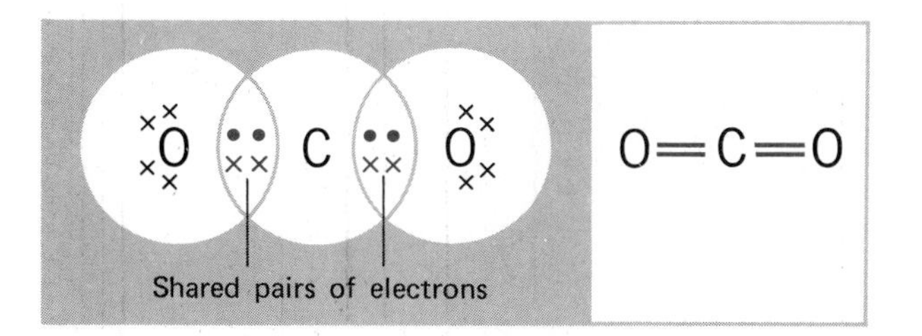

Figure 8.5 The carbon dioxide molecule contains two double bonds, resulting in an octet of electrons around each atom.

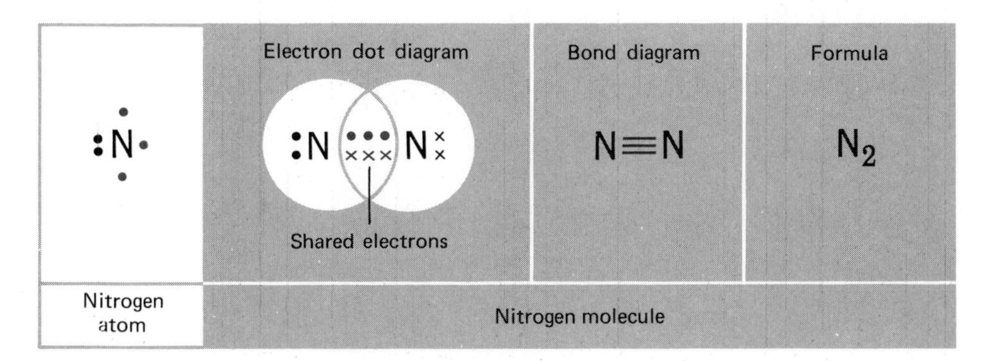

Figure 8.6 The diatomic nitrogen molecule contains a triple covalent bond.

is able to force atmospheric nitrogen to form chemical combinations with other atoms, which then makes it usable by other forms of life. These organisms are bacteria that live in the soil or in the roots of plants such as peas and alfalfa (Figure 8.7). The bacteria convert gaseous nitrogen into compounds that can be used by the plants, and animals then get the nitrogen they need by eating the plants.

EXAMPLE 8-3

Write the electron dot diagram and the bond diagram for the following covalent compounds:

(a) *Water, H_2O.* Oxygen is in group VI, so it has six valence

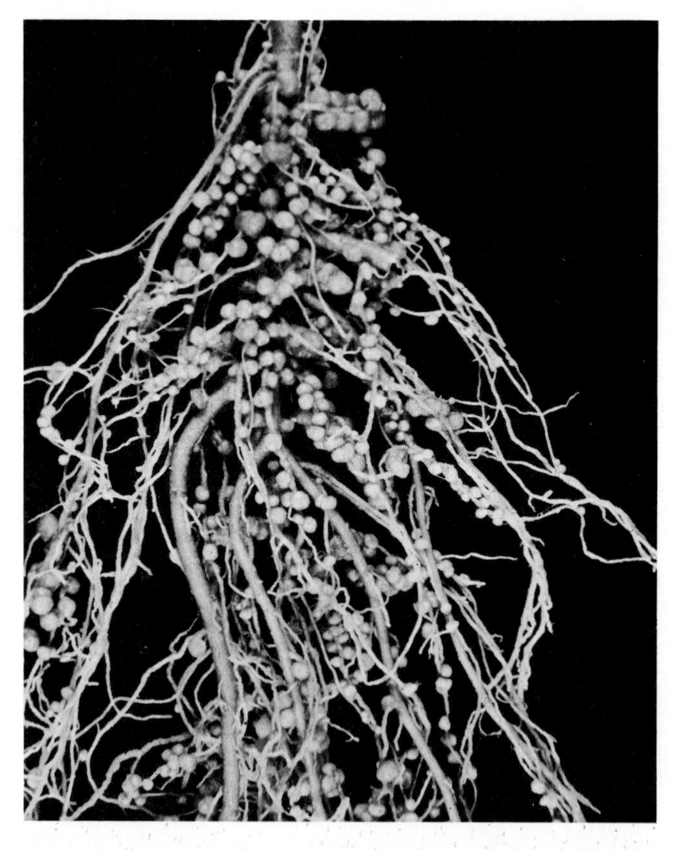

Figure 8.7 Roots of this leguminous plant have nodules containing nitrogen-fixing bacteria. Using nutrients provided through the root, these bacteria convert the nitrogen in the air to ammonia (NH_3) or nitrate (NO_3^-). (Courtesy of The Nitrogen Co., Inc.)

electrons. This means that oxygen needs to share two electrons to become stable. Hydrogen is in group I; it has one valence electron and thus needs to share one electron to have the electron configuration of helium. Therefore, the oxygen atom will share one electron with each of the two hydrogen atoms. Using electron dot diagrams to illustrate this, we have

H· :Ö: ·H ⟶ H
H:Ö:

Therefore, the bond diagram would be

H
|
H—O

(Note, however, that these diagrams are not meant to show the actual shape of the molecule.)

(b) *The compound that is formed between carbon and bromine.* Carbon is in Group IV; it has four valence electrons and needs to share four additional electrons to reach a stable octet. Bromine is in group VII, has seven valence electrons, and needs to share one electron. Therefore, carbon must share one electron with each of four bromine atoms. The electron dot diagram and bond diagram are

Exercise 8-3

Write the electron dot diagrams and bond diagrams for the covalent compounds formed between the following elements:

(a) Hydrogen and chlorine
(b) Hydrogen and nitrogen
(c) Carbon and sulfur

8.7 Electronegativity

An atom of each element has a specific tendency to attract a shared pair of electrons, and the strength of this attraction is known as **electronegativity.** In each period of the periodic table the element with the smallest radius is the most electronegative; among all the elements, fluorine is the most electronegative. The electronegativity of the elements increases to the

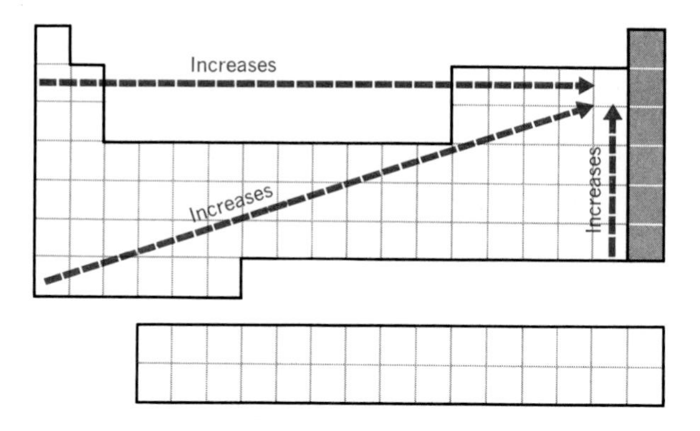

Figure 8.8 A periodic table showing the trends in electronegativity.

right along a period and increases from bottom to top of each group as shown in Figure 8.8. Imagine two identical atoms sharing a pair of electrons in a covalent bond—for example, Cl_2. The two atoms have the same electronegativity, so we would expect the electrons to be shared equally between the two nuclei. This will result in the center of positive charge and the center of negative charge occurring midway along the bond between the two nuclei. Such a bond is called a **nonpolar covalent bond** (Figure 8.9a).

Now imagine that two different atoms are sharing electrons. We would expect that one atom might attract the electrons more strongly than the other, so the electrons will be found closer to the more electronegative (or electron-attracting) atom. In such a bond, therefore, the center of positive charge is not found in the same place as the center of negative charge, resulting in an unequal distribution of charge (Figure 8.9b). Such an unequal distribution of charge results in the formation of an electric dipole; the molecule has two electric poles, just as a magnet has two magnetic poles. This kind of bond is called a **polar covalent bond.**

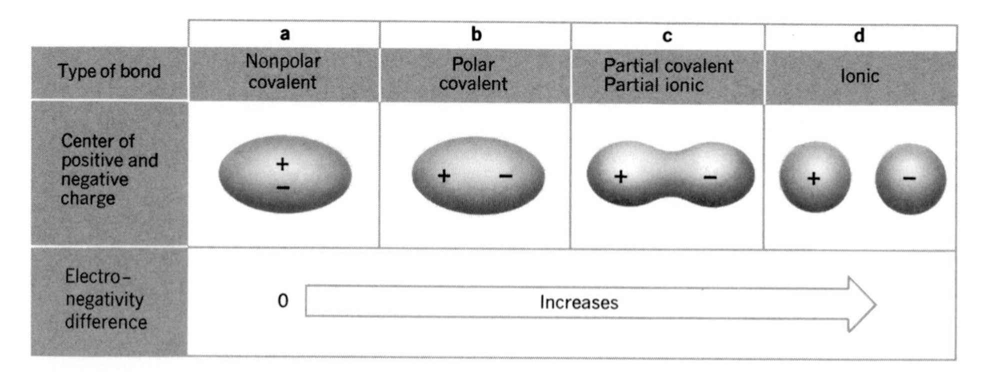

Figure 8.9 The bonding continuum. There is no absolute break between ionic and covalent bonding, but rather a continuous range from nonpolar covalent bonding to ionic bonding.

For example, consider hydrogen chloride, HCl. Chlorine has a greater electron affinity than hydrogen and is the more electronegative element, so the shared electrons will be pulled closer to the chlorine nucleus in the hydrogen–chlorine bond. This will give the chlorine end of the bond a partial negative charge, and the hydrogen end a partial positive charge (Figure 8.10). In general, then, we would expect the negative part of a polar covalent bond to be in the region of the more electronegative atom, and the positive part to be in the region of the less electronegative atom.

Atoms of the same element will form nonpolar covalent bonds, with centers of positive and negative charges coinciding. Atoms of unlike elements that have differences in electronegativity will form bonds with the centers of electrical charge separated. When the difference in electronegativity becomes very large, the electrical charges will be completely separated to the two opposite ends of the bond, and an ionic bond will be formed. That is, we can imagine a continuous range from nonpolar covalent bonding to complete ionic bonding, with different compounds appearing at specific places along this continuum (Figure 8.9). Even though there is no clear break between ionic bonding and polar covalent bonding, it is still convenient to try to classify compounds as either ionic or covalent. In general, bonds formed between metals and nonmetals will be ionic, and bonds formed between nonmetals will be covalent (they will be nonpolar covalent when the electronegativity of the atoms is the same, and polar covalent when one atom has greater electronegativity than the other). As we will see, the type of bonding in a compound will determine many of the compound's important properties, such as boiling point and solubility.

8.8 Polyatomic Ions

A **polyatomic ion** is a group of covalently bonded atoms that, as a group, carries an electrical charge, but which is so stable that it will go through most chemical reactions as a unit and won't come apart. Some polyatomic ions and their formulas are listed in Table 8.6. Compounds containing polyatomic ions are so common and play such a central role in our daily lives that it is important for you to become familiar with their names and formulas. Examples of such familiar compounds are sodium bicarbonate (sodium hydrogen carbonate) $NaHCO_3$, which is baking soda, and magnesium hydroxide $Mg(OH)_2$, which is a common antacid.

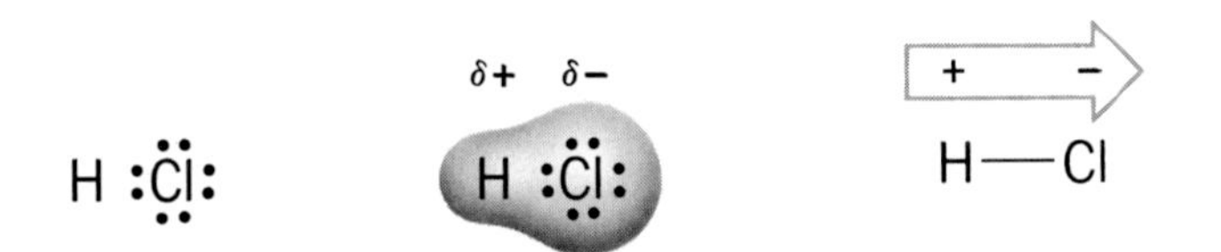

Figure 8.10 Schematic diagrams of the polar covalent bond in a hydrogen chloride molecule. The Greek letter delta (δ) indicates a partial charge.

Table 8.6 Some Common Polyatomic Ions

Name of the Ion (Common Name)	Formula	Name of the Ion (Common Name)	Formula
Ammonium ion	NH_4^+	Permanganate ion	MnO_4^-
Acetate ion	$C_2H_3O_2^-$	Peroxide ion	O_2^{2-}
Carbonate ion	CO_3^{2-}	Phosphate ion	PO_4^{3-}
Hydrogen carbonate ion (bicarbonate)	HCO_3^-	Monohydrogen phosphate ion	HPO_4^{2-}
		Dihydrogen phosphate ion	$H_2PO_4^-$
Chlorate ion	ClO_3^-	Sulfate ion	SO_4^{2-}
Chromate ion	CrO_4^{2-}	Hydrogen sulfate ion (bisulfate)	HSO_4^-
Dichromate ion	$Cr_2O_7^{2-}$		
Cyanide ion	CN^-	Sulfite ion	SO_3^{2-}
Hydroxide ion	OH^-	Hydrogen sulfite ion (bisulfite)	HSO_3^-
Nitrate ion	NO_3^-		
Nitrite ion	NO_2^-		

8.9 Oxidation Number

Oxidation numbers give scientists a way to keep track of the transfer of electrons in chemical reactions. The **oxidation number,** or **oxidation state,** is defined as the charge an atom would have if the electrons in a chemical bond were transferred completely to the more electronegative atom. In an ionic bond, electrons are transferred completely; thus, the oxidation number will equal the charge on the ion formed. When a covalent bond is formed, the atoms have only a partial positive or negative charge, but oxidation numbers still are assigned as if the bond were ionic. If we use the periodic table and the following rules, assigning oxidation numbers is fairly easy and very helpful when writing chemical formulas.

Rule 1. Elements have an oxidation number or oxidation state of zero when they are not combined with atoms of a different element.

Elements can exist in nature as free atoms or as nonpolar, covalently bonded molecules such as N_2, H_2, P_4, or S_8. In either case the oxidation state of the element is zero.

Rule 2. In a covalent bond, oxidation numbers are assigned as if electrons were transferred to the more electronegative atom.

Let's use as our example hydrogen chloride (look back at Figure 8.10). The chlorine is the more electronegative atom and would, therefore, have an oxidation number of 1−. Hydrogen, as a result, would have an oxidation number of 1+.

Rule 3. An ion containing only one atom (a monatomic ion) has an oxidation number equal to the charge on the ion.

We have been discussing the ionic compound sodium chloride, NaCl (look back at Figure 8.1). In sodium chloride, the sodium ion has a 1+ charge and, therefore, an oxidation number of 1+. The chloride ion has a 1− charge and an oxidation number of 1−.

Rule 4. The sum of all the oxidation numbers in a chemical compound is zero.

Both ionic and covalently bonded compounds are neutral. That is, the sum of all the electrical charges in a chemical compound is zero. Therefore, the sum of the oxidation numbers of all the atoms in a compound must be zero. In Example 8-1, we diagrammed the reaction between magnesium and chlorine. The magnesium ion, Mg^{2+}, that is formed has an oxidation number of 2+, and the chloride ion, Cl^-, an oxidation number of 1−. To form an electrically neutral compound, we found we needed two chloride ions for each magnesium ion.

$$
\begin{aligned}
Mg^{2+}&: \text{oxidation number } (2+) \times 1 \text{ atom} = 2+ \\
Cl^{-}&: \text{oxidation number } (1-) \times 2 \text{ atoms} = \underline{2-} \\
&\text{The sum of the oxidation numbers} = 0
\end{aligned}
$$

We can extend this rule for ions containing more than one atom (polyatomic ions) to say that the sum of the oxidation numbers in a polyatomic ion must equal the charge on the ion. For example, the sum of the oxidation numbers of the atoms in the carbonate ion, CO_3^{2-}, must equal 2−, and the sum of the oxidation numbers of the atoms in the ammonium ion, NH_4^+, must equal 1+.

We can make some generalizations when assigning oxidation numbers to atoms. The metals in group I, for example, will always have an oxidation number of 1+; the metals of group II will have an oxidation number of 2+. We call compounds that contain atoms of only two elements **binary compounds.** When nonmetals form binary ionic compounds with metals, group VII elements (the halogens) will have an oxidation number of 1−, group VI elements an oxidation number of 2−, and group V elements an oxidation number of 3− (Table 8.2).

Oxidation numbers are not quite as easy to predict when nonmetals form bonds with other nonmetals in molecules or polyatomic ions. Since fluorine is the most electronegative element, it will always have an oxidation number of 1−. Oxygen is the second most electronegative element and is almost always given the oxidation number of 2− (although there are occasional exceptions). Hydrogen will have an oxidation number of 1+, except when it forms a binary compound with a metal (for example, LiH), and then it has an oxidation number of 1−.

EXAMPLE 8-4

Determine the oxidation number of each element in the following examples.

(a) $KClO_3$ We begin by assigning oxidation numbers to the elements we know and then determine the oxidation number of the remaining element. We have said that oxygen has an oxidation number of 2−, and the group I metals 1+. Therefore,

$$\begin{aligned} \text{K: } (1+) \times 1 \text{ atom} &= 1+ \\ \text{O: } (2-) \times 3 \text{ atoms} &= 6- \\ \text{Cl: } (X) \times 1 \text{ atom} &= \underline{X} \\ \text{The sum of the oxidation numbers} &= 0 \end{aligned}$$

For the sum of the oxidation numbers to be 0, X must equal 5+; therefore, the oxidation number of chlorine in this compound is 5+.

(b) $K_2Cr_2O_7$ Using the same procedure as we did for part (a), we obtain

$$\begin{aligned} \text{K: } (1+) \times 2 \text{ atoms} &= 2+ \\ \text{O: } (2-) \times 7 \text{ atoms} &= 14- \\ \text{Cr: } (X) \times 2 \text{ atoms} &= \underline{2X} \\ \text{The sum of the oxidation numbers} &= 0 \end{aligned}$$

For the sum of the oxidation numbers to be 0, $2X$ must equal 12+; therefore, the oxidation number of each Cr must equal 6+.

(c) CO_3^{2-} This is a polyatomic ion, so the sum of the oxidation numbers must equal the charge on the ion.

$$\begin{aligned} \text{O: } (2-) \times 3 \text{ atoms} &= 6- \\ \text{C: } (X) \times 1 \text{ atom} &= \underline{X} \\ \text{The sum of the oxidation numbers} &= 2- \end{aligned}$$

For the sum of the oxidation numbers to be 2−, X must equal 4+; therefore, the oxidation number of carbon in the carbonate ion is 4+.

Exercise 8-4

Give the oxidation numbers for each element in the following:

(a) $CaCl_2$
(b) H_2SO_4
(c) OCl^-
(d) HCO_3^-
(e) MnO_2
(f) PO_4^{3-}

CHEMICAL FORMULAS AND NOMENCLATURE

8.10 Naming Chemical Compounds

Binary ionic compounds (those containing only two elements) are named for the ions from which they are formed. Positive ions carry the same name as their parent element; for example, Na^{+} is the sodium ion and Ca^{2+} is the calcium ion. If a metal forms more than one positive ion (see Table 8.3), the oxidation number of the ion is indicated in Roman numerals after the name of the element. For example, Fe^{3+} is the iron(III) ion and Cu^{+} is the copper(I) ion. This system using Roman numerals to indicate the oxidation number is called the **Stock system** of nomenclature. An older practice of naming ions gives an *-ic* ending to the name of the ion with the higher oxidation number and an *-ous* ending to the name of the ion with the lower oxidation number. The names of the iron and copper ions in both systems are:

Fe^{2+}: iron(II) ion, or ferrous ion — Cu^{+}: copper(I) ion, or cuprous ion

Fe^{3+}: iron(III) ion, or ferric ion — Cu^{2+}: copper(II) ion, or cupric ion

The negative ion is named by using a prefix taken from the name of the nonmetal element and then adding the suffix *-ide*. For example, Br^{-} is the bromide ion and O^{2-} is the oxide ion. When the name of a binary ionic compound is written, the name of the positive ion appears first. For example, NaBr is sodium bromide, and $CaCl_2$ is calcium chloride. Because copper forms two ions, we must know the oxidation number of copper before we can name a compound such as CuO. Oxygen has an oxidation number of 2−. Copper, therefore, must have an oxidation number of 2+ if the sum of the oxidation numbers is to be zero. The name of the compound is copper(II) oxide (or cupric oxide).

Binary covalent compounds are named in much the same way. The name of the element with the lower electronegativity (the more positive oxidation number) is placed first. The second element is named by adding the suffix *-ide* to the name of the parent element. Prefixes are also added to the name of each element to indicate the number of atoms of that element in the molecule, as follows:

Mono—1   Tetra—4
Di—2   Penta—5
Tri—3   Hexa—6

The prefix "mono" is usually left out, except when it helps distinguish between two different compounds. For example, CO is carbon monoxide, and CO_2 is carbon dioxide.

Many ionic compounds consist of a metal and a polyatomic ion. These compounds are named by placing the name of the metal first, followed by the name of the polyatomic ion. For example, Na_2CO_3 is sodium carbonate and $KHSO_4$ is potassium hydrogen sulfate or potassium bisulfate. Note that when the name of a compound ends in *-ide,* it is a binary compound except when it contains either the hydroxide (OH^-) or cyanide (CN^-) ion. We will learn how to name acids and organic carbon compounds in later chapters.

EXAMPLE 8-5

Name the following compounds:

(a) *KBr.* This ionic compound is formed between the metal potassium and the nonmetal bromine. Placing the name of the metal ion first, we have potassium bromide.

(b) *Cu_2O.* To name this compound we must know the oxidation number of the copper ion. The oxidation number of the oxygen is 2−. Remembering that the sum of the oxidation numbers in a chemical compound must equal zero, we can determine the oxidation number of the copper ion.

$$\begin{aligned} \text{O: } (2-) \times 1 \text{ atom} &= 2- \\ \text{Cu: } (X) \times 2 \text{ atoms} &= \underline{2X} \\ \text{The sum of the oxidation numbers} &= 0 \end{aligned}$$

For the sum of the oxidation numbers to be zero, $2X$ must equal 2+; therefore, the oxidation number of one copper ion must equal 1+. The name is copper(I) oxide, or cuprous oxide.

(c) *HBr.* This is a binary covalent compound. The hydrogen has the lower electronegativity, so it is named first. There is just one atom of each element in the molecule, so the name is hydrogen bromide.

(d) *CCl_4.* This is a binary covalent compound in which carbon has the lower electronegativity. There are four chlorine atoms, so the prefix *tetra-* must be added to the chloride. The name is carbon tetrachloride.

(e) *$CaCO_3$.* This ionic compound contains the positive calcium ion and the polyatomic ion carbonate (Table 8.7). The name is calcium carbonate.

Exercise 8-5

Name the following compounds:

(a) MgI_2 (d) H_2S
(b) FeO (e) $NaHSO_4$
(c) SO_2 (f) $K_2Cr_2O_7$

8.11 Writing Chemical Formulas

In Section 8.4 we saw that a chemical formula is a shorthand way of representing the ratio of atoms of each element found in a chemical compound. When we write a chemical formula of an ionic compound, we write the lowest whole number ratio that will give electrical neutrality. The symbol of the element with the more positive oxidation number is always written first. When a compound contains more than one of the same polyatomic ion, the symbol of the polyatomic ion is put in parentheses, with the subscript following the parentheses, for example, $Ca(OH)_2$.

EXAMPLE 8-6

Write chemical formulas for the following compounds:

(a) *Sodium fluoride.* In Section 8.9 we learned that sodium, a group I metal, will form a positive ion with an oxidation number of 1+, and that fluorine, a halogen, will form a negative ion with an oxidation number of 1−. Therefore, we need one sodium for every fluorine; the ratio will be 1:1. The correct formula is NaF (notice that the subscript "1" is not written).

(b) *Iron(III) oxide.* The name iron(III) means that the iron ion is Fe^{3+} and has an oxidation number of 3+. The oxide ion is O^{2-} and has an oxidation number of 2−. In order to have the sum of the oxidation numbers be zero, we need two iron(III) ions for every three oxide ions:

$$
\begin{aligned}
Fe^{3+}&: (3+) \times 2 \text{ atoms} = 6+ \\
O^{2-}&: (2-) \times 3 \text{ atoms} = \underline{6-} \\
&\qquad\qquad\quad \text{Total} = 0
\end{aligned}
$$

Hint: An easy method to use when writing formulas involving 2 ions is to "crisscross" oxidation numbers, as follows:

$$Fe^{③+} \; O^{②-} \longrightarrow Fe_2O_3$$

(c) *Magnesium hydroxide.* Magnesium, a group II metal, forms the Mg^{2+} ion with an oxidation number of 2+. The polyatomic hydroxide ion, OH^-, has an oxidation number of 1− (Table 8.6). Therefore, we need one magnesium ion for every two hydroxide ions. The correct formula is $Mg(OH)_2$. (Remember that when there is more than one polyatomic ion in a formula, the polyatomic ion is enclosed in parentheses.) We could also have used the crisscross method for this compound.

$$Mg^{②+} \; OH^{①-} \longrightarrow Mg(OH)_2$$

(d) *Hydrogen iodide.* This is a covalent compound formed between two nonmetals, hydrogen and iodine. The name indicates that there is one atom of each element in the molecule. The formula, therefore, is HI.

(e) *Diphosphorus pentoxide.* The prefixes in the name indicate that there are two atoms of phosphorus and five atoms of oxygen in the molecule. The formula, therefore, is P_2O_5.

Exercise 8-6

Write the formula for each of the following compounds (referring, when necessary, to Tables 8.2, 8.3, and 8.7):

(a) Magnesium oxide
(b) Nickel chloride
(c) Potassium sulfide
(d) Sulfur trioxide
(e) Silicon tetrafluoride
(f) Dinitrogen pentoxide
(g) Tin(IV) chloride
(h) Cupric bisulfate
(i) Calcium phosphate

MOLECULAR SHAPE AND HYDROGEN BONDING

8.12 Shapes of Molecules

The properties and behavior of a molecular compound depend not only on the combination of atoms that make it up, but also on the shape or arrangement of these atoms in the molecule. The properties affected by the shape of the molecule range from the odor of the compound to the role that the molecule plays in regulating the chemical reactions in living organisms. For example, all psychedelic drugs seem to affect one particular site in the brain. One current theory is that this site is sensitive to, or will "recognize," a specific three-dimensional shape. According to the theory, the molecules of the psychedelic drugs each have particular regions that closely resemble this three-dimensional shape and, therefore, will each trigger hallucinogenic symptoms.

We have seen before that elements within the same group or family will have similar chemical properties. Such similarities are also true of the arrangement of atoms that are covalently bonded to nonmetals in groups IV through VII. If all bonds to the central atom are single bonds, the arrangement around atoms of elements in group IV will be tetrahedral in shape, group V will be triangular pyramidal, group VI bent, and group VII linear (Figure 8.11).

8.13 Polar and Nonpolar Molecules

A molecule containing many different polar bonds can itself be polar or nonpolar, depending upon the shape of the molecule. As is the case with molecular shape, the polarity or nonpolarity of a molecule also plays a large part in its behavior and role in living organisms. We will see that polar molecules will be found with other polar molecules, and nonpolar molecules will be found with other nonpolar molecules in living systems.

As with covalent bonds, the centers of positive and negative charges will coincide in a nonpolar molecule and will not coincide in a polar molecule. It is important to realize that a polar molecule is still electrically neutral, even though there will be a separation of charge in the molecule resulting in regions of positive and negative charge.

Several examples will illustrate how the shape of a molecule can determine whether the molecule will be polar or nonpolar. Note that in each of our examples, however, the bonds that are formed are polar (Table 8.7). Hydrogen chloride (HCl), as we have already seen, is a linear molecule containing one polar bond. The molecule is polar, with the hydrogen at the positive end and the chlorine at the negative end. Carbon dioxide (CO_2) is

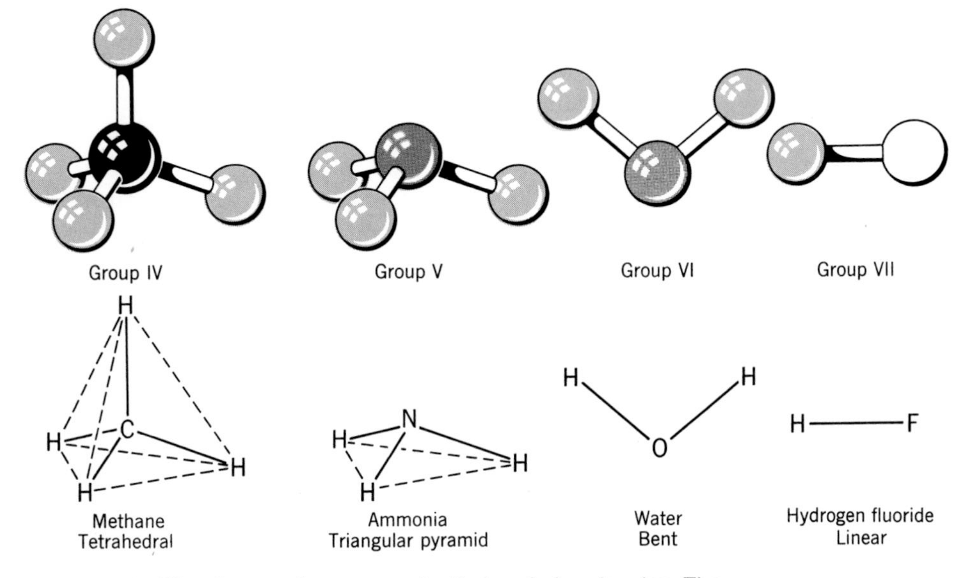

Figure 8.11 The shapes of some covalently bonded molecules. The dotted line is an outline of the shape formed by each molecule.

Table 8.7 Molecular Shape and the Polarity of Molecules

Compound	Formula	Shape	Bond Diagram	Center of Charge	Type of Molecule
Hydrogen chloride	HCl	Linear	$\overset{\delta+}{H}-\overset{\delta-}{Cl}$ (arrow →)	(+ −)	Polar
Carbon dioxide	CO_2	Linear	$\overset{\delta-}{O}=\overset{\delta+}{C}=\overset{\delta-}{O}$ (arrows ← →)	(±)	Nonpolar
Water	H_2O	Bent	$\overset{\delta-}{O}$ bonded to $\overset{\delta+}{H}$ and $\overset{\delta+}{H}$ (arrows toward O)	− above, + below	Polar

also a linear molecule, in this case containing two polar double bonds. However, the linear arrangement of atoms in the molecule makes the centers of positive and negative charge coincide, so the entire carbon dioxide molecule is nonpolar. A molecule of water contains three atoms and two polar bonds just as we found in carbon dioxide. But the water molecule has a bent (rather than linear) shape, and the centers of positive and negative charge do not coincide. Therefore, a water molecule will be polar (Figure 8.12). You can see that for molecules with more than two atoms, we must know the molecular geometry or shape in order to predict whether the molecule is polar or nonpolar. We will have to leave the study of more complex molecular geometries to other courses in chemistry.

8.14 Hydrogen Bonding

Molecules containing hydrogen attached to a highly electronegative atom such as fluorine, oxygen, or nitrogen will have an intermolecular (between molecules) type of bonding called **hydrogen bonding.** The partially positive hydrogen region on one molecule will be attracted to the partially negative fluorine, oxygen, or nitrogen region on another molecule. Such a hydrogen bond has a strength about one-tenth that of an ordinary covalent bond. In some very large biological molecules, hydrogen bonding may occur between different parts of the same molecule, causing the molecule to bend back on itself (intramolecular hydrogen bonding).

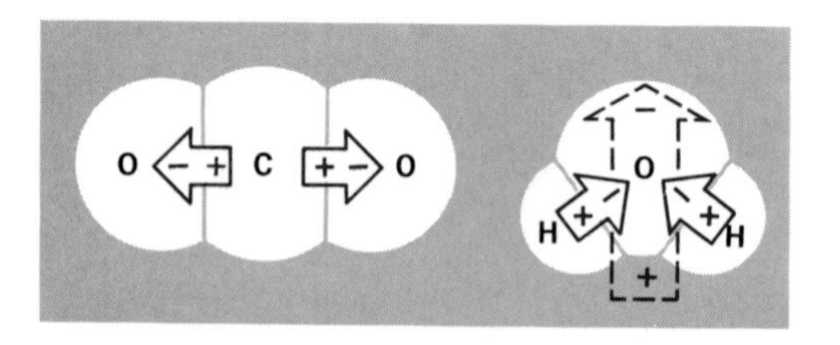

Figure 8.12 Both CO_2 and H_2O contain two polar covalent bonds. But the different shape of their molecules causes a molecule of CO_2 to be nonpolar, and a molecule of H_2O to be polar.

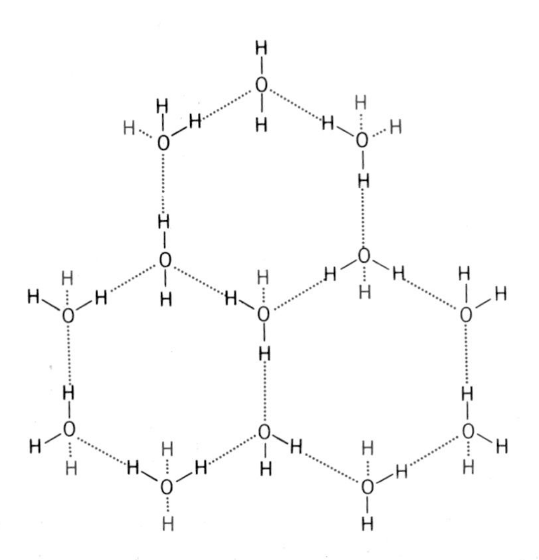

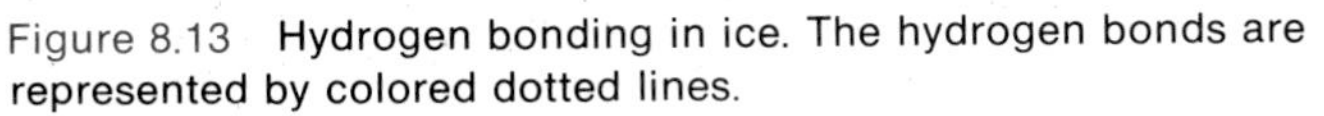
Figure 8.13 Hydrogen bonding in ice. The hydrogen bonds are represented by colored dotted lines.

Hydrogen bonding plays an important role in nature. Hydrogen bonds are responsible for the unusual properties of water (such as high melting and boiling points) which make this fluid so important to all living organisms (Figure 8.13). These properties will be discussed in detail in Chapter 11. Hydrogen bonds also determine the shapes of large biological molecules in living substances and, on a more familiar level, account for the hand-softening property of the lanolin in skin creams, for hard candy getting sticky, and for cotton fabrics taking longer to dry than synthetic fabrics.

CHAPTER SUMMARY

Elements become more stable by entering into reactions in which they lose, gain, or share electrons to attain eight valence electrons. When electrons are transferred from one atom to another, ions are formed. The force of attraction between these ions is called an ionic bond. An ionic compound is a group of ions combined in an orderly fashion called a crystal lattice. In general, ionic bonds form between metals and nonmetals. A covalent bond is formed when electrons are shared between two atoms, creating electrically neutral units called molecules. Two atoms can share two electrons in a single covalent bond, or can share four electrons in a double covalent bond, or six electrons in a triple covalent bond.

The tendency for an atom of an element to attract the electrons in a covalent bond is called the electronegativity of that element. When electrons are shared equally between two atoms, a nonpolar bond is formed. If there is a difference in the electronegativity of the two atoms, a polar bond will be formed. Molecules can be polar or nonpolar depending upon the shape of the molecule and the types of bonds in the molecule. Polyatomic ions are groups of covalently bonded atoms that, as a group, carry an electrical charge and stay together as a unit through most chemical reactions. Another type of bonding that is extremely important in living organisms is hydrogen bonding. A hydrogen bond is formed between a hydrogen attached to a highly electronegative atom, and a fluorine, oxygen, or nitrogen atom on another molecule or on another part of the same molecule.

Oxidation numbers are a way of keeping track of electron transfers in chemical reactions and are very helpful when writing chemical formulas. Oxidation numbers are assigned according to the following rules:

1. Any uncombined element has an oxidation number of zero.
2. For any neutral compound the sum of the oxidation numbers is zero.
3. For any monatomic ion, the oxidation number equals the ionic charge.
4. For any polyatomic ion, the sum of the oxidation numbers of the atoms in the ion equals the ionic charge.
5. Hydrogen has an oxidation number of 1+ (except in metal hydrides, where it has an oxidation number of 1−).
6. Fluorine has an oxidation number of 1−.
7. Oxygen has an oxidation number of 2− (except in peroxides (1−) and a few other compounds).
8. Group I elements always have an oxidation number of 1+, group II elements an oxidation number of 2+, and group III almost always an oxidation number of 3+.

EXERCISES AND PROBLEMS

1. What is the difference between an ionic bond and a covalent bond?
2. Which groups of elements tend to form ionic bonds? Which groups tend to form covalent bonds?
3. What is the difference between the units that make up an ionic compound and the units that make up a covalent compound?
4. In general, what conditions will cause two atoms to combine so as to form (a) an ionic bond, (b) a covalent bond?
5. State the difference between a single, double, and triple covalent bond. Give examples of each.

6. State the difference between the following:
   (a) an atom and an element
   (b) an atom and an ion
   (c) an atom and a diatomic molecule
   (d) an atom and a polyatomic ion

7. Why is it unlikely that (a) calcium would form an ion with a 1+ charge? (b) potassium would form an ion with a 2+ charge?

8. Draw the electron dot diagrams for NaF and MgO. Show that the ions formed in these compounds have the same electron configurations.

9. Draw the electron dot diagrams for each of the following atoms.

   (a) Cesium
   (b) Germanium
   (c) Calcium
   (d) Neon
   (e) Arsenic
   (f) Aluminum
   (g) Sulfur
   (h) Iodine

10. Predict whether the bonds formed between the following atoms will be polar or nonpolar. If the bond is polar, indicate which is the more electronegative atom:

   (a) Cl and Cl
   (b) H and Br
   (c) C and N
   (d) S and Cl
   (e) N and O
   (f) P and O
   (g) F and Br
   (h) O and O

11. Predict whether the following molecules will be polar or nonpolar.

   (a) Oxygen difluoride, OF_2
   (b) Fluorine, F_2
   (c) Hydrogen iodide, HI
   (d) Methane, CH_4
   (e) Chloromethane, CH_3Cl
   (f) Trichloromethane, $CHCl_3$

12. Would you expect hydrogen bonds to form between molecules of hydrogen fluoride, HF? Why or why not?

13. Determine the oxidation number of each element in the following compounds:

   (a) Na_2S
   (b) Cl_2
   (c) $HClO_2$
   (d) N_2O_4
   (e) $HBrO_3$
   (f) $KMnO_4$
   (g) $NaHSO_4$
   (h) $Zn(NO_3)_2$
   (i) $Fe_2(PO_4)_3$
   (j) SO_4^{2-}
   (k) SO_3^{2-}
   (l) $Cr_2O_7^{2-}$

14. Write the correct formula for the compounds composed of the following pairs of ions.

   (a) K^+, CO_3^{2-}
   (b) Na^+, S^{2-}
   (c) Ca^{2+}, NO_3^-
   (d) Sr^{2+}, S^{2-}
   (e) Cr^{3+}, Cl^-
   (f) Fe^{3+}, HPO_4^{2-}
   (g) Cu^{2+}, $C_2H_3O_2^-$
   (h) Ba^{2+}, SO_4^{2-}
   (i) Al^{3+}, SO_3^{2-}
   (j) Sn^{4+}, NO_3^-
   (k) Be^{2+}, F^-
   (l) Cs^+, Br^-

15. Name each of the compounds for which you wrote the formula in question 14.

16. Write the correct formulas for each of the following sets of compounds.

	Set 1	Set 2	Set 3
(a)	Lithium fluoride	Ammonium carbonate	Iodine pentafluoride
(b)	Potassium sulfide	Calcium bisulfate	Hydrogen selenide
(c)	Magnesium bromide	Magnesium bicarbonate	Dinitrogen tetroxide
(d)	Silver chloride	Lithium phosphate	Hydrogen bromide
(e)	Ferric sulfide	Barium nitrite	Boron trichloride
(f)	Barium iodide	Magnesium dihydrogen phosphate	Sulfur dioxide
(g)	Aluminum oxide	Iron(III) sulfate	Carbon disulfide
(h)	Cuprous oxide	Potassium cyanide	Sulfur dichloride
(i)	Mercury(II) nitride	Strontium chromate	Disilicon hexabromide
(j)	Lead(IV) chloride	Beryllium nitrate	Phosphorus trifluoride
(k)	Cesium iodide	Chromium(III) sulfite	Oxygen fluoride
(l)	Gallium nitride	Nickel acetate	Selenium trioxide

17. Name the following compounds:

(a) NH_4I
(b) PCl_3
(c) $Ca(OH)_2$
(d) CBr_4
(e) $FeCl_2$
(f) $(NH_4)_2Cr_2O_7$
(g) O_2F_2
(h) $Zn(NO_2)_2$
(i) P_2O_5
(j) NaH
(k) HI
(l) $NaMnO_4$
(m) Li_2S
(n) K_2SO_3
(o) CaC_2
(p) $Al_2(SO_4)_3$

18. Draw the electron dot diagrams for the following molecules:

(a) HI
(b) F_2
(c) CH_3Cl
(d) HCN
(e) OF_2
(f) PCl_3

19. Draw the bond diagrams for each of the following compounds:

(a) NH_3
(b) CCl_4
(c) HCl
(d) C_2Br_2
(e) SF_2
(f) I_2

20. What is the total number of each type of atom in one unit of each of the following?

(a) $(NH_4)_3PO_4$
(b) $Al_2(HPO_4)_3$
(c) $Ca(C_2H_3O_2)_2$
(d) $Fe_4[Fe(CN)_6]_3$

CHAPTER NINE
CHEMICAL EQUATIONS AND THE MOLE

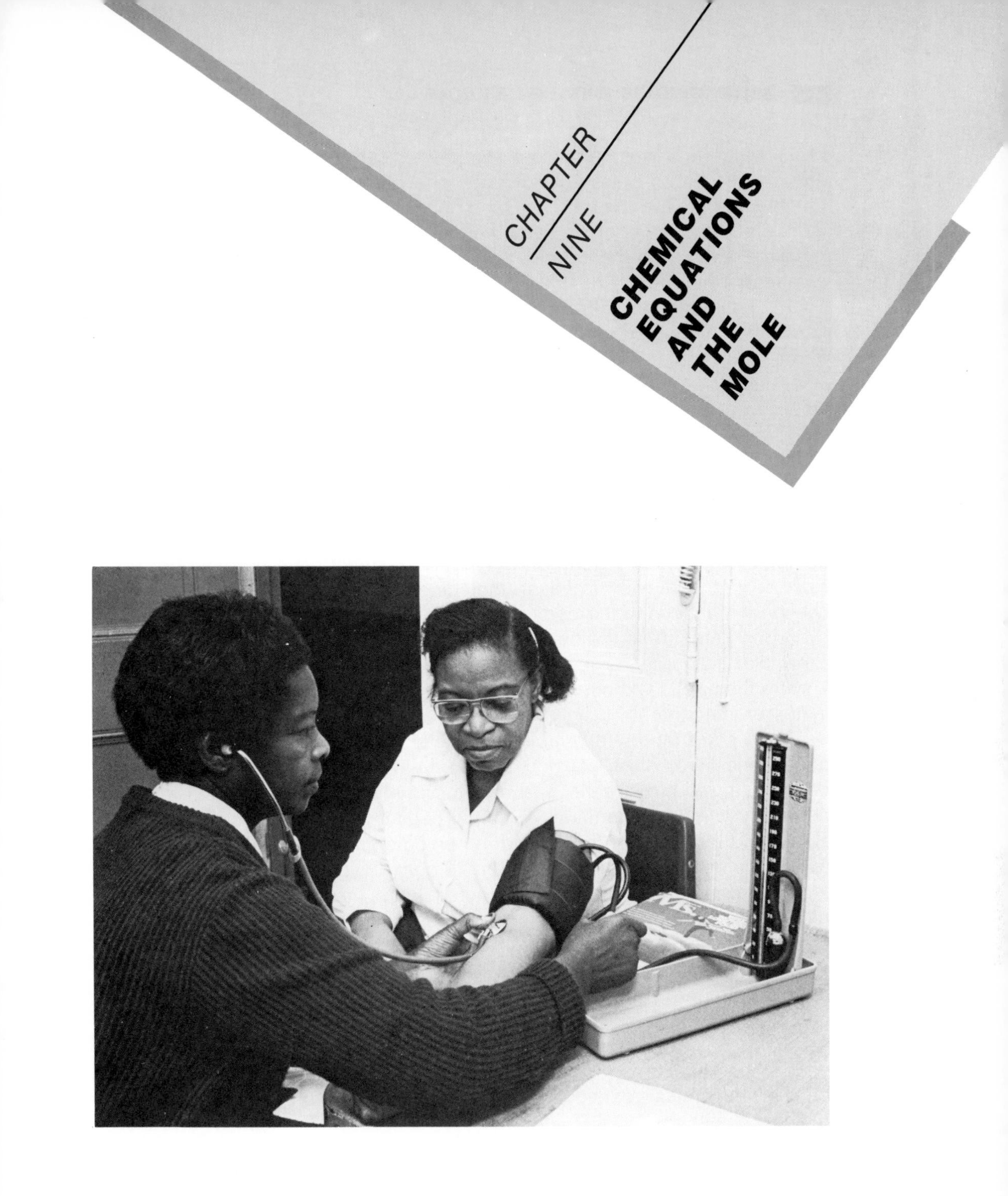

At 45 years of age, Betty Johns felt that she was in perfect health. However, it had been eight years since her last physical examination, and she felt that now was a good time to schedule another checkup. At Betty's exam the doctor measured her blood pressure at the very high level of 180/120. Although she had not suspected it, this meant that Betty was suffering from hypertension, or high blood pressure.

Because initial treatment with a diuretic failed to lower Betty's blood pressure to normal levels, the doctor next prescribed the drug alpha-methyl dopa, which is a muscle relaxant that dilates the small arteries and thereby reduces the blood pressure. However, when Betty returned to the doctor a month later for a follow-up exam, she complained of a constant feeling of fatigue and lack of energy. This time the doctor found that Betty's blood pressure was within the normal range at 140/90, but that her hemoglobin level had fallen from a normal level of 13.5 to a very low level of 10. Further tests confirmed the diagnosis that Betty was now suffering from hemolytic anemia, a destruction of red blood cells caused by the drug alpha-methyl dopa. The doctor immediately stopped the use of this drug and prescribed a different smooth-muscle relaxant for Betty. Within three weeks her hemoglobin level had returned to normal.

It is an unfortunate fact of modern medicine that the treatment given for one disease may very well end up causing another. However, for us the important question at this moment is what exactly caused the destruction of Betty's red blood cells? The alpha-methyl dopa that Betty was taking for her high blood pressure reacts with oxygen in the body to form the compound hydrogen peroxide, H_2O_2. This is usually not a problem because most people have two enzymes—molecules that control the rate of chemical reactions in the body—that protect against the buildup of hydrogen peroxide in the cells by breaking down this compound to form water. Some people, however, are missing one or both of these enzymes, leaving their cells without protection against the harmful effects of hydrogen peroxide.

Betty's red blood cells lacked the selenium-containing enzyme called glutathione peroxidase, which is the main enzyme that destroys hydrogen peroxide in red blood cells. This resulted in a buildup of hydrogen peroxide, which can damage red blood cells in two ways. First, this compound changes the iron found in hemoglobin from Fe^{2+} to Fe^{3+}, which makes the hemoglobin lose its ability to carry oxygen in the blood. Second, the hydrogen peroxide reacts with molecules in the membrane of the red blood cell, resulting in the destruction (hemolysis) of these cells and causing the hemolytic anemia from which Betty suffered.

Each of the processes that we have just described—the breakdown of hydrogen peroxide to water, the change in the iron ion, and the destruction of red blood cells—takes place by means of chemical reactions. But in order for us to study and understand such processes we need to be familiar with the vocabulary and symbols that are used to represent chemical reactions.

LEARNING OBJECTIVES

By the time you have finished this chapter, you should be able to:

1. Write a balanced chemical equation given the names of the reactants and products.
2. Define a *mole*.

3. Calculate the formula weight of a chemical compound.
4. Calculate the number of moles in a given mass of a substance.
5. Calculate the number of grams in a given number of moles of a substance.
6. Use a balanced chemical equation to determine the amount of reactants necessary to produce a given amount of product.

CHEMICAL EQUATIONS

9.1 Writing Chemical Equations

Chemical equations are a shorthand way of representing what occurs in a chemical reaction. A chemical equation contains the formulas of the starting materials, or **reactants,** separated by an arrow from the resulting materials, or **products.**

$$\underset{\text{Reactants}}{A + B} \longrightarrow \underset{\text{Products}}{C + D}$$

The substances that form the reactants and products may be atoms, ions, molecules, or ion groups. In this chapter we will be concerned mainly with reactions between atoms and molecules. We will study reactions between ions in greater detail in Chapter 11.

Atoms are neither created nor destroyed in a chemical reaction, so the chemical equation representing the reaction must be **balanced.** That is, for each element involved in the reaction the equation must show the same number of atoms on the product side as are found on the reactant side. This means that the total mass of the products will equal the total mass of the reactants, which is known as the **Law of Conservation of Mass.**

For example, either hydrogen peroxide or water can be formed from hydrogen and oxygen, but, as we have seen, our tissues are greatly affected by which one is formed. Hydrogen peroxide can badly damage cells, whereas water is an essential part of all cells. In writing a chemical equation for the formation of either hydrogen peroxide or water, the reactants or starting materials are oxygen (O_2) and hydrogen (H_2). Although the hydrogen needed for this reaction in our tissues does not come from the elemental hydrogen molecule H_2, but rather comes from other molecules containing hydrogen, we will just use elemental hydrogen to simplify this introductory discussion:

$$H_2 + O_2 \longrightarrow$$

To complete either equation we must write the correct chemical formulas for the products, water or hydrogen peroxide. We know that water is H_2O. From Table 8.6 we can find that the peroxide ion is O_2^{2-} and, therefore, to maintain electrical neutrality we need two hydrogen ions, H^+. This gives us the correct formula for hydrogen peroxide, H_2O_2. Now we can write the reactants and products of the two equations we have discussed:

(Hydrogen peroxide)	$H_2 + O_2 \longrightarrow H_2O_2$
(Water)	$H_2 + O_2 \longrightarrow H_2O$

Let's check to see if these two equations are balanced. Starting with the equation for the formation of hydrogen peroxide, we must check the number of atoms of each element on both sides of the equation. You can see that there are two hydrogen atoms on the reactant side and two on the product side. There are two oxygen atoms on the reactant side and two on the product side. Therefore, this equation is balanced.

$$H_2 + O_2 \longrightarrow H_2O_2 \qquad (1)$$

Now look at the chemical equation for the formation of water. There are two hydrogen atoms on each side of the arrow, but there are two oxygen atoms on the reactant side and only one on the product side. However, to balance an equation the only numbers that can be changed (after the correct chemical formulas have been written) are the numbers, or coefficients, that appear in front of the formulas of the substances involved. So, to obtain two oxygen atoms on the product side of the equation we must place a coefficient of 2 in front of the formula for water.

$$H_2 + O_2 \longrightarrow 2H_2O$$

Now the oxygen atoms balance, but the hydrogen atoms are unbalanced—there are four hydrogen atoms on the product side and only two on the reactant side. Placing a coefficient of 2 in front of the formula for hydrogen on the reactant side will balance the hydrogens. Now you can check to see that the equation is balanced (Figure 9.1).

$$2H_2 + O_2 \longrightarrow 2H_2O \qquad (2)$$

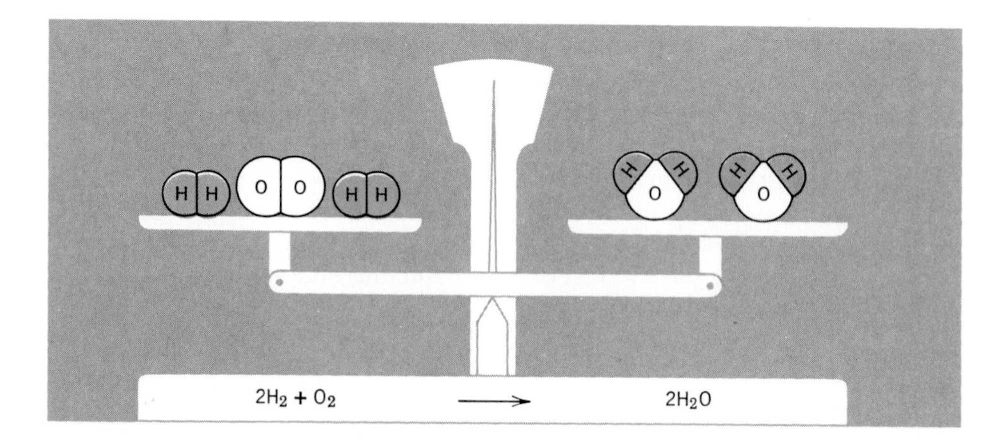

Figure 9.1 When an equation is balanced, the number of atoms of each element on the reactant side will equal the number of atoms of each element on the product side. Therefore, the total mass of the reactants will equal the total mass of the products.

By comparing these two balanced equations, you can now see that the reactions for the formation of water and hydrogen peroxide are actually different: in one reaction 1 molecule of hydrogen combines with 1 molecule of oxygen to form 1 molecule of hydrogen peroxide, and in the other reaction 2 molecules of hydrogen combine with 1 molecule of oxygen to form 2 molecules of water.

9.2 Balancing Chemical Equations

Balancing equations is not really a complicated procedure. The following steps should help you quickly master this skill:

1. First, write the correct chemical formula for each reactant and each product. Once this is done, you should never change the formula of a substance; only the coefficients can be changed. Any change of subscript would change the nature of the compound. For example, CO and 2CO represent one and two molecules of the compound carbon monoxide. But CO and CO_2 are entirely different substances, and you wouldn't want to confuse the two. Carbon monoxide is deadly, but carbon dioxide is produced in, and exhaled from, our bodies.
2. Because the balancing of equations involves juggling coefficients, it is often easiest to begin by giving the coefficient 1 to the compound having the most complicated formula.
3. You must then start balancing each of the elements. You will often find it easiest if you leave oxygen until last.
4. Treat polyatomic ions as one unit (that is, just as though they were a single element) if they remain unchanged in the reaction.
5. When you think the equation is balanced, it is a good idea to check each element again to make sure that it really is balanced.

The following examples illustrate these five steps for balancing equations. After you have read through the examples, try them yourself to see if you can get the same answer. You will find that skill in balancing equations comes with lots of practice.

EXAMPLE 9-1

1. Balance the following equation:

$$Na + Cl_2 \longrightarrow NaCl$$

Step 1: Already completed.
Steps 2, 3, 4: This is not a balanced equation; there are two chlorine atoms on the reactant side of the equation and only

one on the product side. Placing a coefficient of 2 in front of the formula for sodium chloride will balance the chlorine atoms in this equation.

$$Na + Cl_2 \longrightarrow 2NaCl$$

But now there are two sodium atoms on the product side, and only one on the reactant side. Placing a coefficient of 2 in front of the sodium on the reactant side will balance the equation.

$$2Na + Cl_2 \longrightarrow 2NaCl$$

Step 5: We have two atoms of sodium and two atoms of chlorine on each side of the equation.

2. A small amount of the pollutant sulfur trioxide is released into the air when sulfur-containing coal or petroleum is burned. Write the balanced chemical equation for this reaction.

   Step 1: When a substance burns, it reacts with oxygen in the air, so the reactants for this chemical reaction are elemental sulfur and oxygen (remember that elemental oxygen exists as a diatomic molecule). The product is sulfur trioxide.

$$S + O_2 \longrightarrow SO_3$$

   Steps 2, 3, 4: When the equation is written as above, the sulfur atoms balance but not the oxygen atoms. In order for the oxygen to balance, we must have the same number of atoms on both sides, and the lowest possible number is six.

$$S + 3O_2 \longrightarrow 2SO_3$$

   Now to balance the sulfur,

$$2S + 3O_2 \longrightarrow 2SO_3$$

   Step 5: We have two sulfur atoms on each side and six oxygen atoms on each side (Figure 9.2).

3. When aluminum reacts with sulfuric acid (H_2SO_4), hydrogen gas and aluminum sulfate are produced.

   Step 1: Using Tables 8.2 and 8.6, we find that the oxidation number of aluminum is 3+, Al^{3+}, and that of sulfate is 2−, SO_4^{2-}. The formula for aluminum sulfate, therefore, is $Al_2(SO_4)_3$.

$$Al + H_2SO_4 \longrightarrow H_{2(g)} + Al_2(SO_4)_3$$

(*g*) = gas

   Steps 2, 3, 4: Assign $Al_2(SO_4)_3$ a coefficient of 1. Then, balance the aluminum.

$$2Al + H_2SO_4 \longrightarrow H_{2(g)} + Al_2(SO_4)_3$$

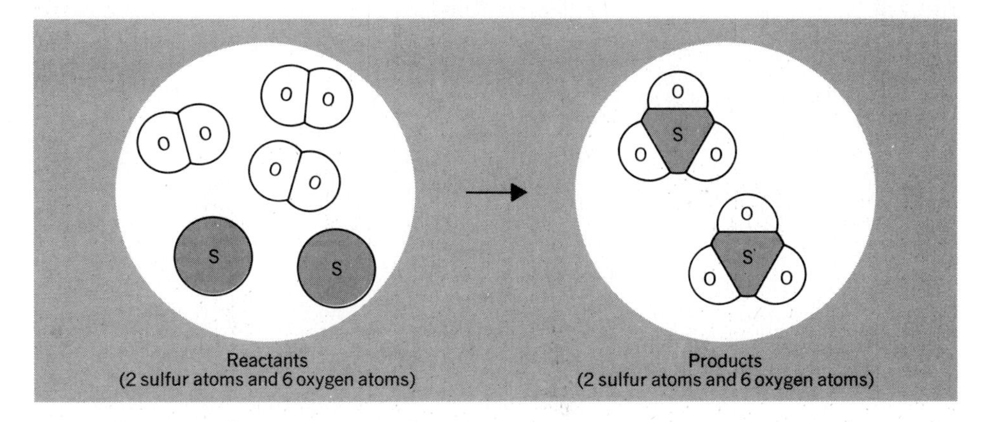

Figure 9.2 The balanced equation for the formation of sulfur trioxide is $2S + 3O_2 \longrightarrow 2SO_3$.

Next balance the sulfate, SO_4^{2-}.

$$2Al + 3H_2SO_4 \longrightarrow H_{2(g)} + Al_2(SO_4)_3$$

And finally, balance the hydrogens.

$$2Al + 3H_2SO_4 \longrightarrow 3H_{2(g)} + Al_2(SO_4)_3$$

Step 5: There are now two atoms of aluminum on each side, six atoms of hydrogen on each side, and three sulfate ions on each side (Table 9.1).

4. Hard water contains dissolved calcium chloride, ($CaCl_2$), that forms a scum with soap. You can soften the water by adding sodium carbonate. The calcium will react with the carbonate to form calcium carbonate, which is insoluble in water. Write a balanced equation for this reaction.

   Step 1: $CaCl_{2(aq)} + Na_2CO_{3(aq)} \longrightarrow CaCO_{3(s)} + NaCl_{(aq)}$

   (*aq*) = aqueous or dissolved in water
   (*s*) = solid

Table 9.1 A Tally of Atoms for the Following Balanced Equation $2Al + 3H_2SO_4 \longrightarrow 3H_2 + Al_2(SO_4)_3$

Symbol of Atom	Number of Atoms on the Reactant Side	Number of Atoms on the Product Side
Al	2	2
H	6	6
S	3	3
O	12	12

Steps 2, 3, 4: Assign $CaCO_3$ a coefficient of one. The calcium and carbonate ions then balance, but we must still balance the sodium ions.

$$CaCl_{2(aq)} + Na_2CO_{3(aq)} \longrightarrow CaCO_{3(s)} + 2NaCl_{(aq)}$$

Checking the chloride ions (Cl^-), we find they are now balanced.

Step 5: There are one calcium ion, two sodium ions, two chloride ions, and one carbonate ion on each side of the equation.

Exercise 9-1

Use the five steps discussed in Section 9.2 to balance the following equations:

(a) $P + O_2 \longrightarrow P_4O_{10}$
(b) $NOCl \longrightarrow NO + Cl_2$
(c) $CH_4 + O_2 \longrightarrow CO_2 + H_2O$
(d) $Ca(OH)_2 + HCl \longrightarrow CaCl_2 + H_2O$
(*Note:* H_2O can also be written as HOH.)
(e) Magnesium reacts with oxygen gas to form magnesium oxide.
(f) Lead(II) sulfide reacts with oxygen gas to form lead(II) oxide and sulfur dioxide gas.
(g) Sodium carbonate reacts with magnesium nitrate to form magnesium carbonate and sodium nitrate.

9.3 Using Oxidation Numbers to Balance Oxidation–Reduction Reactions*

Many chemical reactions involve the transfer of electrons, either completely or partially, from one atom to another. These reactions are called **oxidation–reduction,** or **redox,** reactions. Your car battery uses redox reactions to produce the energy that powers your headlights, and food is broken down within the cells of your body to produce energy through a series of redox reactions. **Oxidation** refers to the loss of electrons by an atom, and **reduction** to the gain of electrons by another atom. Oxidation and reduction reactions are complementary processes. They always occur simultaneously and in equal amounts: the number of electrons lost by a substance must equal the number of electrons gained

* *Optional: This section may be omitted without loss of continuity.*

by another substance. In the formation of sodium chloride, for example, the sodium atom loses an electron while the chlorine atom gains an electron; the sodium atom is oxidized and the chlorine atom reduced.

$$Na \longrightarrow Na^+ + e^- \quad \text{oxidation reaction}$$
$$Cl + e^- \longrightarrow Cl^- \quad \text{reduction reaction}$$

We call the element that gains electrons the **oxidizing agent,** and the element that loses the electrons the **reducing agent.**

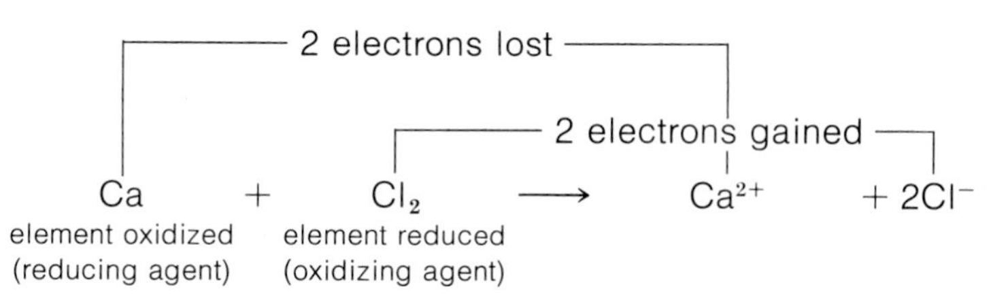

Because the total number of electrons gained must equal the total number of electrons lost, we can use oxidation numbers to help balance complex redox reactions. To calculate the number of electrons gained or lost, we use oxidation numbers to indicate charges on the atoms (even if the compound formed is covalent). Before working through the following examples, you might want to review the discussion of oxidation numbers in Section 8.9.

Let's begin by balancing an equation that we could easily balance by inspection: the reaction of magnesium with oxygen gas to form magnesium oxide.

Step 1: Write the formulas for the reactants and products.

$$Mg + O_2 \longrightarrow MgO$$

Step 2: Write the oxidation number for each element, then determine which elements changed oxidation number, and show the electrons lost or gained.

$$\begin{array}{lccccc} & Mg & + & O_2 & \longrightarrow & Mg\ O \\ \text{oxidation number:} & 0 & & 0 & & 2+\ 2- \end{array}$$

Remember that uncombined elements have oxidation numbers of zero. To go from an oxidation number of 0 to 2–, each oxygen atom had to gain two electrons. (Reduction, or electron gain, causes the oxidation number to decrease.) To go from an oxidation number of 0 to 2+, each magnesium atom had to lose two electrons. (Oxidation, or electron loss, causes the oxidation number to increase.)

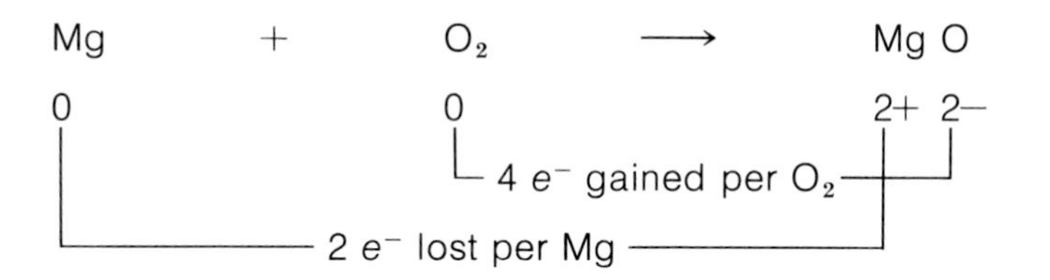

Step 3: Use coefficients to make the number of electrons gained equal to the number of electrons lost.

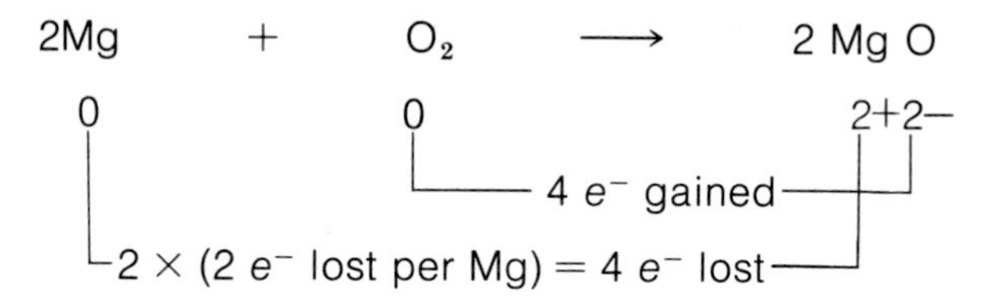

Step 4: Balance the remaining parts of the equation. In this case the equation is now already properly balanced.

$$2Mg + O_2 \longrightarrow 2MgO$$

EXAMPLE 9-2

Balance the following redox equation. This equation is more complicated than the example shown above, but can be solved in the same way.

$$SO_2 + HNO_3 + H_2O \longrightarrow H_2SO_4 + NO$$

Step 1: Completed.
Step 2: When figuring out the oxidation numbers, remember that the sum of the oxidation numbers for a compound is zero. We will start by assigning oxidation numbers to hydrogen and oxygen.

SO_2	O:	(2−) × 2 atoms = 4−	
	S:	(X) × 1 atom = X	
		Total = 0	So, X = 4+
HNO_3	H:	(1+) × 1 atom = 1+	
	O:	(2−) × 3 atoms = 6−	
	N:	(X) × 1 atom = X	
		Total = 0	So, X = 5+
H_2SO_4	H:	(1+) × 2 atoms = 2+	
	O:	(2−) × 4 atoms = 8−	
	S:	(X) × 1 atom = X	
		Total = 0	So, X = 6+
NO	O:	(2−) × 1 atom = 2−	
	N:	(X) × 1 atom = X	
		Total = 0	So, X = 2+

Therefore, we have

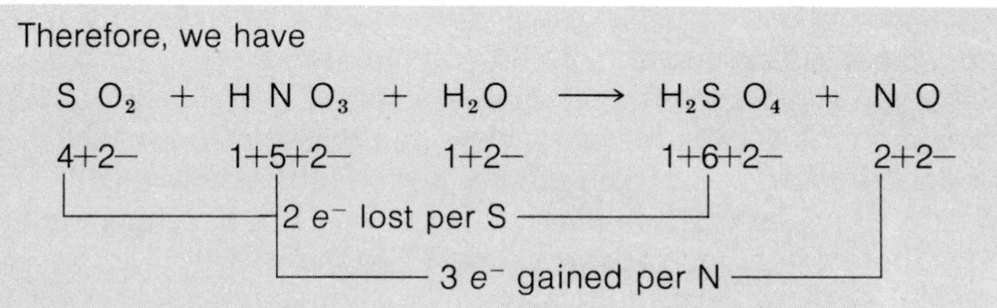

Step 3: The lowest number of electrons that can be equally lost and gained is 3 × 2, or 6.

$$3SO_2 + 2HNO_3 + H_2O \longrightarrow 3H_2SO_4 + 2NO$$

$4+$ (S), $5+$ (N), $6+$ (S), $2+$ (N)

$3 \times (2\ e^- \text{ lost per S}) = 6\ e^-$

$2 \times (3\ e^- \text{ gained per N}) = 6\ e^-$

Step 4: Balance the remaining elements, hydrogen and oxygen.

$$3SO_2 + 2HNO_3 + 2H_2O \longrightarrow 3H_2SO_4 + 2NO$$

Exercise 9-2

For each of the following redox reactions, (1) identify the element being oxidized and the element being reduced, (2) identify the oxidizing and reducing agents, and (3) balance the equations using oxidation numbers.

1. $Mg + Br_2 \longrightarrow MgBr_2$
2. $KCl + MnO_2 + H_2SO_4 \longrightarrow K_2SO_4 + MnSO_4 + Cl_2 + H_2O$
3. $Cu + HNO_3 \longrightarrow Cu(NO_3)_2 + NO + H_2O$
   (*Hint:* In this equation some of the NO_3^- ions change and some remain unchanged. It is helpful to write the HNO_3 twice to take care of both conditions.)

9.4 The Mole

When sodium (Na) is placed in a container of chlorine gas (Cl_2), sodium atoms and chlorine atoms will undergo a chemical reaction to produce the compound sodium chloride, NaCl. One sodium atom will react with each chlorine atom—or more precisely, two sodium atoms will react with each chlorine molecule to produce two sodium chloride ion groups (remember that ionic compounds do not exist as single molecules). If a chemist wanted to produce a certain exact amount of sodium chloride, however, it would be impossible to measure out the sodium and chlorine atoms individually. Atoms are too small to count out one at a time, even millions at a time. Therefore, a unit of measurement had to be invented that would allow us to measure out equal numbers of sodium and chlorine atoms: this unit is the mole. The term "mole" was introduced by Wilhelm Ostwald in 1896 from the Latin word *moles,* meaning heap or pile. A **mole (mol)** is defined to be the number of atoms in 12.0000 grams of carbon-12. Because atoms are so small, this number of atoms is very large: 6.02×10^{23}. It is hard to imagine just how large this number is, but another way to write 6.02×10^{23} is $602{,}000 \times 1$ million $\times$ 1 million $\times$ 1 million. This number is called **Avogadro's number,** named for the nineteenth-century scientist Amadeo Avogadro, who contributed a great deal to our knowledge of atomic weights. A mole, then, contains 6×10^{23} units; these units can be atoms, molecules, ions, electrons, or anything else. A mole of sodium atoms contains 6×10^{23} sodium atoms; a mole of chlorine molecules contains 6×10^{23} chlorine molecules; a mole of marshmallows contains 6×10^{23} marshmallows (enough to cover all 50 of the United States with a blanket of marshmallows 60 miles deep!).

How much does one mole weigh? That depends on the nature of the particles, just as a dozen lemons, grapefruit, or pumpkins each has a different total weight. However, *the weight of one mole of atoms of any element will be exactly equal to the atomic weight in grams of that element.* For example, the atomic weight of one atom of sodium is 23 amu, and we saw in Chapter 5 that one amu weighs 1.66×10^{-24} g. If a mole of sodium contains 6.02×10^{23} atoms of sodium, then the weight of that mole can be calculated as follows:

$$1\ \cancel{\text{mol}}\ \text{Na} \times \frac{6.02 \times 10^{23}\ \cancel{\text{atoms}}\ \text{Na}}{1\ \cancel{\text{mol}}\ \text{Na}} \times \frac{23\ \cancel{\text{amu}}}{1\ \cancel{\text{atom}}\ \text{Na}} \times \frac{1.66 \times 10^{-24}\ \text{g}}{1\ \cancel{\text{amu}}} = 23\ \text{g}$$

Figure 9.3 Although they do not weigh the same, the number of atoms of copper, mercury, lead, and iron in these samples is the same.

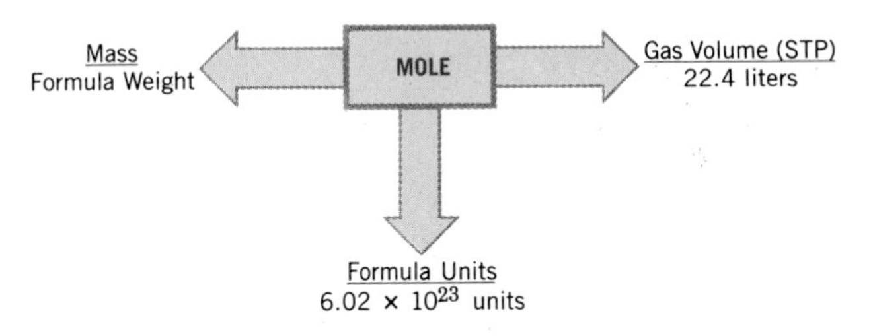

Figure 9.4 **A mole is a specific amount of a substance.**

Aluminum has an atomic weight of 27 amu, so 27 grams of aluminum will contain one mole, or 6×10^{23} atoms, of aluminum; likewise, 35.5 grams of chlorine will contain one mole of chlorine atoms (Figure 9.3). It has also been shown that if a substance is in the gaseous state, one mole of this gas will occupy 22.4 liters at standard temperature and pressure (0°C and 1 atm). The molar volume of a gas at STP, therefore, is 22.4 liters (Figure 9.4).

It has become common to use the term *mole* to refer not only to Avogadro's number of particles, but also to the number of grams of a substance that will contain 6×10^{23} particles. Therefore, we will refer to 23 grams of sodium as one mole of sodium and think of one mole of aluminum as being 27 grams of aluminum.

EXAMPLE 9-3

1. Determine the weight of each of the following:

(a) *One mole of magnesium atoms:* From the table on the back inside cover of this book, we find that the atomic weight of magnesium is 24 amu. Therefore, the weight of one mole of magnesium atoms is 24 g.

(b) *3.50 moles of iron:* The atomic weight of iron is 55.9 amu, so we know that one mole of iron = 55.9 g. This equality allows us to write two unit factors showing the relationship between moles and grams of iron:

$$\frac{1 \text{ mol Fe}}{55.9 \text{ g}} \quad \text{and} \quad \frac{55.9 \text{ g}}{1 \text{ mol Fe}}$$

Therefore, 3.50 moles of iron will weigh

$$3.50 \cancel{\text{mol Fe}} \times \frac{55.9 \text{ g}}{1 \cancel{\text{mol Fe}}} = 196 \text{ g}$$

2. Determine the number of moles in each of the following:

(a) *1.8×10^{24} atoms of lead:* We know that one mole of lead will contain 6.02×10^{23} atoms, so we can construct a unit factor that will allow us to convert atoms of Pb into moles of Pb.

$$1.8 \times 10^{24} \text{ atoms Pb} \times \frac{1 \text{ mol Pb}}{6.02 \times 10^{23} \text{ atoms Pb}} = \frac{1.8}{6.02} \times \frac{10^{24}}{10^{23}} \text{ mol Pb}$$

$$= 0.299 \times 10 \text{ mol Pb} = 3.0 \text{ mol Pb}$$ (rounded off to 2 significant figures)

(If you are unsure of how this calculation was done, see Appendix 1.)

(b) *275 g boron:* The atomic weight of boron is 10.8 amu, so one mole of boron weighs 10.8 g. Using the appropriate unit factor, therefore, we get

$$275 \text{ g} \times \frac{1 \text{ mol B}}{10.8 \text{ g}} = 25.5 \text{ mol B}$$

Exercise 9-3

1. Determine the weight of each of the following:

   (a) 0.050 mol gold
   (b) 2.0 mol zinc
   (c) 0.10 mol sulfur
   (d) 2.5×10^{20} atoms of magnesium

2. Determine the number of moles in each of the following:

   (a) 32.4 g silver
   (b) 980 g silicon
   (c) 0.02 g neon
   (d) 1.2×10^{25} atoms of uranium

9.5 Formula Weight

When atoms react to form compounds, there is no net gain or loss of weight. The particle that forms, whether it be a molecule or ion group, will have a **formula weight** that is equal to the sum of the atomic weights of all the atoms appearing in its chemical formula.

For example, the formula weight for sodium chloride, NaCl, is 58.5 amu—the atomic weight of one sodium atom (23.0 amu) plus the atomic weight of one chlorine atom (35.5 amu). The formula weight of carbon tetrachloride, CCl_4, equals the sum of the atomic weight of one carbon atom plus the atomic weight of four chlorine atoms.

$$C + (4 \times Cl) = CCl_4$$
$$12 + (4 \times 35.5) = 154$$

One mole of any substance will have a mass equal to the formula weight of that substance expressed in grams. For example, one mole of chlorine molecules, Cl_2, will weigh 2×35.5 or 71.0 grams. One mole of the compound sodium chloride, NaCl, will weigh $23.0 + 35.5$ or 58.5 grams. If you weigh out 58.5 grams of NaCl, which is one mole of table salt, your table salt will contain 6×10^{23} NaCl ion groups.

EXAMPLE 9-4

1. Calculate the formula weight for each of the following:

   (a) *KBr:* The atomic weight of potassium is 39 amu, and that of bromine is 80 amu. Therefore, the formula weight of KBr is $39 + 80 = 119$ amu.

   (b) *$Ca(OH)_2$:* The formula weight will be the sum of the atomic weight of one calcium atom (40 amu) plus two times the weight of the polyatomic ion OH^- ($16 + 1 = 17$ amu).

   $$Ca + (2 \times OH) = Ca(OH)_2$$
   $$40 + (2 \times 17) = 74 \text{ amu}$$

   (c) *$Mg_3(PO_4)_2$:* The formula weight can be calculated as follows:

   $$3 \times Mg + \{2 \times [P + (4 \times O)]\}$$
   $$3 \times 24 + \{2 \times [31 + (4 \times 16)]\}$$
   $$72 + (2 \times 95)$$
   $$72 + 190 = 262 \text{ amu}$$

2. What is the weight of one mole of $Ca(OH)_2$?

   One mole of $Ca(OH)_2$ will have a mass equal to the formula weight in grams. Therefore, one mole of $Ca(OH)_2$ will weigh 74 g.

3. How many KBr ion groups are there in 119 g KBr?

   The weight of one mole of KBr is 119 g. Because one mole contains 6×10^{23} units, 119 g or one mole of KBr will contain 6×10^{23} ion groups.

Exercise 9-4

1. Calculate the formula weight for each of the following:

(a) I_2
(b) HF
(c) PbS
(d) $KClO_3$
(e) $Al_2(SO_4)_3$
(f) C_4H_{10}

2. What is the weight of one mole of each of the following?

(a) Hydrogen fluoride
(b) Potassium chlorate
(c) Aluminum sulfate

3. What is the weight in grams of 6×10^{23} molecules of I_2?

9.6 Using the Mole in Problem Solving

As we saw in Section 9.4, problems involving mole calculations often require the use of unit factors to convert from one unit of measure to another. Suppose, for example, a chemist needs 0.45 mole of potassium chlorate, $KClO_3$. How many grams must be weighed out? This problem requires that we convert from moles of $KClO_3$ to grams, so we must construct an appropriate unit factor. We know how to determine the weight of one mole of $KClO_3$.

$$\text{Formula weight of } KClO_3 = 39.1 + 35.5 + (3 \times 16.0) = 122.6 \text{ amu}$$

Therefore,

$$1 \text{ mol } KClO_3 = 122.6 \text{ g}$$

From this equality, we can write the unit factor that will allow us to convert from moles to grams

$$0.45 \text{ mol } \cancel{KClO_3} \times \frac{122.6 \text{ g}}{1 \text{ mol } \cancel{KClO_3}} = 55 \text{ g}$$

Study carefully the next two examples and then work through Exercise 9-5. We will use the following three steps in solving these problems.

Step 1: Determine what the problem is asking for.
Step 2: Set up the unit factors necessary to make the conversions.
Step 3: Do the calculation, making certain that the units of measure cancel so that your answer is in the desired units.

EXAMPLE 9-5

1. If you have a flask containing 9 grams of water, how many moles of water are in the flask?

Step 1: Our problem asks,

$$9 \text{ grams} = (?) \text{ moles of water}$$

Step 2: We need to find the relationship between grams and moles of water.

$$\text{Formula weight of } H_2O = (2 \times 1) + 16 = 18$$

Therefore,

$$1 \text{ mole of } H_2O = 18 \text{ g}$$

Step 3: From the above equality we can construct the unit factor necessary to give us an answer in moles.

$$9 \cancel{\text{g}} \times \frac{1 \text{ mol } H_2O}{18 \cancel{\text{g}}} = 0.5 \text{ mol } H_2O$$

2. How many molecules of chlorine gas will there be in a tank containing 7.10 grams of chlorine?

Step 1: Our problem asks,

$$7.10 \text{ grams} = (?) \text{ molecules of } Cl_2$$

Step 2: To solve this problem, we need to establish a relationship between grams of chlorine and molecules of chlorine. We know that

$$1 \text{ mol } Cl_2 = 2 \times 35.5 = 71.0 \text{ g}$$

and

$$1 \text{ mol } Cl_2 = 6.02 \times 10^{23} \text{ molecules}$$

Therefore,

$$71.0 \text{ g } Cl_2 = 6.02 \times 10^{23} \text{ molecules}$$

Step 3: Using the above equality, we can write a unit factor that will give us an answer in molecules of chlorine.

$$7.10 \text{ g } \cancel{Cl_2} \times \frac{6.02 \times 10^{23} \text{ molecules}}{71.0 \text{ g } \cancel{Cl_2}} = 6.02 \times 10^{22} \text{ molecules}$$

Exercise 9-5

1. Calculate the weight in grams of each of the following (using the formula weights you calculated in Exercise 9-4):

   (a) 0.500 mol I_2
   (b) 2.8 mol PbS
   (c) 0.035 mol $KClO_3$
   (d) 4.00 mol $Al_2(SO_4)_3$

2. How many moles are in each of the following?

   (a) 500 g HF
   (b) 17.4 g C_4H_{10}
   (c) 1.76 g $Ca(NO_3)_2$

3. How many molecules are there in 1.45 g C_4H_{10}?

9.7 Calculations Using Balanced Equations

A balanced chemical equation contains a great deal of information for the chemist. Not only does it indicate the identity of the reactants and products, but it also tells the relative number of these substances involved in the reaction. Consider the following reaction:

$$2H_2 + O_2 \longrightarrow 2H_2O$$

This equation can be read: "Two hydrogen molecules will react with one oxygen molecule to produce two water molecules," or in quantities with which we can work: "Two moles of hydrogen will react with one mole of oxygen to produce two moles of water." This, in turn, tells us that 4 grams of hydrogen will react with 32 grams of oxygen to produce 36 grams of water.

EXAMPLE 9-6

The following equation gives three sets of information. Write out each in sentence form.

$$2Al + 3H_2SO_4 \longrightarrow 3H_2 + Al_2(SO_4)_3$$

(1) Two atoms of aluminum will react with three molecules of hydrogen sulfate (sulfuric acid) to produce three molecules of hydrogen and one aluminum sulfate ion group.

(2) Two moles of aluminum will react with three moles of sulfuric acid to produce two moles of hydrogen and one mole of aluminum sulfate.

(3) 54 grams of aluminum will react with 294 grams of sulfuric acid to produce 6 grams of hydrogen and 342 grams of aluminum sulfate.

Exercise 9-6

Write out in sentence form the three sets of information given us by the following equation (CH_4 is methane):

$$CH_4 + 2O_2 \longrightarrow CO_2 + 2H_2O$$

Now we can put together both of the procedures that we have learned in this chapter—balancing chemical equations and doing calculations involving the mole—to analyze many different types of problems concerning chemical reactions As an example, let's determine how many moles of hydrogen will combine with 7.0 moles of oxygen to form water. We have just looked at the balanced equation for the formation of water, which we know gives us the relationship between moles of reactants and products in this reaction. The coefficients in this balanced equation tell us the ratio of moles that must hold among the reactants and products, which allows us to construct unit factors for solving the problem.

$$2H_2 + O_2 \longrightarrow 2H_2O$$

The unit factors giving the relationship between moles of hydrogen and moles of oxygen in this reaction are

$$\frac{2 \text{ mol } H_2}{1 \text{ mol } O_2} \quad \text{and} \quad \frac{1 \text{ mol } O_2}{2 \text{ mol } H_2}$$

We want to know how many moles of H_2 will react with 7.0 moles of O_2, so the first unit factor is the one that will allow our units to cancel properly.

$$7.0 \cancel{\text{mol } O_2} \times \frac{2 \text{ mol } H_2}{1 \cancel{\text{mol } O_2}} = 14 \text{ mol } H_2$$

Assume next that we want to know how many grams of water will be formed when 6.4 grams of oxygen are available to react. In this case we are being asked to convert from grams of oxygen to grams of water, but the unit factors that come from the balanced chemical reaction give us the relationship between moles of oxygen and moles of water. Therefore, we must use a series of unit factors to solve the problem, going through the following conversions:

$$\text{grams of } O_2 \xrightarrow{(1)} \text{moles of } O_2 \xrightarrow{(2)} \text{moles of } H_2O \xrightarrow{(3)} \text{grams of } H_2O$$

Solving the problem by means of these three steps, we have

(1) 6.4 g O_2 = (?) mol O_2

We know that

$$1 \text{ mol } O_2 = 32 \text{ g}$$

This allows us to construct a unit factor that will give us a result in moles.

$$6.4 \text{ g } O_2 \times \frac{1 \text{ mol } O_2}{32 \text{ g } O_2} = 0.20 \text{ mol } O_2$$

(2) Now we must determine how many moles of water will be formed from the reaction of 0.20 mole of oxygen. The conversion factors that we can derive from the ratio of coefficients in our balanced equation are

$$\frac{1 \text{ mol } O_2}{2 \text{ mol } H_2O} \quad \text{and} \quad \frac{2 \text{ mol } H_2O}{1 \text{ mol } O_2}$$

Therefore,

$$0.20 \text{ mol } O_2 \times \frac{2 \text{ mol } H_2O}{1 \text{ mol } O_2} = 0.40 \text{ mol } H_2O$$

(3) Now we need to convert our answer from moles of H_2O to grams H_2O. We know that

$$1 \text{ mol } H_2O = (2 \times 1) + 16 = 18 \text{ g}$$

Constructing the appropriate unit factor from this equality so that our units cancel properly, we can now write

$$0.40 \text{ mol } H_2O \times \frac{18 \text{ g}}{1 \text{ mol } H_2O} = 7.2 \text{ g}$$

To save time and extra calculations, this problem could have been solved in just one step by using a series of conversion factors. The trick is to be careful that you have arranged the conversion factors in such a way that all the units cancel out except the unit desired for the final answer. In this case, we would have done the following calculation:

$$6.4 \text{ g } O_2 \times \frac{1 \text{ mol } O_2}{32 \text{ g } O_2} \times \frac{2 \text{ mol } H_2O}{1 \text{ mol } O_2} \times \frac{18 \text{ g } H_2O}{1 \text{ mol } H_2O} = 7.2 \text{ g } H_2O$$

You will want to study carefully the following examples before trying to solve the problems in Exercise 9-7. Remember especially that you cannot convert directly from grams of one substance to grams of another substance in analyzing a chemical reaction; you must first convert to moles.

$$\text{grams } A \longrightarrow \text{moles } A \longrightarrow \text{moles } B \longrightarrow \text{grams } B$$

EXAMPLE 9-7

1. Imagine that you are working in a laboratory and want to obtain 2.40 moles of magnesium chloride, a compound that is produced (along with water) when magnesium oxide is heated with hydrochloric acid, HCl. How many grams of MgO would you need for this reaction?

(a) First you need to write a balanced chemical equation for this reaction.

Magnesium oxide + hydrochloric acid $\longrightarrow$ magnesium chloride + water

$$\text{Unbalanced: } MgO + HCl \longrightarrow MgCl_2 + H_2O$$
$$\text{Balanced: } MgO + 2HCl \longrightarrow MgCl_2 + H_2O$$

(b) Second, our problem asks how many grams of MgO are necessary to produce 2.40 moles of $MgCl_2$. We need to write the unit factors necessary to make the following conversions:

$$\text{Moles } MgCl_2 \xrightarrow{(1)} \text{Moles MgO} \xrightarrow{(2)} \text{Grams MgO}$$

(1) Looking at the ratio of coefficients in our balanced equation, we obtain the following conversion factors:

$$\frac{1 \text{ mol MgO}}{1 \text{ mol } MgCl_2} \quad \text{and} \quad \frac{1 \text{ mol } MgCl_2}{1 \text{ mol MgO}}$$

(2) To obtain the conversion factor for the second step, we must know the weight of one mole of MgO.

$$1 \text{ mol MgO} = 24.3 + 16.0 = 40.3 \text{ g}$$

Therefore, we have

$$\frac{1 \text{ mol MgO}}{\mathbf{40.3 \text{ g}}} \quad \text{and} \quad \frac{40.3 \text{ g}}{1 \text{ mol MgO}}$$

(c) Let's now solve the problem in a one-step calculation, making certain that we select our unit factors in such a way that all the units cancel except grams of MgO.

$$\mathbf{2.40} \text{ m}\cancel{\text{ol MgC}}\text{l}_2 \times \frac{1 \text{ m}\cancel{\text{ol Mg}}\text{O}}{1 \text{ m}\cancel{\text{ol MgC}}\text{l}_2} \times \frac{40.3 \text{ g MgO}}{1 \text{ m}\cancel{\text{ol Mg}}\text{O}} = 96.7 \text{ g MgO}$$

2. Antacid tablets are taken by millions of persons to reduce the discomfort of an upset stomach. The active ingredient in some commercial antacid tablets is magnesium hydroxide, $Mg(OH)_2$, which will react with stomach acid (HCl) to produce magnesium chloride ($MgCl_2$) and water. One popular tablet

contains 0.10 gram of $Mg(OH)_2$. How many grams of stomach acid will this tablet neutralize?

(a) First, write the balanced equation for this reaction.

$$\text{Unbalanced: } Mg(OH)_2 + HCl \longrightarrow MgCl_2 + H_2O$$
$$\text{Balanced: } Mg(OH)_2 + 2HCl \longrightarrow MgCl_2 + 2H_2O$$

(b) Second, write the conversion factors necessary for the following steps:

$$\text{g } Mg(OH)_2 \xrightarrow{(1)} \text{mol } Mg(OH)_2 \xrightarrow{(2)} \text{mol HCl} \xrightarrow{(3)} \text{g HCl}$$

(1) $1 \text{ mol } Mg(OH)_2 = 24 + 2 \times (16 + 1) = 58 \text{ g}$

$$\frac{1 \text{ mol } Mg(OH)_2}{58 \text{ g}} \quad \text{and} \quad \frac{58 \text{ g}}{1 \text{ mol } Mg(OH)_2}$$

(2) From the coefficients in our balanced equation,

$$\frac{1 \text{ mol } Mg(OH)_2}{2 \text{ mol HCl}} \quad \text{and} \quad \frac{2 \text{ mol HCl}}{1 \text{ mol } Mg(OH)_2}$$

(3) $1 \text{ mol HCl} = 1.0 + 35.5 = 36.5 \text{ g}$

$$\frac{1 \text{ mol HCl}}{36.5 \text{ g}} \quad \text{and} \quad \frac{36.5 \text{ g}}{1 \text{ mol HCl}}$$

(c) To now solve the problem in one step, we select the appropriate conversion factors from steps (1), (2), and (3) above.

$$0.10 \text{ g } \cancel{Mg(OH)_2} \times \frac{1 \cancel{\text{mol } Mg(OH)_2}}{58 \text{ g } \cancel{Mg(OH)_2}} \times \frac{2 \cancel{\text{mol HCl}}}{1 \cancel{\text{mol } Mg(OH)_2}} \times \frac{36.5 \text{ g HCl}}{1 \cancel{\text{mol HCl}}}$$
$$= 0.13 \text{ g HCl}$$

Exercise 9-7

1. How many grams of MgO are needed to produce 418 grams of $MgCl_2$ using the reaction described in Example 9-7 (1)?
2. In Example 9-7 (2), how many grams of hydrochloric acid would have been neutralized if 0.70 mole of H_2O were produced? If 0.19 gram of $MgCl_2$ were produced?

CHAPTER SUMMARY

A chemical equation is a shorthand way of indicating what occurs in a chemical reaction. The formulas of the reactants (or starting substances) are separated from the formulas of the products (or resulting substances) by an arrow. A balanced chemical equation is one in which the number of atoms of each element on the product side of the equation equals the number of atoms of that element on the reactant side.

A mole is a unit of measure that allows equal numbers of particles of different substances to be weighed out. A mole contains Avogadro's number, or 6×10^{23}, particles. The atomic weight in grams of an element will contain one mole of atoms of that element. The formula weight of a compound is equal to the sum of the atomic weights of each atom in the formula of that compound. One mole of a compound will have a mass equal to the formula weight of that compound expressed in grams.

Important Equations

$$\text{number of moles} = \frac{\text{grams of substance}}{\text{formula weight of substance}}$$

1. Calculate the formula weight for each of the following compounds:

Group 1	Group 2	Group 3
(a) Sodium hydroxide	KCl	$Ca(OH)_2$
(b) Calcium chloride	H_2SO_4	$(NH_4)_2SO_4$
(c) Sulfur dioxide	$Mg_3(PO_4)_2$	CH_4
(d) Sodium phosphate	CO_2	HCl
(e) Barium sulfate	$NaHCO_3$	$NaNO_3$
(f) Hydrogen bromide	$KMnO_4$	$AgNO_3$
(g) Boron trifluoride	HNO_3	H_2CO_3
(h) Water	$CuSO_4$	$CaCO_3$

2. What is the weight in grams of each of the following? (Use the formula weights calculated in question 1.)

Group 1	Group 2	Group 3
(a) 0.15 mole of sodium hydroxide	4.20 moles of KCl	0.95 mole of $Ca(OH)_2$
(b) 2.50 moles of calcium chloride	1.50 moles of H_2SO_4	4.60 moles of $(NH_4)_2SO_4$
(c) 0.80 mole of sulfur dioxide	0.015 mole of $Mg_3(PO_4)_2$	12.5 moles of CH_4

(d) 0.50 mole of sodium phosphate	5.50 moles of CO_2	0.025 mole of HCl
(e) 3.60 moles of barium sulfate	0.32 mole of $NaHCO_3$	0.62 mole of $NaNO_3$
(f) 0.020 mole of hydrogen bromide	0.0750 mole of $KMnO_4$	0.50 mole of $AgNO_3$
(g) 0.125 mole of boron trifluoride	0.10 mole of HNO_3	6.50 moles of H_2CO_3
(h) 0.0010 mole of water	0.060 mole of $CuSO_4$	0.0001 mole of $CaCO_3$

3. Calculate the number of moles in each of the following samples. (Use the formula weights that you calculated in question 1.)

Group 1	Group 2	Group 3
(a) 0.012 g of sodium hydroxide	89.4 g of KCl	407 g of $Ca(OH)_2$
(b) 33.3 g of calcium chloride	7.35 g of H_2SO_4	3.3 g of $(NH_4)_2SO_4$
(c) 2.48 g of sulfur dioxide	131 g of $Mg_3(PO_4)_2$	148.8 g of CH_4
(d) 0.82 g of sodium phosphate	198 g of CO_2	2.19 g of HCl
(e) 2.33 g of barium sulfate	67.2 g of $NaHCO_3$	64.6 g of $NaNO_3$
(f) 52.65 g of hydrogen bromide	498.8 g of $KMnO_4$	0.68 g of $AgNO_3$
(g) 122.4 g of boron trifluoride	94.5 g of HNO_3	80.6 g of H_2CO_3
(h) 64.8 g of water	3.19 g of $CuSO_4$	1.25 g of $CaCO_3$

4. Calculate the following:

(a) the number of moles in 3.6×10^{23} molecules of oxygen
(b) the weight in grams of 1.5×10^{23} molecules of chlorine

5. Balance the following equations:

(a) $Na + H_2O \longrightarrow NaOH + H_{2(g)}$
(b) $KClO_3 \longrightarrow KCl + O_{2(g)}$
(c) $MnO_2 + HCl \longrightarrow Cl_{2(g)} + MnCl_2 + H_2O$
(d) $C_3H_8 + O_2 \longrightarrow CO_2 + H_2O$
(e) $NH_3 + O_2 \longrightarrow NO + H_2O$
(f) $CO_3^{2-} + H^+ \longrightarrow CO_2 + H_2O$

6. Balance the following equations:

   (a) $HCl + Cr \longrightarrow CrCl_3 + H_{2(g)}$
   (b) $FeCl_3 + Na_2CO_3 \longrightarrow Fe_2(CO_3)_3 + NaCl$
   (c) $PbS + H_2O_2 \longrightarrow PbSO_4 + H_2O$
   (d) $C_4H_{10} + O_2 \longrightarrow CO_2 + H_2O$
   (e) $CrO_4^{2-} + H^+ \longrightarrow Cr_2O_7^{2-} + H_2O$

7. Balance the following equations:

   (a) $AgNO_3 + CaCl_2 \longrightarrow Ca(NO_3)_2 + AgCl$
   (b) $Fe_2O_3 + H_2 \longrightarrow Fe + H_2O$
   (c) $Al(NO_3)_3 + H_2SO_4 \longrightarrow Al_2(SO_4)_3 + HNO_3$
   (d) $C_3H_8O + O_2 \longrightarrow CO_2 + H_2O$
   (e) $Al + H_2SO_4 \longrightarrow Al_2(SO_4)_3 + H_2$
   (f) $Fe_2O_3 + C \longrightarrow Fe + CO_2$

8. Write a balanced equation for each of the following reactions:

   (a) Hydrogen reacts with bromine to form hydrogen bromide.
   (b) Calcium bicarbonate, when heated, breaks apart to form calcium carbonate, water, and carbon dioxide.
   (c) Silver nitrate will react with copper to form copper(II) nitrate and silver.
   (d) Hydrogen reacts with nitrogen to form ammonia, NH_3.
   (e) Methane (CH_4) and chlorine gases will react to form carbon tetrachloride and hydrogen chloride.

9. Use this balanced equation

$$Mg_3N_2 + 6H_2O \longrightarrow 3Mg(OH)_2 + 2NH_3$$

   to answer the following questions:

   (a) How many moles of $Mg(OH)_2$ would be produced from the reaction of 0.10 mole of Mg_3N_2?
   (b) How many moles of $Mg(OH)_2$ and of NH_3 would be produced from the reaction of 500 g of Mg_3N_2?
   (c) How many grams of Mg_3N_2 and H_2O must react to produce 0.060 mole of $Mg(OH)_2$?
   (d) How many grams of Mg_3N_2 are needed to produce 52.2 g of $Mg(OH)_2$?
   (e) What is the maximum number of grams of $Mg(OH)_2$ that can be produced by the reaction of 10.0 g of Mg_3N_2 and 14.4 g of H_2O?

10. Wine is produced by the process of fermentation, in which the sugar in grapes is converted by the action of yeast into ethyl alcohol and carbon dioxide.

$$\underset{\text{Sugar}}{C_6H_{12}O_6} \xrightarrow{\text{yeast}} \underset{\text{Ethyl alcohol}}{2C_2H_5OH} + 2CO_{2(g)}$$

   How many kilograms of ethyl alcohol would be produced if all of the 5.00 kilograms of sugar in a batch of grapes were fermented?

11 Aspirin is produced by the reaction of salicylic acid with acetic anhydride.

$$\underset{\text{Salicylic acid}}{C_7H_6O_3} + \underset{\text{Acetic anhydride}}{C_4H_6O_3} \longrightarrow \underset{\text{Aspirin}}{C_9H_8O_4} + C_2H_4O_2$$

How many grams of salicylic acid are required to produce an aspirin tablet that contains 0.33 gram of aspirin?

12. Oxygen gas is produced by the decomposition of potassium chlorate, $KClO_3$.

$$2KClO_3 \longrightarrow 2KCl + 3O_2$$

If 1.00 gram of $KClO_3$ is completely decomposed by this reaction, how many grams of oxygen are formed? How many grams of KCl?

13. How many grams of water are formed when 100 grams of natural gas (CH_4) is burned completely? The other product of this reaction is carbon dioxide.

14. Ammonia is used in the production of fertilizers. The fertilizer urea, $(NH_2)_2CO$, is formed by reacting ammonia with carbon dioxide.
    (a) Write the balanced equation for this reaction. (Water is the other product of the reaction.)
    (b) How many tons of urea are produced by the reaction of 6.00 metric tons (1 metric ton = 10^3 kg, or 2200 lb) of ammonia?

15. The overall reaction for the refining of aluminum metal from aluminum ore involves the reaction of aluminum oxide with carbon, producing aluminum and carbon dioxide.
    (a) Balance the equation for the refining of aluminum.
    (b) If 5.4 metric tons of aluminum were produced in the United States each year, how many metric tons of aluminum oxide were used?
    (c) How many metric tons of carbon are consumed in the production of 1.0 metric ton of aluminum?

16. Sodium lauryl sulfate, a detergent, can be prepared by the following reactions:

1. $$\underset{\text{Lauryl alcohol}}{C_{12}H_{25}OH} + H_2SO_4 \longrightarrow \underset{\text{Lauryl sulfonic acid}}{C_{12}H_{25}OSO_3H} + H_2O$$

2. $$\underset{\text{Lauryl sulfonic acid}}{C_{12}H_{25}OSO_3H} + NaOH \longrightarrow \underset{\text{Detergent}}{C_{12}H_{25}OSO_3Na} + H_2O$$

If a day's production of detergent is 11 metric tons, how many metric tons of lauryl alcohol will be required?

17.* Balance the following redox equations using oxidation numbers:
    (a) $Fe_2O_3 + C \longrightarrow Fe + CO_2$
    (b) $KCl + MnO_2 + H_2SO_4 \longrightarrow K_2SO_4 + MnSO_4 + Cl_2 + H_2O$
    (c) $I_2 + HNO_3 \longrightarrow HIO_3 + NO_2 + H_2O$
    (d) $SO_2 + HNO_3 + H_2O \longrightarrow H_2SO_4 + NO$
    (e) $K_2Cr_2O_7 + HCl \longrightarrow KCl + CrCl_3 + Cl_2 + H_2O$

---

* *This problem is optional.*

CHAPTER TEN

REACTION RATES AND CHEMICAL EQUILIBRIUM

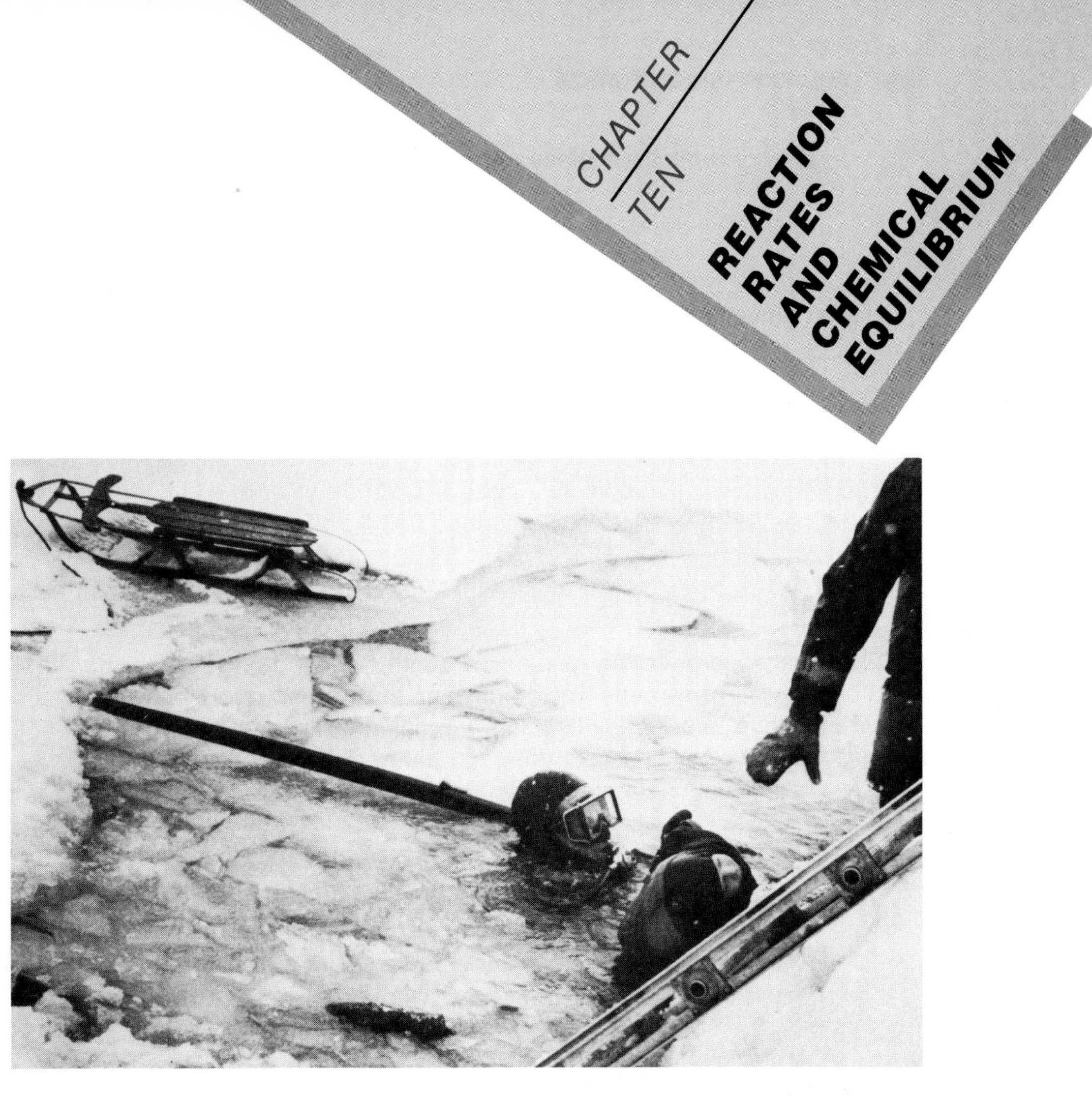

The day was cold and clear. Four-year-old Jimmy Tontlewicz had been spending an enjoyable winter afternoon with his father sledding on his favorite hill beside Lake Michigan. His father chuckled as he noticed Jimmy fall off his sled, but his chuckle turned to cries of horror when he saw Jimmy running after the sled as it continued out onto the frozen lake. Jimmy heard his father's screams just as he reached the sled, but by that time the ice was already breaking beneath him. It was only a second before he disappeared into the icy waters of the lake.

His father jumped into the water but was unable to find Jimmy under the ice. It wasn't until twenty minutes later that fire department scuba divers found Jimmy and brought his lifeless body to the surface. He was clinically dead: his skin was ashen, his pupils fixed and dilated, and there was no sign of a pulse or breathing.

How could it be possible, then, that Jimmy is alive today? When he fell into the water, Jimmy instinctively held his breath. But no one can hold his breath for much more than a minute under such conditions. Eventually Jimmy was forced to breathe in the cold lake water. Upon entering the lungs, this water was absorbed into the circulatory system and immediately made its way through the heart and into the brain. The icy water quickly cooled—to 85°F (or 29.4°C) in Jimmy's case—all the tissues it contacted.

This drastically lowered the tissues' metabolic rate and their need for oxygen. Such a rapid lowering of the rate of reaction in tissue cells can delay tissue death for 45 minutes or more after breathing stops.

The special treatment that Jimmy received after being pulled from the water was critical to his recovery. Paramedics immediately began heart massage and other reviving techniques, but did not try to raise his body temperature. In the emergency room, Jimmy's heart was started with electric shocks, and he was put on a respirator. Heat lamps and warm intravenous fluids were used to very slowly warm his body to 91°F (32.8°C) to prevent further tissue damage. Meanwhile, doctors gave Jimmy massive doses of barbiturates to maintain his coma. Such drugs reduce the brain's need for oxygen and glucose and lessen the chance of lethal swelling. Only after three days did the doctors stop giving Jimmy barbiturates. His brain function slowly began to return over the next several weeks.

Under normal conditions the brain undergoes irreparable damage if it is cut off from its oxygen supply for as little as three minutes. Jimmy survived because his brain tissue was cooled by the icy water he had inhaled, slowing down all the chemical reactions in the brain. This allowed the brain tissue to survive without its normal supply of oxygen.

The rate of reactions can mean the difference between life and death to a young child such as Jimmy. On a much grander scale, it can also mean the difference between a nuclear disaster and the peaceful use of nuclear power to produce electricity. To understand how scientists control reaction rates for our benefit, it is important to study the various factors that influence the rate at which chemical reactions occur.

LEARNING OBJECTIVES

By the time you have finished this chapter, you should be able to:

1. Define *activation energy* and *activated complex.*
2. Describe an endothermic and exothermic reaction and draw a potential energy diagram for each.
3. List four factors that will affect the rate of a chemical reaction.
4. Define *catalyst* and explain how a catalyst affects the rate of a chemical reaction.
5. Define *chemical equilibrium* and give two examples.
6. State two ways in which a chemical equilibrium can be disrupted.
7. State Le Chatelier's Principle and predict the changes that will occur in an equilibrium when a given stress is applied.

RATES OF CHEMICAL REACTIONS

10.1 Activation Energy

For a reaction to occur between two particles, they must be brought close enough together for their outer-shell electrons to interact. In fact, they must collide. Not only must they collide, however, but they must do so with enough energy to overcome the repelling forces between the electrons surrounding the two nuclei. The amount of energy necessary for a successful collision is called the **activation energy, E_{act}.** In order for a reaction to occur, molecules must collide with energy at least equal to the activation energy. When such a collision occurs, the molecules form a reactive group called the **activated complex.** This complex of atoms is neither the reactants nor the products. Rather, it is a highly unstable combination of atoms that represents a state of change between the reactants and the products.

It may help to imagine the activation energy as a hill the particles must get over to complete the reaction. Each reaction is associated with a hill of a different height, or different activation energy. Imagine a person trying to roll a bowling ball up such a hill; in most cases the bowling ball will roll only part way up the hill and then roll down again. This is true of a reaction—most collisions between molecules do not occur with enough energy to overcome the activation barrier and to form the activated complex; the molecules simply bounce off each other unreacted. Just as the bowler can only occasionally give the ball enough energy to get up over the hill, so it is only on occasion that the particles in a reaction collide with sufficient energy to overcome the activation energy barrier and eventually form products (Figure 10.1).

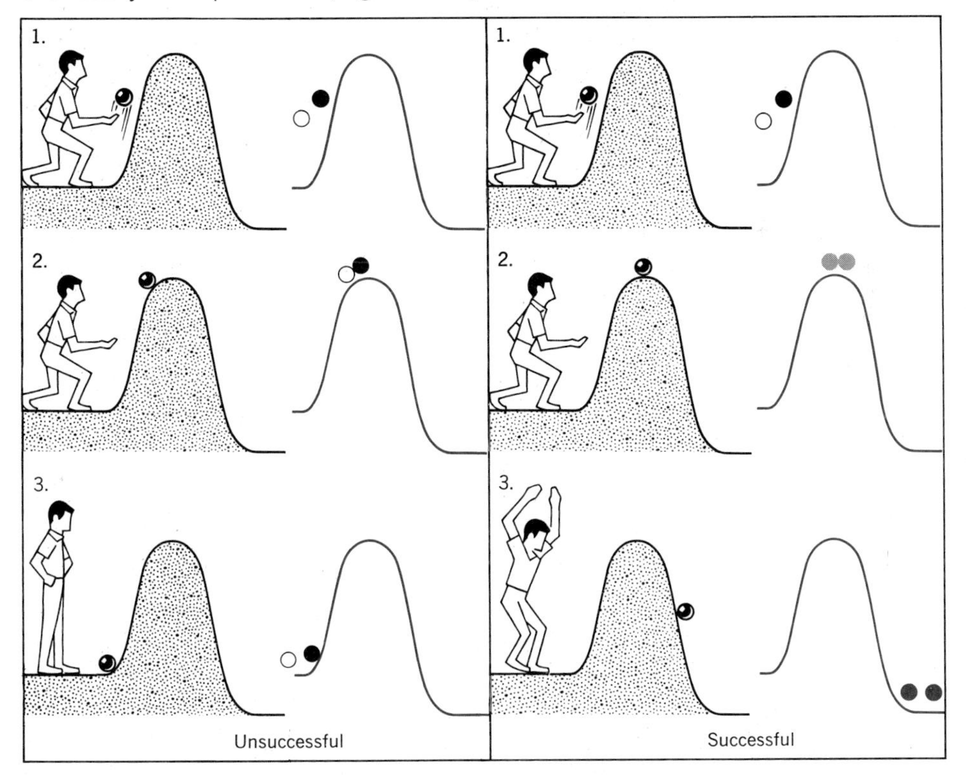

Figure 10.1 Just as a bowling ball will roll over the hill only when it has been given enough energy, so a reaction will occur between two molecules only when they collide with sufficient energy to overcome the activation energy barrier.

10.2 Exothermic and Endothermic Reactions

You know that a candle will not spontaneously burst into flame at room temperature, but you may wonder what allows it to keep burning once it is lit. The reaction between candle wax and oxygen has a high activation energy, and very few molecules have enough energy at room temperature to react. The heat of a match is required if a large number of molecules are to overcome the activation energy barrier and cause the wax to burn. Now, the reaction between candle wax and oxygen is **exothermic,** meaning that energy is released as the reaction occurs. Exothermic reactions result when the products of the reaction have less potential energy than the reactants. Therefore, once these molecules begin to react, this reaction releases enough energy to boost additional molecules over the activation energy barrier. Hence, the reaction becomes self-sustaining, and the candle continues to burn (Figure 10.2).

Some highly exothermic reactions can be explosive. Dynamite, for example, must be set off with a percussion cap. The very small explosion of the percussion cap releases enough energy for a small number of dynamite molecules to react, and the energy released by these molecules is enough to cause the remainder of the molecules to all react at once, releasing a tremendous amount of energy.

Other reactions, such as the decomposition of water into hydrogen and oxygen, or the formation of sugar molecules in the process of photosynthesis, require the addition of energy for the reaction to continue. Reactions requiring a continuous input of energy are called **endothermic.** In such reactions the potential energy of the products is greater than that of the reactants. In effect, energy has been absorbed by the reaction. In endothermic reactions, reactants require an initial input of energy to get over the activation energy barrier, and then a continuing supply of energy to keep the reaction going. Stop the supply of electricity, and the decomposition of water will come to a halt; keep a green plant in the dark, and photosynthesis will stop, causing the plant to die (Figures 10.3 and 10.4).

Figure 10.5 contains potential energy diagrams for an exothermic and an endothermic reaction. The energy required or released by a reaction is

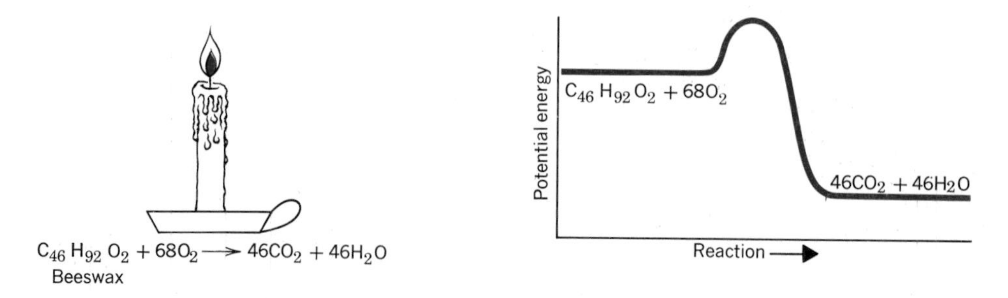

Figure 10.2 The burning of candle wax is an exothermic reaction in which the products have less potential energy than the reactants.

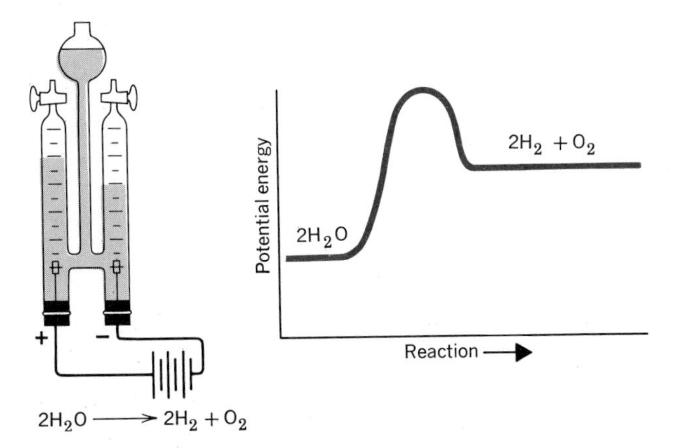

Figure 10.3 Electrolysis (the breaking apart of water molecules with electrical energy) is an endothermic reaction in which the products (hydrogen and oxygen) have more potential energy than the reactant (water).

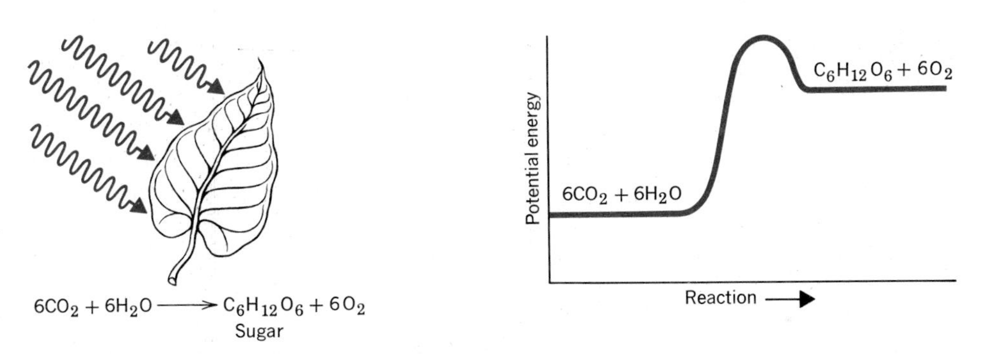

Figure 10.4 Photosynthesis is an endothermic process requiring a continuous supply of energy from the sun.

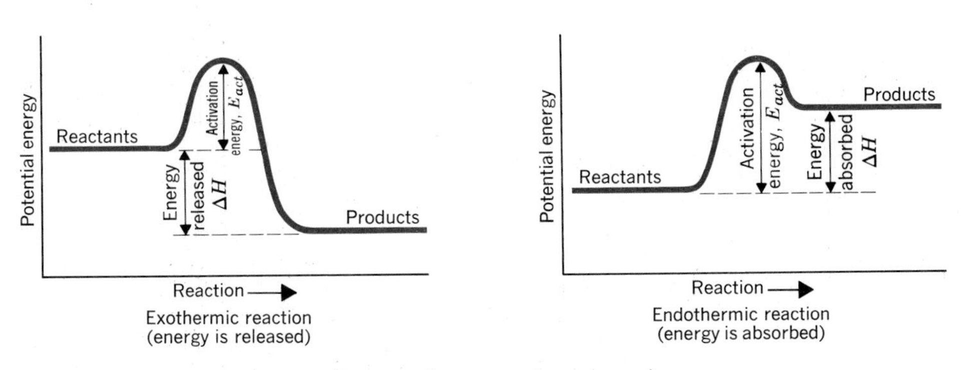

Figure 10.5 Potential energy diagrams for an exothermic and endothermic reaction. E_{act} = activation energy. ΔH = heat of reaction (the energy released or absorbed).

called the **heat of reaction, ΔH.** This change in heat content (or enthalpy) is expressed in kilocalories per mole of reactant. The value of ΔH is the difference between the potential energy of the products and the potential energy of the reactants. The heat of reaction will have a positive value for an endothermic reaction, and a negative value for an exothermic reaction. Each reaction has its own characteristic activation energy and heat of reaction.

EXAMPLE 10-1

1. Are the following reactions endothermic or exothermic?

   (a) $2Na_2O_2 + 2H_2O \longrightarrow 4NaOH + O_2 + 30.2$ kcal
   (b) $Si + 2H_2 + 8.2$ kcal $\longrightarrow SiH_4$

   Reaction (a) is exothermic because energy is given off as a product of the reaction. Reaction (b) is endothermic because energy is required by the reaction; that is, energy is one of the reactants.

2. What is the value of the heat of reaction, ΔH, for the reactions in the previous problem?

   (a) $\Delta H = -30.2$ kcal. The heat of reaction for an exothermic reaction has a negative value.
   (b) $\Delta H = +8.2$ kcal. The heat of reaction for an endothermic reaction has a positive value.

3. Draw the potential energy diagram for the following reaction, whose activation energy $E_{act} = 39$ kcal:

$$2ICl + H_2 \longrightarrow I_2 + 2HCl + 53 \text{ kcal}$$

   To draw the potential energy diagram, we must first determine if the reaction is endothermic or exothermic. Because energy is given off by this reaction, it is exothermic. This means that the potential energy of the reactants is greater than the potential energy of the products—in this case, 53 kcal greater. There are several ways to draw the diagram, but to simplify things we can arbitrarily assign the substances with the least potential energy (in this reaction, the products) a relative potential energy of 100 kcal. The reactants, therefore, will have a potential energy of 153 kcal in relation to the products. The activation energy is 39 kcal, so the top of the curve will be at 153 kcal (the potential energy of the reactants) + 39 kcal (the activation energy) = 192 kcal.

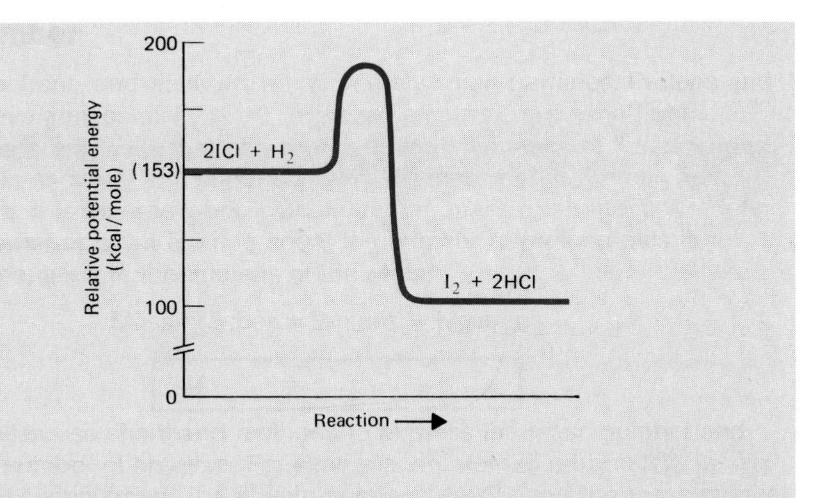

Exercise 10-1

1. The heat of reaction for the following reaction is $\Delta H = +31.4$ kcal.

$$C_{(s)} + H_2O_{(g)} \longrightarrow CO_{(g)} + H_{2(g)}$$

   (a) Is this an exothermic or endothermic reaction?
   (b) Will energy be absorbed or released by this reaction?
   (c) Rewrite the equation so that the heat of reaction appears in the equation.

2. The following reaction has a heat of reaction $\Delta H = +3$ kcal, and an activation energy $E_{act} = 43.8$ kcal. Draw the potential energy diagram for this reaction, labeling the reactants, products, activated complex, activation energy, and heat of reaction.

$$2HI \longrightarrow H_2 + I_2$$

10.3 Factors Affecting Reaction Rates

The Nature of the Reactants

The nature of the reactants influences the rate of a chemical reaction. For example, when colorless nitric oxide escapes from a test tube into the air, reddish-brown nitrogen dioxide forms very quickly. But when carbon monoxide from automobile exhaust fumes is released into the air, the reaction with oxygen to form carbon dioxide is quite slow (unfortunately for the residents of large urban areas).

$$2NO + O_2 \longrightarrow 2NO_2 \qquad \text{Fast reaction at 25°C}$$
$$2CO + O_2 \longrightarrow 2CO_2 \qquad \text{Very slow reaction at 25°C}$$

These two balanced equations look quite similar, but the large difference in their rates of reaction comes from the nature of the carbon monoxide and nitric oxide molecules themselves.

The Concentration of Reactants

The concentration of reactants can greatly influence the rate of a chemical reaction. We have seen that particles must collide in order for them to react, and it seems reasonable that the more reactant particles there are in a given space, the more collisions will occur. (We see the same thing on our highways; the more crowded the highways, the higher the probability of a fatal collision. This is especially apparent when you consider the high death rates on holiday weekends.) Increasing the concentration of the reactants will increase the number of collisions and, therefore, will increase the reaction rate (Figure 10.6).

Physicians make use of this principle when prescribing medicines to treat specific diseases. The appropriate treatment for tonsillitis or pneumonia may involve a dose of 250 mg of penicillin, but meningitis would require two to five times that dosage. This higher concentration of penicillin increases the rate of absorption into the bloodstream and, therefore, increases the effective concentration in the blood.

The Surface Area of a Solid

Increasing the surface area of a solid reactant will increase the rate of reaction. An increase in the surface area increases the number of solid particles that are exposed, thereby increasing the number of collisions possible between the reactants. Increasing the surface area can sometimes increase the rate of reaction to explosive levels. For example, lumber mills seldom worry about their log piles spontaneously catching fire, but sawdust piles can burst into flame and, therefore, must be kept wet. You probably don't think of flour as being potentially dangerous, but finely divided flour dust can explode (Figure 10.7). As another example a crushed aspirin tablet, because of its increased surface area, will relieve your headache faster than a whole tablet.

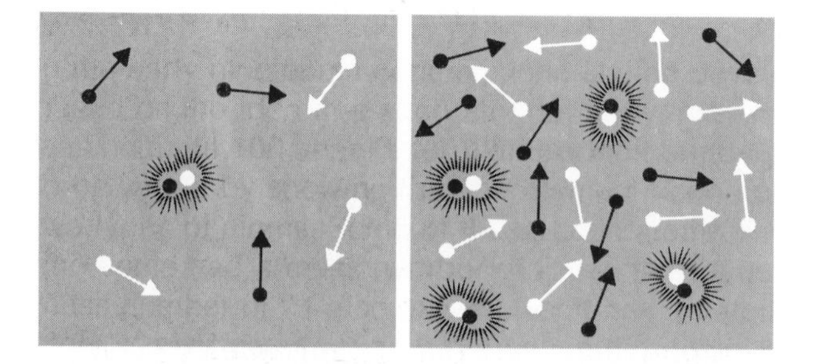

Figure 10.6 Increasing the concentration of reactants will increase the frequency of collisions, thereby increasing the rate of the reaction.

Figure 10.7 These grain elevators were destroyed by the force of an explosion of grain dust. (Wide World Photos.)

The Temperature of the Reaction

Increasing the temperature of the reactants will increase the kinetic energy of the particles. This will not only increase the frequency of collisions, but will also increase the likelihood that the colliding particles will have enough energy to overcome the activation energy barrier. (This is similar to the idea that increasing speed limits on highways will not only increase the number of collisions, but will also increase the likelihood that any collision will be a fatal one.) (See Figure 10.8.)

Changes in temperature have important effects on living organisms.

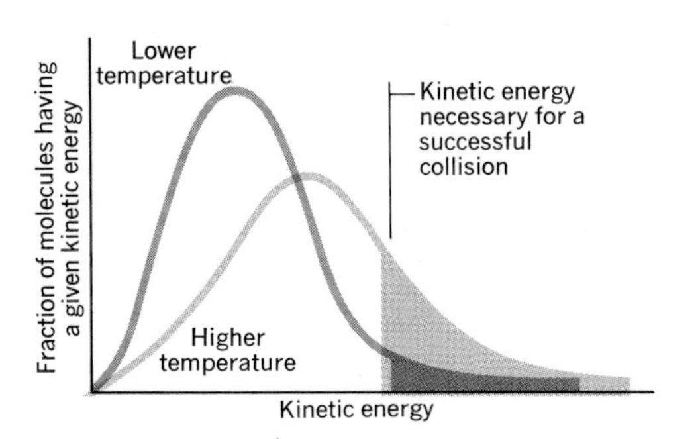

Figure 10.8 Increasing the temperature of a reaction increases the average kinetic energy of the molecules, thus increasing the number of molecules having enough energy to collide successfully and produce products.

A fever increases the rate of chemical reactions in the body, as can be seen by the increased pulse rate, increased breathing rate, and abnormalities in digestive and nervous systems. When you run a fever, your basal metabolism rate goes up by 5% for each degree rise in body temperature. The effects on living organisms from an increase in external temperatures is illustrated by what happens when warm water from power plants is released into streams and lakes, causing an increase in the metabolic rate in the aquatic life. In order to support their new higher rate of metabolism, these organisms require more oxygen. But the warming of the water also decreases the concentration of oxygen in the water, contributing to the death of fish in these streams and lakes. Another detrimental effect of these higher water temperatures is a resulting increase in the sensitivity of fish to pollutants in the water (Figure 10.9).

Decreasing the body's temperature slows down bodily reaction rates. During open heart surgery, a patient's body temperature is usually lowered four to five degrees Fahrenheit to decrease the metabolism rate and oxygen requirements. At the beginning of this chapter we saw an extreme example of the decrease in bodily reactions from a lowering of body temperatures. As another example, an athletic trainer might spray a surface anesthetic called ethyl chloride on the injured knee of a basketball player to relieve the pain. The rapid evaporation of the ethyl chloride will lower the temperature of the tissue enough to greatly retard the reactions responsible for the transmission of nerve impulses.

Many animals hibernate during the cold winter months. Their body

Figure 10.9 This fish kill resulted when warm waste water from a nuclear power plant was discharged into a nearby stream. (U.P.I.)

temperatures fall to a few degrees above freezing, and all of the chemical reactions in their bodies slow down. Their breathing and heart rates become very slow, and they require much less energy to sustain life. Therefore, the animals can live on stored body fat alone. To illustrate the extent of this slowdown, consider that a woodchuck's heart beats about 80 times a minute while he is active, but when hibernating his heart will beat only four times a minute (Figure 10.10). The interesting idea of keeping human beings in suspended animation by lowering their body temperatures has been the subject of many science fiction stories. In 1974 two scientists from the Darwin Research Institute discovered that bacteria that had been frozen in the cold rock of Antarctica 10,000 to 1 million years ago could be revived in the warmth of the laboratory and, incredibly, could even reproduce.

A Catalyst

Many reactions that proceed slowly can be made to occur at a more rapid rate by the introduction of substances called catalysts. A **catalyst** is a substance that increases the rate of a chemical reaction without being consumed in the reaction. The effect of the catalyst is to lower the activation energy required for the reaction. In our illustration of the bowler and the bowling ball, the action of a catalyst would resemble that of a bulldozer cutting a much lower path over the hill, allowing many more balls

Figure 10.10 The body temperature of this hibernating woodchuck is only a few degrees above freezing, slowing down all chemical reactions in its body. (Wilford L. Miller/Photo Researchers.)

rolled by the bowler to reach the top and roll down the other side (Figure 10.11).

Industry makes wide use of catalysts, especially those catalysts which allow companies to produce large amounts of a product at a lower temperature, thereby saving on energy costs. An example of this occurs in the production of sulfuric acid, H_2SO_4, the so-called "king of chemicals." Over 20 million tons of sulfuric acid are used each year in the United States, mostly by the steel, fertilizer, and petroleum industries. One step in the production of sulfuric acid involves the production of sulfur trioxide (SO_3) from sulfur dioxide (SO_2) and oxygen (O_2). This reaction has a very high activation energy and is quite slow even at high temperatures. But the reaction has been made economically worthwhile by the introduction of a catalyst, such as finely divided platinum, which greatly increases the reaction rate (Figure 10.12). Similarly, the cost of producing gasoline from much larger molecules of crude petroleum is kept low through the use of catalysts in the cracking process—the process by which large molecules are broken into small fractions.

Most of the chemical reactions that occur in our bodies would not ordinarily occur at body temperature at a high enough rate to sustain life. The body, however, produces special compounds called **enzymes,** which act as biological catalysts, permitting these reactions to occur readily at body temperature. For example, carbon dioxide, which is produced as a waste product by our tissues, is converted to carbonic acid in red blood cells before being transported to the lungs to be exhaled.

$$CO_2 + H_2O \longrightarrow H_2CO_3$$

This reaction is very slow at body temperature, but red blood cells contain an enzyme called carbonic anhydrase that increases the rate of the uncatalyzed reaction—making it ten million times faster!

Damaging or removing body enzymes can have disastrous effects. For example, the fish poison rotenone, which is used to clear lakes and ponds of undesirable species, kills the fish by preventing enzymes from catalyzing

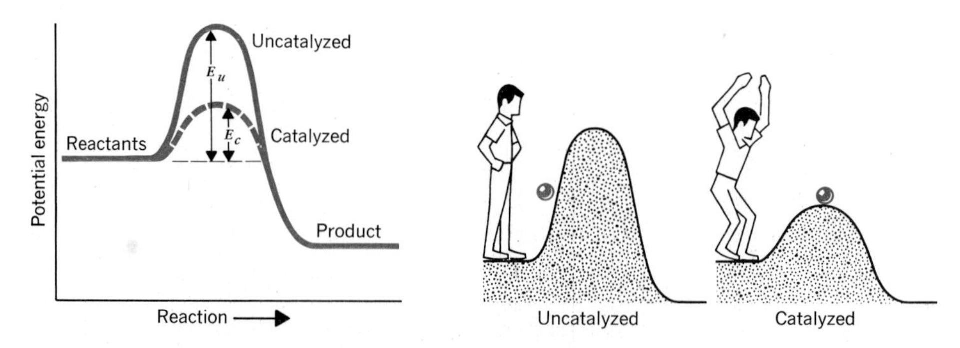

Figure 10.11 A catalyst lowers the activation energy, increasing the chances for a successful collision between reactant particles.

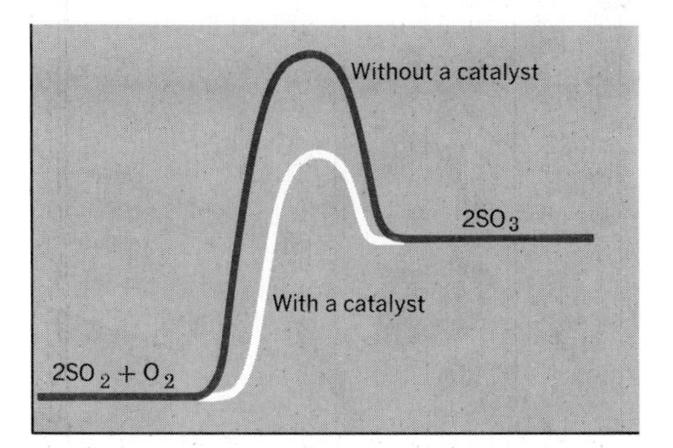

Figure 10.12 The reaction between sulfur dioxide and oxygen to produce sulfur trioxide has a very high activation energy. The addition of platinum as a catalyst lowers the activation energy enough to make the process economically feasible.

reactions essential to the production of cellular energy. The digestive tracts of 90% of non-Caucasian adults lack the enzyme lactase, which catalyzes the breakdown of lactose (the sugar found in milk). Because these individuals cannot digest the lactose, drinking milk results in nausea, gas, and diarrhea.

CHEMICAL EQUILIBRIUM

10.4 What Is Chemical Equilibrium?

At one time or another you may have added sugar to a glass of iced tea, stirred, and watched the sugar dissolve. You may have noticed that additional sugar added to this solution continued to dissolve until you reached a certain point, after which any more sugar added just settled to the bottom of the glass. Although you may have continued stirring, or let the glass sit for a period of time, neither the amount of sugar on the bottom of the glass nor the amount of sugar dissolved in the iced tea changed.

Under these circumstances the iced tea contained all the sugar molecules it could possibly hold, and we will see that the molecules of sugar dissolved in the iced tea and the molecules of sugar in the crystals on the bottom of the glass were in equilibrium. A molecular equilibrium such as this is not a static, motionless condition such as the equilibrium reached when two children are equally balanced on a seesaw. Rather, a chemical equilibrium is a dynamic state in which events are constantly occurring, but at exactly equal rates. In that glass of iced tea, for example, new sugar molecules were constantly dissolving and other sugar molecules were constantly crystallizing out of solution, but the rate of each of these processes was the same. Therefore, there was no overall change in the number of sugar molecules dissolved in the iced tea or lying on the bottom of the glass. In other words, for every sugar molecule that dissolved, one sugar molecule crystallized out of solution (Figure 10.13).

Let's examine this iced tea example in chemical terms. We said that when we first add sugar to the iced tea, it dissolves. We can represent this process by the following equation.

$$\text{Sugar}_{(s)} \longrightarrow \text{Sugar}_{(aq)}$$

As more and more sugar molecules are added to the iced tea, the number dissolved in the tea will increase to the point that the reverse reaction (that is, the crystallization of sugar from the tea) begins to occur.

$$\text{Sugar}_{(s)} \rightleftharpoons \text{Sugar}_{(aq)}$$

Notice that in this equation we use two arrows pointing in different directions to indicate that this reaction is reversible. As the number of sugar molecules in the tea continues to increase, the rate of the reverse reaction will increase until the rate of the reverse reaction equals the rate of the forward reaction. If we denote the rate of the forward reaction by rate_f and the rate of the reverse reaction by rate_r, we then have $\text{rate}_f = \text{rate}_r$.

$$\text{Sugar}_{(s)} \underset{\text{Rate}_r}{\overset{\text{Rate}_f}{\rightleftharpoons}} \text{Sugar}_{(aq)}$$

At this point, a state of equilibrium has been reached and no net change will occur in our glass of sugared iced tea. But it is again important to emphasize that the forward reaction and reverse reaction still continue to occur, only at exactly equal rates. We can define **chemical equilibrium,** then, as a dynamic state in which the rate of the forward reaction equals the rate of the reverse reaction.

10.5 Altering the Equilibrium

Altering a chemical equilibrium is an important technique in industrial or

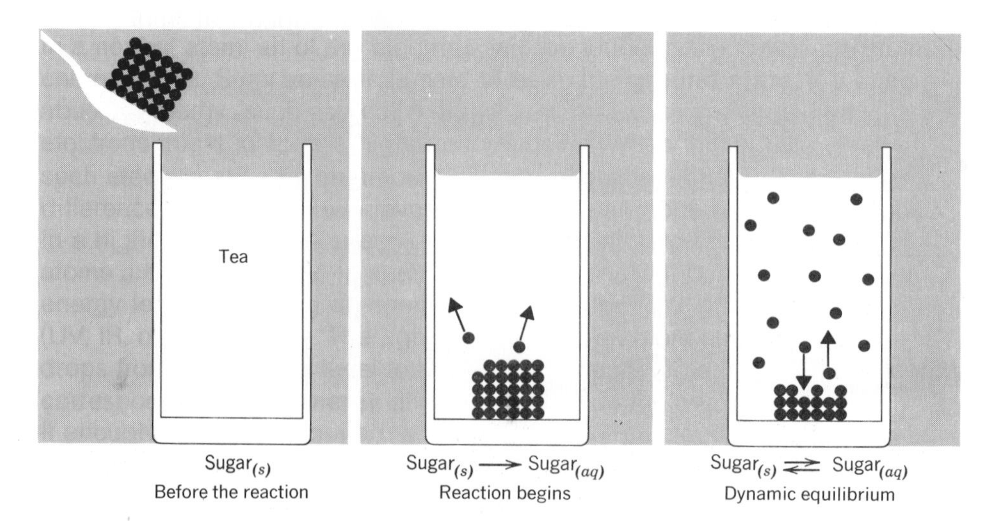

Figure 10.13 A dynamic equilibrium is established when the rate of the forward reaction equals the rate of the reverse reaction.

biological chemistry, especially when it is desirable for a reaction to go to completion even though the equilibrium state may be reached very early in the reaction. Various factors affecting the rate of a chemical reaction can alter a chemical equilibrium. Let's review each of the factors that we learned would cause a change in the rate of chemical reactions.

Changes in Concentration

Changing the concentration of a reactant or product in a reaction that is at equilibrium can affect the rate of the forward or reverse reaction. Increasing the concentration of a reactant increases the number of collisions between reactant molecules, thereby increasing the rate of the forward reaction. When this occurs, the system is no longer at equilibrium.

$$\text{Rate}_f > \text{Rate}_r$$

The forward reaction will proceed at a higher rate, producing more product molecules until the rate of the reverse reaction again increases enough to equal the rate of the forward reaction. At this point a new equilibrium is established.

One colorful example of the effect on a chemical equilibrium of changing reactant concentrations can be seen in the equilibrium between the yellow chromate ion, CrO_4^{2-}, and the orange dichromate ion, $Cr_2O_7^{2-}$.

$$\underset{\textit{Yellow}}{2CrO_4^{2-}(aq)} + 2H^+(aq) \underset{\text{Rate}_r}{\overset{\text{Rate}_f}{\rightleftharpoons}} \underset{\textit{Orange}}{Cr_2O_7^{2-}(aq)} + H_2O$$

Increasing the concentration of hydrogen ions on the reactant side of the equation will increase the number of collisions between the chromate ions and the hydrogen ions, thus causing an increase in the rate of the forward reaction. As more dichromate forms, the rate of the reverse reaction also increases until it again equals the rate of the forward reaction. At that point a new equilibrium will be established. This new equilibrium will have a higher concentration of dichromate ions and a lower concentration of chromate ions than the previous equilibrium. Therefore, the solution will appear more orange (Figure 10.14). It is a common procedure for chemists

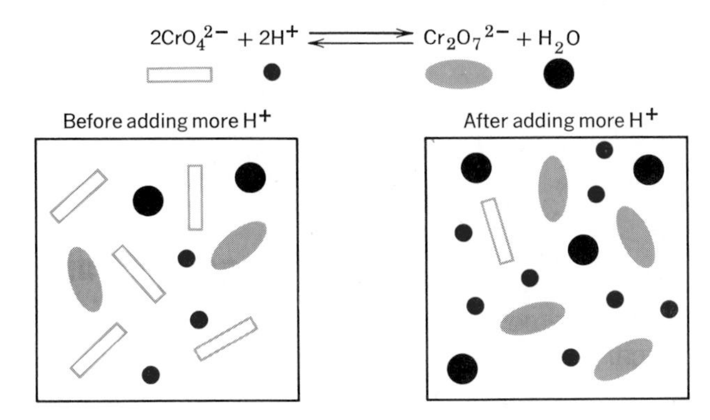

Figure 10.14 A molecular view of the chromate–dichromate equilibrium.

to add more reactant, or to remove a product in a reaction, in order to upset a chemical equilibrium and to drive a reaction toward completion.

In the blood, an equilibrium exists between carbon dioxide and carbonic acid.

$$\underset{\text{Carbon dioxide}}{CO_2} + H_2O \underset{\text{Rate}_r}{\overset{\text{Rate}_f}{\rightleftharpoons}} \underset{\text{Carbonic acid}}{H_2CO_3}$$

In the tissues, carbon dioxide is a waste product that enters the blood. As the concentration of carbon dioxide in the blood increases, the above reaction is driven to the right ($\text{rate}_f > \text{rate}_r$), and carbonic acid, which can be transported by the bloodstream, is formed. When the blood reaches the lungs, we exhale carbon dioxide. This lowers the concentration of carbon dioxide in the blood, driving the reaction to the left and allowing us to rid the bloodstream of the waste product carbonic acid (Figure 10.15).

Changes in Temperature

Changing the temperature of a system in equilibrium will alter the equilibrium. Increasing the temperature will favor the endothermic reaction (the reaction requiring energy), and lowering the temperature will favor the exothermic reaction (the reaction in which energy is released). For example, let's return to our iced tea example. The process of dissolving is an endothermic reaction. Therefore, increasing the temperature will increase the number of sugar molecules that will dissolve.

$$\text{Sugar}_{(s)} + \text{Energy} \underset{\text{Rate}_r}{\overset{\text{Rate}_f}{\rightleftharpoons}} \text{Sugar}_{(aq)}$$

Warming our tea will increase rate_f, because that is the endothermic

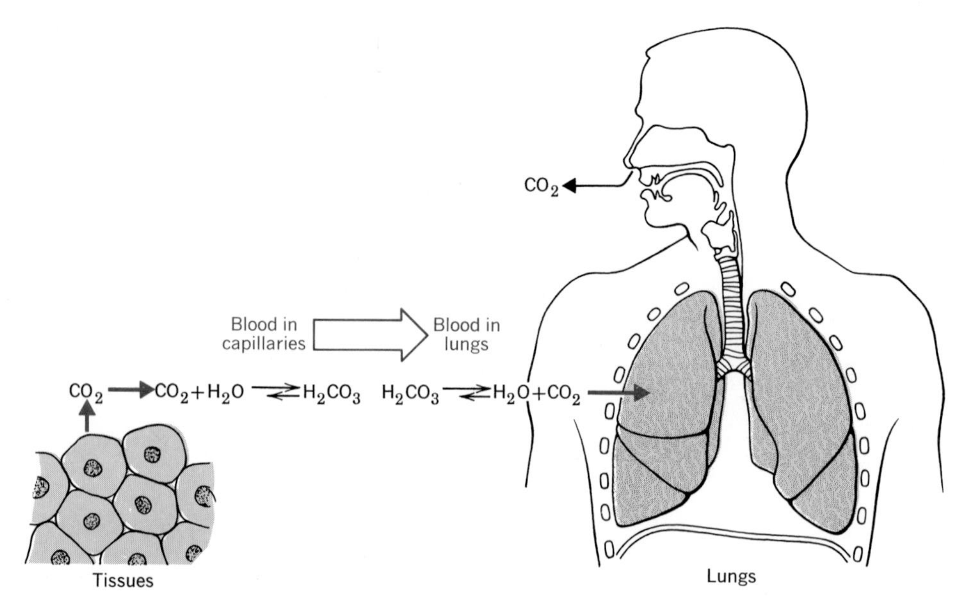

Figure 10.15 Changes in the concentration of carbon dioxide in the body affect the carbon dioxide–carbonic acid equilibrium.

reaction. Therefore, we will have $rate_f > rate_r$, and more sugar will be dissolving than is crystallizing. If we raise the temperature and hold it constant, $rate_r$ will increase until it equals $rate_f$, thus establishing a new equilibrium with a greater number of sugar molecules dissolved in the tea (Figure 10.16). Obviously, if you like your tea sweet you should drink it warm!

Catalysts

We have discussed the fact that a catalyst will increase the rate of a chemical reaction, but you may be surprised to learn that a catalyst will have no effect on the equilibrium concentrations of the reactants and products in an equilibrium system. Why is this? Remember that the way a catalyst increases the rate of a chemical reaction is by lowering the activation energy of the reaction (Figure 10.11). At the same time, however, the activation energy for the reverse reaction will be lowered by an equal amount, so the rate of the reverse reaction will also be increased. A catalyst, then, increases the rate of both the forward and the reverse reactions by the same amount and, therefore, will have no effect on the equilibrium system.

10.6 Le Chatelier's Principle

We have seen that a chemical system is in equilibrium when the rates of the forward and reverse reactions are equal; moreover, changes made in such systems can disrupt the equilibrium, thereby changing the equilibrium concentrations of the reactants and products. After studying a great number of equilibrium systems, the French chemist Henri Louis Le Chatelier was able to state a principle that helps predict whether the reactants or products will be favored in a given change. **Le Chatelier's Principle** states that a system at equilibrium will resist attempts to change its temperature or pressure, or the concentration of a reactant or product. In other words,

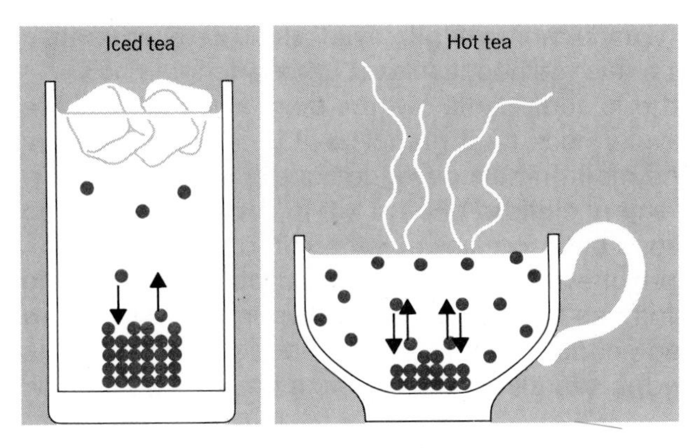

$Sugar_{(s)} + Heat \rightleftharpoons Sugar_{(aq)}$

Figure 10.16 Increasing the temperature of a system at equilibrium will increase the rate of the endothermic reaction. As the temperature of the iced tea is increased, more sugar will dissolve.

when upset by a change in temperature, pressure, or concentration, an equilibrium system will shift in the direction necessary to resist the change and to reestablish an equilibrium. Le Chatelier's Principle tells us, for example, that if we increase the concentration of a reactant, the system will move in a direction to remove that increase; that is, $rate_f$ will be greater than $rate_r$. If we increase the temperature, the reaction which uses up that increase in energy will be favored; that is, the rate of the endothermic reaction will be greater than the rate of the exothermic reaction, until a new equilibrium has been established.

Let's consider the effect of various changes on the following reaction at equilibrium.

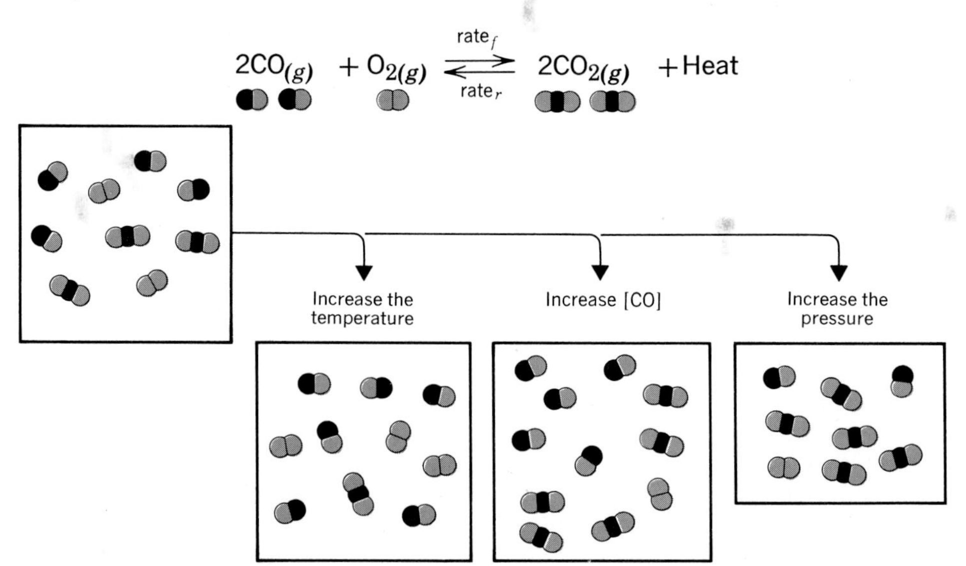

1. If we increase the temperature, the system will shift in a direction to resist the change; that is, the direction that removes heat. This means that the endothermic, or reverse, reaction will be favored. Therefore, $rate_r$ will be greater than $rate_f$.

2. If we increase the concentration of CO, the system will shift to lower the CO concentration: the forward reaction will be favored.

3. If we increase the pressure, the reaction that will be favored is the one that produces a system exerting less pressure; that is, a system with fewer gas molecules. The forward reaction produces a system with two gas molecules, and the reverse reaction produces a system with three gas molecules. Therefore, the forward reaction will be favored, and $rate_f$ will be greater than $rate_r$.

Le Chatelier's Principle can be very helpful in explaining the results of stresses on equilibrium systems in living organisms. For example, *calculus* is a word used in medicine to describe an abnormal build-up of ionic solids, called mineral salts, in a framework or matrix of other substances, such as proteins. Kidney stones, bladder stones, and gall stones are examples of calculi. The mineral salts found in kidney and bladder stones are formed from ions circulating in body fluids; the positive ions are calcium (Ca^{2+}) and magnesium (Mg^{2+}), and the negative ions are PO_4^{3-}, HPO_4^{2-}, or $H_2PO_4^-$. One of the equilibriums involved is

$$3Ca^{2+}{}_{(aq)} + 2PO_4^{3-}{}_{(aq)} \underset{Rate_r}{\overset{Rate_f}{\rightleftharpoons}} Ca_3(PO_4)_{2(s)}$$

When a person develops kidney stones, something has happened to disrupt the system that controls the amount of calcium and magnesium ions in body fluids. The concentration of calcium ions goes up, and this stress increases the rate of the forward reaction—the formation of the solid $Ca_3(PO_4)_2$. The stone will grow slowly as this solid, as well as other salts, are deposited in the matrix (Figure 10.17).

As another example, aspirin, which is taken by millions of people for relief of headaches, is a very safe drug. In some people, however, aspirin can disrupt the normal functioning of the stomach lining and can cause bleeding. The manner in which aspirin enters the stomach lining involves various stresses on the equilibrium that exists between the neutral aspirin molecule and the aspirin ion that is formed when the aspirin molecule is dissolved in water.

$$\underset{\text{(nonpolar)}}{\text{Aspirin}} \underset{Rate_r}{\overset{Rate_f}{\rightleftharpoons}} \underset{\text{(polar)}}{\text{Aspirin}^-} + H^+$$

When aspirin is dissolved in water, an equilibrium is formed between the

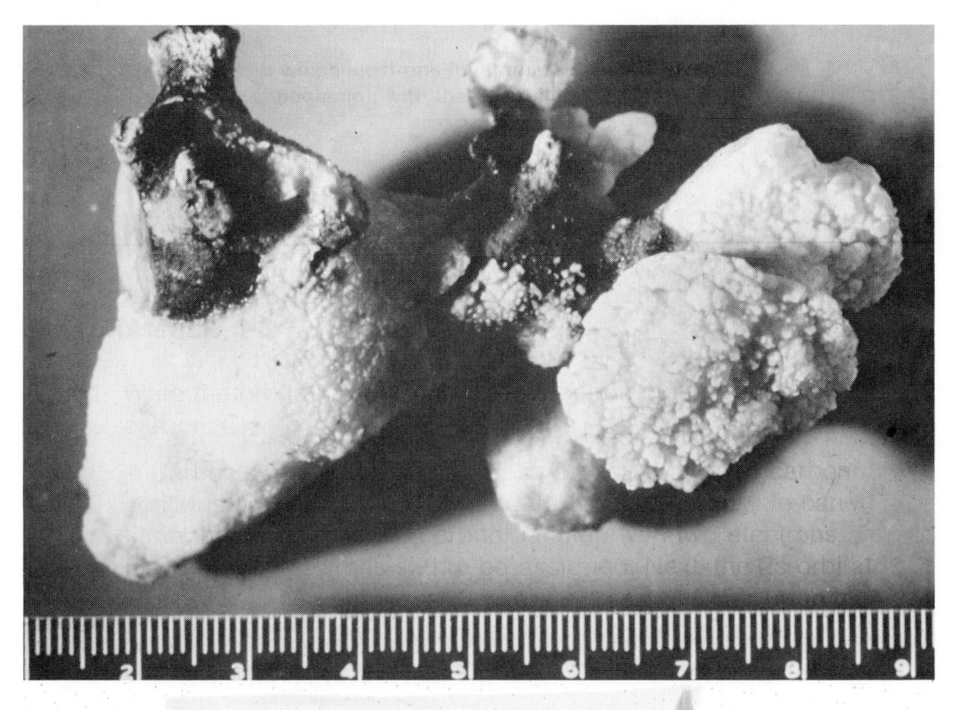

Figure 10.17 These kidney stones, composed mainly of calcium phosphate, were formed in the kidneys of a person with a urinary tract infection. (Courtesy William P. Mulvaney, M.D.)

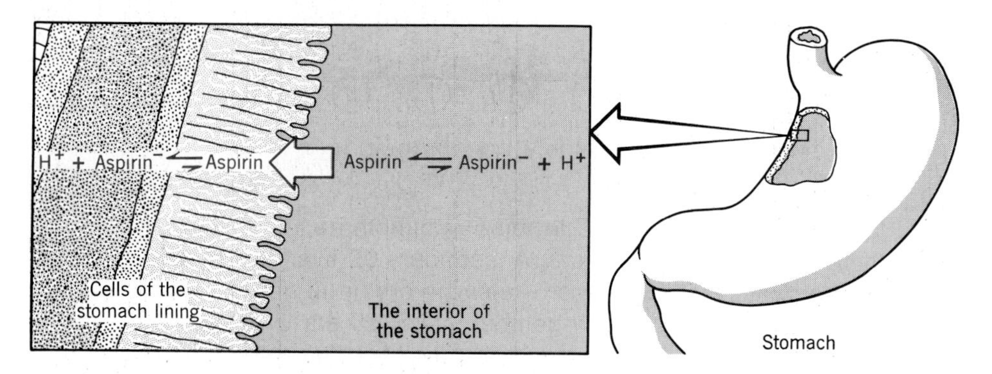

Figure 10.18 The passage of aspirin molecules through the stomach wall is accelerated by the stress that is applied to the aspirin-aspirin ion equilibrium by the hydrogen ions in stomach fluid.

polar and nonpolar forms of this compound. Therefore, when you swallow an aspirin with some water, a significant amount of the aspirin will be found in the polar ion form, which cannot pass through the protective lining of the stomach. But the contents of the stomach are very high in hydrogen ions, H^+. This puts a stress on the equilibrium, increasing the rate of the reverse reaction and forming more nonpolar aspirin molecules, which can pass through the protective lining. Once these aspirin molecules have passed into the cells of the stomach, the concentration of hydrogen ions becomes very low, and the equilibrium shifts to the right. This again produces the polar aspirin ions, which cannot pass back into the stomach and, therefore, become trapped in the cells, where they produce the damage that causes the bleeding (Figure 10.18).

EXAMPLE 10-2

Suppose that a glass cylinder with a movable piston at one end contains the brown gas nitrogen dioxide, NO_2, in equilibrium with the colorless gas dinitrogen tetroxide. N_2O_4.

$$\underset{\text{brown}}{2NO_{2(g)}} \rightleftharpoons \underset{\text{colorless}}{N_2O_{4(g)}} + 13.6 \text{ kcal}$$

What effect would the following changes have on the color of the gas in the cylinder (that is, on the equilibrium concentration of NO_2)?

(a) *Removing N_2O_4 from the flask.* The brown color will decrease. The equil brium system will res st the removal of the N_2O_4 by increasing the forward reaction, thereby reducing the number of NO_2 molecules in the flask and decreasing the color.

(b) *Putting the flask in an ice-water bath.* The brown color will decrease. Putting the flask in an ice-water bath will decrease the temperature of the flask. The system will resist this change by shifting in the direction of the exothermic reaction—the forward reaction—thereby lowering the concentration of NO_2.

(c) *Putting the flask in a boiling water bath.* The brown color will increase. In this case, the temperature is increased, so the system will shift in the direction of the endothermic reaction.

(d) *Decreasing the volume of the cylinder.* The brown color will decrease. Decreasing the volume increases the pressure within the cylinder. The system, therefore, will shift in the direction of the forward reaction, which is the direction that produces fewer gas molecules and lowers the pressure. The concentration of NO_2 will decrease.

Exercise 10-2

What would be the effect of each of the following on the equilibrium concentration of O_2 in this equilibrium system?

$$2N_2O_{5(g)} \rightleftharpoons 4NO_{2(g)} + O_{2(g)} + \text{heat}$$

(a) Adding NO_2
(b) Removing N_2O_5
(c) Increasing the temperature
(d) Decreasing the pressure
(e) Adding a catalyst

CHAPTER SUMMARY

For a chemical reaction to occur, the reactants must collide with enough energy to overcome an energy barrier called the activation energy, E_{act}. When this occurs, the particles first form a reactive group called the activated complex (a transition state between reactants and products) and then go on to form the products. A reaction that gives off energy is called exothermic, and one that absorbs energy is called endothermic. The amount of energy given off or absorbed by a reaction is called the heat of reaction, ΔH. The following factors will all affect the rate at which a reaction occurs: the nature of the reactants, the concentration of the reactants, the temperature at which the reaction takes place, and the addition of a catalyst.

A chemical reaction is in a state of equilibrium when the rate of the forward reaction equals the rate of the reverse reaction, and there is no net change in the concentrations of reactants and products. According to Le Chatelier's Principle, when a system at equilibrium is put under stress (by changes in temperature or pressure, or in the concentration of reactants or products), the system will change in a direction that will tend to remove the stress.

EXERCISES AND PROBLEMS

EXERCISES AND PROBLEMS

1. Using the concept of activation energy, explain why a dropped glass might remain intact on one occasion and shatter on another.

2. Draw potential energy diagrams for a general endothermic and exothermic reaction. Label the reactants, the products, the activation energy E_{act}, and the heat of reaction ΔH.

3. Which of the following reactions are endothermic and which are exothermic?

   (a) $2H_{2(g)} + O_{2(g)} \longrightarrow 2H_2O_{(g)} + 115.6$ kcal
   (b) $N_{2(g)} + 2O_{2(g)} + 16.2$ kcal $\longrightarrow 2NO_{2(g)}$
   (c) $2NH_{3(g)} + 22$ kcal $\longrightarrow N_{2(g)} + 3H_{2(g)}$
   (d) $3C_{(s)} + 2Fe_2O_{3(s)} + 110.8$ kcal $\longrightarrow 4Fe_{(s)} + 3CO_{2(g)}$

4. What is a catalyst?

5. The activation energy for the following reaction is 43.8 kcal. Adding platinum to the reaction mixture lowers the activation energy to 29 kcal.

$$2HI + 3 \text{ kcal} \longrightarrow I_2 + H_2$$

   (a) What effect will the platinum have on the rate of the reaction? Why?
   (b) Draw the potential energy diagrams for the catalyzed and uncatalyzed reactions.
   (c) What is the activation energy of the reverse reaction before adding the platinum? After adding the platinum?
   (d) What is the heat of reaction, ΔH, for the reverse reaction before adding the platinum? After adding the platinum?

6. Explain why special industrial procedures must be designed for handling large amounts of finely divided, dry, combustible materials.

7. The browning of cut peaches is caused by the reaction of molecules on the surface of the peach with oxygen in the air. Explain on the molecular level why covering the cut peaches with plastic wrap and placing them in the refrigerator will slow the browning process.

8. Why do labels of certain antibiotic drugs state that the drugs must be kept under refrigeration?

9. Many animals, such as snakes, are cold-blooded; that is, their body temperature is the same as the temperature of their surroundings. Using your knowledge of reaction rates, explain why snakes are very sluggish on cold mornings, and often sun themselves on rocks before searching for food.

10. What is *equal* in a reaction at equilibrium?

11. Why is a chemical equilibrium described as dynamic rather than static?

12. Is an equilibrium established in a can of ether with the lid off; with the lid on? Give a reason for your answers.

13. What is the effect of a catalyst on a reaction at equilibrium?

14. State Le Chatelier's Principle in your own words.

15. State three ways to increase the concentration of NOCl in the following reaction:

$$2NO_{(g)} + Cl_{2(g)} \rightleftharpoons 2NOCl_{(g)}$$

16. The heat of reaction for the conversion of graphite to diamond is quite small, and yet this reaction will take place only under extremely high temperatures and pressure. Why is this? Draw a potential energy diagram for the reaction

$$0.45 \text{ kcal} + C_{(\text{graphite})} \longrightarrow C_{(\text{diamond})}$$

17. (a) Compare the difference in reaction rates for the following reaction if the carbon is supplied (1) in chunks and (2) as a finely divided powder.

$$C_{(s)} + O_{2(g)} \longrightarrow CO_{2(g)}$$

(b) Draw the potential energy diagrams for the reactions in (1) and (2).

18. The following is the potential energy curve for the reaction

$$CO + NO_2 \longrightarrow CO_2 + NO$$

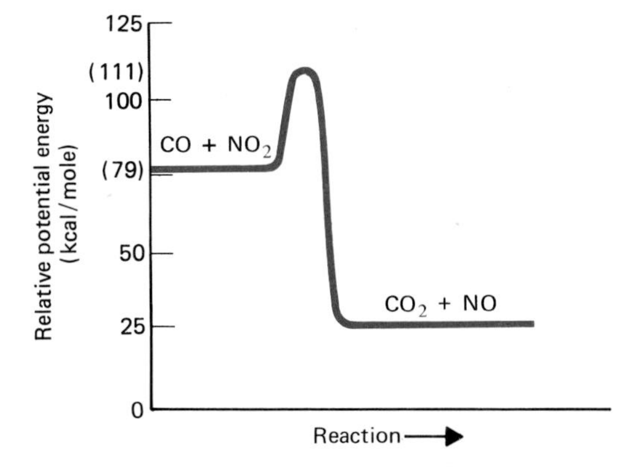

(a) Is this reaction endothermic or exothermic?
(b) What is the heat of reaction, ΔH?
(c) What is the activation energy E_{act} for the forward reaction?
(d) What is the activation energy E_{act} for the reverse reaction,

$$CO_2 + NO \longrightarrow CO + NO_2$$

19. An important industrial process is the production of ammonia through the following reaction.

$$N_{2(g)} + 3H_{2(g)} \rightleftharpoons 2NH_{3(g)} + 22 \text{ kcal}$$

This process uses a catalyst.

(a) Draw the potential energy diagram for the uncatalyzed and the catalyzed reaction.
(b) Describe three steps that manufacturers could take to maximize the yield of ammonia in this reaction.

20. What effect would the following changes have on the equilibrium concentration of $O_{2(g)}$ in the following reaction? Give a reason for each of your answers.

$$4HCl_{(g)} + O_{2(g)} \rightleftharpoons 2H_2O_{(g)} + 2Cl_{2(g)} + 27 \text{ kcal}$$

(a) Increasing the temperature of the reaction.
(b) Increasing the pressure.
(c) Decreasing the concentration of Cl_2.
(d) Increasing the concentration of HCl.
(e) Adding a catalyst.

21. Nitrogen dioxide, NO_2, causes the reddish-brown haze associated with air pollution. It is formed by the reaction of nitric oxide, NO, with oxygen.

$$2NO_{(g)} + O_{2(g)} \rightleftharpoons 2NO_{2(g)}$$

What would be the effect of each of the following on this system at equilibrium?

(a) Doubling the volume of the reaction flask.
(b) Adding more oxygen to the reaction flask.
(c) Increasing the pressure in the flask.

22. What would be the effect of each of the following changes on the equilibrium concentration of PCl_3 in the system:

$$PCl_{3(g)} + Cl_{2(g)} \rightleftharpoons PCl_{5(g)}, \qquad \Delta H = -21 \text{ kcal}$$

(a) Decreasing the concentration of PCl_5.
(b) Increasing the concentration of Cl_2.
(c) Increasing the temperature.
(d) Decreasing the volume of the reaction container.

23. An industrially important reaction for the production of hydrogen gas is

$$CO_{(g)} + H_2O_{(g)} \rightleftharpoons H_{2(g)} + CO_{2(g)} + 9.9 \text{ kcal}$$

Describe how each of the following changes would affect the equilibrium concentration of H_2.

(a) Addition of H_2O.
(b) Addition of CO_2.
(c) Removal of CO.
(d) Increasing the temperature of the reaction.
(e) Decreasing the volume of the reaction container.

INTEGRATED PROBLEMS

1. On August 29, 1982, element 109 was produced by bombarding bismuth-209 with iron-58.
   (a) Element 109 would be expected to have chemical properties similar to what other elements?
   (b) State the atomic number, mass number, number of protons, and number of neutrons in an atom of bismuth-209 and iron-58.
2. Low-density lipoprotein (LDL), the major cholesterol carrier in the blood, is a spherical particle with a mass of 3 million daltons and a diameter of 22 nanometers.
   (a) What is the mass of an LDL particle in μg?
   (b) What is the diameter of the particle in mm?
3. You are running a carefully controlled experiment that keeps track of the use of sodium ions by a specific one-celled organism. You discover that your culture water has been contaminated by trace amounts of the following ions: potassium ion, barium ion, iodide ion. Do you think any of these ions would interfere with your data? Why or why not? Design an experiment that would support your conclusion.
4. Strontium-90, a beta emitter, is a radioactive waste product of nuclear testing. When emitted into the atmosphere, it settles back to earth and is ingested by grazing animals such as cows. Why would finding strontium-90 in cows' milk pose a serious health risk?
5. In 1983 in the Mexican city of Juarez, an obsolete cancer therapy machine containing 6000 pellets of cobalt-60 was mistakenly broken up as scrap and refined into metal table legs. Six men who were breaking up the machine could have been exposed to as much as 500 rem whole-body radiation.
   (a) If the exposure was that high, how many of these men would be expected to be alive after one month?
   (b) If the activity of the cobalt-60 source was 400 Ci, what is the rate of decay of this source in disintegrations per second? in Bq?
   (c) Several children played in the bed of the truck that had carried the machine. When identified, the truck was giving off 50 rem/hr of radiation. What would you expect an examination of the chromosomes in the children's blood cells to reveal?
6. Iodine-131 is used in the treatment of thyroid cancer. A patient admitted to the hospital for treatment is placed in a special room in which the bed is surrounded by lead shields 1 cm thick. The patient is given an oral dose of 150 millicurie of NaI^{131} and remains in the room for about 4 days until the radiation given off by his body is below a minimum level.
   (a) I-131 is produced in the laboratory by bombarding tellurium-130 with neutrons. Write the balanced equation for this transmutation.
   (b) I-131 decays by giving off beta and gamma radiation. Write the balanced equation for this decay.

(c) What is the rate of disintegration in becquerel of the treatment dose given to the patient?
(d) If a hospital received a shipment of 20.0 g of iodine-131, how much I-131 would remain in the laboratory after 20 days?
(e) Why is it important that the nurses caring for this patient wear film badges and not go inside the shielding unless absolutely necessary?
(f) Must extra care be taken in handling the urine and blood of this patient while he is under treatment? Why?
(g) If the half-life of I-131 is 8 days, why is it that the patient can go home in 4 days?

7. Sulfur trioxide can be produced by the reaction of the air pollutant sulfur dioxide with oxygen in the air.
   (a) Write the balanced equation for this reaction.
   (b) How many grams of sulfur dioxide must be released into the air to produce 1.00 kilogram of sulfur trioxide?

8. Sulfur trioxide is removed from the atmosphere by rain, which reacts with the sulfur trioxide to form sulfuric acid, H_2SO_4. Along major freeways, rainwater has been found to be quite acidic and damaging to plant life. For every 1.00 kilogram of sulfur trioxide that reacts with rain, how many grams of sulfuric acid will be formed?

9. The hemoglobin molecule in red blood cells carries oxygen from the lungs to the tissues. This results in the following equilibrium:

$$\underset{\text{Hemoglobin}}{HHb^+} + O_2 \rightleftharpoons \underset{\text{Oxyhemoglobin}}{HbO_2} + H^+$$

A second equilibrium system that is present in the blood involves carbon dioxide, carbonic acid, and the bicarbonate ion:

$$\underset{\text{Carbon dioxide}}{CO_2} + H_2O \rightleftharpoons \underset{\text{Carbonic acid}}{H_2CO_3} \rightleftharpoons \underset{\text{Bicarbonate ion}}{HCO_3^-} + H^+$$

Use Le Chatelier's Principle and Figure 4.18 to explain how the production of the waste product carbon dioxide by the tissues and the concentration of oxygen in the lungs cause these two equilibrium systems to shift in such a way as to aid in the transport of oxygen to the tissues and the exhalation of carbon dioxide from the lungs.

CHAPTER ELEVEN
WATER, SOLUTIONS, AND COLLOIDS

Judy examined Amaresh and shook her head in wonder and frustration. It had taken Amaresh and her three children 30 days to make the terrifying journey from their Ethiopian village in Tigre to the refugee camp in the desert across the Sudanese border. They had walked the whole way, hiding during the day and traveling only at night to avoid the bombs and bullets of the Ethiopian military planes. They started their journey with only a small amount of food, which had been used up long before they reached the Sudanese border.

The trucks they met at the border took them directly to the refugee camp. Judy, as one of the camp's volunteer health workers, had examined many such families as they arrived weak with starvation and fatigue. She assigned Amaresh's two youngest children to a special feeding program to combat their severe malnutrition and then showed the family their new home—a single 10 foot by 10 foot tent that already housed 15 of their relatives.

The camp routine never varied. Each day Amaresh went to collect her family's allotment of raw wheat, beans, and cooking oil. She mixed these together and boiled them in the blackish-green water drawn from the camp's water tanks. Until recently water had been taken directly from a nearby irrigation canal, and diarrhea had been rampant in the camp. Unfortunately, many of the refugees drinking this water were so weakened by malnutrition that even such simple diarrhea was life-threatening—especially for the children and the elderly. The health workers had put a high priority on construction of the water tanks, which allowed the canal water to be chlorinated and to settle before being used in the camp.

The water was not the only source of worry for the health workers. The camp had no sanitary facilities. All of the camp's 12,000 refugees used the open field next to the camp. The danger of this arrangement became obvious a week after Amaresh's arrival at the camp, when a fierce sandstorm destroyed most of the tents and the camp's water system. The only available water was that pumped directly from the canal, without any chlorination or settling. By the next morning, with the temperature rising to 110°F, everyone in the camp was suffering from terrible diarrhea.

Judy was not surprised when Amaresh came to her that evening carrying her youngest son. He was very pale and unresponsive and was obviously suffering from severe dehydration from his diarrhea. Since there were no antibiotics available, she gave Amaresh a yellow plastic jug and several small sealed packets containing oral rehydration salts (ORS). Judy instructed Amaresh to pour the white powder from a packet into the jug and then to fill the jug with chlorinated water up to a clearly marked one liter line. She was to feed as much of this liquid to her children as they could drink. Judy predicted that the children would be out playing with their friends in only a few days.

The oral rehydration salts that Judy distributed have become almost a miracle treatment for refugees suffering from dehydration caused by diarrhea. It was remarkable that in the six weeks Judy had worked at the camp not one child had died from dehydration with diarrhea! Of the 27.5 grams of white powder in each packet of ORS, 7.5 grams is made up of sodium chloride, sodium bicarbonate, and potassium chloride. When added to water, these salts help restore the fluids and electrolytes lost from the body during diarrhea. The remaining 20 grams is glucose, a sugar that gives the children the calories they need to regain their strength. The additional water also helps control the body's temperature, an extremely important concern in the sub-Saharan heat of the refugee camp. By restoring normal fluid and ion balance, this ORS therapy allows the body's immune system to take over and fight the cause of the diarrhea.

Why is dehydration so potentially dangerous? Water is critical to the survival of all living organisms and is second only to oxygen in importance for human survival. We can survive for several weeks without food, but only for a few days without water (and only a few minutes without oxygen). The water content of living organisms varies from less than 50% in some bacteria cells to 96% and 97% in some marine invertebrates. It is by far the most abundant chemical compound in the human body, making up 60% to 70% of the adult body weight. This water is found inside the cells, in the extracellular (or interstitial) fluid that bathes the cells, and in the blood plasma (Figure 11.1).

Water performs many biological functions. It is the fluid found throughout living organisms, transporting food and oxygen to the cells and

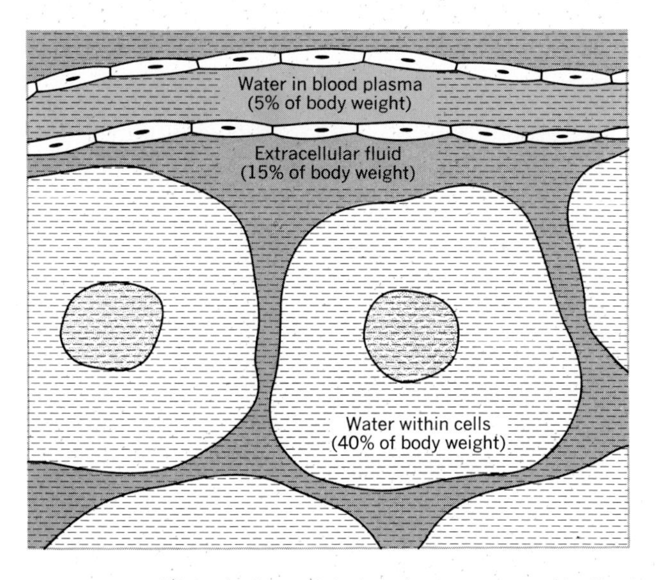

Figure 11.1 Water is the most abundant compound in the human body. It is distributed within the cells, in the extracellular fluid that bathes the cells, and in the blood plasma.

carrying away wastes. It is the fluid in which digestion takes place and is a lubricant for the cells and tissues. Water also plays a major role in the regulation of body temperature and the acid–base balance of body fluids. Water is an important reactant and product in many chemical reactions that take place in living organisms.

Every day you lose between 1500 and 3000 milliliters of water in the form of urine, perspiration, water vapor in your breath, and feces. This lost water must be replaced through the liquids you drink or foods that you eat. Dehydration can be caused by diarrhea, high fever, bleeding, burns, or ulcers—all of which can disrupt the normal water balance in the body. For adults, a 10% loss in total body fluids causes serious dehydration and disruption of normal body chemistry; a 20% loss can be fatal.

To understand the chemistry of living organisms, we need to examine more closely the properties of water that make it so essential for life.

LEARNING OBJECTIVES

By the time you have finished this chapter, you should be able to:

1. List three special properties of water and explain each in terms of the hydrogen bonding in water.
2. Describe and compare suspensions, colloids, and solutions and state three properties that are unique to colloids.
3. Define *solute, solvent,* and *aqueous solution.*
4. Describe in your own words the process by which sodium chloride dissolves in water.

5. Define *electrolyte* and *nonelectrolyte* and state the difference between a strong and weak electrolyte.
6. Predict whether a precipitate will form when two ion-containing solutions are combined.

PROPERTIES OF WATER

11.1 Solvent Properties

Water is the most abundant liquid in the world, yet knowledge of its structure is far from complete. It is the special properties of water that make it so vital to life. Water, H_2O, is a molecule containing an oxygen atom covalently bonded to two hydrogen atoms. The shape of the molecule is bent.

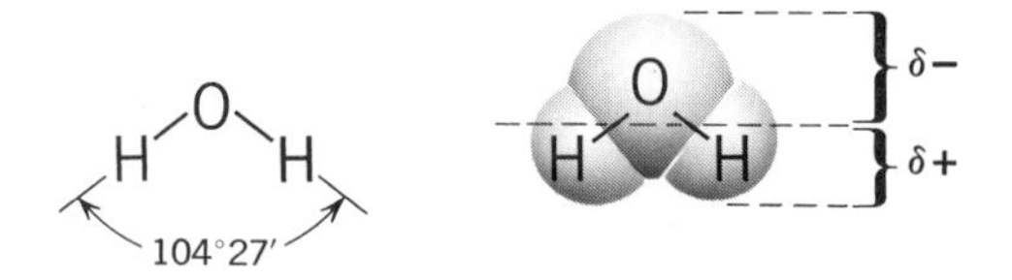

The oxygen atom is much more electronegative than the hydrogen atom, and each covalent bond is polar. In addition, the bent shape of the molecule makes the entire molecule polar, with the oxygen at the negative end and the hydrogens at the positive end. The polarity of water allows the formation of hydrogen bonds and gives this molecule its solvent properties. Water is called the *universal solvent* because it is a better solvent than most other liquids. It will dissolve ionic compounds and molecular compounds that are polar or that contain polar groups.

11.2 High Melting and Boiling Points

Other properties of water can be explained by the hydrogen bonding that exists between water molecules (Figure 8.13). Water has a high melting point and a high boiling point when compared with other molecules of comparable formula weight. In fact, most compounds of comparable formula weight are gases at the temperature that water is a liquid. For example, water has a formula weight of 18 and a boiling point of 100°C, whereas ammonia (with a formula weight of 17) and methane (with a formula weight of 16) have boiling points of −33° and −164°C, respectively. The high melting point and boiling point result from the hydrogen bonding between the molecules of water, thus requiring more energy to pull these molecules apart.

11.3 Density of Ice

You will recall that the density of a substance is a measure of the substance's mass per unit volume (density = mass/volume). When heated, nearly all substances will increase in volume. Because the mass of the substance remains constant, its density will therefore decrease as the temperature increases (Figure 11.2a). As water cools, it contracts until it reaches its maximum density at 3.98°C. As it continues to cool, however, and then freeze at 0°C, it expands by 9% (Figure 11.2b). It is the open lattice structure of ice, in which each molecule of water is hydrogen bonded to four other water molecules, that makes ice less dense than the more compact liquid water. This is why pipes break when the water in them freezes. The fact that ice is less dense than water is critical to aquatic life. When the surface of a lake freezes, the ice floats and the most dense water (at 4°C) sinks to the bottom, giving plants and animals a place to survive. If ice were more dense than water, lakes and oceans would freeze from the bottom up and would probably never completely thaw in summer.

11.4 Surface Tension

We mentioned in Chapter 4 that water has a high surface tension. Water molecules beneath the surface are strongly attracted in all directions to other water molecules by hydrogen bonding. But the molecules on the surface are not attracted to the nonpolar air molecules. Therefore, rather than being attracted in all directions, these surface molecules are attracted only downward and inward toward the water. This surface tension pulls the surface molecules together, creating the effect of a thin, elastic membrane on the surface. This is why, for example, dust particles or water spiders can remain on the water's surface even though they are more dense than water. Also, you have probably noticed that water, when dropped onto waxed paper, forms small beads of liquid. This occurs because the surface water molecules are attracted neither to the nonpolar wax nor to the surrounding air, and thus the attraction between water molecules draws the liquid up to

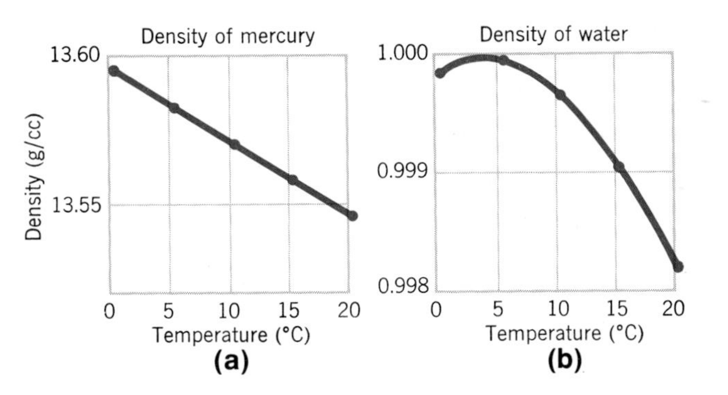

Figure 11.2 The density of most substances, such as mercury, gradually increases as the temperature decreases. The density of water, however, increases to a maximum at 4°C and then decreases as the temperature decreases.

form a bead. Glass, on the other hand, is more polar than water. Thus, water is attracted to glass and, when placed on a glass surface, will spread out rather than form beads. Substances that reduce the surface tension of water are called **surface active agents,** or **surfactants.** In order for water to penetrate the dirt and oil in clothes in the laundry, or to aid in the digestion of fats, surfactants such as soaps, detergents, or bile must be present.

11.5 High Heat of Vaporization

Water has a high heat of vaporization (539 cal/g). The strong attractive forces between water molecules cause this liquid to boil away very slowly. It takes nearly seven times as much energy to boil away one liter of water at 100°C as it does to heat it from 21°C (room temperature) to 100°C. It is important to realize that evaporation is essentially the same process as boiling; it involves changing liquid water to gas, only at a slower rate. As with boiling, evaporation uses up a great deal of energy or heat. Animals make use of this principle to rid their bodies of excess heat through the evaporation of sweat.

11.6 High Heat of Fusion

Water also has a high heat of fusion (80 cal/g). That is, a great deal of heat must be released before ice can be formed. To convert one kilogram of water at 0°C to one kilogram of ice requires almost four times as much refrigeration as is needed to cool this much water from 21 to 0°C. This explains why ice forms slowly in the winter and melts slowly in the spring. Water stored as snow in the mountains will, therefore, run off relatively slowly, preventing disastrous floods in the spring.

11.7 High Specific Heat

The specific heat of a liquid is the amount of heat required to raise the temperature of one gram of that liquid from 15 to 16°C. Water has a high specific heat compared with other liquids (Table 11.1). The higher the specific heat of a substance, the less its temperature will change when it absorbs a given amount of heat. The high specific heat of water enables this fluid to keep the temperature of an organism relatively constant in the face of fluctuating internal or external heat levels. On a larger scale, the water in lakes and oceans will absorb and store large quantities of solar energy, explaining why such large bodies of water have moderating effects on local climates.

COLLOIDS AND SUSPENSIONS

11.8 Three Important Mixtures

In Chapter 2 we stated that matter could be divided into pure substances and mixtures. Mixtures make up most of the things we see around us. The

Table 11.1 Specific Heat of Several Compounds

Compound	Specific Heat (cal/g)	Compound	Specific Heat (cal/g)
Water	1.0	Chloroform	0.23
Ethanol	0.58	Ethyl acetate	0.46
Methanol	0.60	Liquid ammonia	1.12
Acetone	0.53		

soil, buildings, lakes, our body's cells—all are made up of mixtures. We will discuss three important kinds of mixtures in this chapter: suspensions, colloids, and solutions. Their properties are compared in Table 11.2; the main property that distinguishes these three types of mixtures is the size of the particles in the mixture.

11.9 Suspensions

Suspensions are heterogeneous mixtures containing relatively large particles that, in time, will settle out. Suspensions made up of solid particles suspended in a liquid can be separated using filter paper or a centrifuge. Whole blood is an example of a suspension; the red and white blood cells will settle out in time, or can be separated from the plasma by centrifugation.

11.10 Colloids

Colloidal dispersions, or **colloids,** are heterogeneous mixtures whose particles are larger than the molecules or ions forming solutions, but

Table 11.2 A Comparison of Some Properties of Solutions, Colloids, and Suspensions

Property	Solutions	Colloids	Suspensions
Particle size	Less than 1 nm[a]	1 to 1000 nm	More than 1000 nm
Filtration	Will pass through filters and membranes	Will pass through filters but not membranes	Stopped by filters and membranes
Visibility	Invisible	Visible in an electron microscope	Visible to the eye or in a light microscope
Motion	Molecular motion	Brownian movement	Movement only by gravity
Passage of light	Transparent, no Tyndall effect	May be transparent, often translucent or opaque. Tyndall effect	Often opaque, may be translucent

[a] 1 nanometer (nm) = 10^{-9} meter.

smaller than particles forming suspensions. The particles in a colloidal dispersion cannot be separated out by ordinary filter paper. Homogenized milk, for example, is a colloidal dispersion of butterfat in water, and it passes through filter paper. Colloids can be classified by the solvent (or dispersing medium) and by the colloidal matter (or the dispersed medium). The eight classes that result are shown in Table 11.3. Colloidal chemistry is quite important in the study of biological systems. Tissues and cells are colloidal dispersions, and reactions occurring within them take place through colloidal chemistry. Food digestion, for example, involves the formation of colloids before the food can be digested. The contraction of muscles can be explained by colloidal chemistry, and the body's proteins are of colloidal size.

11.11 Properties of Colloids

Have you ever watched particles of dust dancing in a sunbeam, or moving randomly about in the light from a movie projector? The random movement of such particles in a colloid is caused by the bombardment of these particles by the solvent molecules. The resulting motion is called **Brownian movement** after the English botanist Robert Brown, who first observed this irregular movement of particles while studying pollen grains suspended in water under a microscope. This constant bombardment by solvent molecules helps keep the colloidal particles from settling out.

When you watch dust dancing in a sunbeam, you are not seeing the actual dust particles—they are too small to be seen with the naked eye. What you are actually seeing is the scattering of light by these particles. The particles in a solution are too small to scatter light well, but colloidal particles will scatter light. A dilute colloidal dispersion that may look the same to you as a true solution can be distinguished from the solution by passing a strong beam of light through the two mixtures. The path of the light beam will be clearly visible in the colloidal dispersion because of the

Table 11.3 Classes of Colloids

Class	Example
1. Solid in solid	Colored glass; certain alloys
2. Solid in liquid	Gelatin in water; protein in water; starch in water
3. Solid in gas	Aerosols; dust in air; smoke
4. Liquid in solid	Water in gems (opals and pearls); jellies; butter; cheese
5. Liquid in liquid	Emulsions; milk; mayonnaise; protoplasm
6. Liquid in gas	Aerosols; fog; mist; clouds
7. Gas in solid	Activated charcoal; styrofoam; marshmallows
8. Gas in liquid	Foam; whipped cream; suds
(Gas in gas)	(A solution, not a colloid)

Figure 11.3 The Tyndall Effect. A laser beam is passed through three beakers. The beakers on the left and right contain a dilute colloidal dispersion of starch in water; the one in the middle contains a solution. Because colloidal particles scatter light, the light beam can be seen passing through the colloidal dispersions but not the solution. (Photo by David Crouch, from T. R. Dickson, *Introduction to Chemistry,* John Wiley & Sons, 1971.)

scattering of the light by the colloidal particles (Figure 11.3). This property of colloids is known as the **Tyndall effect.** Light rays can be seen streaming through a forest, for example, as a result of the scattering of light by colloidal particles suspended in the forest air (Figure 11.4).

Colloidal particles range in size from 1 to 1000 nanometers (1 nm =

Figure 11.4 Sunbeams decorate the forest as a result of the Tyndall effect. (U.S. Forest Service.)

1×10^{-9} m) in diameter. Particles of this size have very large surface areas compared to their volume. This large surface area gives colloidal particles the ability to take up, or **adsorb,** substances on their surface. Powdered charcoal is an example of a substance that has many practical uses because of the colloidal size of its particles. It is put in gas masks to adsorb poisonous gases in the air and is used to remove gases and odors from city water supplies. It is also used to remove colored impurities from solutions in the laboratory and in industry, and even serves as an antidote for swallowed poisons. A person who has swallowed poison or taken an overdose of drugs can often be saved by passing their blood through a machine containing a charcoal filter that removes the toxic substance from the blood.

SOLUTIONS

11.12 Solutions

A **solution** is a homogeneous, or uniform, mixture of two or more substances (called crystalloids) whose particles are of atomic or molecular size. Club soda (carbon dioxide in water), air (gas in gas), rubbing alcohol (alcohol in water), and steel (carbon in iron) are all examples of solutions. We call the substance being dissolved the **solute,** whereas the **solvent** is the substance in which the solute is being dissolved. Our major interest in this text will be **aqueous** solutions: solutions in which water is the solvent.

Water is often called the universal solvent because its highly polar nature allows a large number of substances to dissolve in it. The process

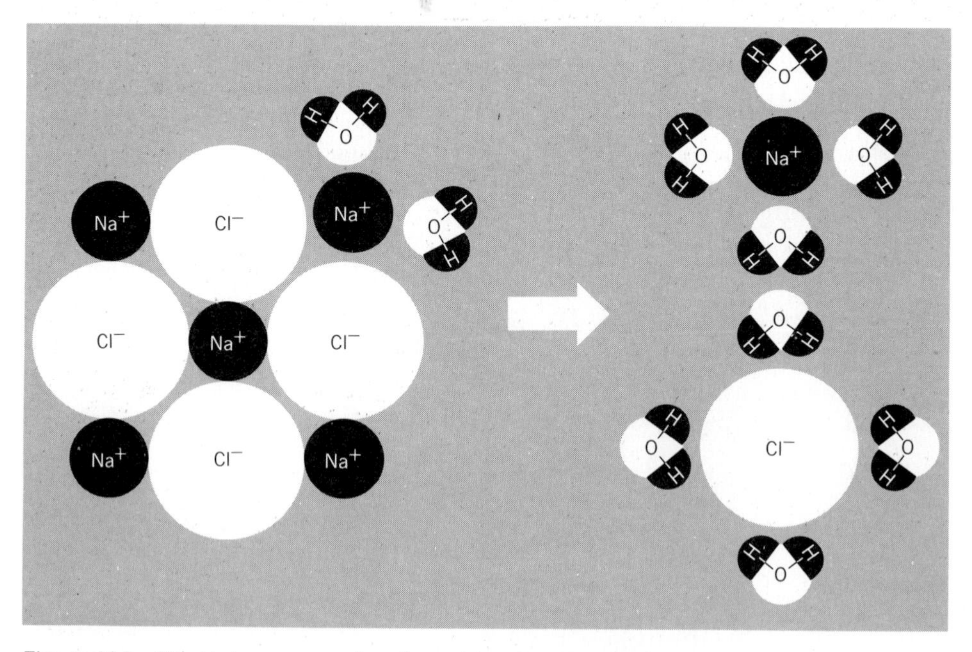

Figure 11.5 The ionic compound sodium chloride will dissolve in water because the attraction of the water molecules for the sodium and chloride ions is greater than the attraction between these ions in the crystal.

of dissolving occurs as follows. When, for example, a sodium chloride crystal is placed in water, the surface of the crystal is bombarded by the water molecules, whose polar ends exert an attractive force on the surface ions of the crystal and pull them loose (Figure 11.5). As each ion is pulled loose, it is surrounded by water molecules. This action prevents the sodium and chloride ions from rejoining. Such an ion surrounded by water molecules is said to be **hydrated.** Note, however, that not all ionic substances dissolve in water. For some crystals, the attraction of the water molecules may not be strong enough to overcome the attraction between the oppositely charged ions within the crystal.

Highly polar molecular substances, such as sugar, are also soluble in water. The hydration of sugar molecules occurs because of the attraction and hydrogen bonding between oppositely charged polar regions on the sugar and water molecules (Figure 11.6).

11.13 Electrolytes and Nonelectrolytes

We can see from our discussion of the dissolving process that solute particles take two forms: they may be charged ionic particles, or uncharged molecular particles. Substances such as sugar, which form uncharged molecular particles when dissolved in a solvent, are called **nonelectrolytes.**

$$\text{Sugar}_{(s)} + H_2O \longrightarrow \text{Sugar}_{(aq)}$$

Ionic compounds such as sodium chloride, which dissolve (or dissociate) to form charged ionic particles in solution, are called **electrolytes.**

$$NaCl_{(s)} + H_2O \longrightarrow Na^+_{(aq)} + Cl^-_{(aq)}$$

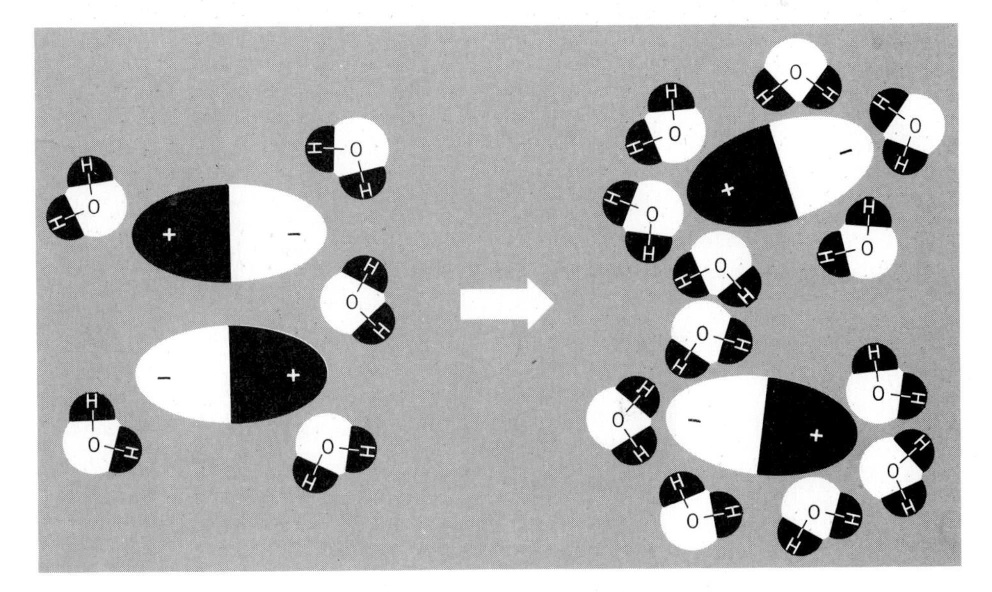

Figure 11.6 Highly polar molecular substances are soluble in water, especially if they contain atoms such as oxygen or nitrogen that can form hydrogen bonds with the water molecules.

Some molecular substances, such as hydrochloric acid, are so highly polar that the molecule will be pulled apart or ionized by water molecules. These substances will also form electrolytes in solution.

$$HCl_{(g)} + H_2O \longrightarrow H_3O^+ + Cl^-_{(aq)}$$

The term *electrolyte* refers to the fact that solutions of electrolytes conduct electricity. Electrolytes may be classified as either strong or weak, depending upon the number of charged ionic particles formed when the substance dissolves. A **strong electrolyte** is a substance that completely ionizes, or dissociates into ions, in solution. A **weak electrolyte** is a substance that only partially ionizes in solution (Figure 11.7).

Electrolytes perform many important regulatory roles in our bodies and are responsible for maintenance of the acid-base and water balance. The major positive ions, or **cations,** found in living tissue are Na^+, K^+, Ca^{2+}, and Mg^{2+}. The major negative ions, or **anions,** are HCO_3^-, Cl^-, HPO_4^{2-}, SO_4^{2-}, organic acids, and proteins (Figure 11.8).

11.14 Factors Affecting the Solubility of a Solute

Many factors affect the solubility of a solute. A major factor is the nature of the solvent and the solute; polar solvents will dissolve polar solutes, and nonpolar solvents will dissolve nonpolar solutes. For most solid substances, increasing the temperature will increase the solubility of the solute. However, there are a few compounds, such as $CaCr_2O_7$, $CaSO_4$, and $Ca(OH)_2$, whose solubility decreases with an increase in temperature.

The solubility of a gas in a liquid decreases with an increase in temperature. For example, we have discussed how the warming of river water caused by heat discharged from a nuclear power plant decreases the supply of oxygen available for the river's fish. Finally, as mentioned in

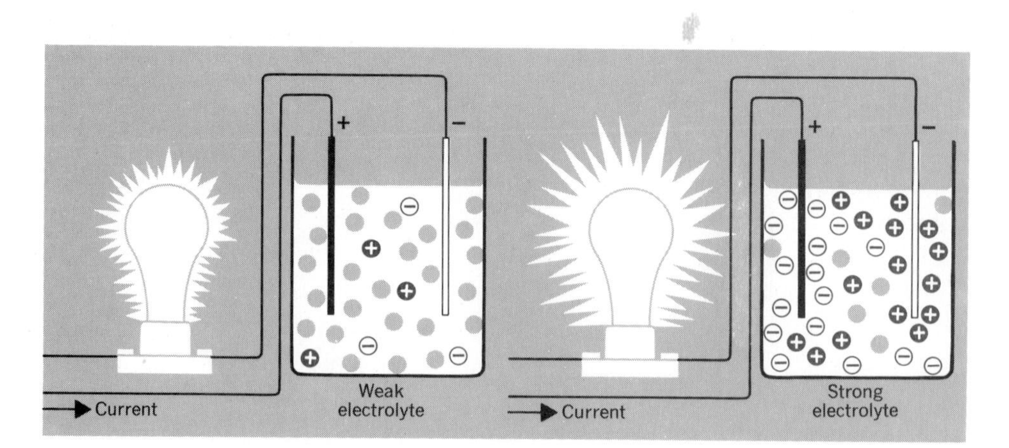

Figure 11.7 Solutions of electrolytes will conduct electricity. The stronger the electrolyte, the better the solution conducts electricity.

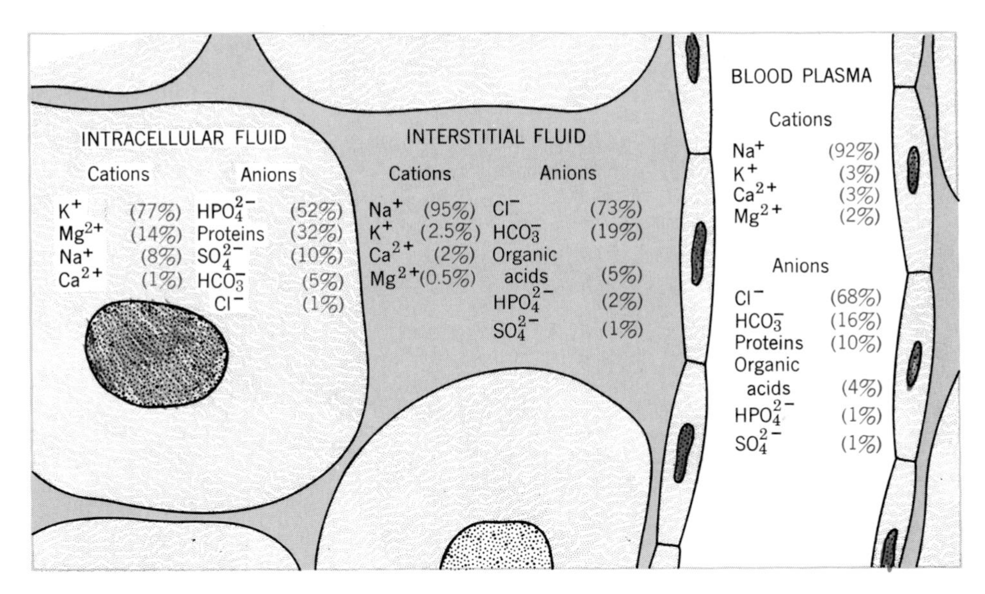

Figure 11.8 The principal electrolytes found in body fluids.

Chapter 4, the solubility of a gas in a liquid increases with an increase in pressure (Henry's law).

11.15 The Solubility of Ionic Solids

We have said that ionic solids vary in their solubility in water. Although the exact solubility of each compound must be determined by experiment, it is possible to make a few general statements about the solubility or insolubility of various solids.

Using Table 11.4, we can predict whether an insoluble solid, called a **precipitate,** will form when two aqueous solutions of soluble ionic compounds are mixed. For example, will a precipitate form when aqueous solutions of sodium chloride (NaCl) and silver nitrate ($AgNO_3$) are mixed? A solution of sodium chloride contains sodium ions (Na^+) and chloride ions (Cl^-), and a solution of silver nitrate contains silver ions (Ag^+) and nitrate ions (NO_3^-). The two new substances that could be formed from this combination are AgCl and $NaNO_3$. We see from Table 11.4 that $NaNO_3$ is soluble, but AgCl is not soluble. Thus, we would expect a precipitate of AgCl to form. We can write these results in equation form:

$$Na^+_{(aq)} + Cl^-_{(aq)} + Ag^+_{(aq)} + NO_3^-{}_{(aq)} \longrightarrow AgCl_{(s)} + Na^+_{(aq)} + NO_3^-{}_{(aq)}$$

Cancelling the ions that appear on both sides of the equation, but that were not involved in the reaction, we get

$$Ag^+_{(aq)} + Cl^-_{(aq)} \longrightarrow AgCl_{(s)}$$

This equation is called the **net-ionic equation,** because it shows only those ions that actually react.

Table 11.4 Some General Rules for the Solubility of Ionic Compounds

I	II	III
Compounds containing these cations are generally soluble:	Compounds containing these anions are generally soluble:	All ionic compounds formed from ions not listed in I or II are generally not soluble or, at most, only slightly soluble.
Li^+ Na^+ K^+ NH_4^+	NO_3^-, ClO_3^-, CH_3COO^-	
	Cl^-, Br^-, and I^- (except with Ag^+, Pb^{2+}, and Hg_2^{2+})	
	SO_4^{2-} (except with Pb^{2+}, Sr^{2+}, and Ba^{2+})	

EXAMPLE 11-1

Kidney stones are formed from the precipitate of the reaction between sodium phosphate and calcium chloride in solution in the kidneys. Write the net-ionic equation for the formation of kidney stones.

Sodium phosphate, Na_3PO_4, will form sodium and phosphate ions in solution, and calcium chloride, $CaCl_2$, will form calcium and chloride ions. The two new substances that can be formed from this reaction are sodium chloride, NaCl, and calcium phosphate, $Ca_3(PO_4)_2$. From Table 11.4 we see that NaCl is soluble, but $Ca_3(PO_4)_2$ is not. It, therefore, forms the precipitate that results in kidney stones.

Step 1: First, we must write a complete balanced equation for this reaction

Unbalanced: $Na_3PO_{4(aq)} + CaCl_{2(aq)} \longrightarrow Ca_3(PO_4)_{2(s)} + NaCl_{(aq)}$

Balanced: $2Na_3PO_{4(aq)} + 3CaCl_{2(aq)} \longrightarrow Ca_3(PO_4)_{2(s)} + 6NaCl_{(aq)}$

Step 2: Now we need to expand this equation to show all the ions that are free to react.

$$6Na^+_{(aq)} + 2PO_4^{3-}{}_{(aq)} + 3Ca^{2+}{}_{(aq)} + 6Cl^-{}_{(aq)} \longrightarrow Ca_3(PO_4)_{2(s)} + 6Na^+_{(aq)} + 6Cl^-{}_{(aq)}$$

Step 3: Cancelling the "spectator" ions on both sides of the equation, we get the following net-ionic equation:

$$2PO_4^{3-}{}_{(aq)} + 3Ca^{2+}{}_{(aq)} \longrightarrow Ca_3(PO_4)_{2(s)}$$

Exercise 11-1

1. Will a precipitate form when aqueous solutions of the following compounds are mixed?

   (a) $NaOH$ and $MgCl_2$
   (b) $NaCl$ and $(NH_4)_2SO_4$
   (c) $NaCl$ and $Pb(NO_3)_2$
   (d) $Ba(NO_3)_2$ and Li_2SO_4

2. Using the three steps discussed in Example 11-1, write the net-ionic equations for those reactions in question 1 that produce a precipitate.

3. Write the net-ionic equation for the reaction described in question 4 of Example 9-1.

CHAPTER SUMMARY

CHAPTER SUMMARY

Water is critical to all living organisms. It is a good solvent for polar substances, has a high freezing and a high boiling point, is less dense as a solid than as a liquid at 0°C, and has a high surface tension, heat of vaporization, heat of fusion, and specific heat. All of these properties make water especially well suited for performing many biological functions.

Three important kinds of mixtures are suspensions, colloids, and solutions. A suspension is a heterogeneous mixture containing relatively large particles that will settle out in time. A colloid is a mixture in which the particles of the dissolved substance are larger than the particles in a solution, but smaller than those in a suspension. Brownian movement, the Tyndall effect, and the ability to adsorb large amounts of other substances on the surface of the colloidal particles are all properties of colloids.

A solution is a homogeneous mixture of two or more substances, called crystalloids, whose particles are of atomic or molecular size. The dissolving medium is called the solvent, and the substance that is dissolved is the solute. An aqueous solution is one in which the solvent is water. The solute may be an electrolyte, which forms charged ions in solution, or a nonelectrolyte, which forms molecular particles in solution. The solubility of a substance in a solvent will depend upon the nature of the solute and solvent, the temperature and, for a gas, the pressure. A precipitate may form when two solutions containing electrolytes are mixed if one or more of the possible combinations of the ions is insoluble in the solvent.

EXERCISES AND PROBLEMS

1. Explain the reason for each of the following properties of water in terms of the hydrogen bonding that exists in water.
   (a) High boiling point
   (b) Density of ice
   (c) High heat of vaporization
   (d) Solubility of molecular substances
2. What is the most important difference in composition between pure substances (elements and compounds) and mixtures?
3. State three properties that allow us to distinguish between a solution, a colloidal dispersion, and a suspension.
4. What test could you use to tell if a clear aqueous fluid contained a colloidal dispersion?
5. What causes the Brownian movement observed in a colloidal dispersion?
6. Which of the following are cations and which are anions?
   (a) PO_4^{3-}
   (b) H^+
   (c) Br^-
   (d) Ba^{2+}
7. State in your own words the steps that occur when sodium chloride is dissolved in water.
8. Which would you expect to be more soluble in water: carbon tetrachloride or ammonium sulfate? Give a reason for your answer.
9. Explain why a soft drink will go "flat" faster at room temperature than in the refrigerator.
10. Urea is a waste product formed in the liver from ammonia that is produced in the breakdown of protein. This urea is toxic in high concentrations and, therefore, is excreted by the body in the urine. Explain why people often feel thirsty after a meal rich in protein.
11. State which of the following are soluble in water:
    (a) $MgCO_3$
    (b) NaOH
    (c) NH_4Br
    (d) $PbSO_4$
    (e) KNO_3
    (f) AgCl
12. Hydrochloric acid, $HCl_{(aq)}$, reacts with sodium hydroxide, $NaOH_{(aq)}$, to produce water, $H_2O_{(liquid)}$ and sodium chloride, $NaCl_{(aq)}$. Write the net-ionic equation for this reaction.
13. Silver bromide (AgBr), the light-sensitive coating on photographic film, is made in a dark room by mixing aqueous solutions of silver nitrate and sodium bromide. Write the net-ionic equation for this reaction.
14. Barium ions, when swallowed, are highly toxic. If a person swallows some barium chloride, the antidote that is given is sodium sulfate. Explain why this is an effective antidote and write the net-ionic equation for the reaction that takes place.

CHAPTER TWELVE
SOLUTION CONCENTRATIONS

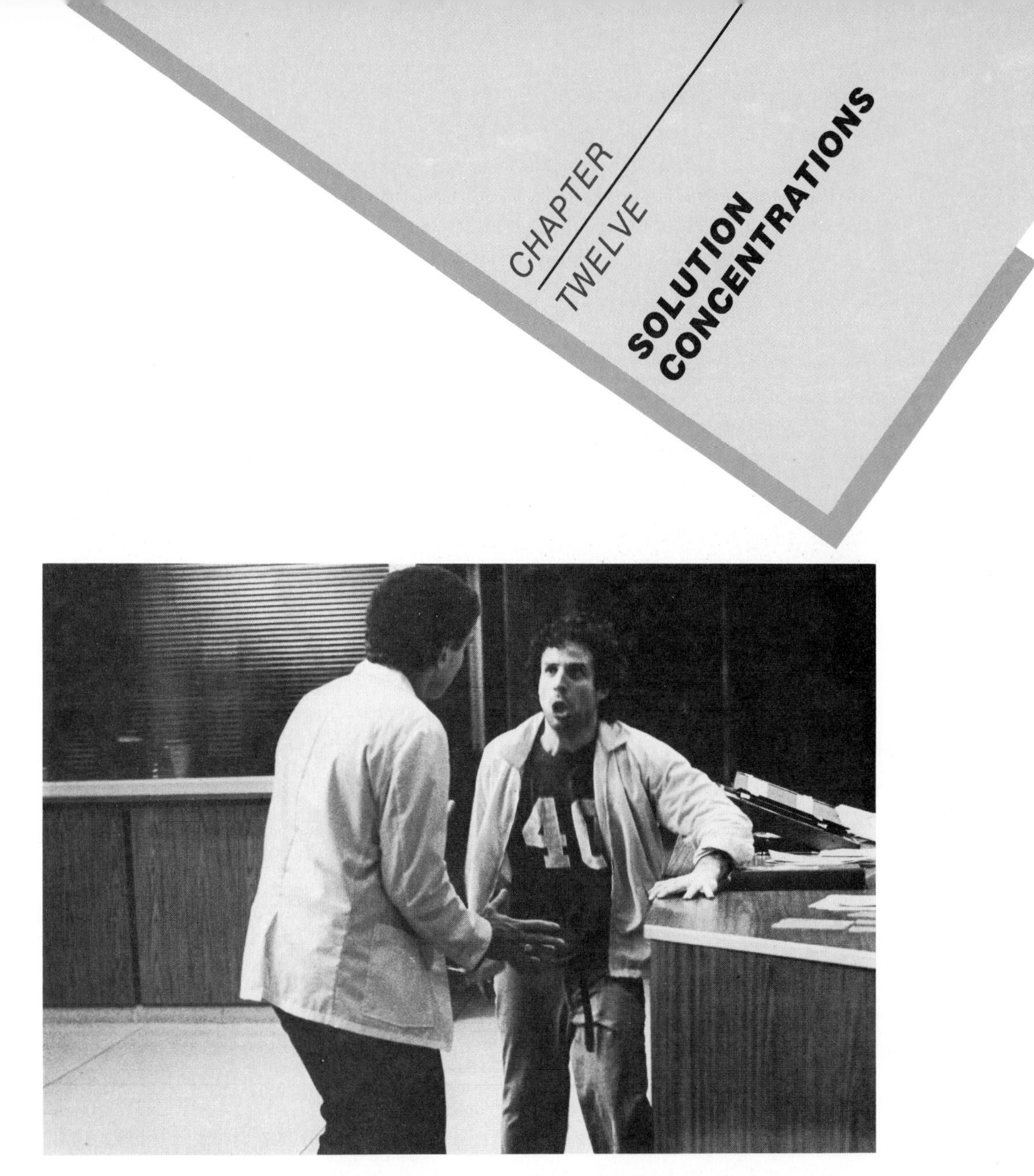

Doug Carson, a 23-year-old college student, was brought to the emergency room of the local hospital by his friends. Doug was in an extremely agitated state: he was unable to sit still, was talking in a rapid-fire, nonstop fashion—insisting to everyone that he had just been appointed secretary to the governor—and seemed confused by even simple questions. His friends told the doctor that Doug had been staying up with little or no sleep for the last three weeks, had been eating little, and had been quite abnormally cheerful and carefree. That evening he had shocked them all by throwing a deskful of books out of his dorm window "to get rid of alien influences."

A physical examination and several blood tests ruled out the possibility that Doug was suffering from a physical disease or from hallucinations that might be caused by drugs. The emergency room doctor then called in a psychiatrist, who diagnosed mania (that is, Doug was suffering from the agitated phase of manic-depressive illness). The psychiatrist recommended that Doug be admitted to the hospital and that he be treated immediately with lithium carbonate (600 mg three times a day). Lithium has been used for many years for its calming effects; it is thought to help regulate the balance between ions inside and outside of cells in the body.

Two days after beginning lithium therapy, Doug became confused and uncoordinated and began vomiting. A blood serum test for lithium was performed immediately. Lithium is effective therapeutically at blood serum levels of 0.6 to 1.5 mEq/liter. Doug's blood, however, was found to have a Li^+ concentration of 1.8 mEq/liter. High levels of lithium can cause toxic effects as severe as kidney damage, coma, and death.

Doug was immediately given extra fluids and sodium to reduce the serum lithium level in his blood, and his new symptoms soon disappeared. When Doug's original agitated symptoms again began, lithium therapy was resumed at 300 mg four times daily. Four days later, a blood test revealed Doug's serum lithium level to be at 0.75 mEq/liter, and Doug was noticeably calmer and more rational. After nine days of treatment in the hospital, Doug was discharged and was able to return to his college studies. He continued to see the psychiatrist for brief weekly visits and, happily, both he and his friends reported that he was now "back to normal." Doug continued to take lithium carbonate for six months. During this time his serum lithium concentration was carefully monitored, and the drug dosage adjusted to maintain the lithium level within the therapeutic range.

As you can see, knowing the exact concentration of a component of the blood, or a drug in a prescription, can be critical to the diagnosis and treatment of disease. In this chapter we will discuss some of the more widely used units for measuring solution concentration.

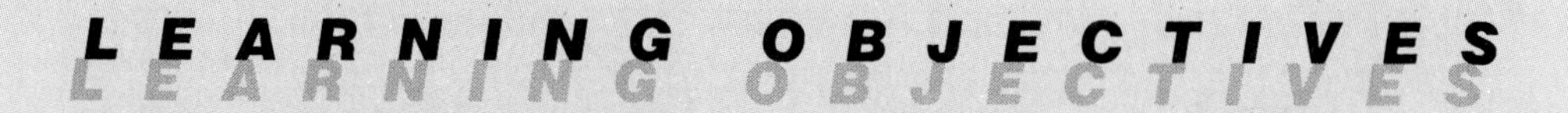

By the time you have finished this chapter, you should be able to:

1. Define *saturated, unsaturated,* and *supersaturated.*
2. Explain how to prepare a solution of a given molarity or a given percentage of solute.
3. Determine the concentration of a solution in parts per million or parts per billion.
4. Given an ion, state its gram-equivalent weight, and calculate the number of milliequivalents of that ion in a given weight/volume percent solution.
5. Describe how to prepare a specific volume of a given concentration from a stock solution.
6. Define *osmosis, osmolarity, osmol, isotonic, hypertonic,* and *hypotonic.*
7. Describe a laboratory procedure for separating cellular colloids such as proteins from the crystalloid contents of the cell.

CONCENTRATION

12.1 Saturated and Unsaturated Solutions

We may say that a solution is **dilute** if there are only a few solute particles dissolved in it, or **concentrated** if there are many solute particles dissolved. But the terms *dilute* and *concentrated* are not particularly precise, so various methods have been developed to more accurately describe the concentration of a solute in solution. A **saturated** solution is one that contains all the solute particles the solvent can normally hold at that temperature (if it contained fewer particles the solution would be **unsaturated**). A state of equilibrium will exist, in a saturated solution, between the dissolved solute and any undissolved solute in the container (Figure 12.1).

$$\text{Undissolved solute} \underset{\text{Rate}_r}{\overset{\text{Rate}_f}{\rightleftharpoons}} \text{Dissolved solute}$$

When a saturated solution is cooled, the solubility of the solute decreases. The rate of the crystallizing reaction (rate_r) will be greater than the rate of the dissolving reaction (rate_f). By Le Chatelier's Principle, the solute will continue to form crystals until a new equilibrium is established with fewer solute particles in solution. Occasionally, however, the solution will initially contain no crystals on which the dissolved particles can deposit. In such a case, when the temperature of the solution is lowered, no crystals will form and the solution is then said to be **supersaturated.** A supersaturated solution, however, is very unstable. If a small crystal is added, onto which the solute can deposit, a very rapid formation of

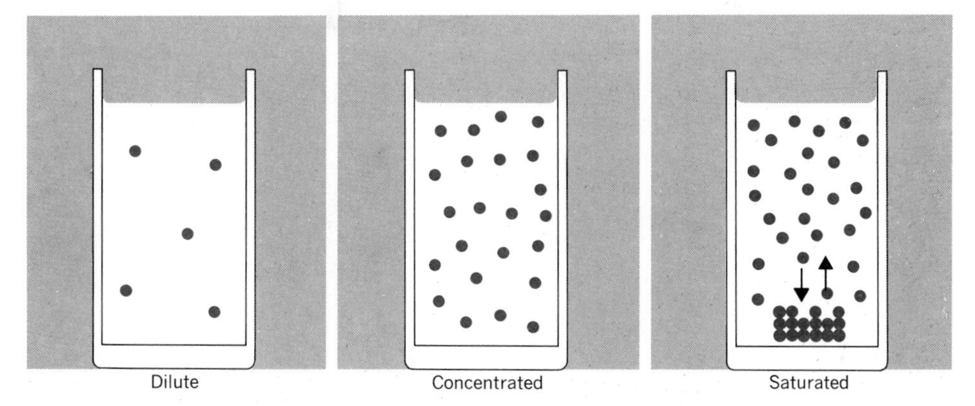

Figure 12.1 A dilute solution has very few solute particles. A concentrated solution has many solute particles. A saturated solution has all the solute particles the solvent can hold at that temperature.

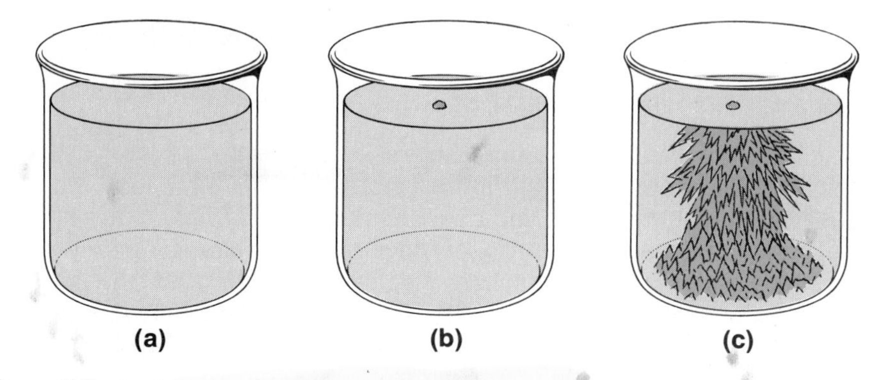

Figure 12.2 (a) A supersaturated solution is very unstable. (b) When a small crystal (called a seed crystal) is added to the solution, (c) solute crystals will rapidly form on the seed crystal. (From Brady and Humiston, *General Chemistry,* copyright © 1976, John Wiley & Sons, New York.)

crystals will often occur (Figure 12.2). Honey and jellies are often supersaturated sugar solutions that, after long storage, may be found to contain sugar crystals.

12.2 Molar Concentration

When working with solutions that are unsaturated, we often need to know exactly how much solute is present. The **concentration** of a solution is a numerical measure of the relative amount of solute in the solution; this measure, therefore, is always expressed as a ratio.

One of the most useful units of concentration is molarity. The **molarity (*M*)** of a solution is defined to be the number of moles of solute in a liter of solution.

$$\text{Molarity } (M) = \frac{\text{mol solute}}{\text{liter solution}}$$

A one molar, or 1 *M*, solution will contain one mole of solute in enough water to make one liter of solution. A bottle labeled 6.0 *M* HCl, therefore, contains 6.0 moles of HCl in each liter of solution.

EXAMPLE 12-1

1. How would you make one liter of a 0.100 *M* NaCl solution?

   (a) First, you would weigh out 0.100 mole of NaCl.

   Formula weight of NaCl = 23.0 + 35.5 = 58.5
   1 mole of NaCl = 58.5 grams

$$0.100 \text{ mole NaCl} \times \frac{58.5 \text{ grams}}{1 \text{ mole NaCl}} = 5.85 \text{ grams}$$

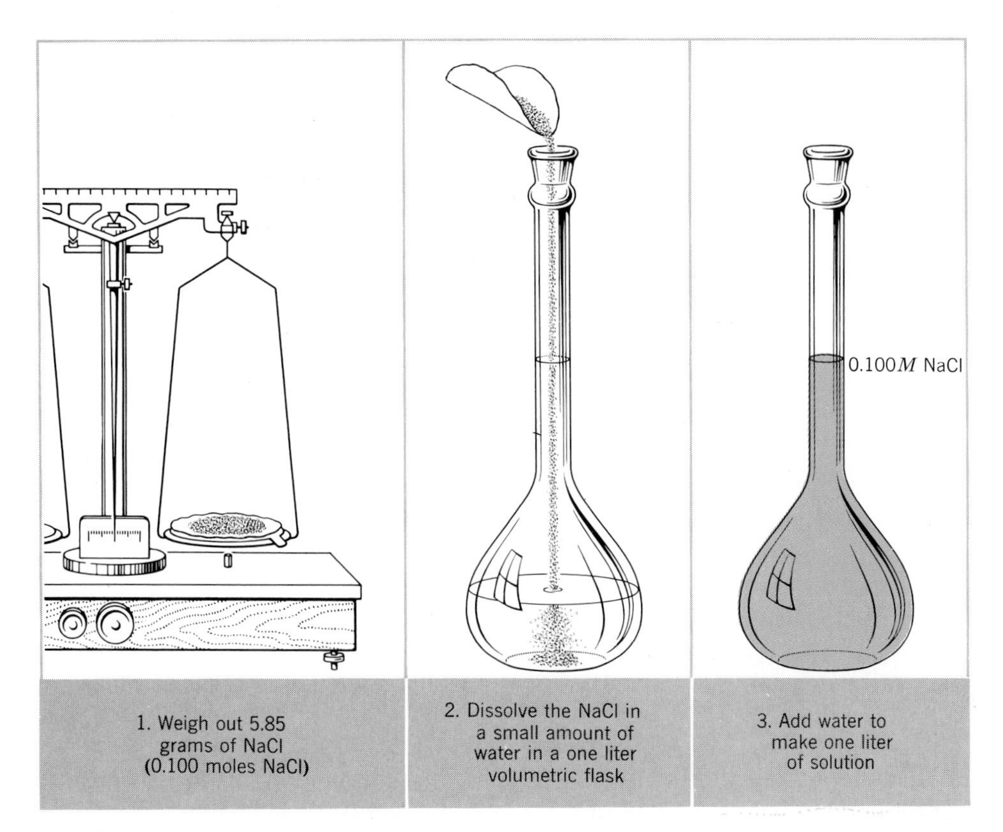

Figure 12.3 The procedure for making one liter of 0.100 M NaCl.

(b) Second, you would dissolve 5.85 grams of NaCl in a small amount of water and then add enough water to make one liter of solution. (Note that it is the total volume of the solution that is measured, not the volume of water added.) (Figure 12.3)

2. To carry out a particular reaction in the laboratory, you need 1 mole of hydrochloric acid dissolved in water. The only solution available to you is a 2 M HCl solution. How many milliliters of this stock solution should you use?

(a) 2 M HCl means $\frac{2 \text{ moles HCl}}{1 \text{ liter}}$ or $\frac{2 \text{ moles HCl}}{1000 \text{ ml}}$

(b) The problem asks how many milliliters contain 1 mole HCl

$$1 \cancel{\text{mole}} \text{ HCl} \times \frac{1000 \text{ ml}}{2 \cancel{\text{moles}} \text{ HCl}} = 500 \text{ ml}$$

Exercise 12-1

1. Describe how you would prepare each of the following solutions:

(a) 250 ml of 0.200 *M* Na_2CO_3
(b) 1.5 liters of 0.75 *M* H_3PO_4
(c) 150 ml of 0.600 *M* $KMnO_4$

2. A student requires 0.250 mole of KOH for a reaction, but the only reagent available is a stock solution marked 0.400 *M* KOH. How many milliliters of this stock solution should she use?

12.3 Percent Concentration

Another way of describing the concentration of a solution—a way that does not take into account the formula weight of the solute—is by means of percent concentration. We will discuss several different types of percent concentration: weight/weight percent, volume/volume percent, weight/volume percent, and milligram percent. The last two types of percent concentrations are used extensively in clinical reports and the biological sciences.

Weight/Weight Percent

When percent concentration is used in chemical laboratories, what is most often meant is weight/weight percent. **Weight/weight (w/w) percent** gives the number of grams of solute per 100 grams of solution.

$$\text{Weight/weight (w/w) \%} = \frac{\text{g of solute}}{\text{100 g of solution}} \times 100\%$$

A 10% (w/w) NaOH solution, therefore, will contain 10 grams of NaOH per 100 g of solution.

Volume/Volume Percent

When both the solute and the solvent are liquids, it is convenient to describe the concentration of the solution in volume/volume percent, also called percent by volume. **Volume/volume (v/v) percent** gives the number of volumes of solute per 100 volumes of solution. The specific unit of volume used can be any unit of liquid measure, but the units must be the same for the solute and solvent.

$$\text{Volume/volume (v/v) \%} = \frac{\text{volume of solute}}{\text{100 volumes of solution}} \times 100\%$$

A 10% (v/v) solution of ethyl alcohol, therefore, would contain 10 ml of ethyl alcohol in 100 ml of solution (or, if the units being used were quarts, 10 quarts of ethanol in 100 quarts of solution).

Weight/Volume Percent

Weight/volume percent is the concentration often found on clinical reports.

Weight/volume (w/v) percent gives the number of grams of solute per 100 ml of solution (Figure 12.4).

$$\text{Weight/volume (w/v) \%} = \frac{\text{g of solute}}{\text{100 ml of solution}} \times 100\%$$

A 10% (w/v) NaOH solution, therefore, will contain 10 g of NaOH in 100 ml of solution.

Milligram Percent

Milligram percent is a unit of concentration often used in clinical reports to describe extremely low solute concentrations (for example, trace minerals in the blood). **Milligram percent (mg%)** gives the number of milligrams of solute per 100 ml of solution.

$$\text{Milligram percent (mg\%)} = \frac{\text{mg solute}}{\text{100 ml solution}} \times 100\%$$

For example, blood urea nitrogen is measured in milligram percent. An infant suffering from dehydration might have a blood urea level of 32 mg%. This would mean that there are 32 mg of urea per 100 ml of blood.

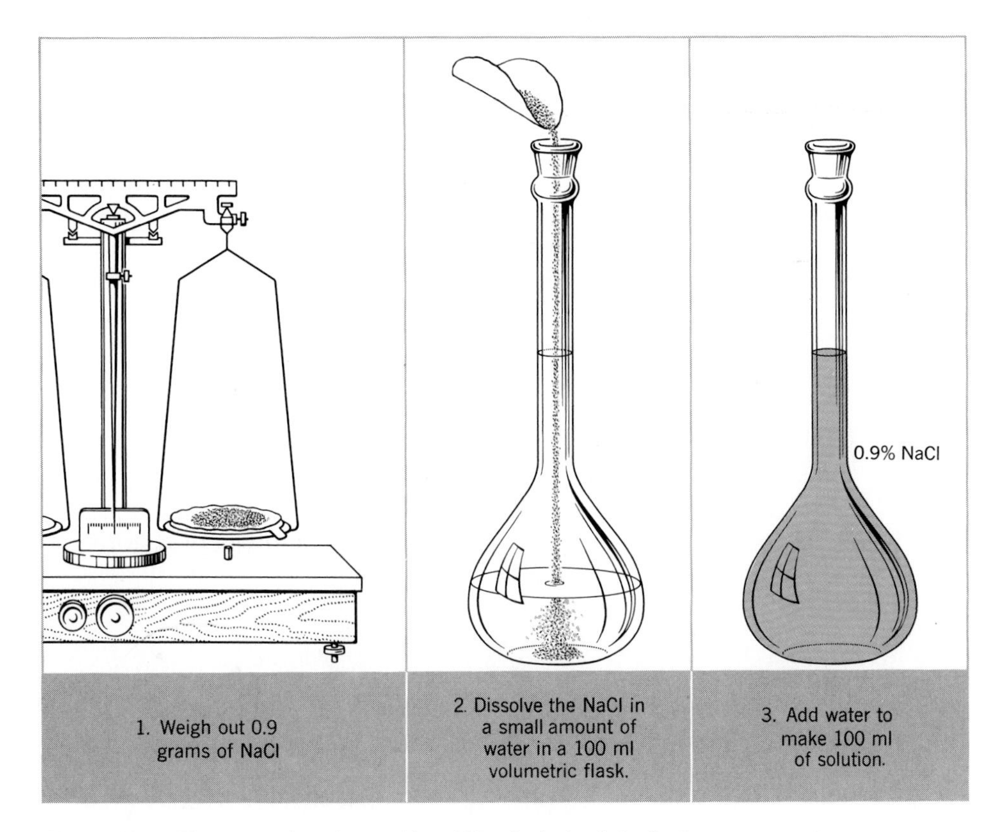

Figure 12.4 The procedure for making 100 ml of physiological saline (0.9% NaCl).

EXAMPLE 12-2

1. How would you prepare each of the following solutions?

(a) *500 g of 10% (w/w) NaOH:* To solve problems involving percent concentrations, we will start by writing the conversion factors given by the definition of that particular unit of concentration. In this case, 10% (w/w) NaOH gives us the conversion factors

$$\frac{10 \text{ g NaOH}}{100 \text{ g solution}} \quad \text{and} \quad \frac{100 \text{ g solution}}{10 \text{ g NaOH}}$$

We want to make 500 g of solution, so we will use the conversion factor that will allow us to calculate how many grams of NaOH are needed.

$$500 \text{ g } \cancel{\text{solution}} \times \frac{10 \text{ g NaOH}}{100 \text{ g } \cancel{\text{solution}}} = 50 \text{ g NaOH}$$

To prepare 500 g of 10% (w/w) NaOH, therefore, we would start with 50 g NaOH and then add enough water to make 500 g of solution.

(b) *650 ml of 25.0% (v/v) methanol:* The volume unit in this example is milliliters, so the conversion factors associated with this concentration unit are

$$\frac{25.0 \text{ ml methanol}}{100 \text{ ml solution}} \quad \text{and} \quad \frac{100 \text{ ml solution}}{25.0 \text{ ml methanol}}$$

We must calculate how many milliliters of methanol are needed to make 650 ml of solution.

$$650 \text{ ml } \cancel{\text{solution}} \times \frac{25.0 \text{ ml methanol}}{100 \text{ ml } \cancel{\text{solution}}} = 163 \text{ ml methanol}$$

To prepare this solution, therefore, we start with 163 ml of methanol and then add enough water to make 650 ml of solution.

(c) *150 ml of 0.4% (w/v) $NaHCO_3$:* The two conversion factors associated with this concentration unit are

$$\frac{0.4 \text{ g } NaHCO_3}{100 \text{ ml solution}} \quad \text{and} \quad \frac{100 \text{ ml solution}}{0.4 \text{ g } NaHCO_3}$$

We must calculate how many grams of $NaHCO_3$ are needed to prepare 150 ml of solution.

$$150 \text{ ml } \cancel{\text{solution}} \times \frac{0.4 \text{ g } NaHCO_3}{100 \text{ ml } \cancel{\text{solution}}} = 0.6 \text{ g } NaHCO_3$$

To prepare 150 ml of 0.4% (w/v) $NaHCO_3$, therefore, we would start with 0.6 g $NaHCO_3$ and then add enough water to make 150 ml of solution.

2. How much of a 15% (w/w) dextrose solution is needed to obtain 165 g of dextrose?

The two conversion factors associated with this unit of concentration are

$$\frac{15 \text{ g dextrose}}{100 \text{ g solution}} \quad \text{and} \quad \frac{100 \text{ g solution}}{15 \text{ g dextrose}}$$

We need to know how many grams of solution will contain 165 g of dextrose.

$$165 \text{ g } \cancel{\text{dextrose}} \times \frac{100 \text{ g solution}}{15 \text{ g } \cancel{\text{dextrose}}} = 1100 \text{ g solution}$$

Exercise 12-2

1. Describe how to prepare each of the following:

   (a) 350 g of 4.60% (w/w) NaCl
   (b) 2.0 liters of 30% (v/v) ethylene glycol
   (c) 55.0 ml of 0.220% (w/v) K_2CO_3
   (d) 25 ml of 80 mg% glucose

2. How many milliliters of a blood sample will contain 5.6 mg of uric acid if the concentration of uric acid in the blood is 7.0 mg%?

3. How many grams of glucose are there in 275 ml of a 12.0% (w/v) glucose solution?

12.4 Parts per Million and Parts per Billion

When a solution is extremely dilute, the concentration units of **parts per million (ppm)** or **parts per billion (ppb)** may be used. One part per million is equivalent to 1 mg of solute per 1 liter of solution. One part per billion is 1 μg of solute per 1 liter of solution.

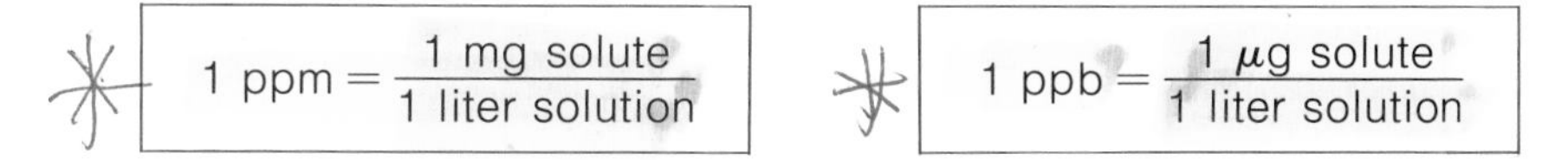

$$1 \text{ ppm} = \frac{1 \text{ mg solute}}{1 \text{ liter solution}} \qquad 1 \text{ ppb} = \frac{1\ \mu\text{g solute}}{1 \text{ liter solution}}$$

It is hard to imagine concentrations this small, but the following comparison might help: If you were to earn a yearly salary of $100,000, then 1 ppm of your salary would be 10 cents, and 1 ppb would be only one one-hundredth of a penny! But it would be wrong to think that such concentrations are too small to even care about. Some of the industrial pollutants that are being released daily into the water we drink and the air we breathe can be extremely harmful in concentrations as small as 1 ppm.

EXAMPLE 12-3

1. Infant formula is often prepared from evaporated milk. In 1972, however, it was found that canned evaporated milk contained up to 3.2 ppm of lead, which had dissolved out of the lead solder used in fabricating the can. At this concentration, how many grams of lead would be present in an 8-oz (470-ml) baby bottle of evaporated milk?

   The two conversion factors associated with this unit of concentration are

$$\frac{3.2 \text{ mg Pb}}{1000 \text{ ml milk}} \quad \text{and} \quad \frac{1000 \text{ ml milk}}{3.2 \text{ mg Pb}}$$

   Therefore, 470 ml of evaporated milk would contain

$$470 \not{\text{ml milk}} \times \frac{3.2 \text{ mg Pb}}{1000 \not{\text{ml milk}}} = 1.5 \text{ mg Pb}$$

   Converting our answer to grams, we have

$$1.5 \not{\text{mg}} \text{ Pb} \times \frac{1 \text{ g}}{1000 \not{\text{mg}}} = 0.0015 \text{ g Pb}$$

2. After four months of drinking evaporated milk containing 1 ppm lead, an infant was found to have a lead concentration in her blood of 55 μg Pb/100 ml blood. What was the lead concentration in the infant's blood in ppb and ppm?

   (a) To calculate the concentration in ppb, we need to know the number of μg of lead in one liter of blood.

$$\frac{55 \ \mu\text{g Pb}}{100 \not{\text{ml}}} \times \frac{1000 \not{\text{ml}}}{1 \text{ liter}} = \frac{550 \ \mu\text{g Pb}}{1 \text{ liter}}$$

   Therefore, the concentration of lead in the infant's blood was 550 ppb.

   (b) To calculate the concentration in ppm, we need to know the number of mg of lead in one liter of blood.

$$55 \not{\mu\text{g}} \text{ Pb} \times \frac{1 \text{ mg}}{1000 \not{\mu\text{g}}} = 0.055 \text{ mg Pb}$$

Then,

$$\frac{0.055 \text{ mg Pb}}{100 \text{ ml}} \times \frac{1000 \text{ ml}}{1 \text{ liter}} = \frac{0.55 \text{ mg Pb}}{1 \text{ liter}}$$

Therefore, the concentration of lead in the infant's blood was 0.55 ppm.

Exercise 12-3

The purity of drinking water is of increasing concern around the world as underground water supplies become contaminated with environmental pollutants. A 250-ml water sample from a private well was tested for the four contaminants listed in the following table. Were any of the contaminants in the well water above the safe limits set by the Environmental Protection Agency (EPA) in 1980?

Contaminant	250-ml sample	EPA Drinking Water Limits
Carbon tetrachloride (CCl_4)	0.40 μg	6.4 ppb
Chlordane	1.3 μg	3.0 ppb
Lead	11 μg	0.050 ppm
Mercury	0.21 μg	2.0 ppb

12.5 Equivalents

Doctors often refer to the important ionic components of the blood in terms of their ionic charge. The unit that is used to describe the concentration of ionic components is the **equivalent (Eq).**

$$1 \text{ Eq} = 1 \text{ mole of charge (either + or −)}$$

For example, one mole of the sodium ion Na^+ contains one mole of positive charge and is equal to one equivalent. For Mg^{2+}, each mole contains two moles of positive charge. Therefore, $\frac{1}{2}$ mole of Mg^{2+} is 1 Eq. For HCO_3^-, one mole contains one mole of negative charge and equals 1 Eq.

The **gram-equivalent weight** of a substance is the weight of the substance, in grams, that will contain one equivalent.

$$\text{Gram-equivalent weight} = \frac{\text{g of substance}}{1 \text{ Eq}}$$

TRIG	8		mg/dl			■
NA +	9	135 - 148	mEq/L	122	L	■
K+	10	3.5 - 5.3	mEq/L	3.2	L	■
CL −	11	96 - 109	mEq/L	83	L	■
BICARB	12		mEq/L			■
CA ++	13	8.1 - 10.7	mg/dl	9.1		■
PHOS	14	2.6 - 4.8	mg/dl	2.9		■

Figure 12.5 A portion of a blood chemistry laboratory report, with normal serum levels on the left and values for the patient on the right. The three L's on the far right signal the doctor that the patient's serum sodium, potassium, and chloride are below normal. (Courtesy Good Samaritan Hospital, Corvallis, Oregon.)

For Na^+, we know that 1 Eq = 1 mole. Therefore, the gram-equivalent weight of Na^+ is 23 g. For Mg^{2+}, 1 Eq = $\frac{1}{2}$ mole. So the gram-equivalent weight is $\frac{24}{2}$ = 12 g. For HCO_3^-, 1 Eq = 1 mole, so the gram-equivalent weight is 61 g.

Because the concentration of ions in the blood is very dilute, medical reports will often state ion concentrations in **milliequivalents.**

$$1000 \text{ milliequivalents (mEq)} = 1 \text{ Eq}$$

For example, it is important to regularly check the concentration of sodium ions (Na^+) and potassium ions (K^+) in the blood serum of a person with congestive heart failure who is also taking diuretics (Figure 12.5). A diuretic is a substance that increases the amount of water released into the urine by the kidneys. However, potassium and sodium ions will also pass into the urine with the water. In such a case, these ions will have to be added back into the person's diet if their concentrations in the blood fall below the normal range of 135 to 148 mEq/liter for Na^+ and 3.5 to 5.3 mEq/liter for K^+.

EXAMPLE 12-4

The calcium ion Ca^{2+} serves many extremely important functions in the body.

(a) What is the gram-equivalent weight of the calcium ion?

The calcium ion is Ca^{2+}.
1 mole of Ca^{2+} contains 2 moles of + (positive charge).
Therefore, $\frac{1}{2}$ mole of Ca^{2+} contains 1 mole of +, or 1 Eq.

1 mole of Ca^{2+} = 40 g.
$\frac{1}{2}$ mole of $Ca^{2+} = \frac{40}{2} = 20$ g.
Therefore, the gram-equivalent weight of Ca^{2+} = 20 g.

(b) How many milliequivalents of calcium ions are present in 100 ml of a 0.1% (w/v) Ca^{2+} solution?

From Section 12.3 we know that

$$0.1\%\ (w/v)\ Ca^{2+} = \frac{0.1\ g\ Ca^{2+}}{100\ ml}$$

Therefore, 100 ml of 0.1% (w/v) Ca^{2+} contains 0.1 g Ca^{2+}.
We now need to know

$$0.1\ g\ Ca^{2+} = ?\ \text{milliequivalents}$$

If we use the result of (a), 1 Eq of Ca^{2+} = 20 g

$$0.1\ g\ Ca^{2+} \times \frac{1\ Eq\ of\ Ca^{2+}}{20\ g} = 0.005\ Eq\ of\ Ca^{2+}$$

We know that 1000 mEq = 1 Eq, so

$$0.005\ Eq\ of\ Ca^{2+} \times \frac{1000\ mEq}{1\ Eq} = 5\ mEq\ of\ Ca^{2+}$$

Exercise 12-4

1. What is the gram-equivalent weight of the following:

   (a) K^+ (c) HPO_4^{2-}
   (b) Cl^- (d) SO_4^{2-}

2. A normal value for the concentration of HPO_4^{2-} in a liter of body fluid is 140 mEq. How many grams of HPO_4^{2-} are there in one liter of body fluid?

3. A patient's chart records his serum chloride ion concentration as 94 mEq/liter. What is his chloride ion concentration in mg%?

12.6 Dilutions

A very common task in a chemical or medical laboratory is preparing a needed solution using a concentrated stock solution. You can use the following relationship to determine the amount of stock solution required for a desired solution, but be sure that you use the same units of concentration and units of volume on both sides of the equation.

(concentration desired) × (volume desired) = (concentration of stock) × (volume of stock) (1)

You might also be asked to make a dilution of a concentrated solution by adding solvent to increase the volume of the solution. If the volume of the solution is doubled, you have made a 1:2 dilution; if the volume is increased by five times, you have made a 1:5 dilution. For example, instructions on frozen lemonade concentrate tell you to add four cans of water to one can of frozen lemonade concentrate. This increases the total volume from one can to five cans, making a 1:5 dilution of the lemonade concentrate. We can rewrite eq. (1) to determine the final concentration of our diluted solution.

(Final concentration) × (Final volume) = (Initial concentration) × (Initial volume) (2)

EXAMPLE 12-5

1. How would you prepare 100 ml of 1.5 *M* HCl from a 6.0 *M* HCl stock solution?

   Substituting the known values in eq. (1), we have
   $(1.5\ M\ \text{HCl}) \times (100\ \text{ml}) = (6.0\ M\ \text{HCl}) \times V$

   $$\frac{1.5\ \cancel{M\ \text{HCl}} \times 100\ \text{ml}}{6.0\ \cancel{M\ \text{HCl}}} = V$$

   $$25\ \text{ml} = V$$

   Therefore, to make 100 ml of 1.5 *M* HCl you would take 25 ml of the 6.0 *M* HCl stock solution and add enough water to make 100 ml of solution (Figure 12.6).

2. What is the final concentration of a 1:5 dilution of 100 ml of 10% NaCl?

   A 1:5 dilution means that the volume has been increased five times by adding more solvent.

   100 ml (Initial volume) × 5 = 500 ml (Final volume)

   Using eq. (2), we have

   (Final concentration) × 500 ml = 10% NaCl × 100 ml

   $$(\text{Final concentration}) = \frac{10\%\ \text{NaCl} \times 100\ \cancel{\text{ml}}}{500\ \cancel{\text{ml}}}$$

   (Final concentration) = 2% NaCl

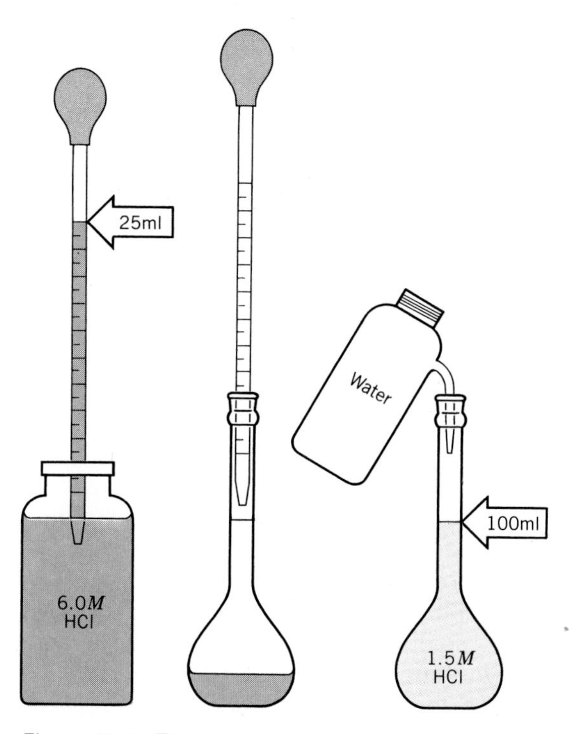

Figure 12.6 The procedure for making 100 ml of 1.5 *M* HCl from a stock solution of 6.0 *M* HCl.

Exercise 12-5

1. How would you prepare 1.65 liters of 2.75 *M* NaCl
   (a) Using solid NaCl.
   (b) Using a 4.12 *M* NaCl stock solution.
2. Describe how you would make a 1:6 dilution of the solution prepared in question 1. What would be the final concentration of this solution?

PROPERTIES OF SOLUTIONS

12.7 Colligative Properties

Solutions have certain properties, called **colligative properties,** that depend only on the number of solute particles that are dissolved, regardless of the chemical nature of the solute. Such properties include the lowering of the freezing point and the raising of the boiling point of water by solute particles. The more solute particles in the solution, the lower the freezing point and the higher the boiling point. For example, automotive antifreeze is added to the water in a car's radiator to lower the freezing point of the water, preventing it from freezing in the winter and cracking the radiator. The antifreeze also raises the boiling point of the water, preventing the radiator from boiling over in the summer. Nature has given some insects their own antifreeze. They have a high concentration of the compound glycerol in their body fluids, which helps them survive the harsh winter temperatures by lowering the freezing point of the water in their bodies.

12.8 Osmosis

Another colligative property is osmotic pressure. **Osmosis** is the flow of water molecules through a differentially permeable membrane (a barrier that will allow the passage of water, but not solute particles) from a region of lower solute concentration to a region of higher solute concentration. For example, if pure water and a sugar solution are separated by a differentially permeable membrane, water molecules will flow through the membrane into the sugar solution at a higher rate then they return. Therefore, the volume of the sugar solution will increase (Figure 12.7).

Osmotic pressure is defined to be the amount of pressure that must be applied to prevent the flow of water, or osmosis, through the membrane from a solution of lower solute concentration to a solution of higher solute

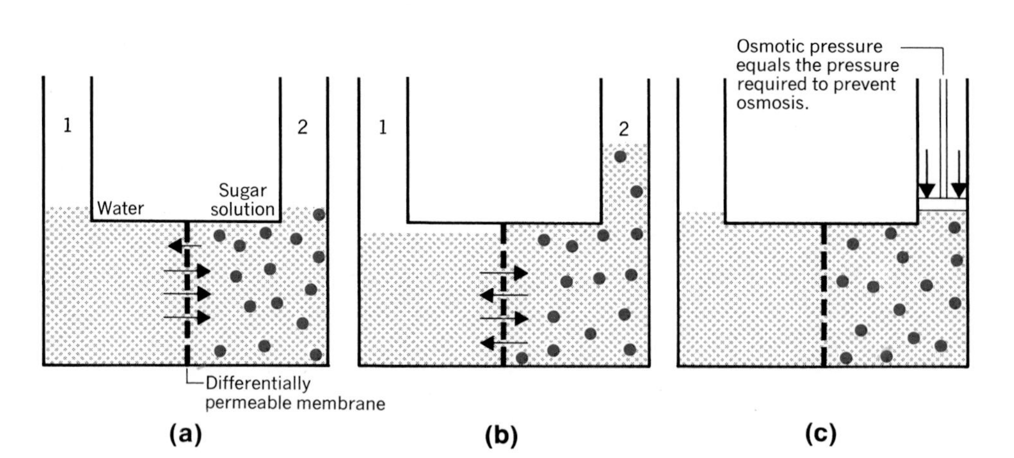

Figure 12.7 Osmosis. (a) A differentially permeable membrane will permit water molecules to pass through, but not sugar molecules. The water will move from the region of lower concentration to the region of higher concentration. (b) When equilibrium is established, the side with the sugar solution ends up with more material. (c) The osmotic pressure equals the pressure that must be applied to the sugar solution to prevent osmosis, the movement of water into the solution.

concentration. Osmotic pressure depends only on the number of solute particles in the solution—the greater the solute concentration difference between the solutions, the greater the pressure that must be applied to prevent osmosis (Figure 12.7).

A good example of the way in which nature uses osmotic pressure is the method by which trees get water to their leaves. A tree loses water (through transpiration) from its leaves to the atmosphere; as a result, the solute concentration within the leaves increases. This generates osmotic pressure within the tree that forces ground water into the roots, up the trunk, and through the branches to the leaves. A coast redwood tree can grow to be 364 feet tall. In order to force water up a pipe as tall as a coast redwood, you would need a pump capable of generating 12 atmospheres of pressure!

12.9 Osmolarity

The total concentration of particles in the blood or urine is often recorded in terms of osmolarity. Because every particle in solution (whether an ion or molecule) contributes to the osmotic pressure of that solution, **osmolarity** describes the total number of moles of all particles in a liter of solution. One **osmol** is defined to be one mole of any combination of particles. For example, one mole of KCl will dissolve in a liter of solution and will produce one mole of K^+ and one mole of Cl^-, or two moles of particles—that is, 2 osmols. Therefore, a liter of 1 *M* KCl will have an osmolarity of 2 osmols. Solutions containing different solute particles, but having the same osmolarity, will have the same osmotic pressure. A 1 *M* KCl solution will have the same osmolarity and, therefore, the same osmotic pressure as a 1 *M* NaBr solution. A comparison of the osmolarity of the blood and urine is often a good indication of kidney function. Normal serum osmolarity is 280–290 osmols/liter, and normal urine osmolarity is 300–1000 osmols/liter.

12.10 Hemolysis

The passage of water in and out of living cells is an important biological process. If the concentration of solute is equal on both sides of the cell membrane, the osmotic pressures will be equal and the solutions are then called **isotonic.** The usual concentration of the salts in blood plasma is isotonic to a 0.9% NaCl solution. This concentration of NaCl is called **normal saline** or **physiological saline.**

A normal saline solution is isotonic to red blood cells. When red blood cells are placed in the saline, no change will be observed. If red blood cells are placed in a solution of lower solute concentration, called a **hypotonic solution** (for example, distilled water), water will move into the cells. The cells will then swell and rupture in a process called **hemolysis.** If red blood cells are placed in a solution of higher solute concentration, called a **hypertonic solution** (for example, blood plus 3% NaCl solution), the water will move from the cells to the solution. The cells will then shrink in a process called **crenation** (Figure 12.8). As you can see, solutions cannot be safely introduced into the bloodstream unless they are isotonic. All intravenous preparations are made with isotonic solutions.

Some of the solutes in the bloodstream are not capable of passing through the walls of the capillaries; this gives the blood a higher solute concentration than the surrounding extracellular fluid. As a result, water moves from the extracellular fluid into the capillaries, keeping the veins full and preventing the collapse of the blood vessels. In some diseases such as malnutrition or kidney failure, the concentration of particles in the blood goes down, decreasing the osmotic pressure and decreasing the flow of water from the tissues into the veins. This results in a condition

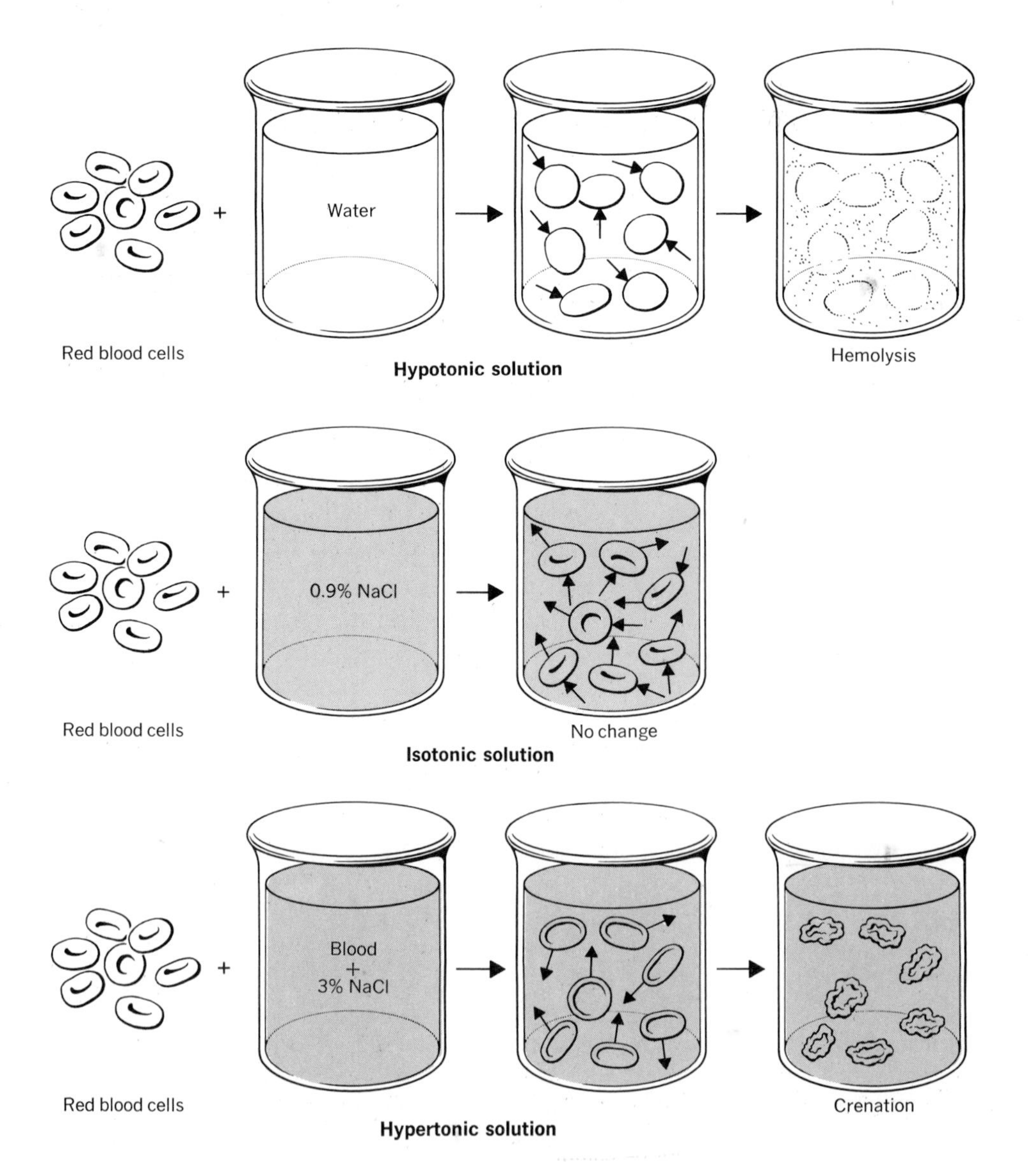

Figure 12.8 Blood cells undergo various changes when placed in solutions of different concentrations.

called **edema,** which is a swelling of the tissues such as those of the hands or legs.

Sometimes the introduction of a hypertonic or hypotonic solution into the body can have therapeutic effects. Saline laxatives such as epsom salts act by forming hypertonic solutions in the intestines. The resulting osmotic pressure draws water through the intestinal wall, moistening the contents of the intestines and making evacuation easier.

12.11 Dialysis

Cellular membranes are not osmotic membranes. In order for a cell to live, the membrane must allow the passage of not only water molecules but also ions, nutrients, and waste products. Membranes that allow crystalloids to pass through them, but not large molecules or colloids, are called **dialyzing membranes. Dialysis** is the movement of ions and small molecules through dialyzing membranes. Most animal membranes are dialyzing membranes.

Dialysis can be a useful laboratory tool for obtaining purified colloids from solutions of cell contents that may contain both crystalloids and colloids (Figure 12.9). For example, biochemists often use dialysis to separate protein molecules from aqueous ions. By controlling the nature of the dialyzing membrane, they can also separate large protein molecules from smaller ones.

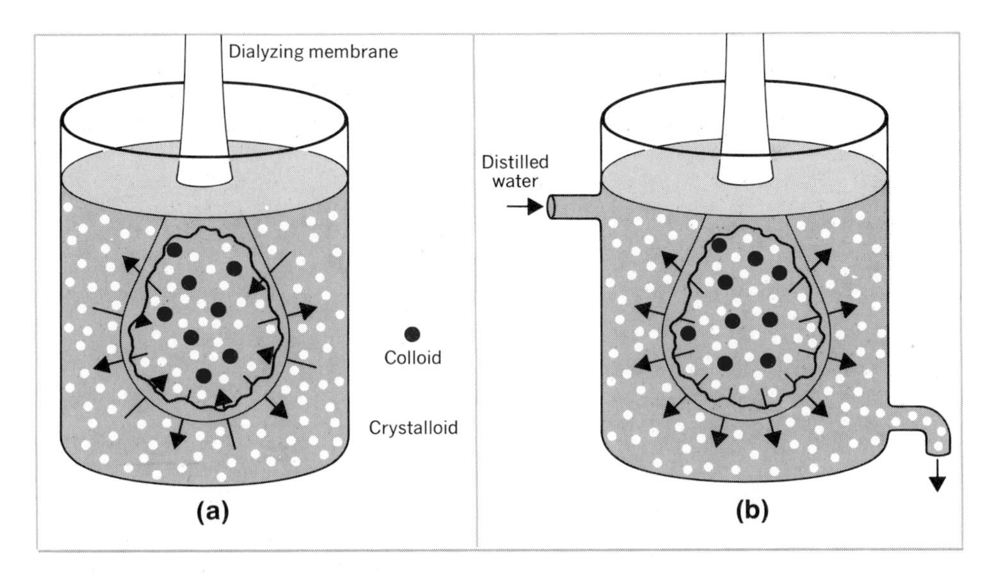

Figure 12.9 A dialysis system. (a) Crystalloid particles readily pass through the membrane, while colloidal particles cannot. An equilibrium will eventually be established between the crystalloid particles inside and outside the bag. (b) Complete separation of the crystalloid particles from the colloidal particles can be obtained if a slow flow of distilled water is maintained through the system. Because an equilibrium cannot be established, the crystalloid particles will continue to diffuse from inside the bag, leaving behind only colloidal particles.

The function of the kidneys is to remove waste products from the blood. If the kidneys are damaged and fail to function, waste products of crystalloid size will continue to build up in the blood, and the patient will die. However, a device called the artificial kidney is now saving the lives of patients with kidney failure. This machine passes blood from an artery in the arm or leg of such patients through long, coiled cellophane tubes that act as dialyzing membranes. Surrounding these tubes is a solution, called the dialysate, that is isotonic for all the components that are to remain in the blood. Therefore, these solutes will pass in and out of the blood at an equal rate. But the dialysate contains no waste products, so these wastes will pass out of the blood at a rate greater than they return. In this way the patient's blood is cleansed. Each week, dialysis patients require two to three treatments that can last 5 to 7 hours in order to remove a sufficient amount of waste products from their blood.

CHAPTER SUMMARY

The concentration of a solution may be expressed in many ways: dilute, concentrated, unsaturated, saturated, or supersaturated—all of which are fairly general terms. To be more precise, solution concentration can be stated in molarity, weight/weight percent, volume/volume percent, weight/volume percent, milligram percent, parts per million or parts per billion of a solute, or equivalents or milliequivalents of a particular ion.

Adding solute particles to a liquid solvent will raise the boiling point and lower the freezing point of the liquid. Osmosis is the flow of water through a membrane from a solution of lower solute concentration to a solution of higher solute concentration. Osmotic pressure is the pressure that must be applied to prevent such a flow. Osmolarity describes the total concentration of all the particles in a solution. Isotonic solutions are solutions with the same osmolarity. A hypotonic solution has fewer solute particles and a hypertonic solution more solute particles than a given solution. Dialysis is the movement of crystalloids, but not colloids, through a membrane from a region of higher to lower crystalloid concentration. Cellular membranes are dialyzing membranes.

Important Equations

$$\text{Molarity } (M) = \frac{\text{mol solute}}{\text{liter solution}}$$

$$\text{ppm} = \frac{\text{mg solute}}{\text{liter solution}}$$

$$\text{Weight/weight \%} = \frac{\text{g of solute}}{\text{100 g solution}} \times 100\%$$

$$\text{ppb} = \frac{\mu\text{g solute}}{\text{liter solution}}$$

$$\text{Volume/volume \%} = \frac{\text{volume solute}}{\text{100 volumes solution}} \times 100\%$$

$$\text{Weight/volume \%} = \frac{\text{g of solute}}{\text{100 ml solution}} \times 100\%$$

$$\text{Osmolarity} = \frac{\text{osmols}}{\text{liter}}$$

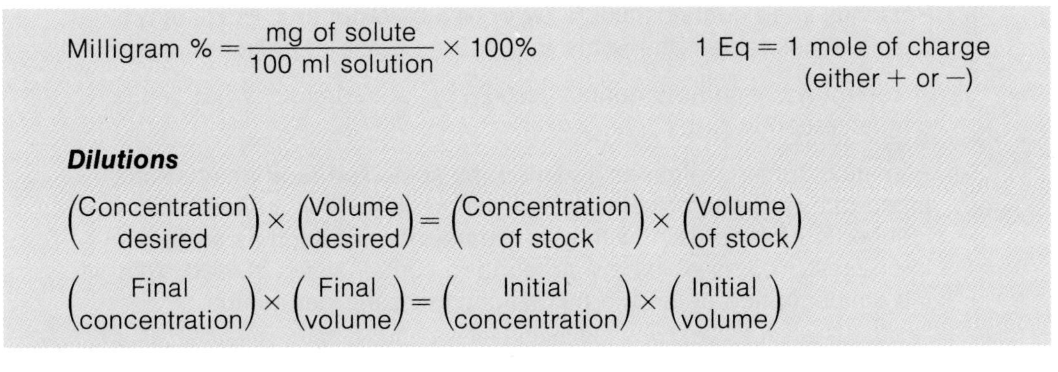

$$\text{Milligram \%} = \frac{\text{mg of solute}}{\text{100 ml solution}} \times 100\%$$

1 Eq = 1 mole of charge (either + or −)

Dilutions

$$\begin{pmatrix}\text{Concentration}\\\text{desired}\end{pmatrix} \times \begin{pmatrix}\text{Volume}\\\text{desired}\end{pmatrix} = \begin{pmatrix}\text{Concentration}\\\text{of stock}\end{pmatrix} \times \begin{pmatrix}\text{Volume}\\\text{of stock}\end{pmatrix}$$

$$\begin{pmatrix}\text{Final}\\\text{concentration}\end{pmatrix} \times \begin{pmatrix}\text{Final}\\\text{volume}\end{pmatrix} = \begin{pmatrix}\text{Initial}\\\text{concentration}\end{pmatrix} \times \begin{pmatrix}\text{Initial}\\\text{volume}\end{pmatrix}$$

EXERCISES AND PROBLEMS

1. State the difference between a dilute, concentrated, saturated, and supersaturated solution.

2. Describe how you would prepare each of the following aqueous solutions:

   (a) 50 ml of 0.20 *M* NaOH
   (b) 250 ml of 0.11 *M* Na_2SO_4
   (c) 0.50 liter of 5.0% (w/v) KOH
   (d) 475 ml of 20% (v/v) ethyl alcohol
   (e) 0.10 liter of 0.50 *M* NH_4Cl
   (f) 125 g of 8.25% (w/w) glucose
   (g) 3.5 liters of 4.0% (v/v) methyl alcohol
   (h) 125 ml of 22.0 mg% lactic acid
   (i) 55 ml of 6.0 *M* HCl
   (j) 10 ml of 0.20 *M* K_2HPO_4
   (k) 0.25 liter of 0.010 *M* $(NH_4)_2SO_4$
   (l) 350 ml of 0.832% (w/v) KCl

3. What is the concentration of each of the following in (1) molarity, (2) mg%, and (3) ppm?

   (a) 11 μg $HgCl_2$ in 10 ml of solution
   (b) 5.0 mg NaF in 250 ml of solution
   (c) 0.039 g KCN in 100 ml of solution
   (d) 0.99 mg SeO_2 in 750 ml of solution

4. If you dissolve 4.9 g of H_2SO_4 in 250 ml of solution, what is the molarity of the solution?

5. What is the molarity of NaCl in normal saline (0.900% (w/v) NaCl)?

6. Dextrose is often added to normal saline for intravenous feeding. How many grams of dextrose and NaCl would you need to prepare 500 ml of 5.0% (w/v) dextrose in normal saline?

7. How many grams of 0.10% (w/w) sodium carbonate would contain 2.5 g of Na_2CO_3?

8. Phisohex soap contains 3.00% (w/v) hexachlorophene. How many grams of hexachlorophene are in a 148-ml bottle of Phisohex soap?

9. A 150-ml water sample contained 0.26 μg of cadmium. What is this concentration in ppb?

10. Recently, drinking glasses given out by some fast food chains were found to have lead in their painted decorations. This lead could be leached out of the paint with mildly acidic drinks such as fruit juice. In one test, 100 ml of juice was found to contain 4 mg of lead. What is the concentration of lead in this solution in parts per million?

11. Nitrosamines are potent carcinogens (chemicals that cause cancer). In 1979, the average level of dimethylnitrosamine in beer was found to be 5.9 ppb.

    (a) How many micrograms of dimethylnitrosamine were in a 12-oz can of beer in 1979?
    (b) By 1981, the average concentration of dimethylnitrosamine in beer had been reduced to 0.2 ppb. By how many micrograms was the nitrosamine content of a 12-oz can of beer reduced?

12. Give the gram-equivalent weight for the following ions.

    (a) PO_4^{3-}
    (b) Br^-
    (c) Ba^{2+}
    (d) CO_3^{2-}

13. The normal range for the concentration of potassium ion in the blood of an adult is 3.5 to 5.3 mEq/liter. Does a person taking diuretics need a potassium supplement if a 50-ml blood sample contains 3.9 mg K^+?

14. The blood of a patient being treated with lithium carbonate was sampled and found to contain 1.4 mg of Li^+ in 100 ml. The blood concentration of Li^+ should not exceed 1.5 mEq/liter. Should the patient stop taking lithium carbonate?

15. If the concentration of Mg^{2+} in interstitial fluid (the fluid found between body cells) is 2.0 mEq/liter, how many milligrams of Mg^{2+} are in 10 ml of interstitial fluid?

16. What is the final concentration of a 1:5 dilution of

    (a) 50 ml of 6.0 *M* H_2SO_4
    (b) 150 ml of 2 ppm Cd^{2+}
    (c) 30 ml of 3% KCl
    (d) 10 ml of 32 mg% urea

17. How would you prepare each of the following aqueous solutions from the given stock solution?

    (a) 250 ml of 8.00% (w/v) NaCl from 10.0% (w/v) NaCl.
    (b) 150 ml of 0.030 *M* NaOH from 0.10 *M* NaOH.
    (c) 1:10 dilution of 75 ml of 2.5 *M* H_2SO_4.
    (d) 4.5 liters of 30% (v/v) isopropyl alcohol from 70% (v/v) isopropyl alcohol.
    (e) 55 ml of 6.4 mg% uric acid from 7.8 mg% uric acid.
    (f) 650 ml of 4.0 ppm Pb^{2+} from 10 ppm Pb^{2+}.

18. You have available three stock solutions to use in performing an experiment: 6.00 *M* HCl, 0.100 *M* NaOH, and 5.00% (w/v) Na_2CO_3. State the amount of each stock solution required to supply the following amounts of reactants needed for the experiment.

    (a) 0.300 mol HCl
    (b) 0.150 mol NaOH
    (c) 2.50 g Na_2CO_3
    (d) 40.0 mg NaOH
    (e) 21.9 g HCl
    (f) 0.250 mol Na_2CO_3

19. Dextran is a substance that is administered in saline solution to help maintain the blood pressure of persons in shock. The recommended dosage for the first 24 hours is 2.00 g of dextran per kilogram of body weight. How many milliliters of 10.0% (w/v) dextran in normal saline should be administered to a 165-lb man during the first 24 hours?

20. Why is covering a sidewalk with salt an effective way of preventing ice from forming on the sidewalk in the winter?

21. Quite often celery stored for long periods in a refrigerator goes limp. It can be made crisp again by putting it in a container of water. Explain the principle behind the celery's return to its original crispness.

22. Referring to Figure 12.7(a), state which side, 1 or 2, would rise for each of the following:
    (a) 0.1 *M* HCl in 1 and 0.01 *M* HCl in 2
    (b) 1% dextrose in 1 and 5% dextrose in 2
    (c) 1 *M* NaCl in 1 and 1% (w/v) NaCl in 2

23. Potassium sulfate (K_2SO_4) is a strong electrolyte, and alcohol is a nonelectrolyte. Explain why the osmotic pressure of a 1 *M* solution of K_2SO_4 is about three times that of a 1 *M* alcohol solution.

24. Which of the following will have the same osmotic pressure (osmolarity) as 0.1 *M* NaCl?

    (a) 0.1 *M* $CaCl_2$
    (b) 0.05 *M* Na_2CO_3
    (c) 0.1 *M* NaOH

25. What would happen to red blood cells if they were placed in the following solutions?

    (a) 1.5% (w/v) NaCl solution
    (b) 0.154 *M* NaCl solution
    (c) 0.15% (w/v) NaCl solution

26. Identify each of the solutions in question 23 as isotonic, hypertonic, or hypotonic to red blood cells.

27. The packet of oral rehydration salts (ORS) given to the Ethiopian refugees discussed at the beginning of Chapter 11 contains the following:

Glucose	20.0 g	Sodium bicarbonate	2.5 g
Sodium chloride	3.5 g	Potassium chloride	1.5 g

The instructions on the packet say to dissolve the contents in enough water to make 1 liter of solution.

(a) What is the concentration in (w/v)% of each of the components of ORS?
(b) What is the concentration in molarity of each of the components of ORS?
(c) What is the osmolarity of the ORS solution?

28. When a house plant is grown in a pot without holes, salt dissolved in the tap water will build up in the soil and will appear as white crystals on the soil surface. Explain why, as time goes on, the plant's growth slows and the plant eventually dies. Why is it that some plants are able to live very successfully in saltwater marshes?

29. Draw a schematic diagram of an artificial kidney, label its parts, and describe what takes place at the molecular level along the cellophane tube.

CHAPTER THIRTEEN

ACIDS AND BASES

Robert had been coming with his father to Darts Lake in the Adirondacks since he was a small boy. Now he was bringing his grandson Jon to fish with him, but somehow things were not the same. The lake was still beautiful and the cabins still rustic, but the fishing had become terrible. He and Jon had been out every day for a week and had caught only two fish—a far cry from the strings of fish he remembered catching as a boy.

Robert listened as the local cafe owner criticized the government for not cracking down on industries that were polluting the air. He was claiming that this caused acid rain, which was killing all the lakes in the region. Wanting to learn more about this theory, Robert decided to attend the program on acid rain being held that night at the main lodge.

Two university biologists had come up that night to debate the causes of acidification of the Adirondack lakes. The first speaker explained the acid-rain theory that had been mentioned by the cafe owner. This involves two air pollutants: nitrogen dioxide (NO_2), which is produced when nitrogen monoxide (NO) from auto and furnace exhausts reacts with oxygen in the air, and sulfur dioxide (SO_2), which is produced from burning sulfur-containing coal and oil. NO_2 and SO_2 react with water to form acids. So, when they are washed out of the atmosphere by rain, these compounds cause the rain-water to become acidic. This water collects in the streams and lakes, causing them slowly to become more acidic. Most living plants and animals can tolerate only a very small range of acidity. The continued acidification of the lakes has been killing the fish and other aquatic life (Figure 13.1).

The second biologist claimed that the acid-rain theory didn't take account of all the evidence. He described data showing that the acidification of the lakes and streams began to occur at the beginning of the 1900s, long before the air pollution from automobiles, power plants, and other industries became a problem in the eastern United States. In fact, scientists have found that lakes in the western United States—where the rain is not particularly acidic—are also showing a slow acidification. But what, then, are other sources of acids that could be affecting our lakes? It turns out that many natural biological processes, such as plant decay and the natural oxidation of organic nitrogen in the soil, produce acids. Such acids will be leached out of the soil as the rainwater passes through on the way to streams and lakes. The biologist thought that these naturally produced acids are probably a significant contributor to the acidification of lakes and streams.

To better understand the arguments of the two biologists, we need to know exactly what acids are and why changes in their concentrations in the environment can have such significant effects on living systems. In this chapter we will discuss the nature of acids and bases and the ways in which their concentrations are regulated by living organisms.

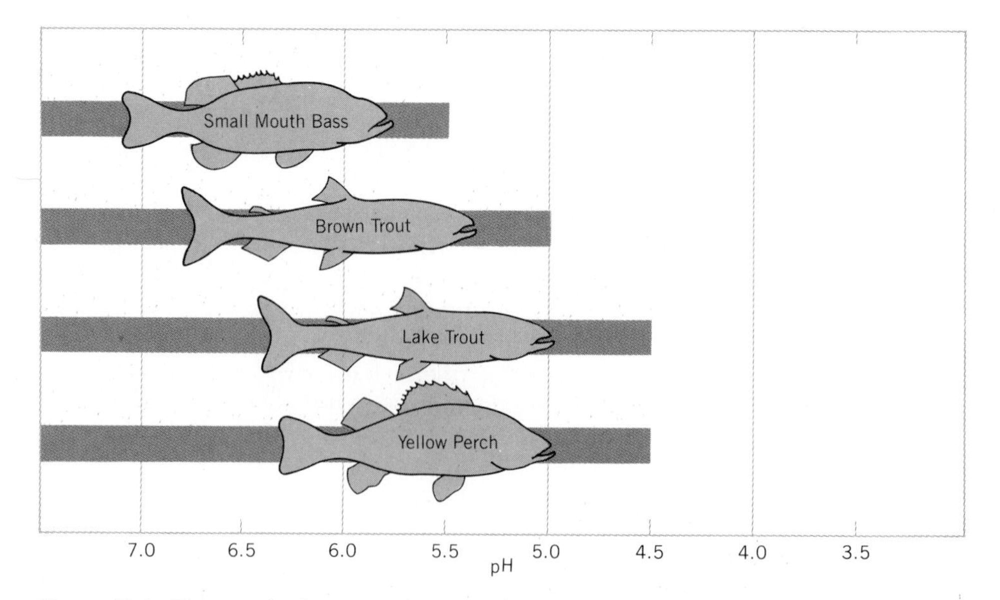

Figure 13.1 Fish vary in their sensitivity to the acidity of the water in which they live. Acid water washing into lakes and ponds decreases the pH of the water—a suspected factor in the death of all fish in 6% of the lakes and ponds in New York's Adirondack Mountains.

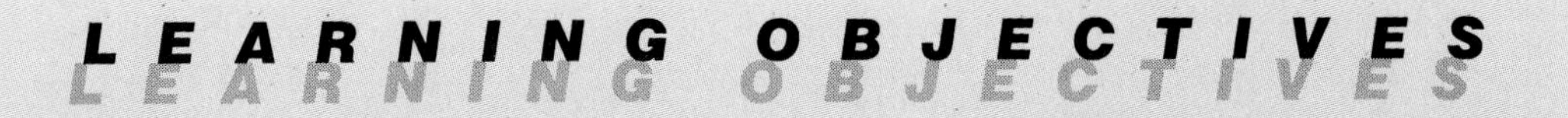

By the time you have finished this chapter, you should be able to:

1. State the Brønsted–Lowry definition of an acid and base and identify the conjugate acid–base pairs in a chemical reaction.
2. State in your own words the difference between a strong and weak acid, and a strong and weak base.
3. Write a balanced equation for a neutralization reaction between an acid and base.
4. Name an acid or acid salt when given its formula.
5. Write the equation for the ionization of water.
6. Given the pH of a solution, calculate the hydrogen ion concentration and hydroxide ion concentration and state whether the solution is acidic, basic, or neutral.
7. Define *normality* and calculate the normality of a specified solution of acid or base.
8. Perform the equivalence point calculations for a titration.
9. Define *buffer* and give an example of the way in which the blood plasma is protected against large changes in pH.
10. Define *acidosis* and *alkalosis* and describe conditions in the body that will cause acidosis to occur.

ACIDS AND BASES

13.1 Acids and Bases: The Brønsted–Lowry Definition

Acids and bases play important roles in living organisms. Strong acids and bases can be quite harmful; they will destroy tissue by dissolving protein material and drawing out water. For example, concentrated sulfuric acid is a strong dehydrating (water-removing) agent and will rapidly injure tissues on contact. Concentrated bases will react with the fats that make up the protective membranes of cells, destroying such membranes and causing even more widespread destruction to tissues than acids. Strong laundry soaps and detergents contain bases. Clothes containing wool and silk (which are animal proteins) cannot be washed in such soaps because the base in the soap will cause the fibers of these materials to shrink and partially dissolve.

In general, **acids** are compounds that, when dissolved in water, produce solutions that conduct electricity, react with metals such as zinc or magnesium to produce hydrogen gas, taste sour, and turn paper containing litmus dye from blue to red. **Bases** also form solutions that conduct electricity; however, they taste bitter, feel slippery to the touch, and turn litmus paper from red to blue. Bases will react with acids to neutralize

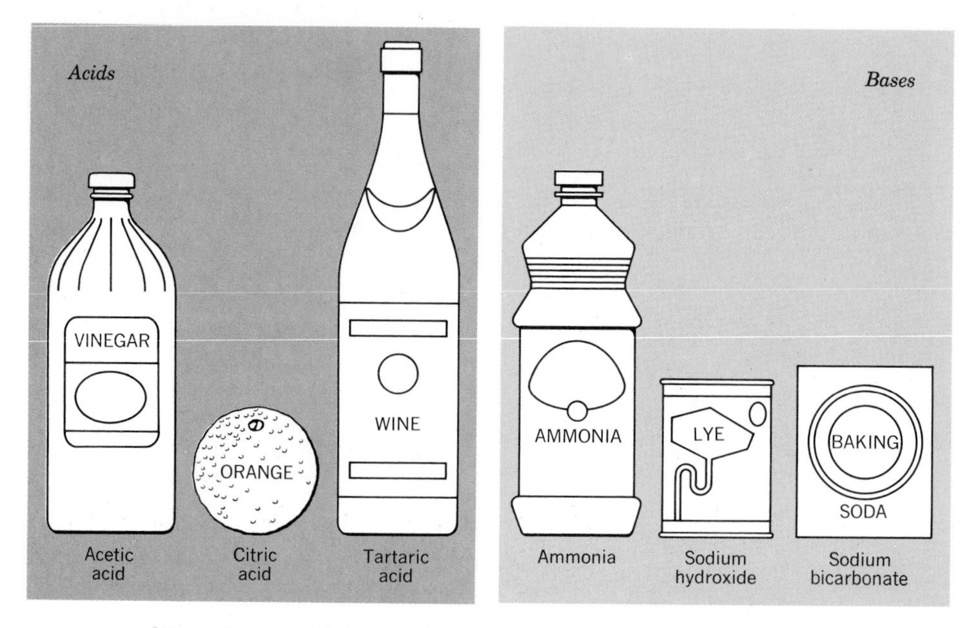

Figure 13.2 Some household acids and bases.

each others' properties. For example, medicines for the relief of pain, indigestion, and constipation often contain the base bicarbonate (HCO_3^-), which will neutralize the hydrochloric acid (HCl) in the stomach (Figure 13.2).

What makes a compound an acid or a base? Although there are several definitions that can be used, we will find it convenient for our study of biochemistry to use the definitions proposed in 1923 by the Danish chemist J. N. Brønsted and the English chemist T. M. Lowry:

An acid is a substance that can donate a proton*, H^+.
A base is a substance that can accept a proton, H^+.

Let's look at some examples. Consider the following reaction:

$$\underset{\text{Nitrous acid}}{HNO_2} + \underset{\text{Water}}{H_2O} \rightleftharpoons \underset{\text{Nitrite ion}}{NO_2^-} + \underset{\text{Hydronium ion}}{H_3O^+} \qquad (1)$$

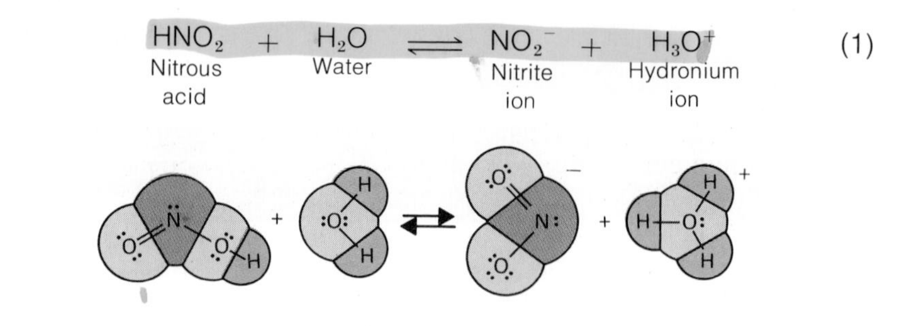

In this reaction, nitrous acid is donating a proton to water. Therefore, nitrous acid is the acid (the proton donor), and water acts as the base (the proton acceptor).

---

* *A hydrogen atom has just one proton in its nucleus, and one electron. Therefore, the hydrogen ion (H^+, a hydrogen atom that has lost one electron) is simply a proton.*

$$\underset{\text{Ammonia}}{NH_3} + \underset{\text{Water}}{H_2O} \rightleftharpoons \underset{\text{Ammonium ion}}{NH_4^+} + \underset{\text{Hydroxide ion}}{OH^-} \qquad (2)$$

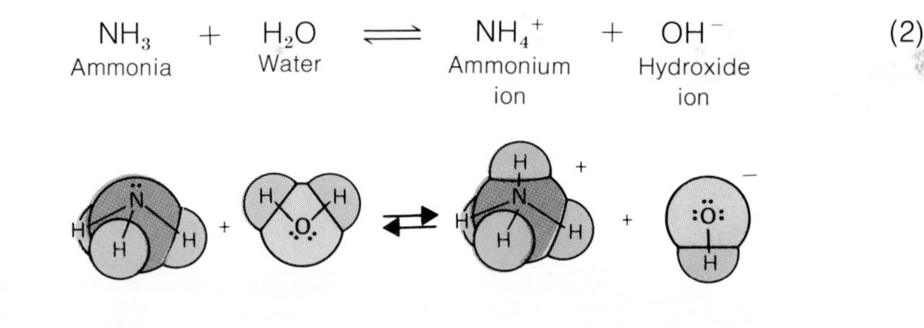

In this reaction, ammonia (NH_3) accepts a proton from a water molecule. Therefore, ammonia is the base, and water (as the proton donor) is the acid. You can see that water may function as either an acid or a base depending upon the substance with which it is reacting.

Consider for a moment the reverse reaction in eq. (1). Here, the hydronium ion (H_3O^+) is the acid and the nitrite ion (NO_2^-) is the base. Similarly, in the reverse of eq. (2), the ammonium ion (NH_4^+) is the acid and the hydroxide ion (OH^-) is the base.

$$\underset{\text{Acid}_1}{HNO_2} + \underset{\text{Base}_2}{H_2O} \rightleftharpoons \underset{\text{Base}_1}{NO_2^-} + \underset{\text{Acid}_2}{H_3O^+}$$

$$\underset{\text{Base}_1}{NH_3} + \underset{\text{Acid}_2}{H_2O} \rightleftharpoons \underset{\text{Acid}_1}{NH_4^+} + \underset{\text{Base}_2}{OH^-}$$

You might notice that HNO_2 and NO_2^-, and H_2O and H_3O^+, differ only by a proton. They are known as **conjugate acid-base pairs.** The nitrite ion is the conjugate base of nitrous acid, and the hydronium ion is the conjugate acid of water.

Some acids, called **polyprotic acids,** can donate more than one proton in a reaction with a base. For example, sulfuric acid (H_2SO_4) and carbonic acid (H_2CO_3) each have two ionizable hydrogens that can be donated in an acid-base reaction. Phosphoric acid (H_3PO_4) has three ionizable hydrogens.

$$H_3PO_4 + H_2O \rightleftharpoons H_3O^+ + H_2PO_4^-$$
$$H_2PO_4^- + H_2O \rightleftharpoons H_3O^+ + HPO_4^{2-}$$
$$HPO_4^{2-} + H_2O \rightleftharpoons H_3O^+ + PO_4^{3-}$$

13.2 The Strength of Acids and Bases

We will define strong and weak acids in much the same way that we defined strong and weak electrolytes. A strong acid is one that completely, or almost completely, ionizes to donate all of its protons. The result of adding a strong acid to water is a large increase in the concentration of hydronium ions. For example, nitric acid is a strong acid. In 0.1 *M* HNO_3, 100% of the nitric acid molecules are ionized to hydronium ions and nitrate ions.

$$HNO_3 + H_2O \rightleftharpoons H_3O^+ + NO_3^-$$

Other strong acids are hydrochloric acid (HCl), hydrobromic acid (HBr), hydroiodic acid (HI), and sulfuric acid (H_2SO_4). A weak acid only partially ionizes in water to donate protons. Therefore, the addition of a weak acid to water results in only a small increase in the concentration of hydronium ions. Acetic acid is an example of a weak acid. In a 0.1 *M* CH_3COOH solution, only 1.3% of the molecules will be ionized.

$$CH_3COOH + H_2O \rightleftharpoons H_3O^+ + CH_3COO^-$$

Other weak acids are nitrous acid (HNO_2), carbonic acid (H_2CO_3), and boric acid (H_3BO_3).

Similarly, a strong base will have a very large attraction for protons. A weak base will have a weak attraction, and only a small percentage of its molecules will accept protons. The hydroxide ion is a strong base, whereas ammonia is a weak base.

If an acid is strong and, therefore, has a strong tendency to donate protons, its conjugate base will be weak and will have a low attraction for protons. The reverse will be true for a weak acid; a weak acid will have a conjugate base with a strong attraction for protons. Table 13.1 lists the relative strengths of some conjugate acid-base pairs. (Note that the number of hydrogens in the formula of an acid is not an indication of its strength as an acid.)

Table 13.1 Relative Strengths of Some Conjugate Acid-Base Pairs

ACID				BASE	
Sulfuric acid	H_2SO_4	Increasing Acid Strength (↑)	Increasing Base Strength (↓)	HSO_4^-	Hydrogen sulfate ion
Hydrochloric acid	HCl			Cl^-	Chloride ion
Nitric acid	HNO_3			NO_3^-	Nitrate ion
Hydronium ion	H_3O^+			H_2O	Water
Sulfurous acid	H_2SO_3			HSO_3^-	Hydrogen sulfite ion
Hydrogen sulfate ion	HSO_4^-			SO_4^{2-}	Sulfate ion
Phosphoric acid	H_3PO_4			$H_2PO_4^-$	Dihydrogen phosphate ion
Nitrous acid	HNO_2			NO_2^-	Nitrite ion
Acetic acid	$HC_2H_3O_2$			$C_2H_3O_2^-$	Acetate ion
Carbonic acid	H_2CO_3			HCO_3^-	Hydrogen carbonate ion
Hydrogen sulfite ion	HSO_3^-			SO_3^-	Sulfite ion
Dihydrogen phosphate ion	$H_2PO_4^-$			HPO_4^{2-}	Hydrogen phosphate ion
Ammonium ion	NH_4^+			NH_3	Ammonia
Hydrogen carbonate ion	HCO_3^-			CO_3^{2-}	Carbonate ion
Hydrogen phosphate ion	HPO_4^{2-}			PO_4^{3-}	Phosphate ion
Water	H_2O			OH^-	Hydroxide ion
Ammonia	NH_3			NH_2^-	Amide ion

From looking at Table 13.1, we see that for polyprotic acids, the neutral acid that donates the first proton is a stronger acid than the negative ion that donates the second proton. For example, H_2CO_3 is a stronger acid than HCO_3^- because it is easier to remove a positive ion (H^+) from a neutral molecule than from a negatively charged ion.

13.3 Naming Acids

When dissolved in water, certain binary compounds made up of hydrogen and a nonmetal form solutions that have acidic properties. These compounds are called **binary acids** and are named by (1) using the prefix *hydro-,* (2) adding an *-ic* ending to the name of the second element, and (3) adding the word *acid.* For example, HBr is hydrobromic acid. However, not all binary hydrogen compounds are acids. When the formula of a binary hydrogen compound has the hydrogen listed first (HCl, HI, H_2S), it is an acid, but compounds such as CH_4 and NH_3 are not acidic.

Salts of these binary acids (formed when the acid reacts with a base) contain the negative ion of the nonmetal; their name always ends in *-ide.* For example, hydrochloric acid, HCl, will produce chloride salts containing the chloride ion, Cl^-.

HCl	+	NaOH	$\longrightarrow$	H_2O	+	NaCl
Hydrochloric acid		Sodium hydroxide		Water		Sodium chloride

Some acids, called **oxoacids,** contain three different elements: hydrogen, oxygen, and another nonmetallic element. The naming of these acids is based on the oxidation number of the third element. If the same nonmetal appears in two oxoacids, the acid in which it has the higher oxidation number is named by adding an *-ic* ending to the name of that element and then adding the word *acid.* The oxoacid in which the nonmetal has the lower oxidation number is given an *-ous* ending. For example, H_2SO_4 (oxidation number of S = 6+) is sulfuric acid, and H_2SO_3 (oxidation number of S = 4+) is sulfurous acid.

Elements in group VII, the halogens, can form as many as four different oxoacids. In this case, the oxoacid with the fewest oxygens is given the prefix *hypo-,* and the acid with the most oxygens carries the prefix *per-.* For example, there are four oxoacids formed from hydrogen, oxygen, and chlorine.

HClO:	Hypochlorous acid
$HClO_2$:	Chlorous acid
$HClO_3$:	Chloric acid
$HClO_4$:	Perchloric acid

Salts formed from oxoacids contain polyatomic ions. The name of the polyatomic ion formed from an *-ic* acid ends in *-ate.* For example, sulfuric acid, H_2SO_4, produces the sulfate ion, SO_4^{2-}. The name of the polyatomic ion formed from an *-ous* acid ends in *-ite.* So sulfurous acid, H_2SO_3, produces the sulfite ion, SO_3^{2-}.

Polyprotic acids can form ions that still contain a hydrogen and are themselves acids. These ions form compounds called **acid salts** and are named by indicating the number of hydrogens still present. For example, phosphoric acid, H_3PO_4, can give ions that form sodium dihydrogen phosphate, NaH_2PO_4, and sodium hydrogen phosphate, Na_2HPO_4.

For acid salts of diprotic acids, the prefix *bi-* is still often used to indicate the presence of one hydrogen. So, for example, $NaHCO_3$ is sodium bicarbonate.

EXAMPLE 13-1

Name the following acids: HF, HNO_2, and HNO_3.

HF: This is a binary acid, so its name is hydrofluoric acid.

HNO_2 and HNO_3: These are oxoacids containing nitrogen. The oxidation number of nitrogen in the first acid is 3+, and in the second is 5+. Therefore, HNO_2 is nitrous acid and HNO_3 is nitric acid.

Exercise 13-1

1. Name the following acids: HI, HIO, HIO_2, HIO_3, and HIO_4.
2. What is the formula of (a) sodium bisulfite, and (b) the periodate ion?

13.4 Neutralization Reactions

Acids will react with bases to form water and the dissolved ions of an ionic compound, commonly called a **salt.** For example,

$$\underset{\text{Acid}}{HCl_{(aq)}} + \underset{\text{Base}}{NaOH_{(aq)}} \longrightarrow \underset{\text{Water}}{H_2O} + \underset{\text{Salt}}{NaCl_{(aq)}}$$

If equal amounts of hydronium ions and hydroxide ions react, the resulting solution will have neither acidic nor basic properties and is said to be **neutral.** A **neutralization reaction,** therefore, is one in which either an acid or basic solution is converted to a neutral solution.

The reaction between, for example, hydrochloric acid and sodium hydroxide actually occurs between aqueous ions. The complete ionic equation for the reaction is

$$H^+_{(aq)} + Cl^-_{(aq)} + Na^+_{(aq)} + OH^-_{(aq)} \longrightarrow H_2O + Na^+_{(aq)} + Cl^-_{(aq)}$$

The net-ionic equation for this reaction is

$$H^+_{(aq)} + OH^-_{(aq)} \longrightarrow H_2O$$

EXAMPLE 13-2

Write the balanced equation for the neutralization of sulfuric acid by potassium hydroxide.

The reactants are H_2SO_4 and KOH. The products are water (the H^+ from H_2SO_4 and the OH^- from KOH) and the salt potassium sulfate (formed by K^+ and SO_4^{2-}). The correct formula for potassium sulfate is K_2SO_4. The unbalanced equation is

$$H_2SO_4 + KOH \longrightarrow H_2O + K_2SO_4$$

If we first balance the potassium atom, we have

$$H_2SO_4 + 2KOH \longrightarrow H_2O + K_2SO_4$$

The sulfur atoms already balance, so next we balance the hydrogen atoms.

$$H_2SO_4 + 2KOH \longrightarrow 2H_2O + K_2SO_4$$

Finally, check the total number of oxygen atoms on both sides. The equation is balanced.

Exercise 13-2

Write the balanced equation and net–ionic equation for the neutralization of potassium hydroxide by nitric acid.

13.5 Ionization of Water

In Chapter 11 we discussed water's ability to dissolve ionic compounds and to ionize some highly polar molecular compounds. But to a very small extent water molecules are also able to ionize other water molecules. At 25°C one water molecule out of every 550 million will be pulled apart to form a hydronium ion, H_3O^+, and hydroxide ion, OH^-. In a liter of water, then, the concentration of H_3O^+ and OH^- is 1×10^{-7} moles.

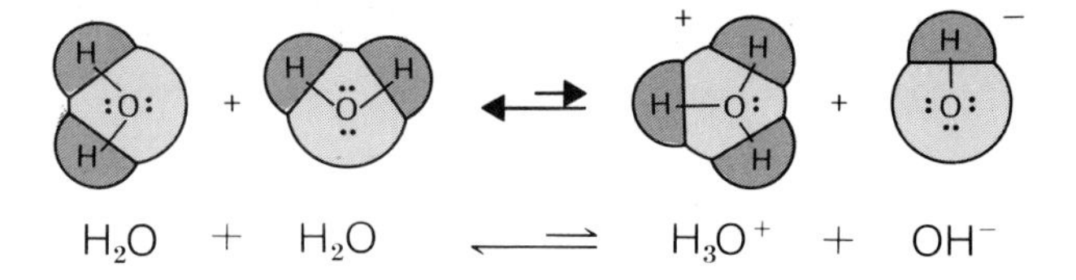

$$H_2O + H_2O \rightleftharpoons H_3O^+ + OH^-$$

For convenience, we often write the equation for the ionization of water as follows:

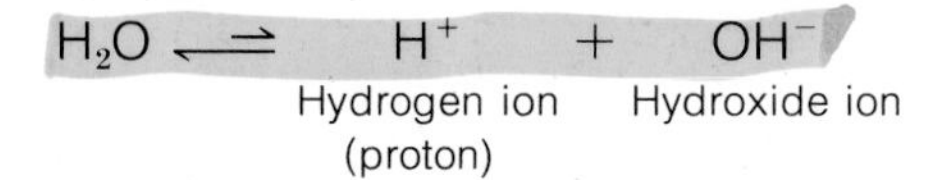

But it is important to understand that although we often use the terms *protons* or *hydrogen ions* when talking about aqueous solutions, single hydrogen ions never exist in water. They are always associated with a water molecule in the form of a hydronium ion (hydrated proton).

$$H^+ + :\ddot{\underset{\cdot\cdot}{O}}:H \longrightarrow H:\ddot{\underset{\cdot\cdot}{O}}:H^+$$

In the covalent bond that is formed between the oxygen and the hydrogen ion, the oxygen molecule donates both of the electrons. This type of bond, in which one atom donates both of the electrons, is known as a **coordinate covalent bond.**

13.6 Ion Product of Water, K_w

We just stated that only a very small number of water molecules ionize to form hydrogen ions* and hydroxide ions. In pure water at room temperature the concentration of hydrogen ions equals only 0.0000001 *M*, or 1×10^{-7} *M*. Because a hydroxide ion is formed for every hydrogen ion that is formed, the concentration of OH^- is also 1×10^{-7} *M*. Therefore,

$$[H^+] \times [OH^-] = [1 \times 10^{-7}][1 \times 10^{-7}] = 1 \times 10^{-14} = K_w$$

(where the brackets indicate "concentration in moles per liter"). This product, known as the **ion product of water, K_w,** always equals 1×10^{-14} at room temperature. From this equation we can calculate the concentration of either H^+ or OH^- if we know the concentration of the other ion.

$$[H^+] = \frac{1 \times 10^{-14}}{[OH^-]} \qquad [OH^-] = \frac{1 \times 10^{-14}}{[H^+]}$$

EXAMPLE 13-3

The concentration of hydrogen ions in a slightly acid solution is 1×10^{-5} *M*. What is the hydroxide ion concentration in this solution?

* *The terms* hydrogen ions *and* hydronium ions *will be used interchangeably in this discussion.*

Substituting in the equation above, we obtain

$$[OH^-] = \frac{1 \times 10^{-14}}{[H^+]} = \frac{1 \times 10^{-14}}{1 \times 10^{-5}} = 1 \times 10^{-9}$$

(If you are unsure about how to divide numbers in exponential form, see Appendix 1.)

Exercise 13-3

Calculate the concentration of hydrogen ions in moles/liter for a solution whose hydroxide ion concentration is:

(a) 1×10^{-3} *M*
(b) 5×10^{-6} *M*
(c) 0.0004 mole in 200 ml
(d) 0.00040 *M*
(e) 0.01 *M*
(f) 8.00×10^{-8} mole in 500 ml

MEASURING THE CONCENTRATION OF ACIDS AND BASES

13.7 The pH Scale

Small changes in hydrogen ion concentration can be of great importance to living cells and are critical in many fields of scientific investigation. As a result, scientists are constantly measuring hydrogen ion concentrations. The pH scale was developed in 1909 by a chemist named Sorenson in order to provide a way of indicating the hydrogen ion concentration more conveniently than by using a negative exponent (10^{-7}) or a decimal fraction (0.0000001). pH is the negative power to which the number 10 must be raised to express the concentration of hydrogen ions in moles/liter. Mathematically,

$$[H^+] = 1 \times 10^{-pH} \qquad \text{or} \qquad pH = -\log [H^+]$$

At room temperature, the $[H^+]$ in pure water is 1×10^{-7}. Therefore, the pH of pure water is 7. Because in pure water we have $[H^+] = [OH^-]$, the water is neutral. This means that the pH of a neutral solution is seven. Acidic solutions are ones in which the hydrogen ion concentration is greater than the hydroxide ion concentration. From Table 13.2 we see that for an acidic solution the exponent of ten will be less negative than −7. Thus, the pH of an acidic solution will be less than seven. A basic solution is one in which the hydrogen ion concentration is less than the hydroxide ion concentration, and the pH will be greater than seven (Figure 13.3).

Table 13.2 The pH Scale and the Corresponding Concentrations of Hydrogen and Hydroxide Ions

	$[H^+]$	pH	$[OH^-] = \frac{1 \times 10^{-14}}{[H^+]}$
Acidic	$10^0 = 1$	0	10^{-14}
	$10^{-1} = 0.1$	1	10^{-13}
	$10^{-2} = 0.01$	2	10^{-12}
	$10^{-3} = 0.001$	3	10^{-11}
	$10^{-4} = 0.0001$	4	10^{-10}
	$10^{-5} = 0.00001$	5	10^{-9}
	$10^{-6} = 0.000001$	6	10^{-8}
Neutral	$10^{-7} = 0.0000001$	7	10^{-7}
Basic	$10^{-8} = 0.00000001$	8	10^{-6}
	$10^{-9} = 0.000000001$	9	10^{-5}
	$10^{-10} = 0.0000000001$	10	10^{-4}
	$10^{-11} = 0.00000000001$	11	10^{-3}
	$10^{-12} = 0.000000000001$	12	10^{-2}
	$10^{-13} = 0.0000000000001$	13	10^{-1}
	$10^{-14} = 0.00000000000001$	14	10^0

EXAMPLE 13-4

We saw at the beginning of this chapter that the air pollutants sulfur dioxide and nitrogen dioxide, when washed out of the air

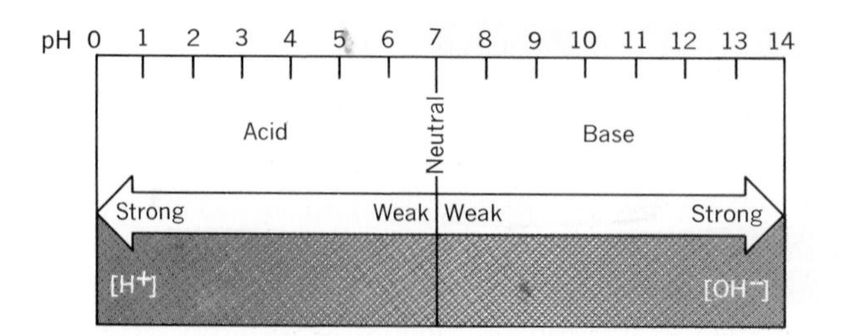

Figure 13.3 The pH scale. Acidic solutions range in pH from 0 to 7. Basic solutions range from 7 to 14. A solution with a pH of 7 is neutral.

by rain, react with the rainwater to form acids. Acid rain is a concern not only in the United States but throughout the world. A sample of rainwater from Sweden was recently found to have a pH of 4.

(a) Is this sample acidic or basic?
(b) Determine the $[H^+]$ and $[OH^-]$ of this sample.

(a) The pH is less than 7, so the solution is acidic.
(b) $[H^+] = 1 \times 10^{-pH} = 1 \times 10^{-4}$

$$[OH^-] = \frac{1 \times 10^{-14}}{[H^+]} = \frac{1 \times 10^{-14}}{1 \times 10^{-4}} = 1 \times 10^{-10}$$

Exercise 13-4

1. For solutions with the following pH values, state whether they are acidic or basic and determine their $[H^+]$ and $[OH^-]$.

   (a) pH 11 (b) pH 2 (c) pH 5 (d) pH 9

2. What is the pH of a sample of saliva if the concentration of hydrogen ions in a 20-ml sample is 2.0×10^{-9} mol?

13.8 Measuring pH

The measurement of pH is an important laboratory procedure because the pH of a solution affects the activity of biological molecules. This means

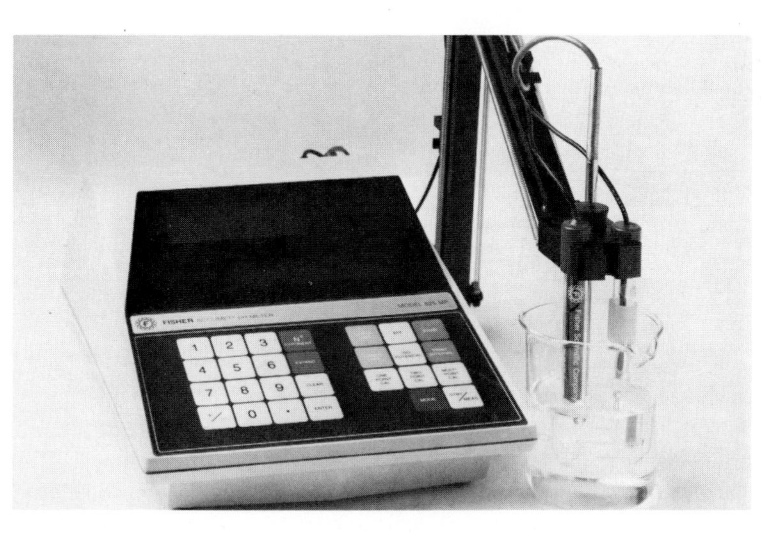

Figure 13.4 A pH meter. (Courtesy Fisher Scientific.)

Table 13.3 The Normal pH Range of Some Body Fluids

Fluid	pH	Fluid	pH
Gastric juice	1.0–3.0	Blood	7.35–7.45
Vaginal secretion	3.8	Intestinal secretions	7.7
Urine	5.5–7.0	Bile	7.8–8.8
Saliva	6.5–7.5	Pancreatic juice	8

that it can influence the behavior of cells and even entire organisms. For example, bacteria will grow best in a very narrow range of pH. Therefore, the pH of culture media must be carefully controlled. Enzymes, the biological catalysts, work best in a very narrow range of pH that can vary from an optimum pH range of 1 to 4 for pepsin (an enzyme in the stomach), to an optimum pH range of 8 to 9 for trypsin (an enzyme in the small intestine). Most body fluids remain in a very narrow range of pH that, if changed, can be toxic to the organism (Table 13.3).

The pH of a solution is best measured using a pH meter (Figure 13.4). These instruments make use of the fact that the voltage of an electric current passing through a solution will change according to the pH of the solution. A second, but less accurate, method of measuring pH is by a colorimetric indicator. Such methods use chemical dyes, called acid-base indicators, which will change color at certain hydrogen ion concentrations (Table 13.4). For example, paper on which the dye nitrazine has been placed is yellow at a pH of 4.5, and blue at a pH of 7.5. Such paper is used to test the pH of urine. Acidic urine—urine with a pH of 4.5 or lower—will turn the paper yellow and is often an indication of a serious disorder.

Table 13.4 Colors of Some Acid–Base Indicators at Various pH Levels

Indicator \ pH	0	1	2	3	4	5	6	7	8	9	10	11	12	13	14
Thymol blue*		Red	Transition	Yellow											
Methyl orange				Red	Transition	Yellow									
Methyl red					Red	Transition		Yellow							
Litmus						Red	Transition		Blue						
Bromothymol blue						Yellow		Transition	Blue						
Metacresol purple								Yellow	Transition		Purple				
Thymol blue*								Yellow		Transition	Blue				
Phenolphthalein								Colorless		Transition	Red				

* Thymol blue indicator undergoes two color changes—one in the acid range and one in the base range.

13.9 Titrations

We can use a neutralization reaction to measure the amount of acid or base in a solution by means of a procedure called **titration** (Figure 13.5). In a titration, a solution of known concentration of acid or base is added from a buret to a solution of unknown concentration of base or acid. The neutralization reaction can be monitored by a pH meter or an acid–base indicator. The point at which an equal number of moles of acid have been added to the base is the **equivalence point** of the titration. The acid–base indicator used in a titration must be carefully chosen so that the color change occurs at the pH of the equivalence point. Some important uses of titrations are to determine the alkali (or basic) constituents of the blood, the acidity of the stomach, or the acidity of the urine. In general, a person is considered to be suffering from hyperacidity when 44 ml of 0.10 *M* NaOH are required to neutralize 100 ml of stomach fluid; the person is suffering from hypoacidity when 10 ml or less are required.

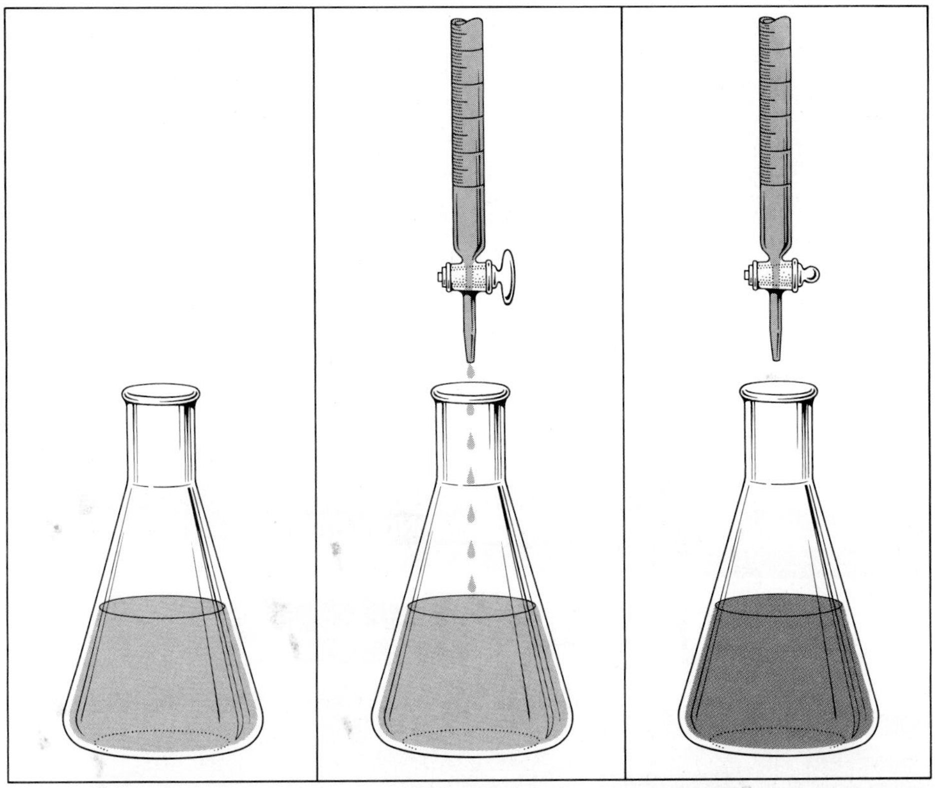

1. An acidic solution of unknown concentration is placed in a flask with an indicator.

2. A basic solution of known concentration is added slowly from the buret.

3. The flow of base from the buret is stopped when the indicator changes color, showing that just enough base has been added to react with all the acid in the flask.

Figure 13.5 A titration.

EXAMPLE 13-5

1. What is the concentration of hydroxide in a solution, if 5.00 ml of 0.0200 *M* HCl completely neutralizes 100 ml of the solution?

   A 0.0200 *M* HCl solution contains 0.0200 mole of H^+ in 1000 ml. Therefore, 5.00 ml will contain

$$5.00 \text{ ml} \times \frac{0.0200 \text{ mole}}{1000 \text{ ml}} = \frac{0.100 \text{ mole}}{1000} = 1.00 \times 10^{-4} \text{ mole of } H^+$$

   Because this many moles of H^+ completely neutralize the base, the 100 ml of solution contains 1.00×10^{-4} mole of OH^-.

Exercise 13-5

Titration of the contents of the stomach can be very useful in medical diagnosis. What is the molar concentration of acid in 100 ml of stomach fluid if complete neutralization occurs when 27.0 ml of 0.10 *M* NaOH have been added?

13.10 Normality

Some acids can donate more than one hydrogen in a neutralization reaction. For doing acid–base titrations, chemists have developed a unit of concentration called normality, which indicates the number of hydrogen ions available in a neutralization reaction. **Normality (*N*)** gives the number of equivalents of acid or base per liter of solution.

$$\text{Normality } (N) = \frac{\text{Eq (acid or base)}}{\text{liter solution}}$$

For a given reaction, an **equivalent (Eq)** of an acid is the amount of acid that will donate one mole of hydrogen ions. One equivalent of base is the amount of base that will neutralize one mole of hydrogen ions in a given reaction. The **gram-equivalent weight** of an acid or base is the weight in grams of acid or base that will contain one equivalent. For example, one mole of NaOH will neutralize one mole of hydrogen ions. Therefore, 1 mol NaOH = 1 Eq NaOH. One mole of NaOH weighs 40 grams, so the gram-equivalent weight of NaOH is 40 grams. This means that a 1 *N* NaOH solution will contain 40 g NaOH in one liter of solution.

look at equation

Similarly, one mole of sulfuric acid (H_2SO_4) can donate two moles of hydrogen ions. Therefore, 1 mol H_2SO_4 = 2 Eq H_2SO_4. One mole of H_2SO_4 weighs 98 grams, so the gram-equivalent weight of H_2SO_4 is 98/2 = 48 grams. This means that a 1 *N* H_2SO_4 solution will contain 48 g H_2SO_4 in one liter of solution.

EXAMPLE 13-6

1. What is the gram-equivalent weight of phosphoric acid, H_3PO_4, if all three of its hydrogens are neutralized?

   One mole of H_3PO_4 will donate three moles of hydrogen ions, so we can write

$$1 \text{ mol } H_3PO_4 = 3 \text{ Eq } H_3PO_4$$

and

$$1 \text{ mol } H_3PO_4 = 98.00 \text{ g}$$

Using these relationships, we can calculate the gram-equivalent weight as follows:

$$\frac{98.00 \text{ g } H_3PO_4}{1 \cancel{\text{mol } H_3PO_4}} \times \frac{1 \cancel{\text{mol } H_3PO_4}}{3 \text{ Eq } H_3PO_4} = \frac{32.68 \text{ g } H_3PO_4}{1 \text{ Eq } H_3PO_4}$$

2. If 4.085 g of H_3PO_4 is dissolved in 250.0 ml of solution, what is the normality of this solution?

   We need to determine the number of equivalents of H_3PO_4 per liter. First, we determine the number of equivalents of H_3PO_4 in 4.085 g.

$$4.085 \text{ g } \cancel{H_3PO_4} \times \frac{1 \text{ Eq } H_3PO_4}{32.67 \text{ g } \cancel{H_3PO_4}} = 0.1250 \text{ Eq } H_3PO_4$$

Second, we determine the number of equivalents per liter, knowing now that 0.1250 Eq H_3PO_4 are dissolved in 250.0 ml.

$$\frac{0.1250 \text{ Eq } H_3PO_4}{250.0 \cancel{\text{ml}}} \times \frac{1000 \cancel{\text{ml}}}{1 \text{ liter}} = \frac{0.5000 \text{ Eq } H_3PO_4}{1 \text{ liter}}$$

The normality of this solution, therefore, is 0.5000 *N* H_3PO_4.

Exercise 13-6

1. What is the gram-equivalent weight of

   (a) HNO_3 (b) $Ca(OH)_2$

2. How would you prepare 200 ml of 0.120 N HCl?

3. How would you prepare 500 ml of 0.0500 N $Ca(OH)_2$?

At the equivalence point of a titration, the number of equivalents of acid equals the number of equivalents of base, or $Eq_a = Eq_b$. When doing a titration, we do not know the equivalents, but we do know the normality or molarity of the stock solution, and the volumes of acid and base. The definition of normality tells us

$$N_a = \frac{Eq_a}{V_a} \qquad N_b = \frac{Eq_b}{V_b}$$

Rearranging these equations, we have

$$N_a \times V_a = Eq_a \qquad N_b \times V_b = Eq_b$$

Since $Eq_a = Eq_b$ at the equivalence point, we have

$$\boxed{N_a \times V_a = N_b \times V_b} \tag{1}$$

Because the units on both sides of this equation cancel, V_a and V_b can be expressed in any unit of volume so long as the same unit is used for both. This equation greatly simplifies titration calculations.

EXAMPLE 13-7

How many milliliters of 0.36 N H_2SO_4 are required to completely neutralize 55 ml of 0.24 N NaOH?

Substituting in eq. (1), at the equivalence point we must have

$$0.36\ N \times V_a = 0.24\ N \times 55\ \text{ml}$$

$$V_a = \frac{0.24\ \cancel{N} \times 55\ \text{ml}}{0.36\ \cancel{N}} = 37\ \text{ml}$$

Exercise 13-7

What is the normality of a nitric acid solution if 1.5 liters of this solution is completely neutralized by 0.75 liter of 0.10 N KOH?

BUFFER SYSTEMS

13.11 What Are Buffers?

Buffers are substances that, when present in solution, resist sudden changes in pH. In particular, they protect against large changes in pH when acids or bases are added to the solution. Living cells are extremely sensitive to even very slight changes in pH. As we stated earlier, the reason for this sensitivity is that the enzymes that catalyze metabolic reactions operate in only a small range of pH. Altering the pH will slow down or stop the action of the enzyme. Fortunately, the contents of cells, the extracellular fluid, and the blood have all developed buffer systems that protect against pH changes.

The best buffer systems consist of a weak acid and its conjugate base, or a weak base and its conjugate acid. Such systems have their highest buffering capacity at a pH where the [concentration of acid] = [concentration of conjugate base], or [concentration of base] = [concentration of conjugate acid]. The following are some acid-base pairs that can be used in buffer systems.

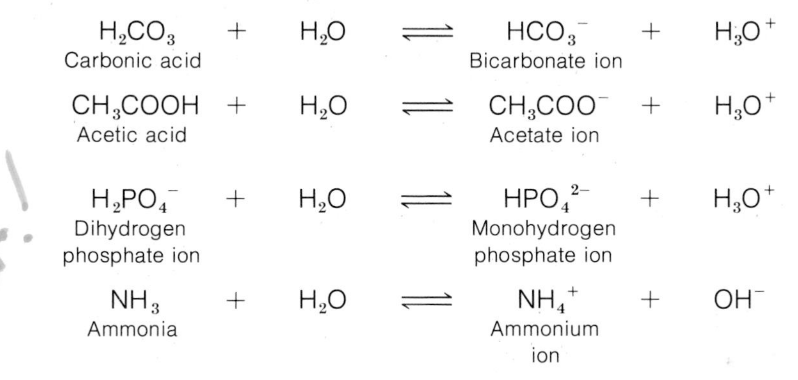

13.12 Control of pH in Body Fluids

In the human body the blood plasma has a normal pH of 7.4. If the pH should fall below 7.0 or rise above 7.8, the results would be fatal. The buffer systems in the blood are very effective in protecting this fluid from large changes in pH. For example, if 1 ml of 10.0 *M* HCl were added to 1 liter of unbuffered physiological saline (0.15 *M* NaCl) at a pH of 7, the pH would fall to 2. But if 1 ml of 10.0 *M* HCl is added to 1 liter of blood plasma at pH 7.4, the pH will drop to only 7.2 (Figure 13.6).

How do such buffer systems protect the blood? The major buffer system in the blood is the carbonic acid-bicarbonate system. Consider the following equilibrium equation:

$$H_2CO_3 \rightleftharpoons HCO_3^- + H^+$$

Adding a strong acid to the system will increase the concentration of H^+, driving the reaction to the left and forming more carbonic acid.

$$H_2CO_3 \leftharpoons HCO_3^- + H^+$$

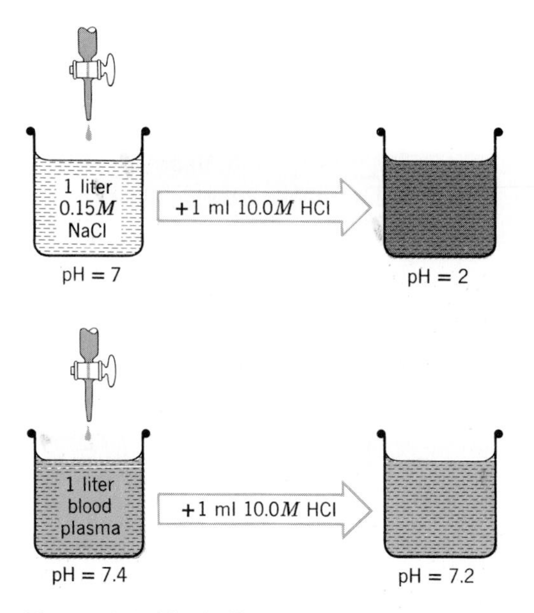

Figure 13.6 The buffer systems in the blood are very effective in preventing large changes in pH, which could be fatal.

But carbonic acid is unstable and will decompose to form carbon dioxide and water.

$$H_2CO_3 \longrightarrow CO_{2(g)} + H_2O$$

The carbon dioxide so formed can be removed from the blood and exhaled by the lungs. This buffer system will continue to protect against the pH change until all the bicarbonate has reacted.

Various factors can cause abnormal increase in acid levels in the blood. Such factors are hypoventilation caused by emphysema, congestive heart failure or bronchopneumonia; an increase in the production of metabolic acids, such as occurs in diabetes mellitus or some low-carbohydrate/high-fat diets; ingestion of excess acids; excess loss of bicarbonate in severe diarrhea; or decreased excretion of hydrogen ions through kidney failure. Each of these conditions will cause an increase in the hydrogen ion level in the blood, and a decrease in the concentration of basic components (such as bicarbonate), known as the alkaline reserves. The pH of the blood can drop to 7.1 or 7.2, resulting in a condition known as **acidosis** (called respiratory acidosis if its origin is in the respiratory system, and metabolic acidosis if the origin is other than respiratory) (Table 13.5). However, the body has ways to restore the blood pH to normal. First, it can expel the excess carbon dioxide, formed from the carbonic acid, through an increase in the rate of breathing. Second, it can increase the excretion of H^+ and the retention of HCO_3^- by the kidneys, resulting in acidic urine (pH about 4).

The bicarbonate buffer system also protects against an addition of strong base to the system. A base will react with the hydrogen ions to

Table 13.5 Acidosis and Alkalosis

ACIDOSIS			
Type	This Condition Results from	Causes	Compensatory Reactions
Respiratory	Retention of CO_2 Blood pH decrease	Hypoventilation Emphysema Congestive heart failure Bronchopneumonia Hyaline membrane disease Drugs that depress brain respiratory center	Increased rate of breathing Kidneys excrete acidic urine
Metabolic	Increase in H^+ Blood pH decrease	Diabetes mellitus Kidney failure Ingestion of acidic drugs such as aspirin Loss of HCO_3^-	Increased rate of breathing Kidneys excrete acidic urine

ALKALOSIS			
Type	This Condition Results from	Causes	Compensatory Reactions
Respiratory	Rapid expulsion of CO_2 Blood pH increase	Hyperventilation High fevers Trauma Hysteria	Slower rate of breathing Kidneys excrete less acid
Metabolic	Increase in basic components in the blood Blood pH increase	Severe vomiting causing loss of stomach acid Excess ingestion of basic substances Kidney disease	Slower rate of breathing Kidneys excrete less acid

produce water, decreasing the concentration of hydrogen ions in the system. This will drive the reaction to the right.

$$H_2CO_3 \rightleftharpoons HCO_3^- + H^+$$

Such an increase in base in the blood can occur in cases of hyperventilation during extreme fevers or hysteria, from excessive ingestion of basic substances such as antacids, and in severe vomiting. The pH of the blood can increase to a pH of 7.5, resulting in a condition known as **alkalosis.**

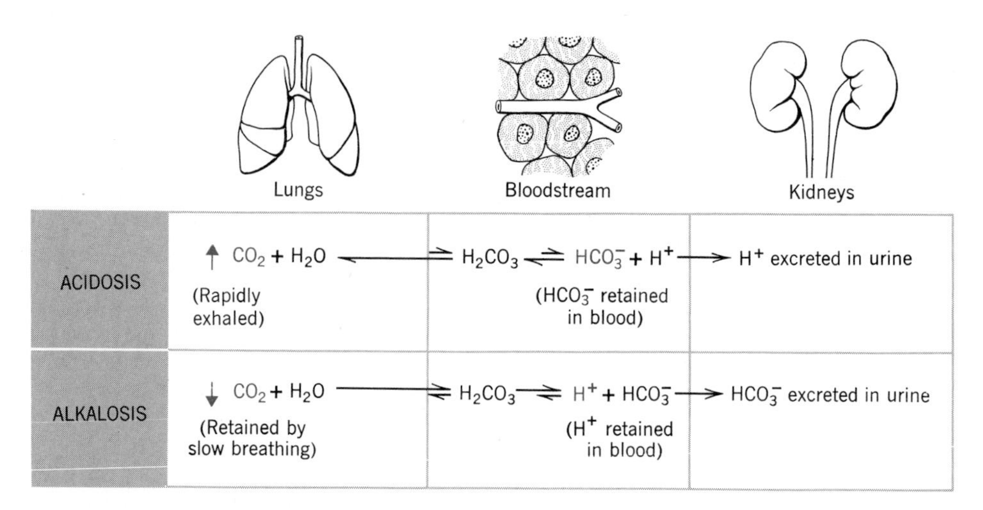

Figure 13.7 The carbonic acid/bicarbonate buffer system acts in the blood to prevent both acidosis and alkalosis.

Alkalosis is not as common as acidosis. The body's means for returning the pH to normal are a decrease in expulsion of carbon dioxide by the lungs and an increase in excretion of HCO_3^- by the kidneys, resulting in an alkaline urine (pH > 7) (Figure 13.7).

Another buffer system, active mainly within the cells, is the phosphate buffer system, which has a maximum buffering action at a pH of 7.2.

$$H_2PO_4^- \rightleftharpoons HPO_4^{2-} + H^+$$

Adding strong acid to this system will drive the reaction to the left, increasing the concentration of $H_2PO_4^-$, which is only weakly acidic. Large amounts of $H_2PO_4^-$ will result in acidosis, but the body will eliminate the

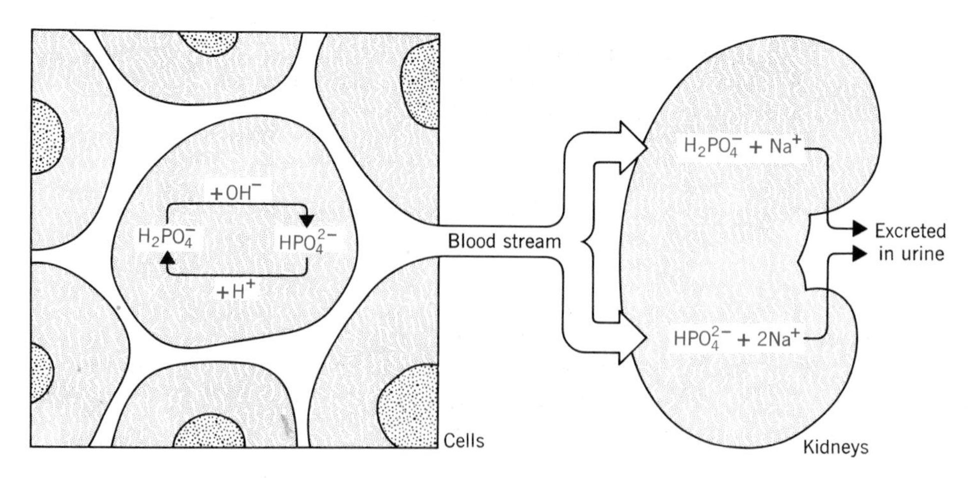

Figure 13.8 The phosphate buffer system acts in the cells to prevent changes in pH. The kidneys remove any excess HPO_4^{2-} or $H_2PO_4^-$ from the blood for excretion in the urine.

excess in the urine. Adding strong base to the system will drive the reaction to the right, as the hydrogen ions react with the base to form water. Large amounts of HPO_4^{2-} would be found in alkalosis, but under normal kidney function the HPO_4^{2-} is also excreted in the urine (Figure 13.8).

Normal metabolic reactions in the body result in the continuous production of acids. Cells produce an average of about 10 to 20 moles of carbonic acid each day, which is equivalent to 1 or 2 liters of concentrated HCl. This acid must be removed from the cells and carried to the organs of excretion without disrupting the pH of the blood. It is through the action of the buffer systems in the cells and extracellular fluid that our bodies are protected from changes in pH that would otherwise be caused by these acids.

CHAPTER SUMMARY

CHAPTER SUMMARY

Acids are compounds that, when dissolved in water, produce solutions that conduct electricity, react with metals to produce hydrogen, taste sour, and turn litmus paper red. Bases are compounds that form solutions which conduct electricity, taste bitter, feel slippery, and turn litmus paper blue. The Brønsted–Lowry definition of acids and bases states that an acid is a substance that can donate protons (H^+), and a base is a substance that can accept protons. A strong acid is one that has a large tendency to donate protons and, when dissolved in water, is almost completely ionized or dissociated. Only some of the molecules of a weak acid will donate their protons, so a weak acid added to water will cause a much smaller increase in the hydrogen ion concentration than the same amount of strong acid. A strong base will have a very large attraction for protons, whereas a weak base will have a weak attraction.

Acids will neutralize bases. If equal amounts of acid and base are reacted, a neutralization reaction occurs. Titration is a procedure that uses a neutralization reaction to determine an unknown concentration of acid or base. Normality is a unit of concentration that gives the number of equivalents of acid or base in one liter of solution.

A few molecules of pure water will ionize to form hydroxide ions and hydrogen ions. In water, the hydrogen ion concentration $[H^+]$ times the hydroxide ion concentration $[OH^-]$ will always equal 1×10^{-14}, or K_w, called the ion product of water. The pH scale is a convenient way to indicate the hydrogen ion concentration of a solution. Pure water, which is neutral, has a pH of 7. Solutions with a pH less than 7 are acidic, and those with a pH greater than 7 are basic.

Buffers are systems containing either a weak acid and its conjugate base, or a weak base and its conjugate acid, that protect against sudden changes in pH caused by the addition of acid or base. Because living organisms are so sensitive to sudden changes in pH, they contain buffer systems within the cells, the extracellular fluid, and the blood to protect against such changes.

Important Equations

Acid + Base ⟶ Salt + Water

$[H^+] \times [OH^-] = 1 \times 10^{-14}$

$[H^+] = 1 \times 10^{-pH}$ or $pH = -\log [H^+]$

$$\text{Normality } (N) = \frac{\text{Eq (acid or base)}}{\text{liter solution}}$$

At the equivalence point, $N_a \times V_a = N_b \times V_b$

EXERCISES AND PROBLEMS

EXERCISES AND PROBLEMS

1. In your own words, state the Brønsted–Lowry definition of (a) an acid and (b) a base.

2. What is a polyprotic acid? Give two examples.

3. Explain the difference between a coordinate covalent bond and a covalent bond.

4. Label the conjugate acid-base pairs in the following reactions:

   (a) $HI + H_2O \rightleftharpoons H_3O^+ + I^-$
   (b) $CO_3^{2-} + H_2O \rightleftharpoons OH^- + HCO_3^-$
   (c) $CH_3COOH + H_2O \rightleftharpoons H_3O^+ + CH_3COO^-$
   (d) $HF + NH_3 \rightleftharpoons NH_4^+ + F^-$
   (e) $O^{2-} + H_2O \rightleftharpoons OH^- + OH^-$
   (f) $NO_2^- + N_2H_5^+ \rightleftharpoons HNO_2 + N_2H_4$
   (g) $HCl + NH_2OH \rightleftharpoons NH_3OH^+ + Cl^-$

5. Using Table 13.1, state which is the stronger acid in each of the following pairs:

   (a) NH_4^+ or H_3O^+
   (b) HCO_3^- or H_2CO_3
   (c) $H_2PO_4^-$ or HPO_4^{2-}
   (d) HCl or H_2SO_3

6. Using Table 13.1, state which is the stronger base in each of the following pairs:

   (a) Cl^- or OH^-
   (b) PO_4^{3-} or HPO_4^{2-}
   (c) NO_3^- or NH_3
   (d) HCO_3^- or CO_3^{2-}

7. Hydrogen sulfide, H_2S, is a stronger acid than phosphine, PH_3. Compare the relative strengths of their conjugate bases, HS^- and PH_2^-.

8. Name the following acids:

   (a) HI
   (b) H_2SO_3
   (c) $HClO_2$
   (d) $HClO_3$

9. Name the salts formed from the neutralization by KOH of the acids in question 8.

10. What are two names for each of the following compounds:

    (a) $NaHCO_3$ (b) $KHSO_4$

11. One of the uses of baking soda ($NaHCO_3$) described on its label is to relieve acid indigestion. Explain how baking soda acts in the relief of indigestion (use words or an equation).

12. Write the balanced equation and the net-ionic equation for the complete neutralization of arsenic acid (H_3AsO_4) by sodium hydroxide. Name the salt formed in this neutralization reaction.

13. For the neutralization reaction between ammonium chloride and sodium hydroxide, write the following:

    (a) a balanced equation for the reaction
    (b) the net-ionic equation for the reaction

    Would the odor of ammonia be greater before or after the sodium hydroxide was added to a solution of ammonium chloride? Give the reason for your answer.

14. In some areas the increasing acidity of lakes has led conservationists to drop lime (calcium hydroxide) into these lakes. Explain how this would help lower the acidity of a lake.

15. You may have noticed that when you add lemon juice to your tea, the tea changes color. Suggest a reason for this color change.

16. Arrange the following pH values in order of increasing acid strength (from the least acidic to the most) and indicate which values represent basic, neutral, and acidic solutions.

    4, 6.3, 9.5, 1.4, 7, 5.5, 8.4, 12, 7.4

17. At room temperature, what is the pH of a solution for which

    (a) $[H^+] = 0.01$
    (b) $[H^+] = 1 \times 10^{-8}$
    (c) $[OH^-] = 1 \times 10^{-4}$

18. At room temperature, what is the $[H^+]$ and the $[OH^-]$ of a solution having a pH of (a) 1, (b) 6, (c) 12?

19. Indicate whether each of the solutions in questions 16 and 17 is acidic or basic.

20. Solution A has a pH of 3 and solution B a pH of 5. Which solution is more acidic, A or B? What is the hydrogen ion concentration in each solution, and by what factor do the two hydrogen ion concentrations differ?

21. Rainwater has a large amount of CO_2 dissolved in it. Explain why the pH of normal rain is 5.6.

22. Two 5-ml samples of fog water collected in Southern California had pHs of 3.0 and 4.0. How many moles of hydrogen ions and hydroxide ions are there in each sample?

23. How many milliliters of a 0.1 *M* NaOH solution are required to completely neutralize 30 ml of a solution of strong acid that has a pH of 2?

24. Four aqueous solutions were prepared as follows:

    1. 36.5 mg HCl in 100 ml of solution
    2. 2.45 g H_2SO_4 in 500 ml of solution
    3. 0.740 mg $Ca(OH)_2$ in 2.00 liters of solution
    4. 0.400 g NaOH in 100 ml of solution

    For each of the above solutions, calculate:

    (a) The molarity of the solution.
    (b) The normality of the solution.
    (c) The hydrogen ion concentration in moles per liter (if we assume that each compound is 100% ionized).
    (d) The pH of the solution.

25. A small private laboratory specializes in analyzing samples of waste water to see if they meet federal pollution standards. The analyst uses titrations to determine the concentration of acid or base in each sample. One set of tests gave the following results:

    Sample 1: 100.0 ml required 52.0 ml of 0.100 *N* NaOH.
    Sample 2: 1.00 liter required 150.0 ml of 0.200 *N* HCl.
    Sample 3: 25.0 cc required 30.0 ml of 0.150 *N* KOH.
    Sample 4: 50.0 ml required 15.0 ml of 0.0900 *N* H_2SO_4.

    (a) Calculate the normality of the samples.
    (b) Indicate for each sample whether the analyst was testing for acid or base.

26. Use words and equations to explain how an acetic acid-acetate ion buffer system can protect against pH changes when (a) acid is added to the system and (b) when base is added.

27. A large amount of lactic acid is produced in the muscles during very hard exercise. This lactic acid must be transported by the blood to the liver, where it is broken down. Why doesn't the pH of the blood change drastically after very hard exercise?

28. What means does the body use to correct the conditions of (a) acidosis and (b) alkalosis?

29. Would you expect a person in a diabetic coma to be breathing faster or slower than normal? Upon analysis would the pH of her urine be higher (more basic) or lower (more acidic) than normal? Give reasons to support both of your answers.

30. The following equilibrium is formed between ammonia and the ammonium ion:

$$NH_3 + H^+ \rightleftharpoons NH_4^+$$

Use the following statements and your knowledge of Le Chatelier's Principle to explain how ammonia is removed from the blood into the kidney tubules that carry urine to the bladder.

1. NH_3 is soluble in, and can pass through, cell membranes, but NH_4^+ cannot.
2. The normal pH of the urine in the kidney tubules is between 5.5 and 6.5.

INTEGRATED PROBLEMS

1. Dextran is a substance that is administered in saline solution to help maintain the blood pressure of persons in shock. The recommended dosage for the first 24 hours is 2.00 g of dextran per kilogram of body weight. How many milliliters of 10.0% (w/v) dextran in normal saline should be administered to a 165-lb man during the first 24 hours?
2. Bile is produced in the liver and stored in the gall bladder, from which it is released after each meal to aid in digestion of fats. The capacity of the gall bladder is only 40 ml. To store sufficient fluid between each meal, the bile produced by the liver is concentrated to 10% of its original volume. The cells of the gall bladder actively transport sodium and chloride ions out of the bile and into the interstitial (between cells) fluid. Explain how this action works to concentrate the bile.
3. The following equilibrium is formed between ammonia and the ammonium ion:

$$NH_3 + H^+ \rightleftharpoons NH_4^+$$

   Use the following statements and your knowledge of Le Châtelier's Principle to explain how ammonia is removed from the blood into the kidney tubules that carry urine to the bladder.
   (a) NH_3 can pass through cell membranes, but NH_4^+ cannot.
   (b) The normal pH of the urine in the kidney tubules is between 5.5 and 6.5.
4. Barium sulfate is used to coat the gastrointestinal tract of patients having CT scans. How many moles of barium sulfate would be produced when 50 ml of 0.1 *M* barium chloride is added to 100 ml of 0.05 *M* sodium sulfate?
5. What is the blood alcohol level of a suspect if 1 cc of her blood contains the same amount of alcohol as 7.5 ml of a 0.020% (w/v) standard alcohol solution? If the legal limit for drunk driving in the state is 0.08% blood alcohol, should the district attorney prosecute this suspect?
6. A 100-ml sample of well water was found to contain the following levels of contaminants: arsenic, 40 ppb; cadmium, 5 ppb; and mercury, 5 ppb. Do any of these contaminants exceed the following maximum contaminant levels for drinking water as set by the EPA: arsenic, 0.05 mg/liter; cadmium, 0.01 mg/liter; and mercury, 0.002 mg/liter?
7. A solution of acetic acid will form the following equilibrium:

$$HC_2H_3O_2 + H_2O \rightleftharpoons H_3O^+{}_{(aq)} + C_2H_3O_2^-{}_{(aq)}$$

   (a) What will happen to this equilibrium if sodium acetate is added to the solution?
   (b) What effect would adding NaOH to the solution have on this equilibrium?

SECTION THREE
THE ELEMENTS NECESSARY FOR LIFE

CHAPTER FOURTEEN

CARBON AND HYDROGEN: SATURATED HYDROCARBONS

According to the Russian chemist A. I. Oparin, the planet earth was born 4.5 billion years ago, having formed from a ball of hot gases in space. Over endless centuries, the surface of the earth slowly cooled and became solid, forming the rocks known as the earth's mantle. As this cooling continued, steam condensed to form water that filled the low places in the mantle, creating the oceans of the earth. This infant earth was a lifeless landscape of rock and water, constantly pounded by storms and heavy rains. The first atmosphere that may have formed, enriched by gases pouring from the interior of the earth in volcanic eruptions, consisted mostly of methane (CH_4) and nitrogen, with hydrogen, ammonia, carbon monoxide, and water vapor in smaller quantities. There was no free oxygen.

In this very earliest, or primordial, atmosphere, chemical reactions were started by ultraviolet radiations, ionizing radiations, and lightning. Many of the products of these reactions were simple carbon compounds and water. These carbon compounds, formed in the atmosphere, were brought to earth with the rains. They settled in the pools and oceans, forming a warm, dilute, aqueous solution of carbon compounds—a "primordial soup" rich in the building blocks for basic life processes.

Oparin suggests that as the centuries passed, the carbon compounds began to clump together to form droplets (or coacervates). These coacervates, which developed a surface layer, or boundary, with properties different from the droplet itself, had the extraordinary ability to absorb only selected materials from the surrounding environment. Thus, various molecules could be held very close to one another within the coacervate, allowing their concentrations to be much higher than in the surrounding environment. This represented the first major step toward the beginning of life.

But where on the earth were conditions just right for this first step toward life? What were the sources of the necessary energy, trace minerals, and compounds of nitrogen and sulfur? American oceanographers John Barross, Sarah Hoffman, and John Corliss believe they have discovered the answer deep in the bottom of the sea. There, in 1979, they observed vents in the ocean floor through which hot water shot up from deep within the earth.

Some of the vents they have studied are at great depths (1500 to 3000 m) and high pressure, and are extremely hot (up to 500°C)—all conditions favorable to a great number of the chemical reactions required to produce the chemicals of life. Additionally, all the needed trace elements, as well as compounds containing carbon, nitrogen, and sulfur, were found in the vicinity of these vents. It appears, therefore, that there were many places on the primitive earth suitable for the development of Oparin's coacervates.

According to Oparin's theory, as the ages passed, increasingly complex coacervate systems appeared that were capable of growth as well as simple maintenance. This capability required the development of complicated chemical pathways to trap chemical energy obtained from other molecules in the primordial soup. Oparin called these coacervate droplets with improved internal organization, protobionts. He suggests that life truly began about 3.5 billion years ago as the protobionts developed methods for controlling their own chemical reactions and their reproduction. The protobionts, therefore, represent the link between the original droplets and the most primitive of living organisms.

Evidence that coacervates could have evolved into primitive life forms near the ocean vents comes from the great diversity of bacteria living close to such vents today. These bacteria are capable of living at surprisingly high temperatures and thrive on chemicals such as methane, hydrogen sulfide, hydrogen, and iron(II) ions as their source of energy.

LEARNING OBJECTIVES

LEARNING OBJECTIVES

By the time you have finished this chapter, you should be able to:

1. Define *organic chemistry.*
2. Describe the difference between molecular and structural formulas of a compound.
3. Define *structural isomer* and, given the molecular formula of a compound, draw the structural formulas of its isomers.
4. Define *alkane* and *saturated hydrocarbon.*
5. Draw the structural formula of an alkane when given its IUPAC name, and state the IUPAC name when shown the alkane's structural formula.
6. Write the chemical equations for the combustion and substitution reactions of a given alkane.

ELEMENTS NECESSARY FOR LIFE

14.1 Elements Abundant in Living Organisms

Living organisms, like all other matter on earth, are composed of atoms of the naturally occurring elements. However, not all 90 of these elements are found in such organisms. The periodic table on the front overleaf of this book shows the 25 elements that have so far been shown to be essential to life. Hydrogen, carbon, nitrogen, and oxygen are the most plentiful, or abundant, elements in the living organism. They make up 99.3% of all the atoms in your body, whereas the remaining 21 elements account for only 0.7%.

14.2 The Role of Carbon

At the beginning of this chapter we described a primordial soup rich in molecules containing carbon atoms. But what is so special about carbon-containing molecules that they should be the building blocks of life?

Look at the periodic table on the front overleaf of this book. Carbon is the first member of group IV, a family also containing silicon (Si), germanium (Ge), tin (Sn), and lead (Pb). Each member of this family has four valence electrons. To attain a stable octet of electrons these elements can lose four electrons or share four electrons. Under ordinary conditions carbon has a strong tendency to share four electrons, and thus to form four covalent bonds.

A carbon atom can form four single bonds with four different atoms or with other carbon atoms. In this second case the other carbon atoms can,

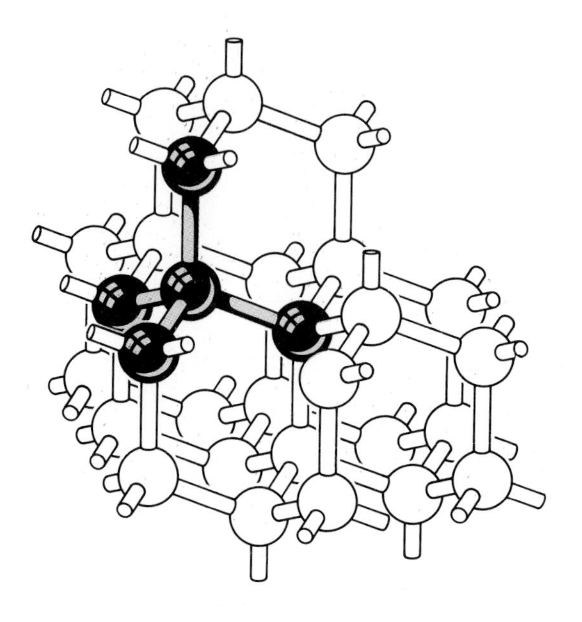

Figure 14.1 The bonding in diamond. Each carbon atom sits at the center of a tetrahedron and is bonded to four other carbon atoms.

in turn, be bonded to up to three more carbon atoms, and so on. The very stable molecule that results when every carbon atom is bonded to four other carbon atoms is diamond (Figure 14.1). A major feature, then, that makes carbon very special among the elements is its ability to form strong stable bonds with up to four other atoms identical to itself.

We have mentioned that diamond is the molecule formed when every carbon is bonded to four other carbon atoms. But all four bonds need not be to other carbon atoms, and the molecules found in living organisms are rarely of this type. Rather, in such molecules we find carbon atoms bound to other carbon atoms in the form of long chains, branched chains, or rings (Figure 14.2). As the number of carbon atoms in a molecule increases, the number of ways in which they can be arranged increases, thus making it possible to form compounds with the same chemical composition (the same number of atoms of each element), but with different structures and with correspondingly different chemical and physical properties. Compounds having the same molecular formulas but with different geometric structures are called **isomers** (Figure 14.3).

14.3 Bonding with Hybrid Orbitals*

In Chapter 8 we learned that two or more atoms can share some of their valence electrons, thereby forming covalent bonds. Earlier, in Chapter 5, we discussed how these valence electrons are located in atomic orbitals (or probability regions) of various shapes, surrounding the nucleus. A covalent bond is formed when two atoms are brought close enough to each other that an atomic orbital of one atom overlaps with an atomic orbital of the second atom. The strength of the covalent bond is determined

---

* *This section is optional and may be omitted without loss of continuity.*

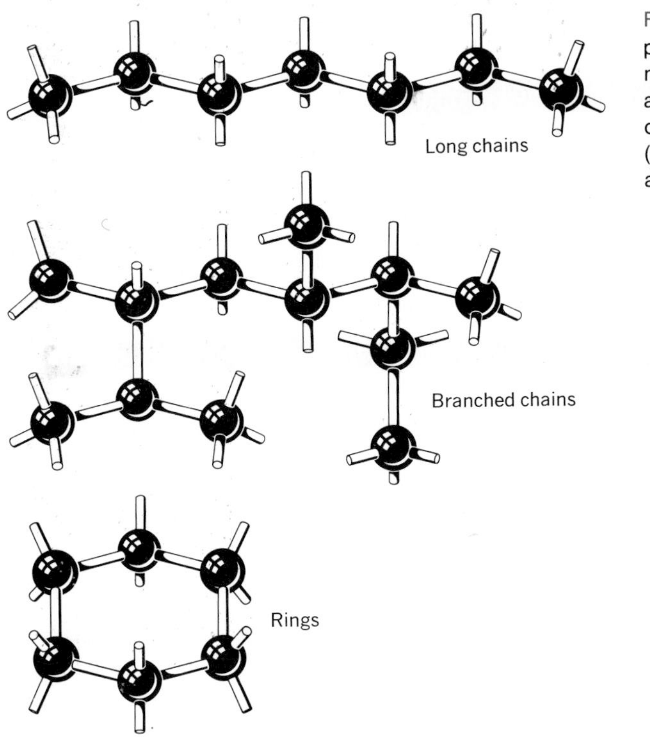

Figure 14.2 Some possible arrangements of carbon atoms found in organic compounds. (Only the carbon atoms are shown.)

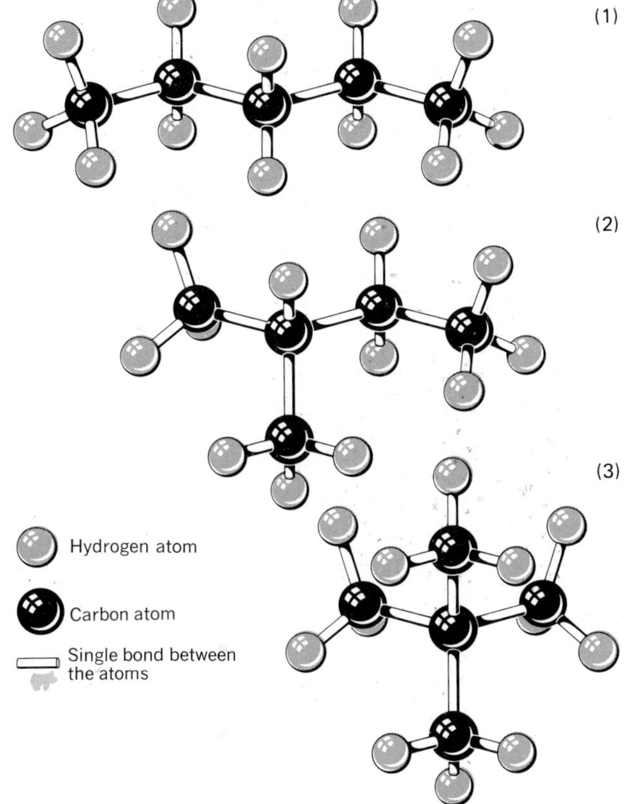

Figure 14.3 These molecules all have the molecular formula C_5H_{12}. They have different structural formulas and, therefore, are isomers.

by the amount of overlap of the two orbitals (different shaped orbitals will allow different amounts of overlap).

This "orbital-overlap" explanation of covalent bonding helps us understand the shapes taken on by the resulting molecules—in particular, it is used to explain the angles formed between different bonds in a molecular compound. However, experimental observations of the bond angles formed in some molecular compounds cannot be explained by the simple overlapping of *s*, *p*, or *d* atomic orbitals. It seems that in forming some bonds these atomic orbitals "mix together" to form **hybrid orbitals,** whose shapes allow for greater overlap—resulting in the formation of stronger bonds (Figure 14.4).

This theory of orbital **hybridization** is used to explain the geometry of carbon compounds. For example, it has been shown experimentally that carbon can form four equal single covalent bonds that are directed toward the four corners of an imaginary tetrahedron surrounding the atom (Figure 14.5). But in Example 5-3 we saw that carbon has a valence electron configuration of $2s^2$, $2p^2$. Given these two different types of orbitals, how could it be that carbon forms four equal covalent bonds?

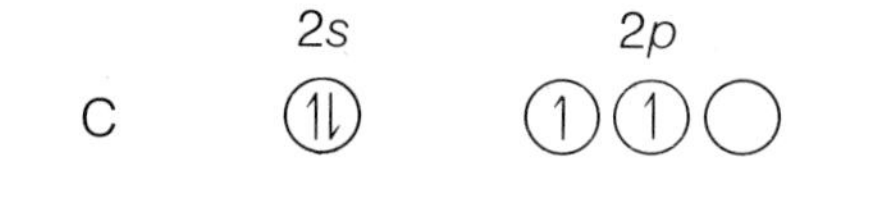

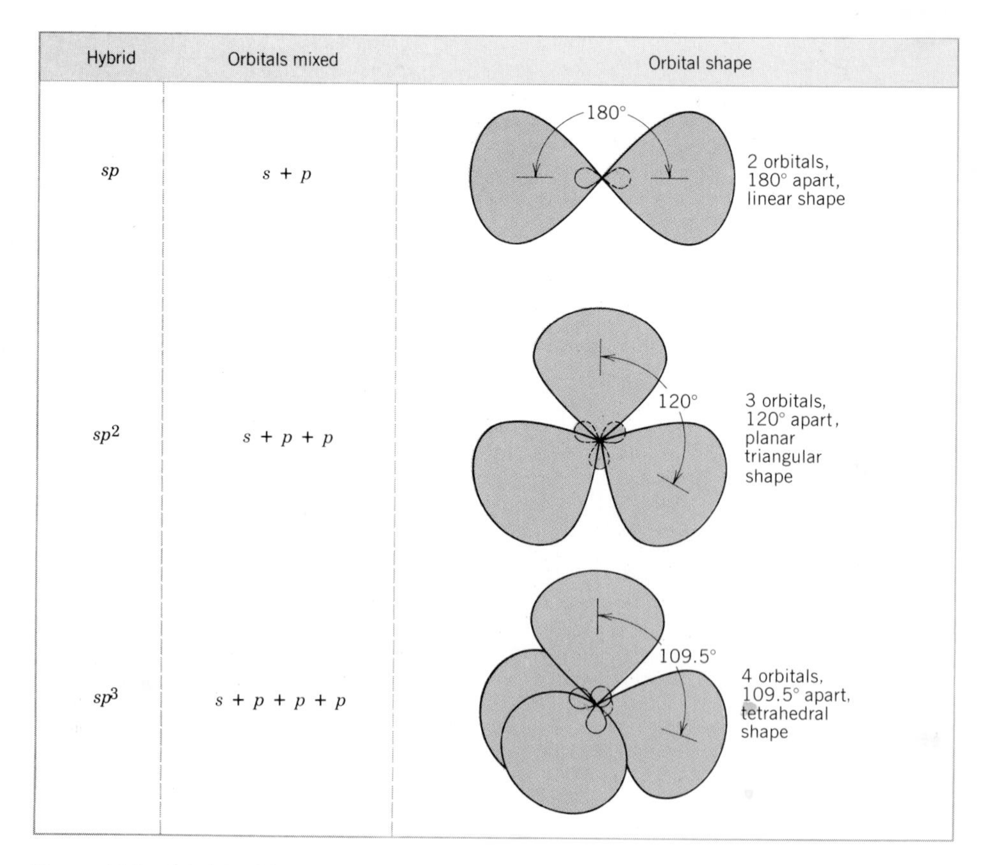

Hybrid	Orbitals mixed	Orbital shape
sp	*s* + *p*	180°; 2 orbitals, 180° apart, linear shape
*sp*2	*s* + *p* + *p*	120°; 3 orbitals, 120° apart, planar triangular shape
*sp*3	*s* + *p* + *p* + *p*	109.5°; 4 orbitals, 109.5° apart, tetrahedral shape

Figure 14.4 **Hybrid orbitals.**

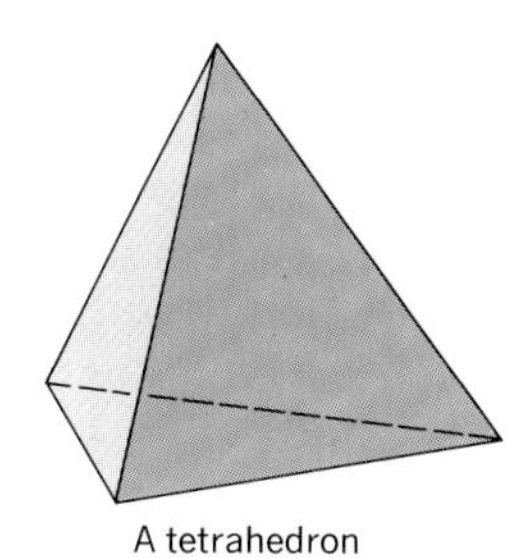

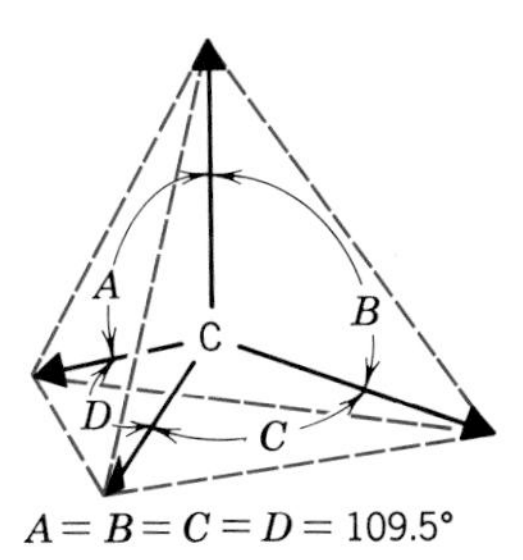

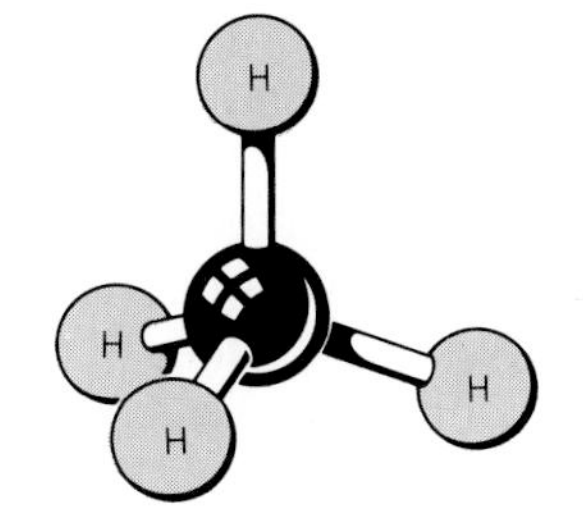

Figure 14.5 The tetrahedral carbon atom.

In order to form four equal covalent bonds, one of the 2*s* electrons moves to the empty 2*p* orbital. These four orbitals "mix," or hybridize, to form four equal sp^3 hybrid orbitals, each containing one electron and each directed toward a corner of the imaginary tetrahedron.

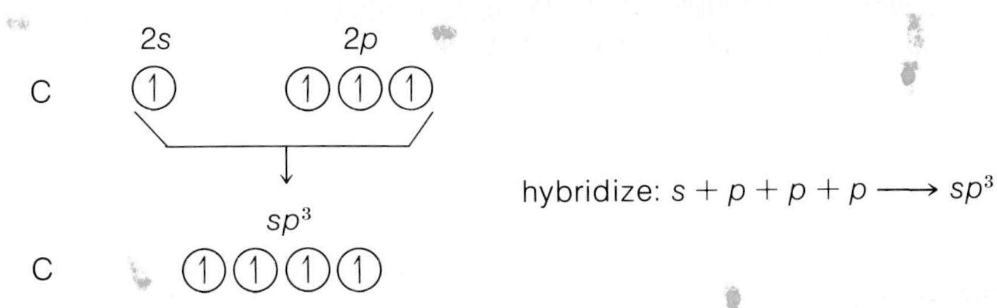

For example, in methane (CH_4) carbon uses four sp^3 orbitals to form four single covalent bonds with hydrogen. The resulting molecule has the expected tetrahedral shape (Figure 14.5). This characteristic tetrahedral bonding of the carbon atom is one of the primary features in the study of carbon-based compounds.

14.4 Organic Chemistry

Carbon's special properties make it unique among the elements. First, it is the only element whose atoms can bond together to form very long straight and branched chains, as well as intricate ring structures. (The atoms of silicon, directly below carbon on the periodic table, can form only short chains.) Second, carbon can also form strong bonds with a number of other elements: hydrogen, oxygen, nitrogen, sulfur, phosphorus, and the halogens. Third, these complex carbon compounds can have more than one arrangement of atoms (or many isomers). The resulting large number of different compounds has given rise to a field of chemistry devoted entirely to the study of just such carbon compounds; this is the field of **organic chemistry.** The number of known organic compounds, both natural and synthetic, is greater than 2 million, whereas the number of known inorganic compounds (those formed from combinations of all the other known elements) is only about 500,000. The reason for this comparatively small number of inorganic compounds is that molecules of covalent inorganic compounds are composed of only a few atoms. Two or three atoms will form a very stable inorganic molecule, but as more atoms are added, the molecule becomes unstable and is more likely to fall apart. Inorganic molecules containing more than 12 atoms are quite rare. However, it is not uncommon for a large organic molecule such as a complex protein to contain more than a million atoms!

You might not look forward to studying a field of chemistry covering more than two million compounds, but fortunately the chemical properties of many of these compounds are similar, allowing them to be conveniently grouped into several classes or series of compounds. In that way, by looking at one or two examples of the chemistry of organic compounds belonging to a given class, we can obtain an understanding of the chemistry of all compounds belonging to that class.

14.5 Hydrocarbons

One large class of organic compounds, whose molecules contain atoms only of carbon and hydrogen, are the **hydrocarbons.** These compounds are the basic building blocks from which all other organic compounds can be formed. A major source of hydrocarbons in commercial quantities is petroleum, which is a mixture of hundreds of different hydrocarbons. Petroleum was formed over a period of millions of years as large regions of plant life died, decayed, and were covered over by the slowly changing earth.

After crude petroleum is pumped out of the ground, it is transported to refineries where the hydrocarbons are separated by a process called **fractional distillation.** This process makes use of the fact that compounds composed of long hydrocarbon chains (high molecular weight) require more heat and higher temperatures to vaporize than do compounds composed of shorter chains (lower molecular weight). In other words, a hydrocarbon with a long carbon chain has a higher boiling point than one with a shorter carbon chain. In fractional distillation, the petroleum is heated slightly; this causes the molecules with short carbon chains to vaporize, and they are drawn off. As the heat is increased, molecules with longer and longer carbon chains vaporize, are drawn off, and are collected. The actual contents of each fraction that is drawn off may vary with the source of the petroleum. Each fraction may then be further processed to produce the particular hydrocarbons that are desired (Figure 14.6).

14.6 Writing Structural Formulas

Previously, we mentioned that in organic chemistry it is common to find several different compounds with the same molecular formula (see again Figure 14.3). We will often find it important, therefore, to be able to draw the **structural formula** of the molecule. Structural formulas are no more than

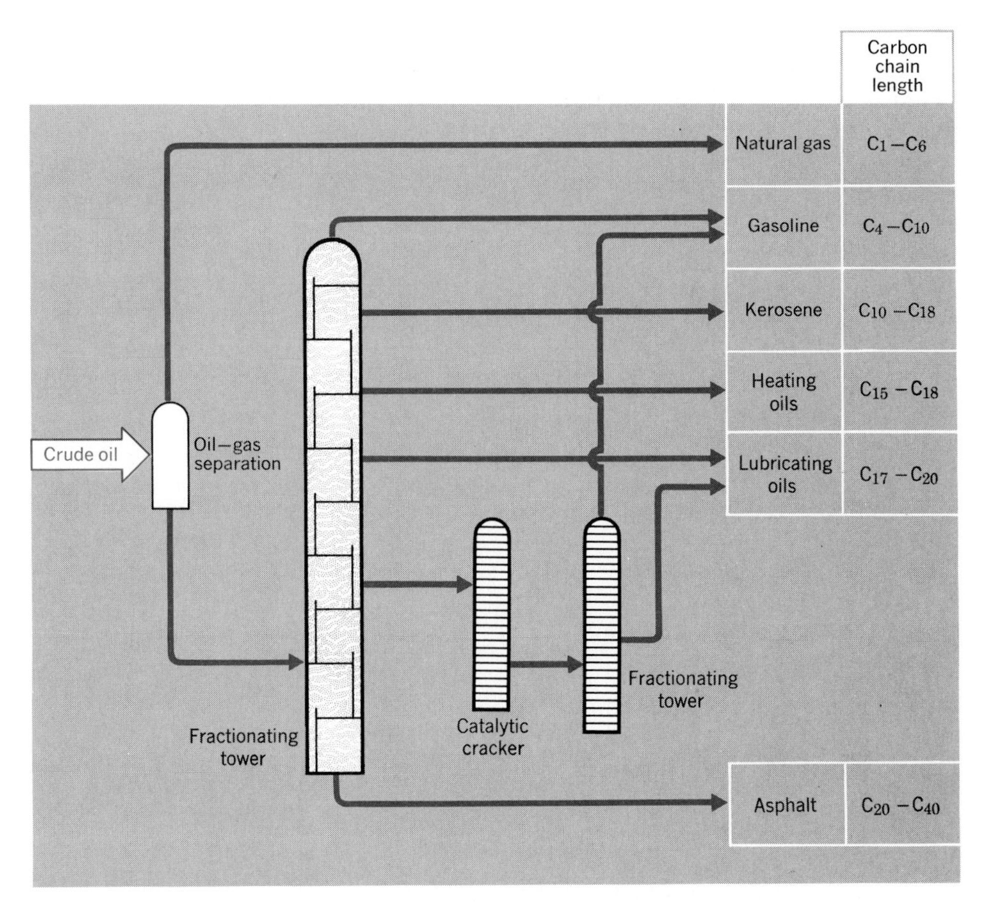

Figure 14.6 Fractional distillation of petroleum. The molecules found in petroleum (crude oil) are separated on the basis of their boiling points. The molecules that remain are called heavy bottoms; they have very long carbon chains and are used to make asphalt.

chemical diagrams. They make organic chemistry easier to follow in much the same way that diagrams can help in assembling a bicycle or in sewing a dress. Although chemical compounds actually have three-dimensional structures, it is often very difficult to draw them in that way. Instead, therefore, we will use two-dimensional diagrams to represent these three-dimensional structures. For example,

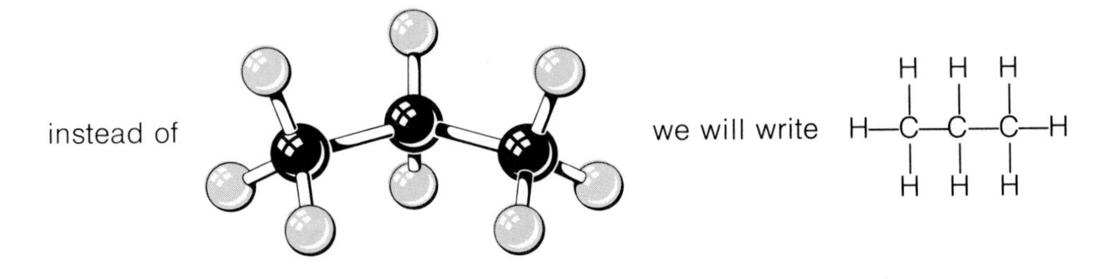

In the structural formula of a compound, the shared pair of electrons in a covalent bond is indicated by a line drawn between the two atoms connected by that bond. Each atom is represented by the symbol of its element. For example, the three isomers of C_5H_{12} shown in Figure 14.3 can be written as follows:

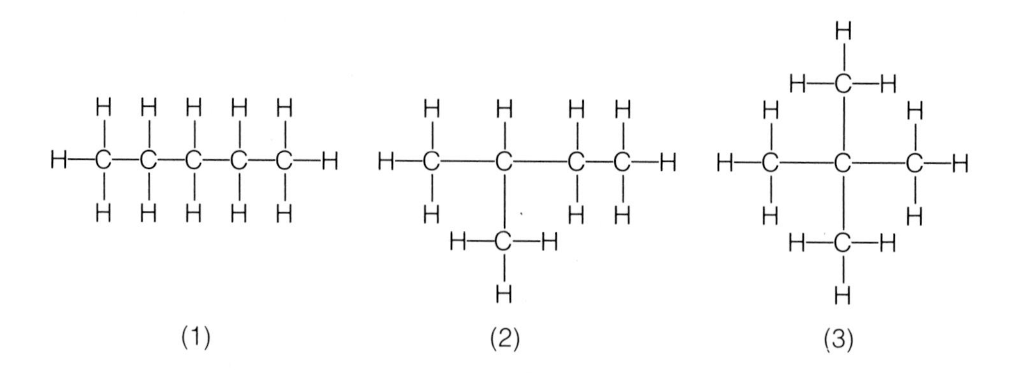

This subject would be fairly straightforward if we could stop our discussion right here, but drawing the structural formulas of large organic molecules can be quite time-consuming. Therefore, organic chemists have devised several ways of shortening or condensing the procedure. One commonly used method is to indicate the atoms that are bonded to the carbon by writing their symbol after the carbon symbol, but without bothering to use a dash for the bonds. For example,

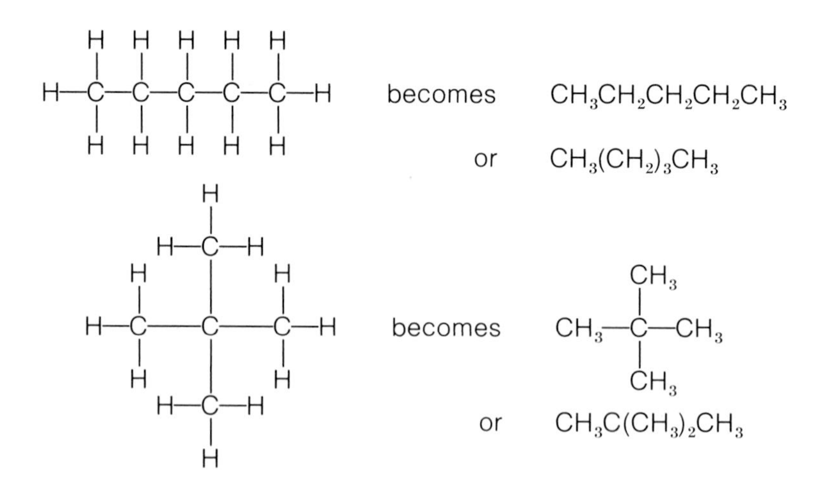

Or, quite often the part of the molecule in which we are interested is drawn out in detail, while the rest of the molecule is either represented in shorthand or simply denoted by the letter R.

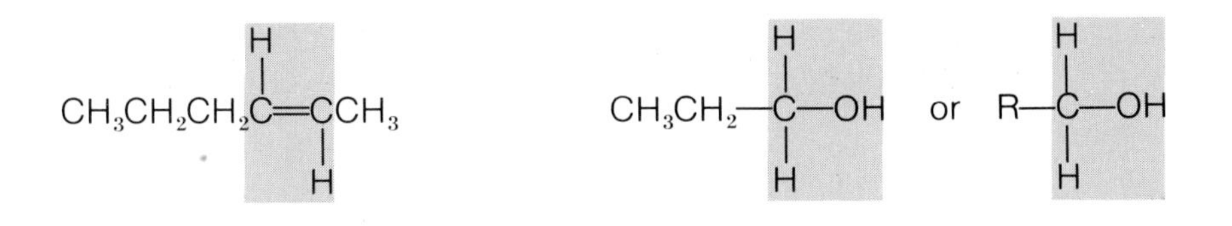

As you read the following chapters and see more structural formulas, you will find that they soon become quite easy to read and help make chemical discussions much easier to follow.

EXAMPLE 14-1

Write the complete structural formula for each of the following compounds:

(a) $CH_3(CH_2)_5CH_3$
(b) $CH_3CH(CH_3)CH_2CH(CH_2CH_3)CH_2CH_3$

When you write the structural formula for a carbon compound, you need to remember that each carbon must have four bonds—that is, four lines must connect to each carbon symbol.

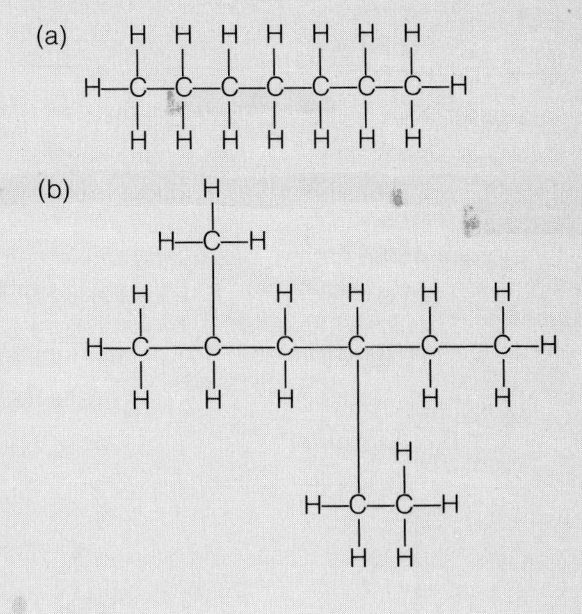

Exercise 14-1

1. Write the condensed structural formula for each of the following compounds:

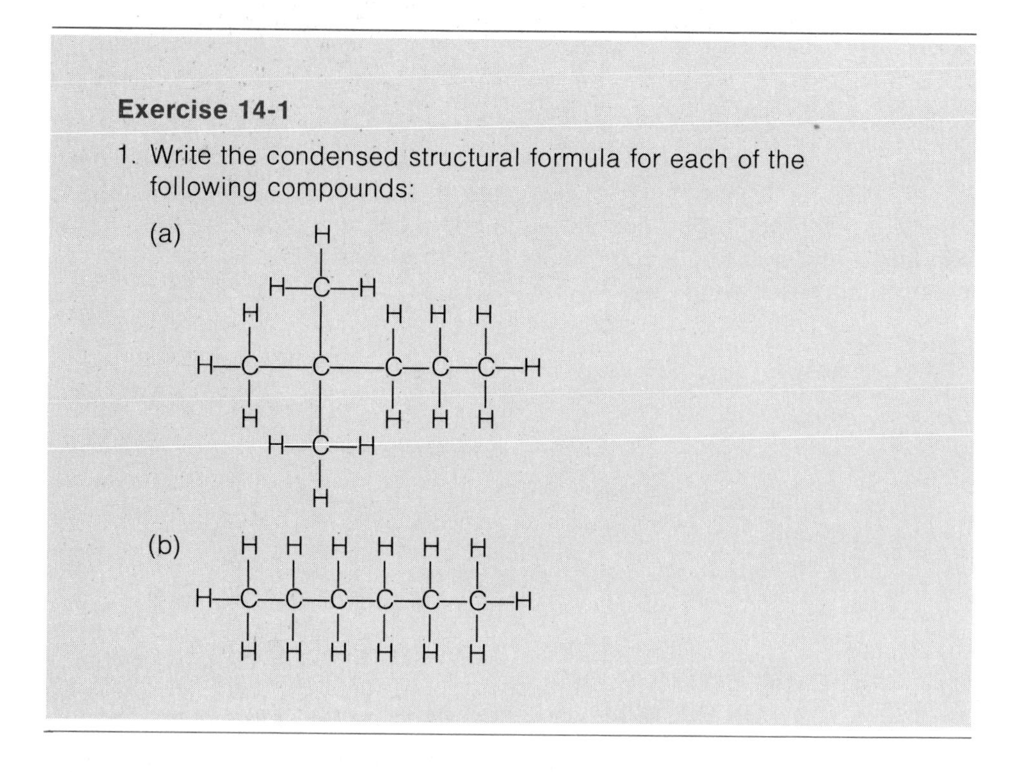

SATURATED HYDROCARBONS

14.7 Alkanes

The simplest of all hydrocarbons is the gas methane, CH_4. In this molecule the carbon is bonded to four hydrogen atoms that are located at the four corners of a tetrahedron. This makes the methane molecule symmetrical and nonpolar. It is not soluble in water (a highly polar solvent) and has a very low boiling point: − 161.5°C (Figure 14.7).

Methane is a major component of the natural gas used to heat homes and cook food. It is occasionally found concentrated in pockets in fields of coal and is one of the causes of explosions in coal mines. Methane can be formed by the action of bacteria on decaying matter and is found among

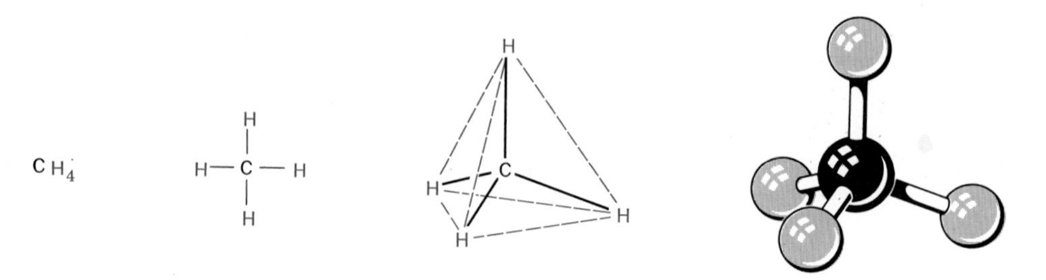

Figure 14.7 Methane.

Table 14.1 Some Alkanes

Number of Carbons	Molecular Formula	IUPAC Prefix	Name	Structural Formula	Boiling Point in °C
1	CH_4	meth-	Methane	CH_4	−162
2	C_2H_6	eth-	Ethane	CH_3CH_3	−89
3	C_3H_8	prop-	Propane	$CH_3CH_2CH_3$	−42
4	C_4H_{10}	but-	Butane	$CH_3CH_2CH_2CH_3$	0
5	C_5H_{12}	pent-	Pentane	$CH_3CH_2CH_2CH_2CH_3$	36
6	C_6H_{14}	hex-	Hexane	$CH_3CH_2CH_2CH_2CH_2CH_3$	69
7	C_7H_{16}	hept-	Heptane	$CH_3CH_2CH_2CH_2CH_2CH_2CH_3$	98
8	C_8H_{18}	oct-	Octane	$CH_3CH_2CH_2CH_2CH_2CH_2CH_2CH_3$	126
9	C_9H_{20}	non-	Nonane	$CH_3CH_2CH_2CH_2CH_2CH_2CH_2CH_2CH_3$	151
10	$C_{10}H_{22}$	dec-	Decane	$CH_3CH_2CH_2CH_2CH_2CH_2CH_2CH_2CH_2CH_3$	174

the gases that bubble out of marshes and swamps (and for that reason was once known as marsh gas). In the Middle Ages, the fires that resulted when these marsh gases were ignited were thought to be the spirits of the dead.

The second member of this group is ethane, CH_3CH_3 or C_2H_6, which is formed when a hydrogen on methane is replaced by a methyl group, —CH_3 (Figure 14.8). Methane and ethane are members of the hydrocarbon class of compounds called the **alkanes,** whose formulas all fit the general pattern of C_nH_{2n+2} (Table 14.1). The identifying characteristic of the alkanes is that in each molecule the carbons are bonded singly to four other atoms. That is, each carbon atom forms four single bonds, so additional atoms cannot be added to the molecule. Molecules having this property are said to be **saturated.** Single bonds between carbon atoms are strong and stable, making the alkanes the least reactive class of hydrocarbons.

Propane ($CH_3CH_2CH_3$ or C_3H_8) and butane ($CH_3CH_2CH_2CH_3$ or C_4H_{10}) are the next two members of the alkane series. They can be liquified in tanks under pressure, allowing them to be stored and transported easily.

C_2H_6

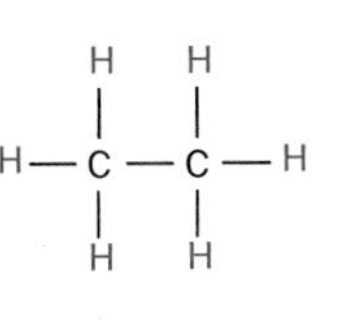

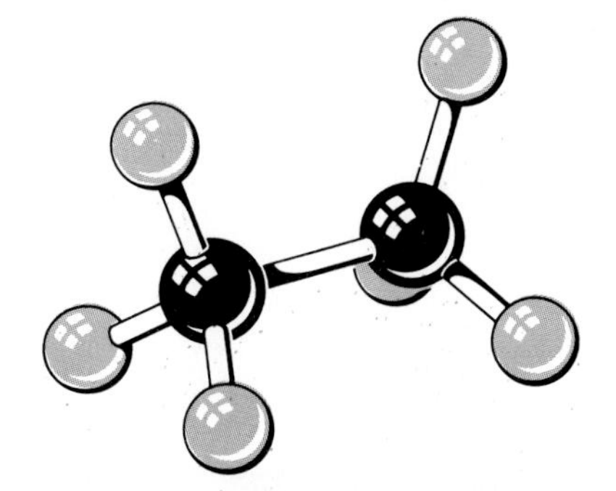

Figure 14.8 Ethane.

This gives them wide use as fuels for lighters, torches, and furnaces in rural homes.

14.8 IUPAC Nomenclature

Before a standard procedure for naming compounds was finally established, most organic compounds were known by common names that usually indicated the source of the compound rather than its chemical structure. For example, methane was called marsh gas. Given that there are more than 2 million known organic compounds, the study of organic chemistry would clearly be impossible without some standardized system of nomenclature that would indicate the structure of the molecule being named. For this reason, the International Union of Pure and Applied Chemistry (IUPAC) began meeting in Geneva in 1892 to establish rules for such a naming system. A complete discussion of the IUPAC rules for naming compounds must be left to other texts, but we will briefly describe some of the basic rules.

Rule 1. Organic compounds can be categorized into groups or classes. Each group is given a suffix that identifies the group.

Suffix	Class of Compounds	Example
-ane	Alkanes	Prop*ane*
-ene	Alkenes[a]	Prop*ene*
-yne	Alkynes[a]	Prop*yne*
-ol	Alcohols[a]	Propan*ol*
-one	Ketones[a]	Propan*one*
-oic acid	Carboxylic acids[a]	Propan*oic acid*

[a] These classes will be discussed in later chapters.

Rule 2. Prefixes are used to indicate the number of carbon atoms in the main carbon chain.

Number of Carbon Atoms	Prefix	Number of Carbon Atoms	Prefix
1	meth-	6	hex-
2	eth-	7	hept-
3	prop-	8	oct-
4	but-	9	non-
5	pent-	10	dec-

Rule 3. Groups of atoms that are attached to the main carbon chain and that are made up solely of carbon and hydrogen

Table 14.2 Some Common Alkyl Groups

Formula	Name	Formula	Name
CH_3—	Methyl	CH_3 | CH_3CH_2CH—	*sec*-Butyl (*s*-Butyl)
CH_3CH_2—	Ethyl		
$CH_3CH_2CH_2$—	Propyl (*n*-Propyl)	CH_3 | CH_3—C— | CH_3	*tert*-Butyl (*t*-Butyl)
CH_3 | CH_3CH—	Isopropyl	$CH_3CH_2CH_2CH_2CH_2$—	Pentyl (*n*-Pentyl)
$CH_3CH_2CH_2CH_2$—	Butyl (*n*-Butyl)		
CH_3 | CH_3CHCH_2—	Isobutyl	CH_3 | $CH_3CHCH_2CH_2$—	Isopentyl
		$CH_3CH_2CH_2CH_2CH_2CH_2$—	Hexyl (*n*-Hexyl)
		CH_3 | $CH_3CHCH_2CH_2CH_2$—	Isohexyl

arranged in straight or branched chains are called **alkyl groups.** These alkyl groups are named using the prefixes shown in Rule 2. When writing condensed or general structural formulas, entire alkyl groups are often represented simply by the letter R. Some common alkyl groups are shown in Table 14.2.

Rule 4. Atoms or groups of atoms (other than hydrogen) attached to the carbon chain are listed before the name of the compound. For example, bromomethane is a compound in which a bromine atom has replaced one of the hydrogen atoms on a molecule of methane; chloroethane is a compound in which a chlorine atom has replaced one of the hydrogens on a molecule of ethane.

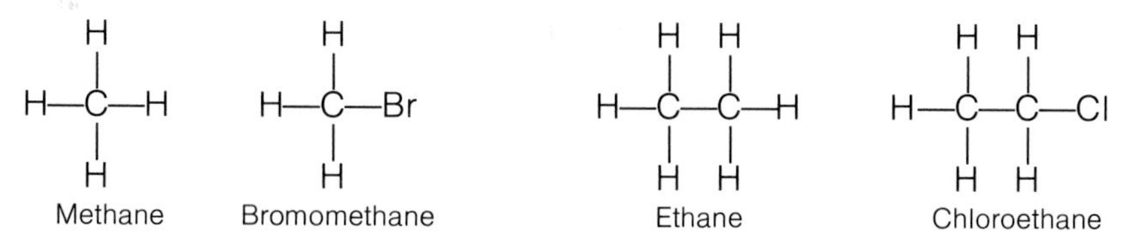

The names of several important groups often found attached to the carbon chain are:

Group	Name	Group	Name
—F	fluoro-	—I	iodo-
—Cl	chloro-	$—NH_2$	amino-
—Br	bromo-	$—NO_2$	nitro-

Rule 5. If the organic compound contains more than one of the same type of these "other" atoms, the number is indicated by a prefix: *di-* indicates two, *tri-* indicates three, and *tetra-* indicates four. For example, dibromomethane is a methane molecule containing two atoms of bromine, and tetrachloromethane is a methane molecule containing four atoms of chlorine.

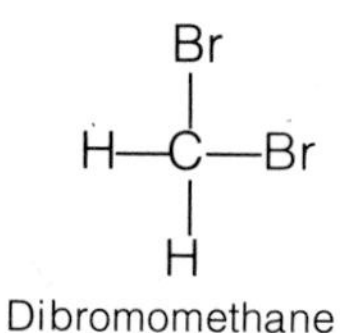

Dibromomethane

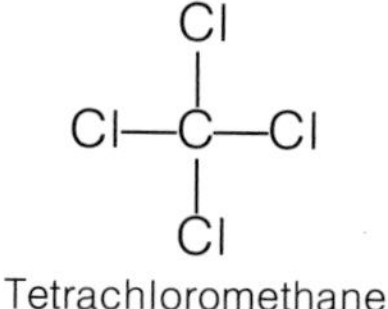

Tetrachloromethane
(Carbon tetrachloride)

Rule 6. If two or more different groups are attached to the carbon chain, they are named either in order of the length of the attached chain (for example, methylethylamine), or simply in alphabetical order (for example, ethylmethylamine).

Rule 7. To indicate the carbon atoms to which these "other" atoms are attached, numbers corresponding to the carbon atoms precede the name of the attached group. The carbon atoms are numbered from the end of the chain nearest the attached group. For example, 1,2-dichloroethane is an ethane molecule containing two atoms of chlorine—one attached to carbon number 1 and one attached to carbon number 2.

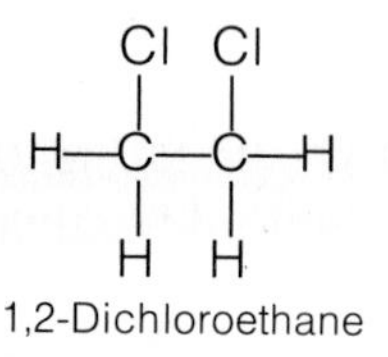

1,2-Dichloroethane

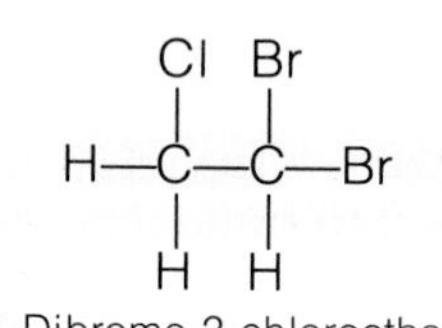

1,1-Dibromo-2-chloroethane

EXAMPLE 14-2

Let's see how these rules work in actual practice by naming the following compound.

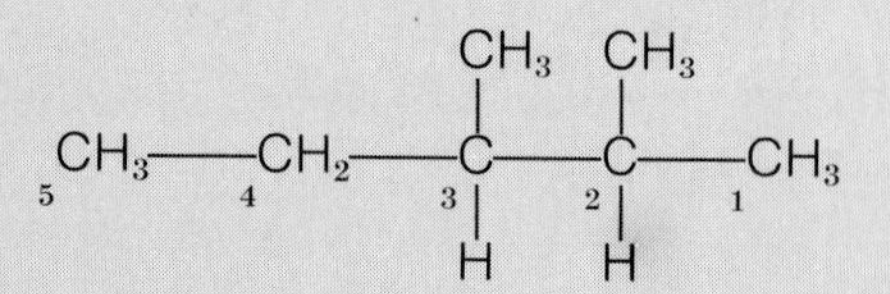

(a) First we must identify the class of hydrocarbons to which this compound belongs. All the bonds are single bonds, so it is an alkane. From Rule 1, we know that the name of all alkanes ends in *-ane*.

(b) Next we count the number of carbons in the main, or longest, chain. In this case, there are five carbons. From Rule 2, therefore, we know that the prefix to use is *pent-;* the compound is a pentane.
(c) There are two —CH_3, or methyl, groups attached to the carbon chain. Rule 5 tells us that the correct prefix in this case is *dimethyl,* making the compound a dimethylpentane.
(d) To indicate the position of the methyl groups, we number the carbons in the main chain, starting with the end that is nearest the attached groups. The methyl groups, therefore, are attached to carbon 2 and carbon 3.
(e) Putting all the above together, the name of the compound is 2,3-dimethylpentane.

Exercise 14-2

Name the compounds shown in Example 14-1.

14.9 Structural Isomers

There are two different compounds that have the molecular formula C_4H_{10}. One is butane (common name, *n*-butane), whose carbons all lie in a straight chain; the other is 2-methylpropane (common name, isobutane), whose carbons are arranged in a branched chain (Figure 14.9). The

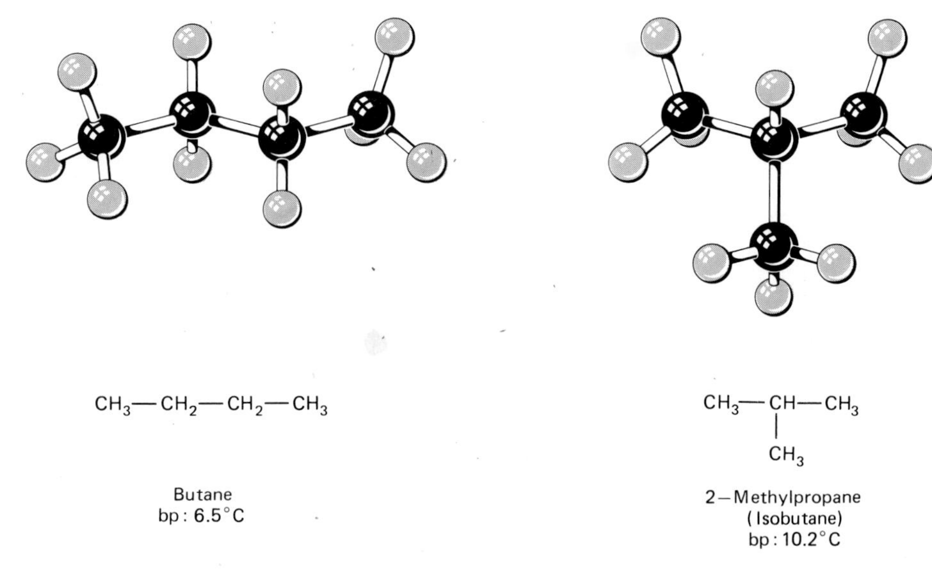

Figure 14.9 Butane and 2-methylpropane (isobutane) are structural isomers having the same molecular formula, C_4H_{10}.

difference between these two compounds, therefore, is that their carbon and hydrogen atoms are arranged in a different order in the three-dimensional structure of the molecule. Such compounds are called **structural isomers.** It is important to note that structural isomers are, in fact, different compounds and will have different physical and chemical properties.

Although there are only two structural isomers of butane, the number of possible isomers increases as the number of carbon atoms in a molecule increases. Octane, which is the eight-carbon alkane, has 18 different structural isomers. Each octane isomer behaves in a slightly different way. One of them, "isooctane," burns very well in car engines and is used as a standard in determining the octane rating of a gasoline.

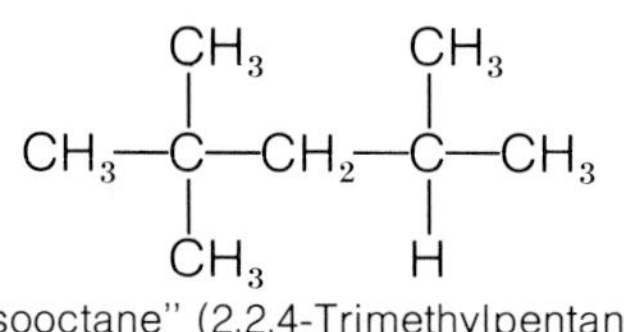

"Isooctane" (2,2,4-Trimethylpentane)

Someone once calculated that a 40-carbon compound would have more than 60 trillion possible isomers!

Exercise 14-3

Draw structural formulas for nine of the structural isomers of octane, C_8H_{18}.

14.10 Reactions of Alkanes

Oxidation

You are probably most familiar with alkanes as the fuels that heat your home, run your car, and power your camping stove and lantern. In general, hydrocarbons are fairly unreactive compounds. One of the reactions that these compounds do undergo, however, is burning, which is an oxidation reaction with oxygen. In this reaction, oxygen atoms from the air combine with the carbon and hydrogen atoms in the hydrocarbon until the carbon and hydrogen atoms are all bonded to oxygens. The reaction is exothermic, and the end products of the reaction are carbon dioxide (CO_2) and water (H_2O). This reaction with oxygen, by the way, is occurring all the time, even at room temperature. However, it usually occurs at a rate too slow to be noticed. If the compound is heated in the presence of air, this process will speed up until a point is reached, called the ignition point, at which the heat that is generated can be seen and felt. It is when we actually see and feel the reaction occurring that we say the substance is burning and refer to the reaction as *combustion.* Although the process of oxidation occurring in a forest fire is easily seen and is impressive in its force, even the results of very slow oxidation can be observed by noticing the rusting of old farm

Figure 14.10 Both of these photographs illustrate oxidation reactions, but the reactions are occurring at vastly different rates. (Top, Courtesy U. S. Forest Service; bottom, Phyllis Lefohn.)

equipment (Figure 14.10). The equation for the oxidation reaction that occurs when methane is burned is

$$CH_4 + 2O_2 \longrightarrow CO_2 + 2H_2O + \text{Energy}$$

Oxidation reactions are occurring continuously in our bodies and, in fact, are the source of energy for our cells. If this oxidation occurred in the same way as methane burns, our cells would be destroyed by heat. Oxidation occurs in the body at a very controlled rate, however, allowing cells to trap and use a portion of the energy that is released. Actually, our bodies do not use alkanes as a fuel supply, but rather use derivatives of alkanes—carbohydrates, fats, and proteins. The oxygen needed for oxidation in the body is supplied to the tissues by the red blood cells, which pick up oxygen in the lungs and distribute it throughout the body, returning the waste product carbon dioxide to the lungs to be exhaled. The water produced in the oxidation reaction either remains in the cells and tissues, or is excreted through sweat and urine. Just as a fire can be smothered by a blanket that cuts off the oxygen supply, we can be smothered by anything that cuts off our supply of oxygen for as short a period as five minutes.

If there is an insufficient amount of oxygen present when alkanes are burned, incomplete combustion (or incomplete oxidation) occurs, producing end products of carbon monoxide (CO) and water. For example, if methane is burned in an oxygen-poor environment, the following reaction occurs.

$$2CH_4 + 3O_2 \longrightarrow 2CO + 4H_2O + \text{Energy}$$

The carbon monoxide produced by incomplete combustion in factories and cars is the largest source of carbon monoxide pollution in our society. Studies have shown that the carbon monoxide from automobile exhaust can speed up the formation of photochemical smog.

Although carbon dioxide is a normal part of the environment of living cells, carbon monoxide is toxic. When inhaled, carbon monoxide greatly reduces the oxygen-carrying ability of the blood by binding very strongly with the hemoglobin molecules in red blood cells. At high levels, this reduction can be so great as to produce coma and death, but even at low levels, disruption of the central nervous system can be measured. Cigarette smokers inhale carbon monoxide into their lungs together with the cigarette smoke, producing elevated levels of carbon monoxide in their bloodstreams. These levels are often sufficient to produce measurable effects on the central nervous system. Even nonsmokers who work in smoke-filled rooms or live with smokers have elevated levels of carbon monoxide in their blood. City drivers who smoke can easily elevate their carbon monoxide level enough to produce headaches, dizziness, and fatigue—the first symptoms of carbon monoxide poisoning (Table 14.3).

Our bodies have several different ways of responding to elevated carbon monoxide levels, all of which result in increased strain on the heart and increased risk of heart disease. Short-term high levels cause the heart to pump faster, and long-term low levels cause the body to increase the number of red blood cells that carry oxygen, thereby thickening the blood and increasing the work load on the heart.

Table 14.3 The Effects of Carboxyhemoglobin[a] Blood Levels on the Human Body

Blood Levels of Carboxyhemoglobin	Effects
2 to 5%	Impairment of the central nervous system
5%	Impairment of perception and psychomotor performance
10%	Oxygen transport significantly impaired
15%	Headaches, dizziness, and lassitude
15 to 40%	Ringing ears, nausea, vomiting, heart palpitations, difficulty breathing, muscular weakness, apathy
40% and above	Collapse, coma, and death

[a] Carboxyhemoglobin is a hemoglobin–carbon monoxide complex that is formed when carbon monoxide enters the blood.

Substitution Reactions

Under the proper chemical conditions, alkanes can also react with nitric acid (HNO_3) and with the elements of group VII (the halogens: F_2, Cl_2, Br_2, I_2) in reactions called **substitution reactions.** These reactions are so named because another atom or group of atoms is substituted for one or more of the hydrogens on the alkane. For example, when methane reacts with nitric acid, nitromethane is formed.

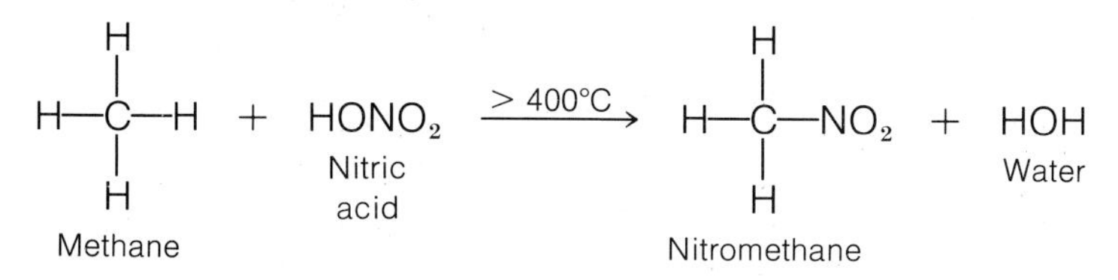

In the above chemical equation you will notice that > 400°C has been written above the arrow. When special conditions are required for a reaction to occur, they are often indicated in this way. In this case, then, the temperature must be greater than 400°C for the reaction to occur. Nitromethane, the product of this reaction, is used as a solvent, as an important chemical in the production of other organic compounds, and as a high-energy fuel for racing cars.

If a mixture of methane and chlorine is kept in the dark at room temperature, the mixture will not react. But when the mixture is exposed to light or heat, a reaction will begin immediately, producing chloromethane and hydrogen chloride.

$$CH_4 + Cl_2 \xrightarrow{\text{Light}} CH_3Cl + HCl$$

EXAMPLE 14-3

If chloromethane from the preceding reaction is allowed to react with additional chlorine, a mixture of chlorination products is formed as additional hydrogens on the methane are replaced by chlorines. Write the equation for each of these reactions and name each of the products.

$$CH_4 + Cl_2 \longrightarrow CH_3Cl + HCl$$

$$CH_3Cl + Cl_2 \longrightarrow CH_2Cl_2 + HCl$$

$$CH_2Cl_2 + Cl_2 \longrightarrow CHCl_3 + HCl$$

$$CHCl_3 + Cl_2 \longrightarrow CCl_4 + HCl$$

With the help of Rules 4 and 5 in Section 14.8, we can determine that

CH_3Cl is chloromethane (also known as methyl chloride).
CH_2Cl_2 is dichloromethane (also known as methylene chloride).
$CHCl_3$ is trichloromethane (also known as chloroform).
CCl_4 is tetrachloromethane (also known as carbon tetrachloride).

Each of these compounds is widely used as a solvent.

Exercise 14-4

1. Write the balanced equation for the complete combustion of hexane.
2. Chloroethane (ethyl chloride) is a fast-acting, topical local anesthetic. Write the equation for its formation from ethane.

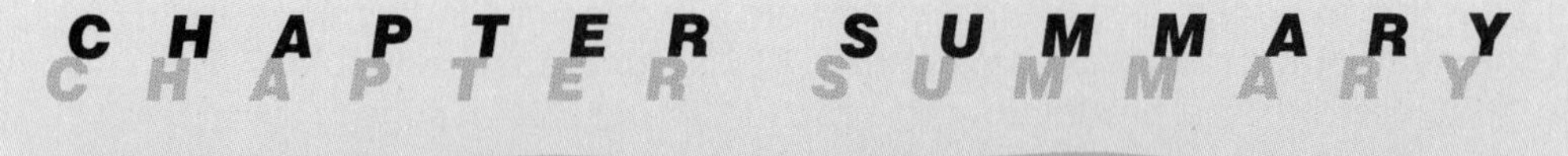

Living organisms are made up of compounds containing only 25 of the 90 naturally occurring elements. Of these elements, the four most abundant are carbon, hydrogen, oxygen, and nitrogen. Carbon-containing compounds are the building blocks of life. The carbon atom is especially suited for this purpose because it can form strong stable bonds with up to four other carbon atoms. Molecules can have carbon atoms connected in long chains, branched chains, or rings—resulting in a great variety of molecules.

The field of chemistry that studies carbon compounds is called organic chemistry.

Hydrocarbons are a large group of organic compounds composed only of carbon and hydrogen. Compounds that have the same molecular formula, but a different ordering of the carbon and hydrogen atoms within the molecule are called structural isomers. Hydrocarbon molecules are nonpolar and will dissolve only in other nonpolar substances. Hydrocarbons are saturated when the molecules contain only carbon-to-carbon single bonds. The alkanes, compounds whose general formula is C_nH_{2n+2}, are saturated hydrocarbons. Alkanes are relatively unreactive, but will enter into oxidation and substitution reactions.

Important Equations

Substitution Reaction of Alkanes

With nitric acid

$$CH_4 + HNO_3 \longrightarrow CH_3NO_2 + H_2O$$

With halogens (halogenation)

$$CH_4 + Cl_2 \xrightarrow{\text{Light}} CH_3Cl + HCl$$

Oxidation of Alkanes

$$CH_4 + 2O_2 \longrightarrow CO_2 + 2H_2O$$

1. Carbon is often said to form the "backbone" of life molecules. What does this mean, and what special properties of carbon make this possible?
2. Define *organic chemistry* in your own words.
3. What is the difference between a molecular and a structural formula of a compound?
4. Write the structural formulas for the five isomers of C_6H_{14}.
5. State whether the following pairs of structures are (a) the same compound, (b) isomers, or (c) unrelated compounds. (Note that to be the same compound, the structural formula must have the identical sequence (or order) of atoms, no matter how they are arranged on the paper. If the molecular formulas are the same, but the sequence of atoms is different, then the compounds are isomers.)

(a) $CH_3—CH_2—CH_3$

```
            H
            |
   H   H    C   H   H
    \  |  / | \  |  /
       C    H    C
       |         |
       H         H
```

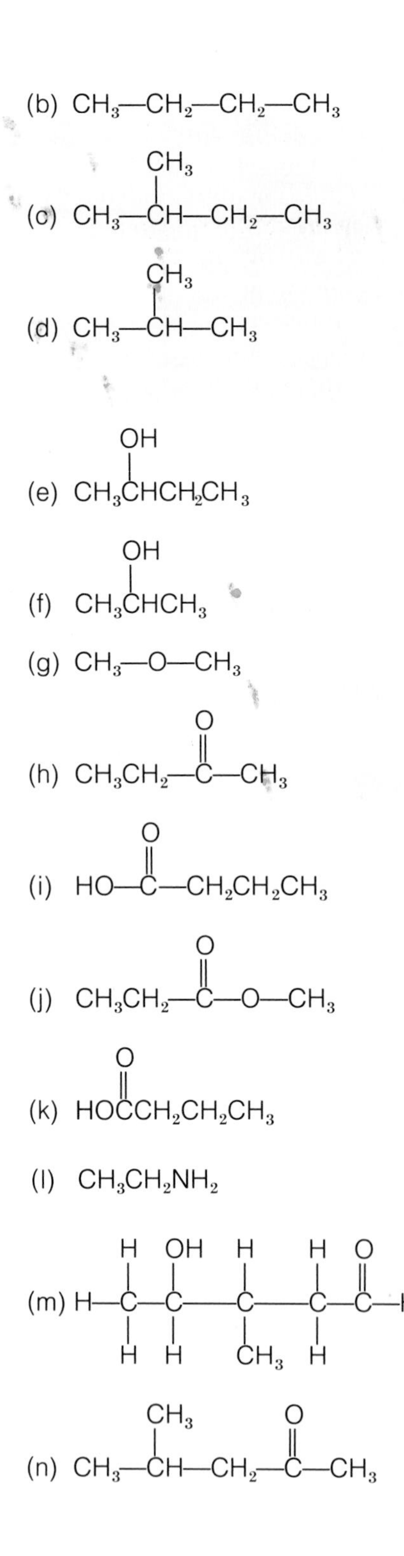
(b) CH3—CH2—CH2—CH3
CH3
(c) CH3—CH—CH2—CH3
CH3
(d) CH3—CH—CH3
OH
(e) CH3CHCH2CH3
OH
(f) CH3CHCH3
(g) CH3—O—CH3
O
(h) CH3CH2—C—CH3
O
(i) HO—C—CH2CH2CH3
O
(j) CH3CH2—C—O—CH3
O
(k) HOCCH2CH2CH3
(l) CH3CH2NH2
H OH H H O
(m) H—C—C—C—C—C—H
H H CH3 H
CH3 O
(n) CH3—CH—CH2—C—CH3

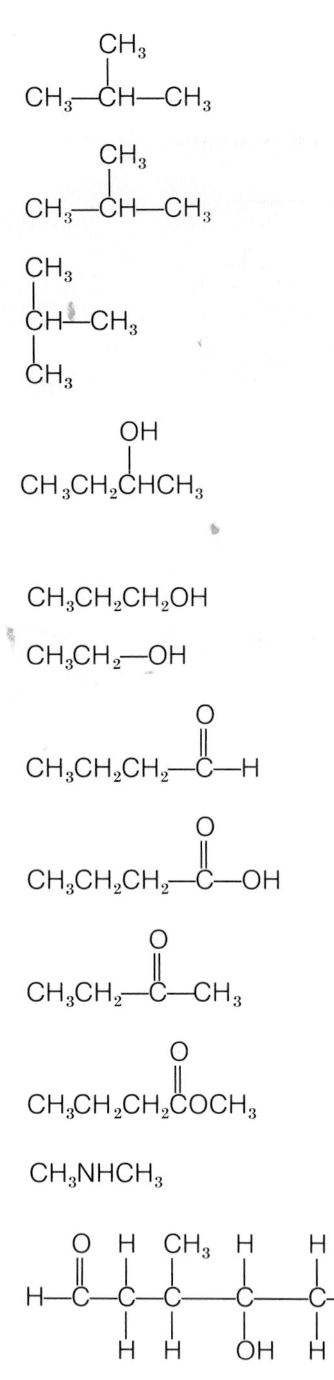
CH3
CH3—CH—CH3
CH3
CH3—CH—CH3
CH3
CH—CH3
CH3
OH
CH3CH2CHCH3
CH3CH2CH2OH
CH3CH2—OH
O
CH3CH2CH2—C—H
O
CH3CH2CH2—C—OH
O
CH3CH2—C—CH3
O
CH3CH2CH2COCH3
CH3NHCH3
O H CH3 H H
H—C—C—C—C—C—H
H H OH H
CH3 OH
CH3—CH—CH2—CH—CH3

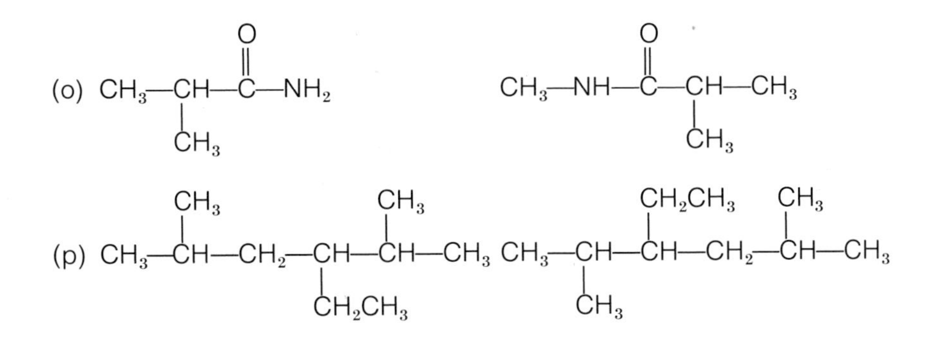

6. Give the IUPAC name for the following alkanes:

(a) $CH_3CH_2CH_3$

(b) $CH_3(CH_2)_7CH_3$

(c) $CH_3CHCH_2CH_3$
$\quad\quad |$
$\quad\; CH_3$

(d) $CH_3-CH-CH_2-CH-CH_3$
$\quad\quad\quad |\quad\quad\quad\quad |$
$\quad\quad CH_3\quad\quad\; CH_3$

(e) $(CH_3)_3CH$

(f)
$\quad\quad\quad CH_2CH_3$
$\quad\quad\quad |$
$CH_3CH_2CHCHCH_2CH_2CH_3$
$\quad\quad |$
$\quad\; CH_3$

(g) $CH_3CH_2-CH-CH_2CH_2CH_3$
$\quad\quad\quad\quad |$
$\quad\quad CH_3CHCH_3$

(h) $CH_3CH_2CH_2CHCH_3$
$\quad\quad\quad\quad |$
$\quad\quad\quad CH_2CH_2CH_3$

(i)
$\quad\quad\quad\quad CH_2CH_2CH_3$
$\quad\quad\quad\quad |$
$CH_3CH_2CH_2-C-CH_2CH_2CH_2CH_3$
$\quad\quad\quad\quad |$
$\quad\quad CH_3CHCH_3$

7. Write the structural formulas for the following compounds:

(a) 5-ethyldecane
(b) 2,2-dimethylpropane
(c) 4-isopropyloctane
(d) 1,1,2-trichlorobutane
(e) 2-iodo-2,4,6-trimethylheptane
(f) 5-butyl-2,4-dimethylnonane

8. State the IUPAC name for each of the following compounds:

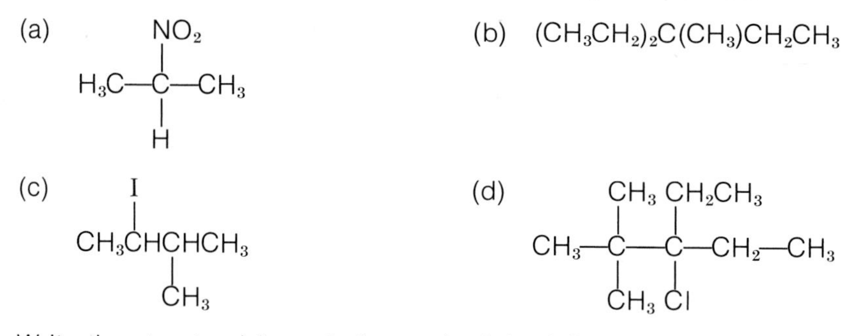

9. Write the structural formula for each of the following compounds:
   (a) 1,1,2,2-tetrabromoethane
   (b) 2-aminoheptane
   (c) 3-chloro-2,3-dimethylhexane
   (d) 3,3-diethylpentane
   (e) 1,4-diaminobutane
   (f) 1-fluoro-4-isobutyloctane

10. Write the balanced equation for the complete combustion of
    (a) propane (b) octane (c) 3-methylpentane

11. Write the equation for the monosubstitution reaction (the reaction in which only one hydrogen on the alkane is replaced) of
    (a) methane with bromine
    (b) ethane with nitric acid
    (c) propane with bromine (there are two different products possible; write the structural formulas for both)

CHAPTER FIFTEEN

CARBON AND HYDROGEN: UNSATURATED HYDROCARBONS

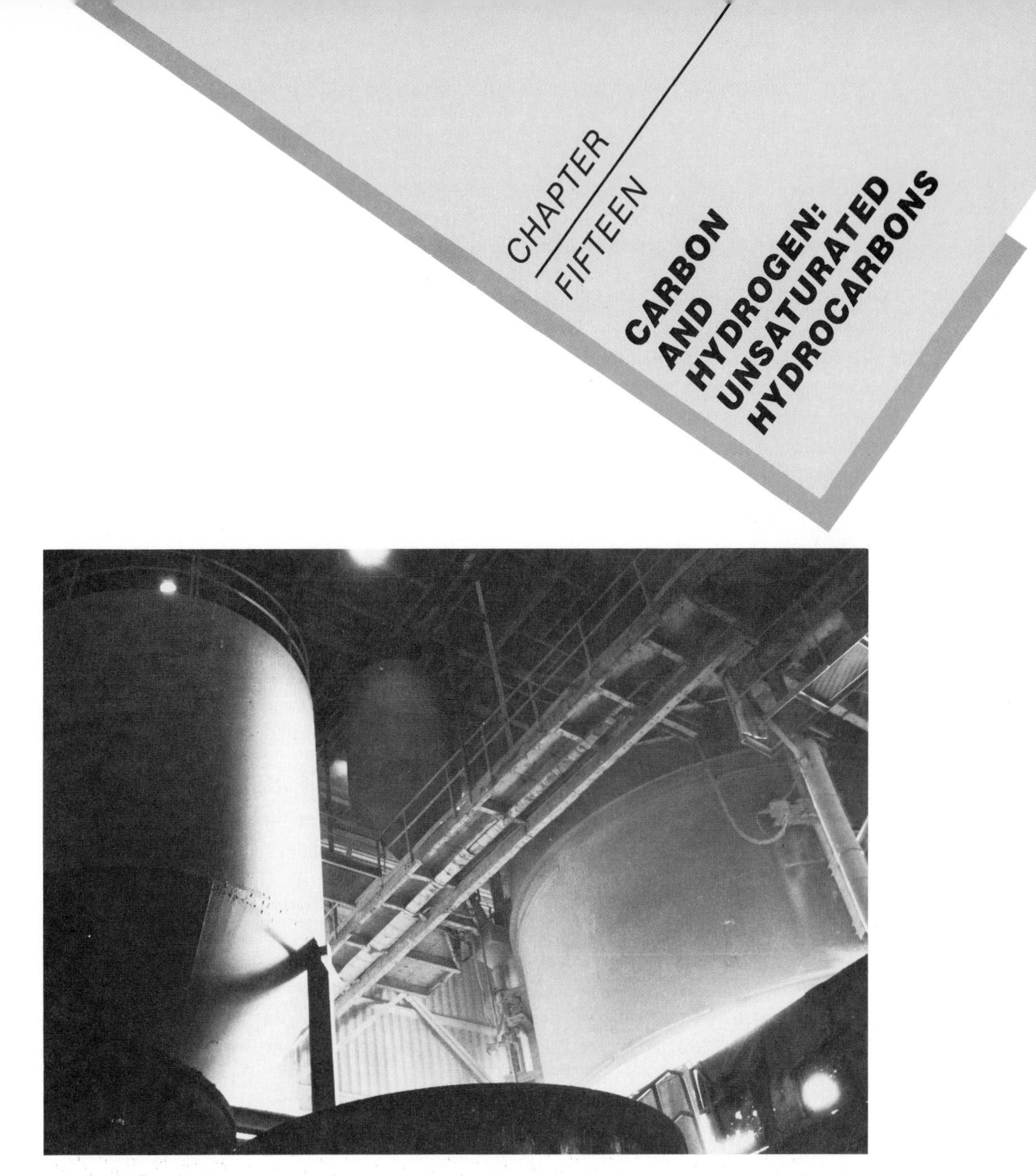

Joe's evening job as a maintenance worker often left a lot to be desired, but this night's repair job was especially difficult. He and his supervisor were repairing the filling mechanism on the top of a large cylindrical tank that usually held trichloroethylene (TCE) for the plant's metal-cleaning process. Standing close to the hatch that they had just opened, Joe could still detect the sweetish odor of TCE, even though the tank had been empty for several days.

His thoughts were suddenly interrupted by a yell and a crash as his supervisor tripped over the hatch and fell to the floor of the tank. After a heart-stopping moment the supervisor called up to Joe for help, saying that one of his legs was badly hurt. Joe quickly clambered down from the tank and ran to call the plant's security office. When he returned, he used a nearby ladder to climb down into the tank to help his supervisor.

Within ten minutes the paramedics had arrived, donned protective clothing and their respirators, and entered the tank. They found Joe and his supervisor sprawled unconscious at the bottom of the tank. Both men were rushed to the emergency room of the nearest hospital. The supervisor had not been breathing and had no pulse when he was pulled from the tank. In spite of immediate cardiopulmonary resuscitation (CPR) and intensive efforts to revive him in the emergency room, the patient was pronounced dead by the emergency room physician.

Joe also was unresponsive when pulled from the tank, but at least he was breathing and had a weak pulse. A series of tests at the hospital revealed that he was suffering from pulmonary edema, a build-up of fluid behind the lungs that prevents the lungs from fully expanding and thus decreases the amount of oxygen getting into the blood. Although Joe regained consciousness a few hours after arriving at the hospital, his head ached terribly, his speech was slurred, and he had difficulty finding the right words. Joe remained in the hospital for a week, but showed only slight improvement in his ability to speak. In fact, Joe had suffered irreversible brain damage and has remained handicapped ever since.

The substance that had been in the tank, and whose heavy vapors still filled the air in the lower part of the tank, was trichloroethylene ($ClCH{=}CCl_2$). TCE is just one of a large class of halogenated hydrocarbons that are used extensively as industrial solvents. As you can see, extreme care must be taken in their use because they can have profound effects on many body systems. In high concentrations, TCE can depress the central nervous system and cause loss of consciousness. It can also cause the heart to beat irregularly, which can lead to a complete stop of the heartbeat. When inhaled, TCE affects the tissues of the lungs. This irritation of lung tissue and the irregular heartbeat both can cause pulmonary edema, which lowers the oxygen in the blood and increases the chances of brain damage from lack of oxygen. Sadly, Joe's pulmonary edema and brain damage could have been prevented if only he had thought to put on protective clothing and a respirator before entering the tank to help his supervisor (Figure 15.1).

In this chapter we will continue our discussion of hydrocarbons by widening our study to include the classes of hydrocarbons that function as organic solvents such as trichloroethylene.

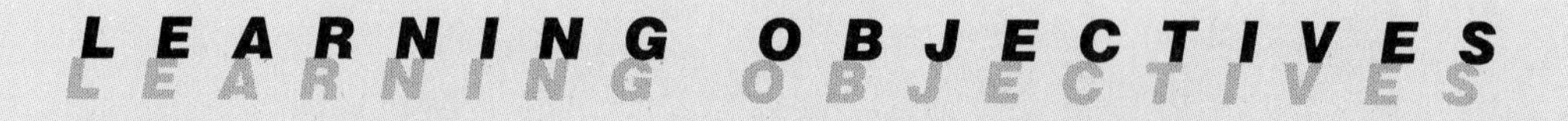

LEARNING OBJECTIVES

By the time you have finished this chapter, you should be able to:

1. Determine whether a hydrocarbon is saturated or unsaturated, given its structural formula.
2. Describe the difference between the alkanes, alkenes, and alkynes, and give examples of compounds found in each class.

3. Draw the structural formula of a hydrocarbon when given its IUPAC name, and state its IUPAC name when given its structural formula.
4. Define *geometric isomer,* and give examples of the *cis* and *trans* structures of an alkene and a cyclic hydrocarbon.
5. Compare the chemical reactivity of the alkanes and alkenes.
6. Write the equations for the combustion, addition, and polymerization reactions of a given alkene.
7. Describe the structure of the benzene molecule, and explain why the molecule is extremely stable.
8. Write the equation for the substitution reaction of benzene with a halogen.

ALKENES AND ALKYNES

15.1 Unsaturated Hydrocarbons

Our study of alkanes in Chapter 14 gave us just a hint of the wide variety of compounds making up the chemistry of carbon. We will now be studying compounds in which carbon can share four electrons with another carbon atom, forming a double bond, or can share six electrons with another carbon atom, forming a triple bond.

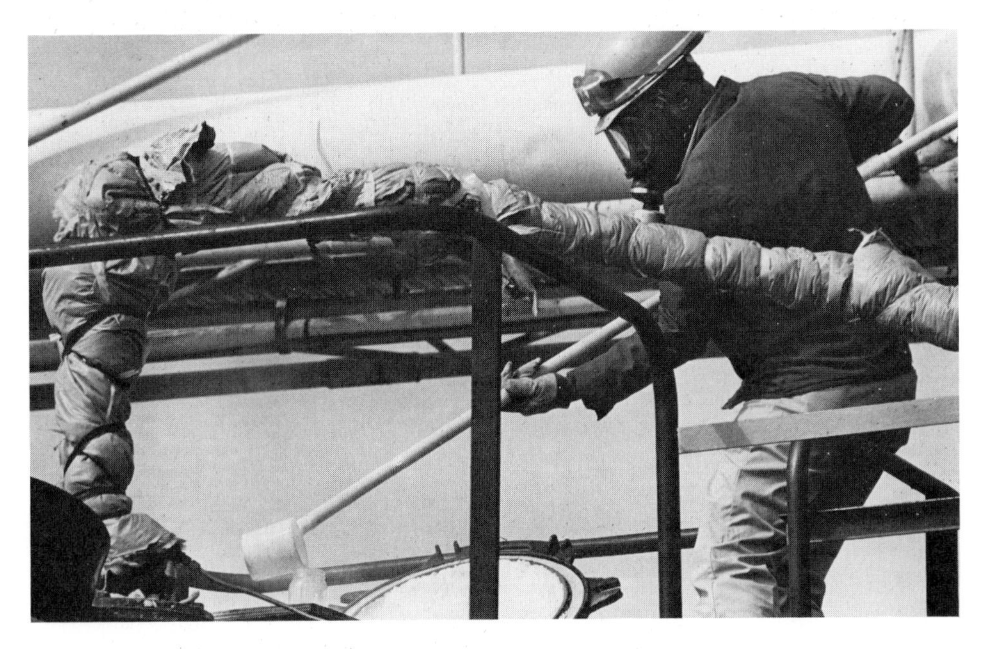

Figure 15.1 Wearing a respirator and protective clothing is critically important when working around potentially toxic chemicals. (John Coletti/Stock, Boston.)

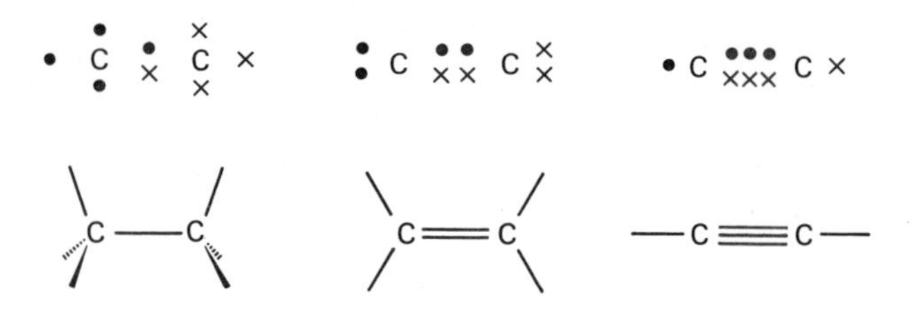

A double bond is less stable, or more reactive, than a single bond. Such multiple bonds, therefore, form an unstable, or reactive, spot in the organic molecule. Compounds having double or triple bonds can add more atoms to their molecules and are said to be **unsaturated.** If a compound has more than one double or triple bond, then it is **polyunsaturated.**

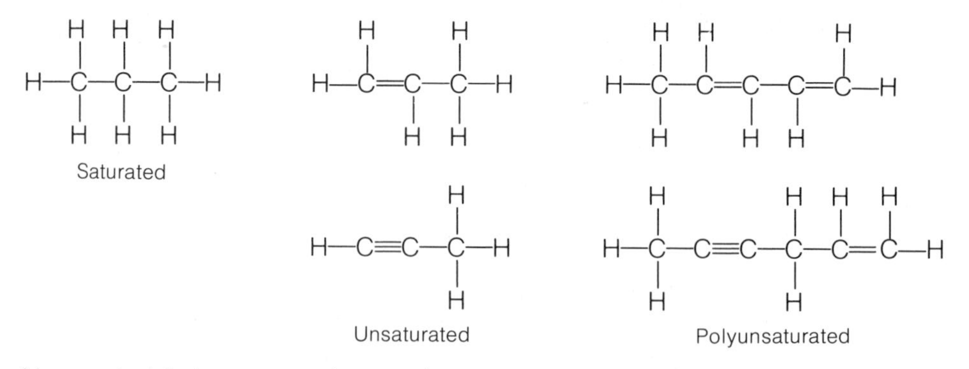

You probably have seen the word *polyunsaturated* before in margarine advertisements, and we will discuss polyunsaturated fats and oils in the chapter on lipids.

15.2 Alkenes

The class of compounds containing double bonds between carbon atoms is known as the **alkenes.** The simplest alkene is ethene, or ethylene, CH_2CH_2 or C_2H_4 (Figure 15.2). Ethylene is a flammable, anesthetic gas that is nontoxic to tissues even in high concentrations. Its anesthetic effects are rapid; a patient is ready for surgery two to four minutes after administration. A commercially useful property of ethylene gas is its ability to shorten the ripening time of citrus fruits. Notice in Figure 15.2 that the structure of ethylene is no longer tetrahedral. The most stable arrangement for this molecule results when all the atoms in the ethylene molecule are located in the same plane.

Larger alkene molecules can have two or more double bonds, which can be located anywhere in the molecule. Large alkene molecules can have dozens of double bonds, leading to millions of possible isomers (Table 15.1).

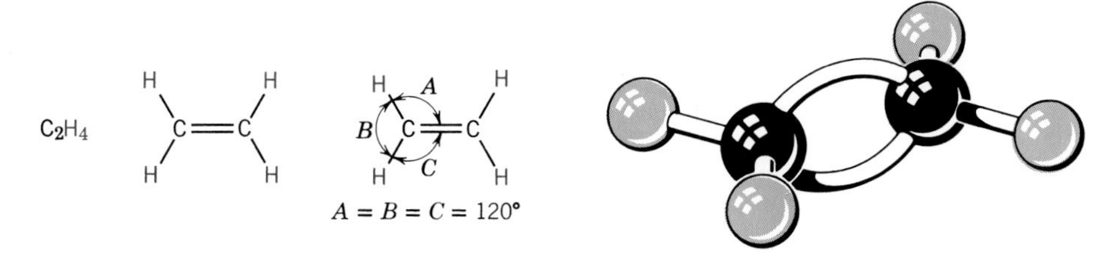

Figure 15.2 Ethene (ethylene).

15.3 Naming Alkenes

The IUPAC rules for naming alkenes are much the same as the rules for naming alkanes, but with the following changes:

1. All the names of alkenes end in **-ene.**
2. The carbon chain chosen for the main carbon chain is the longest carbon chain containing the double bond.
3. The carbons in the carbon chain are numbered starting at the end closest to the first double bond. A number prefix is used to indicate the carbon on each double bond that is nearest the beginning of the chain. For example, $CH_3CH{=}CHCH_2CH_3$ is 2-pentene, not 3-pentene or 2,3-pentene.
4. If two or more double bonds are found in the molecule, the prefixes *di-*, *tri-*, and so on are added to the name of the alkene.

Carefully study the names and structures given in Table 15.1 and in

Table 15.1 Some Alkenes

Number of Carbons	Number of Double Bonds	Molecular Formula	Name	Structural Formula
3	1	C_3H_6	Propene	$CH_2{=}CHCH_3$
4	1	C_4H_8	1-Butene	$CH_2{=}CHCH_2CH_3$
4	1	C_4H_8	2-Butene	$CH_3CH{=}CHCH_3$
4	1	C_4H_8	2-Methylpropene	$CH_2{=}C(CH_3){-}CH_3$
5	2	C_5H_8	1,3-Pentadiene	$CH_2{=}CHCH{=}CHCH_3$
5	2	C_5H_8	2-Methyl-1,3-butadiene	$CH_2{=}C(CH_3){-}CH{=}CH_2$

the following examples to see how these rules are used in the naming of alkenes. (Be sure to notice the way in which commas and hyphens are used in the names.)

EXAMPLE 15-1

1. Name the following compound.

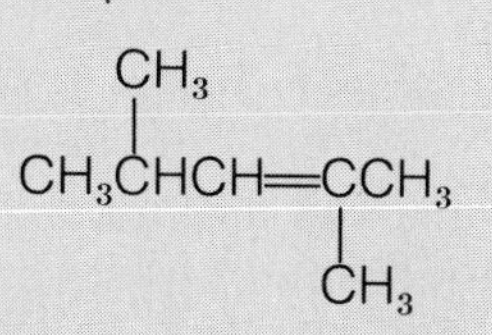

(a) The longest chain containing the double bond has five carbons. If you number the carbons from the end closest to the double bond, you will see that the last part of the name is 2-pentene.

(b) There are two methyl groups attached to the chain, one on carbon 2 and the other on carbon 4.

(c) The name is 2,4-dimethyl-2-pentene.

2. Draw the structural formula for 2-methyl-1,3-hexadiene.

(a) 1,3-Hexadiene tells us that the main carbon chain has six carbons and two double bonds: one starting on carbon 1 and the other starting on carbon 3.

C=C—C=C—C—C

(b) 2-Methyl tells us that there is one group attached to the chain: a methyl group on carbon 2. The rest of the bonds on the carbon atoms will be attached to hydrogen atoms. (But make sure that each carbon atom has only four bonds connected to it.)

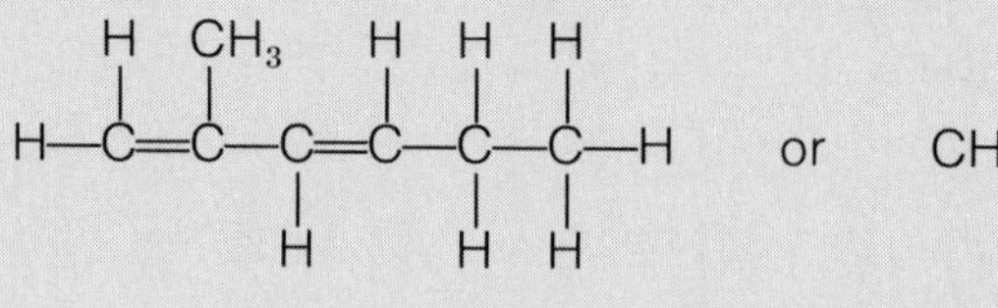

or

$$CH_2{=}C(CH_3)CH{=}CHCH_2CH_3$$

Exercise 15-1

1. Name the following compounds:

(a) $CH_3CH_2CH_2CH{=}CHCH_3$

(b)
$$\begin{array}{l} \quad\;\; CH_3 \\ \quad\;\;\, | \\ CH_2{=}C{-}CH_2{-}CH_3 \end{array}$$

(c)
$$\begin{array}{l} \quad\;\; CH_3 \\ \quad\;\;\, | \\ CH_2{=}C{-}CH{=}CHCH_3 \end{array}$$

(d)
$$\begin{array}{l} \qquad\qquad\quad CH_2CH_3 \\ \qquad\qquad\quad | \\ CH_3{-}C{=}CH{-}CH{-}CH_2{-}CH_2{-}CH_3 \\ \qquad\;\; | \\ \qquad CH_3 \end{array}$$

2. Draw the structural formula for each of the following compounds:

   (a) 2-methyl-3-hexene
   (b) 4-chloro-2-pentene
   (c) 2,3-dimethyl-2-heptene
   (d) 1,3-butadiene

3. Draw five of the structural isomers having the molecular formula C_5H_{10}, and name each compound.

15.4 The Pi Bond*

The double bond in an alkene is actually made up of two kinds of bonds. The first type is the **sigma bond** or **σ bond.** A sigma bond is formed by the overlap of atomic orbitals (*s*, *p*, or hybrid) directly along the bond axis (the imaginary line that connects the two nuclei; see Figure 15.3).

The second type of bond is a **pi bond** or **π bond,** which is formed by the overlap of a *p* orbital above and below the bond axis. That is, the *p* orbital overlaps sideways rather than end to end as in a σ bond (Figure 15.4). Note that a pi bond consists of two parts: two regions that lie on opposite sides of the bond axis.

The formation of pi bonds allows atoms to share more than one pair of electrons between their nuclei. Let's look at ethene.

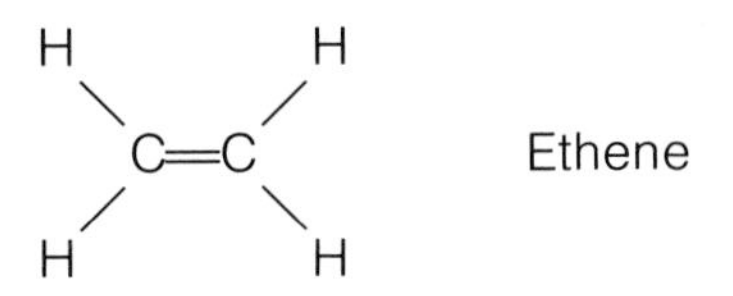

First, the atoms of this molecule must be joined together by σ bonds. Each carbon atom will form three sigma bonds from sp^2 hybrid orbitals, as follows:

* *This section is optional and may be omitted without loss of continuity.*

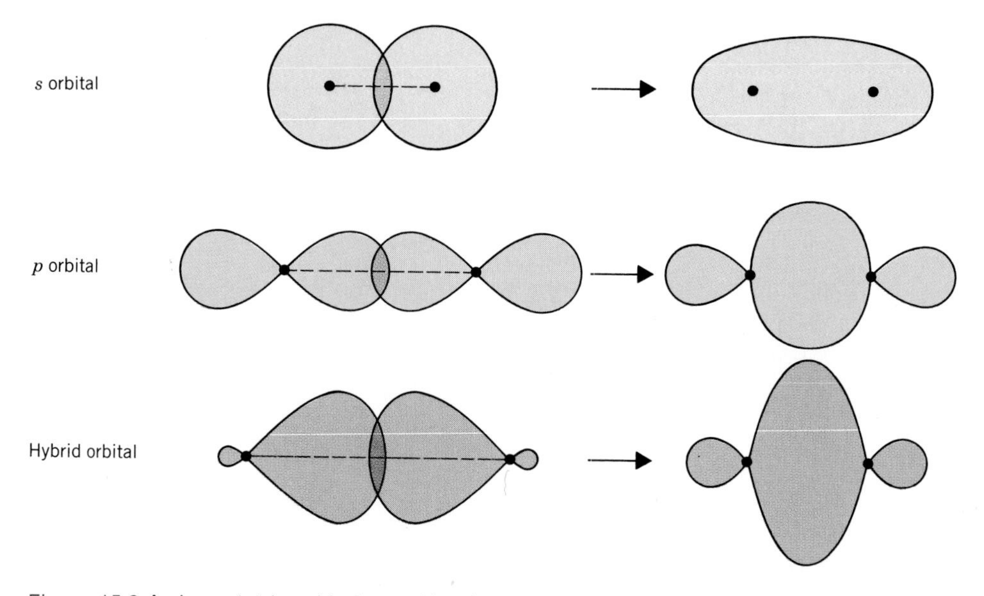

Figure 15.3 A sigma (σ) bond is formed by the overlap of atomic orbitals along the bond axis.

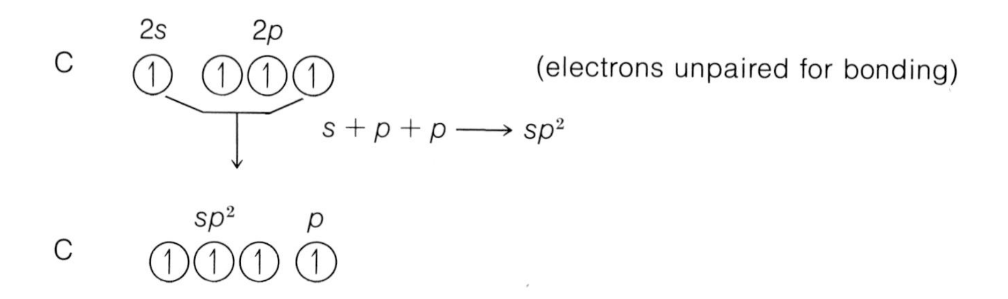

We see that the three σ bonds are formed from sp^2 hybrid orbitals, leaving one p orbital free to form a π bond between the carbons (Figure 14.4). The

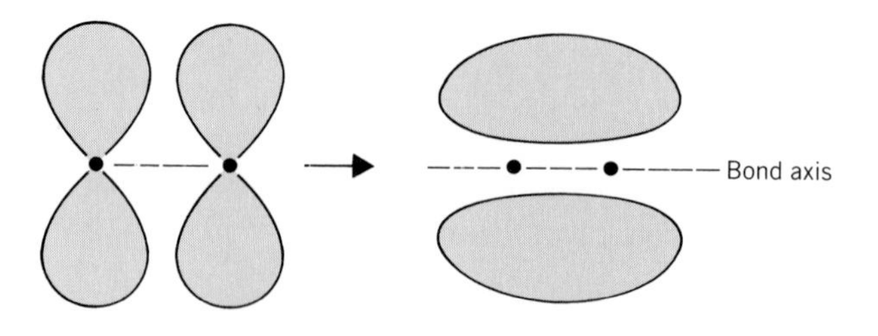

Figure 15.4 A pi (π) bond is formed when a p orbital overlaps on opposite sides of the bond axis.

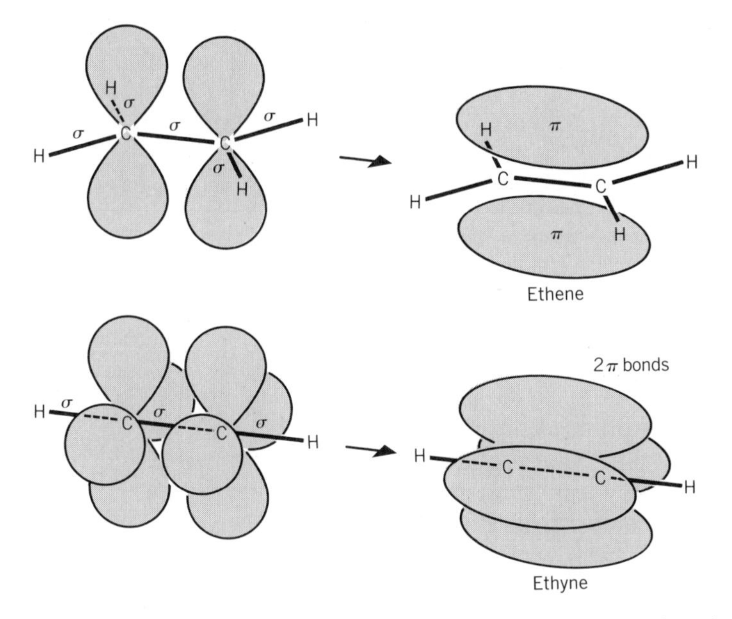

Figure 15.5 Ethene contains one π bond formed by the overlap of one p orbital from each carbon atom. Ethyne contains two π bonds formed by the overlap of two p orbitals from each carbon atom.

sp^2 hybrid orbitals form a planar triangular shape (see Figure 14.4). This means that all the atoms in an ethene molecule lie in the same plane, with the pi bond forming above and below that plane.

A triple bond consists of one sigma bond and two pi bonds. For example, look at ethyne.

$$H{-}C{\equiv}C{-}H \qquad \text{Ethyne}$$

Here, each carbon will form two sigma bonds using sp hybrid orbitals, leaving two p orbitals to form the two pi bonds around the bond axis between the carbon atoms.

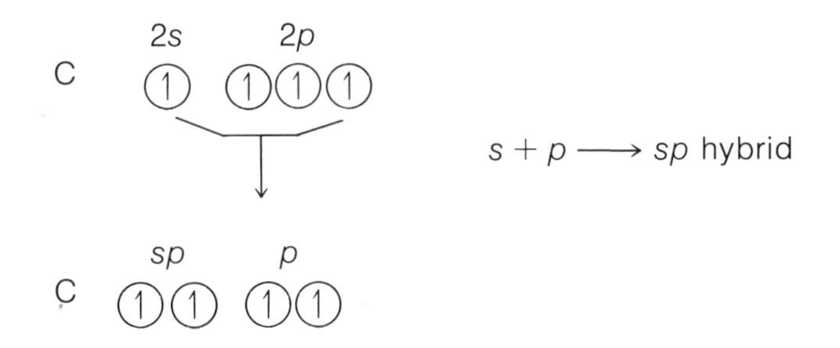

$s + p \longrightarrow sp$ hybrid

Because of the shape of the sp hybrid orbitals (see Figure 14.4), the atoms in a molecule of ethyne lie in a straight line, with the two pi bonds perpendicular to one another along the bond axis between the two carbon atoms (Figure 15.5).

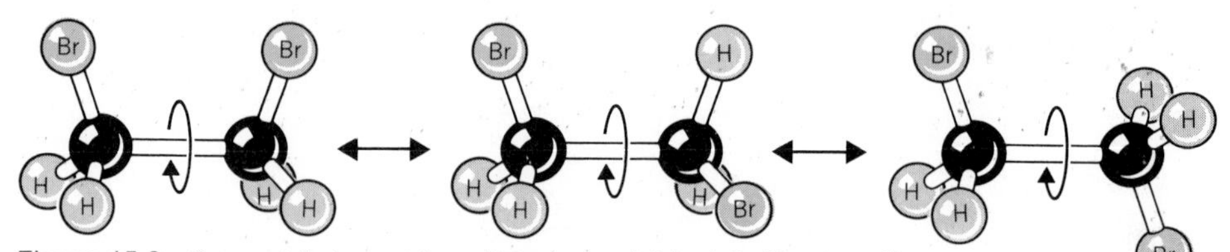

Figure 15.6 Geometric isomerism does not exist in 1,2-dibromoethane because the carbon-to-carbon single bond is free to rotate. As a result, the atoms bonded to the carbons can be in any position.

15.5 Geometric Isomers

A carbon-to-carbon double bond makes another type of isomerism possible, **geometric isomerism.** A single bond between carbon atoms, as in the alkanes, does not restrict the rotation of atoms around that bond. The carbon atoms can twist freely around their single bonds just as two balls can twist freely when connected by a string (Figure 15.6). A double bond between two carbon atoms, however, is structurally rigid, preventing free rotation of the carbon atoms. This results in two possible arrangements of atoms on either side of the unsaturated bond (Figure 15.7). In terms of our comparison, replacing a single bond between two carbon atoms by a double bond is much like replacing the string between the two balls with a rigid pole.

When some specified atoms or groups of atoms attached to the doubly bonded carbons appear on the same side of the bond, the molecule is called a ***cis* isomer.**

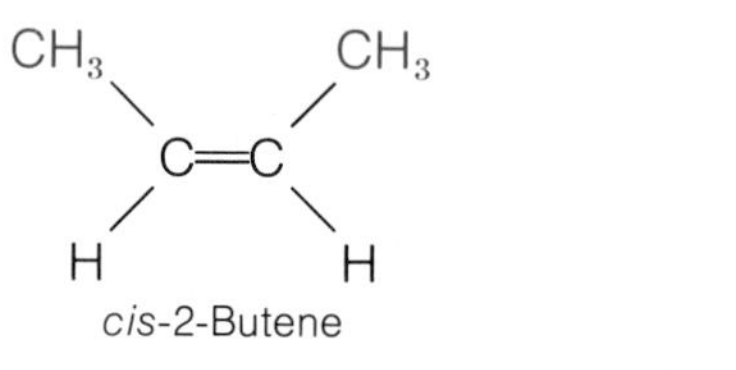

cis-2-Butene

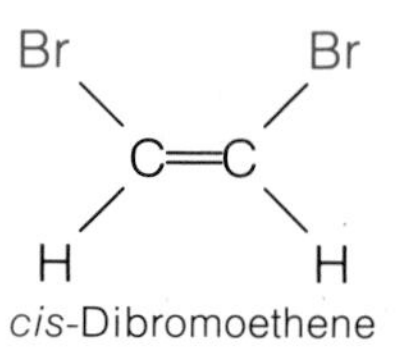

cis-Dibromoethene

A ***trans* isomer** is formed when specified atoms or groups of atoms appear on opposite sides of the double bond.

CH_3 H
C=C
H CH_3

trans-2-Butene

Br H
C=C
H Br

trans-Dibromoethene

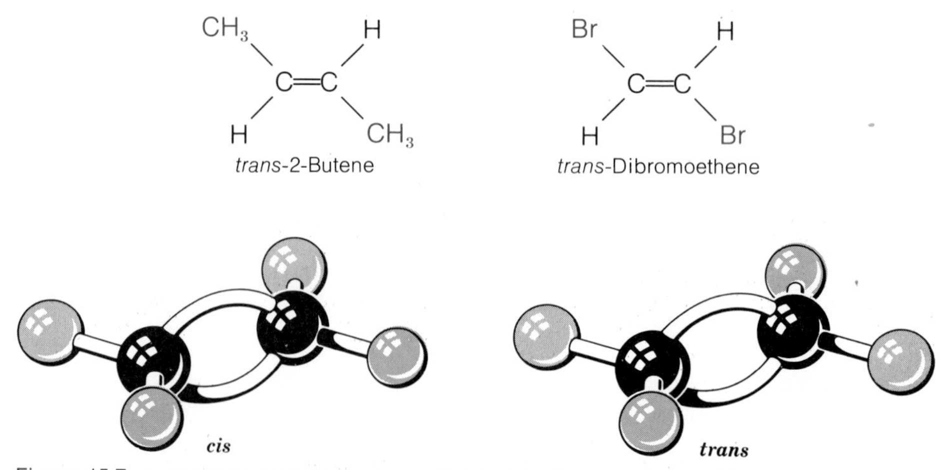

Figure 15.7 Geometric isomerism does exist in 1,2-dibromoethene. The double bond is structurally rigid, preventing free rotation of the carbon atoms.

You may wonder why we are bothering to point out these different types of isomers. A major reason is that isomers are critically important to the chemical reactions of living cells. For example, maleic acid and fumaric acid are *cis* and *trans* isomers.

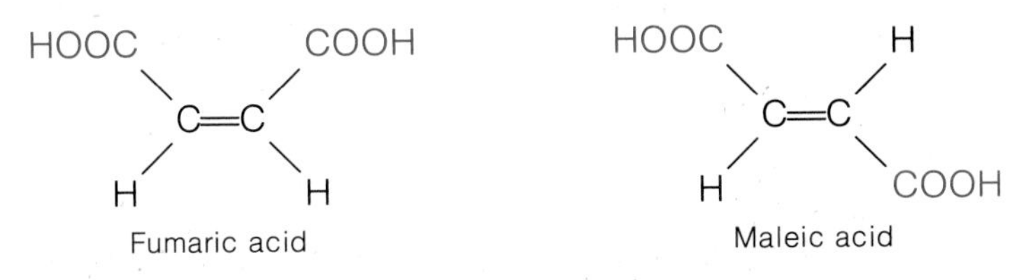

Maleic acid is poisonous to living cells, but fumaric acid is formed as an essential part of the citric acid cycle, the series of reactions that produces energy in the cell. Although you might find it difficult to tell the difference between the *cis* and *trans* isomer of the following complicated compound, the cells in the retina of your eye certainly can.

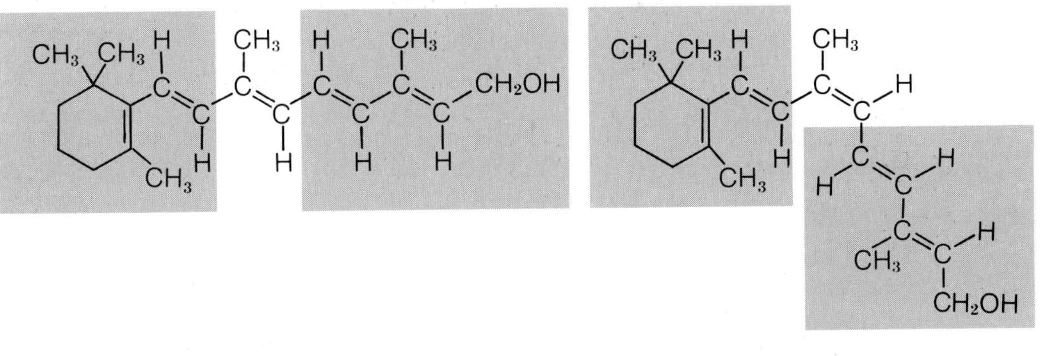

Vitamin A.
(all *trans*—100% activity)

9-*cis* isomer
(22% activity)

The *trans* isomer is Vitamin A, a compound necessary for us to see in dim light. The corresponding *cis* isomer can also be used by the retina cells. These isomers cannot be used as efficiently as the all *trans* isomer, however, and they are much less effective in reducing night blindness.

Exercise 15-2

Draw the structural formulas of the *cis* and *trans* isomers of 2-pentene.

15.6 Reactions of Alkenes

Because a carbon-to-carbon double bond is more reactive than a single bond, alkenes are used commercially in a wide variety of reactions. They are especially important in the chemical industry, where they are used in the production of many other compounds. Three important types of reactions for which alkenes are well suited are oxidation reactions, addition reactions, and polymerization reactions.

Oxidation

In Chapter 14 we discussed the oxidation of alkanes by describing the combustion reaction. But compounds do not have to be burned in order to be oxidized. There is actually a wide range of reactions that are oxidation reactions, or more precisely, **oxidation–reduction (redox)** reactions. Because an oxidation reaction involves the loss of electrons by a reactant, and a reduction reaction involves the gain of electrons by another reactant, oxidation and reduction reactions always occur together. We don't always have to examine oxidation numbers in order to identify the oxidation and reduction reactions that are occurring. For the purposes of our study of organic chemistry, we will simply consider the oxidation reaction to be the one in which a reactant molecule gains oxygen atoms or loses hydrogen atoms (called dehydrogenation). The reduction reaction is the one in which a reactant molecule loses oxygen atoms or gains hydrogen atoms (called hydrogenation).

As was the case with the alkanes, alkenes can be oxidized completely to produce carbon dioxide (CO_2) and water.

$$\underset{\text{Ethene}}{CH_2{=}CH_2} + \underset{\text{Oxygen}}{3O_2} \longrightarrow \underset{\text{Carbon dioxide}}{2CO_2} + \underset{\text{Water}}{2H_2O}$$

Notice that in this reaction the molecule of ethene undergoes oxidation, and the oxygen gas undergoes reduction. Under certain conditions the oxidation of ethene may be incomplete, resulting in the formation of glycol (a type of alcohol) rather than carbon dioxide.

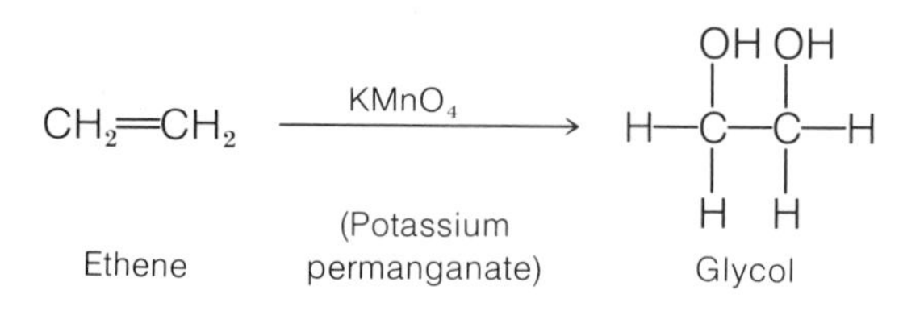

Addition Reactions

An **addition reaction** takes place when atoms react with a double bond, causing the double bond to become a single bond.

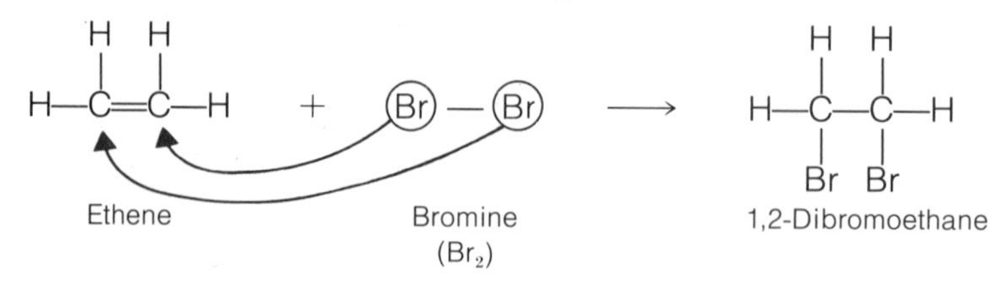

This bromine reaction is often used to test for the presence of unsaturated bonds (that is, double or triple bonds) in a molecule. Bromine in water or in carbon tetrachloride forms a reddish-brown solution, but the dibromides formed from the addition reaction are colorless. Therefore, if the reaction of bromine water with a hydrocarbon results in a colorless solution, the hydrocarbon may have contained unsaturated bonds.

Compounds other than the halogens can also be used in addition reactions. Here are a few examples:

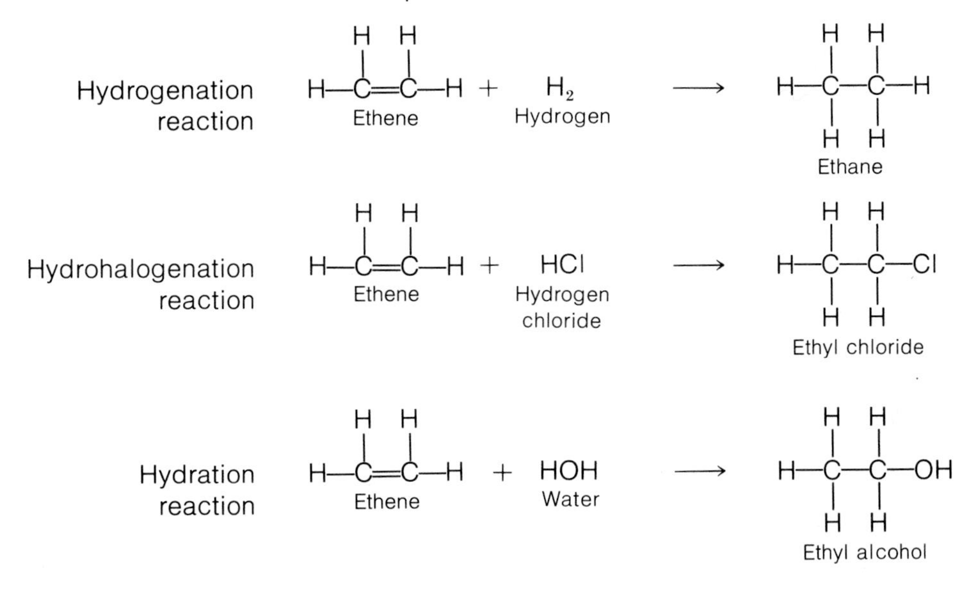

Polymerization

You are probably familiar with the word *ethylene,* having seen it in the name of a type of plastic, polyethylene. Polyethylene is formed by an addition reaction involving thousands of ethylene units. This is how three ethylene units would bond together:

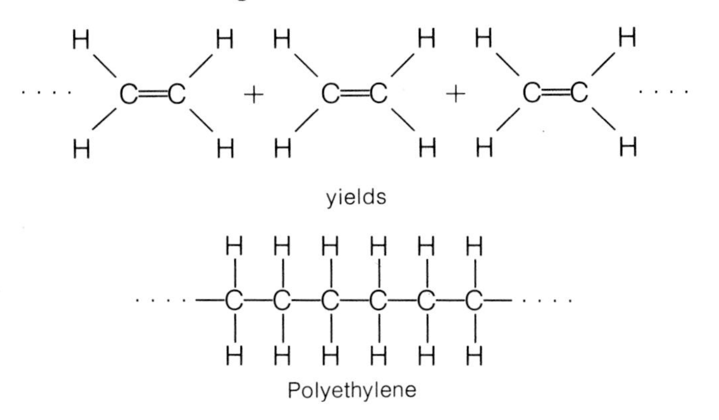

The process of joining many simple units, called **monomers,** together to form very large molecules, called **polymers,** is **polymerization.** Thus, polyethylene is a polymer of the monomer ethylene. Synthetic (that is, artificial) polymers appear constantly in our daily lives, from the plastic containers, bags, and wrapping we use, to the synthetic fibers such as orlon, rayon, and acrylics that we wear, to the synthetic rubber that we use for tires and for parts in appliances (Table 15.2 and Figure 15.8). However, don't think that polymers are the result of human ingenuity alone; the synthetic polymer industry resulted from the attempt to imitate natural polymers. For example, tropical plants produce natural rubber in the form of a milky sap called latex, which is a large molecule composed of 4500 isoprene units. Isoprene is a five-carbon compound with two double bonds.

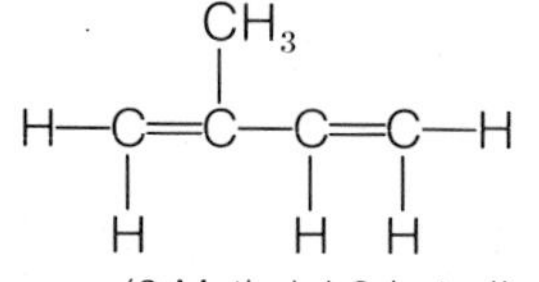

Isoprene (2-Methyl-1,3-butadiene)

A molecule of isoprene has carbon-to-carbon double bonds alternating with single bonds. This alternating arrangement is given the name **conjugated double bonds.** Any compound having conjugated double bonds will be more stable than a similar compound having double bonds arranged in other patterns.

15.7 Alkynes

Alkynes are the class of hydrocarbons that contain triple bonds between carbon atoms. These triple bonds, having three pairs of electrons shared

Table 15.2 Some Commonly Used Polymers

Monomer	Polymer	Uses
Propylene $H_2C{=}CHCH_3$	Polypropylene $-CH_2CHCH_2CHCH_2CH-$ (each CH bearing CH_3: CH_3 CH_3 CH_3)	Film and molded parts
Styrene $H_2C{=}CHC_6H_5$	Polystyrene $-CH_2CH-CH_2CH-CH_2CH-$ (each CH bearing C_6H_5: C_6H_5 C_6H_5 C_6H_5)	Molded objects, insulation, and foam plastics
Vinyl chloride $H_2C{=}CHCl$	Polyvinyl chloride (PVC) $-CH_2CHCH_2CHCH_2CH-$ (each CH bearing Cl: Cl Cl Cl)	Plastic bottles and containers, plastic pipe, and insulation
Acrylonitrile $H_2C{=}CHC{\equiv}N$	Polyacrilonitrile $-CH_2CHCH_2CHCH_2CH-$ (each CH bearing $C{\equiv}N$: $C{\equiv}N$ $C{\equiv}N$ $C{\equiv}N$)	Orlon and clothing fibers
Tetrafluoroethylene $F_2C{=}CF_2$	Polytetrafluoroethylene $-CF_2CF_2CF_2CF_2CF_2CF_2-$	Teflon and lubricating films

Figure 15.8 Polyvinyl chloride (PVC) is used to manufacture all types of plastic pipe including this drainage pipe.

between the two carbon atoms, put quite a strain on the molecule, making such bonds very reactive. Nevertheless, it is possible for a molecule to have more than one triple bond, and there are some molecules that contain both double and triple bonds.

Alkynes are named following the same rules given for alkenes in Section 15.3, except that the names of alkynes end in **-yne.** For example, $CH_3CH_2C{\equiv}CCH_3$ is 2-pentyne.

The simplest alkyne is ethyne, or acetylene, CHCH or C_2H_2 (Figure 15.9). Acetylene is a flammable and explosive gas that burns with a bright flame, making it useful for lighting. When acetylene is burned together with oxygen in an oxyacetylene welding torch, the resulting reaction is highly exothermic, providing enough energy to cut and weld metals. Acetylene is so reactive that it forms a convenient starting material for the industrial production of almost every simple organic compound. However, compounds with triple bonds do not play a part in the chemistry of living organisms. In fact, the human body does not naturally contain any molecules with triple bonds in their structures.

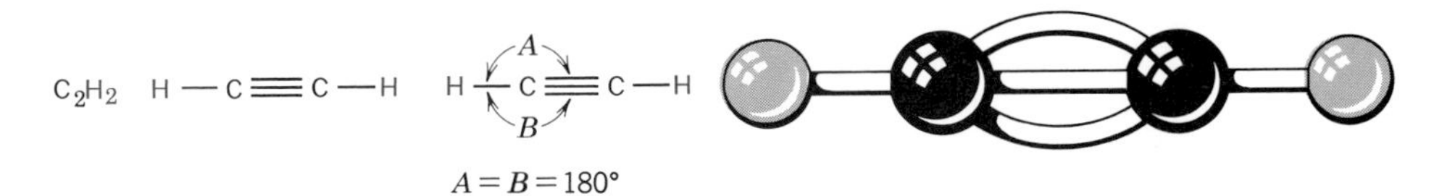

Figure 15.9 Ethyne.

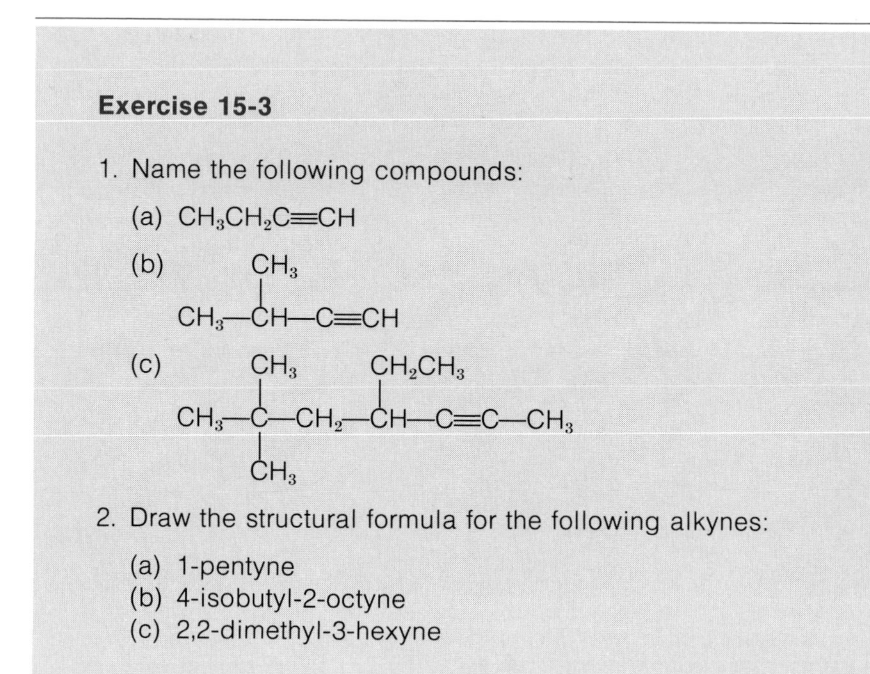

Exercise 15-3

1. Name the following compounds:

   (a) $CH_3CH_2C{\equiv}CH$

   (b)
   $$\begin{array}{l} \qquad\quad CH_3 \\ \qquad\quad\;\, | \\ CH_3{-}CH{-}C{\equiv}CH \end{array}$$

   (c)
   $$\begin{array}{l} \qquad\quad CH_3 \qquad\quad CH_2CH_3 \\ \qquad\quad\;\, | \qquad\qquad\quad | \\ CH_3{-}C{-}CH_2{-}CH{-}C{\equiv}C{-}CH_3 \\ \qquad\quad\;\, | \\ \qquad\quad CH_3 \end{array}$$

2. Draw the structural formula for the following alkynes:

   (a) 1-pentyne
   (b) 4-isobutyl-2-octyne
   (c) 2,2-dimethyl-3-hexyne

CARBON IN RINGS

15.8 Cyclic Hydrocarbons

The great variety and complexity of carbon chemistry becomes even more obvious when we realize that carbon atoms not only form straight chain molecules, but also form stable rings. These compounds, called **cyclic hydrocarbons,** may be found in various sizes and often contain both single and double bonds (Table 15.3). The most common cyclic hydrocarbons have rings composed of five or six carbon atoms, but both larger and smaller rings are possible. The cyclic hydrocarbons are named following the same rules given for the straight chain hydrocarbons, except that the unsubstituted name begins with **cyclo-.**

The saturated cyclic hydrocarbon with the smallest number of carbon atoms is cyclopropane. Cyclopropane is a highly potent anesthetic, but care must be taken in its use because it is both inflammable and explosive. Many hydrocarbons, in addition to cyclopropane, act as anesthetics when inhaled. (Among the others are methane, acetylene, ethylene, and cyclobutane.) An anesthetic is a compound that decreases a person's sensitivity to pain and, in most cases, also causes the person to lose consciousness. The anesthetic property of the hydrocarbons comes from the nonpolar nature of the molecules and the absence of a polar hydrogen atom for hydrogen bonding.

An anesthetic compound acts on the nerves, preventing nerve impulses from moving along the nerve fibers. This happens because each nerve is surrounded, in part, by a protective coating composed of molecules that are nonpolar. When an anesthetic is inhaled, it enters the bloodstream from the lungs and travels to these tissues. There it dissolves in the nonpolar protective coating around the nerves. As the anesthetic accumulates in this coating, it disrupts the transmission of nerve impulses. Because the nerve impulses no longer reach the brain, there is no sensation of pain. The many deaths caused by inhaling methane, or sniffing the solvents in glue, have resulted from these nonpolar hydrocarbon compounds dissolving in the nonpolar protective coatings of the nerves and disrupting the critical nerve impulses from the brain to the lungs and heart.

15.9 *Cis* and *Trans* Isomers

When carbon atoms are arranged in a ring, free rotation around even the carbon-to-carbon single bond is restricted, similar to the restricted rotation

Table 15.3 Some Cyclic Hydrocarbons

Molecular Formula	Name	Structural Formula
C_3H_6	Cyclopropane	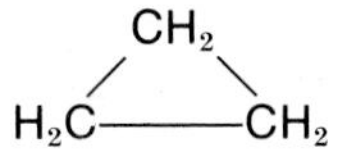 ring: CH_2 / H_2C—CH_2
C_4H_8	Cyclobutane	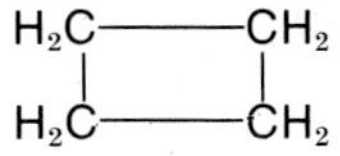 ring: H_2C—CH_2 / H_2C—CH_2
C_5H_{10}	Cyclopentane	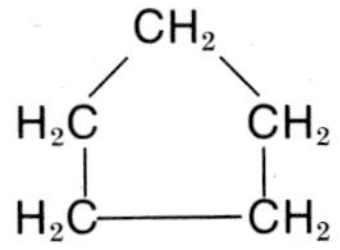 ring: CH_2 / H_2C CH_2 / H_2C—CH_2
C_5H_6	1,3-Cyclopentadiene	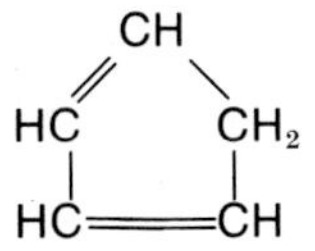 ring: CH / HC CH_2 / HC═CH (with HC═CH double bond at top left)
C_6H_{12}	Cyclohexane	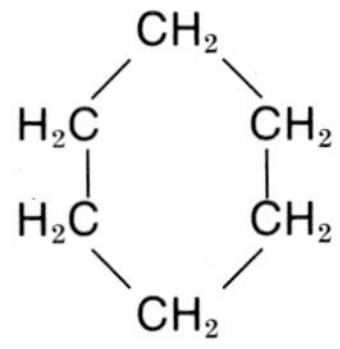 ring: CH_2 / H_2C CH_2 / H_2C CH_2 / CH_2

we have seen in the alkenes. This results in *cis* and *trans* isomers in ring compounds. A *cis* isomer of a ring compound will have specified atoms or groups of atoms lying on the same side of the carbon ring.

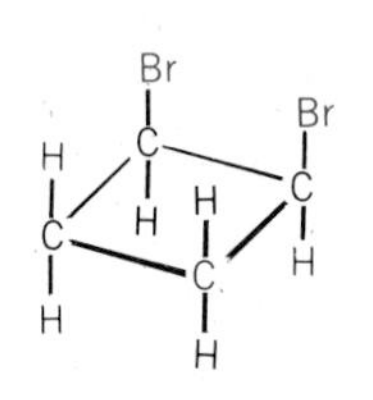

cis–1,2–Dibromocyclobutane

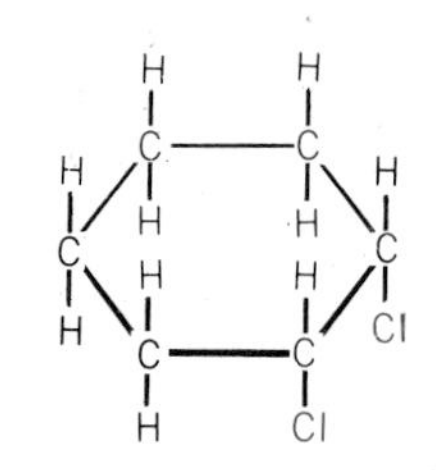

cis–1,2–Dichlorocyclohexane

Similarly, a *trans* isomer will have these atoms or groups of atoms located on opposite sides of the carbon ring.

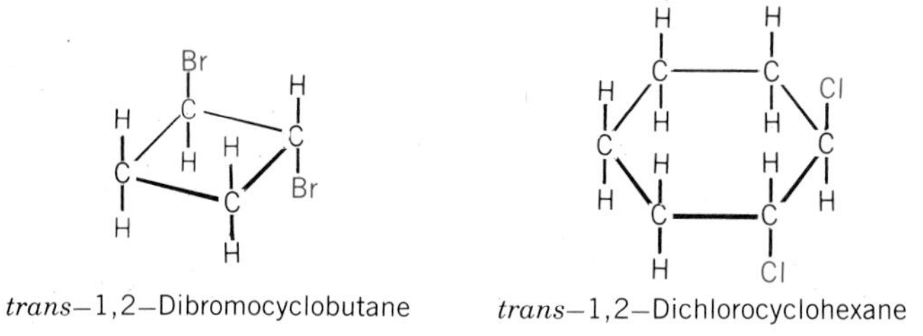

trans–1,2–Dibromocyclobutane   *trans*–1,2–Dichlorocyclohexane

15.10 Benzene and Its Derivatives: The Aromatic Hydrocarbons

The most common and important of the cyclic hydrocarbons is benzene. Benzene and its derivatives make up the class of compounds called the **aromatic hydrocarbons.** In spite of their name, the aromatic hydrocarbons are no more "smelly" than other hydrocarbons. However, the first natural compounds identified as part of this class had a definite and fairly pleasant odor—from which came the name *aromatic.* Benzene has the molecular formula C_6H_6, but its structural formula puzzled scientists for many years. The six carbon atoms in a molecule of benzene are arranged in a ring and are often drawn with alternating single and double bonds.

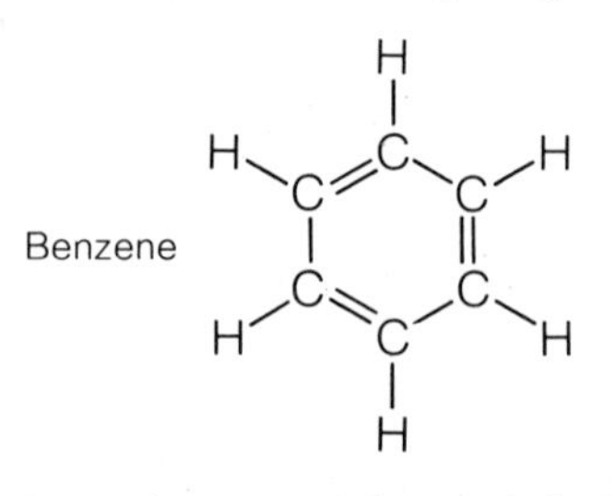

However, this structure does not accurately explain the properties of benzene. The actual benzene ring is much more stable than would be expected from this conjugated alkene structure and does not break open

during reactions. Moreover, each carbon-to-carbon bond is the same strength, and the distance between each of the carbon atoms is equal.

Two-dimensional structural formulas do not adequately represent benzene's structure. It is

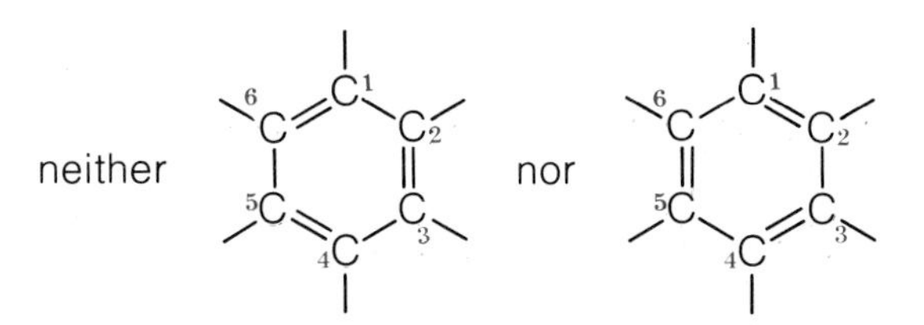

but rather a structure in which the electrons from the three double bonds seem to be delocalized, or spread out, among all six of the carbon atoms. The name *resonance hybrid* is used for this structure (*resonance* is another term used to describe delocalization of electrons). Symbolically, we can represent this resonance structure by drawing a circle in the center of the benzene ring.

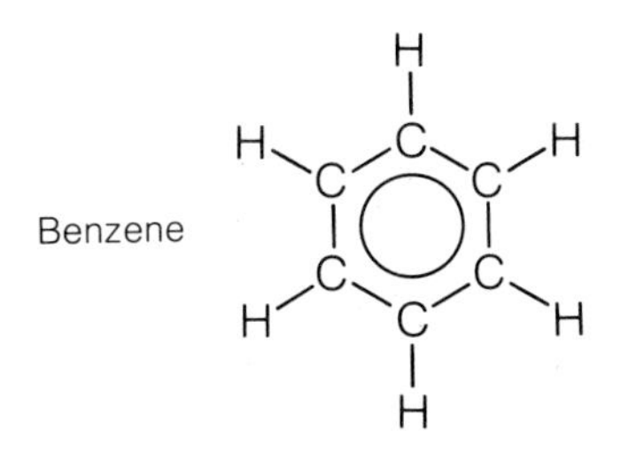

All of the atoms in a molecule of benzene are located in one plane (with bond angles of 120°). As a result, *cis* and *trans* isomers do not exist. Its resonance structure makes the benzene molecule extremely stable, and this molecule forms part of the structure of many complex natural compounds.

To simplify the drawing of the structural formulas of such complex molecules, organic chemists have devised a simplified representation of a hydrocarbon or benzene ring.

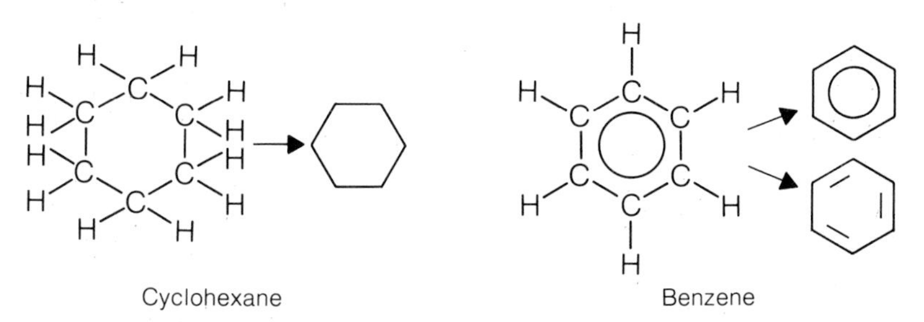

To translate such a diagram back to the full structural formula, place a carbon atom at each angle of the figure and then draw hydrogen atoms on the bonds not involved in the ring. If any elements other than carbon or hydrogen are involved in the compound, they will be separately shown. Here are some examples:

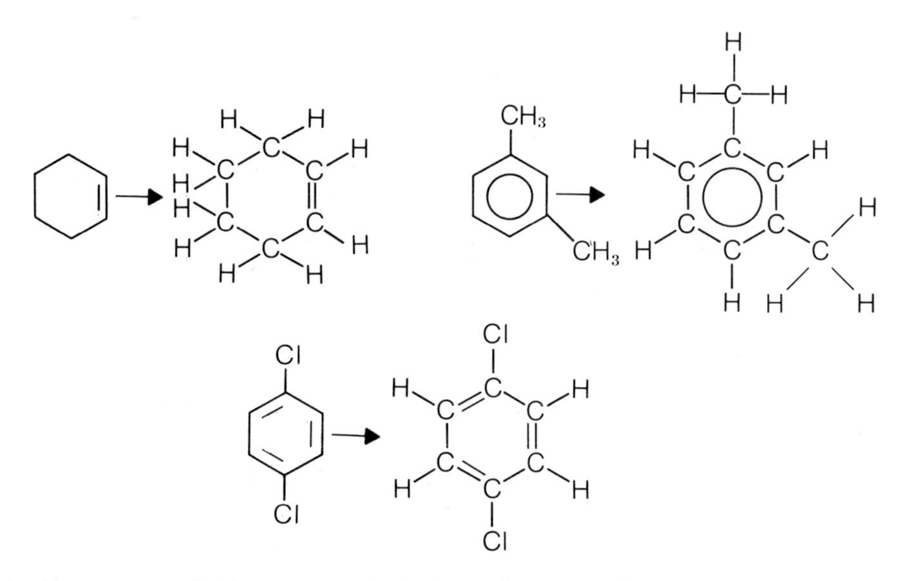

Benzene and other aromatic hydrocarbons can be recovered from coal tar, which is produced by heating bituminous (or soft) coal in the absence of air. Benzene is used by the chemical industry as a solvent and as a starting material in the production of a variety of compounds. Its fumes are toxic and, when inhaled, can cause nausea or death from respiratory and heart failure. Again, let us stress that the danger of these nonpolar compounds, many of which are used as solvents in such common products as paints, glues, and cleaners, is their ability to act on the central nervous system and to interfere with nerve action. Some important aromatic hydrocarbons are listed in Table 15.4.

Table 15.4 **Some Aromatic Compounds**

Name	Structural Formula	Name	Structural Formula
Benzene		Toluene	CH_3
Phenol	OH	Nitrobenzene	NO_2
Benzoic acid	O ‖ C—OH	Aniline	NH_2

The benzene ring is found in many compounds that are important to the life processes of the living organism. Phenylalanine, an amino acid whose importance was illustrated in Chapter 1, is just one example.

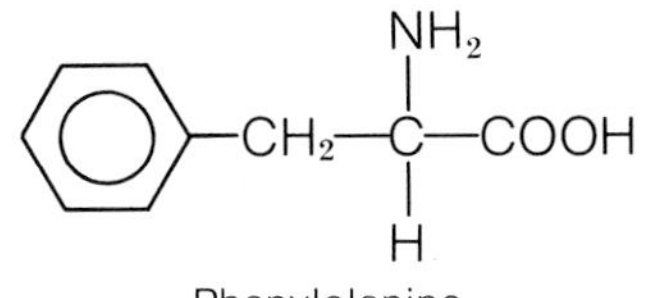

Phenylalanine

Plants are able to synthesize benzene rings from carbon dioxide, water, and inorganic materials. Animals, however, cannot synthesize these aromatic rings, and their survival depends upon obtaining the essential aromatic compounds through their diets.

5.11 Naming Aromatic Hydrocarbons

Aromatic hydrocarbons can be named according to the following general rules.

1. *Benzene Rings with One Substituted Group*

   (a) When a group is substituted onto a benzene ring, the resulting compound is named by adding the name of the substituted group as a prefix before the name benzene.

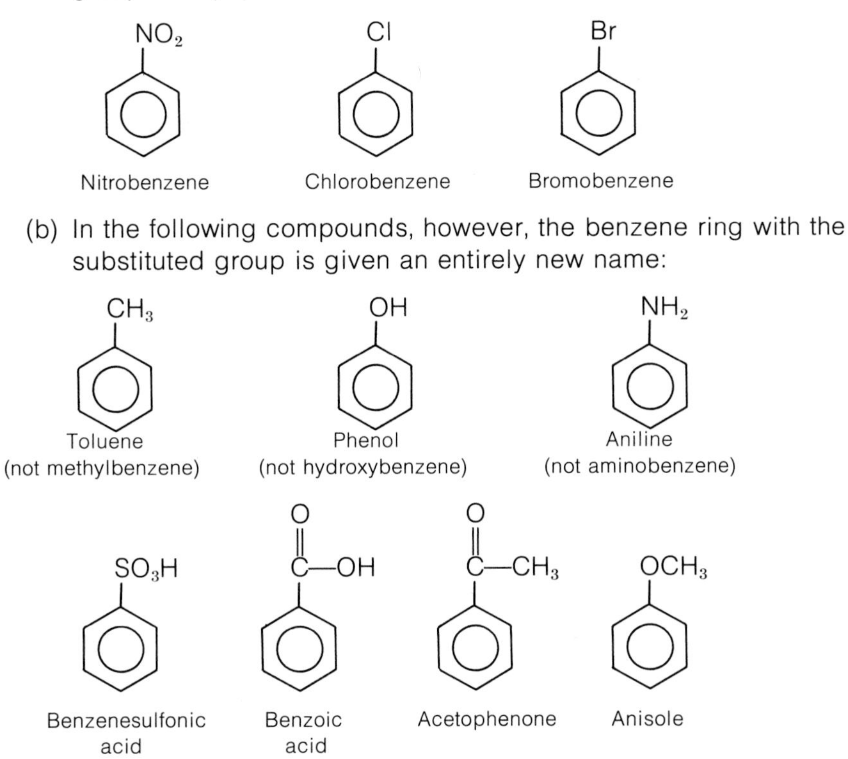

   (b) In the following compounds, however, the benzene ring with the substituted group is given an entirely new name:

2. *Benzene Rings with Two Substituted Groups*

   The position of the second group on the ring is indicated by the prefixes *ortho (o), meta (m),* and *para (p).*

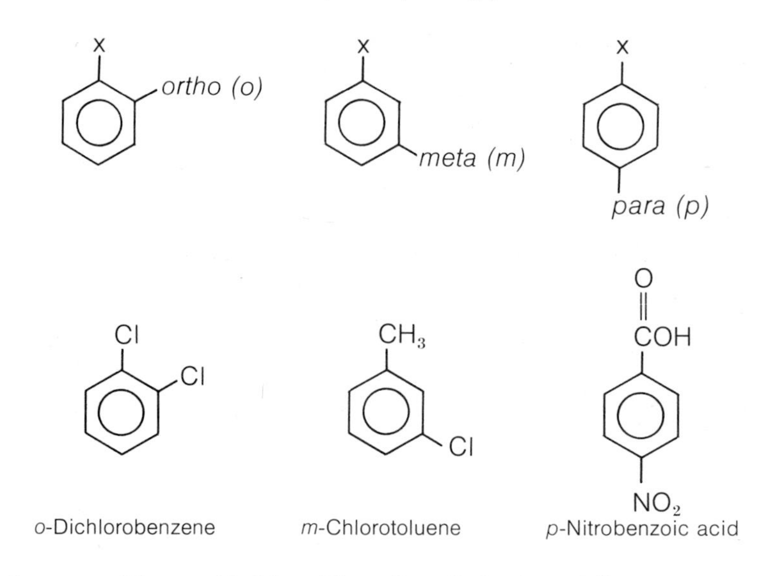

o-Dichlorobenzene   *m*-Chlorotoluene   *p*-Nitrobenzoic acid

3. *Benzene Rings with More Than Two Substituted Groups*

   If more than two substituted groups are present on the benzene ring, their positions are indicated by numbers. The ring is numbered to give the lowest possible number to the substituted groups. Benzene rings with two substituted groups also can be named using numbers. For example,

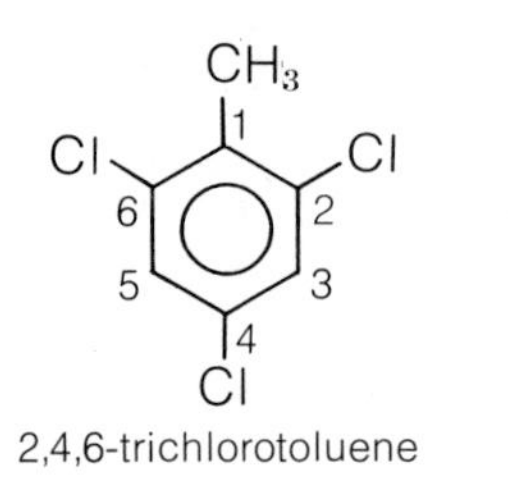

2,4,6-trichlorotoluene

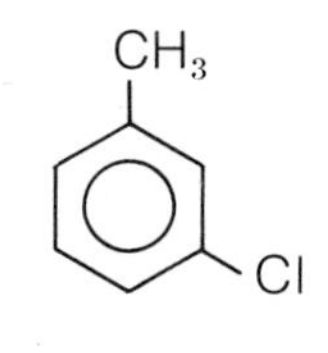

3-chlorotoluene
(*m*-chlorotoluene)

4. *Benzene Ring as a Substituted Group or R-Group*

   When a benzene ring is a substituted group on another hydrocarbon, it is called a **phenyl** group. For example,

Phenyl group

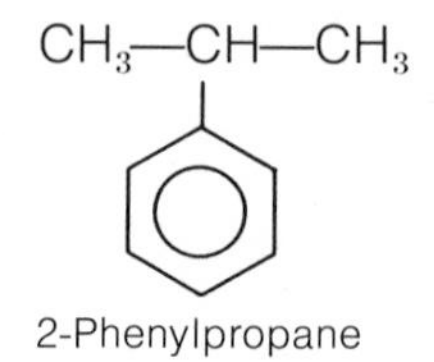

2-Phenylpropane

When the substituted group is derived from toluene, it is called a **benzyl** group.

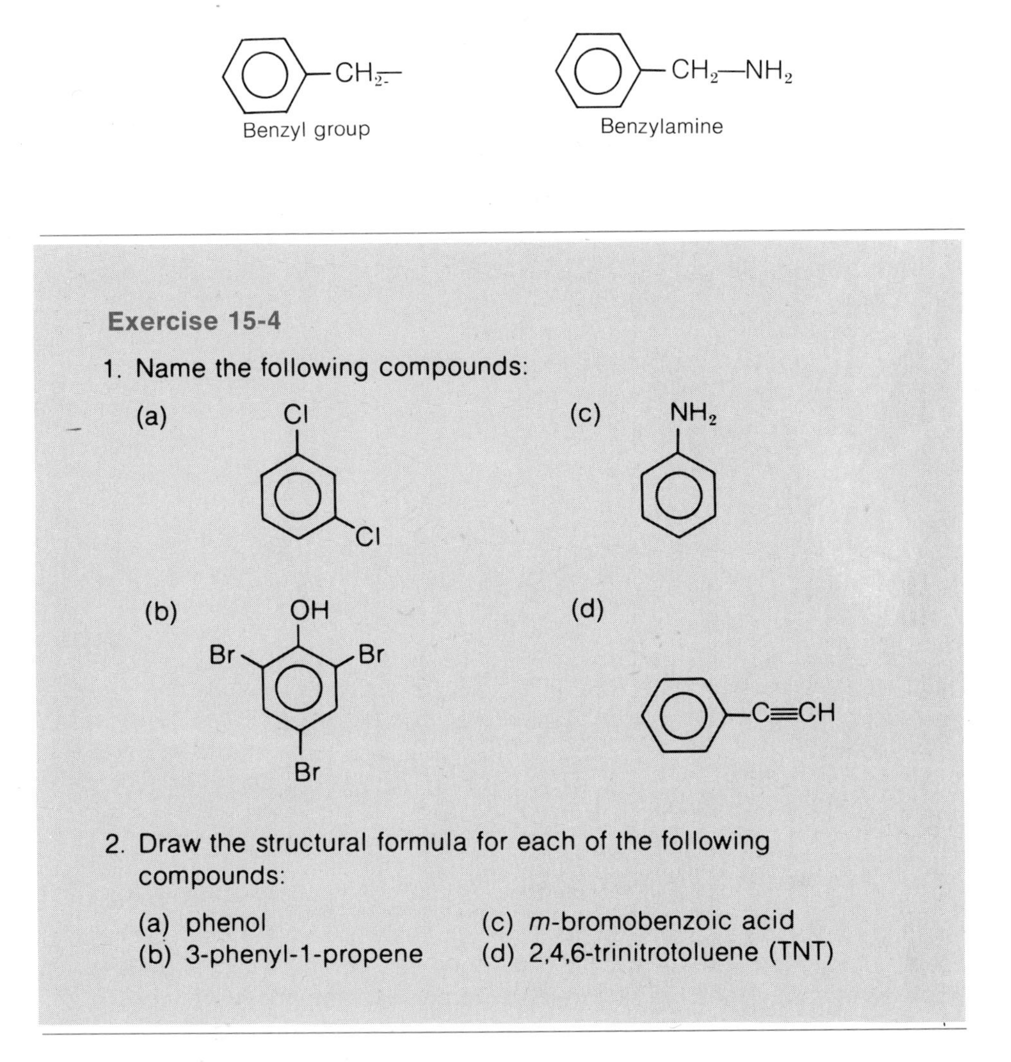

15.12 Reactions of Benzene

The ring structure of benzene is very stable and remains intact during most chemical reactions. Many aromatic compounds are produced both naturally and synthetically by **substitution reactions,** in which one or more hydrogens on the aromatic ring are replaced with other atoms or groups of atoms.

We saw in Section 15.6 that alkenes react readily with bromine, adding bromine atoms across the double bond. However, very special conditions are required for benzene to react with bromine and, when this occurs, the benzene ring remains intact and one bromine atom is substituted for a hydrogen atom.

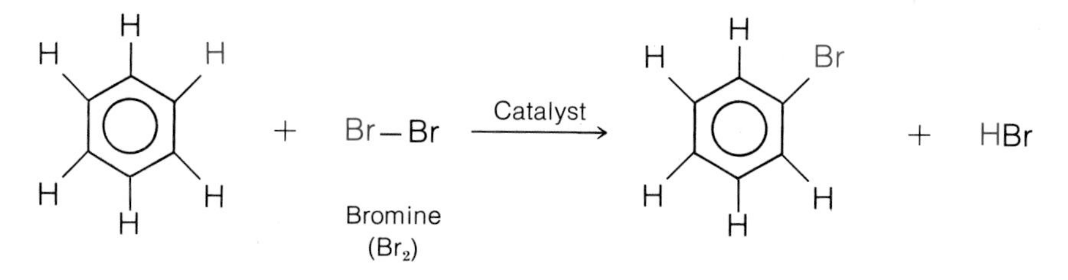

Under certain conditions, benzene can be forced to undergo a substitution reaction with concentrated acids.

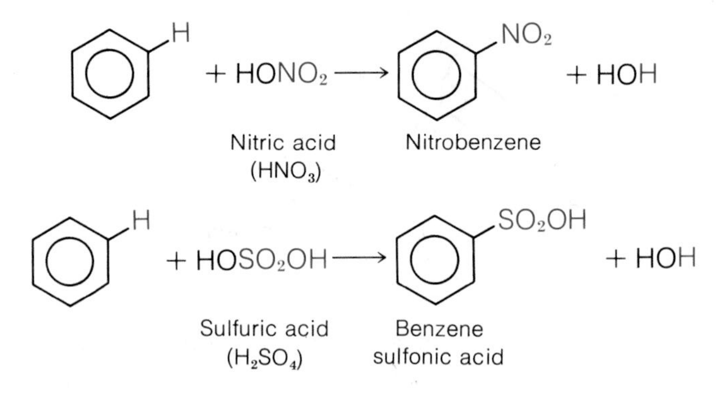

15.13 Other Aromatic Hydrocarbons

Benzene rings may be found in large molecules whose chemical diagrams resemble honeycombs. As a simple example, the compound naphthalene, which gives moth balls their characteristic odor, is composed of two benzene rings joined or fused together. Three benzene rings joined together form anthracene, an important starting material in the production of dyes.

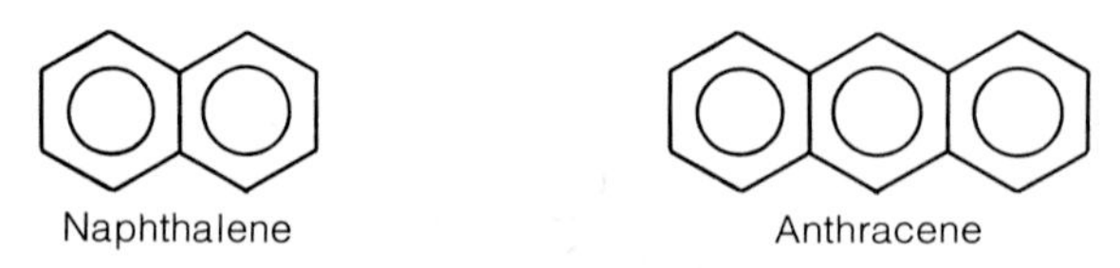

Compounds containing multiple benzene rings are obtained in the production of coal tar. Their harmful effects to humans became evident when workers in European coal tar factories developed skin cancer. A later study of the chemical components of coal tar determined that several aromatic fused-ring compounds were capable of causing cancer in mice (Figure 15.10). Chemicals that cause cancer in animals are known as **carcinogens.** It was found that the carcinogenic hydrocarbons in coal tar

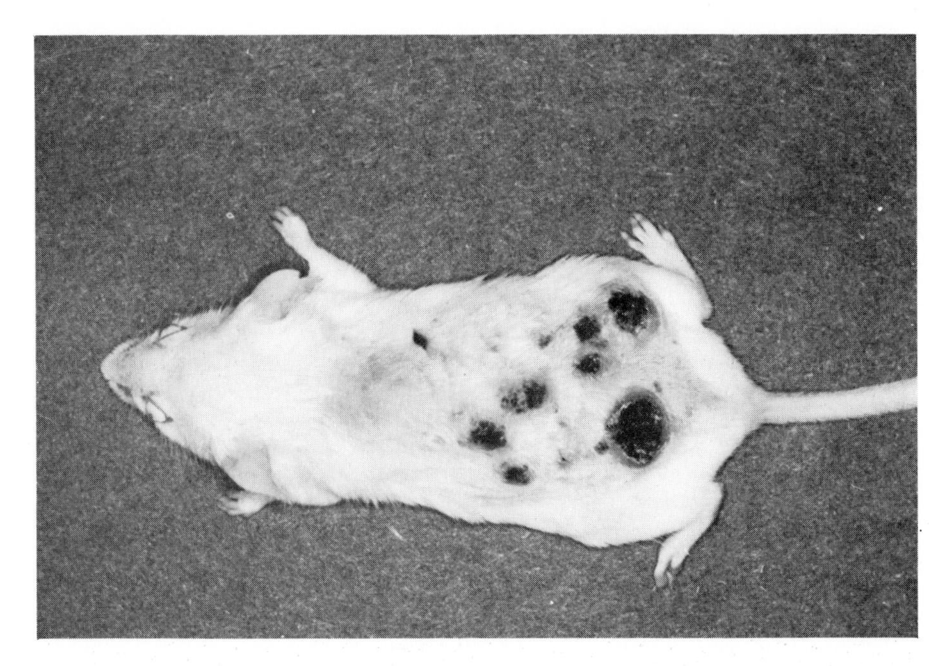

Figure 15.10 Skin tumors were produced by applying a solution containing 7,12-dimethylbenzanthracene to the shaved skin on the back of this mouse. (Courtesy Kanematsu Suguira, Sloan Kettering Institute for Cancer Research.)

all had a similar arrangement of fused benzene rings, as shown in Figure 15.11. These compounds are formed in the partial combustion of many large organic molecules. One of the most active carcinogens, 3,4-benzpyrene, is discharged each year, in very large quantities, into the atmosphere of industrial nations. It is also one of the major carcinogens found in cigarette smoke. Only a few milligrams of 3,4-benzpyrene are enough to produce cancer in experimental animals.

The way in which these rather inert compounds produce cancer in humans and animals has puzzled scientists, but recent studies are beginning to shed light on this process. Our bodies contain a certain set of enzymes located mostly in the liver and kidneys, but also found in the lungs and other tissues, whose function is to detoxify foreign chemicals that

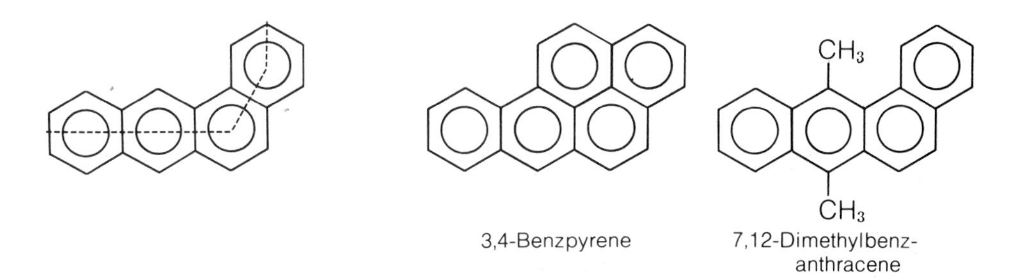

Figure 15.11 These carcinogenic hydrocarbons found in coal tar all have a similar arrangement of fused benzene rings.

enter the body. One way in which this is done is by making nonpolar compounds (such as 3,4-benzpyrene) more polar. This makes these compounds more soluble in water (through hydrogen bonding) and more easily excreted by the kidneys. Most of the products of such detoxifying reactions are less harmful to the cells than were the original reactants, but in some cases the products turn out to be very carcinogenic. This is the way in which these enzymes (some of which are located in the lungs) can contribute to the synthesis of carcinogenic hydrocarbons from cigarette smoke. The way that such carcinogens turn a normal cell into a cancer cell is still an area of active research.

Polychlorinated biphenyls (PCB) and polybrominated biphenyls (PBB) are derivatives of biphenyl in which from 1 to 10 of the hydrogens on the benzene rings have been replaced by chlorine or bromine.

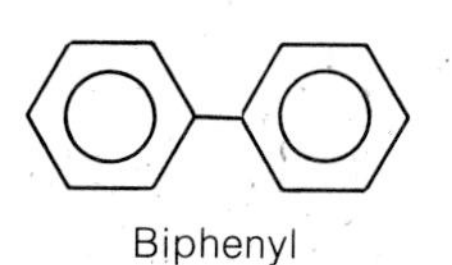

Biphenyl

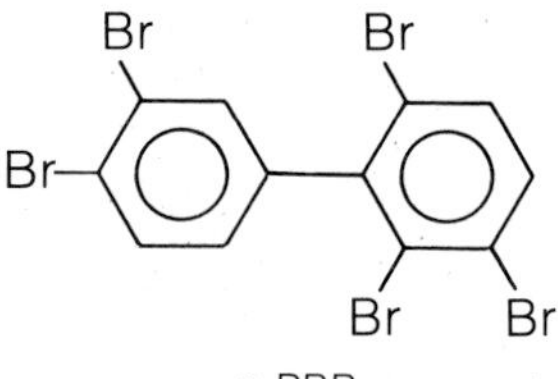

a PBB

These molecules are extremely stable; they are not broken down by reactions with moisture or with the air and are not metabolized by microorganisms. Their chemical stability and solubility in nonpolar solvents have made them very useful as fire retardants, as insulators in electrical equipment, and as components of plastics. However, the toxicity and persistence of the PCBs and PBBs also make them quite dangerous to the natural environment should they be accidentally spilled or enter the food chain as contaminants in feed.

Many complex chlorinated hydrocarbons have been manufactured and used as insecticides, the best known of which is dichlorodiphenyltrichloroethane, or DDT.

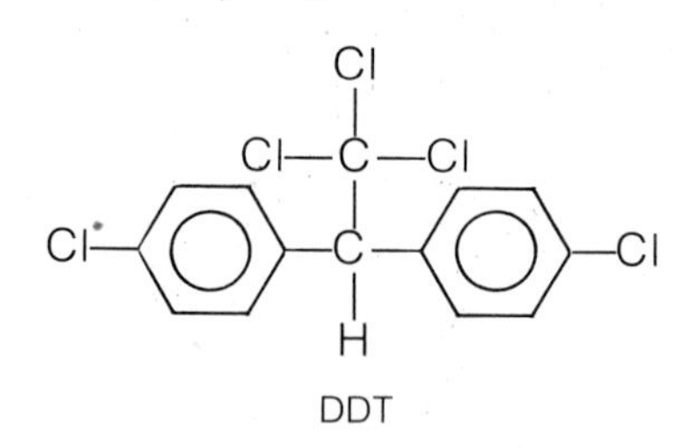

DDT

The widespread and successful use of DDT in the 1940s and 1950s to combat the malaria mosquito has led to the development of resistance to DDT in many insect species. DDT has accumulated in the food chain and has had especially damaging effects on several species of birds. Because of the chemical stability and potential harm to animals of polychlorinated hydrocarbon insecticides, their use is now carefully regulated. Scientists are presently working to develop other methods of insect control that will be safer to the environment.

Unsaturated hydrocarbons are compounds of carbon and hydrogen that contain one or more carbon-to-carbon double or triple bonds. Alkenes are the class of hydrocarbons containing carbon-to-carbon double bonds, and alkynes are the class containing triple bonds. Alkenes and alkynes are more reactive than alkanes and will undergo oxidation, addition, and polymerization reactions. Cyclic hydrocarbons are compounds that contain carbon atoms bonded together in rings.

Because of the restricted rotation around carbon-to-carbon double bonds, alkene compounds may exhibit geometric isomerism. Geometric isomers are molecules having the same ordering of carbon atoms, but with a different arrangement of those atoms in space. Because a ring structure restricts rotation around the carbon-to-carbon bond, *cis* and *trans* isomers are also found in cyclic hydrocarbons.

Benzene and its derivatives make up a large class of compounds called aromatic hydrocarbons. The ring structure of benzene has unusual stability and remains intact through most chemical reactions. Derivatives of benzene are formed through substitution reactions, in which other atoms or groups of atoms are substituted for a hydrogen on a benzene ring.

Important Equations

Oxidation of Alkenes

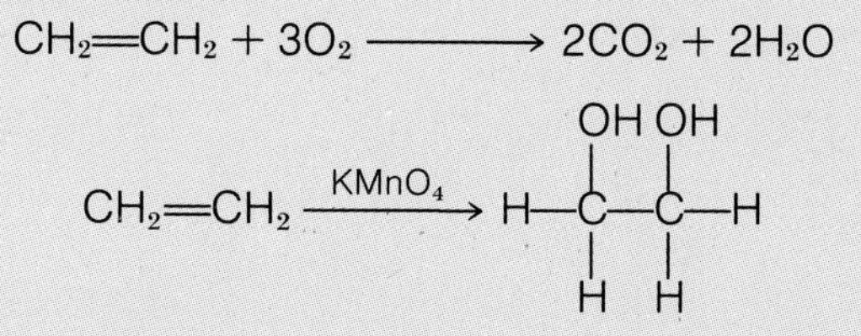

Addition Reactions of Alkenes

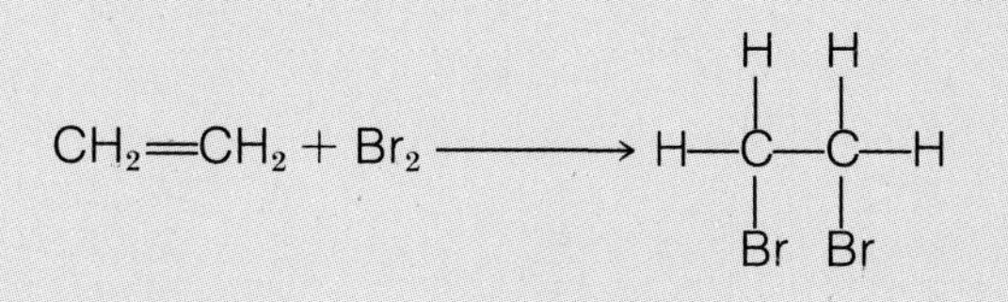

Polymerization of Alkenes

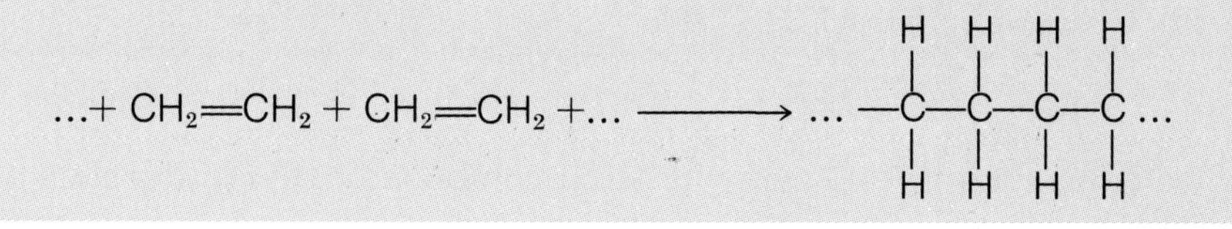

Substitution Reactions of Benzene

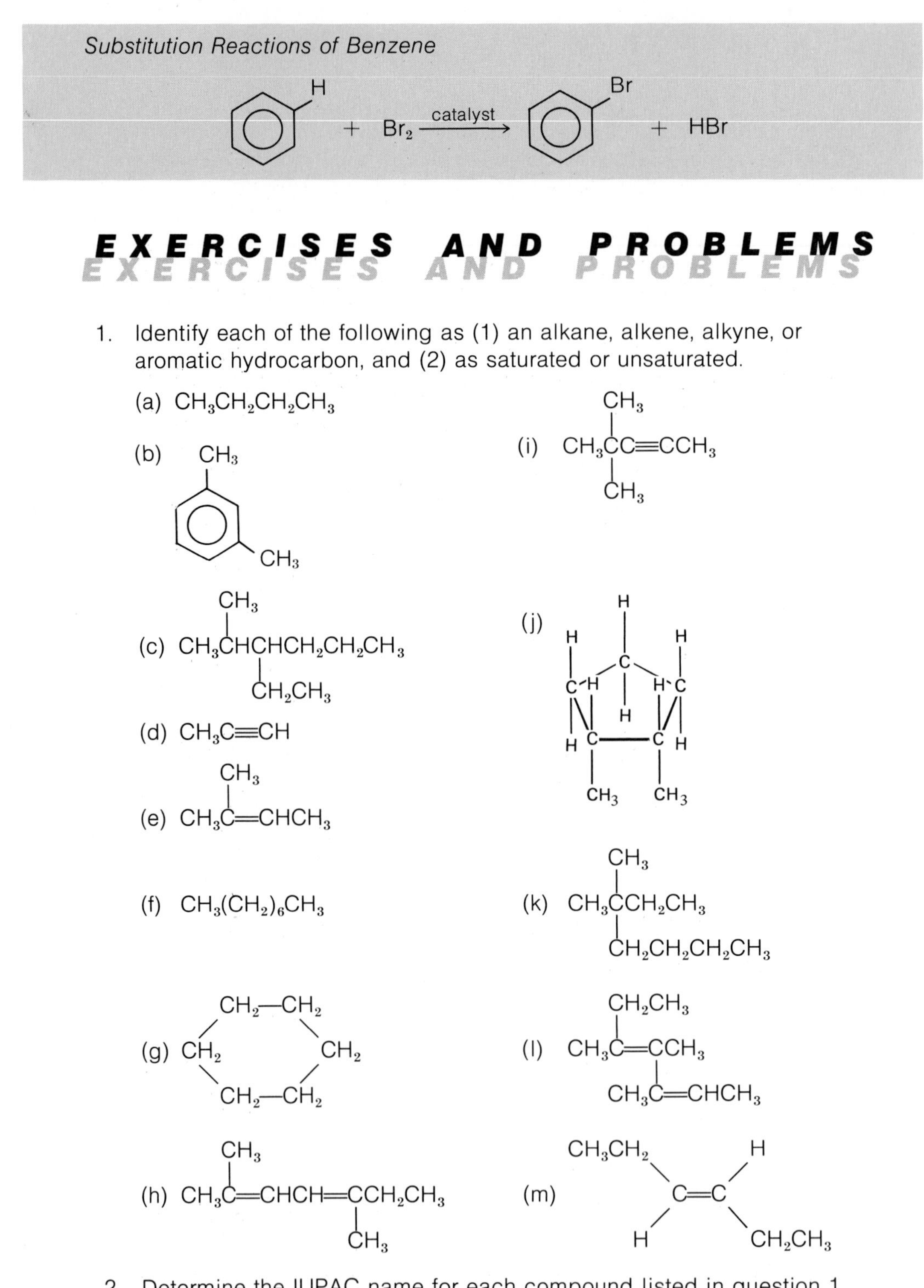

$$C_6H_5\text{—}H + Br_2 \xrightarrow{\text{catalyst}} C_6H_5\text{—}Br + HBr$$

EXERCISES AND PROBLEMS

1. Identify each of the following as (1) an alkane, alkene, alkyne, or aromatic hydrocarbon, and (2) as saturated or unsaturated.

   (a) $CH_3CH_2CH_2CH_3$

   (b) benzene ring with two CH_3 groups (1,3-dimethylbenzene)

   (c) $CH_3CH(CH_3)CH(CH_2CH_3)CH_2CH_2CH_3$

   (d) $CH_3C{\equiv}CH$

   (e) $CH_3C(CH_3){=}CHCH_3$

   (f) $CH_3(CH_2)_6CH_3$

   (g) six CH_2 groups joined in a ring

   (h) $CH_3C(CH_3){=}CHCH{=}C(CH_3)CH_2CH_3$

   (i) $CH_3C(CH_3)_2C{\equiv}CCH_3$

   (j) five-membered ring of C atoms with H atoms and two CH_3 groups

   (k) $CH_3C(CH_3)(CH_2CH_2CH_2CH_3)CH_2CH_3$

   (l) $CH_3C(CH_2CH_3){=}C(CH_3)\text{—}C(CH_3){=}CHCH_3$

   (m) $CH_3CH_2(H)C{=}C(H)CH_2CH_3$

2. Determine the IUPAC name for each compound listed in question 1. (*Hint:* For parts (k) and (l), be sure that you identify the longest carbon chain.)

3. Write the structural formula for each of the following compounds:

   (a) 2-hexyne
   (b) 2-methyl-3-heptene
   (c) *cis*-1,2-dibromocyclopentane
   (d) *cis*-1,2-dichloropropene
   (e) 2,3-dimethyl-2-butene
   (f) *o*-methylaniline
   (g) 1,3,5-octatriene
   (h) 4,5-diethyl-2-heptyne
   (i) 3,3,4-trimethyl-1-pentyne
   (j) 2,2,3,3-tetrabromohexane
   (k) *trans*-4-phenyl-2-pentene
   (l) *p*-diiodobenzene
   (m) *trans*-4,5-diethyl-6-butyl-4-decene
   (n) 2,4,6-trinitrophenol

4. Write the structural formula for each of the following compounds:

   (a) 3-ethylcyclohexane
   (b) 5,5-dimethyl-2-hexene
   (c) 2,3,5-trimethyl-4-propyl-1-heptene
   (d) 4-methyl-2-pentyne
   (e) 2-phenyl-2-butene
   (f) 2,4-difluorobenzenesulfonic acid

5. State the IUPAC name for each of the following compounds:

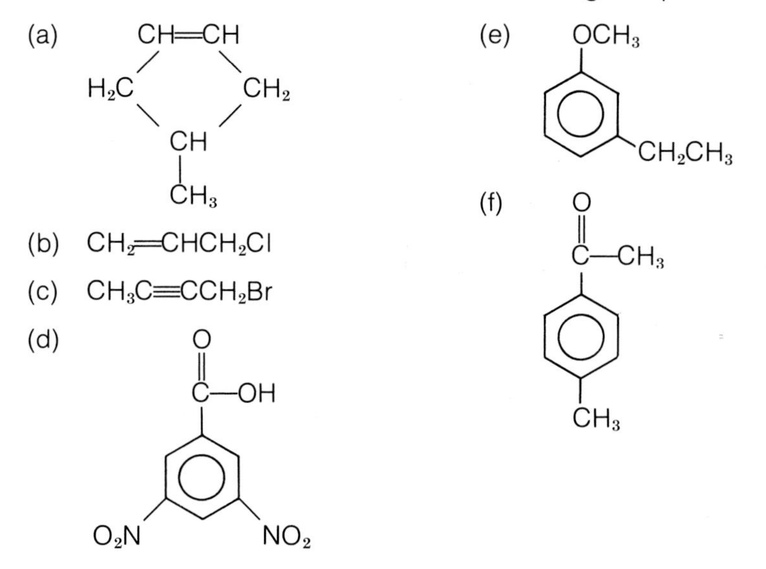

6. State the IUPAC name for each of the alkyne isomers of C_5H_8.

7. Which of the following compounds are structural isomers of each other?

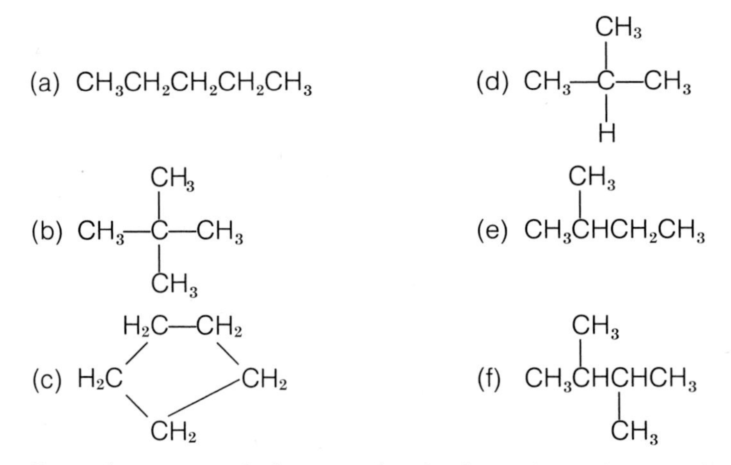

8. Draw the geometric isomers for the following compounds.

   (a) 2-pentene (b) 1,2-dichloroethylene (c) 1,2-dichlorocyclopentane

9. Can dibromoacetylene exist as *cis* and *trans* isomers? Why or why not?

10. For each of the following pairs of molecules, identify the more reactive compound. Give the reason for your choices.

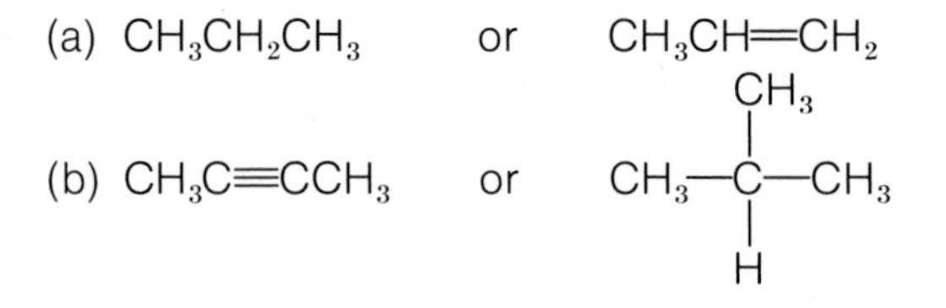

11. Which of the following are aromatic compounds?

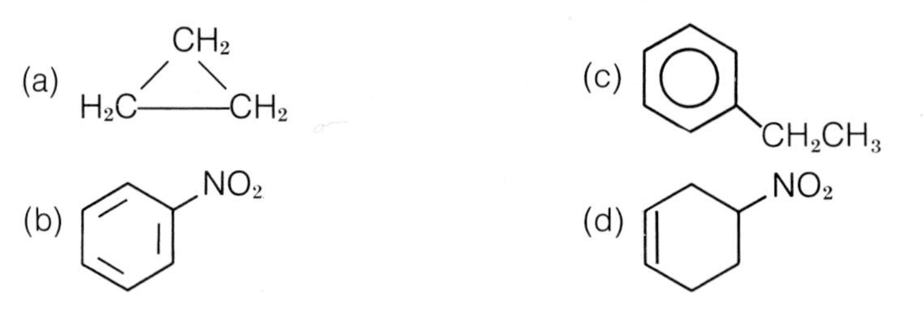

12. Electric switches in a hospital operating room may not be turned on or off while gaseous anesthetic is being administered. Suggest a reason for this regulation.

13. Write the equation for the following reactions:
    (a) the complete oxidation of butene.
    (b) the substitution reaction between benzene and nitric acid.
    (c) the substitution reaction between benzene and chlorine.
    (d) the addition reaction between 2-pentene and hydrogen.

14. Propylene (propene) can undergo a reaction to form the polymer called polypropylene. Show the polymerization of three propylene molecules.

CHAPTER SIXTEEN

ORGANIC COMPOUNDS CONTAINING OXYGEN

There were no classes that Friday, and eighteen-year-old Jane was bored. She felt a bit guilty about wasting the day at her friend's apartment, drinking beer and watching MTV, but after the beer ran out and they switched to vodka and Seven-Up the time seemed to pass more quickly. As

4:30 P.M. came around, Jane knew she had to leave. She realized she was a little drunk, but decided it would be alright if only she was a bit more cautious than usual when driving home.

The young boy on the skateboard never knew what hit him when Jane's car rounded the corner and swerved right into his path. Incredibly, although the boy was thrown onto the hood and bounced off the car's windshield, Jane never realized that she had hit anything. Her only memory was a fuzzy impression of being pulled from the car by a policeman after hitting a tree a bit farther down the road.

Jane's blood alcohol level was measured to be 0.22%, way above the legal limit for driving (0.10% in most states). She was arrested for both drunk driving and hit-and-run.

The specific substance that the police measured in Jane's blood—the chemical responsible for Jane's lack of judgment and her poor muscle coordination—was ethanol. Ethanol, or ethyl alcohol, is just one of many alcohols. These are all substances whose molecules contain a special group of atoms called the hydroxyl group, which consists of an oxygen atom and a hydrogen atom connected by a single bond, —O—H.

What happened as Jane drank her beer and vodka? Even as ethanol is swallowed, it is being absorbed by the mucous membranes in the nose and throat and its vapors absorbed through the lungs. Further absorption of ethanol into the body begins immediately after it is swallowed, with about 20% of the alcohol absorbed through the stomach and the remainder through the small intestines. It takes only about one hour for the alcohol in a single drink to be absorbed when the stomach is empty, but can take up to six hours when the stomach is full.

Ethanol is completely soluble in water. Once in the bloodstream it moves rapidly into the tissues—especially into organs with large blood supplies such as the brain. This movement into the tissues continues until the concentration of ethanol in the tissues equals the concentration of ethanol in the blood. Because the ethanol becomes equally distributed throughout the tissues of the body, the concentration of ethanol in one's breath or urine is a fairly accurate indicator of the level of ethanol in the blood. That is why a breath test can be used to determine the blood alcohol level, and the state of intoxication, of a driver.

A small amount of ethanol in the blood will act as a stimulant to most organs and body systems. But as the level of ethanol increases, it acts as a depressant. This is especially true in the brain, where ethanol has a disruptive effect on the nerve cell membranes, making the brain less responsive to stimuli. As the blood alcohol level rises toward 0.10%, a person is generally in a mood of pleasant relaxation; tension and anxieties are eased. At this level, inhibitions become decreased as the control center in the brain becomes less active. At higher concentrations, the increased disruption of the nervous system results in a lack of muscular coordination, slurred speech, and difficulty in understanding what is seen and heard. A concentration of alcohol in the blood above 0.36% can result in delirium, anesthesia, coma, and even death.

In addition to being found in alcohols, atoms of oxygen appear in many other organic molecules. In this chapter we will study the various classes of oxygen-containing compounds.

LEARNING OBJECTIVES

By the time you have finished this chapter, you should be able to:

1. Classify the organic compounds having oxygen-containing functional groups, and given the structural formula of a compound, state its IUPAC name.
2. Compare the polarity and water solubility of alcohols, ethers, aldehydes, ketones, and carboxylic acids.
3. State the difference between a primary, secondary, and tertiary alcohol.
4. Describe two methods for preparing alcohols in the laboratory.
5. Write the equations for the dehydration and oxidation reactions of an alcohol.
6. State the difference in structure between aldehydes and ketones.
7. Describe one method of preparing aldehydes and ketones.
8. Compare the ease with which aldehydes and ketones can be oxidized.
9. Write the equations for the production of a carboxylic acid and an ester.
10. Define *esterification, hydrolysis,* and *saponification,* and give an example of each reaction.

ALCOHOLS AND ETHERS

16.1 Functional Groups Containing Oxygen

You are probably most familiar with oxygen as the component of air that we breathe to sustain life. Without oxygen, we would die. But oxygen plays a further role in living systems, not just as free oxygen (O_2) or as part of water (H_2O), but as an element in the structure of many molecules that are extremely important in life processes. We have seen that alcohols, for example, are a class of compounds identified by the presence of a hydroxyl group, —OH. Such special groups of atoms, only some of which contain oxygen, are called **functional groups.** Functional groups create a reactive area on an organic molecule, giving the molecule specific chemical properties. In Chapter 14 we mentioned that the study of organic chemistry is simplified by categorizing compounds having similar properties into classes, and then studying several examples of each class. In this chapter we will examine classes of organic compounds that have functional groups containing oxygen (Table 16.1).

Before beginning this discussion, it is important to note again that the letter R is often used when writing general formulas for organic compounds (see the first column in Table 16.1). General formulas are useful in describing properties and reactions of a whole class of compounds. The letter R is used to represent any alkyl group. That is, it could be a methyl group, ethyl group, or propyl group, and so on. In these general formulas, R′ and R″ are used to represent alkyl groups that are different from R.

16.2 Alcohols

To repeat, **alcohols** are organic compounds whose molecules contain the **hydroxyl** function group, —OH. Thus, the general formula for any alcohol is ROH. (The hydroxyl group on an alcohol has no basic (alkaline) properties and should not be confused with the basic hydroxide ion, OH^-, which we discussed in Chapter 13.) This hydroxyl group forms a reactive spot on the molecule, and alcohols are an important class of organic compounds because they are easily formed, are quite reactive, and are good starting materials for the synthesis of many other compounds.

Alcohols are named by changing the *-e* name ending of the parent alkane to **-ol.** The position of the hydroxyl group is indicated by adding a number in front of the name. The presence of more than one hydroxyl group is indicated by placing a prefix before the *-ol* ending. For example, 1,2-ethanediol is a two-carbon alcohol with two —OH groups, one on each carbon.

Table 16.1 Functional Groups Containing Oxygen

Functional group	Class of Compound	Typical Compound	
R—OH	Alcohol	CH_3CH_2—OH	Ethanol
R—O—R′	Ether	CH_3—O—CH_2CH_3	Ethyl methyl ether
R—C(=O)—H	Aldehyde	CH_3CH_2—C(=O)—H	Propanal
R—C(=O)—R′	Ketone	CH_3—C(=O)—CH_3	Propanone (Acetone)
R—C(=O)—OH	Carboxylic acid	CH_3CH_2—C(=O)—OH	Propanoic acid
R—C(=O)—O—R′	Ester	CH_3CH_2—C(=O)—O—CH_3	Methyl propanoate

Alcohols can be divided into three categories according to the placement of the hydroxyl group on the molecule. **Primary alcohols** have the hydroxyl group attached to a carbon atom that is bonded at most to one other carbon atom.

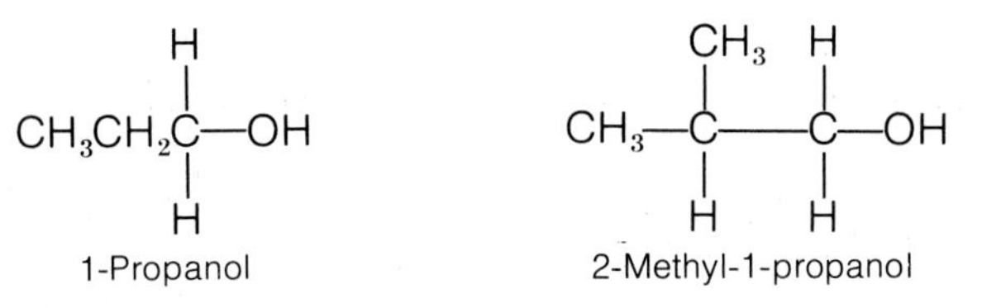

Secondary alcohols have the hydroxyl group attached to a carbon atom that is bonded to two other carbon atoms.

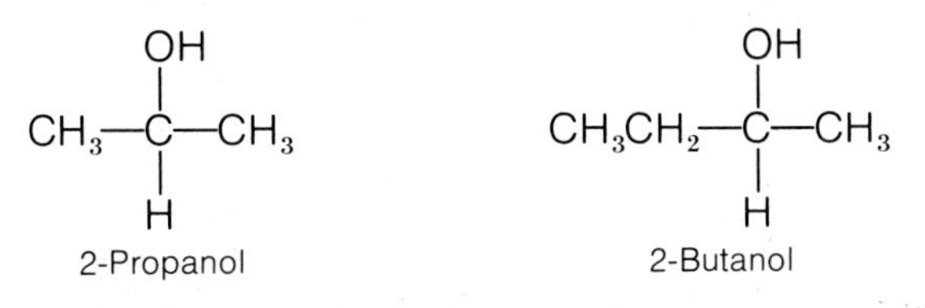

Tertiary alcohols have the hydroxyl group attached to a carbon atom that is bonded to three other carbon atoms.

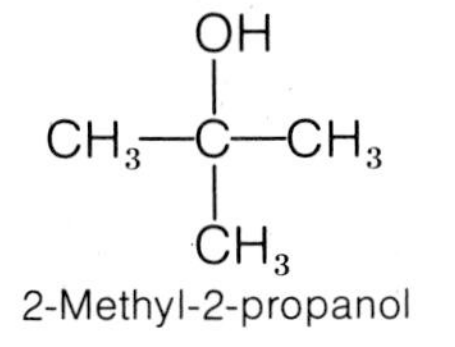

EXAMPLE 16-1

1. Name the following compounds:

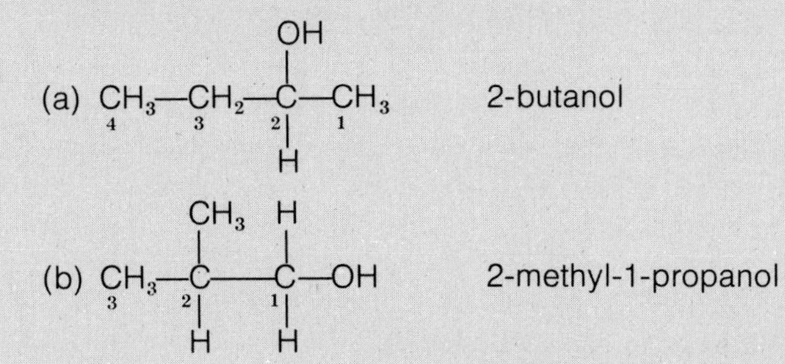

2. Write the structural formula for 2-methyl-2-pentanol.

   This compound is an alcohol with five carbons in its main chain (the chain containing the —OH functional group). Both the —OH group and the methyl group are on carbon 2.

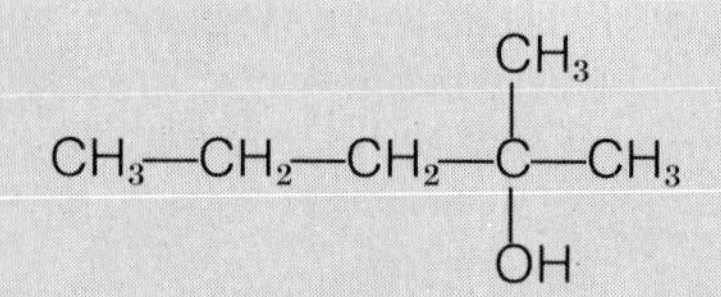

3. State whether the alcohols in problems 1 and 2 above are primary, secondary, or tertiary.

   (1a) 2-Butanol is a secondary alcohol. The hydroxyl group is attached to a carbon atom that is bonded to two other carbon atoms.

   (1b) 2-Methyl-1-propanol is a primary alcohol. The hydroxyl group is attached to a carbon that is bonded to only one other carbon atom.

   (2) 2-Methyl-2-pentanol is a tertiary alcohol. The hydroxyl group is attached to a carbon that is bonded to three other carbon atoms.

Exercise 16-1

1. Name the following alcohols and identify them as primary, secondary, or tertiary:

   (a)

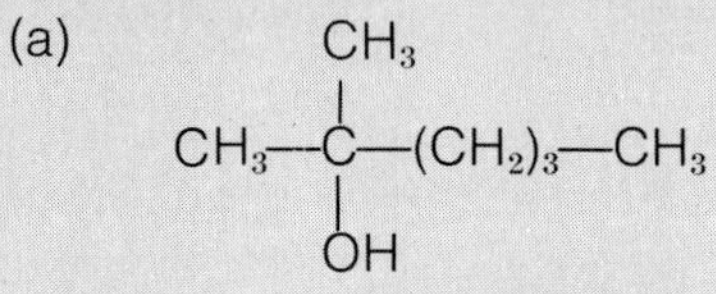

   (b)

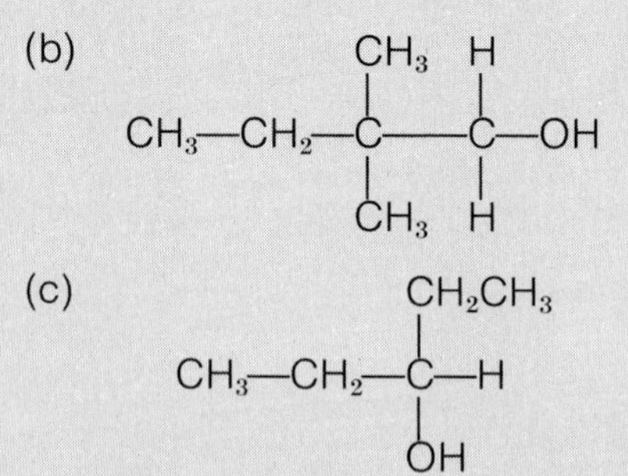

   (c)

$$\begin{array}{c} \quad\quad CH_2CH_3 \\ \quad\quad | \\ CH_3—CH_2—C—H \\ \quad\quad | \\ \quad\quad OH \end{array}$$

2. Draw the structural formula for each of the following compounds:

   (a) 3,3-dimethyl-2-butanol
   (b) 1,2-propanediol
   (c) 5-methyl-3-isopropyl-1-heptanol

The hydroxyl group on an alcohol molecule forms a polar area on the otherwise nonpolar carbon chain. The oxygen atom in the hydroxyl group has a stronger attraction for additional electrons than does the hydrogen, and thus it will form an unequal, or polar, covalent bond with the hydrogen. This hydrogen, then, is capable of hydrogen bonding with other alcohol molecules or with water molecules (Figure 16.1).

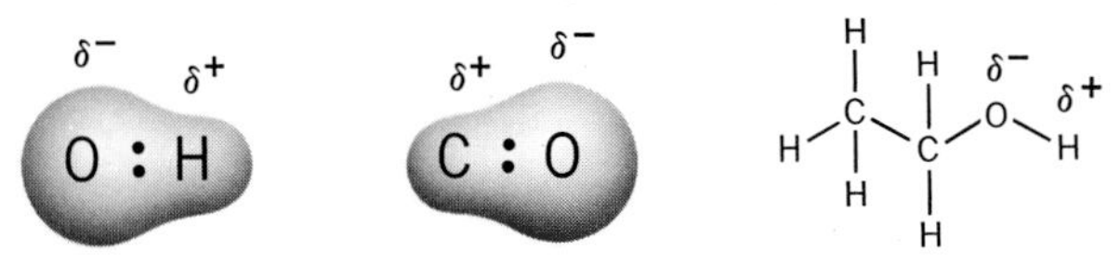

Because of this hydrogen bonding, alcohols with short carbon chains are soluble in polar solvents such as water. As the length of the carbon chain increases, however, the nonpolar nature of the carbon chain becomes more important than the attraction of the hydroxyl group for the water. This makes larger molecules less and less soluble in polar solvents and more soluble in nonpolar solvents such as fats, benzene, and carbon tetrachloride. Increasing the number of hydroxyl groups on a long alcohol molecule increases the number of areas available for hydrogen bonding, making such molecules again more soluble in water (Table 16.2).

The polar hydroxyl group also affects the melting point and the boiling point of the alcohol molecule. The hydrogen bonding that occurs between alcohol molecules increases the amount of energy necessary to pull the molecules apart, thereby increasing the melting and boiling points.

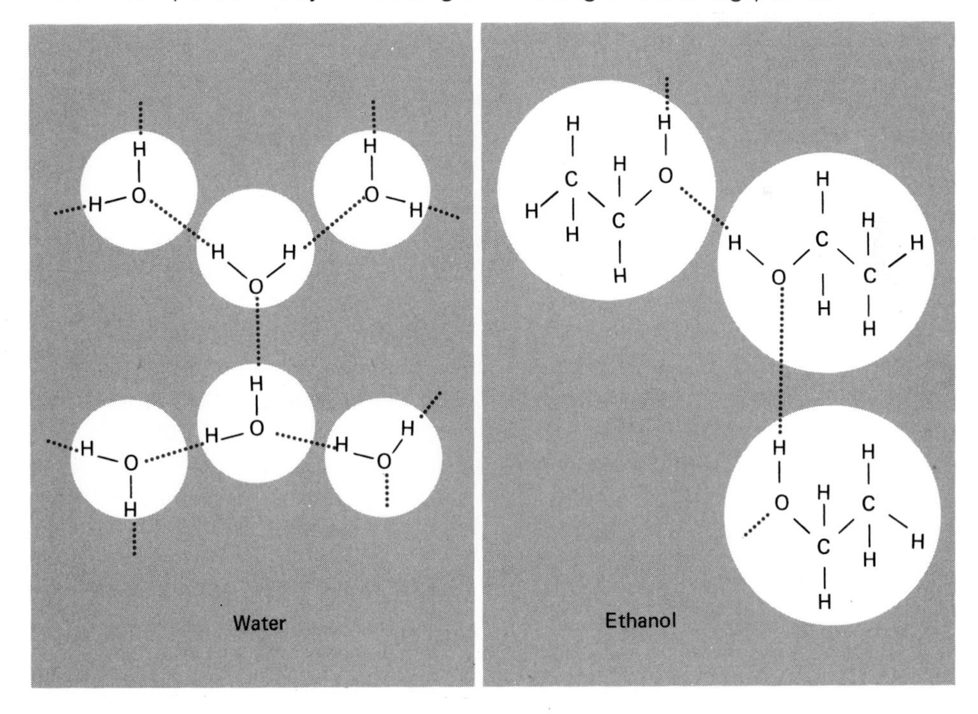

Figure 16.1 Hydrogen bonding in water and ethanol. The polar hydroxyl group on alcohols can enter into hydrogen bonding in much the same way as the —OH group in water. The hydrogen bond is indicated by the colored dotted line.

Table 16.2 Solubility of Some Alcohols (in Water)

Compound	Formula	Solubility
Ethanol	CH_3CH_2OH	Completely soluble
1-Pentanol	$CH_3CH_2CH_2CH_2CH_2OH$	Slightly soluble
1,2-Pentanediol	$CH_3CH_2CH_2CHOHCH_2OH$	Soluble
2-Hexanol	$CH_3(CH_2)_3CHOHCH_3$	Very slightly soluble
2,3-Hexanediol	$CH_3(CH_2)_2CHOHCHOHCH_3$	Soluble
1-Decanol	$CH_3(CH_2)_8CH_2OH$	Insoluble

Alcohols have much higher boiling points than alkanes with comparable molecular weights. For example, ethanol has a molecular weight of 46 and propane a molecular weight of 44. Yet ethanol boils at 78°C, whereas propane boils at −45°C. Similarly, 1-butanol (molecular weight 74) boils at 117°C, whereas pentane (molecular weight 72) boils at 36°C.

16.3 Straight Chain Alcohols

Methanol

The alcohol that is derived from methane is methanol, CH_3OH. Its common name, wood alcohol, comes from the fact that it was first obtained by heating wood in the absence of air. Methanol is very poisonous; less than 10 cc (2 teaspoons) can cause blindness, and 30 cc (2 tablespoons) can cause death.

Ethanol

Ethanol has been known for thousands of years. It was originally produced by mixing honey, fruits, berries, cereals, or other plant materials with water and leaving them in the sun. This created a liquid prized as food, as a ceremonial or religious potion, and as medicine. The production of ethanol through this process known as fermentation became quite an art long before the actual chemistry was understood. In fermentation, yeast cells in the fruit or cereal mixture use the nutrients found there to supply themselves with energy. The waste products of this energy-producing reaction are ethanol and carbon dioxide.

$$\underset{\substack{\text{Glucose}\\\text{(a sugar)}}}{C_6H_{12}O_6} \xrightarrow{\text{Yeast}} \underset{\text{Ethanol}}{2CH_3CH_2OH} + 2CO_2 + \text{Energy}$$

The ethanol produced by the yeast cells dissolves in the surrounding liquid. When the concentration of ethanol rises above 12 to 18% (depending upon the type of yeast), however, it becomes toxic to the yeast. The cells die and settle to the bottom of the container, leaving behind an alcoholic beverage.

Polyhydric Alcohols

Alcohols can have more than one hydroxyl group on their molecules. We have already mentioned that the addition of more polar hydroxyl groups will make the molecule more water soluble and will also raise the boiling point (remember, this is due to more hydrogen bonding). Multiple hydroxyl groups, for a reason still unknown, also make a compound taste sweet.

Ethylene glycol, a dihydric alcohol (that is, an alcohol containing two hydroxyl groups) is a colorless, sweet liquid that is water soluble. It is just as toxic as methanol when taken internally. However, ethylene glycol has great commercial value as the basic ingredient in permanent antifreeze for automobiles, and also in the production of synthetic fabrics.

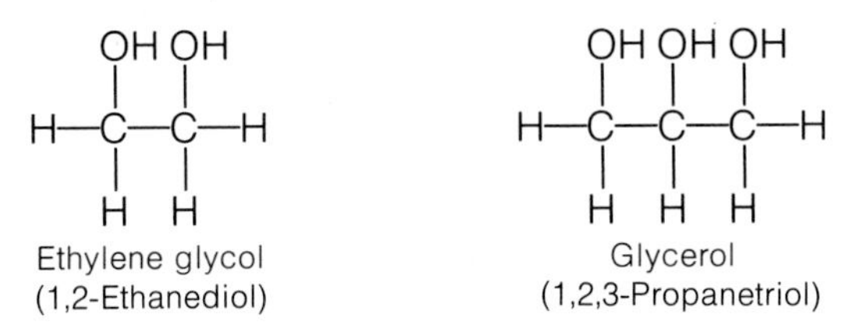

Ethylene glycol (1,2-Ethanediol)   Glycerol (1,2,3-Propanetriol)

Glycerol or glycerin, which has three hydroxyl groups, is a thick, sweet-tasting liquid that is not toxic and that is a component of all natural fats and oils. These properties make it suitable for wide commercial use. It protects the skin and, therefore, is used in hand lotions and cosmetics. Glycerol is also used in inks, tobaccos, cream-filled candies, and plastic clays to prevent loss of water (or dehydration). It is a sweetening agent and a solvent for medicines, a lubricant used in chemical laboratories, and a component of plastics, surface coatings, and synthetic fabrics.

16.4 Cyclic and Aromatic Alcohols

Cyclic Alcohols

Hydroxyl groups can be found on all types of carbon chains. For example, menthol, which is found in peppermint oil, is a cyclic alcohol containing a cyclohexane ring.

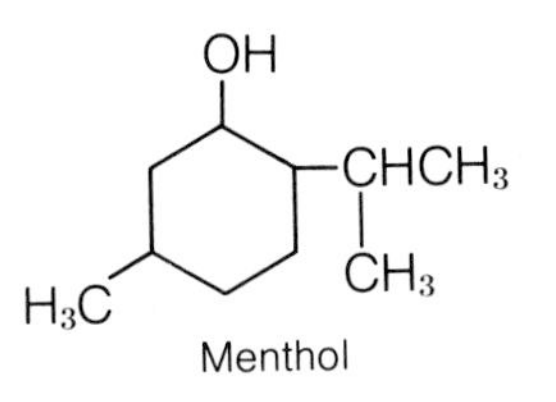

Menthol

The properties of menthol have been known for thousands of years. It causes an unusual cooling and refreshing sensation when rubbed on the skin, leading to its use in aftershave lotions and cosmetics. Menthol soothes inflamed mucous tissues, as in the mucous membranes of the nose and throat, leading to its use in nose and throat sprays, in cough drops, and in cigarettes.

Aromatic Alcohols

Aromatic alcohols are produced when one or more of the hydrogen atoms on a benzene ring is replaced by a hydroxyl group. When one hydrogen is replaced, the compound phenol is produced. Phenol is a powerful germicide, as are most of its derivatives. In 1865, the English surgeon

Lister was the first to apply chemicals to a wound to prevent infection. For this purpose he used a dilute solution of phenol. As other antiseptic chemicals were found, the use of phenol was discontinued because of its damaging effects on the tissues to which it was applied, its absorption through the skin, and its extreme toxicity. However, it is still used by the drug industry as a standard for measuring the germicidal activity of other antiseptics.

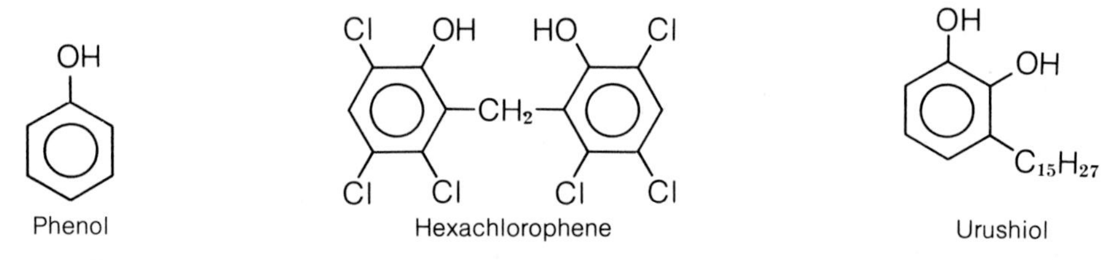

Phenol Hexachlorophene Urushiol

Substituted phenols (phenols having other groups on the benzene ring) are also used as antiseptics. Some soaps, which now can be obtained only by prescription, contain the strong antiseptic hexachlorophene. Another complicated phenol that everyone would like to avoid is urushiol, which is one of the irritants in poison ivy.

16.5 Preparation of Alcohols

Alcohols can be prepared by the addition of water to the double bond of an alkene in the presence of an acid (hydration reaction).

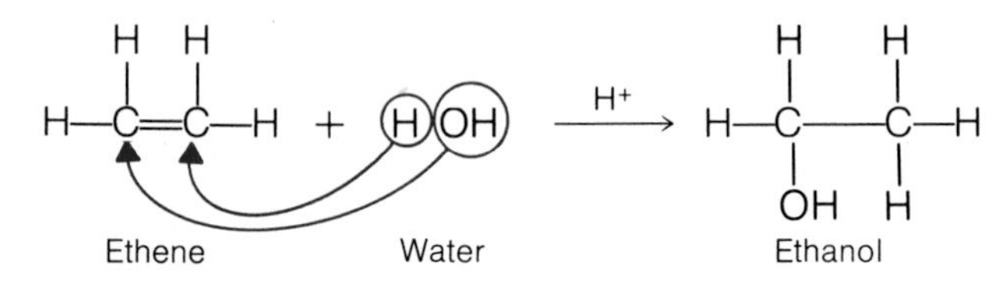

Ethene Water Ethanol

Approximately half of the ethanol used in the United States is produced in this way. Alcohols can also be produced by the addition of hydrogen atoms to aldehydes or ketones in the presence of a catalyst (which we will discuss shortly).

16.6 Reactions of Alcohols

Dehydration

A dehydration reaction of an alcohol is the removal of a water molecule from an alcohol, thereby producing an unsaturated hydrocarbon. Some alcohols undergo dehydration more easily than others. Tertiary alcohols are the easiest to dehydrate and primary alcohols are the most difficult. The dehydration process can be accomplished by heating, but in most cases a dehydrating agent such as sulfuric acid is required.

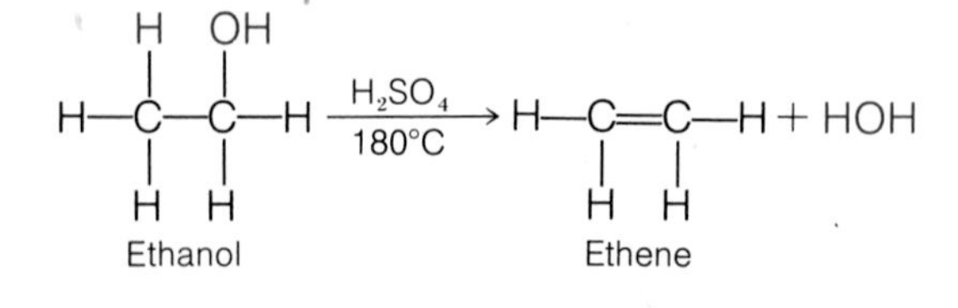

Ethanol Ethene

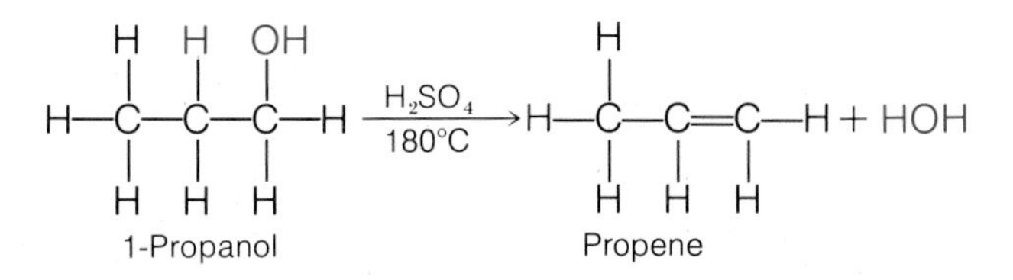

If the above reaction is carried out at a lower temperature, the dehydration reaction takes place between two molecules of the alcohol, thereby producing an ether.

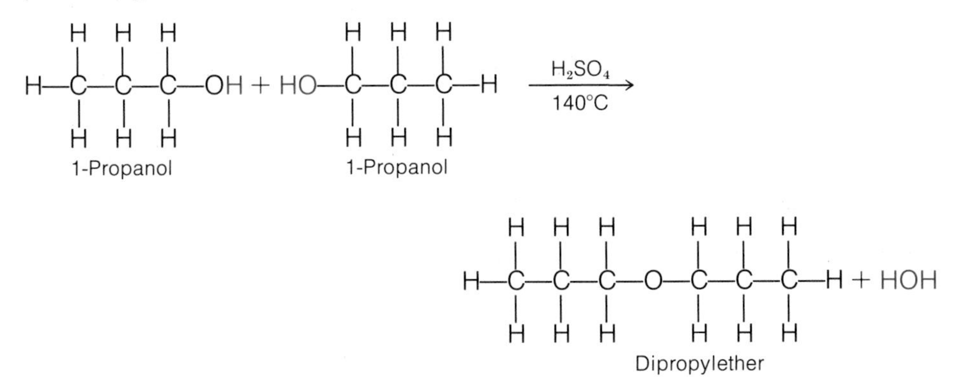

Oxidation

The oxidation of an alcohol involves dehydrogenation, the loss of two hydrogen atoms—one from the hydroxyl group and one from the carbon to which the hydroxyl group is attached. These hydrogen fragments form water by a hydrogenation (reduction) reaction with the oxygen from the oxidizing agent. Potassium permanganate ($KMnO_4$), potassium dichromate ($K_2Cr_2O_7$), and chromic oxide (CrO_3) are commonly used as oxidizing agents for alcohols.

Primary Alcohols. Under controlled experimental conditions, the oxidation (dehydrogenation) of primary alcohols produces members of the class of organic compounds called aldehydes.

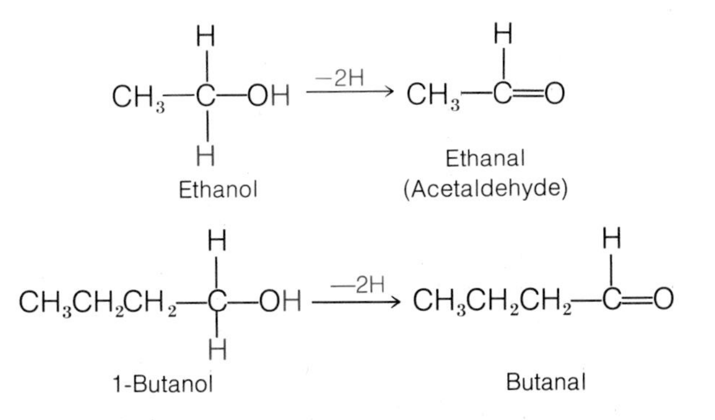

Secondary Alcohols. The oxidation (dehydrogenation) of secondary alcohols produces members of the class of organic compounds called ketones.

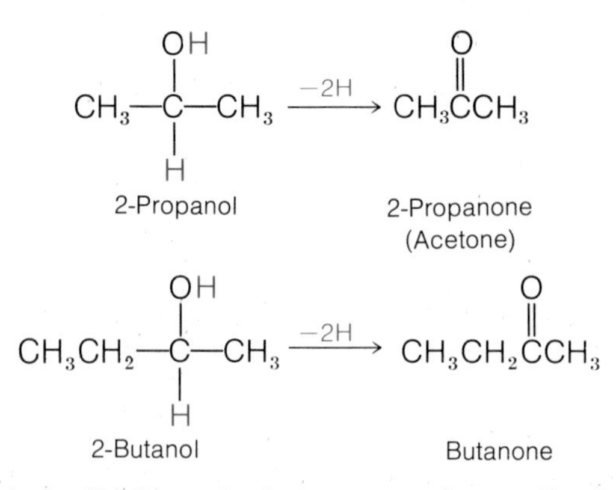

Tertiary Alcohols. Tertiary alcohols cannot be easily dehydrogenated because the carbon attached to the hydroxyl group does not have a hydrogen atom attached to it.

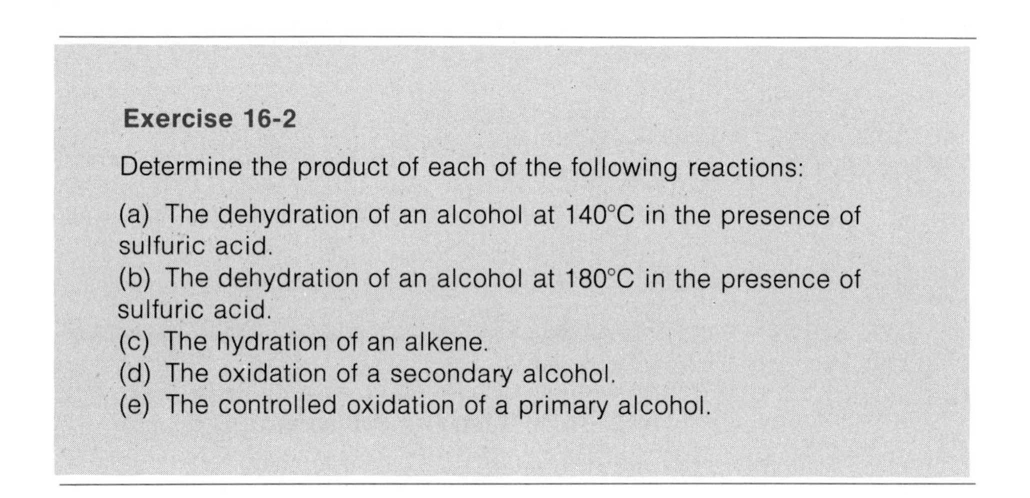

Exercise 16-2

Determine the product of each of the following reactions:

(a) The dehydration of an alcohol at 140°C in the presence of sulfuric acid.
(b) The dehydration of an alcohol at 180°C in the presence of sulfuric acid.
(c) The hydration of an alkene.
(d) The oxidation of a secondary alcohol.
(e) The controlled oxidation of a primary alcohol.

16.7 Ethers

An organic molecule having an oxygen atom bonded to two carbon atoms belongs to the class of compounds called **ethers,** —C—O—C— (general formulas ROR or ROR′). The placement of the oxygen atom between the carbon atoms eliminates the possibility of hydrogen bonding. Therefore, ethers are only slightly more soluble in water than are alkanes, and they are generally soluble in nonpolar solvents. Because there is no hydrogen bonding, ethers boil at much lower temperatures than do alcohols with comparable molecular weights. For example, dimethyl ether (molecular weight 46) boils at −23°C, but ethanol (molecular weight 46) boils at 78°C. Similarly, ethyl methyl ether (molecular weight 60) boils at 11°C, but 1-propanol (molecular weight 60) boils at 97°C. Ethers are extremely flammable, and great care must be taken in their use.

Simple ethers are frequently given common names that just list both alkyl groups attached to the oxygen. For example,

CH_3OCH_3 is dimethyl ether, and

$CH_3OCH_2CH_3$ is ethyl methyl ether.

IUPAC nomenclature is used mostly for complicated ethers that we will not discuss in this book.

Most ethers have some anesthetic properties. Diethyl ether is the anesthetic compound most commonly referred to as "ether" (Figure 16.2). Although it takes a long time for diethyl ether to have full effect on a patient (10 to 15 minutes), it is quite useful in long operations because there is a large safety margin in its use. That is, there is a large difference between the concentration that causes anesthesia and the concentration that will kill the patient. Often a chemical having a more rapid anesthetic effect is used first, and then the patient is switched to diethyl ether.

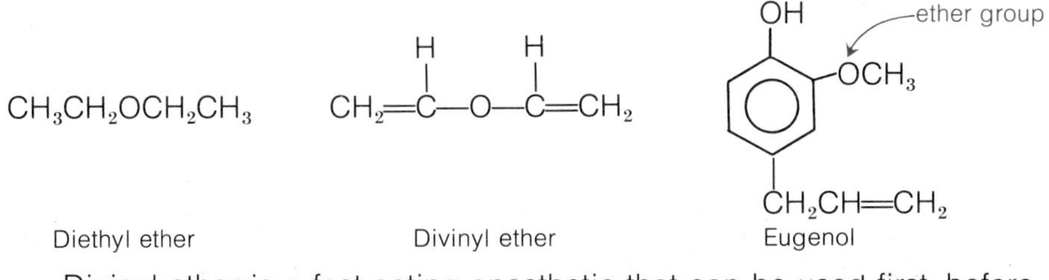

Divinyl ether is a fast-acting anesthetic that can be used first, before diethyl ether. But the double bonds in its structure make divinyl ether highly reactive, and this compound must be handled with care to prevent decomposition. Eugenol, which is obtained from cloves, is a mild local anesthetic used by dentists to lessen pain when filling cavities in teeth.

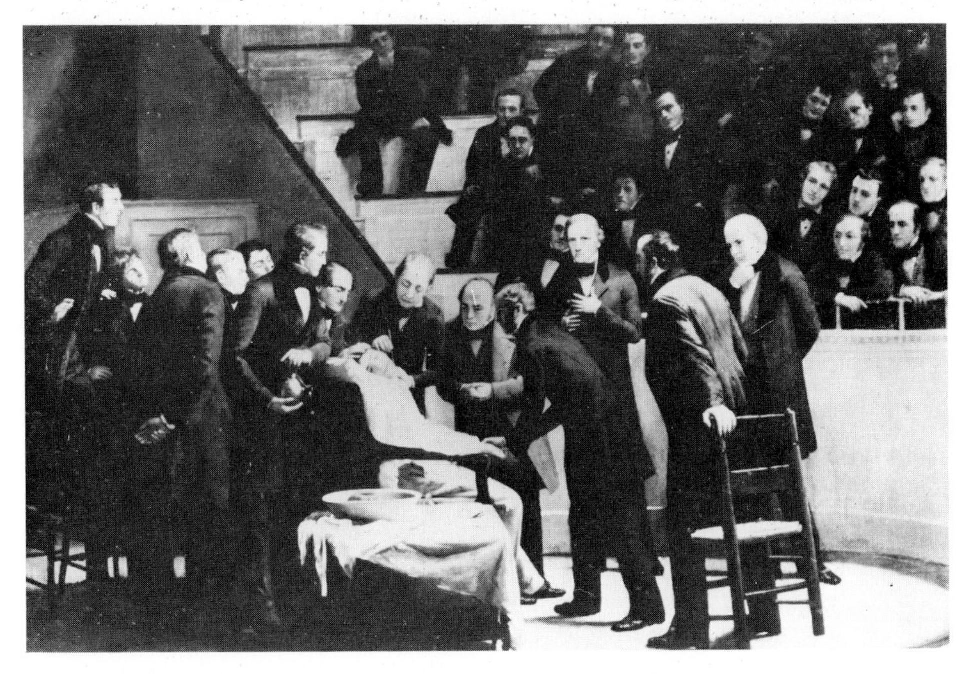

Figure 16.2 The first public demonstration of the use of ether as a surgical anesthetic took place at Massachusetts General Hospital in Boston on October 16, 1846. (Courtesy Massachusetts General Hospital.)

Exercise 16-3

Name the following ethers:

(a) $CH_3OCH_2CH_2CH_2CH_3$

(b) $CH_3CH_2OCH(CH_3)CH_3$

(c) $C_6H_{11}{-}O{-}CH_3$ (cyclohexyl–O–CH_3)

ALDEHYDES, KETONES, CARBOXYLIC ACIDS, AND ESTERS

16.8 Aldehydes and Ketones

Two classes of compounds, **aldehydes** and **ketones,** contain the **carbonyl** group: an oxygen doubly bonded to a carbon atom, $-\overset{\overset{\displaystyle O}{\|}}{C}-$. The difference in properties between the aldehydes and the ketones is due to the position of the carbonyl group. Aldehydes have a terminal carbonyl group—that is, a carbonyl group that is bonded to at most one other carbon atom $-\overset{|}{\underset{|}{C}}-\overset{\overset{\displaystyle O}{\|}}{C}-H$ (general formula $R\overset{\overset{\displaystyle O}{\|}}{C}H$ or RCHO). The carbonyl group in a ketone, by contrast, will appear somewhere in the middle of the molecule. It will be bonded to two other carbon atoms, $-\overset{|}{\underset{|}{C}}-\overset{\overset{\displaystyle O}{\|}}{C}-\overset{|}{\underset{|}{C}}-$ (general formula $R\overset{\overset{\displaystyle O}{\|}}{C}R'$ or RCOR′).

Aldehydes and ketones are both highly reactive compounds; aldehydes are even more reactive than ketones. The presence of the carbonyl group creates a polar region on the aldehyde and ketone molecule. Lower molecular weight aldehydes and ketones are soluble in water because of hydrogen bonding between the carbonyl group and water.

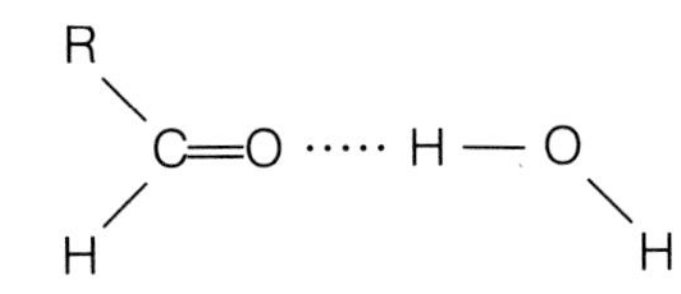

The large class of foods called carbohydrates contains aldehyde and ketone functional groups and will be the subject of a later chapter.

Aldehydes are named by selecting the longest carbon chain that contains the carbonyl group and then replacing the final *-e* on the name of the corresponding alkane with an **-al.** (However, we will use the common name formaldehyde for methanal and acetaldehyde for ethanal because they are so often referred to by these names.) The carbonyl group in aldehydes is always located on the end carbon, so there is no need to indicate its position. When the carbonyl group is attached to a benzene ring, the compound is called benzaldehyde (see Section 16.9 for the structure of this compound).

Ketones are named by selecting the longest carbon chain containing the carbonyl group and then replacing the final *-e* on the name of the corresponding alkane with **-one.** The position of the carbonyl group is indicated by a number appearing before the name.

EXAMPLE 16-2

Name the following compounds:

(a)

$$CH_3CH_2\overset{\overset{\displaystyle O}{\|}}{C}H$$

This compound is an aldehyde with three carbons. Its name therefore, is propanal.

(b)

$$CH_3CH_2\overset{\overset{\displaystyle O}{\|}}{C}CH_2CH_3$$

This compound is a ketone with five carbons. The carbonyl group is located on carbon 3. Its name, therefore, is 3-pentanone.

(c)

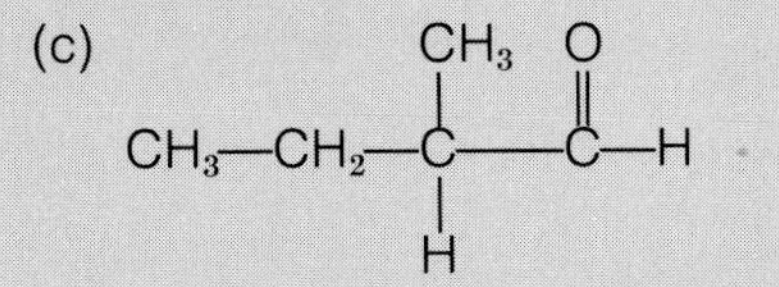

This compound is an aldehyde, and the longest carbon chain containing the carbonyl group has four carbons. There is a methyl group attached to carbon 2. Its name, therefore, is 2-methylbutanal.

(d)

$$\begin{array}{c} \quad O \\ \quad \| \\ CH_3CCH_2CHCH_2CH_3 \\ \qquad | \\ \qquad CH_3 \end{array}$$

This compound is a ketone with six carbons in the longest carbon chain containing the carbonyl group. The carbonyl group is on the second carbon, and there is a methyl group on carbon 4. Its name, therefore, is 4-methyl-2-hexanone.

Exercise 16-4

1. Name the following compounds:

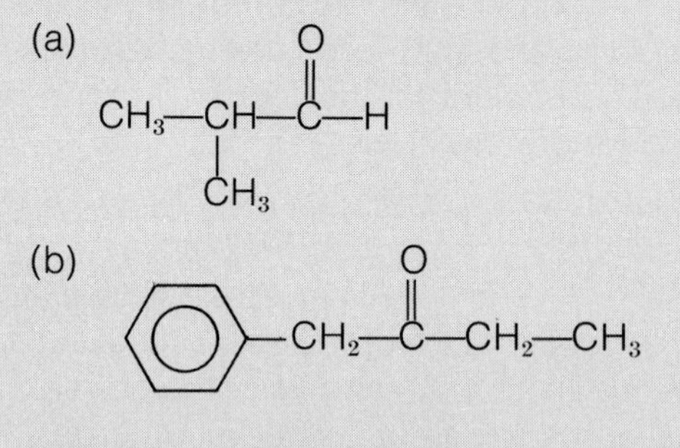

2. Write the structural formula for each of the following compounds:

   (a) 4-ethylheptanal
   (b) 4-ethyl-2-methyl-3-octanone

16.9 Important Aldehydes and Ketones

Formaldehyde is the simplest of the aldehydes.

$$\begin{array}{c} O \\ \| \\ H—C—H \end{array}$$

Formaldehyde

You may be aware of the use of formaldehyde as a tissue preservative. It combines easily with proteins, killing microorganisms and hardening tissues. If you have ever come in contact with formaldehyde, you know that it can irritate the membranes of your eyes, nose, and throat. In our earlier discussion of alcohols we mentioned that even very small amounts of methanol in the body can cause blindness. This happens because the liver, in attempting to rid the body of methanol, changes this compound into formaldehyde. But formaldehyde in the bloodstream will build up in the retina of the eye, destroying the cells and causing blindness. When formaldehyde is polymerized with the alcohol phenol, substances called resins, which are the gummy saps of evergreens, are formed.

Many aldehydes have pleasant odors or tastes and are used in perfumes and artificial flavorings.

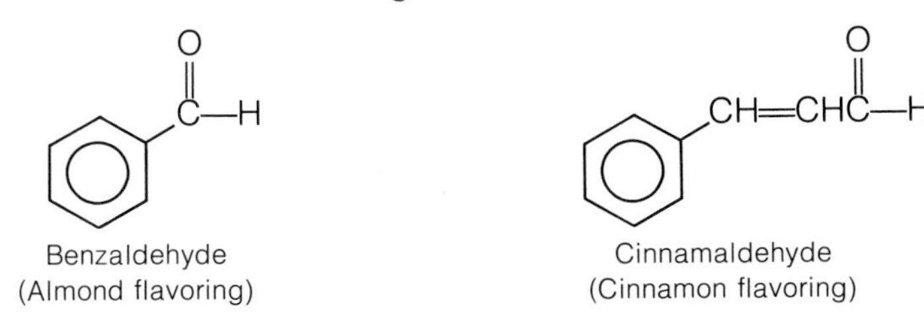

Benzaldehyde (Almond flavoring) Cinnamaldehyde (Cinnamon flavoring)

The aldehydes used in perfumes and flavorings are relatively nonpolar and hence only slightly soluble in water, but will readily dissolve in ethanol. As a result, many perfumes and flavorings use ethanol as their solvent.

Ketones are widely used in the chemical industry as solvents. One of the more common ketone compounds is acetone (propanone).

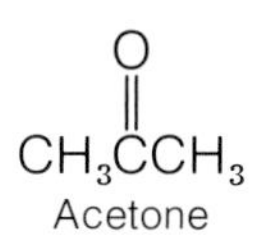

Acetone

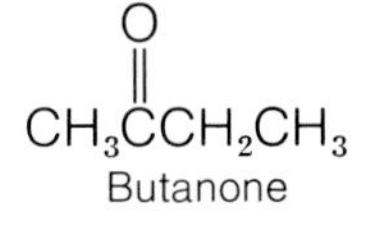

Butanone

Acetone can be produced in the body as the result of a metabolic side reaction, not normally occurring except when there is some metabolic disorder. In the disease diabetes mellitus, for example, some normal metabolic reactions do not occur and acetone is produced in large quantities. This compound builds up in the tissues and appears in the urine and breath of untreated diabetics. Butanone (methyl ethyl ketone) is a solvent with a characteristic odor and is commonly used in nail polish remover.

Some unusual cyclic ketones, isolated from animals, have now become very popular in the cosmetic industry. Are you familiar with musk oil? Civetone, isolated from the African civet cat, and muscone, isolated from the male musk deer, have very unpleasant odors in high concentrations. When used in extremely small quantities in perfumes, however, they intensify and give long-lasting qualities to the scent.

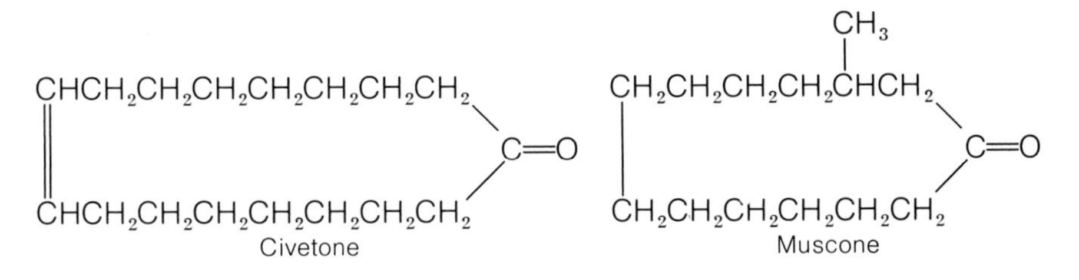

Civetone Muscone

16.10 Preparation of Aldehydes and Ketones

Aldehydes and ketones are produced by the oxidation (dehydrogenation) of primary and secondary alcohols in the presence of oxidizing agents such as potassium permanganate or potassium dichromate.

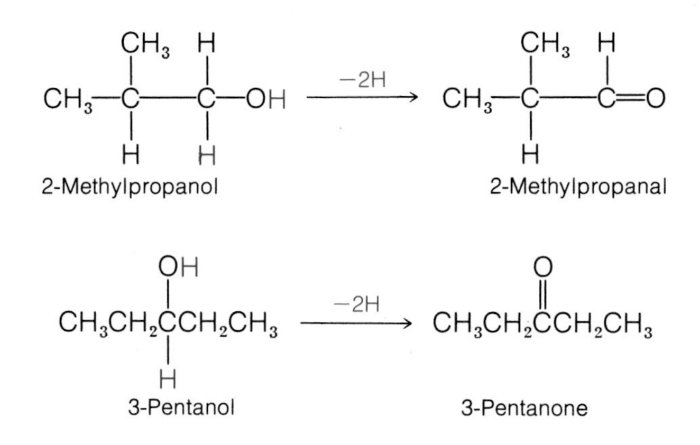

16.11 Reactions of Aldehydes and Ketones

Reduction

Primary alcohols can be produced from aldehydes, and secondary alcohols can be produced from ketones, by the reduction (hydrogenation) of these compounds in the presence of a proper catalyst such as platinum (Pt).

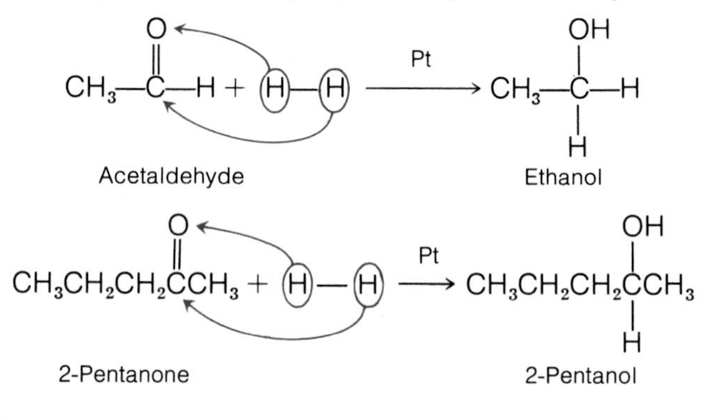

Oxidation

Aldehydes are very easy to oxidize, and such oxidation produces carboxylic acids (which we will discuss shortly). Ketones, however, can be oxidized only under extreme conditions. This, therefore, gives us a convenient way to distinguish between aldehydes and ketones in the laboratory. Notice that in this case the oxidation reaction results in the gain of oxygen atoms by the aldehyde molecule.

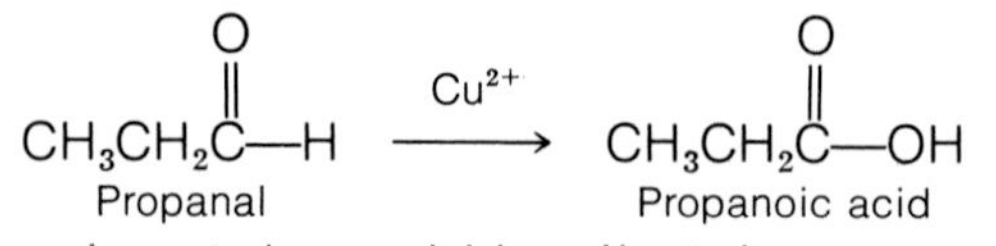

Formation of Hemiacetals and Hemiketals

Dissolving an aldehyde in an alcohol results in the formation of an equilibrium between the aldehyde and a product called a **hemiacetal.**

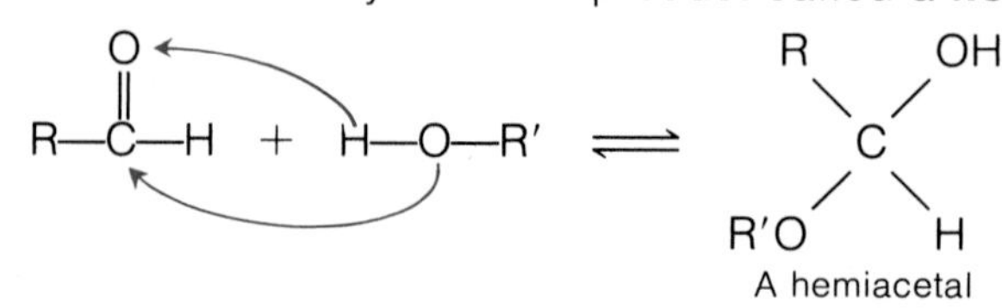

Hemiacetals can form between an aldehyde group and an alcohol group on the same molecule, resulting in the formation of a stable five- or six-member ring structure

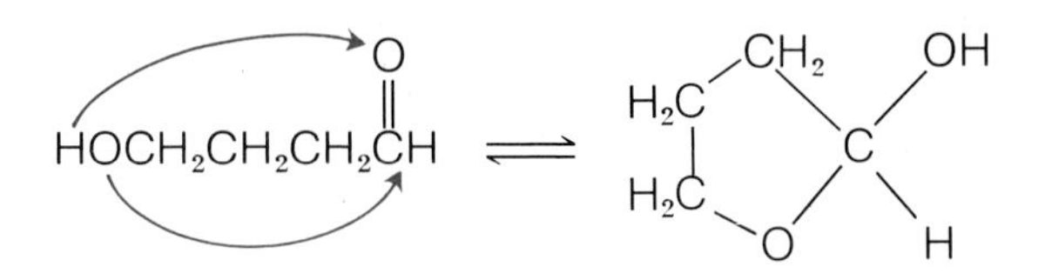

A similar reaction results when a ketone is dissolved in an alcohol, forming a **hemiketal.**

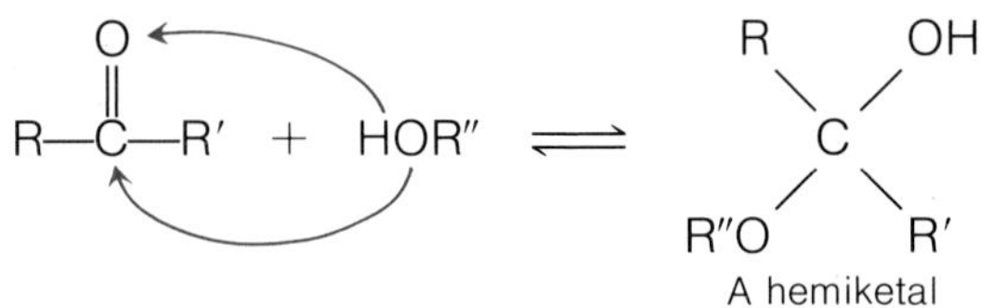

Exercise 16-5

Write the structural formula for the product of each of the following reactions:

(a) The reduction of pentanal
(b) The oxidation of pentanal
(c) The reduction of 3-hexanone
(d) The formation of a hemiketal between acetone and 2-propanol
(e) The formation of a hemiacetal between propanal and 1-butanol

16.12 Carboxylic Acids

Carboxylic acids (general formula $R\overset{O}{\overset{\|}{C}}OH$, or RCOOH) are organic acids that contain a hydroxyl group attached to a carbonyl group. This combined group is called the **carboxyl** or **carboxylic acid** functional group, $-\overset{O}{\overset{\|}{C}}-OH$. The bond between the oxygen atom and the hydrogen atom in the carboxyl group is extremely polar, so much so that when placed in water, some of the molecules ionize as follows:

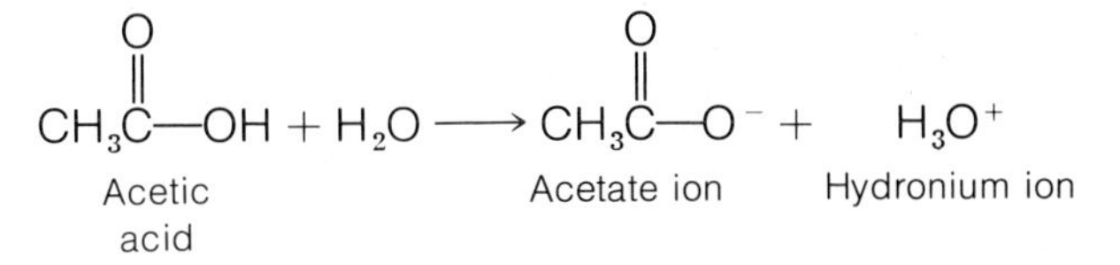

The stronger the acid, the more easily it gives up its hydrogen to form the hydronium ion. But organic acids are very weak compared with inorganic acids such as nitric or sulfuric acid. Because some carboxylic acids are found in fats, these acids are also referred to as fatty acids.

The presence of the polar carboxyl group in carboxylic acids allows the formation of hydrogen bonds between acid molecules and also between acid molecules and water.

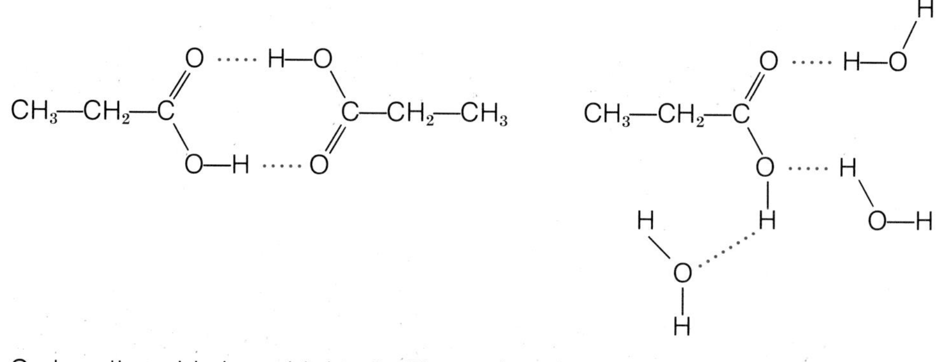

Carboxylic acids have higher boiling points than alcohols with comparable molecular weights, indicating a higher degree of hydrogen bonding. Formic, acetic, propanoic, and butanoic acids are all completely soluble in water, and higher molecular weight carboxylic acids are more soluble than alcohols, aldehydes, or ketones with comparable molecular weights.

Carboxylic acids are named by removing the ending letter *-e* from the alkane having the same length carbon chain and adding **-oic acid.** For example,

$HCOOH$ (H—C(=O)—OH) is methanoic acid, and CH_3CH_2COOH ($CH_3CH_2C(=O)OH$) is propanoic acid.

The carbon atoms in the carboxylic acid molecule are always numbered beginning with the carboxyl group. The names of dicarboxylic acids (organic acids with carboxyl groups on each end of the molecule) end in **-dioic acid.** For example,

HO—C(=O)—C(=O)—OH is ethanedioic acid (common name: oxalic acid)

EXAMPLE 16-3

Name the following compound:

$$CH_3CH_2CH_2CH(CH_3)C(=O)OH$$

This compound has a carboxyl group, so it is a carboxylic acid. There are five carbon atoms in the longest carbon chain, with a methyl group on carbon 2. The name, therefore, is 2-methylpentanoic acid.

Exercise 16-6

1. Name the following compounds:

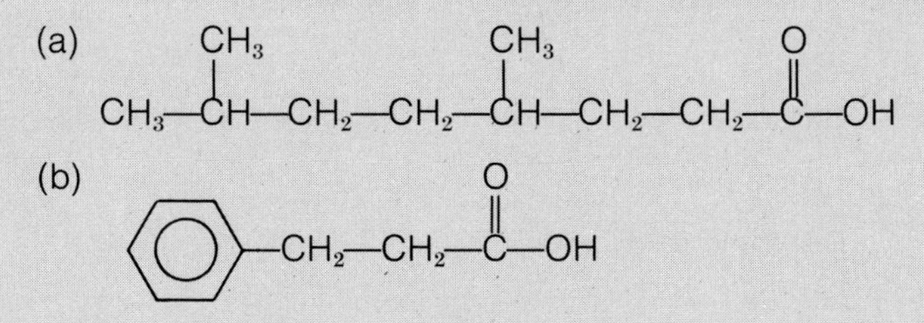

2. Draw the structural formula for adipic acid (hexanedioic acid).

16.13 Important Carboxylic Acids

Formic acid is very irritating to tissues; this is the compound released in bee and ant stings, which causes skin inflammation. (Table 16.3 shows the structure of each of the carboxylic acids that we will discuss.) Formic acid, like formaldehyde, is formed by the liver in the oxidation of methanol. When it enters the bloodstream, it can cause severe acidosis, a condition that disrupts the blood's ability to carry oxygen.

Acetic acid has been known as long as ethyl alcohol, for it is the substance that is formed when wine turns sour. This souring occurs when certain bacteria oxidize ethanol to form acetic acid, producing the familiar taste and smell of vinegar. Acetic acid is the compound used by living cells to produce the longer-chain fatty acids, all of which have an unmistakably strong smell. For example, lactic acid is formed when milk sours. Butyric acid is the unpleasant odor of rancid butter and is one of the compounds that causes "body odor." Capric acid gives limburger cheese its characteristic smell.

Benzoic acid and its sodium salt, sodium benzoate, are used in the making of bread and cheese to prevent spoilage from the growth of mold or bacteria (Figure 16.3). Salicylic acid, whose structure is similar to that of benzoic acid, cannot be used in foods because it is so irritating to tissues. It can destroy horny growths such as corns and warts and is used in products that treat such conditions.

Table 16.3 Some Common Carboxylic Acids

Common Name	IUPAC Name	Formula
Formic acid	Methanoic acid	$H-\overset{\overset{O}{\|}}{C}-OH$
Acetic acid	Ethanoic acid	$CH_3-\overset{\overset{O}{\|}}{C}-OH$
Lactic acid	2-Hydroxypropanoic acid	$CH_3\overset{\overset{OH}{\vert}}{C}H-\overset{\overset{O}{\|}}{C}-OH$
Butyric acid	Butanoic acid	$CH_3CH_2CH_2-\overset{\overset{O}{\|}}{C}-OH$
Capric acid	Decanoic acid	$CH_3(CH_2)_8-\overset{\overset{O}{\|}}{C}-OH$
Tartaric acid	2,3-Dihydroxybutanedioic acid	$HO-\overset{\overset{O}{\|}}{C}-\overset{\overset{OH}{\vert}}{C}H-\overset{\overset{OH}{\vert}}{C}H-\overset{\overset{O}{\|}}{C}-OH$
Benzoic acid	Benzoic acid	$C_6H_5-\overset{\overset{O}{\|}}{C}-OH$ (benzene ring)
Salicylic acid	*o*-Hydroxybenzoic acid	$C_6H_4(OH)-\overset{\overset{O}{\|}}{C}-OH$ (benzene ring with OH ortho)

Figure 16.3 These are some of the many foods containing sodium benzoate as a preservative. (Ron Nelson.)

Salts of carboxylic acids have many commercial uses. The monopotassium salt of tartaric acid is cream of tartar. Salts of long-chain carboxylic acids such as palmitic acid and stearic acid are soaps.

Oxalic acid, a dicarboxylic acid, ionizes to form the oxalate ion, which is a normal product of the body's metabolism.

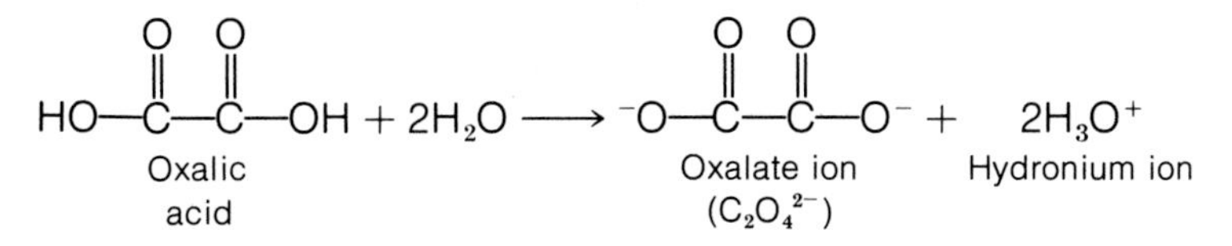

Oxalate ions bond strongly with calcium ions to form an insoluble salt, calcium oxalate (CaC_2O_4). This reaction occurs in the urine of humans, but large crystals of calcium oxalate normally are kept from forming. In some cases, however, such crystals can build up to form small stones known as kidney or bladder stones (urinary calculi).

6.14 Preparation of Carboxylic Acids

Carboxylic acids are prepared by the oxidation of primary alcohols or aldehydes in the presence of a strong oxidizing agent.

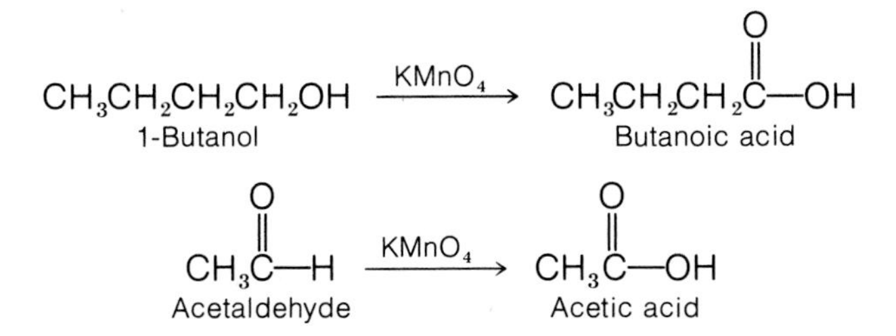

We can summarize the steps in the complete oxidation of a primary alcohol as follows:

$$\text{Alcohol} \xrightarrow{(O)} \text{Aldehyde} \xrightarrow{(O)} \text{Carboxylic acid} \xrightarrow{(O)} CO_2 \text{ and } H_2O$$

where (O) means oxidation by some oxidizing agent (such as $KMnO_4$). Aromatic acids can be prepared by oxidizing a carbon side-chain on a benzene compound with a strong oxidizing agent. The carbon atom attached to the ring becomes the carbon atom in the carboxylic acid group, and the remaining carbons on the side-chain form carbon dioxide.

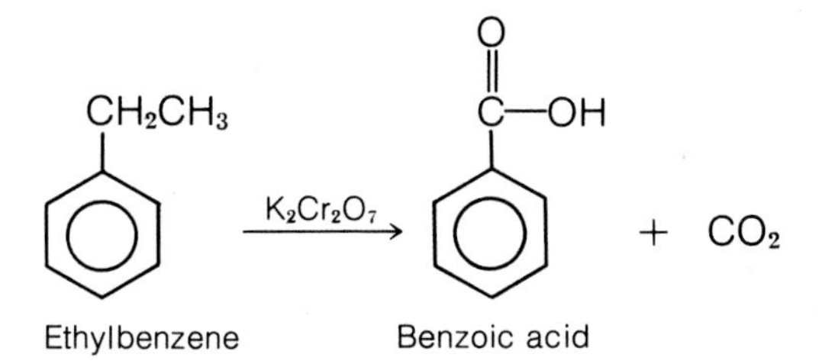

16.15 Esters

An important class of compounds derived from carboxylic acids is the **esters,** whose molecules contain the ester functional group.

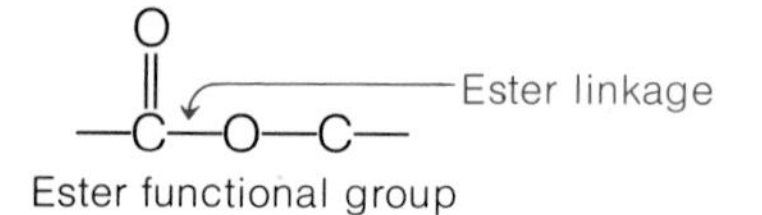

Esters are noted for their pleasant aromas and are the source of the familiar smells of flowers and the tastes of fruits (Table 16.4).

Esters are produced by the reaction of an alcohol and a carboxylic acid in the presence of an acid catalyst. This reaction is called **esterification.** Esterification involves the removal of a molecule of water from a molecule of a carboxylic acid and a molecule of an alcohol.

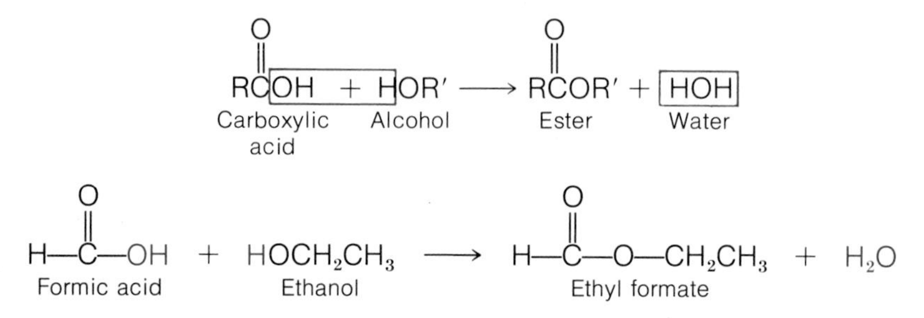

The general name for such a reaction, in which a molecule of water is removed from two reactant molecules with the resulting formation of one product molecule, is a **condensation reaction.** This type of reaction accounts for many of the chemical reactions occurring in biological systems.

Esters are named by first listing the alcohol portion and replacing the *-ol* ending of the alcohol name by a **-yl** ending. Then, the acid portion is listed, with the *-ic* ending of the acid replaced by an **-ate** ending. If the alcohol portion contains a branched chain, the name used is that shown for the alkyl group as listed in Table 14.2.

Table 16.4 **Some Esters Used as Flavorings**

IUPAC Name (Common Name)	Formula	Flavor
Ethyl methanoate (ethyl formate)	$CH_3CH_2{-}O{-}C(=O){-}H$	Rum
Pentyl ethanoate (amyl acetate)	$CH_3(CH_2)_4{-}O{-}C(=O){-}CH_3$	Banana
Octyl ethanoate (octyl acetate)	$CH_3(CH_2)_7{-}O{-}C(=O){-}CH_3$	Orange
Pentyl butanoate (amyl butyrate)	$CH_3(CH_2)_4{-}O{-}C(=O){-}CH_2CH_2CH_3$	Apricot

For example,

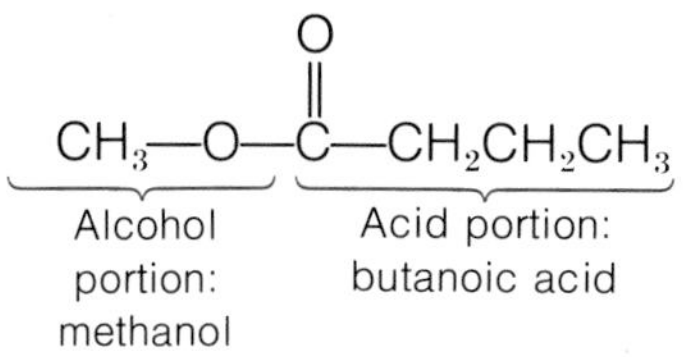

$$\underbrace{CH_3{-}O}_{\text{Alcohol portion: methanol}}{-}\underbrace{\overset{\overset{\displaystyle O}{\|}}{C}{-}CH_2CH_2CH_3}_{\text{Acid portion: butanoic acid}}$$ is methyl butanoate, and

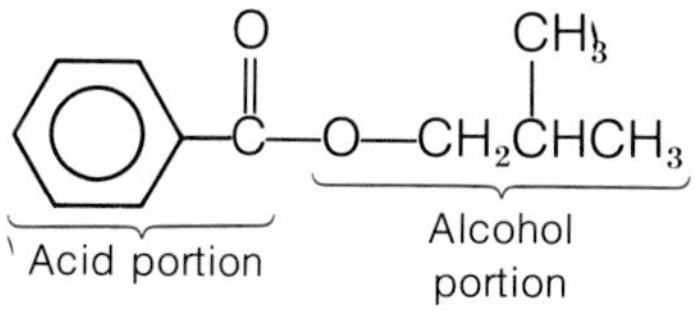

$$\underbrace{C_6H_5{-}\overset{\overset{\displaystyle O}{\|}}{C}}_{\text{Acid portion}}{-}\underbrace{O{-}CH_2\overset{\overset{\displaystyle CH_3}{|}}{C}HCH_3}_{\text{Alcohol portion}}$$ is isobutyl benzoate

EXAMPLE 16-4

Write the structural formula and the name of the ester formed in the esterification of octanoic acid and 2-propanol (isopropyl alcohol).

$$CH_3CH_2CH_2CH_2CH_2CH_2CH_2\overset{\overset{\displaystyle O}{\|}}{C}{-}OH + HO{-}\overset{\overset{\displaystyle CH_3}{|}}{C}HCH_3$$

$$\longrightarrow CH_3(CH_2)_6\overset{\overset{\displaystyle O}{\|}}{C}{-}O{-}\overset{\overset{\displaystyle CH_3}{|}}{C}HCH_3 + H_2O$$

The name of the ester produced in this reaction is isopropyl octanoate.

Exercise 16-7

Write the structural formula and the name of the ester formed from the esterification of

(a) 2-methyl-2-propanol and pentanoic acid
(b) ethanol and *p*-aminobenzoic acid

16.16 Important Esters

Nitroglycerin

Nitroglycerin, which is the active substance in dynamite, is an ester formed by the condensation reaction between an inorganic acid (nitric acid) and an alcohol (glycerol). In addition to its use in dynamite, nitroglycerin is used to relieve pain in the heart disorder angina pectoris.

This compound acts to enlarge (or dilate) smaller blood vessels and to relax smooth muscles found in the arteries, thereby lowering high blood pressure and the accompanying pain.

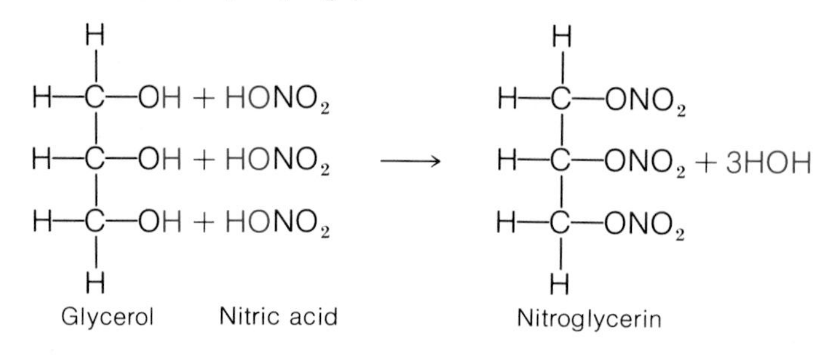

Salicylic Acid Esters

Salicylic acid is found in the bark of willow trees. It is a molecule containing a carboxylic acid group and an alcohol group, both of which can undergo condensation reactions to form esters. We have already mentioned that salicylic acid is a very irritating substance with a disagreeable taste and is not to be taken internally. But the products of condensation reactions with this compound are not as irritating and have a wide range of uses. For example, if the acid group on salicylic acid is condensed with methanol, methyl salicylate is formed. This compound is more commonly known as oil of wintergreen and is used in perfumes, candies, and in ointments that cause a mild burning sensation on your skin in an attempt to take your mind off your sore muscles.

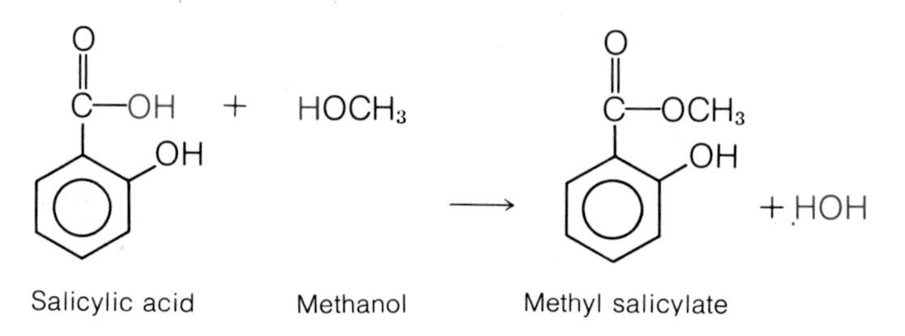

If acetic acid reacts with the alcohol group on salicylic acid, acetylsalicylic acid, one of the most widely used drugs in the world, is formed.

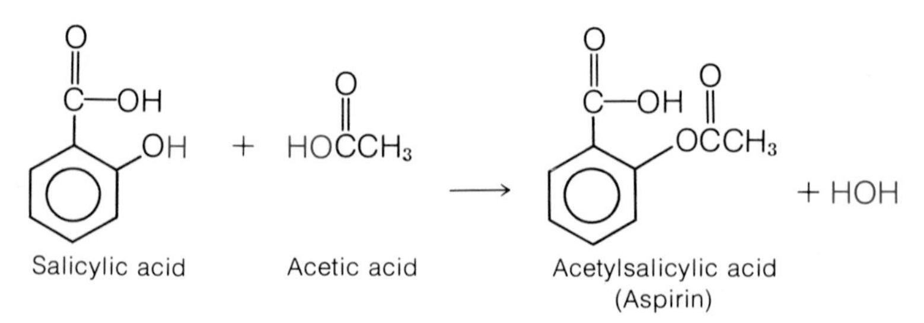

Aspirin, as acetylsalicylic acid is more commonly known, is an analgesic (pain reliever), antipyretic (fever reducer), and antiinflammatory agent. Despite the claims of various aspirin producers, repeated testing has shown that there is no difference among competing brands of aspirin except for their price. Adult aspirin tablets contain only 5 grains (0.33 g) of acetylsalicylic acid. The rest of the tablet is mostly a binder that the manufacturer uses to put the aspirin in tablet form. The difference in size between various brands of aspirin tablets is most often due to the amount of binder used, and not the aspirin itself.

Aspirin is by far the most widely used medicine in the United States, with an annual production of 20 billion aspirin tablets (an average of about 100 tablets per year for every man, woman, and child!). Aspirin has been used since 1899, but only recently have scientists begun to understand how it works. Its major effects come from blocking the production of a family of chemicals called prostaglandins (which we will discuss in Chapter 20). Although aspirin is a relatively safe drug, its continuous use in high doses can cause stomach ulcers and gastrointestinal bleeding. Also, the use of aspirin by children suffering from the flu or chicken pox has been connected with development of the rare, but often fatal, disorder called Reye's syndrome. For this reason, children suffering from these diseases should be given one of the nonaspirin pain relievers instead of aspirin.

Polyesters

Esters have become extremely important in the production of synthetic fibers. Ester polymers called **polyesters,** formed through condensation reactions, give synthetic fabrics many desirable properties. These advantages include the ability to set the fabric into permanent creases and pleats, and the resistance of the fabric to turning gray or yellow with long use. The familiar material Dacron is a polyester made of esters of ethylene glycol and terephthalic acid (Figure 16.4). Notice that both

Figure 16.4 Dacron is used to make the sails on these sailboards because it is durable, light weight, and fast drying. (Don Ivers/ Jeroboam.)

ethylene glycol and terephthalic acid have two positions where ester bonds can be formed.

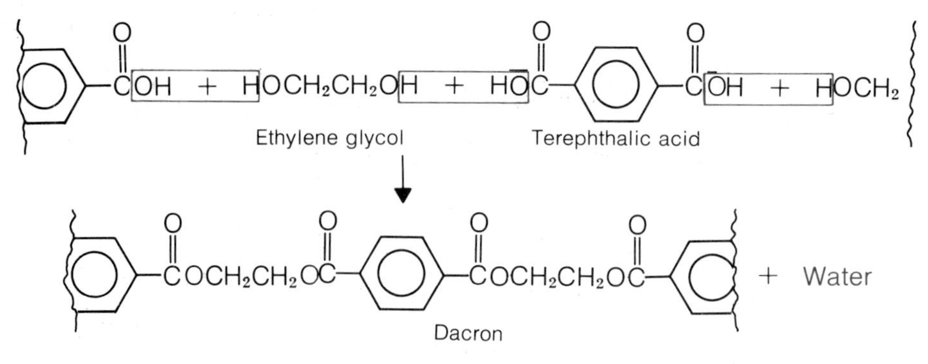

16.17 Reactions of Esters

Hydrolysis

Esters can be broken apart by an addition reaction with water to form the carboxylic acid and the alcohol. This reaction, called **hydrolysis,** is the exact reverse of the condensation reaction by which the ester may have been formed. It can be made to occur by heating the ester in a water solution of a strong inorganic acid.

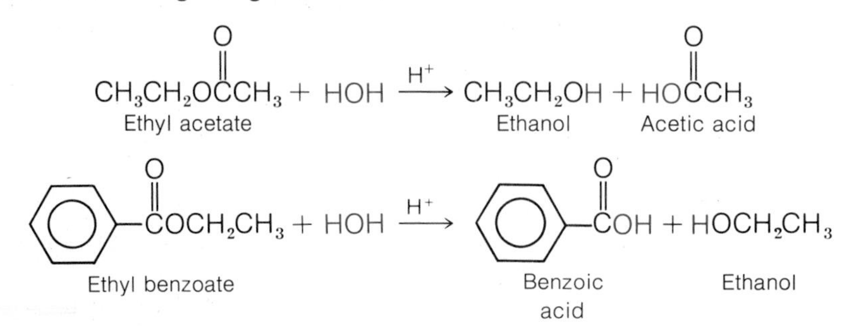

Saponification

Esters can also undergo hydrolysis when heated in an aqueous solution of strong base, such as NaOH or KOH. This reaction, called **saponification,** produces an alcohol and the sodium or potassium salt of the carboxylic acid.

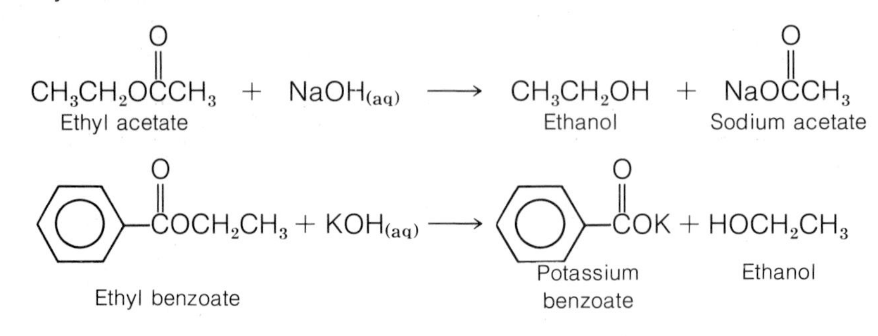

CHAPTER SUMMARY

CHAPTER SUMMARY

When substituted on a hydrocarbon molecule, functional groups containing oxygen create a reactive spot on the molecule. Each oxygen-containing functional group adds its own special properties to the molecule. The hydroxyl group, the carbonyl group, and the carboxylic acid group are all polar and can enter into hydrogen bonding and cause molecules with short carbon chains to be soluble in water. The following are the general formulas for the different classes of compounds with oxygen-containing functional groups.

Alcohol	ROH		Ketone	RC(=O)R′	or RCOR′
Ether	ROR	or ROR′	Carboxylic Acid	RC(=O)OH	or RCOOH
Aldehyde	RC(=O)H	or RCHO	Ester	RC(=O)OR′	or RCOOR′

Important Equations

The chemical reactions that we have discussed in this chapter can be summarized as follows:

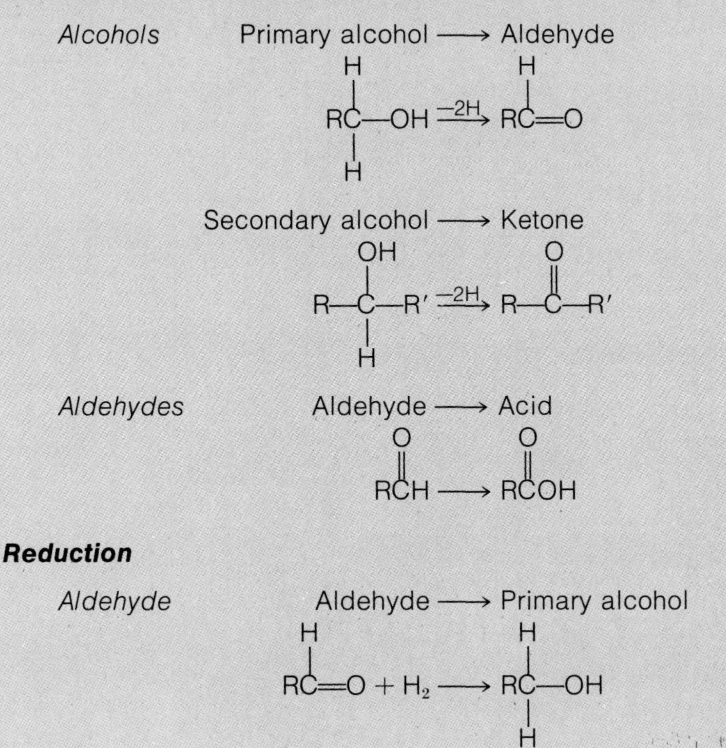

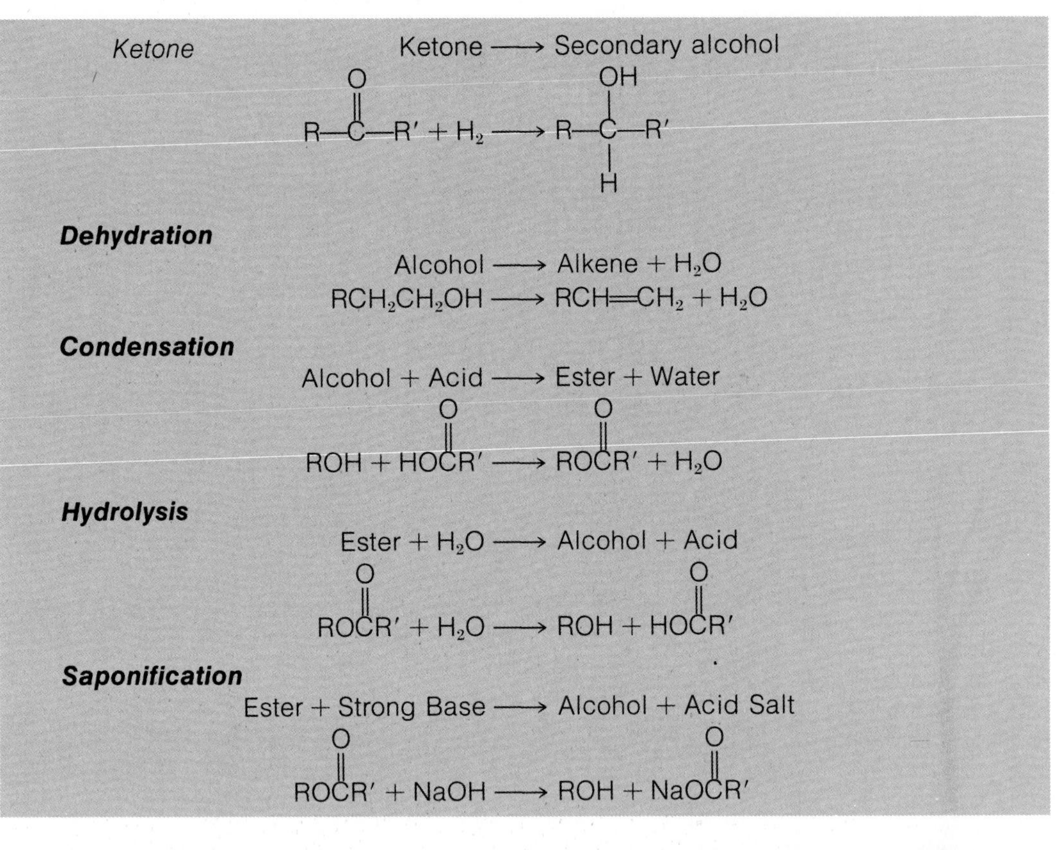

EXERCISES AND PROBLEMS

1. Identify each of the following as either an alcohol, ether, carboxylic acid, ester, aldehyde, or ketone:

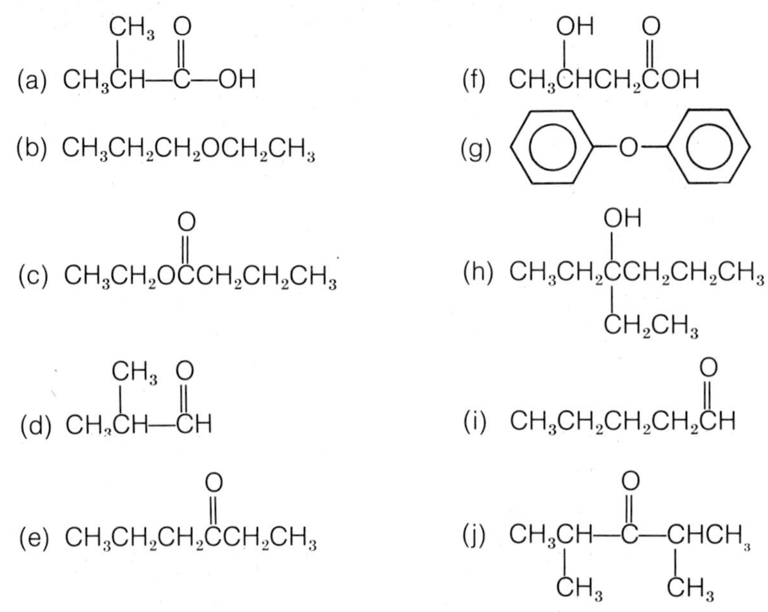

(k) C_6H_5—$OCCH_2CH_2CH_2CH_3$ (with C=O)

(l) $CH_3CH_2CH(CH(CH_3)CH_3)CH(CH_3)CH(OH)CH_2CH_2CH_3$

2. Name each of the compounds shown in question 1.

3. Write the structural formula for the following compounds:

   (a) 4,4-diethyl-2,3-hexanediol
   (b) isobutyl propyl ether
   (c) 3-hydroxypentanal
   (d) 1,3-dichloro-2-butanone
   (e) 2-octyldecanoic acid
   (f) *sec*-butyl benzoate

4. Why is it that propane boils at −45°C and methyl ether at −23°C, but ethanol boils at 78°C?

5. Explain why 2-hexanol is only very slightly soluble in water, whereas 2,3-hexanediol is soluble in water.

6. State the IUPAC name for each of the following compounds, and identify each as either a primary, secondary, or tertiary alcohol.

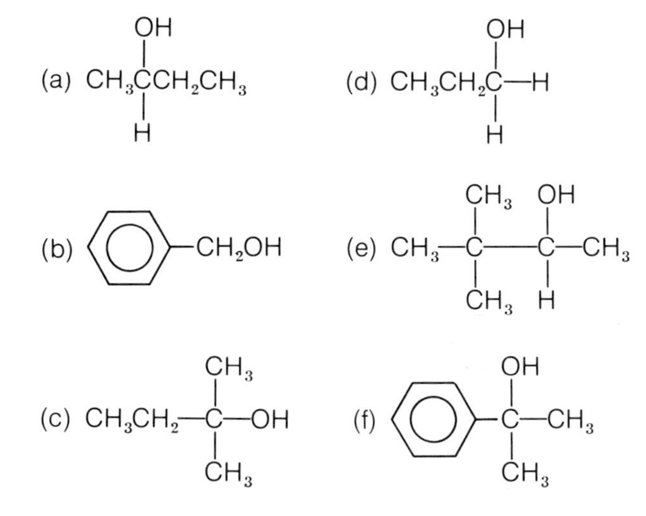

7. Assume you are making chocolate-covered cherries as a fund-raising project. Why might you want to add some glycerol to the cherry filling?

8. Which products used in your home might contain aldehydes and ketones?

9. Name the products of the following reactions and write the general equation for the reaction:

   (a) oxidation of a primary alcohol
   (b) oxidation of a secondary alcohol
   (c) oxidation of an aldehyde
   (d) reduction of an aldehyde
   (e) reduction of a ketone
   (f) dehydration of an alcohol
   (g) esterification of a carboxylic acid and an alcohol
   (h) hydrolysis of an ester
   (i) saponification of an ester

10. Write the structure of the hemiacetal or hemiketal that will form between the following pairs of compounds:

    (a) acetaldehyde and methanol
    (b) acetone and ethanol
    (c) butanone and methanol
    (d) formaldehyde and ethanol
    (e) propanal and methanol
    (f) acetone and 1-propanol

11. 5-Hydroxyhexanal will readily form a stable six-membered cyclic hemiacetal. Draw the structural formula for this cyclic hemiacetal.

12. Write the structure of the aldehyde or ketone that will form from the oxidation of the following alcohols:

    (a) 1-propanol
    (b) 2-propanol
    (c) 3-methyl-1-butanol
    (d) 3-methyl-2-butanol
    (e) cyclopentanol
    (f) ethanol
    (g) 1-pentanol
    (h) 2-hexanol

13. Write the structure of the carboxylic acid that will form from the oxidation of the following aldehydes:

    (a) acetaldehyde
    (b) butanal
    (c) hexanal
    (d) benzaldehyde

14. Write the structural formula of the alcohol that will result from the reduction of the following aldehydes and ketones:

    (a) butanone
    (b) butanal
    (c) benzaldehyde
    (d) 2-pentanone
    (e) acetone
    (f) hexanal

15. Write the structure of the product of the dehydration of

    (a) 1-butanol
    (b) 3-hexanol

16. Write the equation for the esterification of

    (a) butanoic acid and methanol
    (b) benzoic acid and ethanol
    (c) formic acid and 1-propanol
    (d) acetic acid and 2-butanol
    (e) salicylic acid and ethanol
    (f) salicylic acid and propanoic acid

17. Write the equation for the hydrolysis of the following esters:

    (a) ethyl benzoate
    (b) isopropyl butanoate
    (c) octyl acetate
    (d) *t*-butyl propanoate

18. Write the equation for the saponification of the following esters by sodium hydroxide:

    (a) methyl acetate
    (b) ethyl propanoate
    (c) *sec*-butyl benzoate
    (d) diethyl oxalate

19. How would you prepare, in a laboratory, the flavoring for wintergreen candies?

"Well, this is it—they've just called for the 100-meter freestyle. Now, the team's hopes for the state championship rest on me, Ed Scott, the kid who barely made the team his senior year! Two minutes to go—I must get control of myself. I was fine until they called the race, now my heart is pounding (can't everyone hear it?). My hands are sweating and I have a sick feeling in the pit of my stomach. Shake it off, Ed—try to relax those muscles! You'll need everything you've got to beat that guy from Crescent Tech. Slow down that breathing, man, you're panting like the race was over! TAKE YOUR MARK! Concentrate, Ed. GET SET! Well, it's now or never—come on, body, give it all you've got. GO!!"

Faced with a situation full of stress, we've all had reactions much like Ed's. A sudden, loud squeal of brakes close by, or a frightening, suspenseful movie can make your heart pound, your body perspire, and your stomach feel as though it were climbing into your throat. These various body reactions are all caused by a chemical compound that is produced in a pair of hat-shaped organs called the adrenals, located on top of the kidneys. The compound is one of many chemical messengers, called hormones, that travel through the bloodstream and cause changes in body cells. This particular compound is called epinephrine, but you are probably more familiar with its common name, adrenalin. Under normal conditions, small amounts of epinephrine are released into the blood to help control blood pressure and to maintain the level of sugar in the blood. But this compound is also the body's way of meeting emergencies and dealing with stress such as emotional excitement, exercise, extreme temperature changes, severe hemorrhaging, and the administration of certain anesthetics.

Epinephrine affects many parts of the body in reacting to stress. It increases the rate and strength of the heart beat, which increases the cardiac output. It raises the blood pressure by causing constriction of blood vessels in all parts of the body, except for the blood vessels in such vital organs as the skeletal muscles, heart, brain, and liver. It relaxes the smooth muscles in the lungs, making epinephrine a very effective drug in the treatment of severe bronchial asthma attacks. It also increases the rate and depth of breathing, enabling more oxygen to get into the lungs. Epinephrine causes an increase in blood sugar, and a general increase in the metabolic activity of the cells. It slows down the action of the digestive tract, and accelerates blood clotting. It delays the fatigue of skeletal muscles, and increases the strength of contractions of these muscles. This last property has allowed people to show amazing strength under great stress; for example, after an auto accident, a man lifted his car to free his son who was trapped underneath. Each of these changes in normal body function caused by epinephrine enables the body to meet the initial challenge of the stress. The effects last only a short time, because the epinephrine released into the bloodstream is inactivated by the liver in about three minutes.

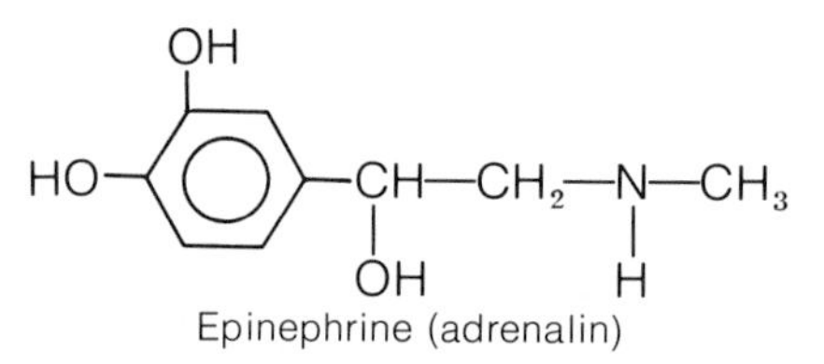

Epinephrine (adrenalin)

Epinephrine is a fairly complex molecule containing a benzene ring, hydroxyl groups, and a functional group that contains nitrogen. In the last few chapters we have discussed important organic compounds composed of carbon, hydrogen, and oxygen. In this chapter, we will discuss those classes of compounds with nitrogen-containing functional groups.

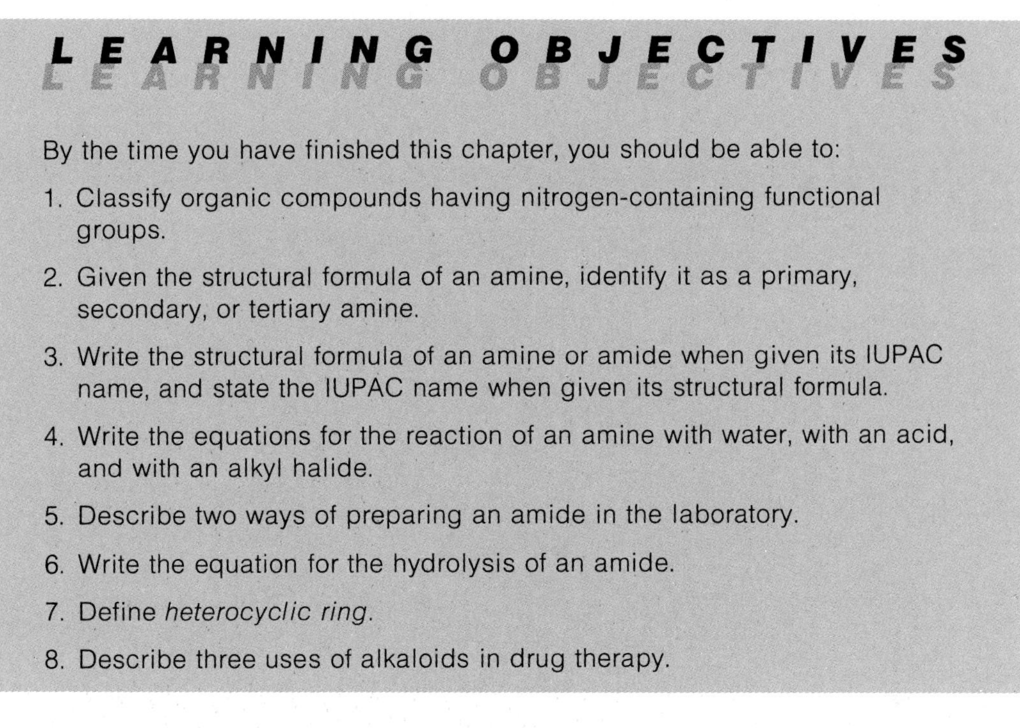

LEARNING OBJECTIVES

By the time you have finished this chapter, you should be able to:

1. Classify organic compounds having nitrogen-containing functional groups.
2. Given the structural formula of an amine, identify it as a primary, secondary, or tertiary amine.
3. Write the structural formula of an amine or amide when given its IUPAC name, and state the IUPAC name when given its structural formula.
4. Write the equations for the reaction of an amine with water, with an acid, and with an alkyl halide.
5. Describe two ways of preparing an amide in the laboratory.
6. Write the equation for the hydrolysis of an amide.
7. Define *heterocyclic ring.*
8. Describe three uses of alkaloids in drug therapy.

AMINES AND AMIDES

17.1 Functional Groups Containing Nitrogen

Nitrogen is the fourth most abundant element in the human body, making up 1.4% of the total number of atoms. Nitrogen is the first member of group V on the periodic table. An atom of nitrogen has five valence electrons and will form three covalent bonds to become stable. Such bonding can occur in several ways: by the formation of three single bonds, a double and a single bond, or one triple bond (Figure 17.1). Nitrogen can form strong bonds with carbon, oxygen, and hydrogen, allowing this element to be

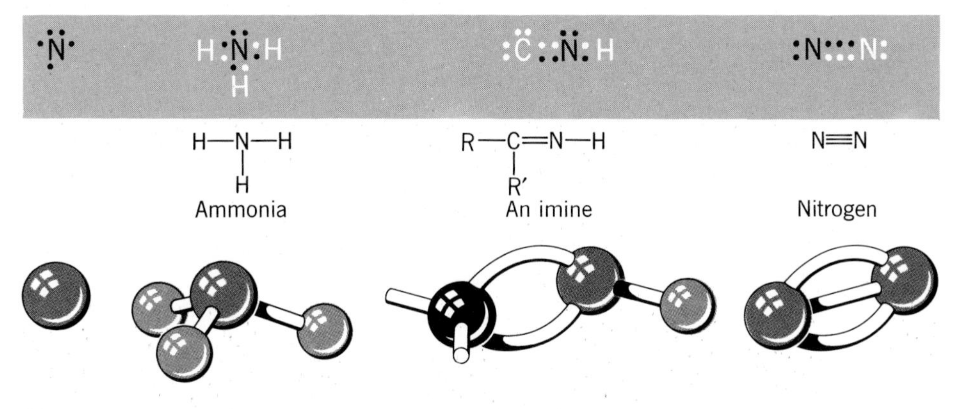

Figure 17.1 Nitrogen can form single, double, and triple covalent bonds.

Table 17.1 Some Nitrogen-Containing Functional Groups

Functional Group	Class of Compound	Typical Compound	
R—NH_2	Amine	CH_3NH_2	Methylamine
R—C(=O)—NH_2	Amide	CH_3C(=O)—NH_2	Acetamide
R—C(R′)=NH	Imine	H_2N—C(NH_2)=NH	Guanidine
R—C≡N	Nitrile	H_2C=CHC≡N	Acrylonitrile

found in many different arrangements in organic molecules. As in the previous chapters, we can categorize these nitrogen-containing molecules by means of the similarities in chemical properties that result from the presence of specific functional groups (Table 17.1).

17.2 Amines

Amines are organic compounds that contain the **amino** functional group, —NH_2 (general formula, RNH_2). They are known for their very strong, pungent odors. For example, methylamine (CH_3NH_2) smells like spoiled fish. Amines are also produced during the natural decay of living organisms, and such amines are called **ptomaines.**

$H_2NCH_2CH_2CH_2CH_2NH_2$
Putrescine

$H_2NCH_2CH_2CH_2CH_2CH_2NH_2$
Cadaverine

The common names of these two ptomaines might give you a hint about their smells. It was once thought that such amines were responsible for the vomiting and diarrhea that results from eating spoiled food, so these symptoms were referred to as ptomaine poisoning. However, it is now known that such body reactions result from more complicated causes.

Amines can be classified as primary, secondary, or tertiary depending upon the number of carbon atoms bonded directly to the nitrogen atom. In a primary amine, the nitrogen atom is bonded to one carbon atom and to two hydrogen atoms. In a secondary amine, the nitrogen is bonded to two carbon atoms and one hydrogen atom. In a tertiary amine, the nitrogen is bonded to three carbon atoms (Figure 17.2 and Table 17.2).

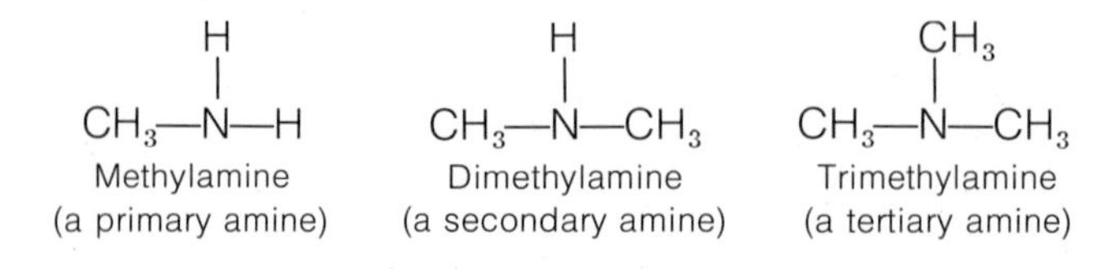

Methylamine (a primary amine) Dimethylamine (a secondary amine) Trimethylamine (a tertiary amine)

Amines have been used successfully to treat patients who suffer from severe depression. Interestingly, it has been found that tertiary amine

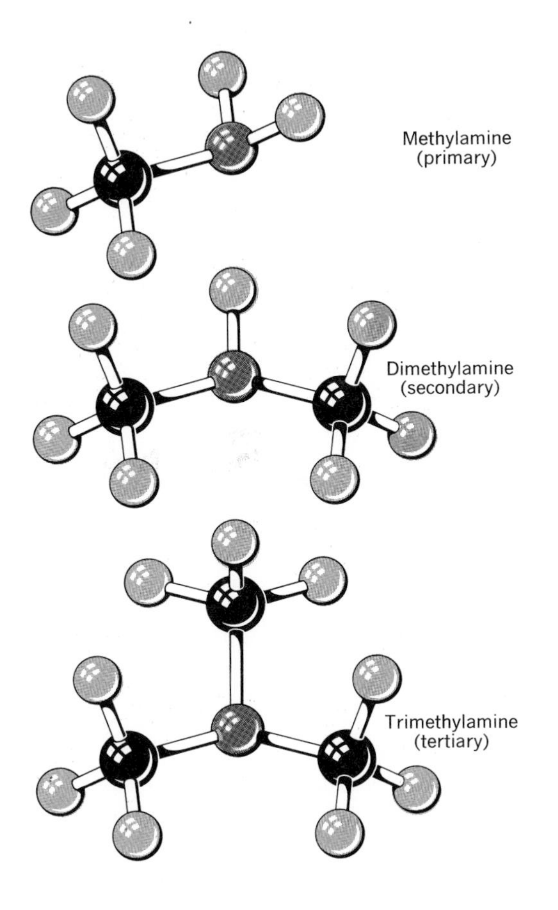

Figure 17.2 Amines can be either primary, secondary, or tertiary.

antidepressants help to calm such patients when they become highly agitated, whereas secondary amine antidepressants stimulate such patients when they are suffering from a depression that renders them totally immobile. The reason for this difference is not entirely clear, but these two classes of amines seem to interact differently with nerve cells in the brain.

Amines are polar compounds, and both primary and secondary amines are able to form hydrogen bonds with other amines and with water.

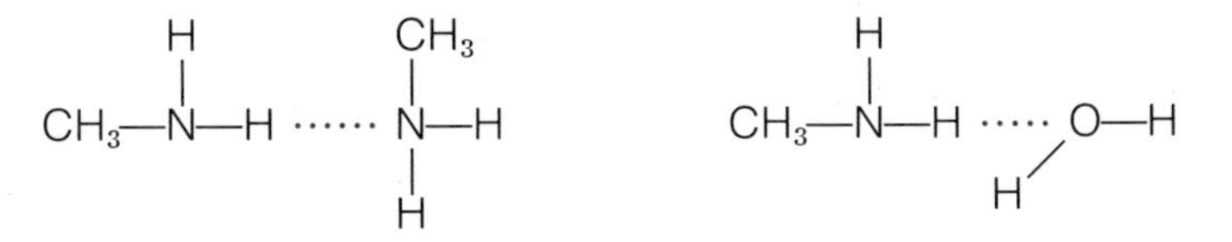

Two results of such hydrogen bonding are that amines have higher boiling points than alkanes with comparable molecular weights, and amines with short carbon chains are soluble in water.

Primary amines are named by first identifying the group attached to the nitrogen and then adding the suffix, **-amine.** For example, $CH_3CH_2NH_2$ is ethylamine, and $CH_3CH_2CH_2CH_2NH_2$ is butylamine. Secondary and tertiary amines are named in the same way: Identify all groups attached to the

Table 17.2 Some Amines

Name	Formula	Type
Propylamine	$CH_3CH_2CH_2NH_2$	Primary
Isopropylamine	$CH_3-CH(CH_3)-NH_2$	Primary
Methylethylamine	$CH_3CH_2-N(CH_3)-H$	Secondary
Dimethylethylamine	$CH_3CH_2-N(CH_3)-CH_3$	Tertiary
Aniline	$C_6H_5-NH_2$	Primary
N-Methylaniline	$C_6H_5-N(H)-CH_3$	Secondary
N,N-Dimethylaniline	$C_6H_5-N(CH_3)-CH_3$	Tertiary

nitrogen if they are different, or use the prefixes *di-* and *tri-*. For example,

$$CH_3CH_2-N(H)-CH_3 \text{ is methylethylamine, and}$$

$$CH_3CH_2-N(CH_2CH_3)-CH_2CH_3 \text{ is triethylamine.}$$

The more complicated amines are often named by using the word *amino-* to identify the $-NH_2$ group in the molecule. For example,

$$H_2NCH_2CH_2C(=O)OH \text{ is 3-aminopropanoic acid, and}$$

$$H_2NCH_2CH_2CH_2CH_2NH_2 \text{ is 1,4-diaminobutane.}$$

And finally, an *N* appearing before the name of a substituted group indicates that the group named after the *N* is attached to the nitrogen of the amino group (and not somewhere else in the molecule) (see Table 17.2).

EXAMPLE 17-1

1. Name the following compound:

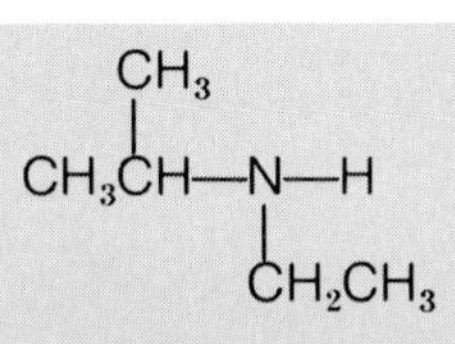

This is a secondary amine with an ethyl group and an isopropyl group attached to the nitrogen. Therefore, the name is ethylisopropylamine.

2. Draw the structural formula for 3-(*N*,*N*-dimethylamino)-butanoic acid.

First draw the carbons for butanoic acid: four carbon atoms.

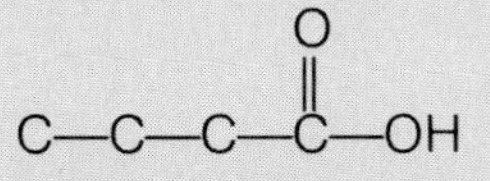

The "3" indicates that on the third carbon we will find the group *N*,*N*-dimethylamino-, which is an amino group with two methyl groups attached to the nitrogen.

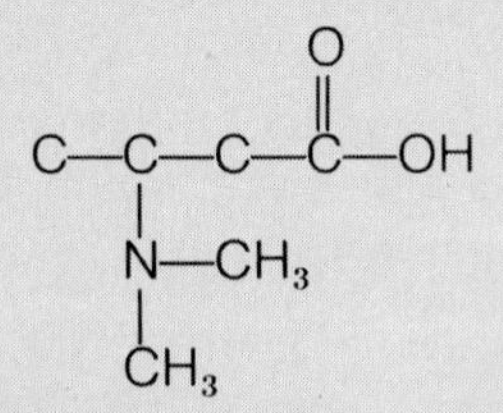

Now all that remains is to add hydrogens so that each carbon has four bonds.

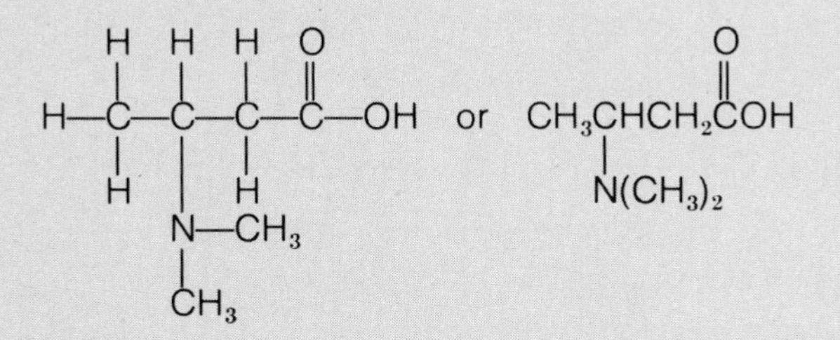

Exercise 17-1

1. Name the following amines:

(a) $CH_3(CH_2)_7NH_2$

(b)
$$\begin{array}{c} \quad\quad\quad CH_2CH_3 \\ \quad\quad\quad | \\ CH_3CH_2—N \\ \quad\quad\quad | \\ \quad\quad\quad CH_2CH_3 \end{array}$$

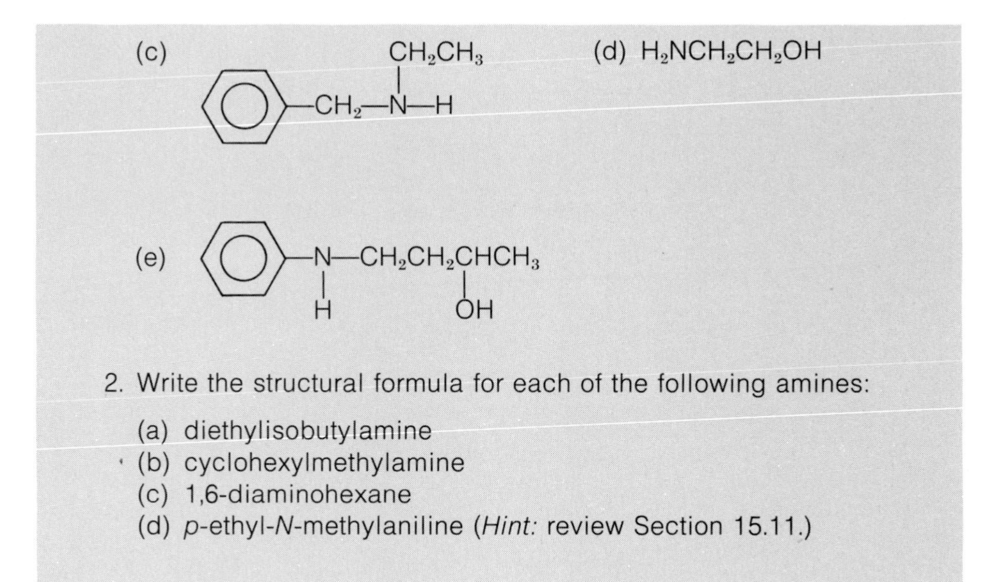

2. Write the structural formula for each of the following amines:
   (a) diethylisobutylamine
   (b) cyclohexylmethylamine
   (c) 1,6-diaminohexane
   (d) *p*-ethyl-*N*-methylaniline (*Hint:* review Section 15.11.)

17.3 Basic Properties of Amines

All amines are derivatives of the weak inorganic base ammonia, NH_3. When ammonia is placed in water, some of the molecules act as hydrogen ion acceptors and form the ammonium ion.

$$\underset{\text{Ammonia}}{:NH_3} + H_2O \longrightarrow \underset{\text{Ammonium ion}}{NH_4^+} + OH^-$$

Similarly, the organic amines are weak bases. For example, when methylamine is placed in water, some of its molecules will accept hydrogen ions and will form the methylammonium ion.

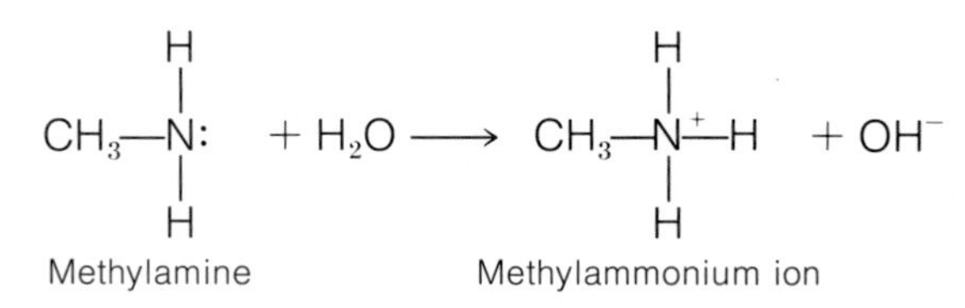

The strength of an amine as a base depends upon the organic groups attached to the nitrogen atom. Carbon chains attached to the nitrogen atom increase the basic properties of the molecule, whereas benzene rings attached to the nitrogen atom greatly decrease the basic properties of the molecule.

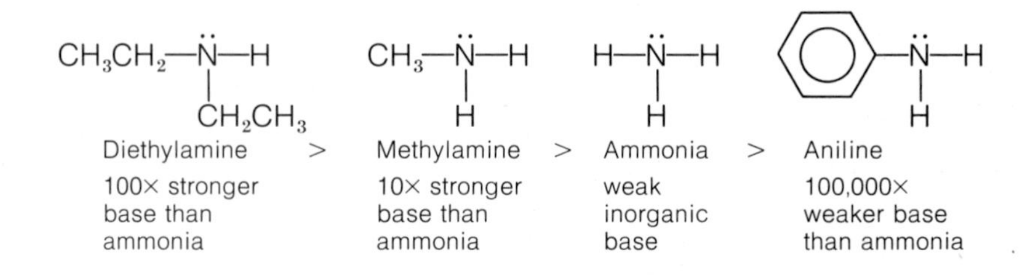

17.4 Reactions of Amines

With Acids

Amines form salts when they react with acids.

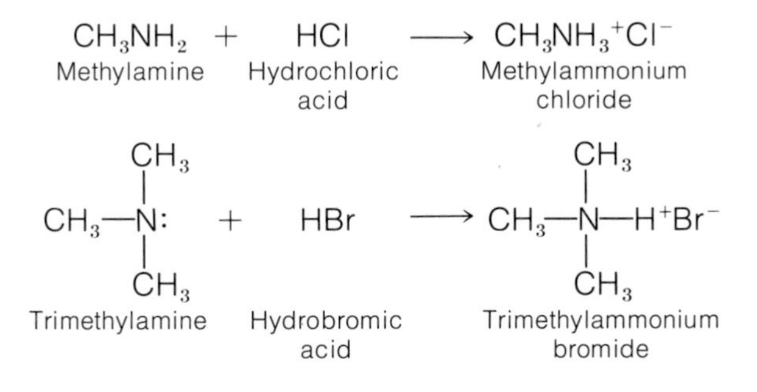

With Alkyl Halides

When heated with alkyl halides such as methyl iodide (CH_3I), amines will form quaternary ammonium salts. After forming three covalent bonds, the nitrogen atom still has a pair of electrons that can be shared with another atom, creating a **coordinate covalent bond** or **donor–acceptor bond.** In the reaction with alkyl halides, the nitrogen on the amine forms a coordinate covalent bond with the methyl group, thereby forming an organic ammonium ion.

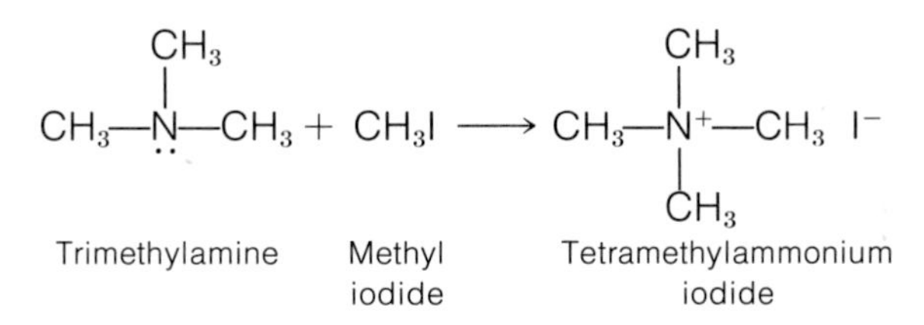

Choline is a compound that contains an organic ammonium ion, and derivatives of the choline molecule are extremely important to the structure of the cell and the transmission of nerve impulses.

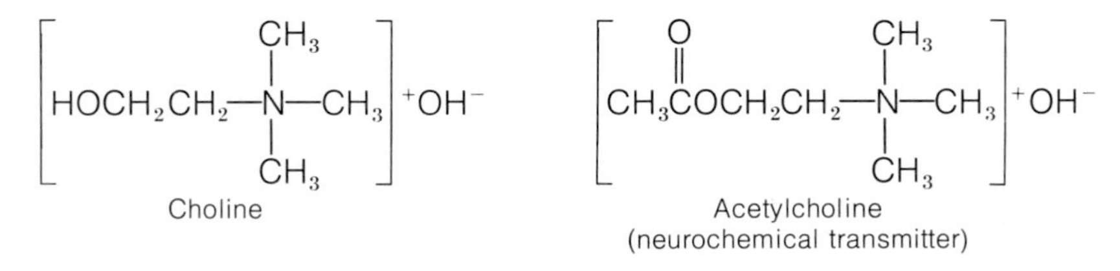

With Nitrites

Secondary amines can react with nitrites (general formula XNO_2) to produce nitrosamines (*N*-nitrosodialkylamines), a class of highly potent carcinogens. These compounds have been shown to cause cancer in all types of

laboratory animals and are suspected of causing human cancers.
In general,

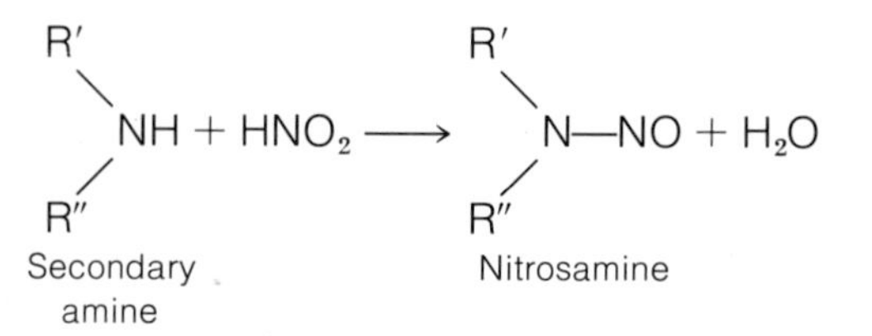

For example,

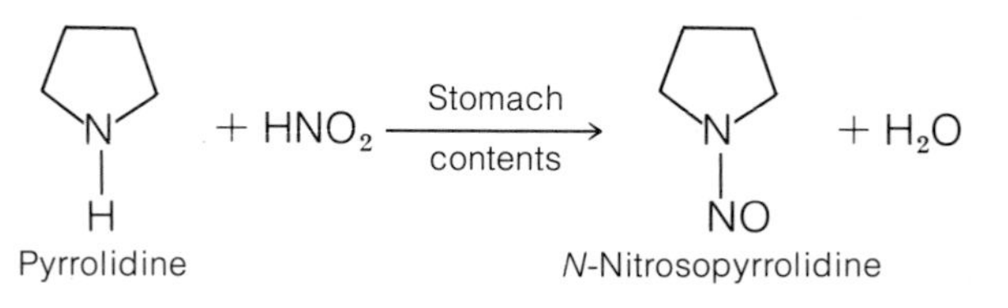

Nitrites are widely used to enhance the color of, and to preserve, meats, poultry, and fish. Also, nitrates, which are naturally present in high levels in such foods as celery and beets, can be converted to nitrites by the bacteria in saliva. No matter what their source, nitrites may react in the stomach with amines naturally present in foods, drugs, and cigarette smoke to produce nitrosamines. The study of nitrosamines as a possible environmental cause of human cancer is currently a very active area of research.

17.5 Amides

Amides are derivatives of carboxylic acids, in which the hydroxyl (—OH) group on the carboxyl group has been replaced by an amino (—NH_2) group. The general formula for the amides, therefore, can be either $\mathrm{R\overset{\overset{O}{\|}}{C}NH_2}$, RCONHR′, or $\mathrm{RCO\overset{\overset{R''}{|}}{N}R'}$ (Table 17.3). The bond between the

Table 17.3 Some Amides

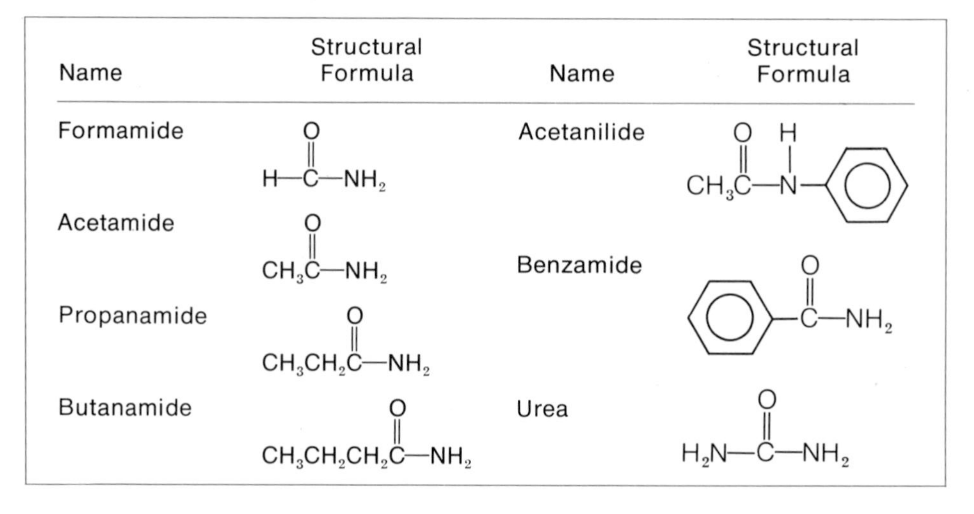

Name	Structural Formula	Name	Structural Formula
Formamide	$\mathrm{H{-}\overset{\overset{O}{\|}}{C}{-}NH_2}$	Acetanilide	$\mathrm{CH_3\overset{\overset{O}{\|}}{C}{-}\overset{\overset{H}{|}}{N}{-}C_6H_5}$
Acetamide	$\mathrm{CH_3\overset{\overset{O}{\|}}{C}{-}NH_2}$	Benzamide	$\mathrm{C_6H_5{-}\overset{\overset{O}{\|}}{C}{-}NH_2}$
Propanamide	$\mathrm{CH_3CH_2\overset{\overset{O}{\|}}{C}{-}NH_2}$		
Butanamide	$\mathrm{CH_3CH_2CH_2\overset{\overset{O}{\|}}{C}{-}NH_2}$	Urea	$\mathrm{H_2N{-}\overset{\overset{O}{\|}}{C}{-}NH_2}$

carbon and the nitrogen atom in an amide molecule is known as the **amide linkage.** It is a very stable bond and is found as a repeating linkage in large molecules such as proteins and industrial polymers such as nylon.

$$\underset{\text{Acetic acid}}{CH_3\overset{\overset{\displaystyle O}{\|}}{C}—OH} \qquad \underset{\text{Acetamide}}{CH_3\overset{\overset{\displaystyle O}{\|}}{C}—NH_2} \leftarrow \text{Amide linkage}$$

Amides with short carbon chains are soluble in water due to the presence of the polar amide group. This solubility decreases as the molecular weight of the amide increases. Amides, like alcohols, have melting and boiling points that are higher than alkanes with comparable molecular weights. These high melting and boiling points result from hydrogen bonding between the polar amide groups on neighboring amide molecules (Figure 17.3).

Amides that have two hydrogens attached to the nitrogen ($R\overset{\overset{\displaystyle O}{\|}}{C}NH_2$) are named by dropping the *-ic* or *-oic* ending from the name of the original acid and adding **-amide.** For example,

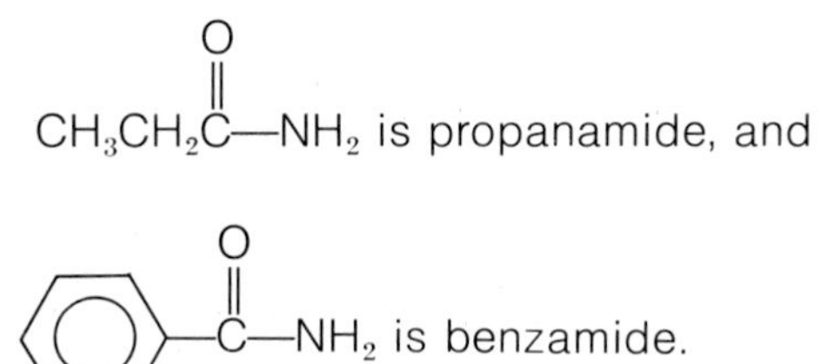

$CH_3CH_2\overset{\overset{\displaystyle O}{\|}}{C}—NH_2$ is propanamide, and

$C_6H_5—\overset{\overset{\displaystyle O}{\|}}{C}—NH_2$ is benzamide.

If an amide contains groups attached to the nitrogen, they are named as alkyl groups, but with an *N* in front of the name to indicate they are

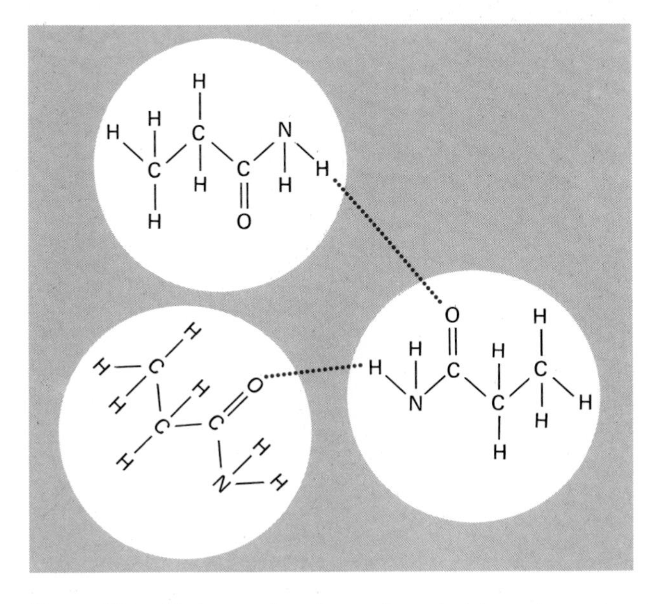

Figure 17.3 Hydrogen bonds can form between the molecules of amides such as propanamide.

attached to the nitrogen. For example,

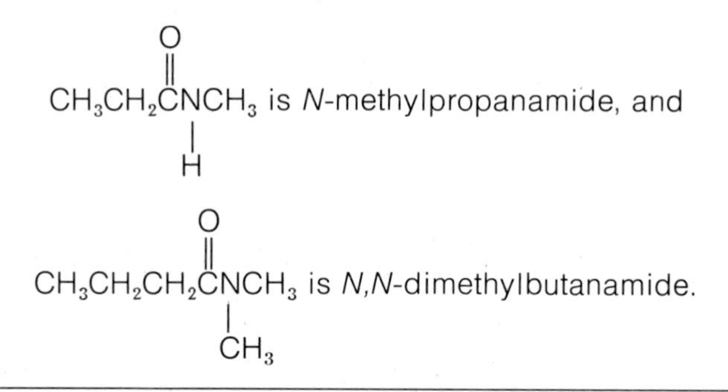

EXAMPLE 17-2

1. Name the following amide:

$$CH_3CH_2CH_2\overset{\overset{\displaystyle O}{\|}}{C}-\underset{\underset{\displaystyle H}{|}}{N}-CH_2CH_3$$

(a) The acid from which this amide is formed has four carbons, so it was butanoic acid. Therefore, the last part of the name is butanamide.

(b) The nitrogen has one substituted group: an ethyl group. The name is *N*-ethylbutanamide.

2. Write the structural formula for *N,N*-dimethylacetamide.

(a) The acid from which the amide is formed is acetic acid.

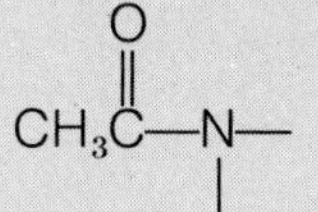

(b) The *N,N*-dimethyl indicates that there are two methyl groups attached to the nitrogen.

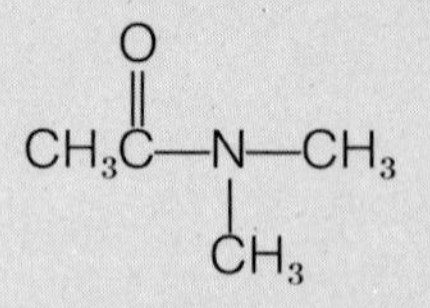

Exercise 17-2

1. Name the following compounds:

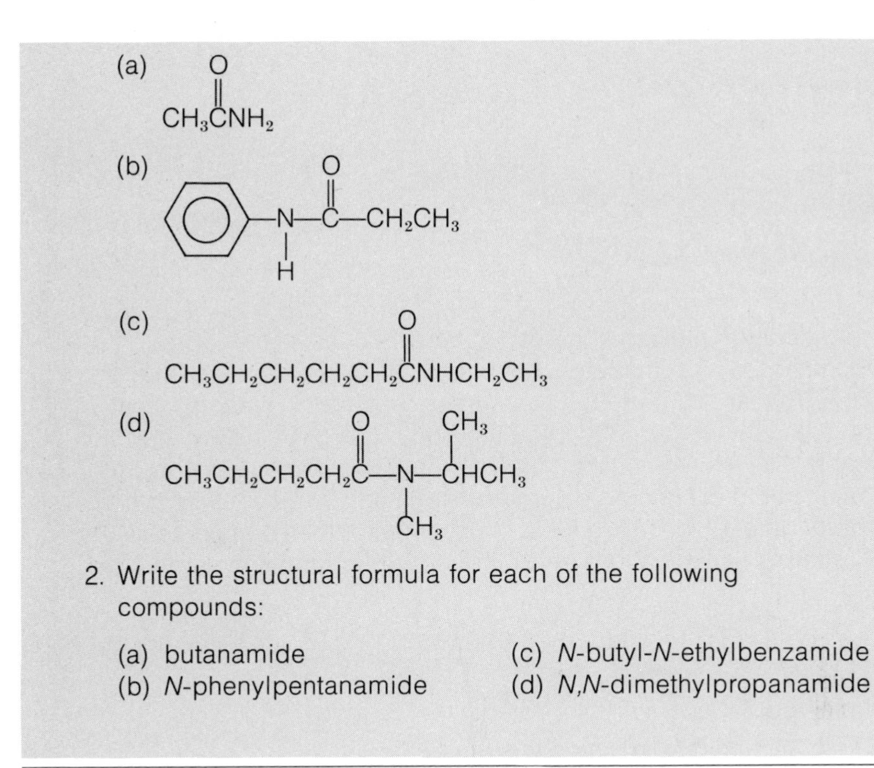

2. Write the structural formula for each of the following compounds:

(a) butanamide
(b) *N*-phenylpentanamide
(c) *N*-butyl-*N*-ethylbenzamide
(d) *N,N*-dimethylpropanamide

17.6 Some Important Amides

Urea is a waste product of the body's metabolism (see Table 17.3). We have previously seen that compounds containing carbon, hydrogen, and oxygen are broken down by the body to produce waste products of carbon dioxide (CO_2) and water (H_2O). The carbon dioxide is then eliminated through the lungs, and water is eliminated through the breath, sweat, and urine. When nitrogen-containing compounds are broken down by the body, the nitrogen is converted into the compound ammonia, NH_3. Ammonia, however, is extremely toxic to living tissues. Aquatic animals are able quickly to rid themselves of ammonia into the surrounding water, but land animals must convert this ammonia to a less toxic substance that can be transported through the body. Such animals convert ammonia into urea, a compound that can accumulate in the blood without harm and that can be eliminated by the kidneys.

We mentioned in Chapter 16 that the continuous use of aspirin in high doses can, in some people, lead to stomach upset, stomach ulcers, and gastrointestinal bleeding. This has led to the development of other analgesics, the most popular of which is acetaminophen, which contains an amide rather than an ester group. Acetaminophen is as effective as aspirin in relieving pain and reducing fever. Unlike aspirin, however, it has no effect in reducing inflammation.

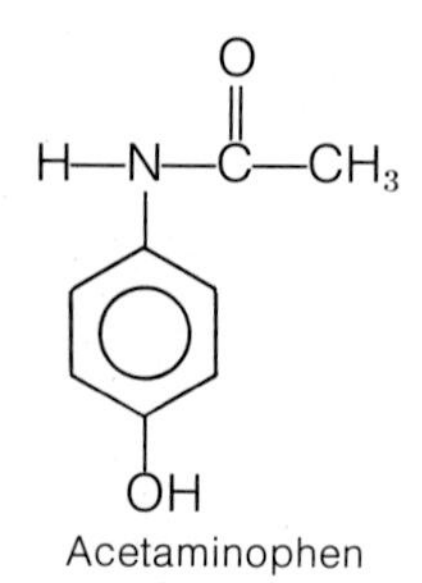

Acetaminophen

Another amide of interest is the hallucinogen LSD, lysergic acid diethylamide. LSD seems to disrupt the transmission of nerve impulses in the brain. The structure of LSD has certain features resembling those of serotonin, a chemical produced by the body to control the transmission of nerve impulses. Nerve cells in the brain probably confuse LSD with serotonin, thus preventing the serotonin from carrying out its control function (Figure 17.4). This results in uncontrolled nerve impulses in the brain, creating hallucinations and other behavioral abnormalities.

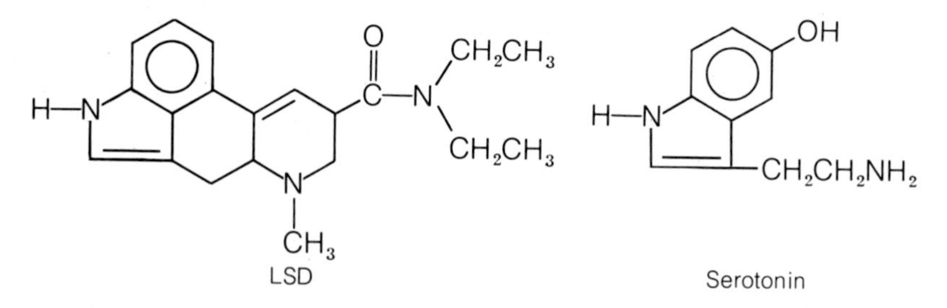

LSD

Serotonin

17.7 Preparation of Amides

Amides are prepared in the laboratory by heating an ammonium salt of a carboxylic acid above its melting point, or by reacting an ester with

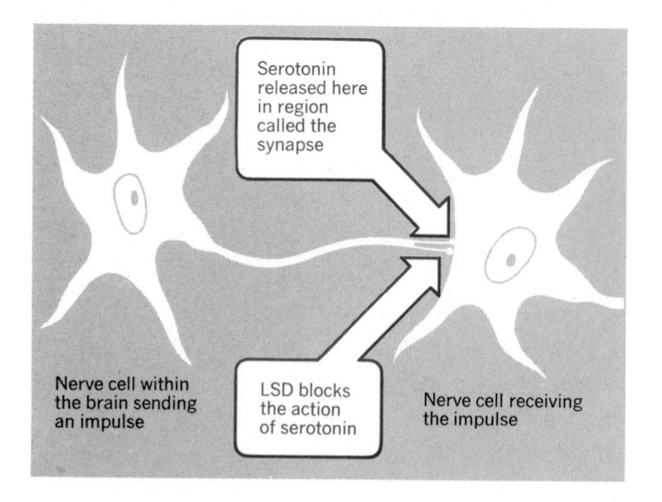

Figure 17.4 The hallucinogen LSD affects brain cells by blocking the action of serotonin.

ammonia or with a derivative of ammonia (such a reaction is called ammonolysis).

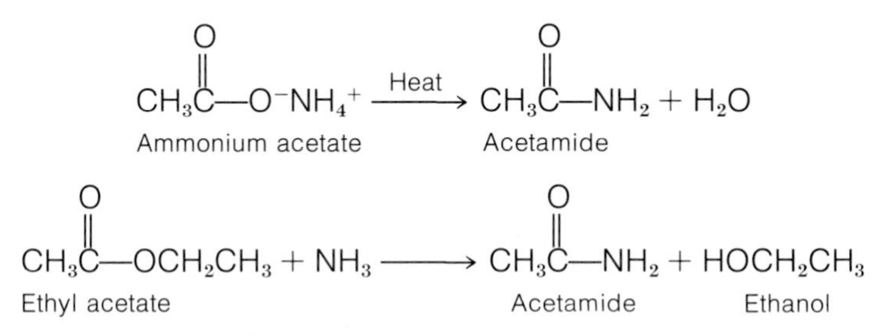

17.8 Reactions of Amides

Hydrolysis

Compounds that contain amide groups are among the most important in our bodies. For example, proteins are complex compounds containing many amide linkages. Amides hydrolyze (that is, are split apart by water) very slowly. The slowness of this reaction explains the stability of the proteins in our bodies. Amides in which the nitrogen is bonded to two hydrogens will hydrolyze to form acids and ammonia. Other amides will form the corresponding acid and an amine. Under certain conditions and in the presence of specific enzymes such as those found in the digestive tract, the hydrolysis of amides will occur in a relatively short time.

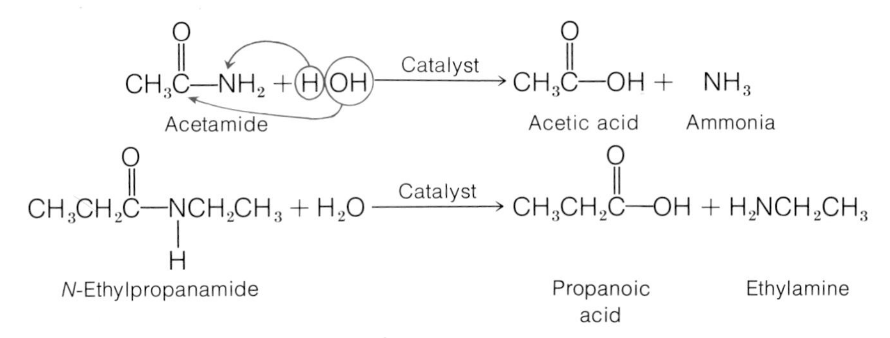

Dehydration

When amides are heated in the presence of a strong dehydrating agent, a molecule of water is removed and a member of the class of compounds called nitriles is formed.

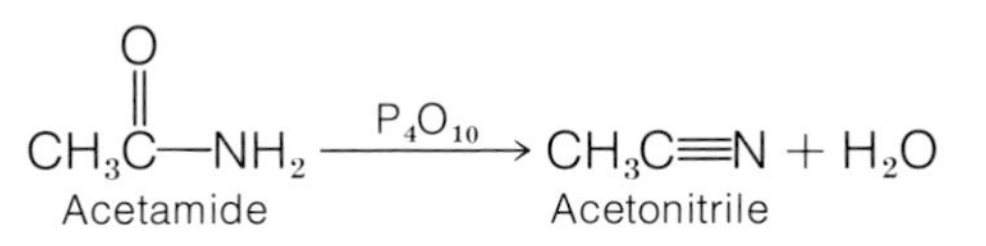

Nitriles are widely used in industrial synthesis. They are polymerized to form many synthetic fibers such as orlon, acrylan, and dynel, which are used in knitwear and carpets (see Table 15.2).

17.9 Nitrogen in Ring Compounds

Our discussion of compounds containing rings has thus far been limited to molecules whose rings are formed by carbon atoms only. However, the elements nitrogen, oxygen, and sulfur can join with carbon atoms to form a ring. The resulting compounds are called **heterocyclic** compounds and have two or more types of atoms making up the ring. Heterocyclic rings are found in an enormous number of naturally occurring compounds and form an entire field of study by themselves. Table 17.4 shows some heterocyclic rings that contain nitrogen. These rings are commonly found in biological systems, and we will study compounds containing such rings in later chapters. Before leaving our study of organic compounds containing nitrogen, however, we will take a brief look at some complex compounds that have important effects on the body.

Table 17.4 Some Heterocyclic Rings Containing Nitrogen

Name	Structural Formula	Found in the Structure of—
Pyrrolidine	N–H	Amino acids
Pyrrole	N–H	Chlorophyll, hemoglobin, and vitamin B_{12}
Imidazole	N, N–H	An amino acid
Indole	N–H	An amino acid
Pyridine	N	The vitamin niacin
Pyrimidine	N, N	Nucleic acids
Purine	N, N, N, N–H	Nucleic acids

17.10 Alkaloids

Alkaloids are a large class of nitrogen-containing compounds with complex structures, often including nitrogen in heterocyclic rings. Most of the complicated alkaloids are found in a specific species of plant and are often part of the plant's defenses. Such alkaloids, in various plant mixtures, have been used as drugs for centuries. Some of the naturally occurring alkaloids have also been synthesized in the laboratory. Moreover, in trying to duplicate the structure of alkaloids, chemists have created compounds with properties far superior to the natural compound when used as a drug. By studying and replacing certain functional groups on these alkaloids, chemists have produced compounds with desirable properties (such as relieving pain or inducing sedation) without the accompanying problems that often result from the use of the natural alkaloid. Such undesirable side effects include dependence (addiction) and tolerance (the need to constantly increase the dosage over time in order to produce the same effect). A complete discussion of the alkaloids and their many uses would be too complicated for this book, but we will briefly mention some interesting examples.

Epinephrine and Norepinephrine

Epinephrine belongs to a class of relatively simple alkaloids containing one benzene ring. We have already discussed the many effects that epinephrine can have on the body. Looking again at the structure of epinephrine, we can recognize that it is a secondary amine.

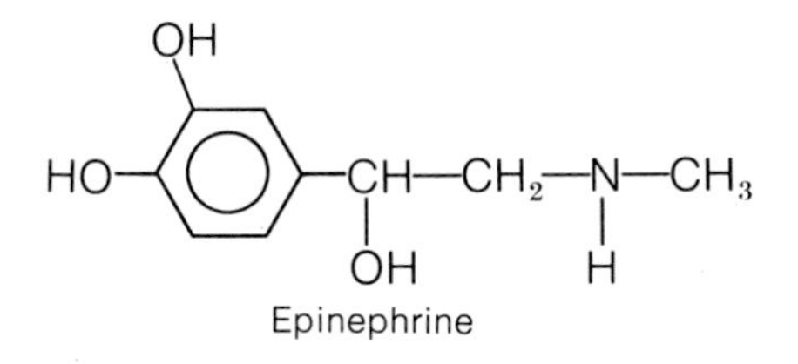

Epinephrine

Another compound with a structure closely related to epinephrine is norepinephrine, also produced by the adrenals. This compound, often called noradrenalin, is a primary amine whose structure differs only slightly from that of epinephrine—it lacks a methyl group on the nitrogen. Just this small change makes a difference in the action of these two compounds in the body.

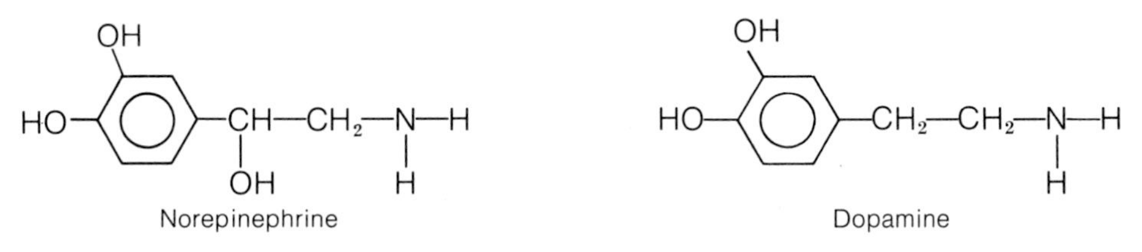

Norepinephrine

Dopamine

Norepinephrine, which is produced by certain nerve endings from dopamine, acts as a chemical messenger between nerve cells. Its main function in the body is to maintain muscle tone in the blood vessels, and in that way to control blood pressure. For example, reserpine (a drug given to help reduce blood pressure in persons suffering from hypertension) works by greatly reducing the amount of norepinephrine in the nerve endings.

Amphetamines

Benzedrine (also called amphetamine) is a synthetic compound with a structure similar to that of epinephrine. Benzedrine stimulates the cortex of the brain, producing a decreased sense of fatigue and an increased alertness. Together with ephedrine, a natural substance extracted from a tree in the pine family, benzedrine has been prescribed to reduce fatigue, overcome drowziness, and suppress the appetite. Benzedrine has also been used to treat bronchial congestion. With continued use of benzedrine for any of these purposes, however, dependence and tolerance can result. Over the last decade the abuse of this drug has increased to such an extent that it is now under strict regulation and control. Benzedrex, a derivative of benzedrine containing a cyclohexane ring, has the same decongestant effects and is now used in inhalers that relieve nasal congestion. This compound does not cause dependence.

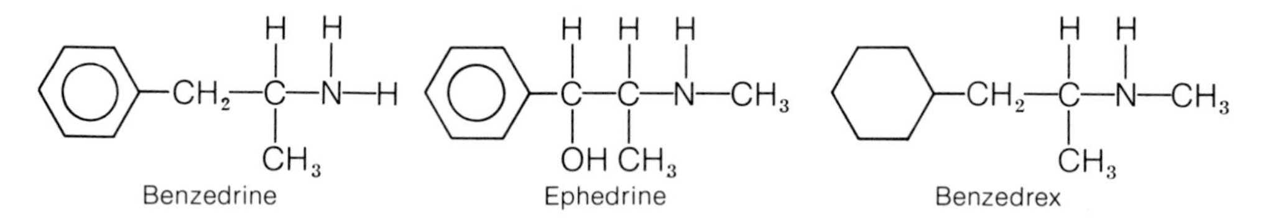

Nicotine

Some alkaloids contain a nitrogen atom as part of the ring. Nicotine, which is found in the leaves of tobacco, has two nitrogen-containing rings.

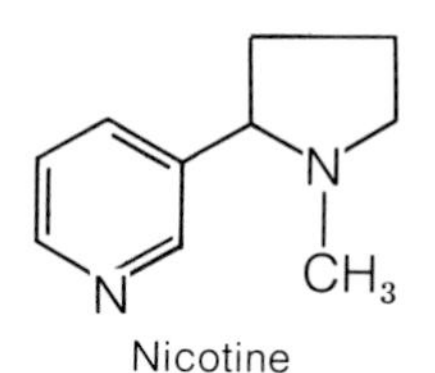

Nicotine

Nicotine in pure form is a rapid acting, extremely toxic drug. There have been reports of gardeners who have died from handling nicotine as an insecticide. In such cases, death from respiratory failure occurred within a few minutes. The action of nicotine in the human body is very complicated. It stimulates the central nervous system (causing irregular heartbeat and blood pressure), induces vomiting and diarrhea, and first stimulates and then inhibits glandular secretions. Nonsmokers can absorb only about 4 mg of nicotine before symptoms of nausea, vomiting, diarrhea, and weakness begin. But a smoker builds up a tolerance to nicotine, and may absorb twice as much without any noticeable effects. The smoke from one cigarette may contain up to 6 mg of nicotine, but only about 0.2 mg is absorbed into the body.

Caffeine

Caffeine, another alkaloid whose structure has two nitrogen-containing rings, is found in coffee beans, tea leaves, and in the seeds of the chocolate tree. In addition to its natural presence in coffee, tea, and chocolate, caffeine is also added to some carbonated beverages and nonprescription medicines (such as some aspirins).

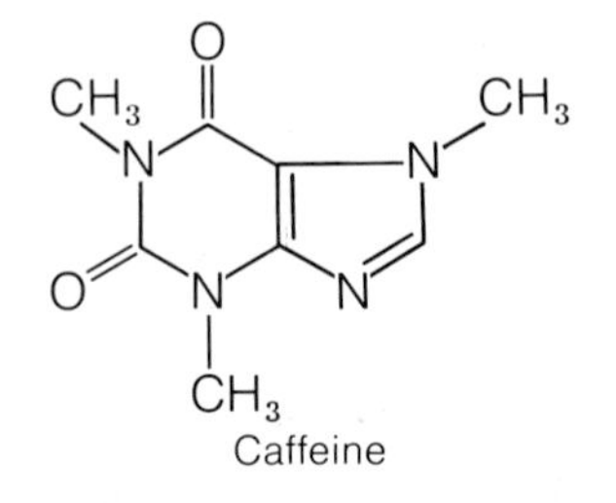

Caffeine

Caffeine is a stimulant to the central nervous system, causing restlessness and mental alertness. Taking caffeine for some people can become a habit—they may become extremely insistent about having their morning coffee—but there is no evidence that caffeine is addictive.

Atropine and Quinine

Some other alkaloids worth mentioning have even more complex nitrogen-containing structures (Table 17.5). Atropine causes dilation of the pupils and is used in eye surgery. Taken internally, it relieves abdominal pain from severe muscle contractions, and it is used as premedication for gas anesthesia because it dries up secretions in the nose and throat. The bark of cinchona trees (found in South America and Java) contains the alkaloid quinine, which is used as an antimalarial drug. An isomer of quinine, called quinidine, is found in a much lower concentration in the bark of these trees and acts as a heart depressant. It is used in the treatment of irregular heart rhythms.

Cocaine

Cocaine is a local anesthetic and is a stimulant to the central nervous system. It has effects similar to, but stronger than, the amphetamines. Because it does not produce tolerance, cocaine has been increasingly popular among drug users. In trying to synthesize compounds that contain the local anesthetic properties of cocaine (but not its side effects), chemists have developed many useful new anesthetics. One such compound is procaine, which has been widely used in dentistry in the form of its derivative Novocaine.

$$H_2N{-}C_6H_4{-}C(=O){-}O{-}CH_2CH_2{-}N(CH_2CH_3){-}CH_2CH_3$$

Procaine

Table 17.5 Alkaloids with Nitrogen-Containing Rings

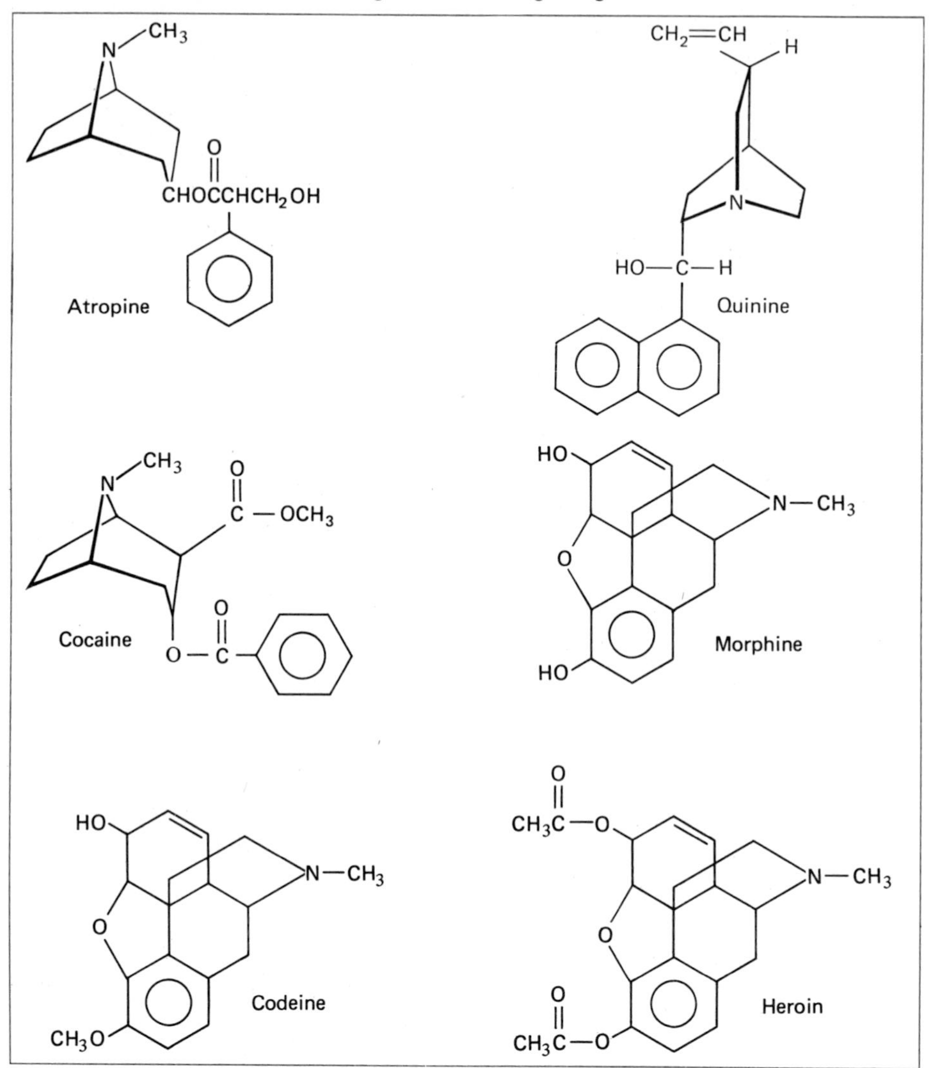

Opiates

Morphine and codeine are alkaloids that can be isolated from the opium poppy. Heroin is a synthetic alkaloid that is derived from morphine. These complicated molecules have a narcotic effect on humans and have benefited society as well as created great problems through their use and abuse. Morphine reduces pain, causes drowsiness and changes in mood, and produces mental fogginess. It has been used to provide pain relief in radical surgery and in cases of wartime injury. However, repeated use of morphine results in addiction and tolerance. Morphine also causes constriction of the smooth muscles such as those found in the intestines and, therefore, can be used in the treatment of diarrhea. Codeine, although less effective than morphine, is also less likely to cause drug addiction. It is used mainly to relieve coughing.

By studying the way in which morphine acts on nerve cells, researchers recently have discovered a group of substances that exist naturally in the brain and that produce the same effects as the opiates. These compounds, called endorphins, are neurotransmitters (chemical substances that pass information between nerve cells). They seem to be important in the body's regulation of pain and emotions.

Barbiturates

Barbiturates are derivatives of barbituric acid (Table 17.6). These compounds act by depressing the central nervous system and are thought to inhibit nerve response centers. The different properties of the various barbiturate compounds depend upon the substituted groups on their molecules. Barbiturates can be used as hypnotics, sedatives, anticonvulsants, and anesthetics. Their most common use (and abuse), however, is as a sleep-inducing drug; excessive use of these compounds can produce physical dependence. One class of barbituric acid derivatives, the thiobarbiturates, contains a sulfur atom in the molecule. The most important member of this class is sodium pentothal, a fast-acting intravenous anesthetic from which recovery is fairly rapid and often without side effects.

In this chapter we have discussed only a few of the nitrogen-containing compounds found in living systems. Other large classes of nitrogen-containing compounds, such as proteins and nucleic acids, will be the subjects of later chapters.

Table 17.6 Some Barbiturates

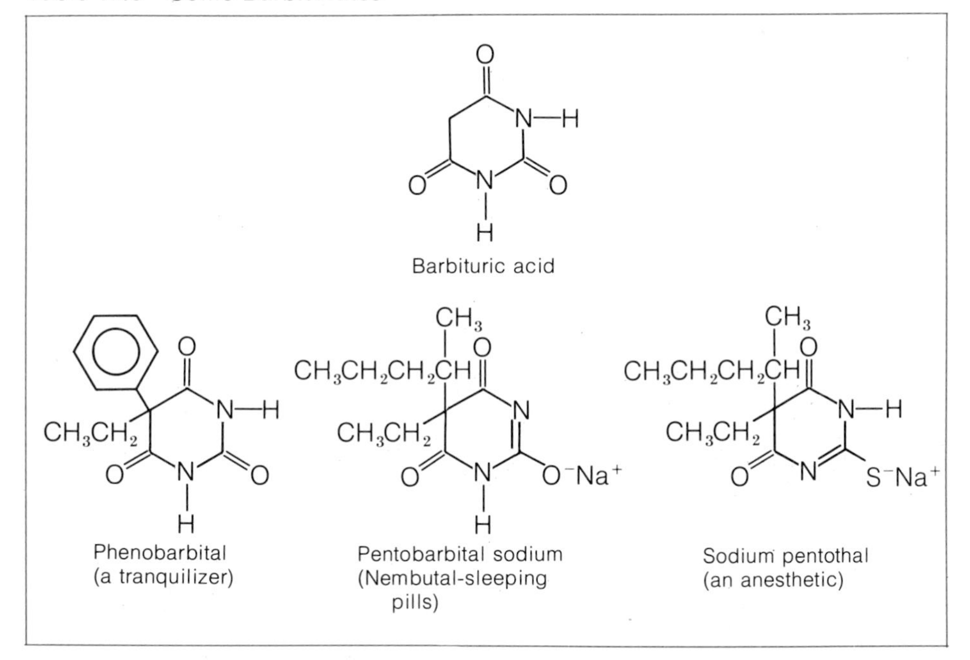

CHAPTER SUMMARY

CHAPTER SUMMARY

Nitrogen is the fourth most common element in the human body and is found in many types of compounds. Amines are weak organic bases that may be classified as primary, secondary, or tertiary according to the groups attached to the nitrogen. Amines form salts in neutralization reactions with acids and when reacted with alkyl halides. Amides are derivatives of carboxylic acids, in which the —OH on the carboxyl group is replaced by $—NH_2$, —NHR, or —NRR′. The bond between the carbon and the nitrogen in these compounds is very stable and is known as an amide linkage. Amides, therefore, undergo hydrolysis very slowly. When dehydrated, amides form nitriles, compounds used to form synthetic fibers. Nitrogen atoms can form ring structures with carbon atoms. Rings containing atoms other than carbon are called heterocyclic rings. Alkaloids are a large class of compounds that often contain nitrogen in ring structures. Most complicated alkaloids occur naturally as part of the defense system of plants, and have been used for centuries as drugs because of their varied physiological effects.

Important Equations

The chemical reactions that we have studied in this chapter can be summarized as follows:

Salt Formation

Amines with acids

$$RNH_2 + HCl \longrightarrow RNH_3^+Cl^-$$

Amines with alkyl halides

$$(CH_3)_3N + CH_3I \longrightarrow (CH_3)_4N^+I^-$$

Ammonolysis

Ester + Ammonia ⟶ Amide + Alcohol

$$R\overset{O}{\overset{\|}{C}}—OR' + NH_3 \longrightarrow R\overset{O}{\overset{\|}{C}}NH_2 + R'OH$$

Hydrolysis

Amide + Water ⟶ Acid + Amine (Ammonia)

$$R\overset{O}{\overset{\|}{C}}—\underset{H}{\underset{|}{N}}—R' + H_2O \longrightarrow R\overset{O}{\overset{\|}{C}}—OH + R'NH_2$$

EXERCISES AND PROBLEMS

EXERCISES AND PROBLEMS

1. Identify each of the following compounds as an amine, amide, imine, or nitrile:

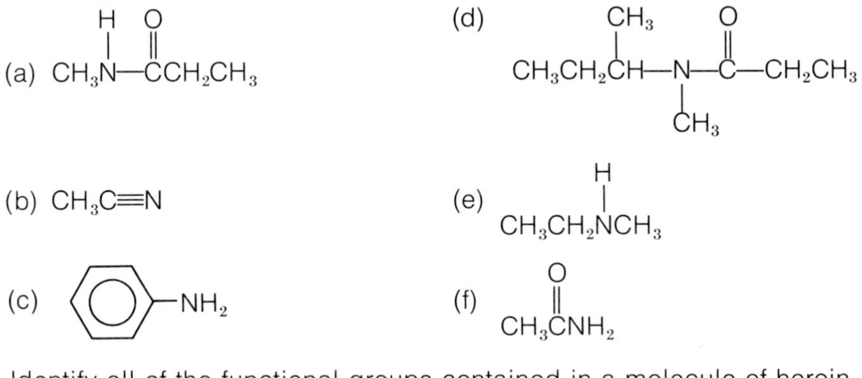

2. Identify all of the functional groups contained in a molecule of heroin.

3. What is a coordinate covalent bond? Why is nitrogen able to form such a bond? List three compounds that contain a coordinate covalent bond.

4. People suffering from Parkinson's disease lack dopamine-producing cells in an area of the brain called the substantia nigra. Is dopamine a primary, secondary, or tertiary amine?

5. Identify each of the following compounds as a primary, secondary, or tertiary amine:

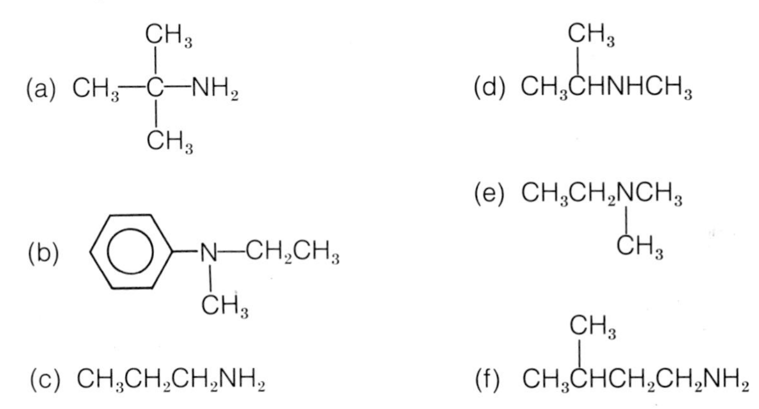

6. Name each of the amines in question 5.

7. What substances can cause the foul smell of decaying organic material?

8. Name each of the following amides:

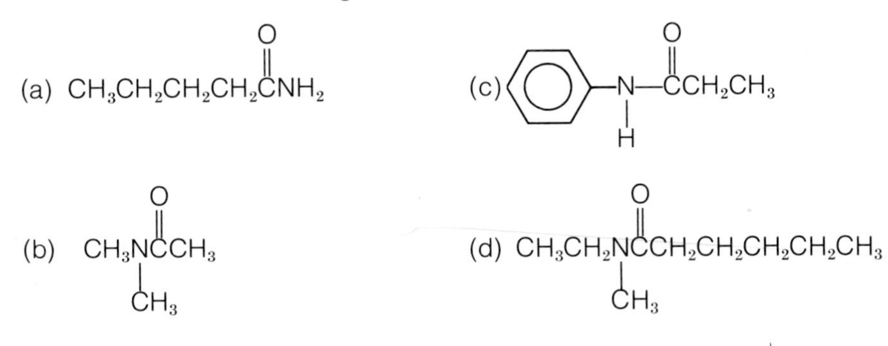

9. Write the structural formula for each of the following:
   (a) methylbenzylamine
   (b) pentanamide
   (c) triisopropylamine
   (d) *N,N*-diethylhexanamide
   (e) 2-(*N*-methylamino)-1-propanol
   (f) *N*-methyl-*N*-phenylpropanamide

10. Draw the structure of two different heterocyclic rings that contain nitrogen.

11. Name the following compounds:
   (a) $(CH_3CH_2CH_2CH_2)_3N$

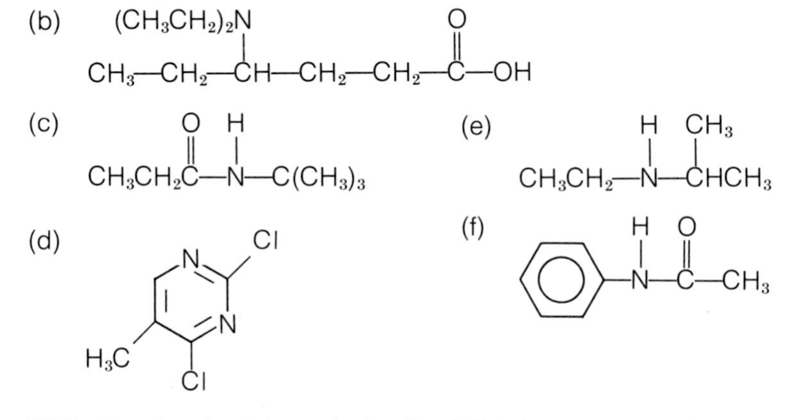

12. Write the structural formula for the following compounds:
   (a) 4-bromo-2,5-dinitroaniline
   (b) 1,5-diaminopentane
   (c) *N*-phenyl-*N*-ethylbutanamide
   (d) indole
   (e) 3-hydroxy-*N*-phenylbenzamide
   (f) ethylhexylamine

13. Morphine is a very effective pain reliever. Why must its continued use in a patient be avoided?

14. What changes in functional groups must take place on the morphine molecule to produce heroin?

15. (a) Describe three specific effects of nicotine on the human body.
    (b) Suggest a reason why smokers can tolerate larger concentrations of nicotine than can nonsmokers before feeling ill effects.

16. Describe three different uses of alkaloids in drug therapy.

17. Write the equation for the following reactions, and name the products.
   (a) isobutylamine + HCl
   (b) triethylamine + HCl
   (c) triethylamine + ethylbromide

18. Write the equation for the hydrolysis of each of the amides in question 8, and name the products of each hydrolysis.

CHAPTER EIGHTEEN

THE REMAINING TWENTY-ONE ELEMENTS

Fred McKay arrived in Colorado Springs, Colorado, in 1901 to begin a practice in dentistry. The young dentist was quite surprised when he discovered that the teeth of people living in that area were badly stained. He was unable to find any mention of such a condition in his dental books, so he named this condition *Colorado Brown Stain.* Sometimes this staining consisted only of small opaque, paper-white areas visible on the surface of normally translucent teeth (a condition called mottling). Other times it was a very noticeable dark brown stain on teeth whose enamel surfaces were often pitted. McKay soon started recording the occurrence and severity of this condition in various parts of the country, and tried to interest other members of the dental profession in joining in a study of Colorado Brown Stain.

By 1916, after a study of 26 communities, McKay had concluded that there was some substance in the drinking water of these communities that was causing the mottled teeth. In addition, he found that this substance had a noticeable effect only during the early stages of permanent tooth development in children. McKay's discovery that Colorado Brown Stain was caused by something in the water prompted communities such as Oakley, Idaho, to switch their source of drinking water. Happily, this change completely eliminated the occurrence of mottled teeth in the children of that town. McKay's investigations continued, and in 1929 he was able to make the important observation that patients suffering from mottled teeth were less likely to have dental caries, commonly known as cavities. In 1931, independent investigations by three different laboratories resulted in the identification of the fluoride ion, F^-, as the substance in drinking water that caused the mottling of children's teeth. Measurements of the fluoride concentration in various water supplies revealed that for half a century the residents of Colorado Springs had been drinking water containing 2 parts per million (2 ppm) of fluoride ion. The original water supply of Oakley, Idaho, was found to contain 6 ppm fluoride (the new water supply for this community contained less than 0.5 ppm fluoride ion). By 1942, it had been established that a level of 1 ppm fluoride in the drinking water would greatly reduce the number of dental caries in a population without producing mottled teeth (Figure 18.1 and Table 18.1). In 1950, the United States Public Health Service officially endorsed the fluoridation of public water supplies to a level of 1 ppm fluoride. Many cities and towns quickly acted to fluoridate their water supplies following studies showing that the rate of decayed and missing teeth in children who used such water was reduced by as much as 60%.

Dental caries are caused by bacteria that are found naturally in the mouth. These bacteria use carbohydrates (the subject of the next chapter) as their source of energy and release carboxylic acids as waste products

Table 18.1 Fluoride Levels in Drinking Water

Parts per Million of Fluoride	Effects
Less than 0.5	High incidence of dental caries; no mottling
1	Low incidence of dental caries; little or no mottling
Greater than 2.5	Disfiguring mottling
10 to 20	Severe mottling or fluorosis

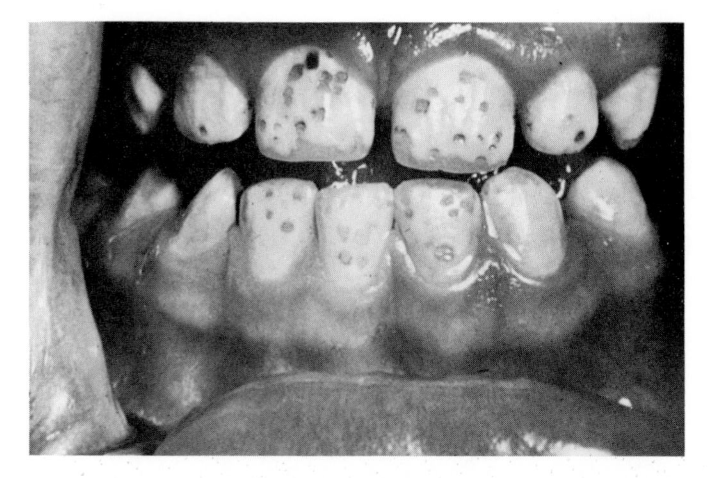

Figure 18.1 These teeth show severe mottling, or fluorosis, caused by the long-term drinking of water containing fluoride in a concentration greater than 10 parts per million. (Courtesy National Institutes of Health.)

of their metabolism. Unfortunately, these bacteria also use carbohydrates (particularly sucrose, or table sugar) to produce plaque, a sticky substance that holds the bacteria and the acids they release against the teeth. These acids dissolve minerals in the teeth, causing the enamel to decay.

The number of dental caries that you have depends on three factors: the particular chemical composition of your teeth, the amount of bacterial plaque on the teeth, and both the type and frequency of your carbohydrate consumption. A person's rate of dental caries can be lowered by removing the plaque with flossing and brushing and by reducing the number of times that sugar is eaten during the day. However, the most effective method of preventing dental caries is the use of drinking water containing 1 ppm fluoride.

Fluoride helps to reduce dental caries in many ways. When children drink fluoridated water, fluoride ions are deposited as a calcium compound, fluorapatite, in the structure of the tooth enamel as the teeth are forming. This compound stabilizes the structure of the enamel and makes it more resistant to the acids produced by mouth bacteria. Fluoride also lowers the bacteria's production of the substance that attaches these bacteria to the teeth. In addition, the bacteria themselves appear to be sensitive to fluoride, and their numbers become reduced in the presence of this chemical. Low levels of fluoride have not been shown to have any adverse effects on humans. But in the presence of very high fluoride concentrations, the body not only deposits fluorides as part of tooth enamel but also deposits these ions in the bones. Over a long period of time, this process can result in enlarged and abnormal bones—a condition called skeletal fluorosis.

Because very small concentrations of fluoride can have such varied effects on our bodies, we might ask if there are other elements that have equally important effects in small concentrations. Such elements are the subject of this chapter.

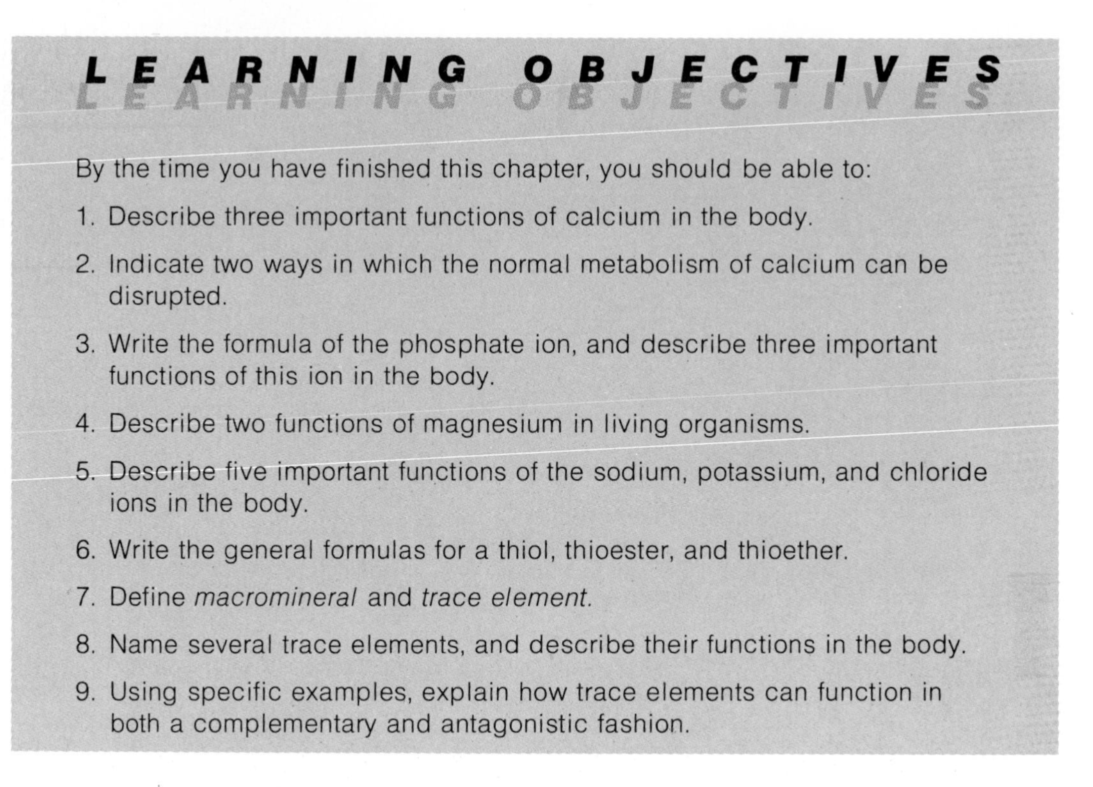

LEARNING OBJECTIVES

By the time you have finished this chapter, you should be able to:

1. Describe three important functions of calcium in the body.
2. Indicate two ways in which the normal metabolism of calcium can be disrupted.
3. Write the formula of the phosphate ion, and describe three important functions of this ion in the body.
4. Describe two functions of magnesium in living organisms.
5. Describe five important functions of the sodium, potassium, and chloride ions in the body.
6. Write the general formulas for a thiol, thioester, and thioether.
7. Define *macromineral* and *trace element.*
8. Name several trace elements, and describe their functions in the body.
9. Using specific examples, explain how trace elements can function in both a complementary and antagonistic fashion.

THE MACROMINERALS

18.1 The Remaining Twenty-One Elements

In the last few chapters we have discussed compounds formed from carbon, hydrogen, oxygen, and nitrogen. These four elements, we have seen, form an enormous number of compounds and make up 99.3% of all the atoms in your body. The identification of other elements that are necessary for life is an area of active research. To date, 21 additional elements have been shown in laboratory experiments to be essential to animal life (Table 18.2). Although these elements make up only 0.7% of the atoms in the human body, they perform a wide variety of functions critical to life. Depriving a living organism of any of these elements will result in disease and possibly death.

As shown in Table 18.2, the remaining 21 elements can be divided into two groups. One group contains seven elements, called the **macrominerals,** which are found in greater concentrations in the body than are the remainder. This group of elements contains potassium, magnesium, sodium, calcium, phosphorus, sulfur, and chlorine. The remaining elements are found in such very small amounts (ppm or ppb) that they are called the **trace elements.** In the past the necessity of these trace elements to good nutrition was largely overlooked, with most emphasis focused on the importance of vitamins. Now, however, it is

Table 18.2 Elements Necessary for Life

Element	Percent of Total Number of Atoms in the Human Body	Number of Grams in a 70-kg Man
Hydrogen	63	7058
Oxygen	25.4	45,348
Carbon	9.5	12,671
Nitrogen	1.4	2180
Calcium	0.31	1381
Phosphorus	0.22	757
Potassium	0.06	261
Sulfur	0.05	181
Chlorine	0.03	118
Sodium	0.03	77
Magnesium	0.01	27
Iron	<0.01	7
Iodine, fluorine, manganese, zinc, molybdenum, copper, cobalt, chromium, selenium, arsenic, nickel, silicon, boron	<0.01	<1

recognized that the trace elements play an equally important nutritional role. All 21 of these elements are found in living systems either as ions or covalently bonded in organic molecules. Their functions are quite varied and may depend on their chemical form or their location in the body's tissues or fluids.

It is important to understand that for each trace element there is a specific range of concentration within which the body functions normally. At less than the minimum concentrations, signs of a deficiency begin to be seen. As the concentration continues to decrease, a deficiency disease develops and death can result. However, every trace element is also potentially toxic if the range of safe and adequate concentration for that element is exceeded. Quite recently, megadoses (very large doses) of dietary mineral supplements, especially zinc and selenium, have been promoted as cures or preventive medicine for a variety of medical problems. What is not mentioned in the advertising is that large doses of trace elements, like large doses of certain vitamins, can be very dangerous to your health. Moreover, the dietary concentration of one trace element can affect the body's ability to absorb other trace elements. Adequate levels of these necessary elements can easily be maintained through good nutrition rather than through mineral supplements (Figure 18.2).

18.2 Calcium

Calcium is the fifth most common element in the body, accounting for 0.31% of the atoms in the body and making up about 2% of an adult's body weight. You've no doubt heard that calcium is important for the formation of strong bones and teeth. Indeed, 99% of the calcium in your body is found in your bones and teeth, in the form of several inorganic salts held in a framework of proteins (Figure 18.3). Maintaining adequate levels of calcium in the blood as we get older is critical to the prevention of osteoporosis, a progressive weakening of the bones that can result in spontaneous fractures and compression of the spine. Although everyone experiences a loss of bone mass after age 35, women lose it at a much faster rate (especially after menopause).

Calcium is a group II element that is found in the body as the stable calcium ion, Ca^{2+}. This is the most closely regulated ion in the body. In the blood, the level of calcium ion follows a complicated relationship involving two hormones (parathormone and calcitonin) and vitamin D. The level of these substances in the blood controls the amount of calcium absorbed in the intestines, the level of calcium ion in the blood, the amount of calcium deposited in the bones and teeth, and the amount excreted by the kidneys.

The calcium ions not found in the skeleton play several other important

Figure 18.2 Eating a balanced diet containing all the essential elements will help you live a long and healthy life. (Phyllis Lefohn.)

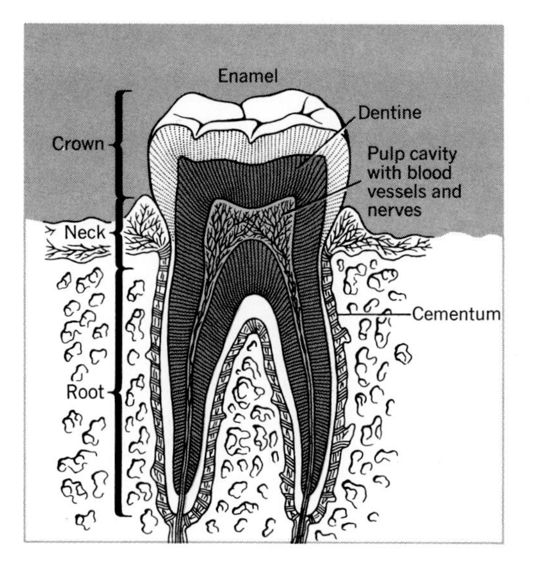

Figure 18.3 **The structure of a tooth. Enamel, dentine, and cementum are composed of the following inorganic materials imbedded in a protein framework:**

Main Inorganic Material

Apatite $Ca(OH)_2 \cdot 3[Ca_3(PO_4)_2]$

Trace Inorganic Material

Fluorapatite $CaF_2 \cdot 3[Ca_3(PO_4)_2]$

Chlorapatite $CaCl_2 \cdot 3[Ca_3(PO_4)_2]$

Dahllite $CaCO_3 \cdot 3[Ca_3(PO_4)_2]$

roles in the body. Calcium ions must be present in the correct concentration to allow the contraction of muscles, and they are especially important in maintaining the rhythmic contraction of the heart muscle. Too low a concentration of calcium can completely stop muscle contraction. Calcium ions affect nerve transmission by their stabilizing effect on the nerve membrane. Too much calcium in the blood results in a deadening of nerve impulses and muscle response, making a person unresponsive to any stimuli. Too little calcium in the blood can result in a high (hyper) irritability of the nerves and muscles. Under such hyperirritability the slightest stimulus, such as a loud noise, cough, or touch, can send a person into convulsive twitching. Such a state is extremely exhausting and will soon result in death.

A specific level of calcium ions in the fluid of the brain is critical in maintaining body temperature. Too high a concentration of calcium will result in the lowering of body temperature. Calcium ions also must be present for the blood to clot. Any condition that removes the calcium ions from the blood will prevent the clotting process from occurring. Leeches, fleas, and other bloodfeeding creatures are able to secrete a substance that reacts with calcium ions in the blood, preventing the blood from clotting while they feed on and digest it. Similarly, citrate ions and oxalate ions will combine with calcium ions in freshly collected blood to form an insoluble precipitate, decreasing the level of calcium ions left in the blood and, therefore, stopping the clotting reactions. For this reason, sodium citrate is used as an anticoagulant in whole blood to be used for transfusions.

$$\underset{\text{Citrate ion}}{2C_6H_5O_7^{3-}} + 3Ca^{2+} \longrightarrow Ca_3(C_6H_5O_7)_{2(s)}$$

Calcium ions, in addition to their role in the clotting process, also activate a variety of enzymes. In fact, the calcium ion may be considered the *coordinator* among the other mineral ions, regulating the flow of these ions in and out of the cell (Figure 18.4).

Dairy products are the major source of calcium in our diets. Three eight-ounce glasses of low-fat milk will provide 1000 mg of calcium, the recommended daily allowance for an adult premenopausal woman (postmenopausal women, 1500 mg; adult males, 800 mg). In addition to milk, good sources of calcium are shellfish and green leafy vegetables. The complex relationships controlling the level of calcium in the blood can be upset by various factors, including low levels of calcium in the diet, diets high in phosphorus or protein (which reduce calcium absorption and retention), or abnormal levels of vitamin D, calcitonin, or parathormone in the blood. Too little vitamin D in the body produces rickets, a disease that causes bones to soften and bend out of shape (Figure 18.5). Although vitamin D is formed in the body when the skin is exposed to sunlight, extra vitamin D has been added to milk supplies in the United States to prevent the occurrence of rickets. However, it has been shown that too much vitamin D can be as harmful as too little; it causes a thickening of bones and a calcification of soft tissues.

Because of their similar charge and ion size, cadmium ions can lower the level of calcium ions in our bodies, causing a disruption of normal calcium metabolism. For example, in the late 1950s residents of Japan's Jinzu River basin were affected by a strange malady called the *Hai–Hai* or *Ouch–Ouch* disease. This disease caused severe and painful decalcification of bones, often resulting in multiple fractures. The malady was found to be caused by the presence of cadmium ions, Cd^{2+}, in rice irrigated by polluted water discharged from industries upstream.

18.3 Phosphorus

We have seen that calcium plays a major role in the formation of bones and teeth. Phosphorus is an important element in the inorganic calcium salts that are found in these bones and teeth (see Figure 18.3). Ninety

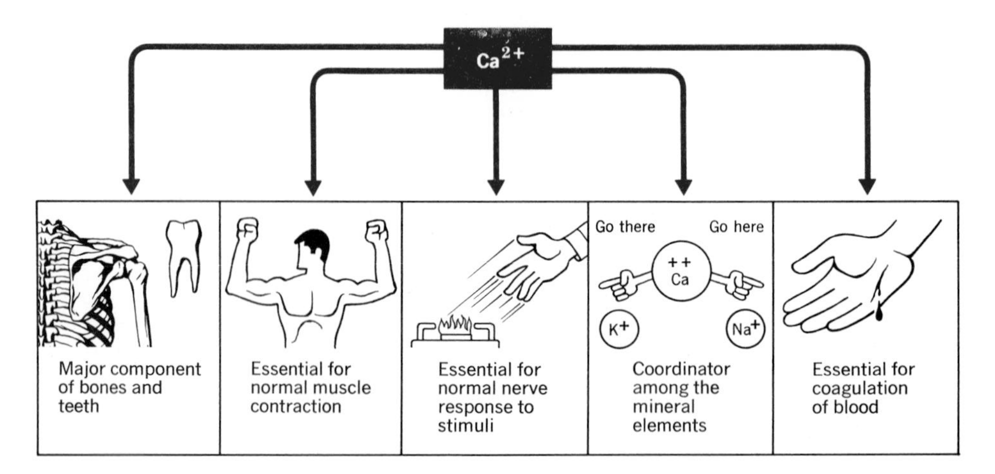

Figure 18.4 The functions of calcium in the body.

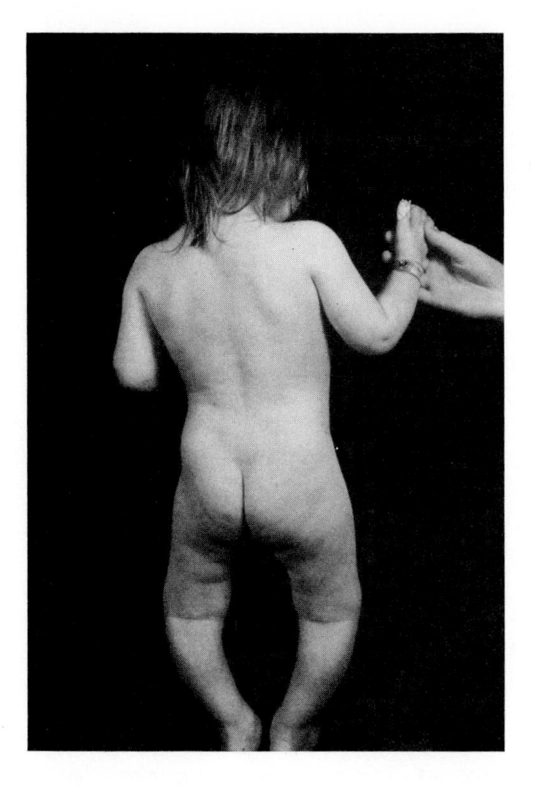

Figure 18.5 This child shows the characteristic deformaties of rickets: bones that become soft from a lack of vitamin D and bend out of shape. (Biophoto Associates/Photo Researchers.)

percent of the phosphorus in the body is found in the bones and teeth in the form of the negative phosphate ion, PO_4^{3-} (in the compound, calcium phosphate).

The phosphate ion is formed when phosphoric acid loses three hydrogen ions.

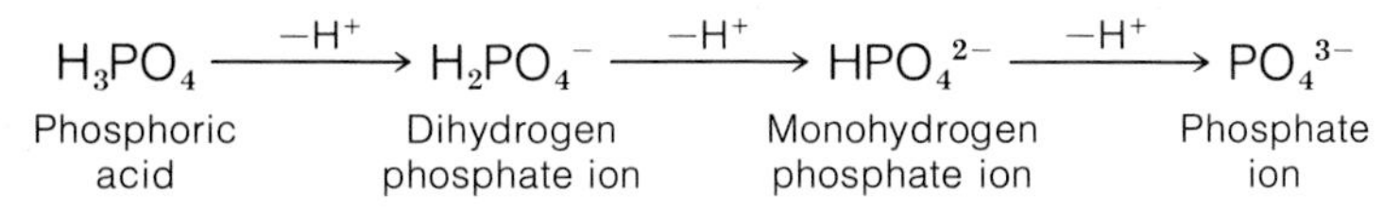

The dihydrogen and monohydrogen phosphate ions act as buffers, which help to maintain the proper pH of body fluids. Phosphoric acid can also react with an alcohol functional group on an organic compound to form a phosphate ester. Organic phosphate esters are found in phospholipids (which make up cell membranes and nerve tissues), DNA and RNA (which control heredity and protein synthesis), and coenzymes (compounds that work with enzymes in the body). We will be discussing each of these compounds in detail in later chapters. When certain organic phosphate esters undergo hydrolysis, considerable chemical energy is liberated. Such phosphates are known as high-energy phosphates and are the compounds that supply the immediate energy needs of the cell. These important functions make phosphorus essential to all body tissues (Figure 18.6). Fortunately, phosphorus is so widely distributed in our daily foods that almost everyone is assured of obtaining at least the 800 mg per day that is required for adults.

18.4 Magnesium

Magnesium ions, Mg^{2+}, make up 0.01% of the atoms in the body. These ions activate many of the enzymes that control the addition and removal of phosphate groups from compounds in the cell. They are also crucial in regulating nerve function and muscle contraction. The magnesium ion forms part of the chlorophyll molecule, which traps the energy from sunlight in the process of photosynthesis and which gives plants their green color (Figure 18.7). Magnesium is found in a wide variety of foods such as green vegetables, nuts, cereals, and seafoods, so we are fairly well assured of enough magnesium in our diets (the recommended daily amount is 300 to 350 mg).

If, however, the magnesium level in the body is lowered, a person may suffer emotional irritability and aggressiveness, muscle spasms, and convulsions. A deficiency in magnesium also affects heart rhythm and general cardiovascular function, and may be responsible for some cases of sudden death from coronary spasms in young people having no history of heart disease. Magnesium deficiencies have been observed in chronic alcoholics, in infants having a protein deficiency disease called Kwashiorkor, in patients with bulemia, and in postsurgical patients on restricted diets. On the other hand, too much magnesium in the body decreases muscle and nerve response, and high levels can produce local or general anesthesia and paralysis.

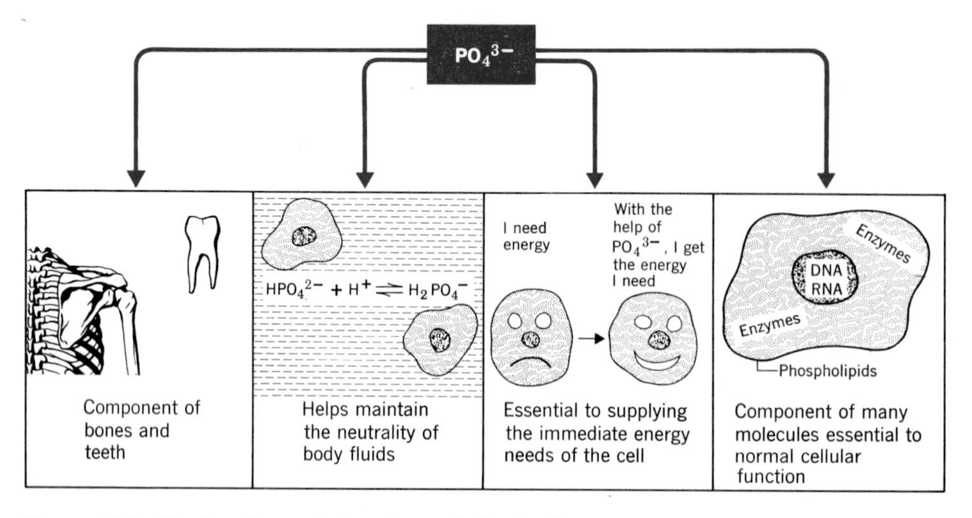

Figure 18.6 The functions of phosphorus in the body.

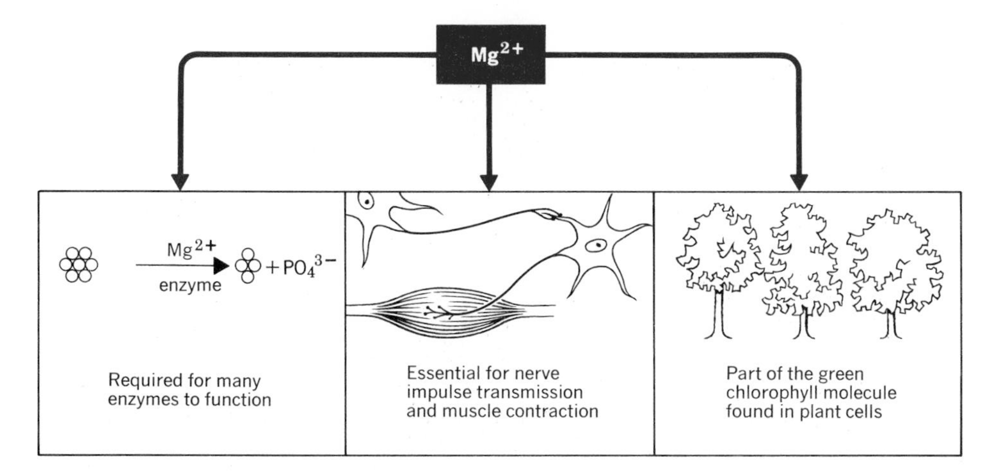

Figure 18.7 The functions of magnesium in living organisms.

18.5 Potassium, Sodium, and Chlorine

The roles of potassium, sodium, and chlorine in the body are intricately interrelated. Sodium and potassium ions are often present in the form of their chloride salts. These ions are the major cations in body fluids; the potassium ion is usually found in the intracellular fluid within the cell, whereas the sodium ion is generally found in the extracellular fluids surrounding the cell. The main function of the potassium, sodium, and chloride ions is to control the cation/anion balance in the cells, tissue fluids, and blood. Such cation/anion balance is necessary to maintain the normal flow of fluids and to control the balance between acids and bases in the body. These three ions also play important roles in the transport of oxygen and carbon dioxide in the blood. Sodium and potassium ions are involved in the transport of sugars across cell membranes and in the breakdown of sugars in the cell. In addition, these two ions (together with calcium and magnesium ions) help maintain the proper level of nerve and muscle response. The effects of sodium and potassium are antagonistic to those caused by calcium and magnesium (that is, they have opposite effects). Therefore, the correct balance in the concentrations of these four ions is critical to normal nerve and muscle performance.

Sodium chloride and potassium chloride act to keep large protein molecules in solution, and thus to regulate the proper viscosity or thickness of the blood. The acid in the stomach that begins the digestion of certain foods is hydrochloric acid, HCl, which is derived from the sodium chloride in the blood. Other digestive compounds found in the gastric juices, pancreatic juices, and bile are likewise formed from the sodium and potassium salts in the blood. The response of the retina of the eye to light impulses is another body process that depends upon the correct concentrations of sodium, potassium, and chloride ions (Figure 18.8).

Having learned the many important body functions that depend upon these three ions, it is easy to imagine that an imbalance in the level of any one of them can have serious effects on the body. Experimental animals have been fed diets deficient in these ions and have experienced slow growth, slow heart rate, muscular atrophy, and sterility. Plant material is high in potassium ions, but high levels of potassium ions in the diet will result in excessive excretion of the sodium ion from the body. This is why herbivorous animals (animals that eat plant material) must be given diets containing a high level of salt or sodium chloride to maintain the proper balance between the sodium and potassium ions in their bodies. Such animals have been known to travel hundreds of miles and to risk their lives in order to reach a salt lick.

You are certainly aware that strenuous exercise, especially in hot weather, results in heavy perspiration. Perspiration is composed mostly of water, but there are also many ions dissolved in this fluid (among which are potassium, sodium, and chloride ions, giving sweat its salty taste). If the concentration of these ions in the body is significantly reduced by heavy perspiration, an imbalance will occur that affects muscle and nerve response. Nausea, vomiting, exhaustion, and muscle cramps can result.

The recommended daily allowance of sodium is 2 to 3 grams, and of potassium and chlorine 2 to 5 grams. Potassium is found in most foods, especially fruits and vegetables; sodium and chloride are found in salt and shellfish. Under normal dietary conditions, humans will not experience a deficiency of these ions. A deficiency of potassium can result, however, in people who suffer from prolonged diarrhea or who use diuretics. A

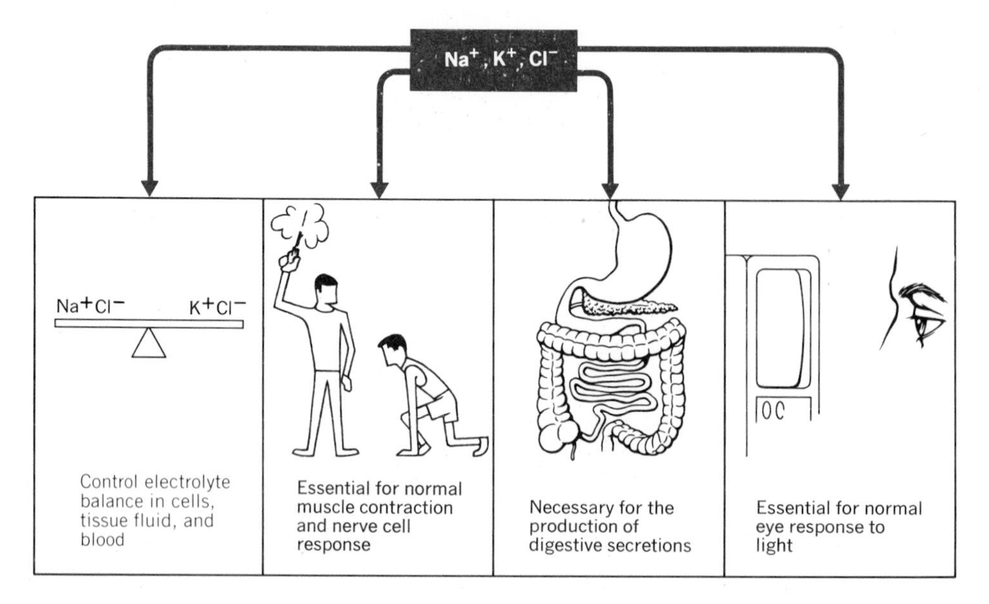

Figure 18.8 The functions of sodium, potassium, and chlorine in the body.

Table 18.3 The Role of Sulfur in the Body

1. Found in proteins and other biological compounds.	
2. Found in several functional groups:	
Thiol or sulfhydryl	—S—H
Disulfide	—S—S—
Thioethers or organic sulfides	—C—S—C—
Thioester	—C(=O)—S—C—

major problem with the American diet at the present time, in fact, is not a deficiency of sodium but rather an excess. We consume 10 to 35 times the recommended daily amount of sodium, mostly in the form of salt added to foods. Long-term high levels of sodium may cause an increase in the amount of extracellular fluid (edema), disruption of fat metabolism, changes in gastric secretions, and high blood pressure (hypertension).

18.6 Sulfur

Sulfur is the last of the seven elements in the group of macrominerals. It is found in group VI on the periodic table; it appears just under oxygen and has properties similar to oxygen. It will form two covalent bonds. One functional group in which sulfur is found is the thiol or sulfhydryl group, which closely resembles the hydroxyl group.

:S̤̈: H —S—H
Thiol or sulfhydryl group

:Ö̤: H —O—H
Hydroxyl group

One class of simple compounds containing the thiol group is called the mercaptans.

CH_3CH_2SH
Ethyl mercaptan

CH_3CH_2OH
Ethanol

These compounds have a terrible smell. In fact, it is mercaptans that produce the strong odor given off by skunks. Because natural gas is odorless and toxic, gas companies add mercaptans to it before it is piped to consumers. This allows leaks to be easily detected.

Other compounds containing sulfhydryl groups need not have any odor. Proteins, for example, are polymers of smaller units called amino acids (which we will discuss in detail in Chapter 21). The amino acid called cysteine has a sulfhydryl group and is very important to the structure and function of protein molecules in which it is found.

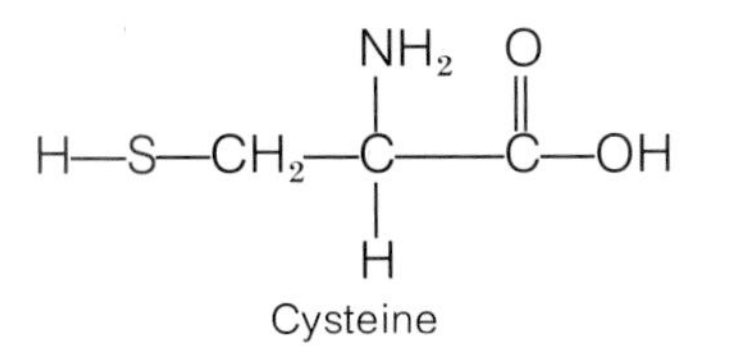

Cysteine

One very important thiol in living cells is coenzyme A, a very complex molecule whose structure is often abbreviated as CoA—SH (showing the sulfhydryl group). Coenzyme A can form an ester with acetic acid. This compound, called acetyl coenzyme A or acetyl CoA, contains the thioester functional group (Table 18.3).

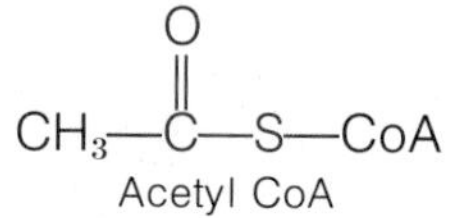

Acetyl CoA

As we will see in the following chapters, acetyl CoA is a very important intermediate in the metabolism of carbohydrates, lipids, and proteins.

Sulfur can replace the oxygen atom in an ether linkage to produce a thioether, or organic sulfide, —C—S—C—. Among the compounds containing this functional group are those giving onions and garlic their familiar flavors.

$H_2C{=}CH{-}S{-}CH{=}CH_2$
Divinyl sulfide
(onion flavor)

$H_2C{=}CHCH_2{-}S{-}CH_2CH{=}CH_2$
Diallyl sulfide
(garlic flavor)

Methionine is an amino acid that contains a thioether group. All body proteins, as well as special protein molecules such as enzymes and hormones, contain this amino acid. The body is able to produce cysteine from methionine, making methionine extremely important in sulfur metabolism.

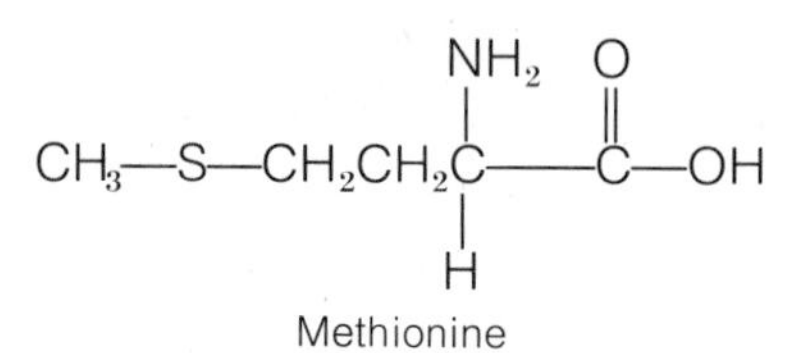

Methionine

We have seen that the roles of the macrominerals in our bodies are varied, complex, and often intricately related. This is why the maintenance of proper concentrations of these elements through good nutrition is so vitally important to assure a healthy body.

THE TRACE ELEMENTS

18.7 The Fourteen Trace Elements

The remaining elements that have been shown to be essential to life are often called the **trace elements** because they are found only in trace (extremely small) amounts in the body. This fact has made it very difficult to determine exactly which elements are essential to life. Recent research, for example, has revealed that essential roles are played by chromium, selenium, and arsenic—elements that had previously been considered only toxic to the body. Safe and adequate daily amounts for some trace elements are fairly well established, but for others there are simply not enough data to determine a recommended daily allowance (Table 18.4).

Table 18.4 The Trace Elements

Element	Daily Requirement[a]	Function
Nonmetals		
Fluorine	1.5–4.0 mg[b]	Found in bones and teeth; important in prevention of dental caries
Iodine	0.15 mg	Required for normal thyroid function; found in thyroid hormones
Selenium	0.05–0.2 mg[b]	Required for the prevention of white muscle and liver disease in some animals and Keshan disease in humans; part of the enzyme glutathione peroxidase
Silicon	Unknown[c]	Required for bone growth and connective tissue development in animals
Arsenic	Unknown[c]	Required for adequate growth and reproduction in animals
Boron	Unknown[c]	Enhances parathormone action and the metabolism of Ca^{2+}, P, and Mg^{2+}
Metals		
Iron	10 mg for men (18 for women)	Found in hemoglobin and many enzymes; important in prevention of anemia
Copper	2.0–3.0 mg[b]	Part of enzymes essential for the formation of hemoglobin, blood vessels, bones, tendons, and the myelin sheath
Zinc	15 mg	Essential for many enzymes, normal liver function, and DNA synthesis
Cobalt	Unknown[c]	Part of vitamin B_{12} molecule
Manganese	2.5–5.0 mg[b]	Essential for several enzymes, bone and cartilage growth, and brain and thyroid function
Chromium	0.05–0.2 mg[b]	Lowers blood sugar level by increasing the effectiveness of insulin
Molybdenum	0.15–0.5 mg[b]	Required for the function of several enzymes
Nickel	Unknown[c]	Iron absorption, needed for optimal growth and reproduction in animals

[a] Recommended daily allowances for men and women (19–22 years) set by the Food and Nutrition Board, National Academy of Sciences–National Research Council, in 1980.
[b] Recommended adequate and safe daily intakes for adults.
[c] Requirements in humans have not yet been established.

The chemical action of many of the trace elements is quite complex, and the exact functions of several are still largely unknown. Many of these elements, however, form important parts of enzymes. Because enzymes can be used over and over again, they can be effective even when found in only very low concentrations in the cells of the body. The transition metals, in particular, have chemical binding properties that make them especially important in enzyme molecules.

18.8 Iodine

Iodine, one of the nonmetal trace elements, has long been known to be essential to life. Seventy to 80% of the iodine in the body is concentrated in the thyroid gland, a small gland in the neck. Iodine is part of two hormones produced by the thyroid gland: thyroxine and triiodothyronine. It is not yet understood why adding iodine atoms to the compound thyronine—three iodines added for triiodothyronine, and four for thyroxine—transforms this otherwise biologically inert compound into powerful hormones that regulate the body's chemical activity and are vital to normal growth.

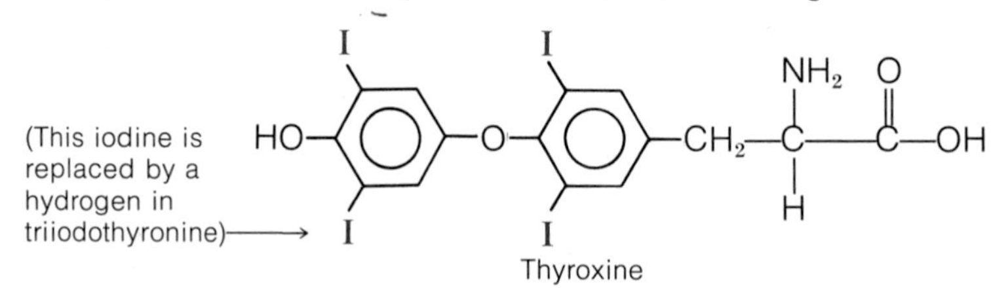

Thyroxine

An insufficient amount of iodine (in the form of the iodide ion) in the diet causes enlargement of the thyroid, a medical condition known as simple goiter (Figure 18.9). Such an increase in the size of the thyroid is a compensatory reaction by the body in response to the low level of iodine. The body attempts to increase the production of thyroid hormone by increasing the number of cells in the thyroid, but this attempt cannot succeed as long as the concentration of iodine remains low. An iodine deficiency during pregnancy results in too little thyroxin production. This can cause a disease in the fetus, called cretinism, which results in mental and physical retardation. Salt-water fish are a rich source of iodine; cases of goiter used to be common in the American Midwest and still are today in many areas of the world where such fish are not commonly available. The amount of goiter in such regions has been greatly reduced by the use of table salt containing the iodide ion, I^-.

18.9 Iron

Iron ions are found in many important enzymes and electron carrier molecules. Four iron(II) ions are found in every molecule of hemoglobin—the molecule in the red cells of our blood which carries oxygen from the lungs to the tissues and which causes the blood to look red (Figure 18.10). If you were able to extract all of the iron from a healthy body, you would have enough to make only two small nails (about 5 to 7 grams). Yet this amount of iron is critical. Only a small reduction in the level of iron in the blood will result in a condition known as anemia, which causes general

Figure 18.9 Some villagers in Paraguay still suffer from endemic goiter. The swelling of the thyroid gland in this woman's neck is the result of a lack of the trace element iodine in her diet. (Courtesy World Health Organization; photo by Paul Almasy.)

body weariness, fatigue, and apathy. Anemias from iron deficiencies result in low hemoglobin levels and often occur in infants at six months and in women from 30 to 50 years of age.

A healthy adult needs between 10 and 18 mg of iron each day. Such dietary iron is found in large amounts in organ meats such as liver, kidney, and heart, as well as in egg yolks, dried vegetables of the pea family, and shellfish. In the body, iron is absorbed in the small intestine in the form of the iron(II) ion, Fe^{2+}. This iron absorption is increased by the presence of vitamin C (ascorbic acid) which reduces the iron(III) ion, Fe^{3+}, in the intestines to the iron(II) ion. Under normal conditions, only 5 to 15%

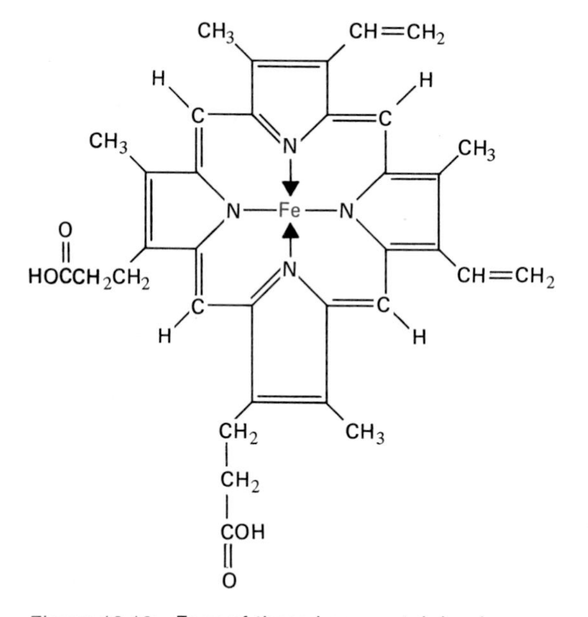

Figure 18.10 Four of these iron-containing heme groups are found in each hemoglobin molecule. Each heme is capable of carrying one oxygen molecule, so one hemoglobin molecule can carry four oxygen molecules.

of the iron in the foods we eat is actually absorbed into our bodies. A deficiency of iron in the diet results in listlessness and fatigue, reduced resistance to disease, and an increase in heart and respiratory rates. Children with iron deficiencies show a decreased growth rate, mental slowness, and abnormal red blood cell growth. High levels of iron, on the other hand, can also be unhealthy. Abnormally high levels will cause cirrhosis of the liver, fibrosis of the pancreas resulting in diabetes mellitus, and congestive heart failure.

The functioning of iron in the body presents a good example of how the concentration of one trace element can be closely dependent upon the concentration of some other trace element. In this case, an enzyme containing copper must be present for a hemoglobin molecule to be formed. Therefore, the concentration of hemoglobin in the body depends not only upon the level of iron, but also upon the concentration of copper. A high concentration of copper will result in a high use of iron. Such relationships between trace elements can also work in an antagonistic fashion. For example, a high concentration of the trace element molybdenum in the diet will cause a decrease in the absorption of copper by the body, resulting in a decrease in the formation of hemoglobin.

18.10 Copper

Since 1818, copper ions have been known to be a component of living tissue. Copper is necessary for the normal functioning of all living cells.

Too much or too little copper will result in malfunction of the cell. Foods high in copper ions include shellfish, dried peas and beans, and cocoa.

Copper performs a variety of functions in the body. It is a component of several important enzymes, one of which helps in the formation of blood vessels, tendons, and bones. Copper-deficient animals will exhibit weakness and fragility of blood vessels and bones (Figure 18.11). The formation of the protective sheath around the nerves is dependent upon copper, and a deficiency of copper plays a role in a degeneration of the nervous system in which nerve impulses are no longer properly transmitted. Copper also helps protect us from the harmful ultraviolet rays of the sun; this element is part of an enzyme that assists in the formation of melanin, the dark pigment of the skin that is our natural protection from ultraviolet radiation. Our cells would not be able to extract energy from foods without another copper-containing compound. Earlier we mentioned that copper must be present for the formation of hemoglobin. And finally, the ability to taste foods may also depend upon the presence of copper.

Because copper is a part of very important enzymes, low levels of this element can cause serious illness, including bone demineralization, cerebral degeneration, depigmentation, and changes in artery walls. But high levels of copper can also be toxic. Because copper ions are very common in all foods and in drinking water, the body has developed complex ways to regulate the absorption and excretion of copper. The copper level in the body depends upon a balance between copper, molybdenum, and sulfate in the diet. A disorder called Wilson's Disease results from a genetic abnormality in which the body's ability to eliminate copper is impaired. The liver, kidneys, and brain of an individual suffering from this disease will show abnormally high levels of copper, possibly leading to mental illness and death. Wilson's Disease can be treated with a chemical that binds with copper ions and, in that way, detoxifies them.

Figure 18.11 These three dogs are from the same litter. The dogs at the center and right were fed diets deficient in copper and have rough hair and deformed legs not found in the normal dog on the left (who was fed a diet containing copper). (Courtesy Baxter and Van Wyk, *Bull. Johns Hopkins Hosp.* **93:**1, 1953.)

18.11 Zinc

Recent research has shown zinc to be an extremely important trace mineral, especially in fetal development and the nutrition of infants. The adult human body contains about 2.3 grams of zinc, which is found in over 100 zinc-containing enzymes. The normal daily requirement for zinc is 15 mg. Foods rich in zinc ions include protein-rich foods such as meat, fish, eggs, and nuts.

Zinc plays a role in carbohydrate, lipid, and protein metabolism, and in the synthesis and breakdown of DNA, the genetic material in the cell. Because of these functions, a zinc deficiency in the fetus will result in retarded growth, malformations of the body, and chromosomal abnormalities. A zinc deficiency after birth may result in dwarfism, retarded sexual development, poor appetite, mental lethargy, loss of hair, and delayed healing of wounds and skin lesions (Figure 18.12).

Human milk contains zinc at concentrations up to 10 times higher than that found in the blood. In addition, this zinc is in a chemical form that promotes increased absorption by the infant. The importance of zinc was demonstrated by a study in which babies who were fed formula diets fortified with zinc showed increased growth rates and were less prone to vomiting and diarrhea than babies fed nonfortified milk.

The zinc ion appears in several liver enzymes, one of which is important in the oxidation of alcohol to less toxic substances. A high level of alcohol in the body may cause this enzyme to break down, which will then produce toxic conditions in the liver and a zinc deficiency. Alcoholics with cirrhosis of the liver show a high level of zinc in the urine, indicating a breakdown of this zinc-containing enzyme. Fetal alcohol syndrome, which is seen in infants born to alcoholic mothers, has characteristics that suggest a zinc deficiency in the fetus.

18.12 Cobalt

We obtain the cobalt that we need from dairy products and meat. A lack of cobalt in the diet results in a disease called pernicious anemia, which produces symptoms of fatigue and general weakness. This disease is not caused by a lack of hemoglobin, but rather results from a lack of erythrocytes (red blood cells that carry the hemoglobin molecules, and thus the oxygen, to the cells). Cobalt is a part of vitamin B_{12}, which is required for the formation of erythrocytes. However, too much vitamin B_{12} in the diet will stimulate the production of too many erythrocytes, producing a condition called polycythemia.

18.13 Manganese

Manganese is necessary for the functioning of several enzymes. This element is abundant in foods such as nuts, whole grain cereals, fruits, and vegetables. Manganese is found in high concentrations in the mitochondria (areas of the cell where cellular energy is produced). Manganese is also essential for normal thyroid function and for cartilage and bone growth. Manganese is required for the normal function of the brain and nervous

Figure 18.12 Both of these boys are 15 years old. The boy on the left was raised in Egypt on a diet deficient in zinc. The boy on the right grew up in Iowa on a diet with adequate zinc. (Reproduced with permission of *Nutrition Today* Magazine, Annapolis, Maryland. © March/April 1981.)

system. In fact, it has been found that as many as one third of the children who suffer from convulsive disorders such as epilepsy have low levels of manganese in their blood. But, as we have seen with other trace elements, high blood concentrations of manganese can be equally dangerous. Miners who dig for manganese often suffer headaches, psychotic behavior, and drowsiness caused by elevated levels of manganese in the blood.

18.14 Selenium

In 1857 a disease that was causing horses to lose hair and hooves was reported in North Dakota. Called "blind staggers," this disease affected other grazing animals as well, causing impaired vision, muscle weakness, and sometimes necrosis of the liver and death from respiratory failure. Not until 1930 was selenium, which is present in large amounts in the region's soils, identified as the element causing the disease. The animals suffering from blind staggers were grazing on plants that accumulated the selenium from the soils. On the other hand, by 1957 it was confirmed that animals living in regions with soils poor in selenium were suffering from other diseases. By supplementing the animals' feed with selenium, ranchers were able to prevent white muscle disease in cattle, horses, and sheep, liver disease in pigs, and pancreatic disease in chickens. It was also found that a selenium deficiency was responsible for reduced reproduction rates in sheep.

Similarly, in a region of China where the people ate only food grown on selenium-deficient soils, as many as 22,000 women and children suffered from a disease of the heart muscle called Keshan disease. This heart disease largely disappeared when the people in the area were treated with selenium salts. Studies have shown that the death rate from heart disease is higher in regions of the United States where soils are deficient in selenium. For this reason, some physicians are suggesting increasing the selenium intake from an average of 150 micrograms per day to 250–350 micrograms per day.

Selenium is part of the enzyme glutathione peroxidase, which protects cellular membranes by preventing the oxidation of lipids in the membranes. This enzyme is the body's main protection against an accumulation of hydrogen peroxide and organic peroxides in the cells. Such organic peroxides are thought to play a role in cancer development. Research has shown that in some special cases selenium can prevent cancer tumor growth; scientists are trying to determine how this occurs. Selenium also counteracts the toxicity of heavy metals such as cadmium and mercury. It causes the heavy metals to bind to high-molecular-weight proteins rather than the low-molecular-weight proteins that are more essential to life processes.

18.15 Chromium

The level of sugar in the blood is critical to the functioning of body tissues, especially the brain. This blood sugar level stays remarkably constant under normal conditions. Many factors help to control the amount of sugar in the blood; one is the compound called insulin. Insulin is secreted by the pancreas and works as a control in lowering the blood sugar level. When the pancreas fails to secrete enough insulin, the disease called diabetes mellitus results. Chromium plays a role in lowering the blood sugar level in the body by increasing the effectiveness of insulin. This may explain why chromium deficiency produces symptoms similar to diabetes mellitus. A series of recent experiments has shown that diets supplemented with inorganic chromium or high-chromium brewer's yeast can significantly lower total serum cholesterol in human subjects. In addition to brewer's yeast, whole grain cereals and liver are good sources of chromium.

18.16 Molybdenum

Molybdenum participates in the energy transfer reactions in the cell, is necessary for the function of certain intestinal enzymes, and is involved in controlling the amount of copper absorbed by the body. Dietary sources of molybdenum include plants in the pea family, cereals, organ meats, and yeast.

18.17 Silicon, Nickel, Arsenic, and Boron

Recent studies have shown that silicon, nickel, arsenic, and boron are essential to animals. Silicon is required for bone growth and connective tissue development. Nickel has been shown to be necessary for iron metabolism. Slowed growth and fetal development result when chicks, goats, pigs, and rats are fed an arsenic-deficient diet. Until 1981 boron was known to be essential for plants but not animals. Since then, boron has been shown to enhance the action of the hormone parathormone and to aid in calcium, phosphorus, and magnesium metabolism.

Trace element research is increasing our understanding of how these elements influence living organisms and the types of diseases that result from changes in their concentrations. Doctors are beginning to recognize trace elements as possible aids in treating disease, monitoring stress, and predicting heart disease. Although the interactions among trace elements are quite complex, a thorough understanding of these interactions is becoming more and more essential in our highly industrialized society. The challenge is to maintain the health of the population in a society that is constantly changing the natural levels of these, as well as other, trace elements in the soil, water, food, and air.

CHAPTER SUMMARY

There are 21 elements that are essential to life, but that together make up only 0.7% of the atoms in the human body. These elements can be divided into two groups: the macrominerals (sodium, potassium, magnesium, calcium, phosphorus, sulfur, and chlorine), required in the diet in amounts ranging from 300 mg to several grams per day, and the trace elements (iodine, iron, copper, zinc, cobalt, manganese, selenium, chromium, molybdenum, fluorine, silicon, boron, nickel, and arsenic), required in much smaller amounts each day. If the normal level of any of these elements in the body is decreased, a deficiency disease can result. These 21 elements play important roles in the control of many bodily processes: the cation/anion balance in cells and body fluids, the contraction of muscles, the transmission of nerve impulses, and the movement of body fluids. They also form a part of digestive secretions, enzymes, hormones, and proteins. The concentrations of these elements in living organisms are closely interrelated and can often have either complementary or antagonistic effects. The trace elements are essential to life in very small amounts, but many are harmful or even toxic to living organisms in more than trace amounts.

EXERCISES AND PROBLEMS

1. (a) How many elements have now been established as being essential to life?
   (b) List the seven elements belonging to the group of elements called the macrominerals.
   (c) What is the name given to the 14 elements required in only minute amounts for good health?
   (d) Describe five functions in the human body performed by elements from part (c).

2. In considering mineral elements necessary for living tissues, calcium and phosphorus are often discussed together. Why? What happens to body growth if there is an inadequate supply of these elements in the diet?

3. (a) List five functions of the calcium ion in the body.
   (b) Name three dietary sources of calcium.

4. What three factors control the amount of calcium in the blood and body fluids?

5. What steps would you take to prevent the clotting of a recently drawn blood sample? Write the equation for the reaction that would take place.

6. Write an equation for the formation of the phosphate ion from phosphoric acid.

7. What are three functions of phosphate ions in the body?

8. What effect will high levels of magnesium and calcium have on nerve function? How does this compare with the effect of high levels of sodium and potassium ions?

9. Why might a doctor inject magnesium salts into the blood of a person who is suffering from convulsions?

10. Sodium, potassium, and chlorine are three essential elements whose varied functions in the human body are intricately related. List five essential functions of these elements.

11. Why does perspiration taste salty? What are several possible harmful effects of excessive perspiration? Why do some marathon runners drink specially formulated drinks or fruit juices during and after a run? Why wouldn't water do just as well?

12. The Indians of South America applied the compound ouabain to their arrowheads to kill prey. At blood concentrations as low as 10^{-4} *M*, ouabain disrupts the regulation of the concentrations of sodium and potassium ions inside and outside the cell. Suggest a reason why this disruption could result in the death of the animal.

13. Identify the sulfur-containing functional groups in each of the following compounds:

    (a) $CH_3CH_2—S—CH_3$

    (b) 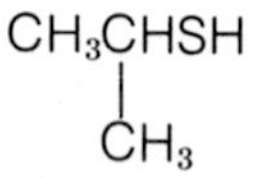 CH_3CHSH with CH_3 on the CH carbon, i.e. $CH_3CH(CH_3)SH$

    (c) $CH_3CH_2—S—S—CH_2CH_3$

    (d) 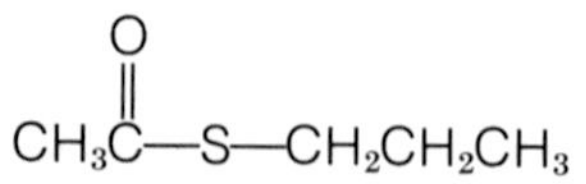 $CH_3C(=O)—S—CH_2CH_2CH_3$

14. Name the functional groups found in a molecule of methionine. What two classes of compounds found in the human body contain this amino acid?

15. Suggest a reason for the substitution of selenium for sulfur in compounds in the cell.

16. In which body tissues is fluoride important?

17. What was the cause of Colorado Brown Stain?

18. (a) What concentration of fluoride ion in drinking water has been shown to prevent dental caries without producing mottling?
    (b) Analysis of a 50-ml sample of water from a well showed a concentration of fluoride of 40 μg. What is the concentration of F^- in ppm? Would fluoride have to be added to this water supply to reach the concentration recommended by the Public Health Service?

19. In what ways can fluoride ions be both beneficial and harmful to living tissue?

20. What is the function of iodide in the human body? What condition results from too little iodide in the diet?

21. What condition results from a lack of iron in the diet?

22. What three trace elements are involved in controlling the rate of synthesis of hemoglobin?

23. Describe the functions of the following trace elements:
    (a) copper
    (b) cobalt
    (c) zinc
    (d) manganese
    (e) selenium
    (f) chromium
    (g) molybdenum

24. Geophagia (clay eating) by children often has harmful effects on their health. When the clay enters the stomach and small intestine, it draws zinc and iron ions from the food being digested. These ions are then excreted with the clay. Suggest some possible adverse health effects of geophagia.

25. It has been reported that General Custer's reinforcements at the Little Bighorn didn't arrive in time from the Dakotas because their horses were suffering from muscle weakness and vision problems. What could have caused this problem?

26. Using specific examples, explain how:
    (a) Trace elements can play complementary roles in living tissues.
    (b) Trace elements can play antagonistic roles in living tissues.

INTEGRATED PROBLEMS

1. A sample group of mothers who smoke were found to have delivered twice as many low-birth-weight infants as a similar group of mothers who did not smoke. Suggest a possible physiological reason for this finding.
2. In laboratory experiments, compounds in several commercially available products have been found to be carcinogenic in mice.
   (a) What does it mean for a compound to be carcinogenic?
   (b) Should humans continue to use such products? Why or why not?
3. A solution containing equal amounts of acetic acid (CH_3COOH) and acetate ion (CH_3COO^-) has a pH that remains quite constant when small amounts of strong acid or base are added. Explain why this happens.
4. Write the structural formula for 1,2-ethanediol. In 1985 this compound was added to some Austrian wines to increase their sweetness. Why should this have resulted in the banning of all sales and export of this wine?
5. In April 1986 all wine exports from Italy were suspended when it was discovered that methanol had been added to some of the wine to increase its alcohol content to 12%. Twenty Italians died after drinking this wine, and other consumers were blinded by the wine. How does methanol act on the body to cause blindness and death?
6. (a) Explain why an amine can act as a base in a chemical reaction.
   (b) Write the equation describing what occurs when propylamine is added to water.
   (c) Which of the following is the strongest base: aniline, dipropylamine, or propylamine?
7. Explain, on the cellular level, why high daily dietary concentrations of sodium can cause edema.
8. Women after menopause often suffer from osteoporosis. Such osteoporosis can be stabilized by the daily consumption of 50 to 100 mg of fluoride. Suggest an explanation for the effectiveness of fluoride in retarding osteoporosis.
9. In the disease bulemia, patients gorge on food and then force themselves to vomit. Why could repeated vomiting cause serious health effects from trace-element deficiency and acid–base imbalance?
10. The trace element fluorine can be absorbed into the body both as the fluoride ion and as hydrofluoric acid. Which form is mainly absorbed in the stomach and which in the small intestine? Explain your answer.

SECTION FOUR
THE COMPOUNDS OF LIFE

CHAPTER NINETEEN
CARBOHYDRATES

Sue hadn't had time to go to bed with this cold—her budget for the advertising department needed to be submitted by Monday! She knew she'd been pushing herself too hard, and now was paying for it with a fever and terrible cough. On the way home she thought she'd better stop in at the immediate care center.

As part of Sue's examination, the doctor ordered a chest X ray, sputum culture, blood count, and routine urine analysis. The chest X ray indicated no signs of pneumonia, so Sue was sent home with a supply of antibiotics, cough medicine, and some strong advice to take it easy for a few days. The next afternoon, however, Sue received a call from the doctor saying that the urine analysis had found sugar in the urine. An appointment was made for her to return for a fasting chemical profile of her blood.

By the time of Sue's appointment, she had recovered completely from her bronchitis and was feeling fine. She couldn't imagine that anything could be wrong with her (although she knew that for a woman of 50, the 225 pounds she carried on her 5 foot-5 inch frame made her excessively overweight). The doctor did a complete physical exam and then asked Sue to join him in his office. Her chemical profile, he said, showed elevated levels of glucose (325 mg%), serum cholesterol (350 mg%), triglycerides (475 mg%), and uric acid (8.2 mg%). When he had examined the retina of her eyes, he saw thickening and irregularities in the arteries. When he had felt the arteries in her feet, the pulses were not as strong as they should be. Both indicated the beginning of atherosclerotic changes unusual in a person as young as she. These findings, and the fact that both Sue's grandmother and aunt had suffered from diabetes suggested a diagnosis of noninsulin-dependent diabetes with early complications of atherosclerosis. Sue was distressed by the diagnosis (since she couldn't even pronounce it!), and wanted more information.

The doctor explained there were two forms of diabetes mellitus: insulin-dependent and noninsulin-dependent. Insulin-dependent diabetes often begins in children, adolescents, or young adults and has a rapid onset that can result in ketosis and diabetic coma. It is caused by a low level or total lack of the hormone insulin, which is critical in controlling the level of the sugar glucose in the blood. The initial treatment requires daily doses of rapid-acting insulin and a strictly controlled diet. Careful regulation of the level of blood glucose in these patients reduces the occurrence of such complications from diabetes as atherosclerosis, visual problems, coronary heart disease, strokes, and circulatory problems in the legs and feet.

The noninsulin-dependent diabetes found in Sue results from the body's inability to use the insulin that it produces. This form of diabetes usually begins later in life, commonly in people who are overweight. It can usually be controlled by a high-fiber, low-cholesterol diet with 50% to 60% carbohydrates and less than 30% total fat. The purpose of the diet is to reduce the patient's weight and to lower the blood sugar to the normal range of 80–120 mg%. If Sue's blood sugar didn't return to normal after 6 to 8 weeks of the diet, the doctor would prescribe an oral antidiabetic drug such as tolbutamide, which stimulates the beta cells of the pancreas to produce more insulin. In addition to the diet, the doctor recommended that Sue attend the hospital's diabetic training course to help her with a weight reduction and exercise program and inform her about oral antidiabetic drugs, insulin therapy, and good foot care. Luckily, Sue was able to lose weight and lower her blood glucose to normal by diet alone.

Diabetes is the most common disease associated with the class of compounds called carbohydrates. In this chapter we will study the structure, properties, and functions of carbohydrates.

LEARNING OBJECTIVES

By the time you have finished this chapter, you should be able to:

1. Define *monosaccharide, disaccharide,* and *polysaccharide.*
2. Given the structure of a monosaccharide, identify the compound as an aldose or ketose.

3. Draw the linear structure and ring structure of glucose.
4. List three hexoses and one pentose that play important roles in human metabolism.
5. Define *reducing sugar* and explain, in terms of their structure, why lactose and maltose are reducing sugars and sucrose is not.
6. Describe the difference in the structures of starch, glycogen, and cellulose and explain why we can digest starch but not cellulose.

CARBOHYDRATES

Carbohydrates are a class of compounds that includes polyhydric aldehydes, polyhydric ketones, and large molecules that can be broken down to form polyhydric aldehydes and ketones. These compounds include sugars, glycogen, starches, cellulose, dextrins, and gums. Carbohydrates are found mainly in plants, where they make up about 75% of the solid plant material. They function both as part of the structure that supports the plant and as storehouses for the plant's energy supply. The carbohydrate

Figure 19.1 This giant sequoia tree is the largest living object in the world. It is as tall as a six-story building, wider at the base than an average city street, and is estimated to be between 2500 and 3000 years old. The support structure of the tree is composed of the carbohydrate cellulose. (Courtesy National Park Service.)

cellulose is the most important component of the supporting tissue of plants (such as the wood in trees), and the carbohydrate starch is the energy storage molecule (Figure 19.1).

19.1 Classification

Carbohydrates are classified according to the size of the molecule. **Monosaccharides** are carbohydrates that cannot be broken into smaller units upon hydrolysis. **Disaccharides** will produce two monosaccharides upon hydrolysis, and **polysaccharides** will produce three or more monosaccharides upon hydrolysis (and can contain as many as 3000 monosaccharide units).

Monosaccharides, also called simple sugars, can be further classified by the number of carbons in the molecule.

3 carbons—triose	5 carbons—pentose
4 carbons—tetrose	6 carbons—hexose, etc.

Monosaccharides may also be classified by the carbonyl functional group found in the molecule.

Aldose—aldehyde functional group
Ketose—ketone functional group

For example, ribose is an aldopentose (a five-carbon sugar molecule containing an aldehyde group) and fructose is a ketohexose (a six-carbon sugar molecule containing a ketone group).

MONOSACCHARIDES

Simple monosaccharides are white crystalline solids that are highly soluble in water as a result of their polar hydroxyl groups (which means, therefore, they are not soluble in nonpolar solvents). Most monosaccharides have a sweet taste (for reasons not totally understood). The most common monosaccharides are the hexoses. Table 19.1 shows the structures of some important monosaccharides.

19.2 Glucose

Glucose (also known as blood sugar, grape sugar, and dextrose) is the most common of the hexoses. It is an aldose that is found in the juices of fruits (especially grape juice), in the saps of plants, and in the blood and tissues of animals. It is the immediate source of energy for energy-requiring cellular reactions such as tissue repair and synthesis, muscle contraction, and nerve transmission. The average adult has 5 to 6 grams of glucose in the blood (about 1 teaspoon). This much glucose will supply the energy needs of the body for only about 15 minutes, so one must continuously replace the glucose in the blood from compounds stored in the liver. The level of glucose in the blood of a normal adult is fairly constant, although it rises after each meal and falls during periods of fasting.

Glucose is a part of many polysaccharides and can be produced by the hydrolysis of these polysaccharides (it is produced commercially by the hydrolysis of cornstarch). Because glucose is found in most living cells, its chemistry is an important part of the carbohydrate chemistry of the body.

Table 19.1 Some Important Monosaccharides

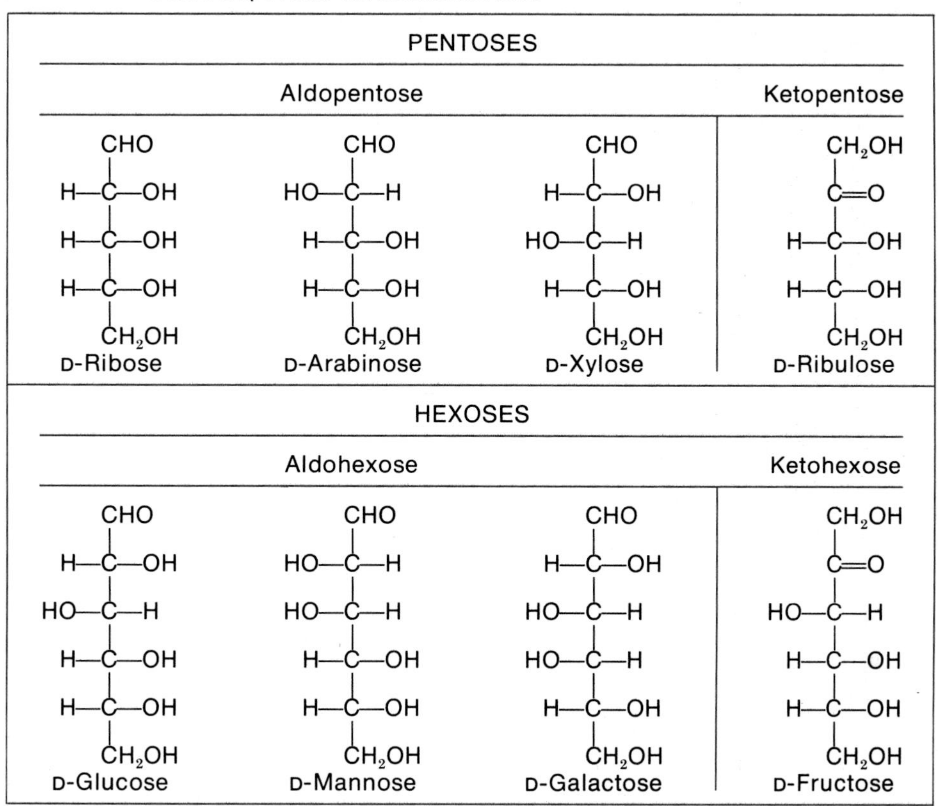

Structure of Glucose

The structure of glucose can be drawn in a straight chain form (Table 19.1). This open chain structure, however, does not explain many of the properties of glucose, which actually is found in three forms in water solution. These three forms exist in equilibrium and are easily converted one into another. The straight chain form of glucose makes up only 0.02% of these molecules. The two other forms are ring compounds that result from the formation of an internal hemiacetal (review Section 16.11). In glucose, the hemiacetal will form between the aldehyde group on carbon 1 and the alcohol group on carbon 5.

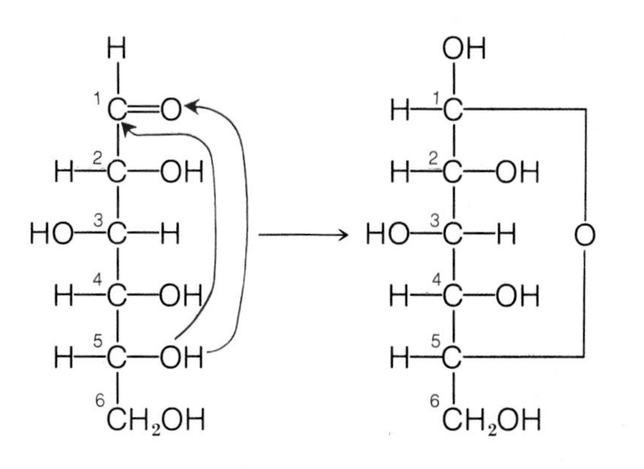

Haworth structural formulas are often used to simplify the drawing of these sugar ring structures. In these formulas the ring formed by the sugar molecule is shown as though we were looking at it from the side (rather than looking down from above the ring). The thickened side of the ring is the one closest to us, and the groups attached to the carbon atoms are then shown either above or below the ring.

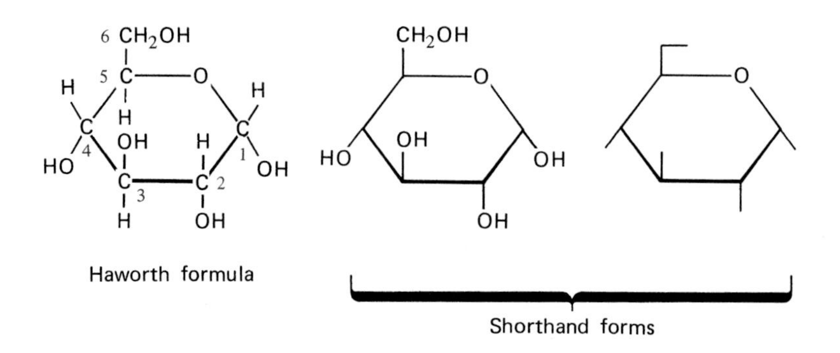

The two ring forms of glucose depend on the placement of the hydrogen and hydroxyl groups on carbon 1. If the hydroxyl group is below the plane of the ring, this is the *cis* or alpha (α) form; if it appears above the ring it is the *trans* or beta (β) form (Figure 19.2).

You may wonder what difference the position of that one hydroxyl group on the ring could possibly make. As we study the metabolism of living organisms, we will see that such small differences can determine whether or not a cell will be able to utilize a molecule. In this case, the difference between starch (a digestible glucose polymer) and cellulose (an indigestible glucose polymer) is the position of the hydroxyl group on carbon 1 of the glucose molecule.

Although Haworth formulas are easy to draw and we will use them throughout the rest of this book, they do not accurately represent the true shape of the ring. In other readings, you might instead see the ring structures drawn in a "puckered" form, as shown for glucose.

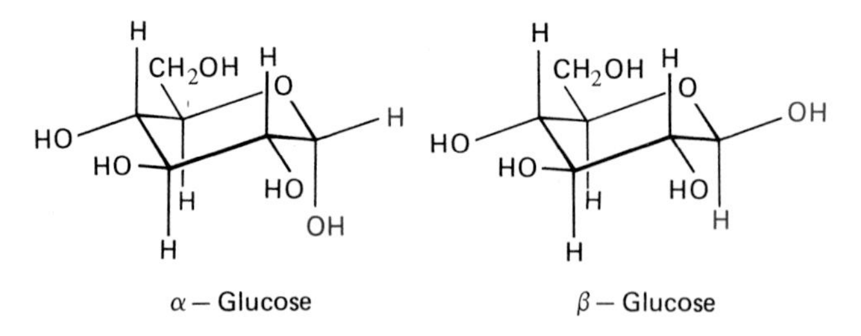

19.3 Optical Isomerism*

In Chapters 14 and 15, we discussed various forms of isomerism in compounds having the same molecular formulas. A different type of

* *This section is optional and may be omitted without loss of continuity.*

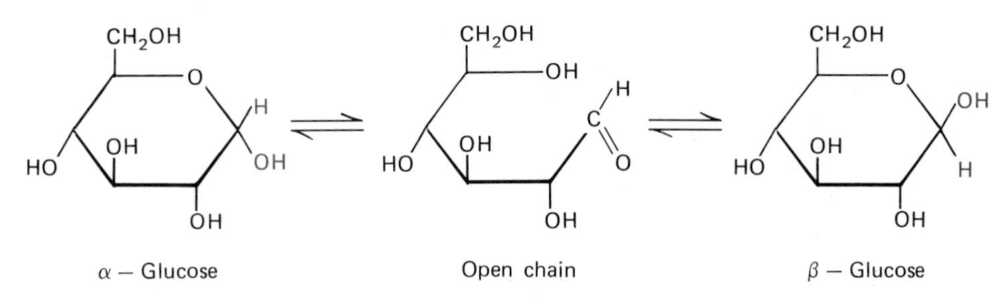

Figure 19.2 Three forms of glucose will exist in equilibrium in water solution: the alpha ring (36%), the open chain (0.02%), and the beta ring (64%).

isomerism, resulting only from the arrangement of atoms in space (not from the order in which they are arranged), is called **optical isomerism.** We have seen that the carbon atom can have four groups attached to it, each directed toward one of the corners of a tetrahedron. If these four groups are all different, the carbon atom will possess a property called **chirality** and is then called a **chiral carbon.** A chiral molecule (one containing a chiral carbon) is a molecule that cannot be superimposed on its mirror image. That is, if you were to take a three-dimensional model of a chiral molecule and a model of its mirror image, there is no way that you could exactly superimpose the two models so that all of the groups matched up (Figure 19.3). Your hand is a familiar example of a chiral or asymmetric object. Your right and left hands cannot be superimposed on one another—there is no way that a right-hand glove will fit correctly on your left hand (except by turning the glove inside out) (Figure 19.4).

What difference will a chiral carbon atom make in a molecule? If molecules are nonsuperimposable mirror images, they will have identical physical properties (such as boiling point, density, and vapor pressure) except for one. They will differ in the way they interact with polarized

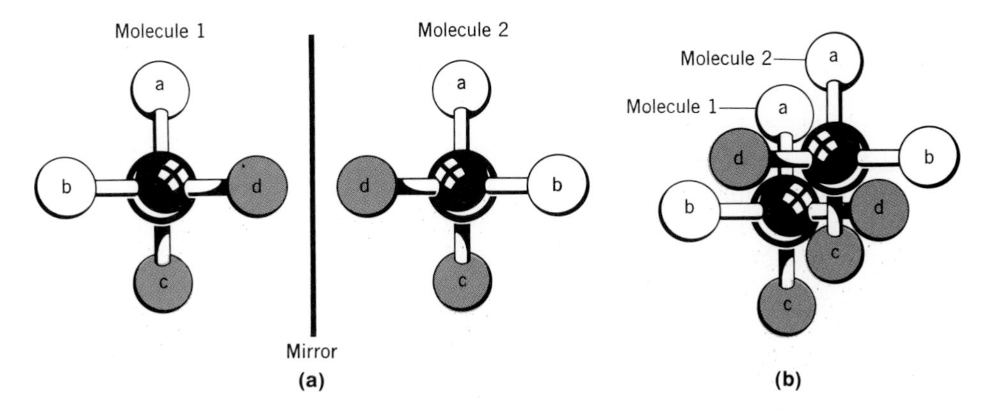

Figure 19.3 Molecules 1 and 2 are optical isomers. (a) Molecule 1 contains a chiral carbon, and molecule 2 is the mirror image of molecule 1. (b) The two molecules cannot be superimposed.

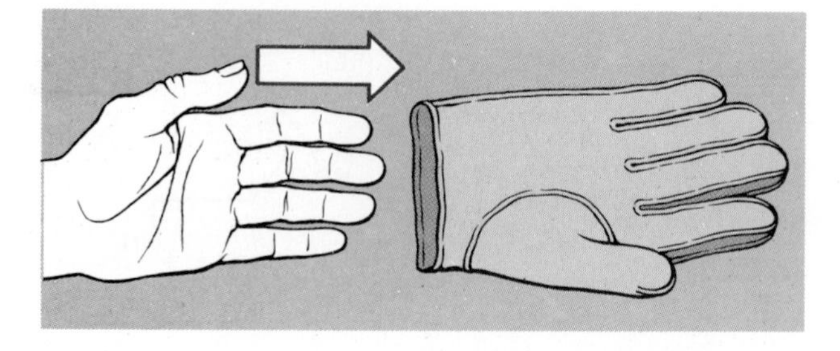

Figure 19.4 Your hand is asymmetric—there is no way that you can put a right-hand glove on your left hand (without turning the glove inside out).

light—light in which the waves vibrate in only a single plane. When placed in solution, chiral molecules will interact with polarized light by rotating the plane of vibration of such light in either a clockwise or counterclockwise direction. Molecules having this property are said to be optically active. **Optical isomers** are optically active compounds that share the same molecular formula, but which have the ability to rotate the plane of polarized light in opposite directions. Optical isomers are critical in the chemistry of the living organism.

The classification system for optical isomers is based on the three-carbon compound glyceraldehyde.

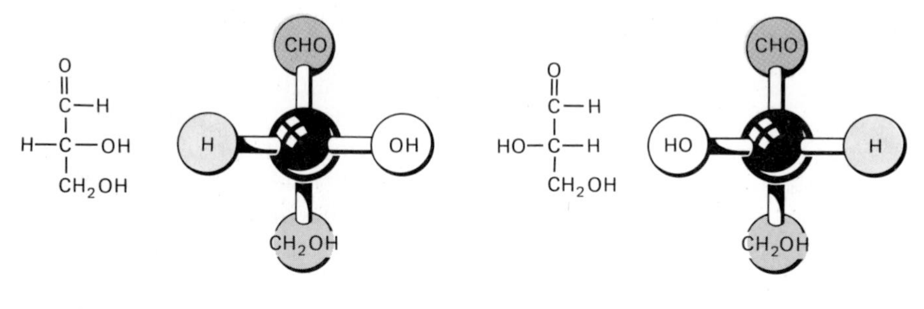

D–Glyceraldehyde L–Glyceraldehyde

The D-family of carbohydrates is represented as having the hydroxyl group located to the right of the chiral carbon farthest from the carbonyl group, and the L-family has the hydroxyl group to the left of the carbon. Most naturally occurring carbohydrates belong to the D-family (Figure 19.5). Of the 16 possible optical isomers in the aldohexoses (8 in the D-family and 8 in the L-family), the three most abundant are D-glucose, D-mannose, and D-galactose.

Again, the small difference in the arrangement of atoms between the D and L isomers may seem unimportant to you, but to your body and its cells it is critical. Cells can recognize this difference and often can use only one of the isomers. For example, yeast can ferment D-glucose to produce alcohol, but not L-glucose. As we will see in Chapter 21, our cells can use only L-amino acids to build proteins. This is because the enzymes that catalyze reactions in cells are asymmetrical compounds themselves, just as your shoes are asymmetrical. To catalyze a reaction, the enzyme must fit the reactant—just as your right shoe can fit only your right foot.

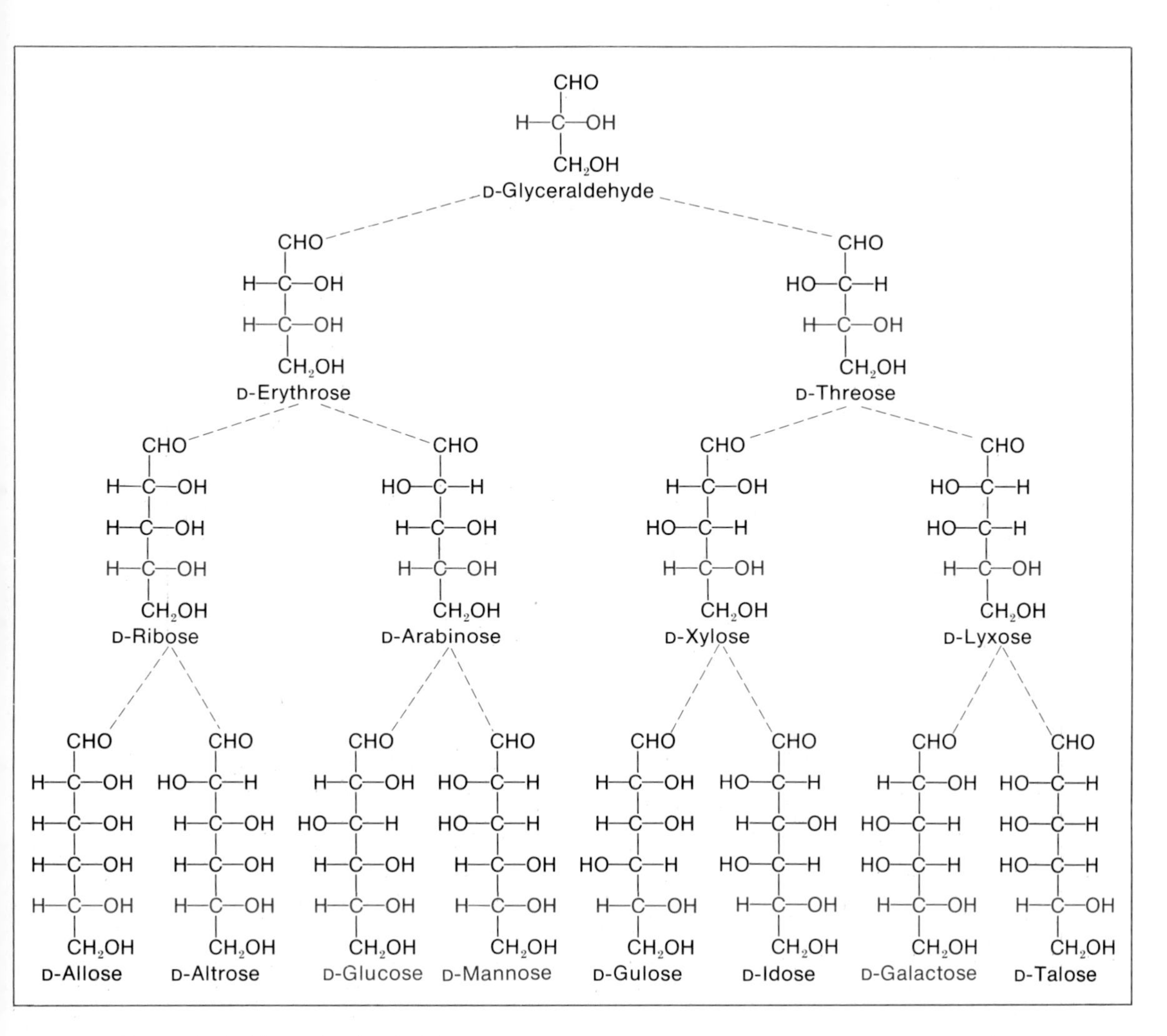

Figure 19.5 The D-family of aldoses.

19.4 Fructose

Fructose (also called levulose or fruit sugar) is a ketohexose found in many fruit juices and in honey (Table 19.1). It is the sweetest sugar known, much sweeter than table sugar. Fructose is a part of the disaccharide sucrose (or cane sugar) and is produced by hydrolysis of the polysaccharide inulin. Fructose molecules form internal hemiketals and exist in 5-member ring structures such as the following.

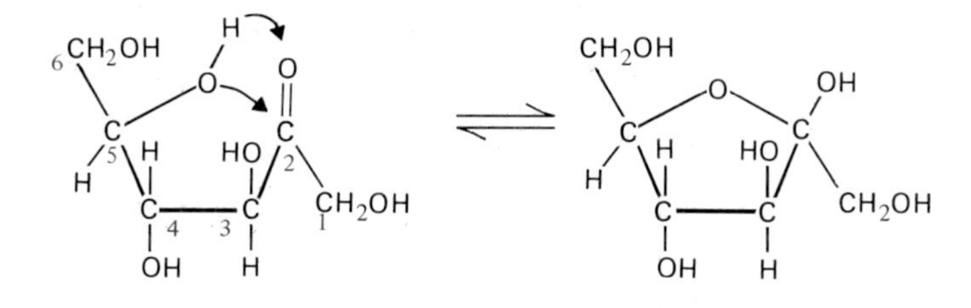

19.5 Galactose

Galactose is not found naturally as a free monosaccharide, but can be formed from the hydrolysis of larger carbohydrates. It is a part of lactose, the sugar found in milk; thus it is found in our diets from birth.

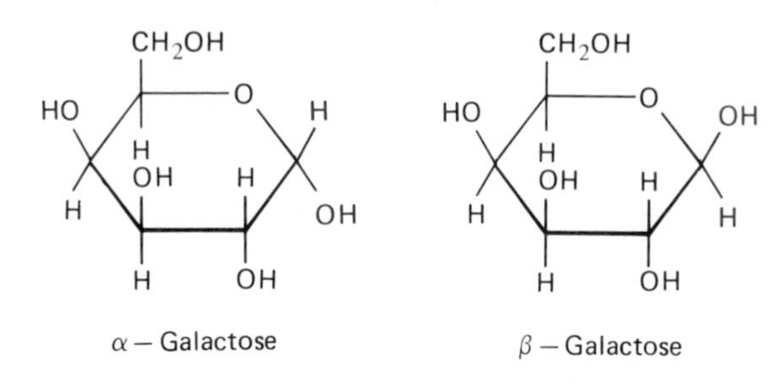

α – Galactose β – Galactose

Galactose is also found in glycolipids, fatlike substances that are components of the brain and nervous system. Agar-agar, a polysaccharide extracted from certain types of seaweed, is a polymer of galactose that humans cannot digest. Nevertheless, it is used as a thickener in sauces and ice creams, as well as in nutrient broths used in microbiology.

19.6 Pentoses

Some important five-carbon sugars are arabinose, which is formed by hydrolysis of gum arabic and the gum of a cherry tree, and xylose, which is a component of wood, straw, corncobs, and bran. (See Figure 19.5.) Pentoses that play a role in human metabolism are ribose and deoxyribose, which are parts of nucleic acids (the subject of Chapter 24). Both the alpha and beta forms of ribose exist in solution, but only the beta form is found in nucleic acids and other metabolically active compounds.

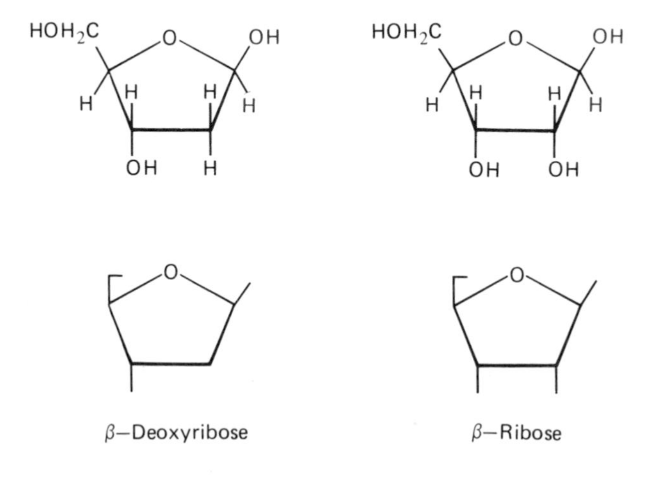

β–Deoxyribose β–Ribose

19.7 Tests for Carbohydrates

Molisch Test

This test is a general test for the presence of a carbohydrate. First, an alcoholic solution of α-napthol is mixed with the unknown solution in a test tube. Then the tube is held at an angle, and cold, concentrated sulfuric acid is poured into the tube. If a red-violet ring forms where the two solutions meet, a carbohydrate is present.

Seliwanoff's Test

This test detects the presence of a polyhydric aldehyde or ketone. The unknown solution is mixed with hot hydrochloric acid and resorcinol. Ketoses will produce a bright red color, whereas aldoses will produce a pink color in the same period of time. This test is frequently used to detect fructose, but sucrose will also give a bright red color because it is hydrolyzed during the test.

Test for Reducing Sugars

Carbohydrates that contain a free, or potentially free, aldehyde or ketone group will reduce alkaline (basic) solutions of mild oxidizing agents such as Cu^{2+} (Benedict's solution, Fehling's solution) or Ag^{+} (Tollens' solution). Benedict's solution is widely used for the detection of reducing sugars. The reagent for this test contains an alkaline solution of copper(II) sulfate, $CuSO_4$. This blue reagent solution is mixed with the unknown solution and heated. If a reducing sugar is present, the Cu^{2+} ions will be reduced to Cu^{+} ions, and a brick-red precipitate of copper(I) oxide (Cu_2O) will form. The amount of precipitate formed indicates the amount of reducing sugar present. Glucose is a reducing sugar, and if a solution of glucose is mixed with Benedict's solution, the following reaction will occur:

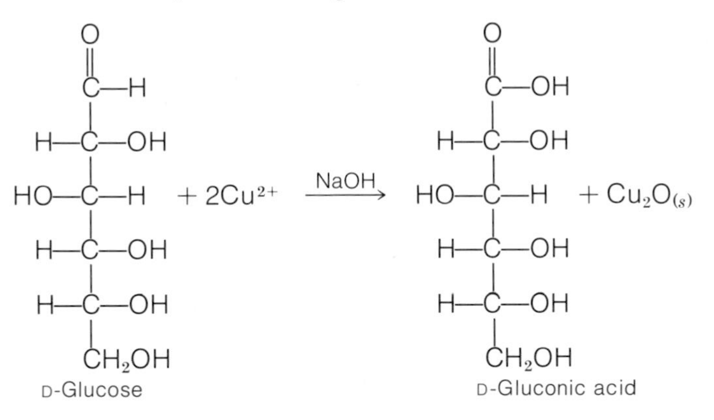

All monosaccharides will yield a positive Benedict's test. Clinitest tablets, which are widely used to test for sugar in the urine, are based on the same principle as Benedict's test. In this case, a green color indicates very little sugar in the urine, whereas a brick-red color indicates more than 2 grams of reducing sugar per 100 ml of urine.

Note, however, that a false positive Benedict's test for glucose (and an erroneous diagnosis of diabetes mellitus) can occur with patients having a rare condition called pentosuria. These persons have the ketopentose xylulose, a reducing sugar, in their urine, but suffer no ill effects from it. To guard against such an erroneous diagnosis, Tollen's pentose test, which identifies pentose sugars, can be used to determine if the sugar found in the urine is glucose or xylulose.

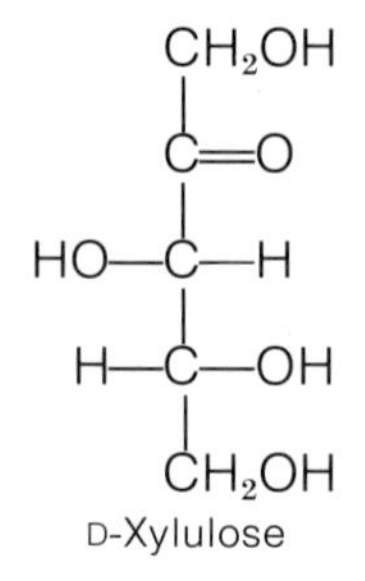

D-Xylulose

Exercise 19-1

Classify each of the following by the number of carbons in the molecule, and the carbonyl functional group. For example, glucose is an aldohexose.

(a) ribose
(b) fructose
(c) threose
(d) galactose
(e) mannose
(f) ribulose

DISACCHARIDES

19.8 Maltose

Maltose (malt sugar) is a disaccharide made up of two glucose units. It is produced by the incomplete hydrolysis of starch, glycogen, or dextrins. Maltose that is produced from grains germinated under controlled conditions is called malt and is used in the manufacture of beer.

Disaccharides are formed by a condensation reaction between two monosaccharides. This reaction involves the formation of an acetal (called a glycoside) from a hemiacetal and an alcohol.

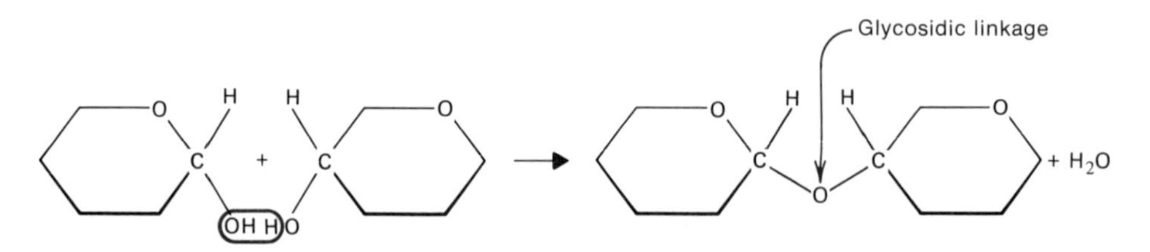

In this condensation reaction, one monosaccharide unit acts as the hemiacetal and the other as the alcohol. The linkage that is formed is called a glycosidic linkage (or acetal linkage) and is more stable than the hemiacetal. This linkage will not react with bases; only acids or specific enzymes are able to break the bond.

The glycosidic linkage in maltose occurs between carbon 1 of a glucose molecule in the alpha form and carbon 4 on the other glucose. Such a bond is called an α,1:4 linkage. (*Note:* A wavy line connecting the OH on the second glucose is used to indicate that this molecule can be in either the alpha or beta form.)

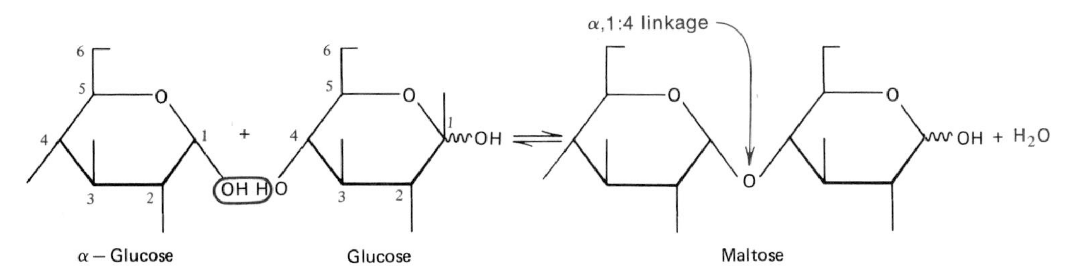

Because the aldehyde group of the second glucose molecule is not involved in the glycosidic linkage, maltose can exist in either an alpha or beta form and is a reducing sugar.

19.9 Lactose

Lactose (milk sugar) is found only in the milk of mammals. Its synthesis within the mammary glands from glucose and galactose is regulated by hormones produced after giving birth. Four to five percent of a cow's milk is lactose, whereas human milk contains 6–8%. Lactose itself is a colorless powder that is nearly tasteless. It can, therefore, be used in large amounts in special high-calorie diets.

Lactose is formed by a condensation reaction between glucose and galactose. The bond is formed between carbon 1 of galactose in the beta form and carbon 4 of glucose, resulting in a β,1:4 linkage.

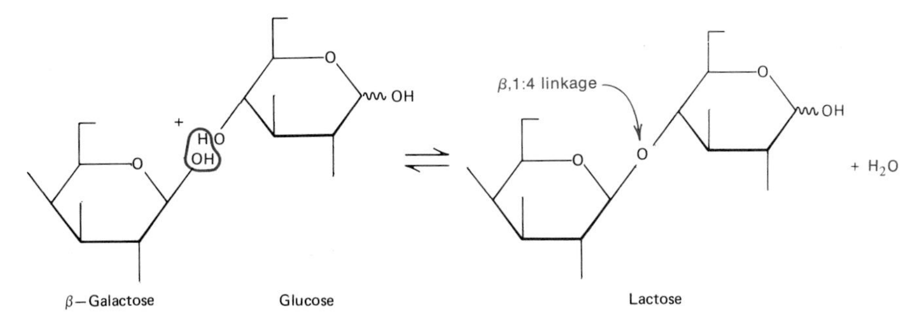

As is the case with maltose, lactose has a potentially free aldehyde group in the glucose unit and is a reducing sugar.

19.10 Sucrose

Sucrose (also called table sugar, cane sugar, and beet sugar) is found in the juices of fruits and vegetables, and in honey. It is produced commercially from sugar cane or sugar beets and is the sugar that we use in cooking. It is estimated that we each consume an average of 100 pounds of sucrose a year.

Sucrose is made up of one unit of glucose and one unit of fructose.

The linkage occurs between the aldehyde group of glucose and the ketone group of fructose.

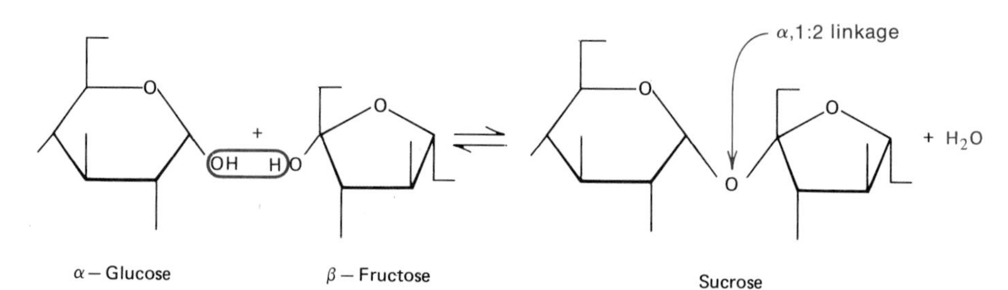

Because both the aldehyde group of glucose and the ketone group of fructose are involved in the linkage, sucrose does not have a potentially free aldehyde or ketone group and is not a reducing sugar. Sucrose can be hydrolyzed by acids or enzymes found in the intestines and yeast. Such a hydrolysis of sucrose produces a mixture of fructose and glucose called invert sugar.

Exercise 19-2

Complete the following chart:

Disaccharide	Source	Monosaccharide Subunits	Reducing Sugar?	Type of Linkage
Sucrose				
Lactose				
Maltose				

POLYSACCHARIDES

19.11 Starch

Polysaccharides (also called complex carbohydrates) are polymers containing three or more monosaccharides. They are used both as a means of storing energy and as part of the structural tissues of the organism. Starch is the storage form of glucose used by plants. It is found in granules in their leaves, roots, and seeds. These granules are insoluble in water; their coating must be broken open for the starch to mix with water. Heat will break open the granules, producing a colloidal suspension whose thickness increases with heating. For this reason, cornstarch is widely used as a thickening agent in cooking.

Natural starches are a mixture of two types of polysaccharides: amylose and amylopectin. Amylose is a large linear polysaccharide (molecular weight of 150,000 to 600,000) whose glucose units are connected by α,1:4 linkages.

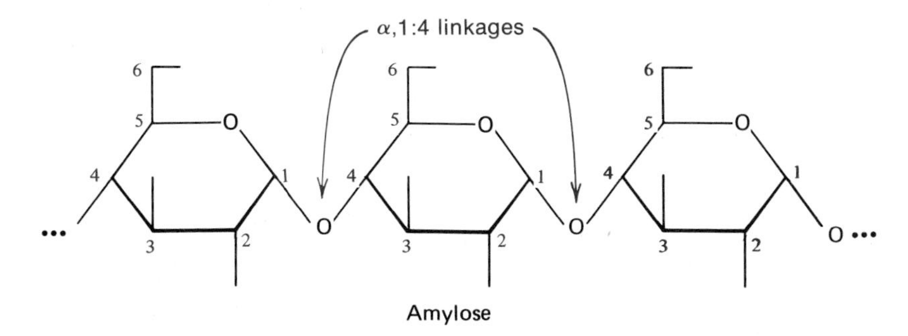

Amylopectin (molecular weight of one to six million) is a highly branched glucose polymer. The nonbranching portion of the molecule consists of glucose units connected by α,1:4 linkages. The branching occurs every 20 to 24 glucose units and is a result of α,1:6 linkages between the glucose units (Figure 19.6).

19.12 Dextrins

Dextrins are polysaccharides formed by the partial hydrolysis of starch by acids, enzymes, or dry heat. The golden color of bread crust results from the formation of dextrins. Dextrins get sticky when wet and, therefore, are used as adhesives on stamps and envelopes and in wallpaper paste.

19.13 Glycogen

Glycogen is a heavily branched molecule that is the storage form of

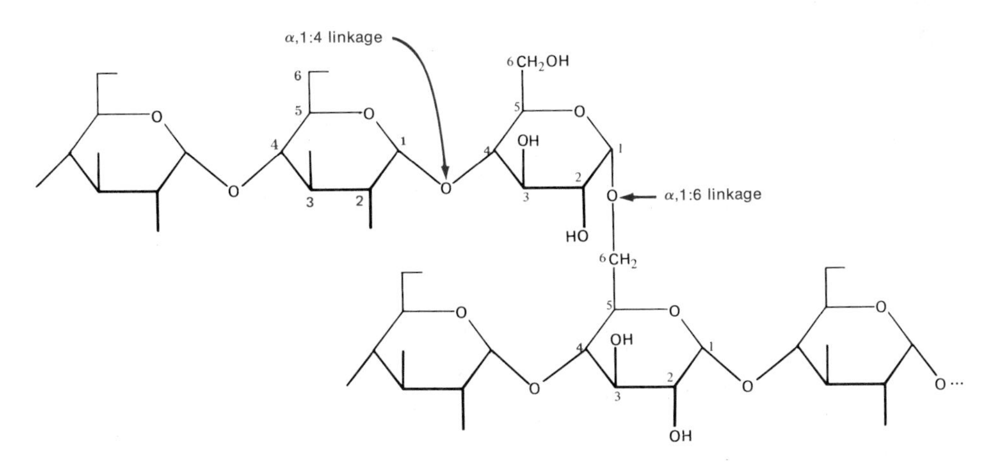

Figure 19.6 The structure of amylopectin and glycogen. The glucose units are connected by α,1:4 and α,1:6 linkages.

glucose in animals (Figure 19.7). It accounts for about 5% of the weight of the liver and 0.5% of the weight of the muscle in the body. There is enough glucose stored in the form of glycogen in a well-nourished body to supply it with energy for about 18 hours.

The structure of a glycogen molecule is similar to that of amylopectin (Figure 19.6). It consists of straight chains of glucose units connected by α,1:4 linkages. The branching that results from α,1:6 linkages between glucose units in a glycogen molecule is more frequent than in amylopectin, occurring every 8 to 12 glucose units.

The ability of the body to form glycogen from glucose (and glucose from glycogen) is extremely important because glucose is the main source of energy for all cells. When we eat a meal, glucose resulting from the breakdown of carbohydrates enters the bloodstream. If this large amount of glucose were to remain in the blood, the osmotic balance between the blood and the extracellular and intracellular fluids would be completely disrupted. This excess glucose, however, does not circulate in the blood, but instead is converted to glycogen in the liver. The large, branched glycogen molecule is ideally suited for storage because it cannot pass through cell membranes. Later, as glucose is used by the cells, the blood glucose is maintained at its normal level by the breakdown of glycogen in the liver and the resulting release of glucose into the blood. In this way, the blood glucose level remains relatively constant even though we eat at widely spaced time intervals during the day.

19.14 Cellulose

Cellulose is a glucose polymer (molecular weight from 150,000 to 1,000,000) produced by plants. It makes up the main structural support for

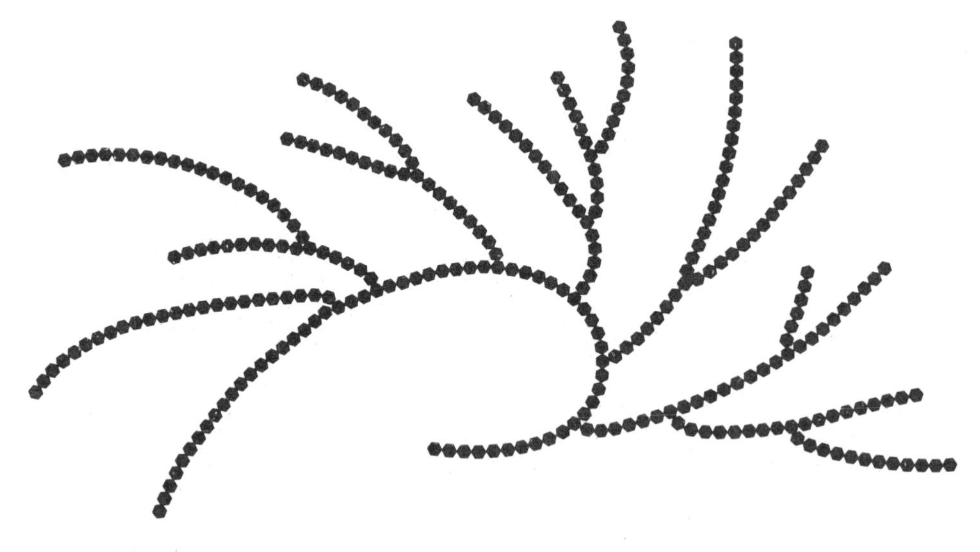

Figure 19.7 The highly branched structure of the amylopectin and glycogen molecules is a result of the α,1:6 linkages that occur about every 20 glucose units. This diagram pictures a larger portion of the molecule than shown in Figure 19.6. Each ● represents a glucose molecule.

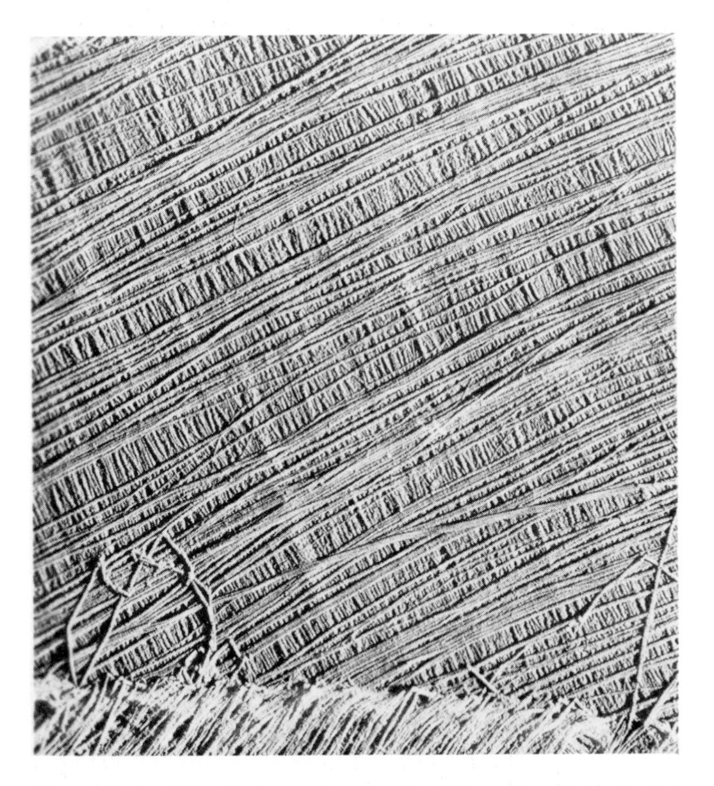

Figure 19.8 This electron micrograph of the cell wall of an alga shows the parallel arrangement of cellulose fibers making up the wall. (Courtesy R.D. Preston.)

plants, whose cells release this compound to form the exterior cell wall. Molecules of cellulose are insoluble in water because of their size and structure. The strength and rigidity that cellulose gives to plants result from hydrogen bonding between the cellulose molecules (Figure 19.8).

The glucose units in cellulose are held together by a glycosidic linkage between carbon 1 in the beta position on the first glucose and carbon 4 on the second glucose. In order to indicate that a cellulose molecule is linear, every second glucose unit in the structural diagram is flipped over.

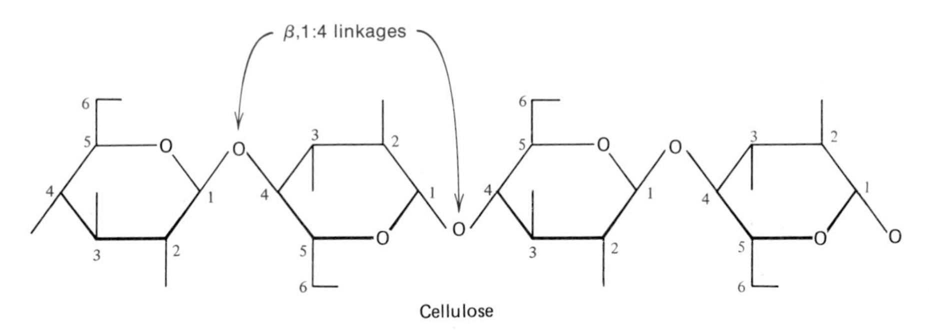

Cellulose

This linkage in cellulose is called a β,1:4 linkage, and human bodies do not possess the enzymes to break this bond. Therefore, any cellulose we eat passes through the digestive tract undigested, supplying the roughage we need for proper elimination. Some microorganisms, however, can digest cellulose. Grass-feeding animals such as cows have extra stomachs to hold the grass for long periods while these microorganisms break down the cellulose into glucose.

Over 50% of the total organic matter in the living world is cellulose. For example, wood is about 50% cellulose, and cotton is almost pure cellulose. When treated with a wide variety of chemicals, cellulose forms many useful products: celluloid; rayon; guncotton (an explosive); cellulose acetate (used in plastics, food wrapping films, and fingernail polish); methyl cellulose (used in fabric sizing, pastes, and cosmetics); and ethyl cellulose (used in plastic coatings and films).

19.15 Dextran

Dextran (molecular weight over 1,000,000) is a glucose polymer produced by bacteria. Partially hydrolyzed dextrans (molecular weight about 70,000) are used as blood plasma substitutes in the treatment of shock from low blood plasma volume. The dextrans used in this way are gradually eliminated through the urine.

In Chapter 18 we discussed the formation of plaque by bacteria in the mouth. Such plaque contains dextrans that act as glue, holding the plaque against the teeth. If the plaque is not removed, calcium compounds slowly deposit in the plaque, turning it into the hard material called tartar.

19.16 Iodine Test

The iodine test is used to detect small amounts of starch in a solution. Starch will produce a dark blue-black color when mixed with the iodine test reagent, a solution of potassium iodide containing iodine. This test can be used to monitor the hydrolysis of starch—the color will slowly change as the starch continues to be broken down into shorter carbon chain products.

Hydrolysis:	Starch	$\xrightarrow{\text{Hydrolysis}}$	Dextrins	$\xrightarrow{\text{Hydrolysis}}$	Maltose
Iodine Test:	*Blue-black*	$\longrightarrow$	*Reddish*	$\longrightarrow$	*Colorless*

19.17 Photosynthesis

We have said that polysaccharides are energy storage molecules. But where does the energy come from? Originally from the sun. Nuclear reactions occurring on the sun produce energy that radiates out into space. This radiant energy is trapped by plants growing on the earth and is used to produce carbohydrates and certain amino acids. **Photosynthesis,** the process by which plants capture and use this energy, is quite complex and not totally understood. We do know that the reactions of photosynthesis require the presence of light and molecules of chlorophyll. The reactions of photosynthesis take place in regions of the plant cell called chloroplasts

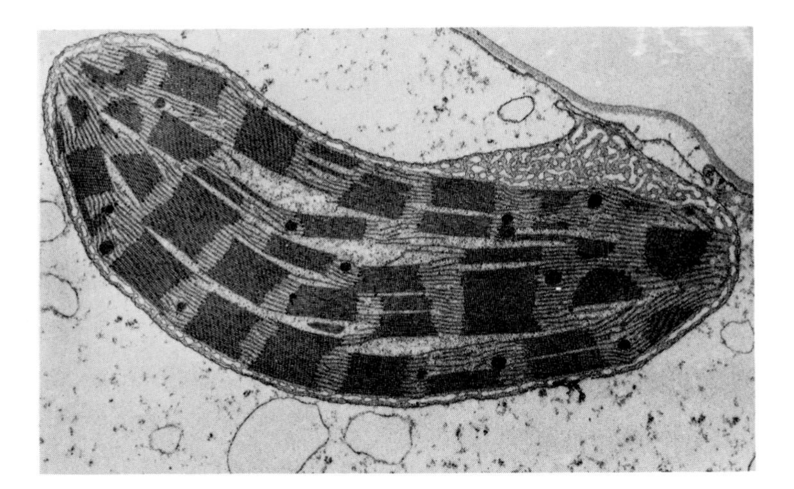

Figure 19.9 An electron micrograph of a chloroplast, the site of the light reactions of photosynthesis. The dense stacks in the chloroplasts are the grana, the main location of the chlorophyll molecules. (Courtesy T. Elliott Weier.)

(Figure 19.9). These reactions can be divided into two categories: the light reactions and the dark reactions. The light reactions require the presence of chlorophyll to absorb radiant light energy and to use this energy in the production of oxygen and energy-rich molecules. The dark reactions then use these energy-rich molecules to reduce carbon dioxide to glucose and other organic products. The overall equation, summarizing the many reactions in this process, shows the formation of glucose from carbon dioxide and water.

$$\text{Energy} + 6CO_2 + 6H_2O \longrightarrow \underset{\text{Glucose}}{C_6H_{12}O_6} + 6O_2$$

The reverse of this equation, in which cells break down glucose to carbon dioxide and water to produce the energy necessary for life, is the subject of Chapter 23.

Exercise 19-3

Identify the following carbohydrates:

(a) Polymer of glucose produced by bacteria in the mouth
(b) Glucose storage molecule in animals
(c) Product of the partial hydrolysis of starch
(d) Glucose storage molecule in plants
(e) Main structural molecule in plants
(f) End product of photosynthesis

CHAPTER SUMMARY

Carbohydrates form a large class of compounds that includes monosaccharides, disaccharides, and polysaccharides. Monosaccharides are polyhydric aldehydes or ketones that cannot be broken down into smaller units upon hydrolysis. Important examples are glucose, fructose, galactose, and ribose. Monosaccharides exist mainly as five- or six-sided ring structures that result from the formation of an internal hemiacetal or hemiketal. These rings have *cis* and *trans* isomers (α and β forms) that make a great difference in the use of the molecule by an organism. All monosaccharides are reducing sugars and will yield a positive Benedict's test.

Disaccharides produce two monosaccharides upon hydrolysis. The linkage formed between the monosaccharide units is a glycosidic linkage and is quite stable. Important examples of disaccharides are sucrose (fructose + glucose), maltose (glucose + glucose), and lactose (glucose + galactose).

Polysaccharides produce more than two monosaccharides upon hydrolysis. Polysaccharides can be linear polymers of glucose, such as amylose (glucose units connected by α,1:4 linkages) or cellulose (glucose units connected by β,1:4 linkages), or can be branched polymers of glucose such as glycogen and amylopectin (glucose units connected by α,1:4 linkages, with branches formed by α,1:6 linkages).

Important Equations

Formation of cyclic hemiacetals

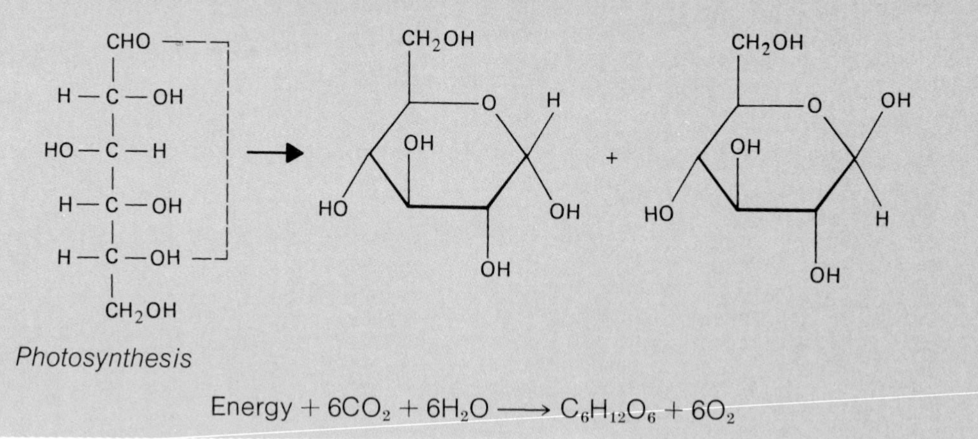

Photosynthesis

$$\text{Energy} + 6CO_2 + 6H_2O \longrightarrow C_6H_{12}O_6 + 6O_2$$

EXERCISES AND PROBLEMS

1. Describe four functions of carbohydrates in living organisms.
2. Describe the differences between monosaccharides, disaccharides, and polysaccharides.

3. Identify each of the following as (a) an aldose or ketose, and (b) a triose, tetrose, pentose, hexose, or heptose.

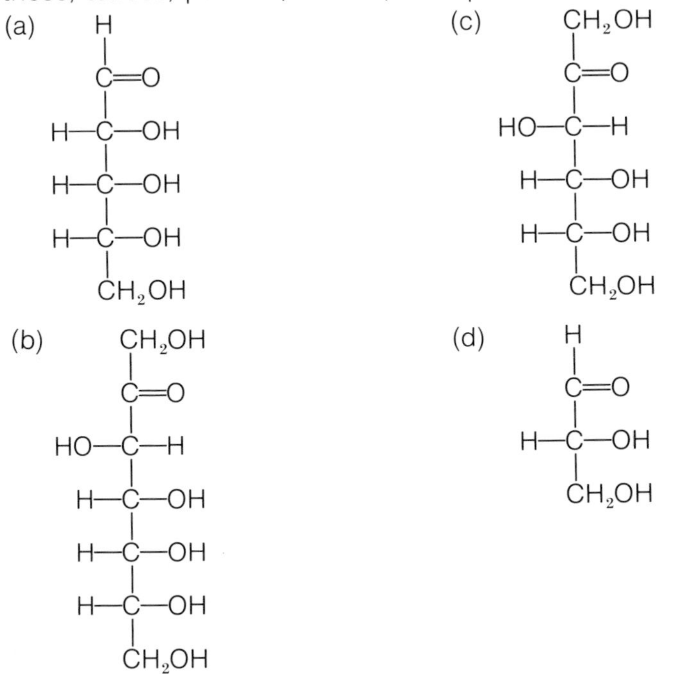

4. Draw the following:
   (a) the linear structure of glucose
   (b) the ring structure of α-glucose
   (c) the ring structure of β-glucose

5. Draw the hemiketal formed by a molecule of fructose.

6. Which of the following sugars will yield a positive Benedict's test? Give the reason for your answer.
   (a) fructose
   (b) ribose
   (c) lactose
   (d) maltose
   (e) sucrose

7. Suppose that a sample of urine from an infant gives a positive test with a Clinitest tablet. Is it correct for the analyst to report glucose in the urine? Why or why not?

8. Why is glucose administered intravenously, whereas sucrose is not?

9. What is the storage form of glucose in animals? How does its structure compare with starch?

10. Write the equation for the hydrolysis of each of the following compounds and name each of the resulting products:
    (a) sucrose (b) maltose (c) lactose

11. (a) Name the two polysaccharides making up starch.
    (b) Describe the difference in their structures.
    (c) Design an experiment for the hydrolysis of starch. How would you check the progress of the hydrolysis?

12. Given three unknown solutions labeled A, B, and C, suggest a method for determining which solution contains starch, which contains glucose, and which contains fructose.
13. Both celery and potato chips are composed of molecules that are polymers of glucose. Explain why celery is a good snack for people on a diet, whereas potato chips are not.
14. The disaccharide called melibiose is found in some plant juices.

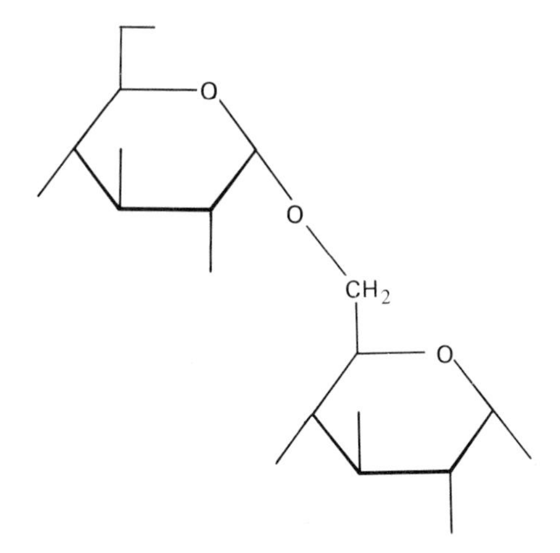

(a) What are the two monosaccharides that make up this disaccharide?
(b) Is this disaccharide a reducing sugar? Why or why not?
(c) What type of linkage connects the two monosaccharides? What polysaccharide also contains this linkage?

15. Raffinose is a trisaccharide that is widely distributed in nature:

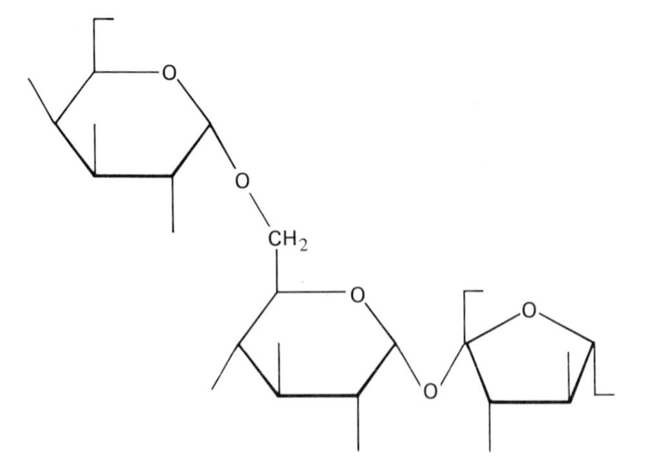

(a) Is raffinose a reducing sugar? Explain your answer.
(b) Identify the types of glycosidic linkages that exist in this trisaccharide.
(c) Identify the monosaccharides produced by the hydrolysis of raffinose, and draw the structure of each.

CHAPTER TWENTY
LIPIDS

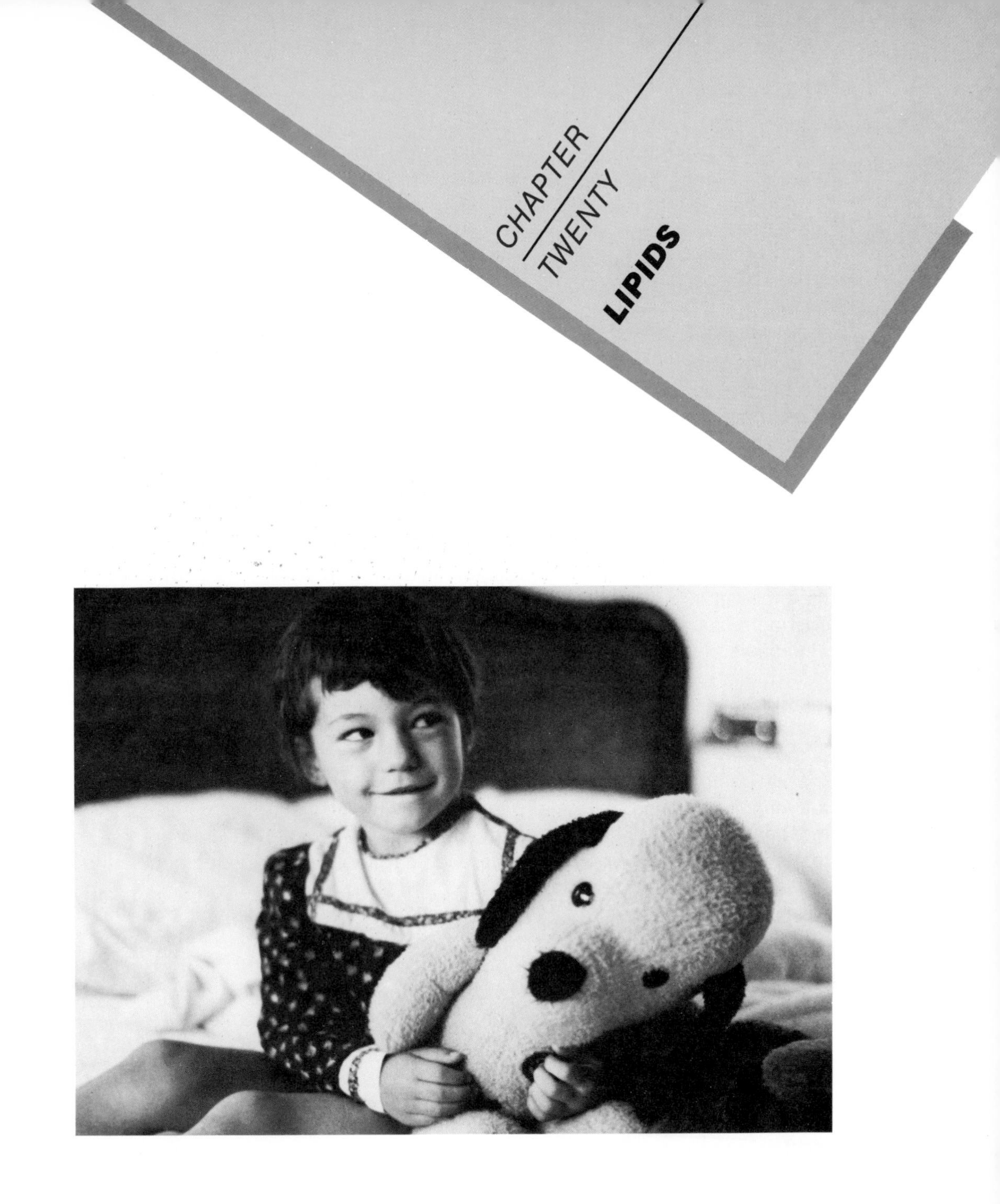

Suzy was found lying in bed. She was snuggled up with her teddy bear and the alarm on her clock radio was ringing. This would have been her sixth birthday, but the morning never came for her. An autopsy the next day revealed that Suzy's arteries were clogged like those of a sixty-year-old man. She had suffered a massive heart attack.

What would make a young child's arteries look like those of a much older person? This was the question that sent Michael Brown and Joseph Goldstein of the University of Texas Health Science Center on a thirteen-year investigation—a scientific journey that ended in Stockholm,

Sweden, where they received the 1985 Nobel Prize in Physiology or Medicine. Along the way they made new biochemical discoveries that not only gave new hope to people like Suzy, but completely revolutionized current thinking about many chemical processes in the human body.

Suzy suffered from a rare and severe form of familial hypercholesterolemia (FH), in which the level of cholesterol in the bloodstream can be as much as ten times higher than that in healthy individuals. Cholesterol is a fascinating molecule. It is absolutely essential to the body for the formation of cells and several body hormones. Yet too much cholesterol in the blood can bring about atherosclerosis, a disease in which fats in the blood (mainly cholesterol) accumulate in the walls of arteries and form bulges, or plaques. Over time such plaques can become so thick that they severely restrict the flow of blood—to the point that a blood clot can eventually block the artery completely. A heart attack occurs when such obstruction takes place in an artery feeding the heart muscle itself (see Figure 20.1).

Because cholesterol is completely insoluble in water, it is carried in the bloodstream by special particles called low-density lipoprotein, or LDL. The more LDL in the bloodstream, the greater the chance of suffering atherosclerosis. It's an unfortunate fact that more than half the people in the industrialized countries of the West have blood LDL levels putting them in the high-risk category. Brown and Goldstein, however, wanted to know why some people have much higher levels of LDL in their bloodstream than do others.

By comparing cells of patients suffering from familial hypercholesterolemia with those of normal people, the two scientists discovered that most normal body cells contain special molecules, called LDL receptors, that stick out from the cell's surface and "capture" LDL particles circulating in the blood. These LDL receptors tend to cluster on the cell wall in indented or dimpled areas called coated pits. Every few minutes these dimpled areas pouch inward and pinch off from the surface to form a sac inside the cell. The LDL particles bound to the LDL receptors in these coated pits are then released and broken down inside the cell to provide the needed cholesterol. The receptors, meanwhile, return to the cell's surface to bind to more LDL particles. Brown and Goldstein showed that the cells of people suffering from FH either have no functional LDL receptors at all or have only very few, and therefore do poorly at removing the cholesterol-carrying LDL from the bloodstream. Individuals having the most severe form of FH, as was the case with Suzy, have virtually no working LDL receptors at all.

But the work of these two scientists went far beyond explaining the cause and developing a treatment for familial hypercholesterolemia. Brown and Goldstein went on to show that cells have a complex feedback mechanism that regulates the amount of cholesterol they contain. As the cholesterol level in the cell rises, the cell stops synthesizing cholesterol and produces fewer LDL receptors to decrease the cholesterol taken in from the blood. Based on this finding, other investigators were able to show that this "receptor" method of incorporating LDL into the body's cells is also used for a variety of other large molecules (such as insulin) that are vital to body function.

The investigative work of Brown and Goldstein has been remarkable in its use of methods from many areas of science: genetics, medicine, cell biology, molecular biology, biochemistry, pharmacology, nuclear medicine, and immunology. But their work is just representative of the investigations of a very large number of scientists who are using their knowledge of chemistry to discover new facts about living organisms.

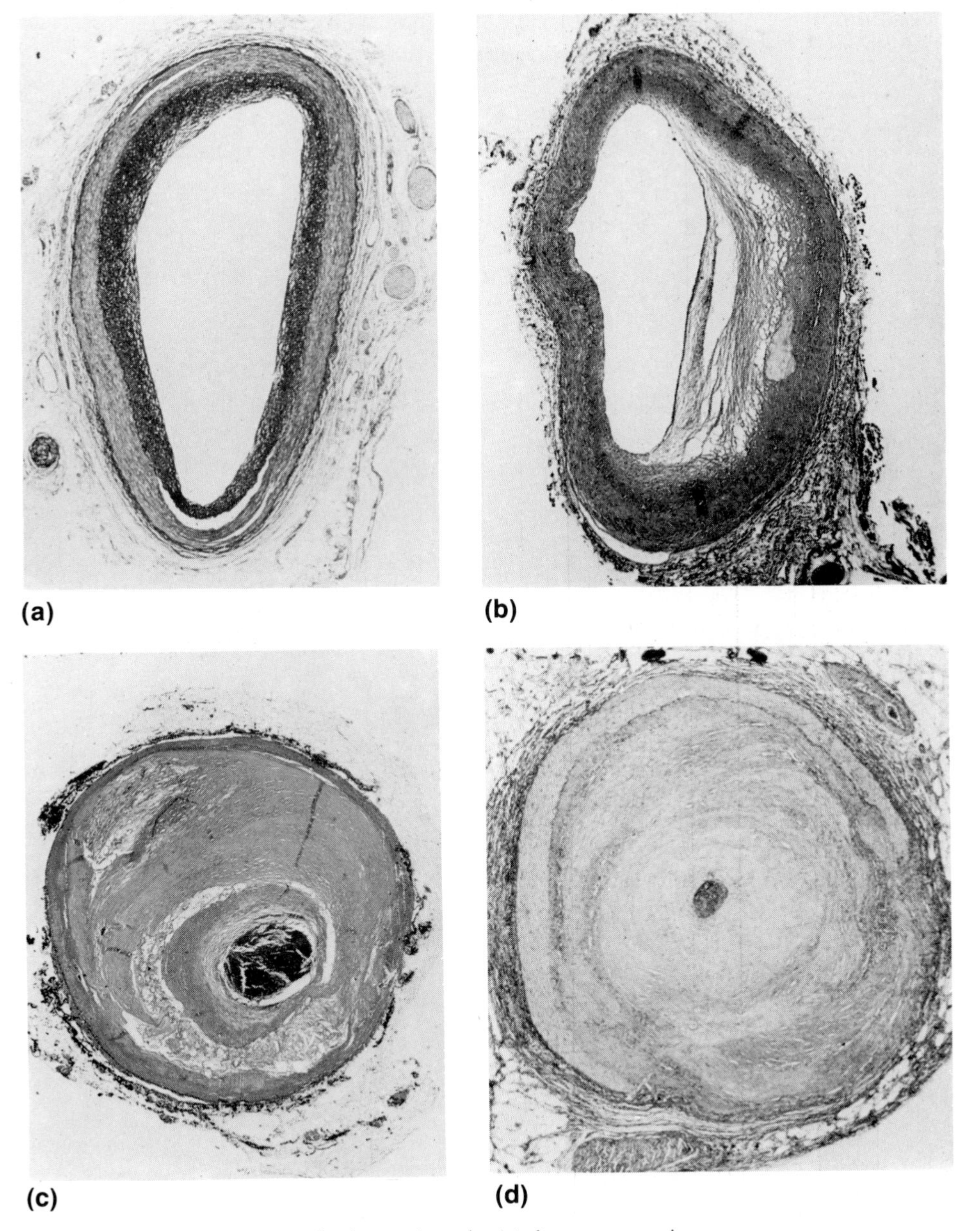

Figure 20.1 The progress of atherosclerosis. (a) A near-normal artery. (b) Plaque forms on the inner lining of the artery. (c) The narrowed channel within the artery is blocked by a blood clot. (d) Atherosclerosis has progressed to the point that this artery is completely blocked. [(a) and (d), Courtesy National Institutes of Health, National Heart & Lung Institute; (b) and (c), Courtesy American Heart Association.]

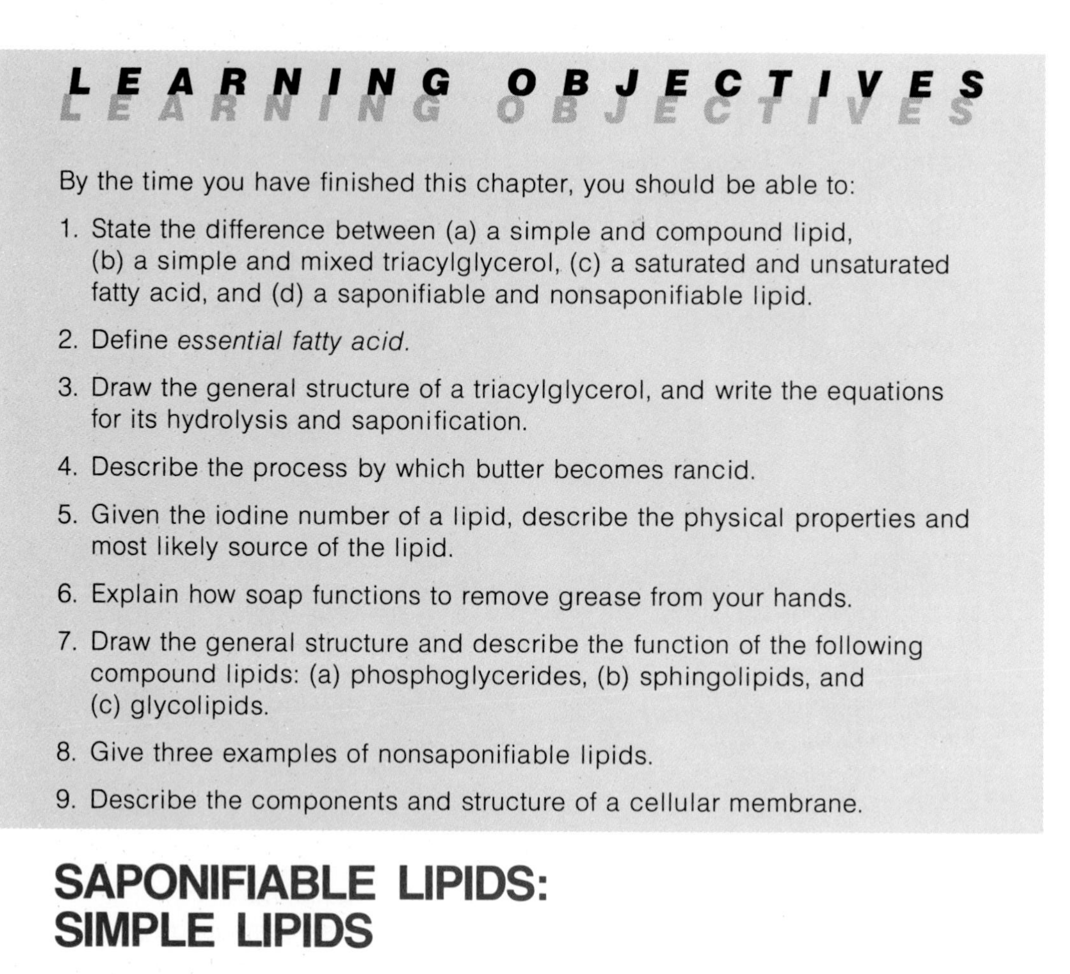

LEARNING OBJECTIVES

By the time you have finished this chapter, you should be able to:

1. State the difference between (a) a simple and compound lipid, (b) a simple and mixed triacylglycerol, (c) a saturated and unsaturated fatty acid, and (d) a saponifiable and nonsaponifiable lipid.
2. Define *essential fatty acid.*
3. Draw the general structure of a triacylglycerol, and write the equations for its hydrolysis and saponification.
4. Describe the process by which butter becomes rancid.
5. Given the iodine number of a lipid, describe the physical properties and most likely source of the lipid.
6. Explain how soap functions to remove grease from your hands.
7. Draw the general structure and describe the function of the following compound lipids: (a) phosphoglycerides, (b) sphingolipids, and (c) glycolipids.
8. Give three examples of nonsaponifiable lipids.
9. Describe the components and structure of a cellular membrane.

SAPONIFIABLE LIPIDS: SIMPLE LIPIDS

20.1 What Are Lipids?

The cholesterol and fats that make up the plaques so important in atherosclerosis belong to a class of compounds called lipids. **Lipids** are substances that contain long-chain hydrocarbon groups in their molecules and are present in, or derived from, living organisms. Lipids are insoluble in water, but are soluble in organic solvents such as chloroform, methanol, ether, or benzene. In living cells they form part of the structure of membranes, store energy for the cell, and are the compounds from which the cell makes prostaglandins.

Table 20.1 The Lipids

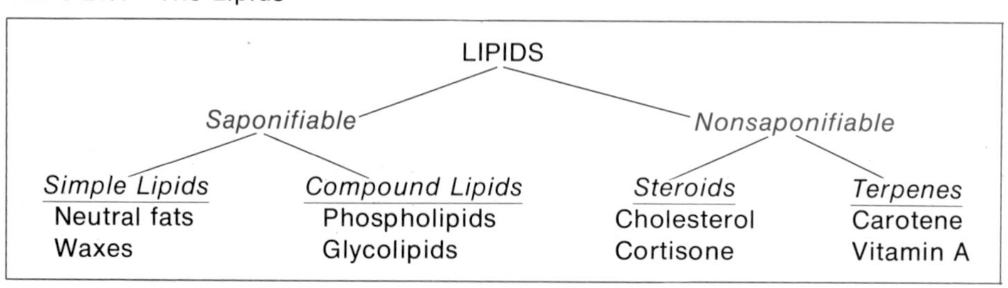

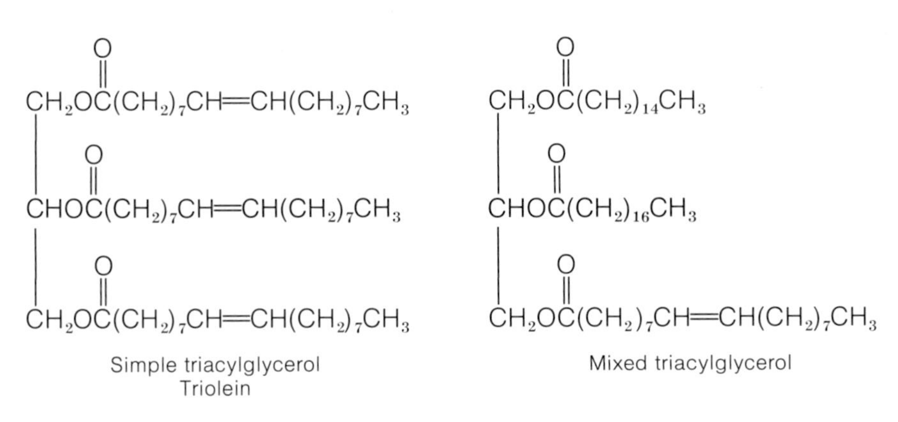

Figure 20.2 Simple triacylglycerols, such as triolein, contain only one type of fatty acid. Mixed triacylglycerols contain two or more different fatty acids.

Lipids are a varied group of compounds that can be categorized in several ways. We can divide the lipids into two major classes: those that can be saponified (hydrolyzed by a base) and those that are nonsaponifiable. The saponifiable lipids can be further subdivided into **simple lipids,** which yield fatty acids and an alcohol upon hydrolysis, and **compound lipids,** which yield fatty acids, alcohol, and some other compounds upon hydrolysis (Table 20.1).

20.2 Fats and Oils

The simplest and most abundant of the lipids are the neutral fats, which are also called **triacylglycerols** or triglycerides.* These compounds are esters of glycerol and three fatty acids, and are the main form of fat storage in plants and in the adipose cells (or fat cells) of animals.

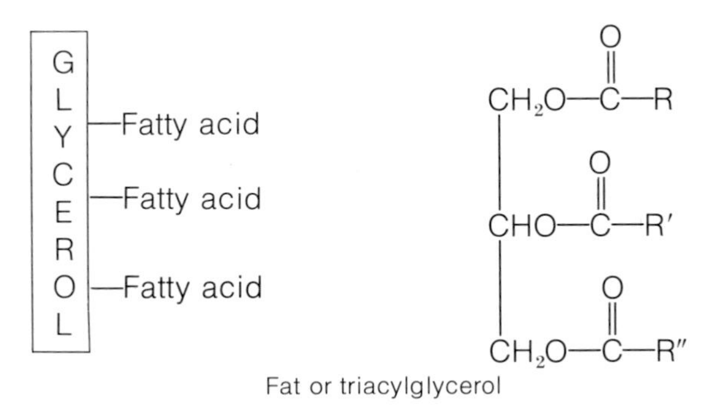

Fat or triacylglycerol

Simple triacylglycerols contain the same fatty acid in all three positions on the glycerol molecule, whereas **mixed triacylglycerols** contain two or more different fatty acids. Natural fats are a mixture of simple and mixed triacylglycerols (Figure 20.2). Triacylglycerols and their

---

* *The IUPAC nomenclature committee has recommended that the commonly used (but chemically inaccurate) term* triglyceride *be replaced by the term* triacylglycerol.

derivatives have very wide commercial use in the production of soaps, varnishes, oilcloth, linoleum, printing inks, ointments, and creams.

20.3 Fatty Acids

Fatty acids are long-chain carboxylic acids that are formed from the hydrolysis of triacylglycerols. The naturally occurring fatty acids have an even number of carbon atoms and are generally nonbranching. The most common fatty acids have 16 or 18 carbon atoms in their chains. These fatty acids are palmitic, stearic, and oleic acids, with oleic acid making up more than half the total fatty acid content of many fats. **Saturated** fatty acids have only carbon-to-carbon single bonds, are unreactive, and are waxy solids at room temperature. **Unsaturated** fatty acids have one or more carbon-to-carbon double bonds and are liquids at room temperature (Figure 20.3 and Table 20.2).

The difference between fats and oils is the number of unsaturated fatty acids that are present. Animal fats, lard, tallow, and butter are mixed fats containing more saturated fatty acids than unsaturated fatty acids. They are waxy, white solids at room temperature. Vegetable oils, olive oil, corn oil, and cottonseed oil contain a higher concentration of unsaturated fatty acids and are liquids at room temperature (Table 20.3).

20.4 Essential Fatty Acids

Our bodies can produce saturated fatty acids, and unsaturated fatty acids containing one double bond. We cannot, however, produce linoleic or linolenic acids, which are called the **essential fatty acids.** Infants lacking these fatty acids in their diets will lose weight and develop eczema. Arachidonic acid is also an essential fatty acid. However, it has been recently shown that animals can synthesize arachidonic acid from linoleic acid, but not in sufficient amounts to supply the body's daily needs.

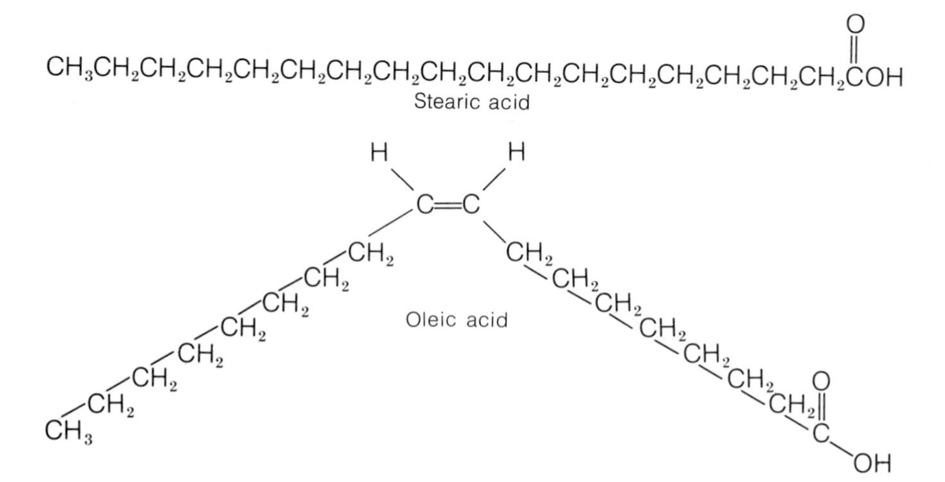

Figure 20.3 Stearic and oleic acids are both fatty acids with 18 carbon atoms. Oleic acid is unsaturated; it contains one double bond in the *cis* configuration, which gives the molecule a rigid bend.

Table 20.2 Some Common Fatty Acids

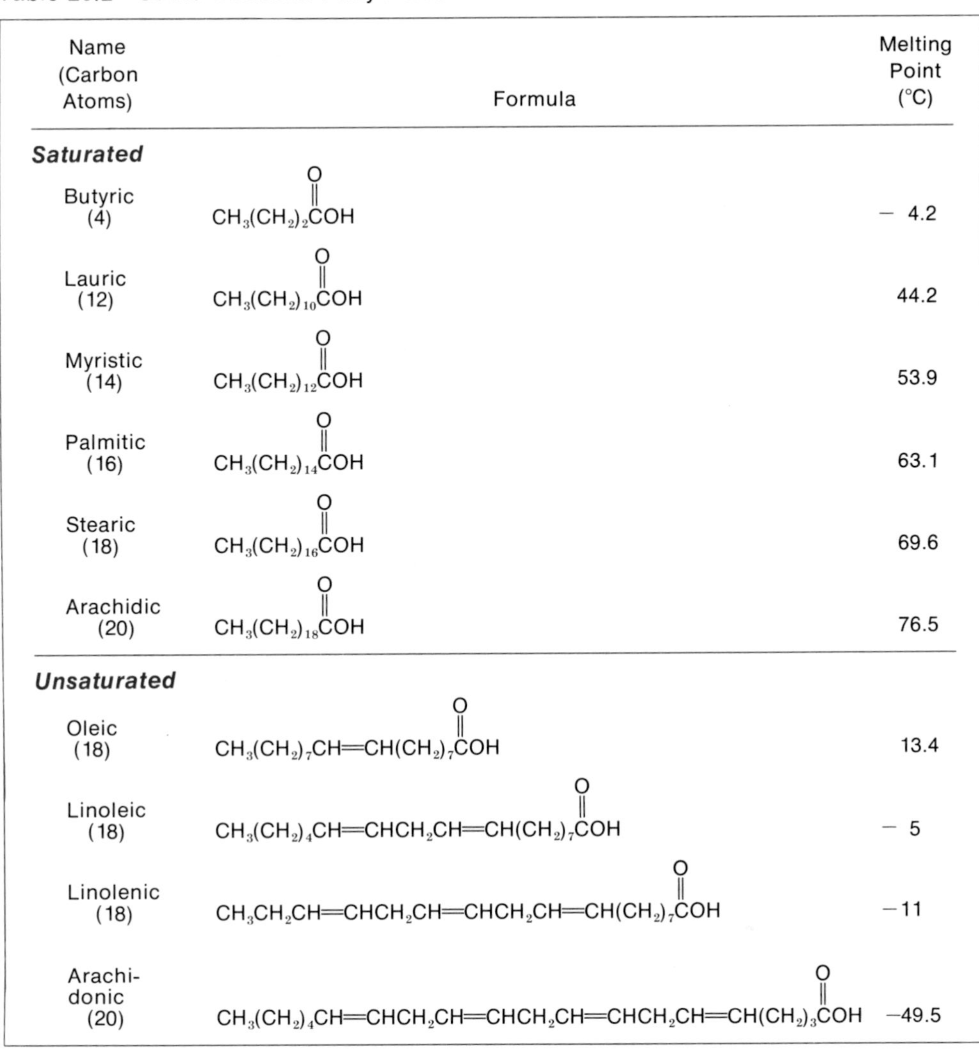

Name (Carbon Atoms)	Formula	Melting Point (°C)
Saturated		
Butyric (4)	$CH_3(CH_2)_2\overset{\overset{O}{\|}}{C}OH$	− 4.2
Lauric (12)	$CH_3(CH_2)_{10}\overset{\overset{O}{\|}}{C}OH$	44.2
Myristic (14)	$CH_3(CH_2)_{12}\overset{\overset{O}{\|}}{C}OH$	53.9
Palmitic (16)	$CH_3(CH_2)_{14}\overset{\overset{O}{\|}}{C}OH$	63.1
Stearic (18)	$CH_3(CH_2)_{16}\overset{\overset{O}{\|}}{C}OH$	69.6
Arachidic (20)	$CH_3(CH_2)_{18}\overset{\overset{O}{\|}}{C}OH$	76.5
Unsaturated		
Oleic (18)	$CH_3(CH_2)_7CH{=}CH(CH_2)_7\overset{\overset{O}{\|}}{C}OH$	13.4
Linoleic (18)	$CH_3(CH_2)_4CH{=}CHCH_2CH{=}CH(CH_2)_7\overset{\overset{O}{\|}}{C}OH$	− 5
Linolenic (18)	$CH_3CH_2CH{=}CHCH_2CH{=}CHCH_2CH{=}CH(CH_2)_7\overset{\overset{O}{\|}}{C}OH$	−11
Arachidonic (20)	$CH_3(CH_2)_4CH{=}CHCH_2CH{=}CHCH_2CH{=}CHCH_2CH{=}CH(CH_2)_3\overset{\overset{O}{\|}}{C}OH$	−49.5

Linoleic, linolenic, and arachidonic acids are used by the body to synthesize prostaglandins, compounds that are found in most mammalian tissues and which have a wide range of physiological effects (see Section 22.15). Prostaglandins play a role in the body's defenses against many sorts of change; in particular, they are powerful inducers of fever and inflammation. Aspirin seems to function as an antipyretic (fever reducer) by regulating the production of prostaglandins in the temperature-regulating cells of the brain.

20.5 Waxes

Waxes are esters of long-chain fatty acids and long-chain monohydric alcohols (alcohols with one hydroxyl group). For example, beeswax is

Table 20.3 Some Common Fats and Oils and Their Fatty Acid Composition[a]

	Melting Point °C	Percent Composition of the Most Abundant Fatty Acids								Iodine Number
		Saturated				Unsaturated				
Animal fats		Myristic	Palmitic	Stearic	Arachidic	Palmitoleic	Oleic	Linoleic	Linolenic	
Butter	32	11	29	9	2	5	27	4	—	36
Lard	30	1	28	12	—	3	48	6	—	59
Tallow	N/A	6	27	14	—	—	50	3	—	50
Human fat	15	3	24	8	—	5	47	10	—	68
Plant oils										
Corn	−20	1	10	3	—	2	50	34	—	123
Cottonseed	− 1	1	23	1	1	2	23	48	—	106
Linseed	−24	—	6	2	1	—	19	24	47	179
Olive	− 6	—	7	2	—	—	84	5	—	81
Peanut	3	—	8	3	2	—	56	26	—	93
Safflower	N/A	←	7		→	—	19	70	3	145
Soybean	−16	—	10	2	—	—	29	51	6	130

[a] Values in this table are averages. Extreme variation may occur in the values depending upon the source, treatment, and age of the fat or oil.

largely an ester of myricyl alcohol ($C_{30}H_{61}OH$) and palmitic acid. Waxes form protective coatings on skin, fur, feathers, leaves, and fruits. They have properties of water insolubility, flexibility, and nonreactivity that make them excellent coatings. Commercially produced waxes are used in cosmetics, floor waxes, furniture and car polishes, ointments, and creams (Table 20.4).

Table 20.4 Some Common Waxes

Name	Melting Point °C	Source	Uses
Beeswax	61–69	Honeycomb	Candles, polishes
Carnauba	83–86	Carnauba Palm	Floor waxes, polishes
Lanolin	36–43	Wool	Cosmetics, skin ointments
Spermaceti	42–50	Sperm whale	Cosmetics, candles

Exercise 20-1

1. Use your own words to define each of the following terms:
   (a) simple and compound lipid
   (b) essential fatty acid
   (c) saturated and unsaturated fatty acid
   (d) wax
   (e) triacylglycerol
2. Write the structural formula for a triacylglycerol formed between glycerol, palmitic acid, stearic acid, and oleic acid. Is this a simple or mixed triacylglycerol?

CHEMICAL PROPERTIES OF SIMPLE LIPIDS

20.6 Iodine Number

The unsaturated bonds in a fatty acid will react to add iodine, giving us a useful tool for determining unsaturation. Thus, chemists have defined the iodine number of a simple lipid to be the number of grams of iodine that will react with 100 grams of fat or oil. The higher the iodine number, the more unsaturated the fat. Fats generally have an iodine number below 70, and oils have an iodine number above 70 (Table 20.3).

20.7 Hydrogenation

Oils can be converted to solid fats by hydrogenation, the addition of hydrogen to the double bonds of the molecule in the presence of a catalyst (see page 351). For example, vegetable shortenings and margarines are commercially produced by the partial hydrogenation of soybean, corn, or cottonseed oil. (The complete hydrogenation of these oils would produce a hard, brittle product.) Natural vegetable oils contain only the *cis* isomers of fatty acids, but in the hydrogenation process a mixture of *cis* and *trans* isomers is produced. The effects on human health from large amounts of *trans* isomers of fatty acids in the diet are largely unknown and this is an area of very active research.

20.8 Hydrolysis

The hydrolysis of fats can occur in the presence of superheated steam, hot mineral acids, or specific enzymes. Hydrolysis under such conditions produces glycerol and three fatty acids.

In general:

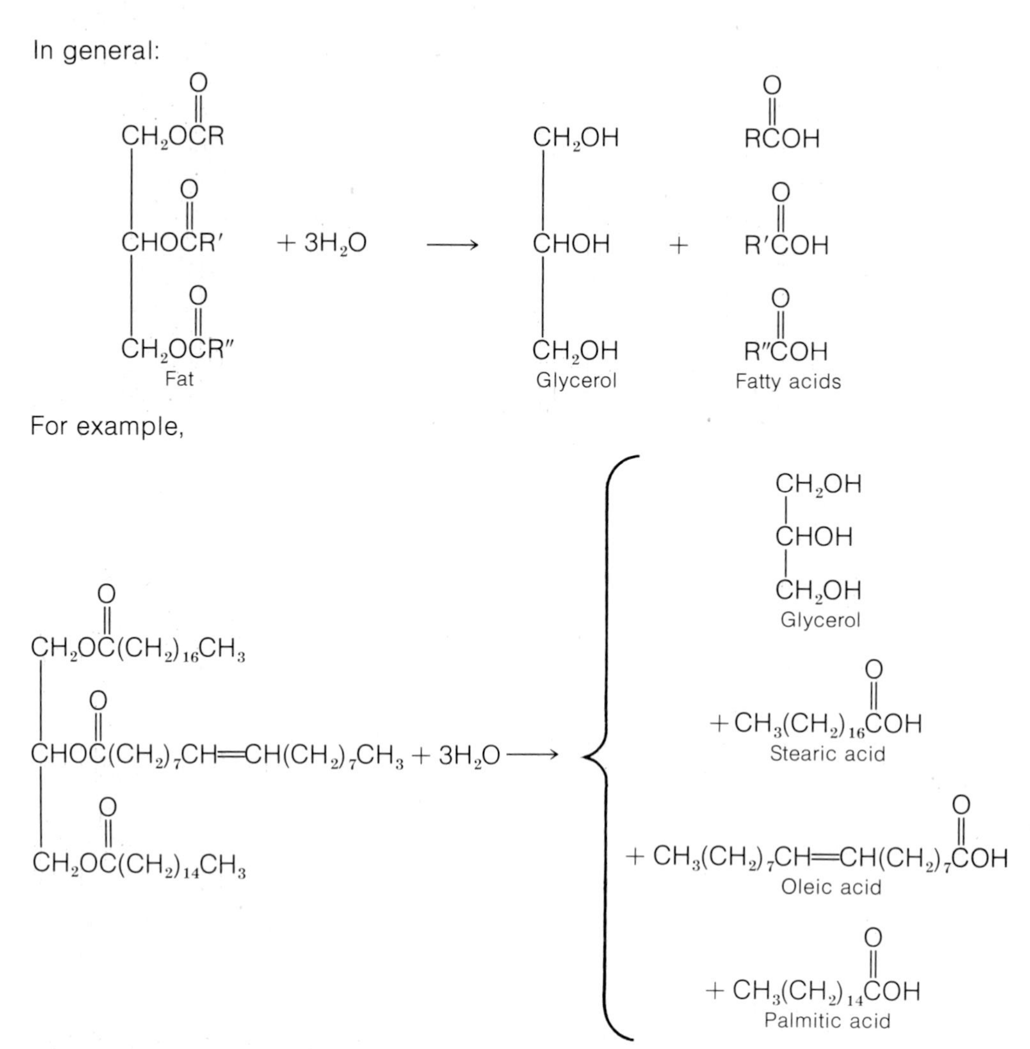

20.9 Formation of Acrolein

When heated to high temperatures, fats will hydrolyze. The glycerol that is produced will react to form acrolein, whose vapors are irritating to the nose and eyes. Acrolein is responsible for the unpleasant odor you recognize when oil or fat is burned. Acrolein is also irritating to the digestive tract and may be responsible for the stomach upset caused occasionally by deep-fried foods.

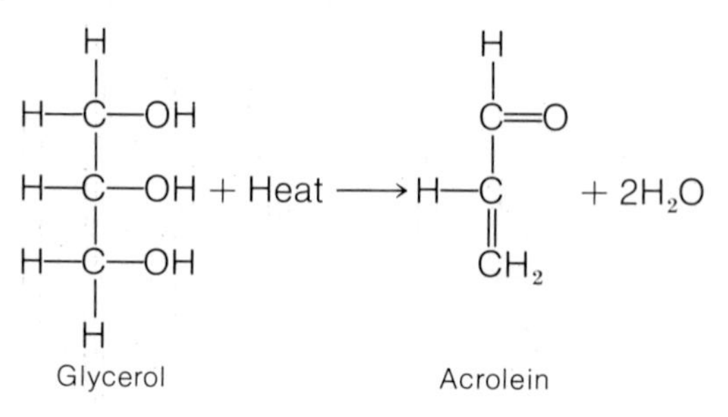

20.10 Rancidity

Fats and oils often develop a disagreeable odor and taste and are then called rancid. There are two causes of rancidity: hydrolysis and oxidation. For example, when left too long at room temperature, butter will become rancid. This occurs because some of the fat in the butter will undergo hydrolysis, accelerated by enzymes produced by microorganisms in the air. This hydrolysis produces the fatty acid butyric acid, which causes the odor of rancid butter. Also, oxygen in the air can oxidize unsaturated fats or oils to produce short-chain acids or aldehydes with disagreeable odors and tastes. Such oxidation is slowed in manufactured products such as crackers, potato chips, and pastries by adding chemicals called antioxidants (such as BHA and BHT).

20.11 Saponification

When hydrolysis is carried out in the presence of a strong base such as sodium hydroxide, glycerol and the sodium salts of the fatty acids are produced. In general:

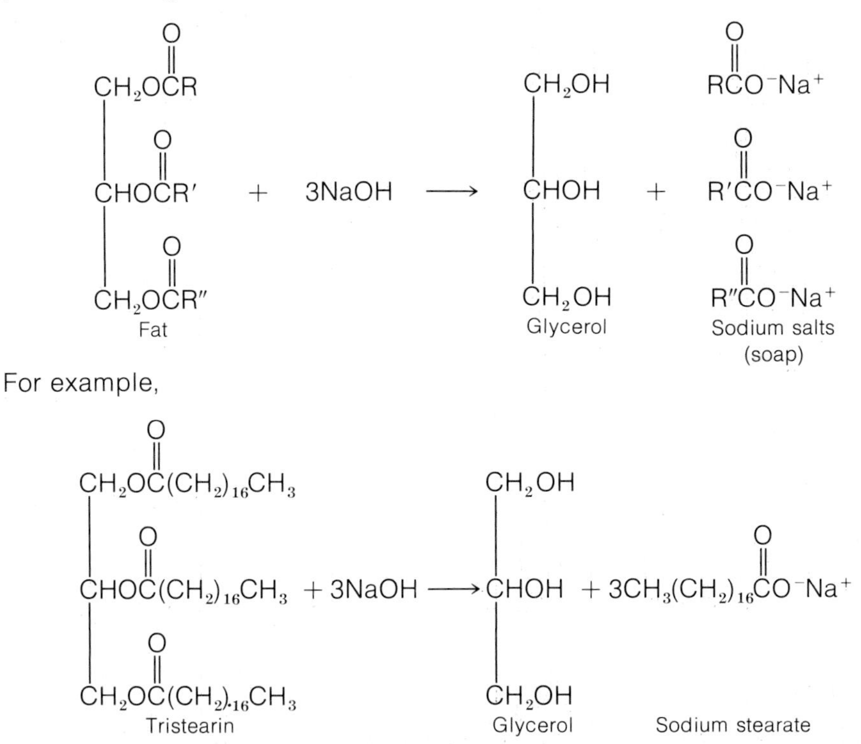

Such salts are called **soaps.** Sodium salts of fatty acids are used in bar soaps, and potassium salts are used in liquid soaps. A hard soap contains a larger number of saturated fatty acids than a soft soap. Various additives are also used to give commercial soaps their colors and scents. In addition, floating soaps contain air bubbles, and scouring soaps contain abrasives.

Cleansing Action of Soap

Water does a poor job removing grease and oil because water molecules tend to stick together rather than penetrate the nonpolar grease. Soap greatly improves the cleansing power of water. A soap molecule has two portions: a nonpolar "tail" formed by the hydrocarbon chain and a polar "head" formed by the carboxyl group.

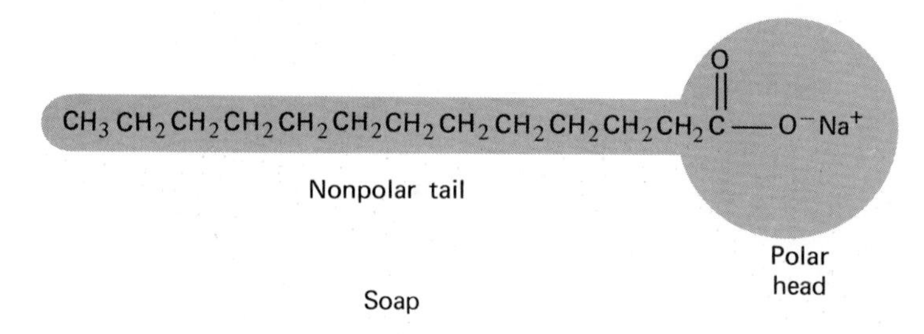

The nonpolar tail will readily dissolve in the nonpolar grease, whereas the polar head tends to remain dissolved in the water. In this way the soap breaks up the grease into small colloidal droplets; that is, it emulsifies the grease, which can then be washed away (Figure 20.4).

The cleansing power of soap is affected by several factors. For example, hard water contains one or more of the metallic ions Ca^{2+}, Mg^{2+}, Fe^{2+}, or Fe^{3+}, which form insoluble salts with soap. These salts precipitate out (forming soap scum and bathtub ring), leaving less soap in the water to do the cleaning. Water softeners work by replacing such metallic ions with other ions, such as sodium, which do not interfere with the action of soap. Lowering the pH of the water will also decrease the cleansing action of soap by neutralizing the charge on the fatty acid ion.

Detergents, which are mixtures of sodium salts of sulfuric acid esters, have cleaning properties similar to soaps, but have important advantages (Figure 20.5). Their calcium and magnesium salts are water soluble, and they are not affected by pH. However, some early synthetic detergents

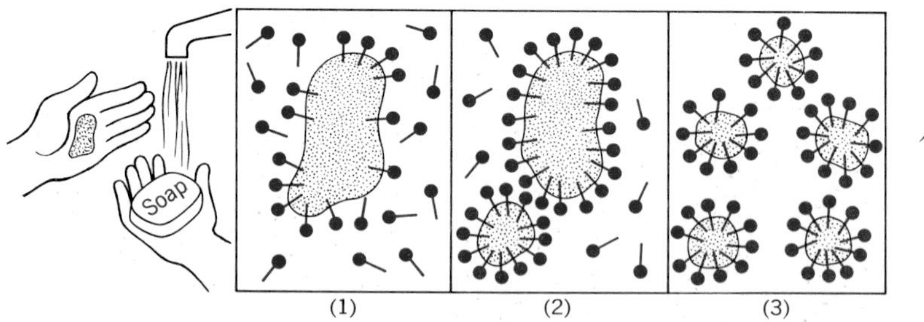

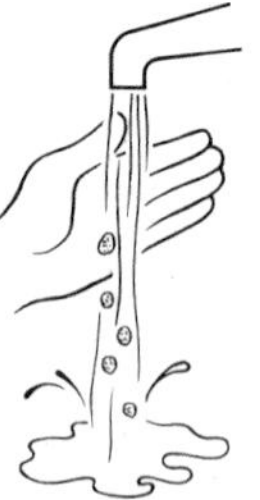

Figure 20.4 The cleansing action of soap. (1) The nonpolar tails of the soap molecules begin to dissolve in the nonpolar grease. (2) As the soap dissolves in the grease, small colloidal grease particles break off and are surrounded by the negatively charged polar heads of the soap molecules. This keeps the grease particles in solution by preventing them from reforming into large droplets. (3) In this manner, the grease can be completely broken up, and the colloidal droplets washed away with water.

Figure 20.5 Two organic salts that possess cleansing properties. Each has a nonpolar, hydrophobic (water-repelling) tail and a polar, hydrophilic (water-attracting) head.

containing branched chains in their hydrocarbon tails could not be naturally broken down by bacteria in sewage (that is, they were not biodegradable) in the same manner as soaps. Therefore, these products could not be removed from the water by sewage treatment plants. This resulted in rivers and streams being covered with foam and suds, creating an alarming pollution problem. Newer detergents are partially or totally biodegradable, and the suds problem has been eliminated.

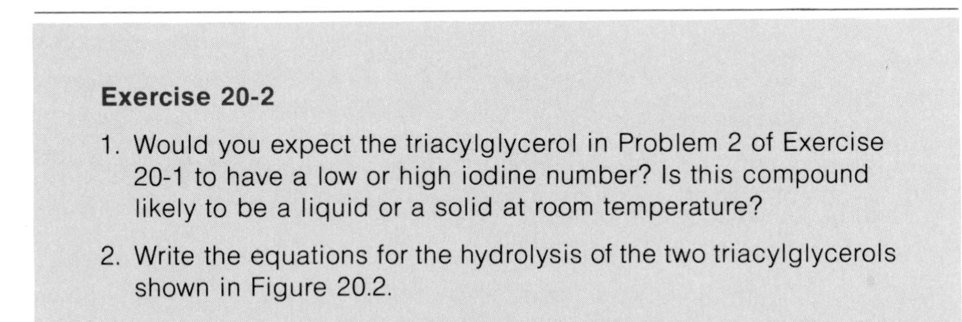

Exercise 20-2

1. Would you expect the triacylglycerol in Problem 2 of Exercise 20-1 to have a low or high iodine number? Is this compound likely to be a liquid or a solid at room temperature?
2. Write the equations for the hydrolysis of the two triacylglycerols shown in Figure 20.2.
3. Write the equations for the saponification by sodium hydroxide of the two triacylglycerols shown in Figure 20.2.

SAPONIFIABLE LIPIDS: COMPOUND LIPIDS

20.12 Glycerol-Based Phospholipids: Phosphoglycerides

Phospholipids are a class of waxy solids that form part of the structure of cell membranes and are important in the transport of lipids in the body. They can be divided into two general categories: glycerol-based phospholipids and sphingosine-based phospholipids.

The glycerol-based phospholipids, or **phosphoglycerides,** are derivatives of phosphatidic acid. They contain glycerol, two fatty acids, a phosphate group, and a nitrogen-containing compound that can be choline, ethanolamine, serine, or inositol.

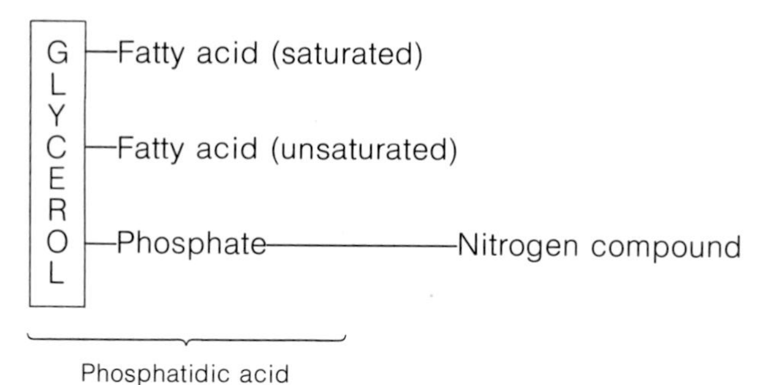

The phosphate group forms a polar, hydrophilic (water-attracting) head on the molecule, and the two fatty acids form two nonpolar, hydrophobic (water-repelling) tails.

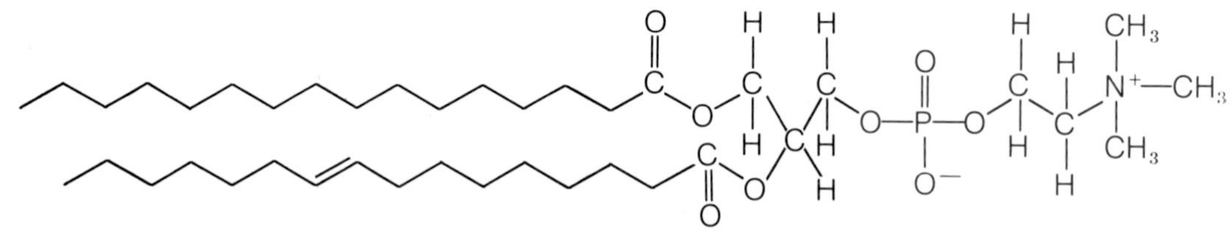

This arrangement gives phospholipids good emulsifying properties and good membrane forming properties.

Phosphatidylcholine (Lecithin)

Phosphatidylcholine (PC), which used to be called lecithin, is a phosphoglyceride in which the nitrogen compound is choline.

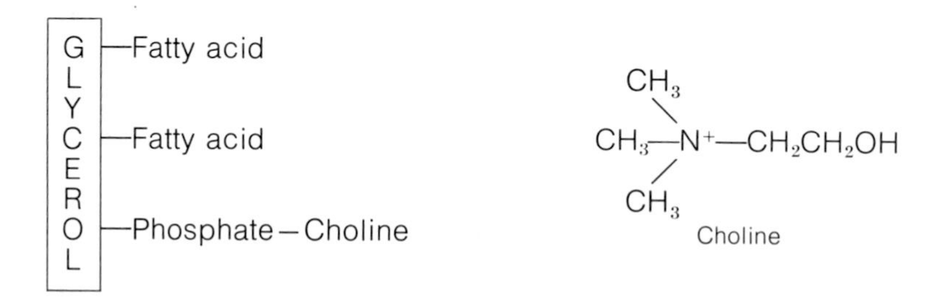

PC plays an important role in the metabolism of fats in the liver and in the transport of fats from one part of the body to another. It serves as a source of inorganic phosphate for tissue formation and is an excellent emulsifying agent. PC finds commercial use as an emulsifying agent in such products as chocolate candies, margarine, and medicines. Egg yolks contain a large amount of PC and are used to emulsify salad oil and vinegar to make mayonnaise. The removal of one fatty acid from PC forms lysolecithin, a compound that causes the destruction of red blood cells and spasmodic muscle contractions. The venom of poisonous snakes contains enzymes to catalyze the formation of lysolecithin from PC.

Premature babies with birth weights of 2 to 3 pounds often suffer from respiratory distress syndrome, or hyaline membrane disease. These tiny infants have lungs that do not yet function properly, mostly because of a lack of dipalmitoyl phosphatidylcholine (DPPC). In the lungs DPPC acts as a surfactant in the air sacs or alveoli, reducing the high surface tension of the water in the alveoli, which otherwise would cause them to collapse. When infants suffering from hyaline membrane disease exhale, their alveoli collapse, preventing oxygen from passing from the lungs into the bloodstream (Figure 20.6). Babies with severe cases of this disease must be put on a ventilator, which forces high concentrations of oxygen into their lungs. If, after 1 or 2 days, these infants do not start producing enough DPPC to be able to breathe on their own, they may suffer blindness from exposure to such high levels of oxygen. After 4 days without producing

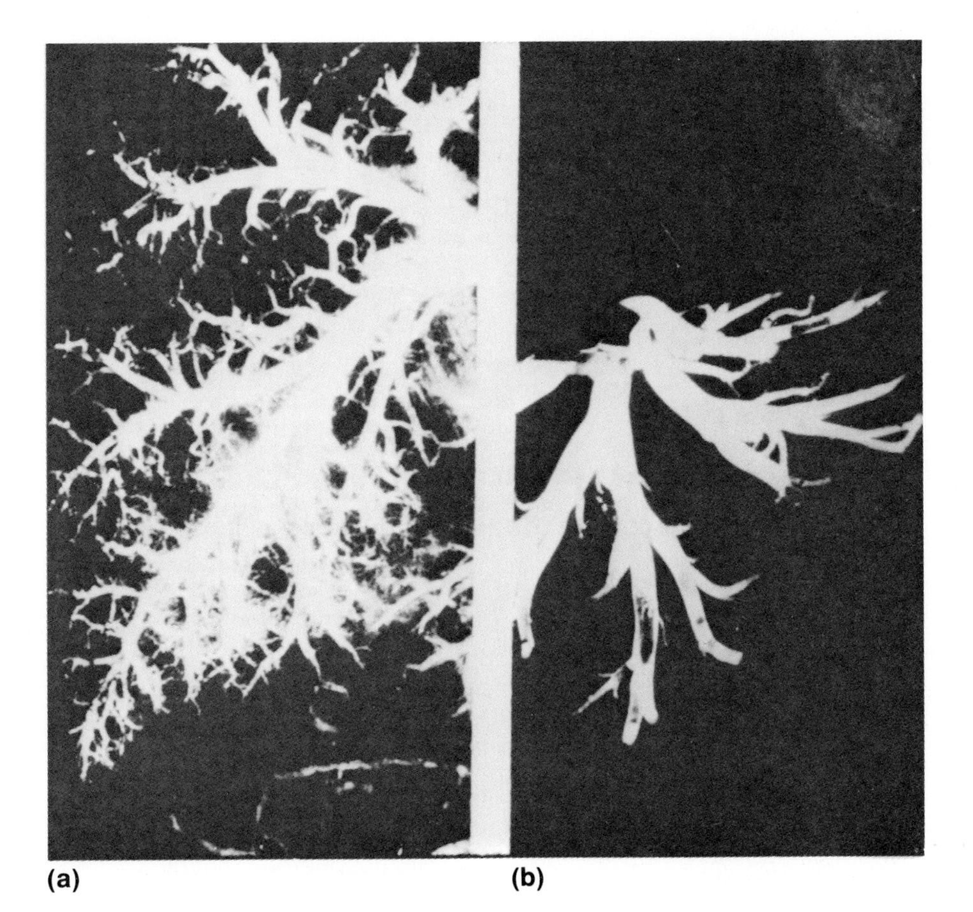

Figure 20.6 (a) A normal lung of an infant. (b) The lung of an infant who died from RDS, respiratory distress syndrome. (Courtesy University of Virginia Neonatal Intensive Care Unit.)

their own DPPC, their chances of recovery are very slim. One treatment used for this disease is to spray DPPC, dissolved in a substance that will help it stick to the lining of the alveoli, into the air of the ventilator. This helps lower the surface tension in the alveoli, allowing the concentration of oxygen being forced into the lungs to be lowered and, therefore, reducing the risk of permanent damage. This same treatment is also currently being used for drowning victims when inhaled water has washed away the surfactant in the alveoli, and for victims of scorched lungs when hot gases have melted the surfactant.

Phosphatidylethanolamine (Cephalins)

Phosphatidylethanolamine (PE), which used to be called cephalin, is a phosphoglyceride in which the nitrogen compound is ethanolamine, $H_2NCH_2CH_2OH$.

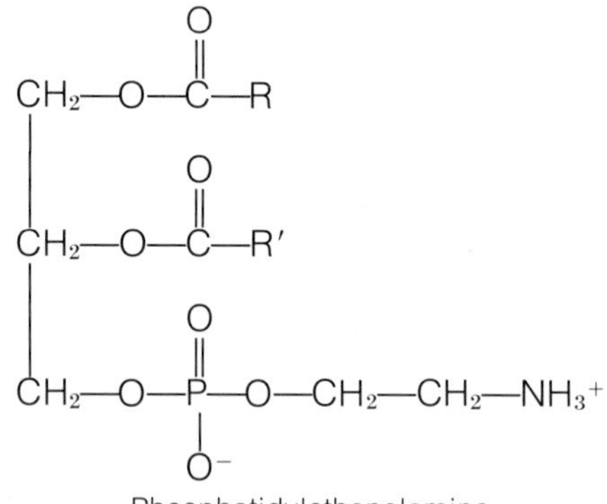

Phosphatidylethanolamine

PE is found in blood platelets and plays an important role in the clotting of blood. It also serves as a source of inorganic phosphate for the formation of new tissue.

20.13 Sphingosine-Based Phospholipids: Sphingolipids

The alcohol of **sphingolipids** is not glycerol, but rather sphingosine. The most common sphingolipid is sphingomyelin.

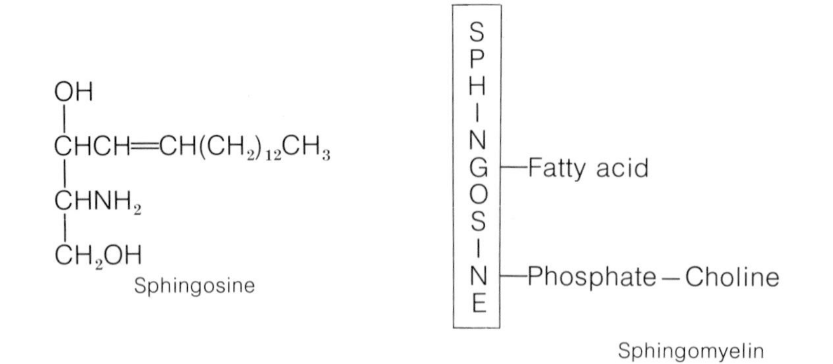

Large amounts of sphingomyelins are found in brain and nervous tissue and form part of the myelin sheath, the protective coating of nerves. The myelin sheath is very stable, due partly to the interlocking of the long fatty acid chains of the sphingomyelins. Certain diseases such as Niemann–Pick disease and multiple sclerosis result in the production of defective myelin sheaths. Niemann–Pick disease is a hereditary disease in which sphingomyelins build up in the brain, liver, and spleen, resulting in mental retardation and early death. Multiple sclerosis (MS) is a disease in which the body's own immune system attacks and destroys areas of myelin sheath. These areas are then replaced by scar tissue, which interrupts or distorts the flow of nerve impulses, causing paralysis, numbness, loss of coordination and balance, and speech difficulties.

20.14 Glycolipids

The main difference between **glycolipids** and phospholipids is that a glycolipid contains a sugar group rather than a phosphate group. The sugar group is usually galactose, but may also be glucose. The alcohol is either glycerol or sphingosine.

Glycolipids

Cerebrosides are glycolipids that contain the base sphingosine. They are found in high concentrations in the brain and nerve cells, especially in the myelin sheath. Several hereditary fat metabolism diseases have been linked to errors in the metabolism of the glycolipids. In Gaucher's disease, the glycolipids contain glucose rather than galactose, and they collect in the spleen and kidneys. In Tay–Sachs disease, the infant lacks an enzyme necessary to break down glycolipids, and they collect in the tissues of the brain and eyes, causing muscular weakness, mental retardation, seizures, blindness, and death by the age of three.

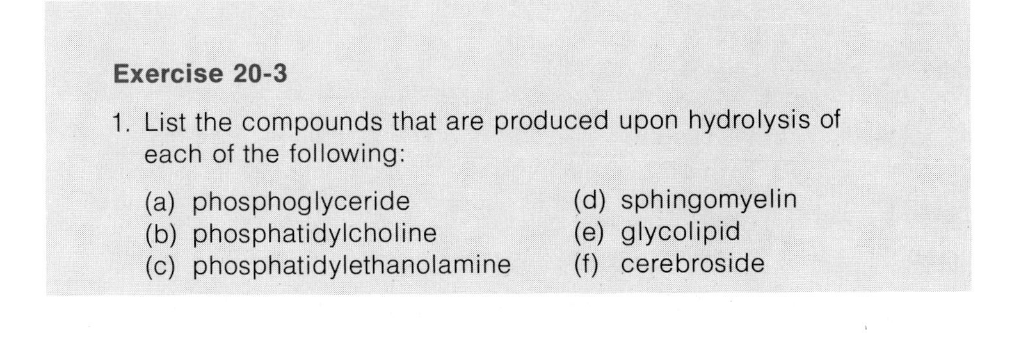

Exercise 20-3

1. List the compounds that are produced upon hydrolysis of each of the following:

   (a) phosphoglyceride
   (b) phosphatidylcholine
   (c) phosphatidylethanolamine
   (d) sphingomyelin
   (e) glycolipid
   (f) cerebroside

2. Describe how the structure of PC allows this compound to form an emulsion between salad oil and vinegar.

NONSAPONIFIABLE LIPIDS

20.15 Steroids

Nonsaponifiable lipids are those that are not broken apart by alkaline hydrolysis. **Steroids** are nonsaponifiable lipids whose structure is based on a complicated four-ring framework consisting of three cyclohexane rings and one cyclopentane ring.

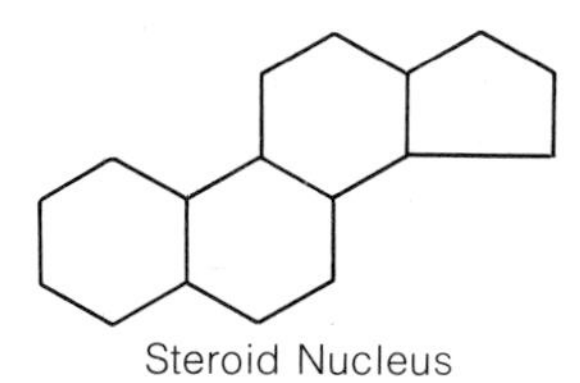

Steroid Nucleus

The steroid nucleus is found in the structure of several vitamins, hormones, drugs, poisons, bile acids, and sterols. The structure and function of some familiar steroids are shown in Table 20.5.

20.16 Cholesterol

Sterols are steroid alcohols. The most common sterol is cholesterol.

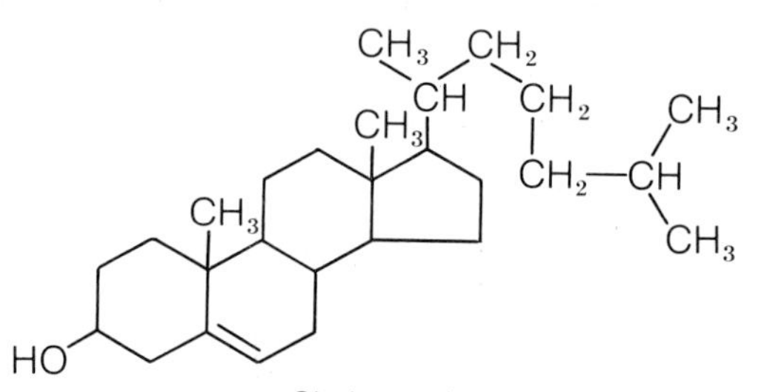

Cholesterol

Cholesterol is synthesized from acetyl CoA by animal cells and forms part of all cell membranes. Specialized cells in the liver, adrenal glands, and ovaries also use cholesterol to synthesize bile acids, steroid hormones, and vitamin D. Ninety percent of the 3 to 5 grams of cholesterol the body makes each day is produced by the liver, so this organ plays a key role in the body's cholesterol balance.

The specific amount of cholesterol produced by the liver at any time, however, is controlled by the amount of cholesterol already circulating in the blood. As we saw at the beginning of this chapter, cholesterol circulates in the blood in several forms, the most important of which is low-density lipoprotein (LDL). When liver cells need cholesterol, they increase the rate of cholesterol synthesis within the cell and also produce more LDL receptors to bring cholesterol-containing LDL particles into the cell. As the

Table 20.5 The Structure and Function of Some Steroids

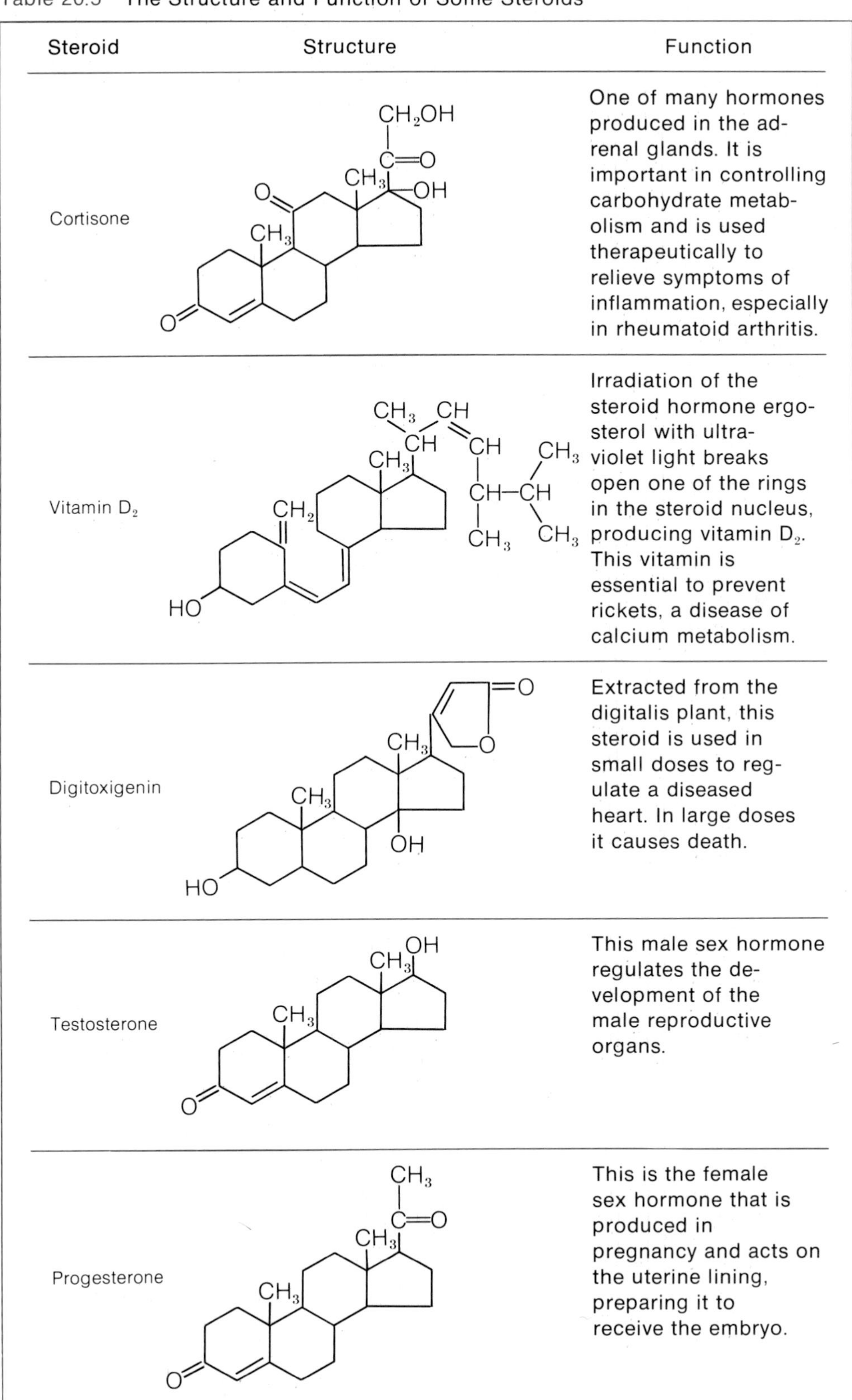

Steroid	Structure	Function
Cortisone		One of many hormones produced in the adrenal glands. It is important in controlling carbohydrate metabolism and is used therapeutically to relieve symptoms of inflammation, especially in rheumatoid arthritis.
Vitamin D_2		Irradiation of the steroid hormone ergosterol with ultraviolet light breaks open one of the rings in the steroid nucleus, producing vitamin D_2. This vitamin is essential to prevent rickets, a disease of calcium metabolism.
Digitoxigenin		Extracted from the digitalis plant, this steroid is used in small doses to regulate a diseased heart. In large doses it causes death.
Testosterone		This male sex hormone regulates the development of the male reproductive organs.
Progesterone		This is the female sex hormone that is produced in pregnancy and acts on the uterine lining, preparing it to receive the embryo.

concentration of cholesterol within the cell increases, the synthesis of both cholesterol and LDL receptors decreases.

Research has shown that a diet rich in saturated fats and cholesterol increases the concentration of circulating LDL and the risk of atherosclerosis. To remove such excess LDL from the blood, the liver would have to increase its production of LDL receptors. But, as we have just seen, the accumulation of this excess dietary cholesterol in the liver actually causes the liver cells to decrease their production of LDL receptors. Thus, the circulating LDL concentration remains high, which increases the rate of plaque formation and the risk of heart attack and stroke.

20.17 Cellular Membranes

Membranes perform many specific functions in living organisms. They control the chemical environment of the space they enclose by keeping out certain compounds and selectively transporting others through the membrane. Cellular membranes maintain the shape of the cell and control cellular movement. The chemical composition of the membrane allows cell-to-cell recognition, and the membrane contains the receptors for many hormones.

The two major components of cellular membranes are lipids and proteins, with the proportion of these two types of components varying among different kinds of membranes. For example, the myelin sheath is about 70% lipid, whereas the nuclear membrane is only about 40% lipid. When viewed through an electron microscope, biological membranes reveal two dark bands on either side of a light band (Figure 20.7). This is due to the fact that the molecular structure of a membrane is a lipid bilayer—two rows of phospholipids, each with their polar hydrophilic heads toward the outside of the membrane (in contact with the aqueous solutions of the cell) and their nonpolar hydrophobic tails toward the water-free interior (Figure 20.8). The protein components of the membrane may be on the surface of, embedded in, or even extending completely

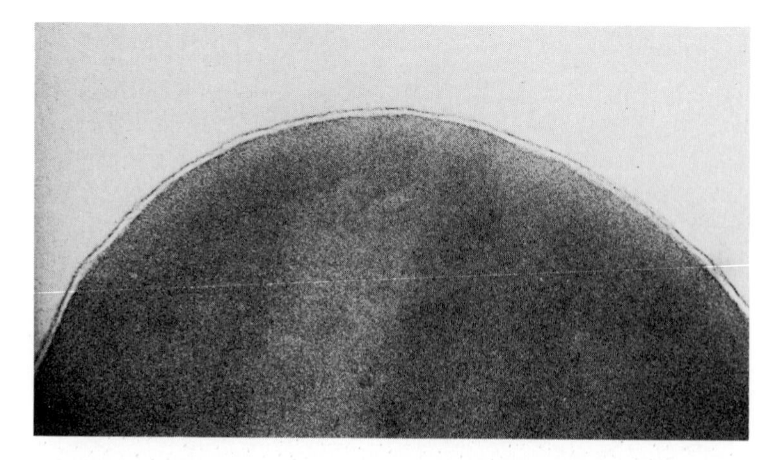

Figure 20.7 Electron micrograph of the plasma membrane of a red blood cell, showing two dark bands on each side of a light band. (Courtesy Dr. J.D. Robertson.)

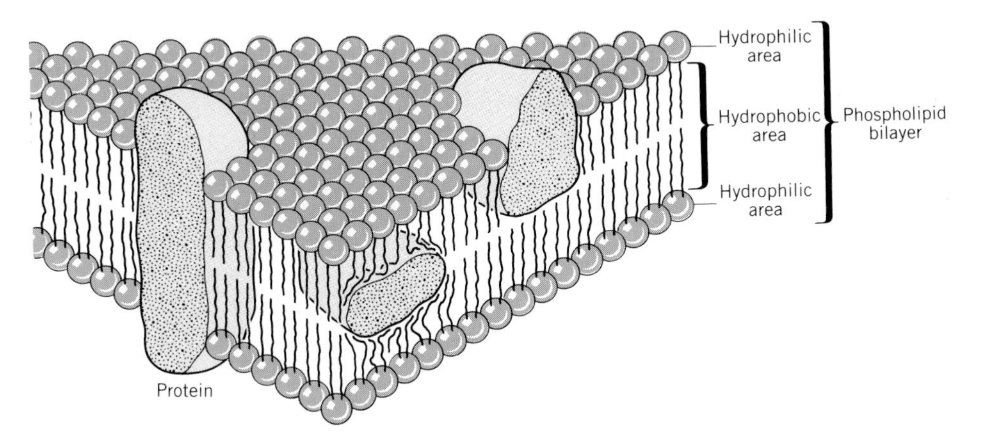

Figure 20.8 **Cellular membranes are composed of phospholipids and proteins. The phospholipids are arranged in a double layer (bilayer) with their polar heads to the outside and their nonpolar tails toward the middle.**

through the lipid bilayer. The lipids found most abundantly in membranes are phosphoglycerides (PE and PC), sphingolipids, and cholesterol. Glycolipids, however, are also found in membranes and are thought to form the structure for receptor sites—areas on the membrane where specific molecules such as hormones attach to the cell.

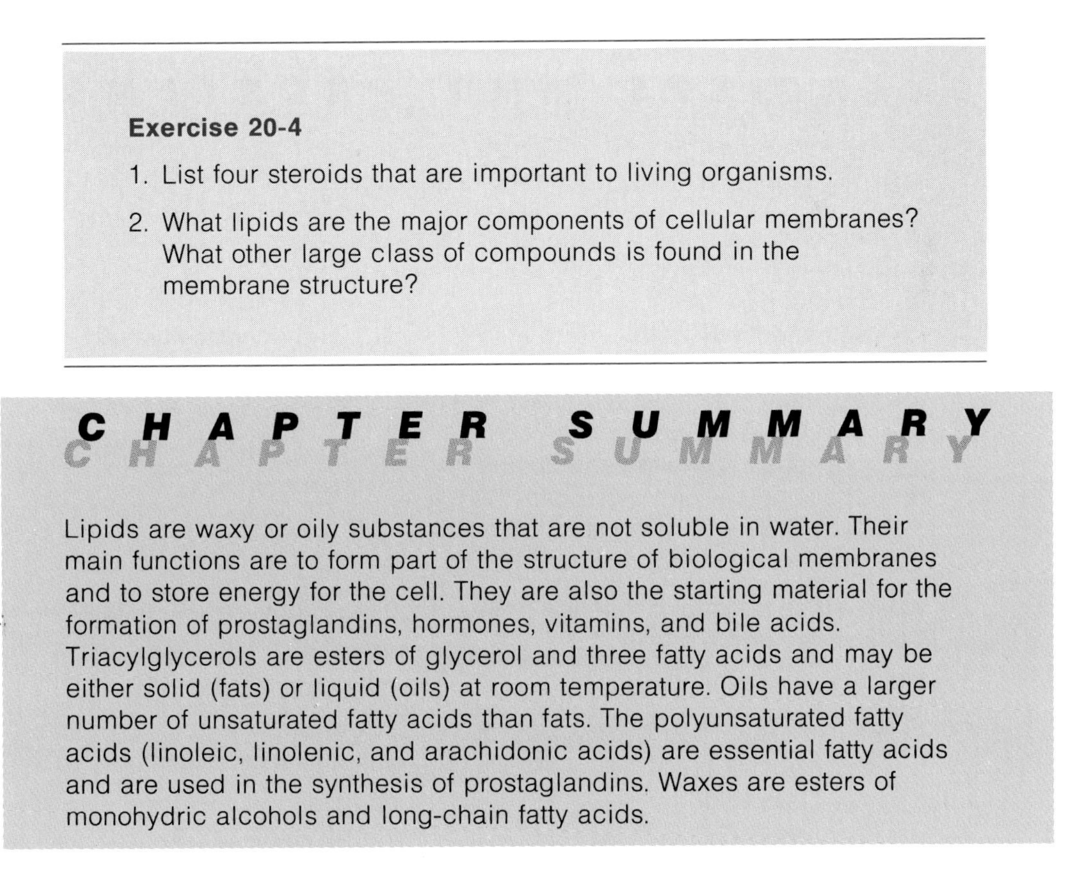

Exercise 20-4

1. List four steroids that are important to living organisms.
2. What lipids are the major components of cellular membranes? What other large class of compounds is found in the membrane structure?

CHAPTER SUMMARY

Lipids are waxy or oily substances that are not soluble in water. Their main functions are to form part of the structure of biological membranes and to store energy for the cell. They are also the starting material for the formation of prostaglandins, hormones, vitamins, and bile acids. Triacylglycerols are esters of glycerol and three fatty acids and may be either solid (fats) or liquid (oils) at room temperature. Oils have a larger number of unsaturated fatty acids than fats. The polyunsaturated fatty acids (linoleic, linolenic, and arachidonic acids) are essential fatty acids and are used in the synthesis of prostaglandins. Waxes are esters of monohydric alcohols and long-chain fatty acids.

Unsaturated fatty acids, or unsaturated fatty acid side chains of triacylglycerols, can be hydrogenated to produce saturated fatty acids or saturated fats. Fats and oils can become rancid by hydrolysis of the triacylglycerol or by oxidation of unsaturated fatty acid side chains. Triacylglycerols can be hydrolyzed to form three fatty acids and glycerol. If the hydrolysis occurs in the presence of a strong base, glycerol and the salts of the fatty acids, or soaps, are formed. Soaps are good emulsifying agents because their molecules contain a nonpolar region that will dissolve in the fat or oil and a polar region that will dissolve in the water.

Phospholipids are compound lipids that form the structure of cell membranes and that help to transport other lipids in the body. All the phospholipids contain a phosphate group, fatty acids, an alcohol (glycerol or sphingosine), and a nitrogen-containing compound (choline, ethanolamine, serine, or inositol). Glycolipids are compound lipids containing a sugar group rather than a phosphate group and are found in high concentrations in brain and nervous tissue. Steroids are nonsaponifiable lipids with a complex ring structure that is found in many hormones, vitamin D, sterols, bile acids, drugs, and poisons. Cholesterol is the most common sterol; it is a component of all cells and is used by the cell as the starting material for the synthesis of many other compounds.

EXERCISES AND PROBLEMS

1. State the difference between each of the following:
   - (a) saponifiable lipid and nonsaponifiable lipid
   - (b) simple triacylglycerol and mixed triacylglycerol
   - (c) fat and oil
   - (d) saturated fatty acid, unsaturated fatty acid, and polyunsaturated fatty acid
   - (e) essential fatty acid and nonessential fatty acid
   - (f) phospholipid and triacylglycerol
   - (g) phosphatidylcholine and sphingomyelin
   - (h) phospholipid and glycolipid
   - (i) hydrolysis and hydrogenation
   - (j) hydrolysis and saponification
2. What is the structural difference between the triacylglycerols in animal fats and vegetable oils?
3. (a) Identify each of the following as a saturated fatty acid or unsaturated fatty acid:
   - (1) myristic acid
   - (2) oleic acid
   - (3) linolenic acid
   - (4) lauric acid

   (b) Which of the above fatty acids would you most likely find in olive oil?

4. Why would a diet lacking in linoleic acid be bad for a person's health?

5. What information about a lipid do you obtain from its iodine number?

6. Describe the process by which margarine is produced from corn oil.

7. Why does butter turn rancid? How does refrigeration slow this process? What causes the odor of rancid butter?

8. Elaidic acid is a *trans* isomer of oleic acid. It is produced in the manufacture of margarine by the partial hydrogenation of corn oil.

   (a) Draw the structural formulas of oleic and elaidic acids.
   (b) Why should there be increasing concern about the production and consumption of *trans* fatty acids in shortenings and margarines?

9. (a) Describe what is happening, on the molecular level, when you wash salad oil from your hands with soap.
   (b) Suggest a possible reason for the antibacterial action of soap, using the fact that bacterial cell membranes are formed by lipids.

10. In what ways are detergents superior to soaps?

11. Why is inhaling a mixture containing DPPC effective in preventing the collapse of alveoli?

12. What steroid is a component of cell membranes?

13. To what class of lipids do the hormones testosterone and progesterone belong? From what compound are they synthesized? What is the difference in the structure of the two compounds?

14. Advertising would lead us to believe that cholesterol is bad for us.

    (a) Give reasons to support this statement.
    (b) Then, explain why this statement is not entirely true.

15. How can companies that produce vegetable oils advertise their products as cholesterol free?

16. Write the general formula for the following compounds, and describe the function of each in the human body:

    (a) phosphoglyceride
    (b) phosphatidylcholine
    (c) phosphatidylethanolamine
    (d) sphingolipid
    (e) sphingomyelin
    (f) glycolipid

17. Write the formula for the following triacylglycerols:

    (a) tripalmitin
    (b) the mixed triacylglycerol containing glycerol, arachidic, stearic, and oleic acids
    (c) triolein

18. Which triacylglycerol in question 12 would:
    (a) have the highest melting point?
    (b) be most likely to be found in animal fat?
    (c) have the highest iodine number?

19. Write the equation for the complete hydrogenation of the triacylglycerols in parts (b) and (c) of question 12.

20. Write the equations for the hydrolysis of the triacylglycerols in question 12.

21. Suppose that hydrolysis of a lipid produced glycerol, stearic acid, oleic acid, and linoleic acid. To what class of lipids does this compound belong? Draw the structural formula of this lipid.

22. If complete hydrolysis of a lipid produces glycerol, stearic acid, oleic acid, phosphate, and choline, what is the name of this lipid? To what class of lipids does this compound belong? Draw the structural formula for this lipid.

23. To what class of lipids does the following compound belong?

$$CH(OH)CH{=}CH(CH_2)_{12}CH_3$$
$$|$$
$$CHNH\overset{O}{\overset{\|}{C}}(CH_2)_{22}CH_3$$
$$|$$
$$CH_2O\overset{O}{\overset{\|}{P}}OCH_2CH_2N^+(CH_3)_3$$
$$\underset{O^-}{|}$$

24. The following triacylglycerol is found in lard:

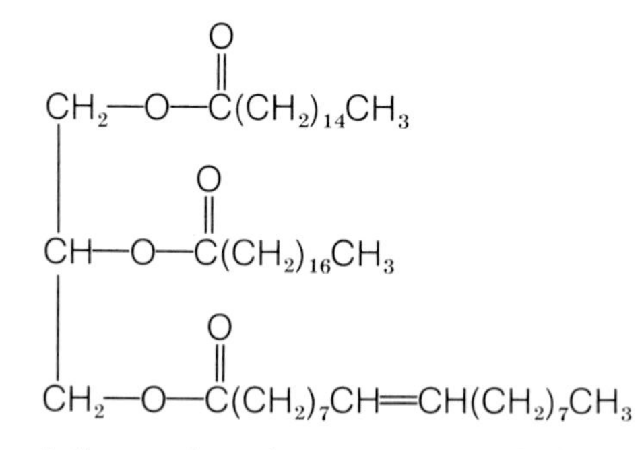

Before commercially produced soap was available, people made soap from lard and lye that was extracted from wood ashes with a small amount of water. This lye solution contained basic substances such as KOH, Na_2CO_3, and K_2CO_3. Write the equation for the formation of soap from this triacylglycerol and KOH.

25. Describe the structure of a cellular membrane, and explain how the structure of phospholipids allows them to form a bilayer.

CHAPTER TWENTY-ONE

PROTEINS

This had not been a good year for Fran Bobo. She had spent a very frustrating winter and spring working with the contractor that was renovating her old house. Now she was faced with a six-year-old daughter suffering from some mysterious ailment. Actually she didn't know whether or not her daughter had something specifically wrong, but she hoped to find out in a few days. It had started when Jackie had come home from school with a bad cold, complaining that she felt tired and had pains in her stomach. Later that evening Jackie had begun vomiting uncontrollably and acting very lethargic. When Fran took her daughter to the town's small clinic, they found nothing requiring immediate attention (although the blood tests did indicate a mild case of anemia).

At the local doctor's suggestion, she took Jackie to the city hospital a few hours away. In the hospital's pediatric clinic, the doctor reviewed the results of the laboratory tests and noted the anemia that had been found the previous week. He also noticed that certain blood cells appeared unusual. This signaled to the physician the possibility of lead poisoning, and a test for blood lead level confirmed this suspicion. Jackie's blood contained a high concentration of lead—70 μg/dl—compared with a normal level of less than 20 μg/dl.

But how could a child living in a small rural town be exposed to dangerously high concentrations of lead? The culprit turned out to be the house renovation. Jackie had been playing in the house as the remodelers were sanding the walls. The dust that she had been breathing contained high concentrations of lead from the old paint. The doctor simply ordered Jackie to stay with friends or relatives until all the house remodeling was finished and to have her blood lead level checked weekly until it returned to the normal range. The blood test would then be repeated at regular intervals for a few years to make certain lead levels didn't increase again.

Lead is just one of many heavy metals that are toxic to the body. Children are much more sensitive to lead exposure than are adults. Lead is absorbed by the body in the form of inorganic lead salts (such as those found in paints) or as tetraethyl lead (found in regular gasoline). The majority of absorbed lead is stored in the bones, but it is the lead in the blood that causes the effects of lead poisoning. At low blood concentrations, the Pb^{2+} ion inhibits several enzymes containing sulfhydryl groups, which affects the synthesis of the heme molecule necessary for the production of hemoglobin. This lowers the concentration of hemoglobin in the blood and causes anemia. At higher concentrations, lead disrupts kidney function and affects the central nervous system. At blood concentrations over 100 μg/dl, lead can cause convulsions, coma, and death. Even if treated, very high lead exposure in children can result in permanent brain damage. Thus it was critical for Jackie to be removed immediately from any further chance of lead exposure.

In this chapter, we will learn about proteins and enzymes and will see why interference with their functions by substances such as the lead ion can have profound effects on living organisms.

LEARNING OBJECTIVES

LEARNING OBJECTIVES

By the time you have finished this chapter, you should be able to:

1. Write the general structure of an amino acid.
2. State the difference between (a) a simple and a conjugated protein and (b) a globular and a fibrous protein.
3. Explain how an amino acid or a protein can act as a buffer.
4. Define *zwitterion* and *isoelectric point.*
5. Describe the differences and the types of bonding found in the primary, secondary, tertiary, and quaternary structure of proteins.
6. Define *native state* and *denaturation.*
7. State five methods for denaturing proteins.

PROTEINS

21.1 Classification of Proteins

Proteins are the most complex and varied class of molecules found in living organisms. They are found in all cells, and their biological importance cannot be overemphasized. This fact was recognized by the German chemist G. T. Mulder in 1839, when he gave this class of compounds the name **protein,** which means "of prime importance."

All proteins are composed of the elements carbon, nitrogen, oxygen, and hydrogen. Most proteins also contain sulfur, and some have phosphorus and other elements such as iron, zinc, or copper. Proteins are large polymers, and upon hydrolysis they will produce monomer units called **amino acids.** The molecular weight of most proteins ranges from 12,000 to 1 million or more, as compared to other groups of compounds whose molecular weights are generally under 1000. This large size gives protein molecules colloidal properties. For example, they cannot pass through differentially permeable membranes such as membranes of the cell. The presence of proteins in the urine, therefore, warns doctors of the possibility of damage to the membranes of the kidneys.

Because of their varied natures, proteins can be classified in several different ways. They may be divided into two major classes: **simple proteins,** which produce only amino acids upon hydrolysis, and **conjugated proteins,** which produce amino acids and other organic or inorganic substances upon hydrolysis. These other substances are called **prosthetic groups** (Table 21.1). A second classification is based on the physical characteristics of the protein molecule. **Globular proteins** are soluble in water, are quite fragile, and have an active function, such as catalyzing reactions (in the case of enzymes) or transporting other substances (as, for example, hemoglobin). **Fibrous proteins** are insoluble in water, are physically tough, and have a structural or protective function. The keratins in hair, skin, and fingernails, the collagen in tendons and

Table 21.1 Conjugated Proteins

Class	Prosthetic Group	Examples
Glycoproteins	Carbohydrate	Connective tissue, mucins, heparin, immunoglobulins
Lipoproteins	Lipid	High and low density lipoproteins in the blood
Nucleoproteins	Nucleic acid	Viruses, chromosomes
Metalloproteins	Metal ions	Ferritin (Fe), alcohol dehydrogenase (Zn)
Chromoproteins	Colored groups: riboflavin, heme, etc.	Hemoglobin, chlorophyll, luciferase, cytochromes

hides, and the silks are examples of such fibrous proteins (Figure 21.1).

The biological importance of proteins results from their wide variety of functions (Table 21.2). Proteins are the body's main dietary source of nitrogen and sulfur. In addition to their catalytic and structural functions, they make up the contractile system of muscles (Figure 21.2). As

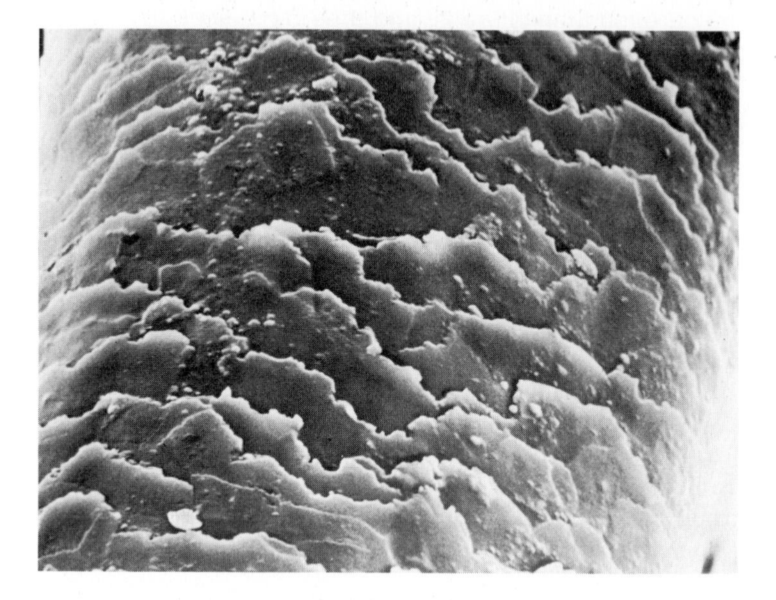

Figure 21.1 A human hair enlarged 500 times by a scanning electron microscope. The outer layer of the hair is composed of flat, dead, scale-like cells filled with keratin. (Dr. Tony Brain/Science Photo Library/Photo Researchers.)

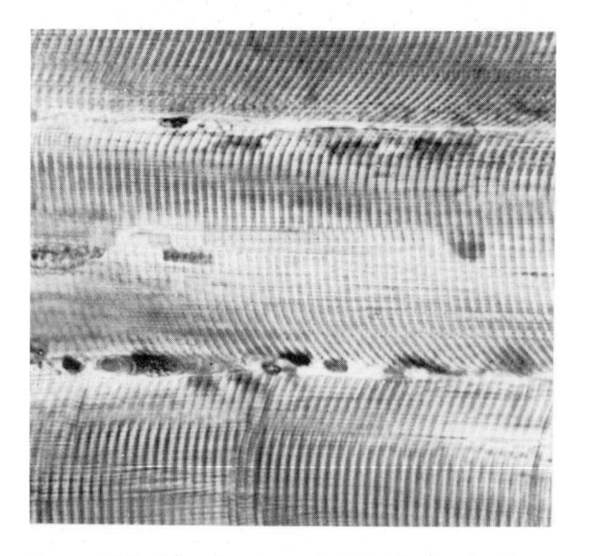

Figure 21.2 The structure of skeletal muscle. The muscle cells contain myofibrils that show cross banding or striations when stained; hence, the name striated muscle. The myofibrils consist of two types of contractile proteins: actin and myosin. (Courtesy J. Robert McClintic.)

Table 21.2 A Classification of Proteins Based on Their Functions in Living Organisms

Class	Function (with an example)
Enzymes	Catalyze biological reactions: pepsin catalyzes the breakdown of proteins in the stomach
Structural Proteins	Provide structural support: collagen is the major extracellular support in tendons and bone
Storage Proteins	Store nutrients: ferritin stores Fe in the spleen
Transport Proteins	Bind and transport specific molecules in the blood: hemoglobin transports oxygen
Hormones	Regulate body metabolism: insulin regulates glucose metabolism
Contractile Proteins	Perform contraction and movement: actin and myosin form the contraction system in muscles
Protective Proteins	Protect against foreign substances: antibodies inactivate foreign proteins in the blood
Toxins	Defend organisms: botulinus toxin is poisonous to organisms other than *Clostridium botulinum*

antibodies, they are the defense system of the body, and as hormones, they regulate the body's glandular activity. In the blood they maintain fluid balance, are part of the clotting process, and transport oxygen and lipids. They can act as poisons, such as the venoms in animal bites and stings, or toxins, such as the bacterial toxin that produces botulism in humans from eating improperly processed foods. Some antibiotics that are secretions of bacteria and fungi also are protein in nature.

21.2 Amino Acids

The particular function of a given protein is determined by the sequence or order of amino acids in the protein molecule. It is necessary, therefore, for any study of proteins to include a thorough discussion of amino acids. **Amino acids** are carboxylic acids that have an amino group on the alpha carbon—the carbon next to the carboxyl group. The general structure of an amino acid is as follows:

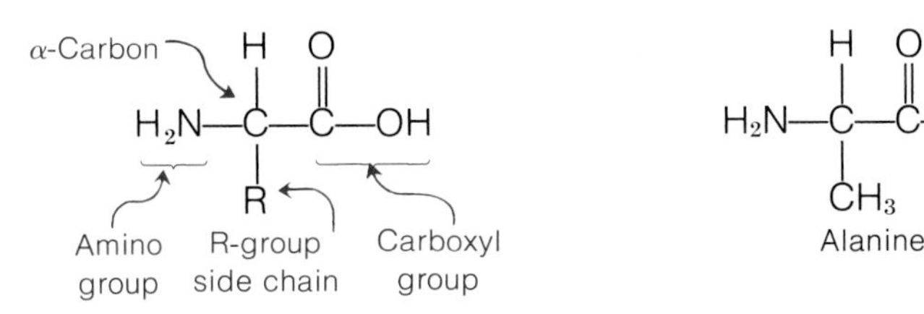

The different R-group side chains on the amino acids make one amino acid different from another. Most naturally occurring proteins are composed of

Figure 21.3 (Opposite page) The structure of the R-group side chains of the 20 most common amino acids. The complete structure of the amino acid can be written by substituting the formula of the R-group in the general formula. For example, the structure of serine can be written

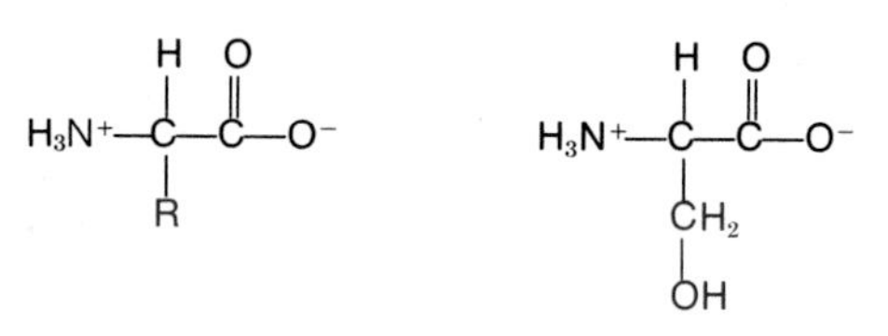

the 20 amino acids shown in Figure 21.3, but there are a few specialized types of proteins that contain other, more rare, amino acids. Still other amino acids not found in proteins exist in a free or combined form; these are, in general, derivatives of the 20 amino acids found in proteins.

21.3 The L-Family of Amino Acids*

All amino acids found in proteins, with the exception of glycine, are optically active and belong to the L-family. D-family isomers of amino acids can be found in nature, but never occur in proteins.

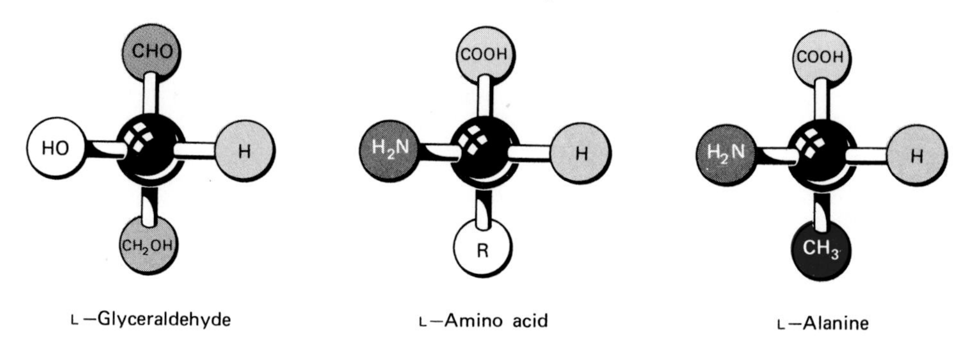

21.4 Essential Amino Acids

The body can synthesize 10 of the 20 amino acids found in proteins. Ten others, however, cannot always be synthesized at a sufficient rate, and so they must be supplied in the diet. Eight of these other amino acids are essential throughout life, whereas the other two (arginine and histidine) are required in the diet in childhood during periods of rapid growth (Table 21.3). These 10 other amino acids are called the **essential amino acids.** Proteins that contain all 10 of the essential amino acids are known as **adequate proteins.** Animal protein and milk are adequate proteins, but many vegetable proteins are missing one or more of the essential amino acids and, therefore, are inadequate proteins. The protein in soybeans is adequate, but the protein in corn is too low in lysine and tryptophan

---

* *This section is optional and may be omitted without loss of continuity.*

	Structure of R in $H_3N^+—CH(R)—C(=O)—O^-$	Name of Amino Acid	Abbreviation
NONPOLAR R–GROUP	$—H$	Glycine	Gly
	$—CH_3$	Alanine	Ala
	$—CH(CH_3)—CH_3$	Valine	Val
	$—CH_2—CH(CH_3)—CH_3$	Leucine	Leu
	$—CH(CH_3)—CH_2—CH_3$	Isoleucine	Ile
	$—CH_2—CH_2—S—CH_3$	Methionine	Met
	$—CH_2—C_6H_5$ (phenyl ring)	Phenylalanine	Phe
	$—CH_2—C_6H_4—OH$ (ring)	Tyrosine	Tyr
	$—CH_2—$(indole ring, N—H)	Tryptophan	Trp
	$^-O—C(=O)—CH—CH_2$, $H_2N^+—CH_2—CH_2$ (ring: CH—H_2N^+—CH_2—CH_2—CH_2) (complete structure)	Proline	Pro
POLAR R–GROUP	$—CH_2—OH$	Serine	Ser
	$—CH(CH_3)—OH$	Threonine	Thr
	$—CH_2—SH$	Cysteine	Cys
	$—CH_2—C(=O)—NH_2$	Asparagine	Asn
	$—CH_2—CH_2—C(=O)—NH_2$	Glutamine	Gln
Acidic R–group	$—CH_2—C(=O)—OH$	Aspartic acid	Asp
	$—CH_2—CH_2—C(=O)—OH$	Glutamic acid	Glu
Basic R–group	$—CH_2—CH_2—CH_2—CH_2—NH_2$	Lysine	Lys
	$—CH_2—CH_2—CH_2—NH—C(=NH)—NH_2$	Arginine	Arg
	$—CH_2—C{=}CH$ (ring: HN, N, C—H)	Histidine	His

Table 21.3 The Essential Amino Acids

Amino Acid	Function in the Human Body
1. Leucine	Associated with digestive enzymes
2. Isoleucine	Associated with digestive enzymes
3. Lysine	Aids in assimilation of other amino acids
4. Methionine	Associated with fat metabolism
5. Phenylalanine	Important in utilization of vitamin C and production of thyroxine
6. Tryptophan	Important in utilization of B vitamins and synthesis of neurotransmitters
7. Threonine	Important in building tissues and utilization of nutrients
8. Valine	Important to functioning of the nervous system

to support growth in young children. Rice is low in lysine and threonine, and wheat is low in lysine. People eating vegetable diets, therefore, must eat a combination of vegetables to obtain all of the essential amino acids. This is why protein deficiency diseases, which cause both slowed growth and reduced resistance to disease in children, are commonly found in parts of the world where a single plant (such as corn) is the major source of dietary protein.

21.5 Acid-Base Properties

Because proteins can, in many cases, be viewed as very large amino acids, a knowledge of the acid-base properties of amino acids can be of great help in understanding some of the properties of proteins. Amino acids contain both an acid group—the carboxyl group—and a basic group—the amino group. In water, amino acids can act as either acids or bases. Molecules having this property are called **amphoteric.**

Amino acids are soluble in water and have very high melting points. This suggests that they do not exist as uncharged molecules, but rather are found in the form of the highly polar **zwitterion** or dipolar ion.

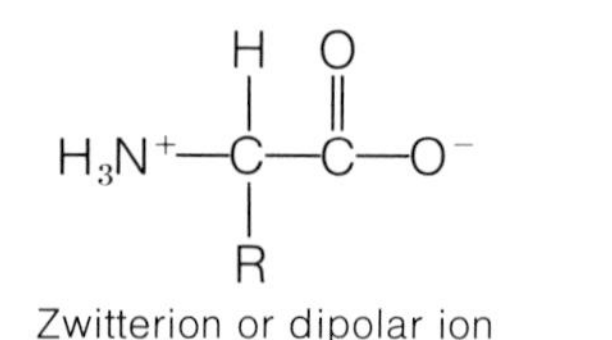

Zwitterion or dipolar ion

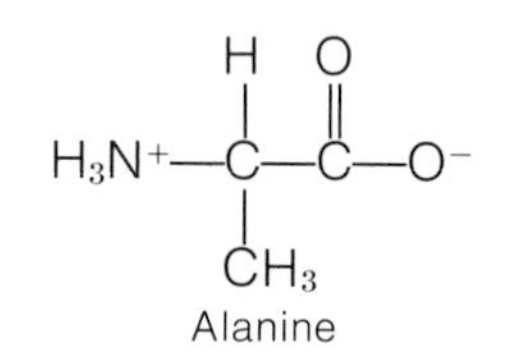

Alanine

The zwitterion is formed when the acidic carboxyl group donates a hydrogen ion to the basic amino group. Although we may write the structure of the amino acid in the uncharged form, keep in mind that it will usually be found as a dipolar ion.

Table 21.4 Isoelectric Points of Some Amino Acids and Proteins

Compound	Isoelectric Point (pI)	Compound	Isoelectric Point (pI)
Amino acid		*Protein*	
Glutamic acid	3.2	Egg albumin	4.6
Phenylalanine	5.5	Urease	5.0
Alanine	6.0	Hemoglobin	6.8
Leucine	6.0	Myoglobin	7.0
Lysine	9.7	Chymotrypsin	9.5
Arginine	10.8	Lysozyme	11.0

21.6 Isoelectric Point

There is a specific pH at which each amino acid and protein will be electrically neutral and will not move in an electric field. This pH is called the **isoelectric point** for that molecule and is indicated by the symbol pI. At a pH more basic than the isoelectric point, the amino acid will have a net negative charge and will move toward the positive electrode. At a pH more acidic than the isoelectric point, the amino acid will carry a net positive charge and will move toward the negative electrode. Because some amino acids have an ionizable R-group, each amino acid and protein has a specific isoelectric point (Table 21.4). Proteins can be separated from one another at different pH levels on the basis of their charge by a process called electrophoresis. In this technique, each protein will migrate toward the positive or negative electrode at a different rate depending upon the pH and voltage applied. Paper electrophoresis is a useful tool in analyzing the proteins in human blood serum (Figure 21.4).

At the isoelectric point, the protein will have minimum solubility. The electrically neutral protein molecules can cluster together and are most

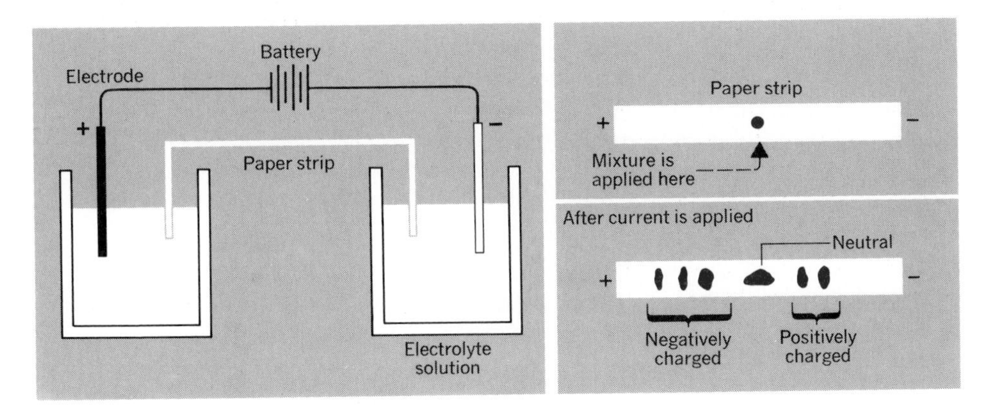

Figure 21.4 Proteins and amino acids can be separated by their rate of migration in an electric field using paper electrophoresis.

easily removed from solution. Casein, for example, is the protein found in cow's milk and has an isoelectric point of pH 4.7. The normal pH of cow's milk is 6.3. In the production of cheese, however, bacteria produce lactic acid that lowers the pH of the milk. This, then, lowers the solubility of the casein, causing the milk to curdle.

21.7 Buffering Properties

Because amino acids (and proteins) can act as either acids or bases, they are effective buffers in an aqueous solution.

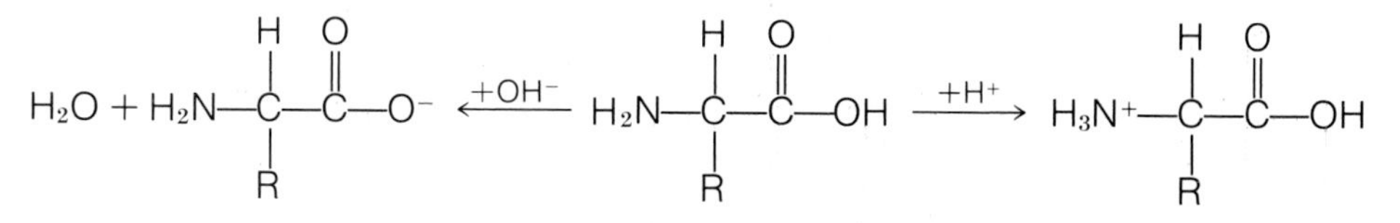

One of the functions served by the proteins in the blood is to act as buffers, helping to keep the blood pH within its very narrow normal range (pH = 7.35 to 7.45).

Exercise 21-1

1. Define the following terms in your own words:
   (a) simple protein
   (b) conjugated protein
   (c) prosthetic group
   (d) globular protein
   (e) fibrous protein
   (f) amino acid
   (g) amphoteric
   (h) zwitterion
   (i) isoelectric point
   (j) electrophoresis
2. (a) How can an amino acid act as a buffer?
   (b) Write the equation for the reaction of leucine with HCl.
   (c) Write the equation for the reaction of phenylalanine with NaOH.
3. (a) What is the pI of phenylalanine?
   (b) If a solution containing phenylalanine at a pH of 7 were placed in an electric field, toward what pole would the phenylalanine migrate?

PROTEIN STRUCTURE

21.8 Primary Structure: The Amino Acid Sequence and the Peptide Bond

The **primary structure** of a protein is given by the sequence of amino acids in the protein molecule. These amino acids are joined together by

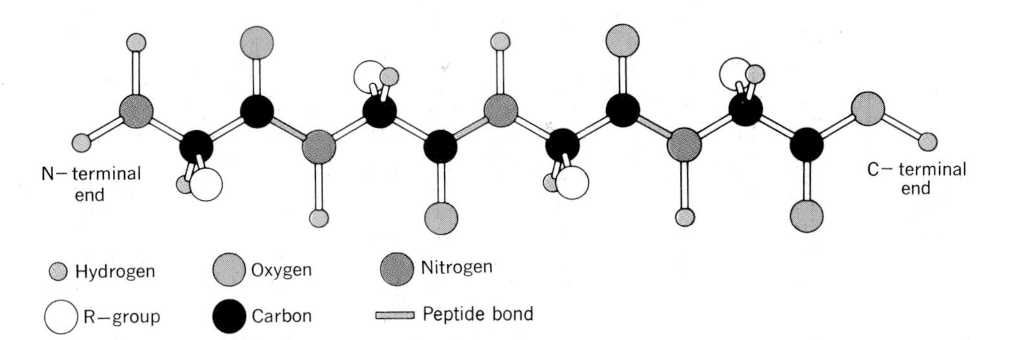

Figure 21.5 Primary structure of proteins. This polypeptide chain contains four amino acids joined by peptide bonds. The end of the molecule with the free amino group is the N-terminal end and the end with the free carboxyl group is the C-terminal end.

covalent bonds called **peptide bonds.** The peptide bond is an amide linkage, formed by joining the carboxyl group of one amino acid to the amino group of a second amino acid through the elimination of water (a condensation reaction) (Figure 21.5).

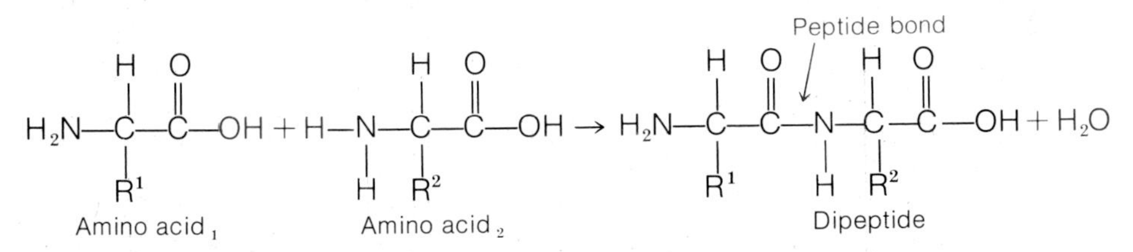

The peptide bond is stable in the face of changes in pH, in solvents, or in salt concentrations. It can be broken only by acid or base hydrolysis, or by specific enzymes. Two amino acids held together by a peptide bond are called a **dipeptide;** three amino acids form a **tripeptide;** and more than three form a **polypeptide.** There is no precise dividing line between polypeptides and proteins. For example, insulin, with 51 amino acids in its primary structure, is a very small protein. Glucagon, with 21 amino acids, however, is considered a large polypeptide. There are many small polypeptides that have important functions in biological systems. Glutathione, which plays a role in oxidation-reduction reactions, is a tripeptide.

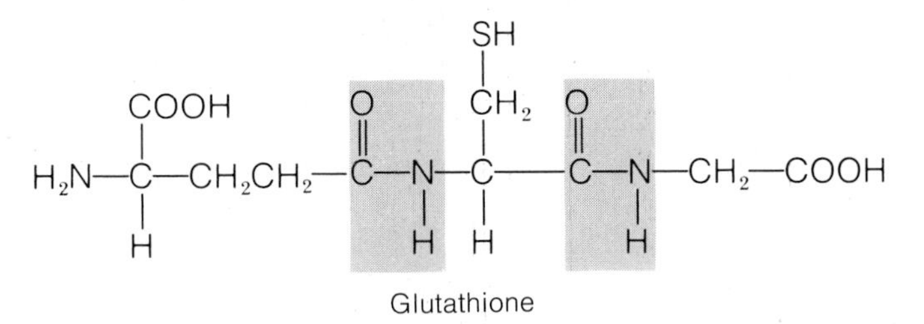

Glutathione

Vasopressin, a diuretic hormone produced by the pituitary, is a nonapeptide, and the antibiotic bacitracin is a polypeptide with 11 amino acids in its sequence.

Two different dipeptides can be formed between the amino acids glycine and alanine.

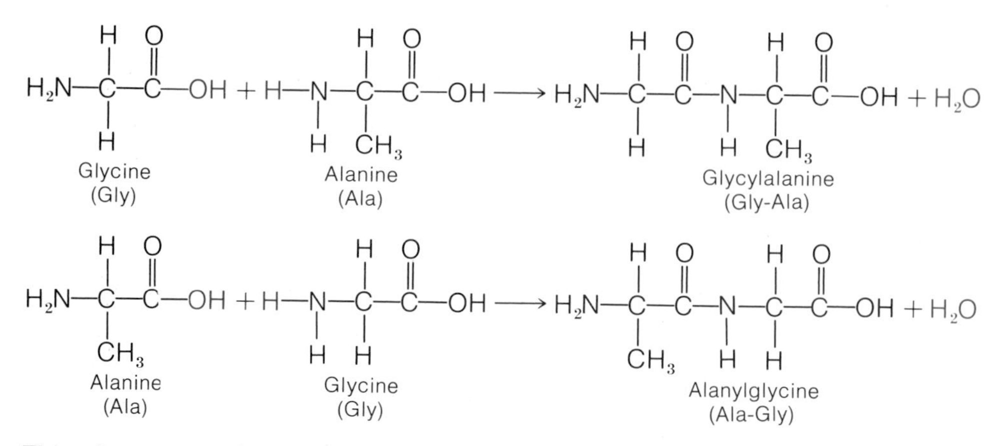

This gives us a good reason for discussing the standard rules for the naming of peptides and proteins. Because proteins contain very large numbers of amino acids, three-letter abbreviations of the amino acid names are used when writing the amino acid sequence (see Figure 21.3 for the abbreviations). The peptide bond is represented by a dash or dot between the amino acid names. One end of the protein will consist of an amino acid having a free amino group. This is called the N-terminal end of the protein, and the N-terminal amino acid is listed first in the sequence of amino acids. At the other end of the protein, there will be an amino acid having a free carboxyl group. This is the C-terminal amino acid, and it is the last amino acid listed in the sequence. The following are two examples of these naming rules:

Lysine–Aspartic acid–Serine–Asparagine–Glutamic acid (I)
Lys–Asp–Ser–Asn–Glu

Valine · Phenylalanine · Alanine · Tryptophan · Leucine (II)
Val · Phe · Ala · Trp · Leu
(N-terminal end) (C-terminal end)

The order of amino acids in a protein will determine its function and is critical to its biological activity. A change of just one amino acid in the sequence can disrupt the entire protein molecule. For example, hemoglobin, the molecule in the blood that carries oxygen, consists of four polypeptide chains with a total of 574 amino acid units. Changing just one specific amino acid in one of the chains results in the defective hemoglobin molecule found in patients with sickle cell anemia.

Adult hemoglobin (Hb-A)	Val–His–Leu–Thr–Pro–Glu–Glu–Lys– . . .
Sickle cell hemoglobin (Hb-S)	Val–His–Leu–Thr–Pro–Val–Glu–Lys– . . .

Similarly, in their study of familial hypercholesterolemia (FH), Brown and Goldstein showed that in the chain of approximately 822 amino acids

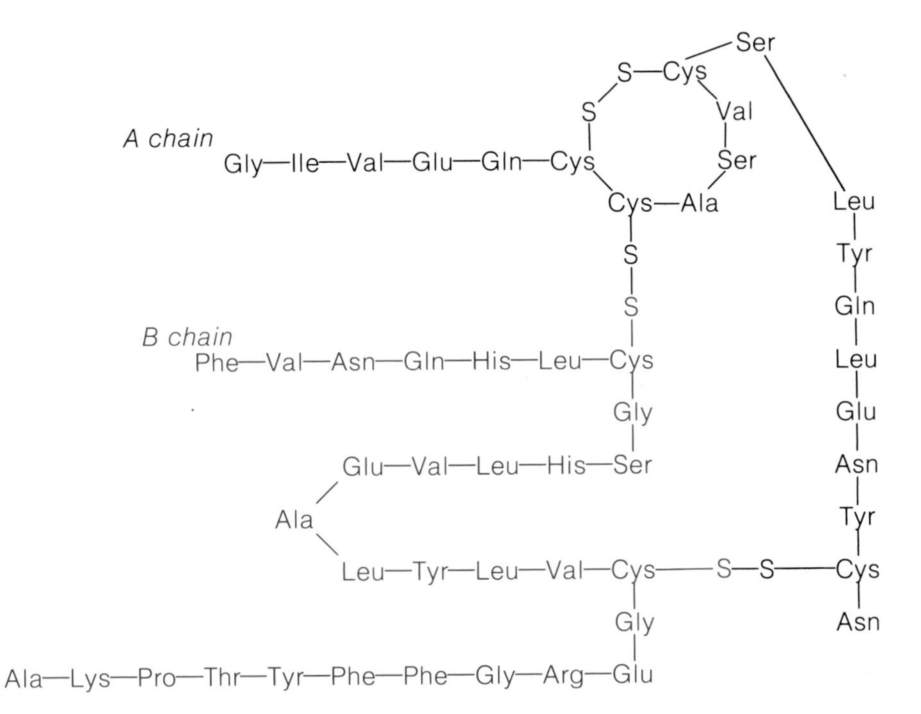

Figure 21.6 The complete amino acid sequence of bovine insulin was first determined by F. Sanger in 1953, for which he was awarded the Nobel prize. The insulin molecule contains 51 amino acids in two polypeptide chains.

making up the LDL receptor molecule, a change of just one amino acid (a cysteine instead of a tyrosine) causes a defect in the receptor. Such defective LDL receptors bind to LDL, but fail to cluster in the coated pits and, therefore, are not able to bring the cholesterol-containing LDL particles into the cell.

The determination of the amino acid sequence in a protein is a complex procedure that was first developed by F. Sanger in 1953. He determined the amino acid sequence of the protein insulin (Figure 21.6).

Both pig and beef insulin are used to treat diabetes mellitus. The amino acid sequence of these insulins, however, differs from human insulin, and some diabetics develop an immune reaction to these insulins because their bodies recognize the insulin as foreign. Through the techniques of genetic engineering, scientists are now producing commercially enough human insulin to replace these animal insulins in the treatment of diabetes.

Exercise 21-2

Enkephalins, produced in the brain, are neurotransmitters that act like opiates. Methionine enkephalin, which acts to inhibit the sense of pain, is a pentapeptide with the following amino acid séquence.

Tyr–Gly–Gly–Phe–Met

Draw the structural formula for methionine enkephalin, boxing in the peptide bonds.

21.9 Secondary Structure: Noncovalent Bonding

The **secondary structure** of a protein is the specific geometric arrangement of the amino acids in space. These arrangements, resulting from hydrogen bonding, were established by Linus Pauling using X-ray diffraction (Figure 21.7).

Keratins

Keratins are the proteins that make up fur, wool, claws, hooves, and feathers. Pauling determined that the polypeptide chains in the protein keratin were curled in an arrangement called an **α helix** (alpha helix). In this arrangement, the amino acids form loops in which the hydrogen on the nitrogen atom in the peptide bond is hydrogen bonded to the oxygen attached to the carbon atom of a peptide bond farther down the chain (Figure 21.8a). There are 3.6 amino acids in each turn of the α helix, and the R-groups on these amino acids extend outward from the helix.

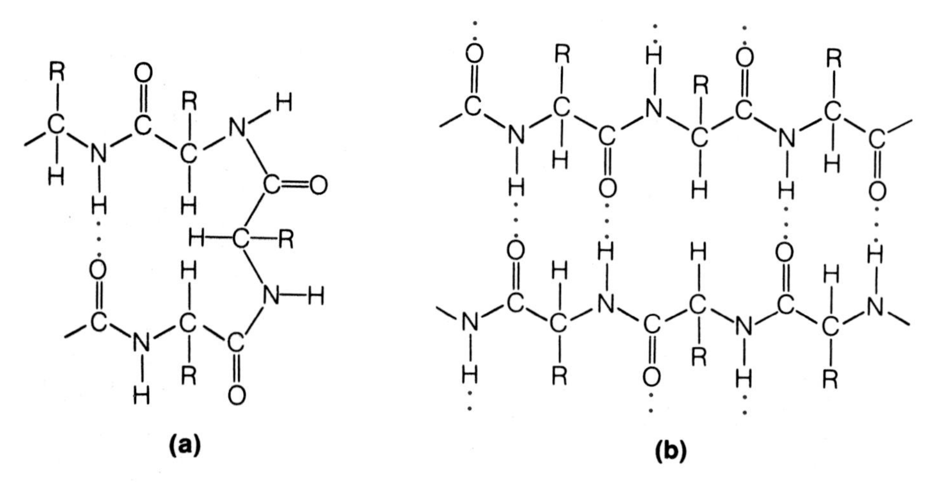

Figure 21.7 Hydrogen bonds can form (a) between amino acids on the same polypeptide chain, forming a loop in the molecule, or (b) between amino acids on different polypeptide chains.

In keratin, three of these α helixes are wound together—much like the fibers in a rope—to form a protofibril that is held together by disulfide bridges (see Section 21.10). The greater the number of disulfide bridges, the less flexible and harder the keratins. Wool fibers are made up of a great number of these fibrils imbedded in an insoluble protein framework.

Collagens

Collagens are the most abundant protein in the body and are found in skin, bones, teeth, tendons, cartilage, blood vessels, and connective tissue. Because of the types of amino acids found in collagen, the polypeptides cannot form an α helix. Instead, three polypeptide chains are twisted

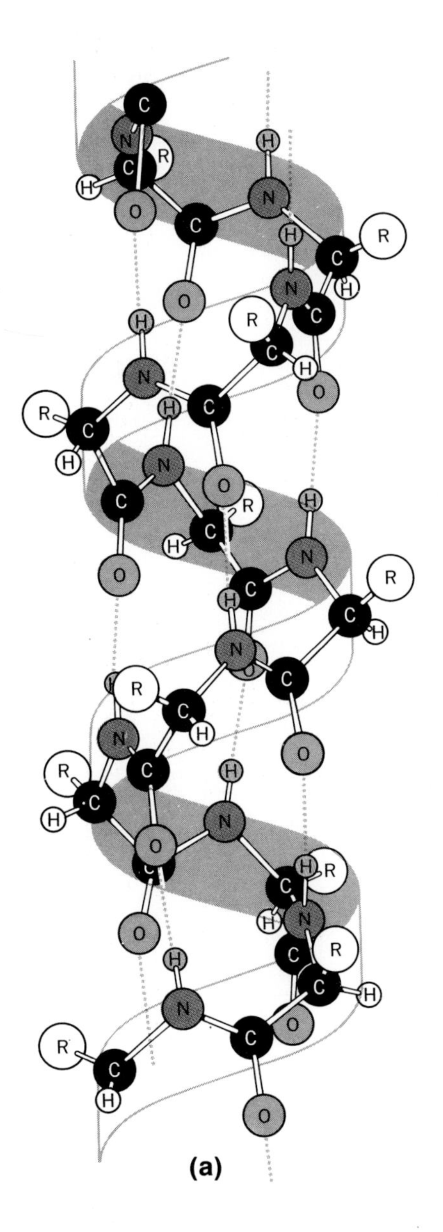

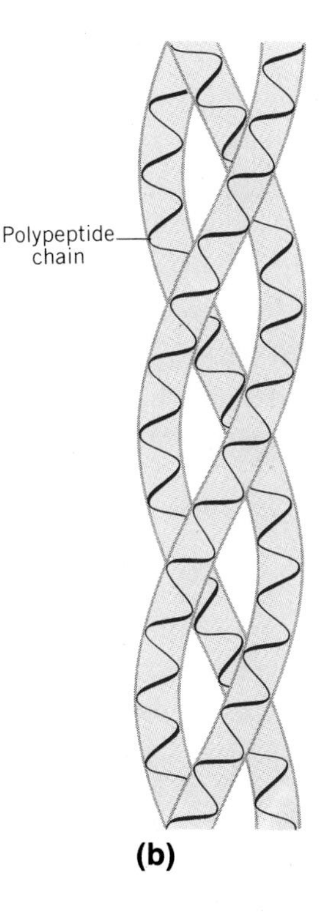

Figure 21.8 (a) Model of a polypeptide chain in an alpha helix configuration. The amino acids are coiled in a manner resembling a circular staircase, with the loops held together by hydrogen bonds. (b) The triple stranded helix of collagen [(a) Adapted from B. Low and E.T. Edsall, *Currents in Biochemical Research,* Wiley-Interscience, New York, 1956. Used by permission.]

together to form a triple helix structure called tropocollagen (Figure 21.8b), which makes collagen fibers extremely strong.

Silks

The fibrous protein of silk has a different secondary structure. In silk, several polypeptide chains oriented in different directions are located next to each other in an arrangement called a ***β* configuration** (beta configuration). This gives the protein a zig-zag appearance, from which we get the name *pleated sheet.* The polypeptide chains in silks are held together only by extensive hydrogen bonding. Note that the R-groups extend above and below the sheet (Figure 21.9).

Each protein will have a specific secondary structure depending upon the amino acid sequence. For example, the α helix is formed when the R-groups are small and uncharged. However, this arrangement will be disrupted by the amino acid proline, which has no amide hydrogen and, therefore, can't form hydrogen bonds. Kinks or bends in the molecule are found at proline positions. In a large protein it is possible to have separate areas of α helix and pleated sheet arrangements (Figure 21.10). Note that

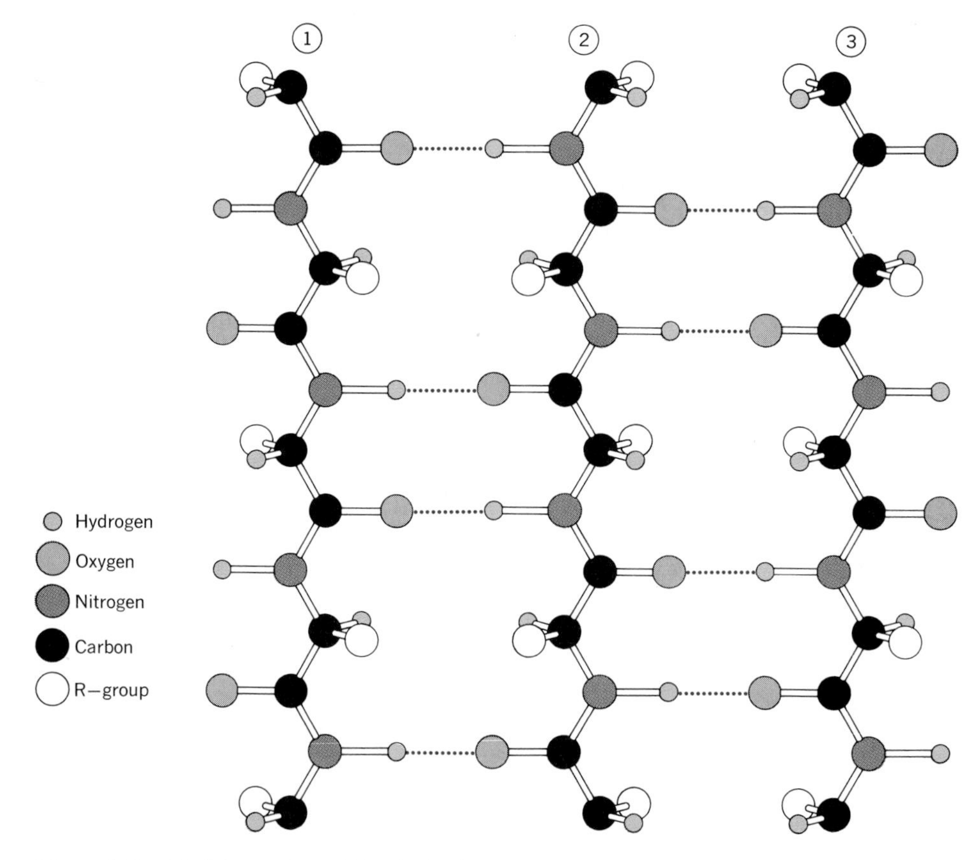

Figure 21.9 Model of polypeptide chains in a pleated sheet structure. In this configuration, several polypeptide chains are held together by hydrogen bonds. The R-groups extend above and below the plane of the sheet.

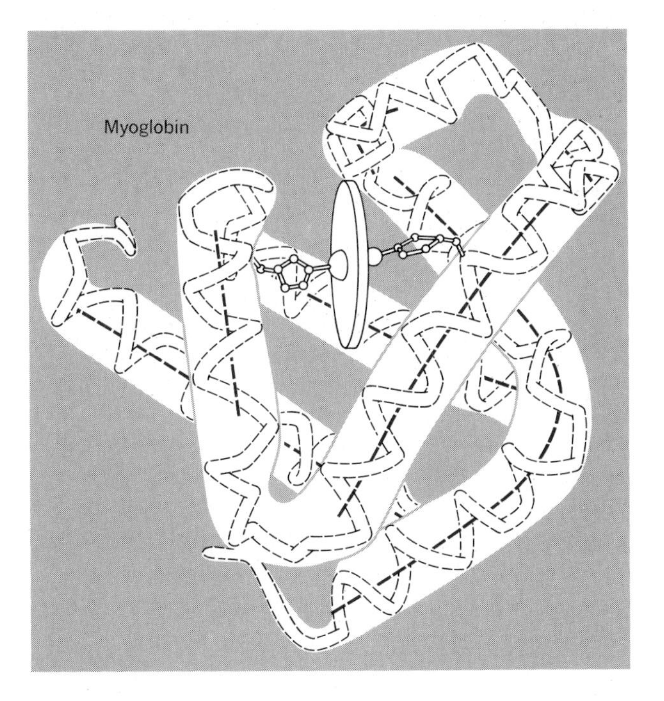

Figure 21.10 The tertiary structure of a molecule of myoglobin consists of eight sections of alpha helix surrounding a heme group (drawn as a flat disk). The main polypeptide chain is shown here without the R-group side chains. (From R.E. Dickerson and I. Geis, *The Structure and Action of Proteins,* W.A. Benjamin, Inc., Menlo Park. Copyright © 1969 by Dickerson and Geis.)

hydrogen bonding is weak, noncovalent bonding and is easily disrupted by changes in pH, temperature, solvents, or salt concentrations.

21.10 Tertiary Structure: Globular Proteins

Tertiary structure is the three-dimensional structure of globular proteins. At normal pH and temperature, each protein will take on a shape that is energetically the most stable given the specific sequence of amino acids and the various types of interactions that may be involved. This shape is called the **native state** or **native configuration** of the protein. In general, globular proteins are very tightly folded into a compact, spherical form (Figure 21.10). This folding results from interactions between the R-group side chains of amino acids and may involve hydrogen bonding and the other following interactions.

Disulfide Bridges

Disulfide bridges are interactions that include disulfide linkages between molecules of the amino acid cysteine. Sulfhydryl groups (—S—H) are easily oxidized to form a disulfide (—S—S—). If this reaction occurs between two cysteines, the amino acid cystine is formed.

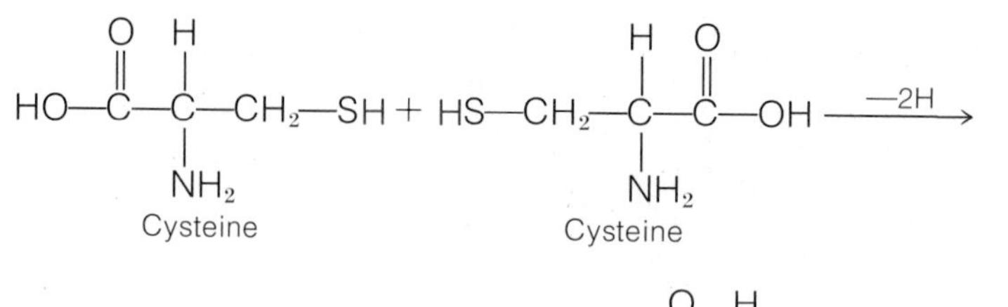

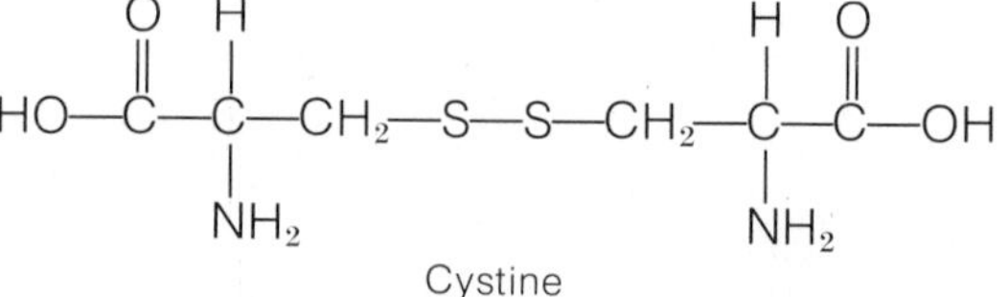

This disulfide linkage can be formed between two cysteines on different amino acid chains, in this way linking the two chains together to form such tough, strong material as keratin. Or, it can be found between cysteines on the same amino acid chain, creating a loop in the chain. The insulin molecule contains examples of both types of disulfide bridges (Figure 21.6). These bridges are covalent linkages that can be broken by reduction, but which are stable in the face of changes in pH, solvents, or salt concentrations.

Salt Bridges

Salt bridges result from ionic interactions between charged carboxyl or amino side chains found on amino acids such as aspartic acid, glutamic acid, lysine, and arginine (see Figure 21.3). These linkages are very easily broken by changes in pH.

Hydrophobic Interactions

Hydrophobic interactions occur between the nonpolar side chains of the amino acids in the protein molecule, and are perhaps the most important interactions in determining protein conformation. Because they are repulsed by water, these nonpolar R-groups tend to be found on the inside of the molecule together with other hydrophobic groups.

21.11 Quaternary Structure: Two or More Polypeptide Chains

The **quaternary structure** of a protein is the manner in which separate polypeptide chains and prosthetic groups fit together in proteins containing more than one chain. Hydrogen bonding, hydrophobic interactions, and salt bridges may be involved in holding the chains in position. Hemoglobin is just one example of a protein that contains more than one polypeptide chain. It consists of four polypeptide chains—two alpha chains and two beta chains—each arranged around an iron-containing heme group (Figure 21.11).

21.12 Denaturation

Various changes in the surroundings of a protein can disrupt the complex secondary, tertiary, or quaternary structure of the molecule. Disruption of the native state of the protein is called **denaturation.** This process involves

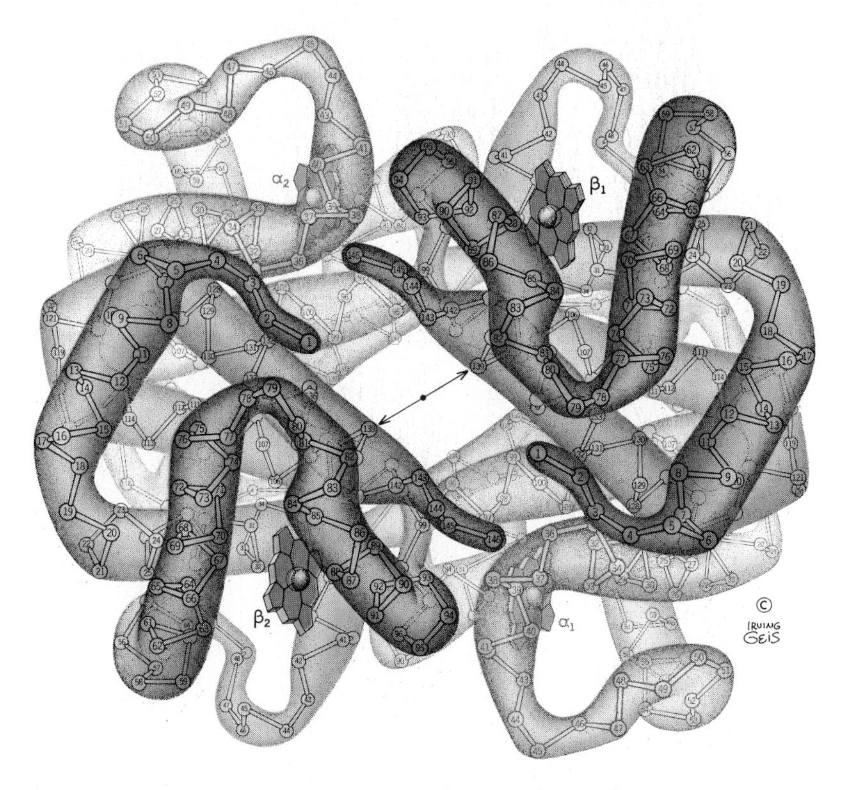

Figure 21.11 The quaternary structure of hemoglobin. The hemoglobin molecule consists of four polypeptide chains, each surrounding a heme group. (Copyright by Irving Geis.)

the uncoiling of the protein molecule into a random state and will cause the protein to lose its biological activity (Figure 21.12). Denaturation may or may not be permanent; in some cases the protein will return to its native state when the denaturing agent is removed. Denaturation may also result in coagulation, with the protein being precipitated from solution. Denaturing agents come in many forms, a few of which are as follows.

pH

Changes in pH have their greatest disruptive effect on hydrogen bonding and salt bridges. For example, the polypeptide polylysine is composed

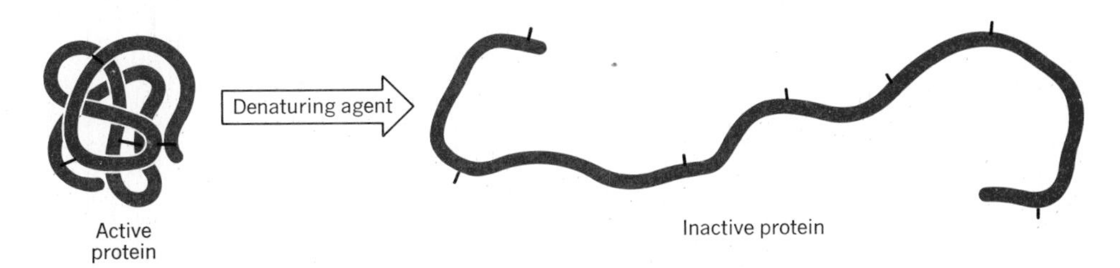

Figure 21.12 Denaturation of a protein disrupts the tertiary structure of the molecule, causing it to uncoil. This results in the loss of protein activity.

entirely of the amino acid lysine, which has an amino group on its side chain. In acidic pH all the side chains will be positively charged and will repel each other, causing the molecule to uncoil. In basic pH, however, the side chains will be neutral. Therefore, they do not repel, and the molecule will coil into an α helix. Exposing proteins to strong acids or bases for long periods of time will lead to the hydrolysis of the peptide chain.

Heat

Heat causes an increase in the thermal vibration of the molecule, disrupting hydrogen bonding and salt bridges. After gentle heating, the protein can usually regain its native state. However, violent heating will result in irreversible denaturation and coagulation of the protein (as, for example, in cooking an egg). Similarly, heat used to sterilize equipment coagulates the protein of microorganisms. Heat can also be used to detect the presence of protein in urine; urine that turns cloudy when heated indicates the presence of protein.

Organic Solvents

Organic solvents such as alcohol or acetone can form hydrogen bonds with protein molecules, which then compete with the hydrogen bonds naturally occurring in the protein. This causes denaturation and coagulation. A 70% alcohol solution is a good disinfectant because it will coagulate the protein in bacteria, thus destroying these organisms.

Heavy Metal Ions

Heavy metal ions such as Pb^{2+}, Hg^{2+}, and Ag^{+} may disrupt the natural salt bridges of a protein by forming salt bridges of their own with the protein. This usually causes coagulation of the protein. Organomercury ions such as methyl mercury (CH_3Hg^+), which are present in mercury pollution, coagulate proteins in this way. Heavy metal ions also bind with sulfhydryl groups, disrupting the disulfide linkages in the protein molecule and denaturing the protein. Such heavy metals, therefore, are toxic to living organisms. As an antidote to heavy metal poisoning, patients are given substances high in protein, such as milk or egg whites. These substances will bond with the metals while they are still in the stomach. The victim must then be made to vomit before the metals are again released through the processes of digestion.

Alkaloidal Reagents

Reagents such as tannic or picric acid affect salt bridges and hydrogen bonding, causing proteins to precipitate. Tannic acid is used to precipitate proteins in animal hides; this is the process of tanning used in the manufacture of leather.

Reducing Agents

Reducing agents disrupt disulfide bridges formed between cysteine molecules. For example, hair permanents work by reducing and disrupting the disulfide bridges in hair. When the hair is curled and oxidizing agents are applied, new disulfide bridges are formed (Figure 21.13).

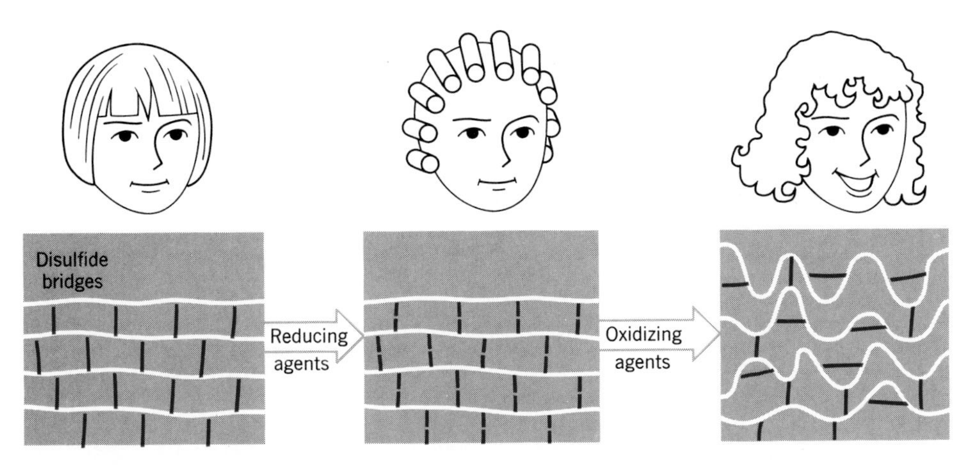

Figure 21.13 Permanent waves work by breaking the disulfide bridges between the cysteines in hair protein and then reforming the bridges in different locations.

Radiation

Nonionizing radiation is similar to heat in its effect on proteins. For example, ultraviolet light will denature proteins in the skin, causing sunburn.

21.13 Identification of Proteins and Amino Acid Sequences

Many color tests and separation procedures are used in protein research. Two commonly used color tests are the biuret test and ninhydrin test.

Biuret Test

The biuret test detects the presence of two or more peptide linkages. The unknown solution is mixed with sodium hydroxide and a few drops of dilute copper sulfate. If a violet color appears, peptides larger than dipeptides are present. This color test is often used to follow the progress of protein hydrolysis.

Ninhydrin Test

The ninhydrin test is widely used to detect the presence of amino acids, peptides, and proteins. The unknown solution is heated with excess ninhydrin, and an intense purple color indicates a positive test (except in the case of the amino acid proline, which gives a yellow color).

Exercise 21-3

1. Describe the types of bonding or attractive interactions that occur in the following protein structures:

(a) the primary structure of a protein
(b) the secondary structure
(c) the tertiary structure
(d) the quaternary structure

2. Define native state and denaturation.
3. Which structure or structures of a protein will be disrupted by

(a) a strong acid (b) sunlight (c) lead (Pb^{2+}) ions

CHAPTER SUMMARY

Proteins are large polymers of amino acids and have many functions in the cell. They may be simple proteins that contain only amino acids, or complex proteins that contain other nonprotein or inorganic components in addition to amino acids. Enzymes, hormones, antibodies, and transport proteins are all globular proteins. Keratin, muscle fibers, and collagen are examples of fibrous proteins. Amino acids contain an amino group and a carboxyl group. As a result, they are amphoteric and can act as buffers. Each amino acid and protein will have a characteristic isoelectric point; this is also the pH at which the protein is the least soluble.

Protein molecules are quite complex. The primary structure of a protein is the sequence of amino acids, held together by peptide bonds, in the polypeptide chain. The secondary structure is the particular shape of the protein, which results from noncovalent interactions between different parts of the peptide. Alpha helix, beta configuration, and triple helix are all examples of different types of secondary structures. Tertiary structure refers to the three-dimensional structure of globular proteins and results from interactions between the R-group side chains of the amino acids. These interactions include hydrogen bonding, disulfide bridges, salt bridges, and hydrophobic interactions. Some proteins contain more than one polypeptide chain, and the way in which these chains fit together is called the quaternary structure of the protein. The native state of a protein is the most stable structural arrangement for that protein. Denaturation is the disruption, whether permanent or temporary, of the native state of a protein. Denaturation can be caused by heat, radiation, changes in pH, organic solvents, heavy metal ions, alkaloidal reagents, or reducing agents.

1. Describe the difference between

(a) a simple protein and a conjugated protein
(b) a globular protein and a fibrous protein
(c) a glycoprotein and a lipoprotein

(d) the primary structure and the secondary structure of a protein
(e) a dipeptide and a polypeptide
(f) alpha helix and beta configuration
(g) tertiary structure and quaternary structure
(h) vegetable protein and animal protein

2. Give the common name for the following amino acids:

   (a) aminoacetic acid
   (b) 2-aminopropanoic acid
   (c) 2-amino-4-methylpentanoic acid
   (d) 2-amino-3-methylbutanoic acid

3. Why does protein in the urine indicate a possible kidney disorder?

4. What explains the difference between the solubility of globular and fibrous proteins?

5. What is meant by an adequate protein? What are the problems associated with eating a strictly vegetarian diet?

6. (a) Is a diet consisting mainly of rice an adequate diet? Why or why not?
   (b) Suggest several ways in which a diet consisting mainly of corn can be supplemented to provide enough adequate protein.

7. Describe how blood proteins help protect against pH changes.

8. If placed in an electric field, how would each of the following proteins migrate if the pH of the solution were 7?

   (a) urease (b) myoglobin (c) chymotrypsin

9. Albumin, the most abundant protein in the blood plasma, transports slightly soluble molecules such as aspirin, bilirubin, and barbiturates through the plasma and extracellular fluid. Albumin has an isoelectric point of 4.8. What charge does this molecule have at physiologic pH (pH 7.4)?

10. Why is a protein the least soluble at its isoelectric point?

11. What fact about the structure of amino acids explains their high melting points?

12. What type of bonding occurs in

    (a) alpha helix configuration
    (b) globular proteins
    (c) beta configuration
    (d) triple helix configuration

13. What structural difference makes the keratins of wool flexible and those of hooves rigid?

14. Which would be more completely disrupted by heat: the structure of keratin or silk? Give the reason for your answer.

15. Milk is pasteurized by heating it to 143°F (61.7°C) for 30 minutes. Why is this process effective in preventing contamination of the milk by microorganisms?

16. Why are eggs a good antidote for lead poisoning? What additional treatment is required when an egg is used as an antidote?

17. Ultraviolet lights are often used in research areas to prevent contamination by foreign bacteria. Why is this an effective method of sterilization?

18. Explain, on the molecular level, the process of hair straightening.

19. (a) Write the formula for leucine in its zwitterion form.
    (b) Write the equation for the addition of acid to the solution in (a).
    (c) Write the equation for the addition of base to the solution in (a).

20. Write the structure and identify each of the products of the acid hydrolysis of the following polypeptide:

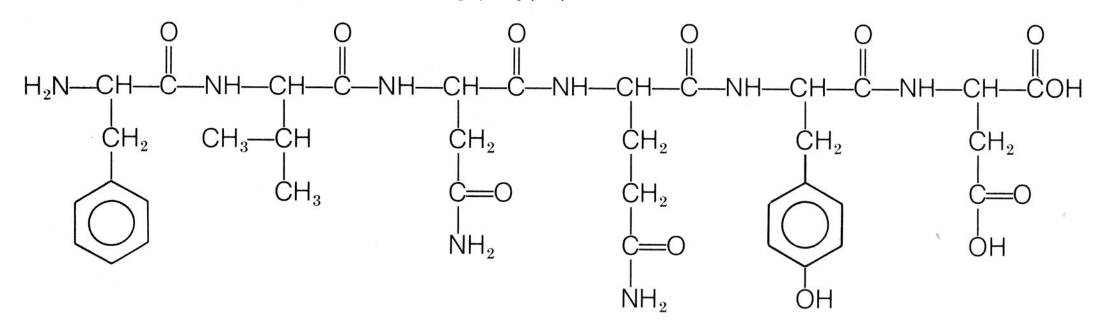

21. Using the three-letter abbreviations for the amino acids, write the amino acid sequence for the polypeptide in question 20.

22. The following is the amino acid sequence of the hormone vasopressin:

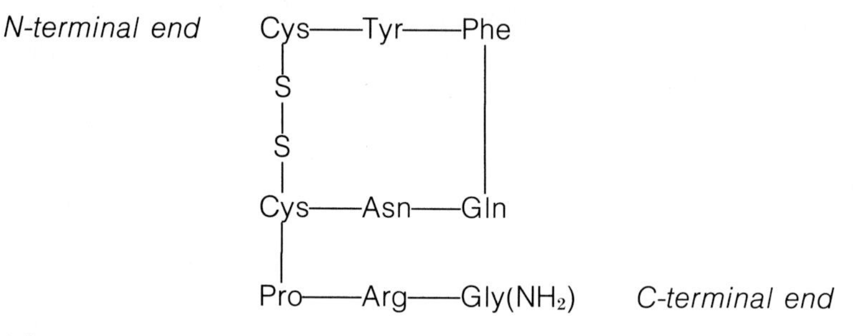

    (a) Draw the structural formula of vasopressin. The notation $Gly(NH_2)$ means that the carboxyl group of Gly has reacted to form an amide $CONH_2$.
    (b) What kind of linkage other than a peptide bond is present in this molecule?

CHAPTER TWENTY-TWO

ENZYMES, VITAMINS, AND HORMONES

Kathy Phipps felt she hadn't a care in the world. She loved her job teaching sixth grade science and math courses, her two children were both doing well in school, and her husband had just been promoted to manager of his department.

Well, she actually did have one little worry. When she had examined her breasts the previous month, she thought she had detected a small lump. But she hadn't been sure, and wasn't always able to find it again. But lately there was no question that the lump was there, and this is what worried Kathy. Her mother had developed breast cancer and had undergone a radical mastectomy about 15 years earlier. Kathy cringed at the thought of losing a breast and looking the way her mother did. She promised herself that if the lump was still there in a month, she would go see her doctor.

Dr. Huntington felt the lump in Kathy's breast and was dismayed to learn that Kathy had first discovered it more than three months ago. He sent Kathy for a mammogram and telephoned her the next day to tell her the lump appeared to be cancerous. He had already scheduled her for a biopsy of the lump and the lymph nodes under her arm. If the pathology reports from the biopsy were positive, he said, then they would discuss the various options for her treatment.

Three days later Kathy learned that the biopsy had found malignant (cancerous) cells both in the lump and in the lymph nodes under her arm. This meant that the malignant cells had begun to spread, or metastasize, into her body. Dr. Huntington explained that she could choose between two courses of treatment: she could undergo surgical removal of the breast (a modified radical mastectomy) followed by chemotherapy to kill the metastatic cells, or she could be treated with chemotherapy followed by irradiation of her breast to kill any cancerous cells remaining in the area of the biopsy incision. The choice was hers—both treatments had shown similar survival rates in clinical trials. Because of her strong negative feelings about mastectomy, Kathy chose to be treated with chemotherapy and radiation.

The following week Kathy began a six-month program of chemotherapy treatment. During this period she was given three drugs in combination: methotrexate, 5-fluorouracil, and cyclophosphamide. Used together, these drugs enhance each others' anticancer properties by killing cells in the body that are rapidly dividing. Unfortunately, cancer cells are not the only cells in a person's body that are rapidly dividing, and Kathy's treatment worked to disrupt and damage these normal tissues as well as the cancer cells.

Dr. Huntington had warned Kathy about these well-known side effects of the chemotherapy. She had been told, for example, that bone marrow cells reproduce very rapidly in order to continually replace the cells in the blood. That is why Kathy needed to have her blood tested weekly to monitor the decrease in her white blood cell count and platelet count. Too severe a decrease in these blood cells would require stopping her chemotherapy. Similarly, in our bodies, the mucous linings of the mouth and intestines are constantly being sloughed off and replaced (the mucosa of the small intestine replaces itself every 48 hours!). The cells producing this lining are among those disrupted by the chemotherapy, commonly resulting in sores in the mouth as well as nausea and vomiting. As a final visible addition to such side effects, cyclophosphamide also causes the hair to fall out. Although she often felt tired and listless, Kathy bore up quite well under the side effects of the chemotherapy and was able to return to her classroom and finish the term.

In this chapter we will learn about enzymes, which are critical to the functioning of living cells. In particular, we will see how the inhibition of enzyme function can lead to disruption and death of the cell. 5-Fluorouracil and methotrexate are antimetabolites, chemicals that obstruct enzyme function. In this case, both compounds inhibit the function of enzymes that are needed for the synthesis of DNA in the cell. But the synthesis of DNA is a necessary step in the division of a cell. By acting to prevent cell reproduction, these two chemicals have especially damaging effects on rapidly dividing cells such as cancer cells. This gives physicians a way to kill cancer cells that have broken off from a tumor and are circulating in the blood.

Extensive research into the structure and function of enzymes has led not only to the development of chemotherapy drugs that inhibit enzyme function, but has also identified ways to monitor specific enzymes that are released into the blood by damaged cells. The level of these enzymes in the blood can tell a physician such things as how much damage has been done to the heart muscle after a heart attack, or whether a patient is suffering from infectious hepatitis. We will start our study of enzymes, however, by learning the language used to describe these important molecules.

LEARNING OBJECTIVES

By the time you have finished this chapter, you should be able to:

1. Define *metabolism* and *anabolic and catabolic reactions*.
2. Identify the parts of an enzyme molecule.
3. Describe the method by which enzymes catalyze reactions.
4. Explain the *lock-and-key* and *induced fit* theories of enzyme action.
5. Explain how changes in pH and temperature will affect enzyme activity.
6. Define *vitamin,* and explain why vitamins are essential for normal cellular function.
7. Describe the function of hormones in the living organism.
8. Define *multienzyme system,* and explain how such systems are regulated.
9. Describe the ways in which enzyme activity can be inhibited, and give examples of each type of inhibition.

ENZYMES: METABOLIC CATALYSTS

22.1 What is Metabolism?

Metabolism can be defined as all the enzyme-catalyzed reactions in the body. These reactions are of two types: **catabolic reactions,** in which molecules are broken down to produce smaller products and cellular energy, and **anabolic reactions,** in which the cell uses energy to produce the molecules it needs for growth and repair (these are also called biosynthetic reactions). Anabolic and catabolic reactions occur continuously in the cell, and each involves a complicated series of carefully controlled enzyme-catalyzed reactions.

22.2 Enzymes

Enzymes are the largest and most highly specialized class of proteins. They function as biological catalysts in the body for the reactions involved in metabolism. Earlier we described catalysts as substances that increase the rate of a chemical reaction by lowering the activation energy of that reaction. The reactions of metabolism would ordinarily occur at extremely slow rates at normal body temperature and pH. Without the enzymes in our digestive tract, for example, it would take us about 50 years to digest a single meal! Enzymes greatly increase this reaction rate, allowing cells to function under normal body conditions. Because a catalyst is not consumed in the reaction, it can be used over and over again. Such compounds, therefore, need to be present in only very small amounts. Enzymes are water-soluble globular proteins that can vary from a molecular weight of 12,000 to over 1 million. The enzyme molecule may consist of a fairly simple single polypeptide chain, or it may be a more complex molecule composed of several polypeptide chains and other nonprotein parts.

For example, ribonuclease is the enzyme that hydrolyzes ribose-containing nucleotides; it consists of a single polypeptide chain of 124 amino acids and has a molecular weight of about 13,700 daltons (remember, 1 dalton = 1 amu). Pyruvate dehydrogenase, on the other hand, is the enzyme that catalyzes the conversion of pyruvate to acetyl CoA; it is a macromolecular complex containing 42 individual molecules and having a molecular weight of about 10 million daltons.

22.3 Some Important Definitions

Enzymes may be simple proteins made up entirely of amino acids, or may be conjugated proteins that require a nonprotein group for their biological activity. Before continuing with our discussion of enzymes, however, it will be helpful for us to define some special terms used in the study of these complex molecules (see Figure 22.1).

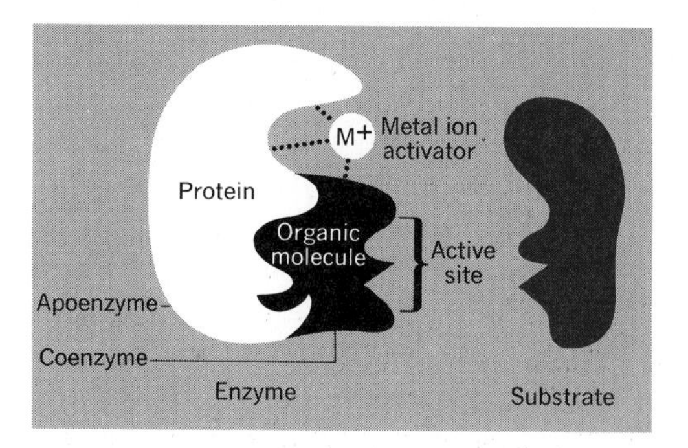

Figure 22.1 Some important terms used in the study of enzymes.

Apoenzyme. Apoenzyme is a term used to describe the protein part of the enzyme molecule.

Cofactor. Cofactors are the additional chemical groups appearing in those enzymes that are conjugated proteins. These cofactors are required for enzyme activity and may consist of metal ions or complex organic molecules. Some enzymes require both types of cofactors.

Activator. When the cofactor of an enzyme is a metal ion (such as the ions of magnesium, zinc, iron, or manganese), the cofactor is called an activator.

Coenzyme. When the cofactor of an enzyme is a complex organic molecule other than a protein, the cofactor is called a coenzyme.

Prosthetic group. In the enzyme molecule the cofactor may be weakly attached to the apoenzyme, or it may be tightly bound to the protein. If it is tightly bound, the cofactor is called a prosthetic group.

Isoenzyme. Enzymes that perform the same catalytic function in different body tissues or different organisms, but which have different sequences of amino acids in various portions of their polypeptide chain are called isoenzymes. Isoenzymes can be separated from one another by electrophoresis.

Proenzyme or zymogen. Proenzyme (or zymogen) is the name given to the inactive form of an enzyme. Enzymes (especially digestive enzymes) are often secreted in their inactive form, transported to the place where activity is desired, and then converted to their active forms.

Substrate. The substrate is the chemical substance or substances upon which the enzyme acts.

Active site. The active site is the specific area of the enzyme to which the substrate attaches during the reaction.

22.4 Enzyme Nomenclature and Classification

Because enzymes are the largest class of proteins, a great number of specific enzymes have been isolated and described. At first, the only general rule of nomenclature was to end the enzyme name with the suffix *-in* to indicate a protein. Names such as trypsin, renin, and pepsin are examples of such nomenclature and are still used today. However, such names give no indication of the reaction being catalyzed or the substrate involved.

In 1961 the Commission on Enzymes of the International Union of

Biochemistry proposed a standard classification of enzymes. They recommended that enzymes be divided into six major classes (each with several subclasses) based on the reactions that are catalyzed (Table 22.1).

Table 22.1 Classes of Enzymes

Hydrolases: Enzymes that catalyze hydrolysis reactions.

Example	Reaction Catalyzed
Carbohydrases:	Polysaccharides and disaccharides $\xrightarrow{+H_2O}$ Monosaccharides
Esterases:	Ester $\xrightarrow{+H_2O}$ Acid + alcohol
Proteases:	Protein $\xrightarrow{+H_2O}$ Peptides and amino acids
Nucleases:	Nucleic acids $\xrightarrow{+H_2O}$ Pyrimidines + purines + sugars + phosphoric acid

Oxido-Reductases: Enzymes that catalyze oxidation-reduction reactions.

Example	Reaction Catalyzed
Oxidases:	Addition of oxygen to a substrate
Dehydrogenases:	Removal of hydrogen from a substrate

Transferases: Enzymes that catalyze reactions involved in the transfer of functional groups.

Example	Reaction Catalyzed
Transaminases:	Transfer of $-NH_2$
Transmethylases:	Transfer of $-CH_3$
Transacylases:	Transfer of $-\overset{\overset{\displaystyle O}{\|}}{C}-R$
Transphosphatases: (Kinases)	Transfer of $-O-\overset{\overset{\displaystyle O}{\|}}{\underset{\underset{\displaystyle OH}{\|}}{P}}-OH$

Lyases: Enzymes that catalyze the addition to double bonds.

Isomerases: Enzymes that catalyze the interconversion of isomers.

Ligases: Enzymes that, in conjunction with ATP, catalyze the formation of new bonds.

22.5 Method of Enzyme Action

We have stated that enzymes catalyze reactions in cells by lowering the activation energy. They do this by forming a complex with the substrate (that is, by attaching to the substrate) that then increases the probability that the reaction will occur. We can outline the steps of an enzyme-catalyzed reaction as follows:

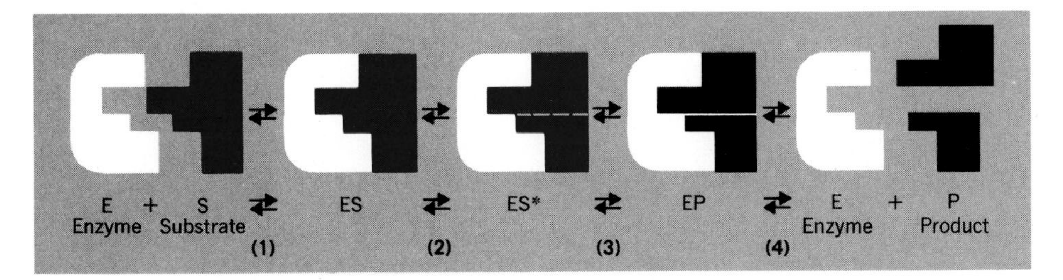

1. The substrate (or substrates) becomes weakly bound to the enzyme surface, forming the enzyme-substrate complex.

$$E + S \rightleftharpoons ES$$

2. The substrate becomes activated; that is, the bonds in the substrate become polarized.

Indicates activated state

$$ES \rightleftharpoons ES^*$$

3. The products of the reaction form on the surface of the enzyme.

$$ES^* \rightleftharpoons EP$$

4. The products are released or "kicked off" the surface of the enzyme, making the enzyme available to catalyze further reactions.

$$EP \rightleftharpoons E + P$$

The rate at which an enzyme catalyzes a reaction will vary with different cellular conditions, and from enzyme to enzyme. Some enzymes, such as carbonic anhydrase, are extremely efficient in catalyzing a reaction. Such efficiency of enzyme action is measured by the **turnover number,** which is the number of substrate molecules transformed per minute by one molecule of enzyme under optimal conditions of temperature and pH (Table 22.2).

EXAMPLE 22-1

Cholinesterase, the enzyme that catalyzes the breakdown of acetylcholine to acetate and choline at nerve endings, has a

Table 22.2 Turnover Numbers of Some Enzymes

Enzyme	Turnover Number (molecules of substrate per minute)
Carbonic anhydrase	36,000,000
Sucrose invertase	1,000,000
Glutamic dehydrogenase	30,000
Phosphoglucomutase	1,240
Chymotrypsin	100
DNA polymerase	15

turnover number of 1.5×10^6 molecules/minute. How many molecules of acetylcholine can one molecule of cholinesterase hydrolyze in a second?

From the problem:

$$1 \text{ molecule hydrolyzes } \frac{1.5 \times 10^6 \text{ molecules}}{1 \text{ minute}}$$

and

$$1 \text{ minute} = 60 \text{ seconds}$$

Therefore, 1 molecule will hydrolyze

$$\frac{1.5 \times 10^6 \text{ molecules}}{\cancel{1 \text{ minute}}} \times \frac{\cancel{1 \text{ minute}}}{60 \text{ seconds}} = \frac{1.5 \times 10^6 \text{ molecules}}{60 \text{ seconds}}$$

$$= 0.025 \times 10^6 \frac{\text{molecules}}{\text{second}} = 2.5 \times 10^4 \frac{\text{molecules}}{\text{second}}$$

Exercise 22-1

1. Define the following terms in your own words:

   (a) apoenzyme
   (b) coenzyme
   (c) activator
   (d) cofactor
   (e) proenzyme
   (f) substrate
   (g) active site
   (h) turnover number
   (i) hydrolases
   (j) oxido-reductases
   (k) transferases
   (l) lyases
   (m) isomerases
   (n) ligases

2. Chymotrypsin catalyzes the hydrolysis of proteins in the small intestine. How many moles of protein can be hydrolyzed in 1 hour by 100 molecules of chymotrypsin?

22.6 Specificity

One of the main differences between enzymes and inorganic catalysts is the specificity of enzymes. For example, platinum will catalyze several different types of reactions. However, each enzyme will catalyze only one type of reaction, and in some cases will limit its activity to only one particular type of reactant molecule.

Pancreatic lipase, for example, will hydrolyze the ester linkage between glycerol and fatty acids in lipids, but will have no effect on the hydrolysis of proteins or carbohydrates. Kidney phosphatase catalyzes the hydrolysis of esters of phosphoric acid, but at a different rate for each substrate. Urease is even more specialized; it will catalyze only the hydrolysis of urea. An extreme example of the specificity of enzyme action is given by the enzyme aspartase, which will catalyze only the following reversible reaction.

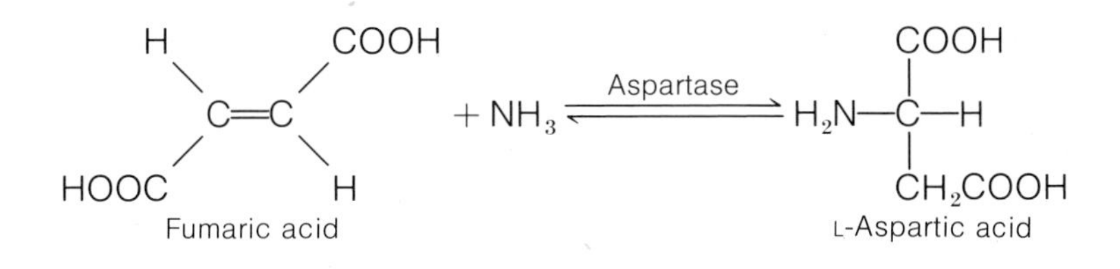

This enzyme will not catalyze the addition of ammonia to any other unsaturated acid—not even maleic acid, which is the *cis*-isomer of fumaric acid.

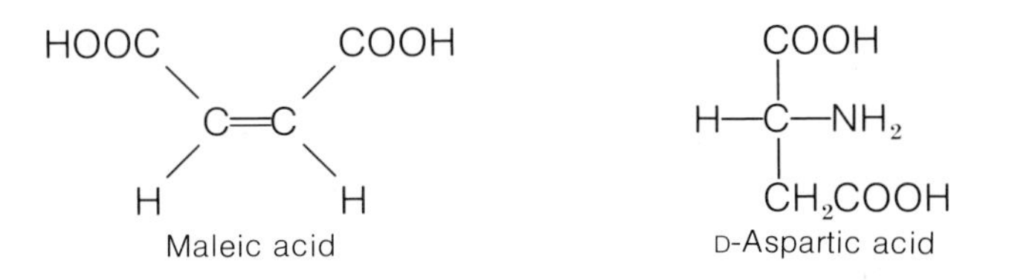

Moreover, aspartase will not even catalyze the removal of ammonia from D-aspartic acid.

22.7 Lock-and-Key Theory

How can we explain this specificity of enzymes? There is a particular area, called the **active site,** that is found on the surface of the enzyme molecule. This is where the substrate attaches during a reaction. The shape, or configuration, of this active site is especially designed for the specific substrate involved. Because the configuration is determined by the amino acid sequence of the enzyme, the native configuration of the entire enzyme molecule must be intact for the active site to have the correct configuration. In such a case, the substrate then fits into the active site of the enzyme in much the same way as a key fits into a lock. The configuration of the lock is specific for only one key; no other keys will turn the lock (Figure 22.2).

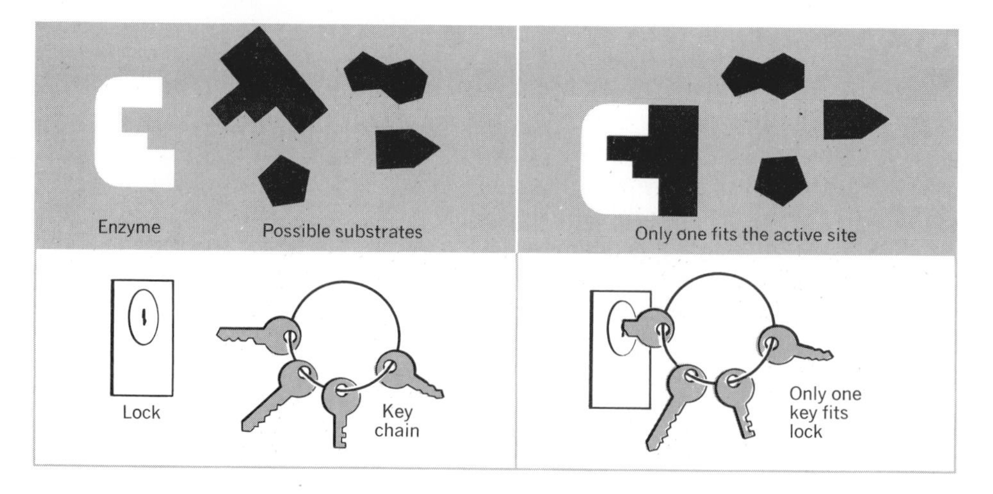

Figure 22.2 Many enzymes are very specific, often limiting their action to only one particular molecule (just as a lock is specific for only one key).

Not only is the geometry of the active site important, but so also is the arrangement of charged R-groups around the active site. There is an electrical attraction between the substrate and the enzyme that pulls them together. The reaction products that then occur, however, will have a different distribution of charges, and the repulsion between the products and the active site may cause the products to be "kicked off" the enzyme.

22.8 Induced-Fit Theory

The model of a lock-and-key fit, which helps to explain the action of some enzymes, must be slightly changed for other enzymes. In these cases a better comparison would be a hand slipping into a glove, which then causes, or induces, a fit. Enzyme molecules are flexible, and the active site of some enzymes may not initially match the substrate. However, the substrate itself, as it is drawn to the enzyme, may induce the enzyme to take on a shape that matches the substrate. In such cases the active site is

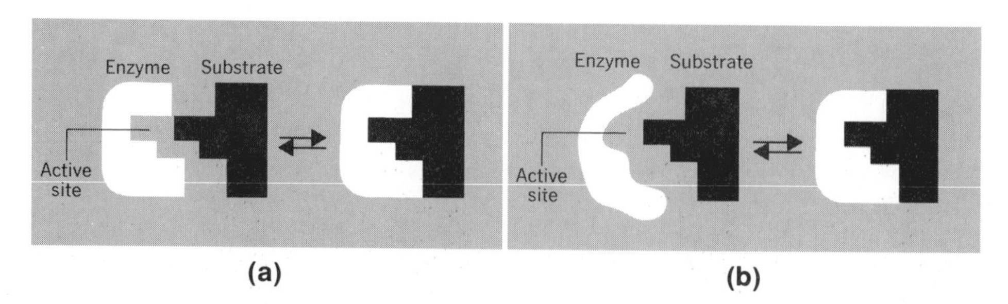

Figure 22.3 Two theories of enzyme action. (a) In the lock-and-key theory, the active site conforms exactly to the substrate molecule. (b) In the induced-fit theory, the substrate induces the active site to take on a shape complementary to the shape of the substrate molecule.

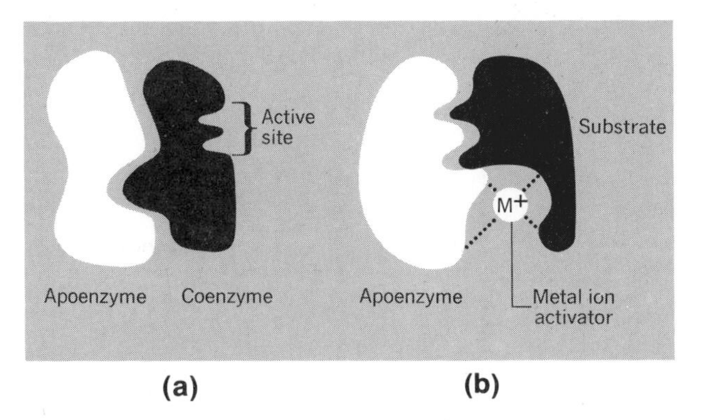

Figure 22.4 Cofactors may contribute to the activity of the enzyme (a) by providing the active site or (b) by forming a bridge between the enzyme and the substrate.

still quite specific to the substrate, just as a left-hand glove will not normally fit a right hand (Figure 22.3).

Enzymes that are secreted as zymogens have their active sites blocked. To activate the enzyme, these sites must be unblocked by the hydrolysis of part of the molecule. Cofactors contribute to the activity of the enzyme either by providing the arrangement of molecules necessary for the active site, or by forming a bridge between the substrate and the enzyme (Figure 22.4).

22.9 Factors Affecting Enzyme Activity

We have seen that factors such as pH, temperature, solvents, and salt concentrations can change the structure of a protein. Such factors, therefore, will have an effect on the activity levels of enzymes.

pH

Changes in the pH of the surrounding medium can change the secondary or tertiary structure of an enzyme. This may alter the geometry of the active site or the surrounding charge distribution. Each enzyme has a certain pH at which it is most active. For example, pepsin has an optimum pH of 1.5, whereas trypsin has its maximum activity at pH 8 (Figure 22.5).

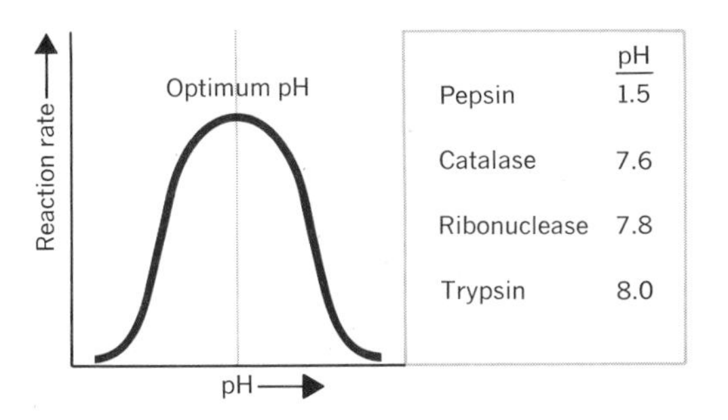

	pH
Pepsin	1.5
Catalase	7.6
Ribonuclease	7.8
Trypsin	8.0

Figure 22.5 Each enzyme has its highest activity at a specific pH.

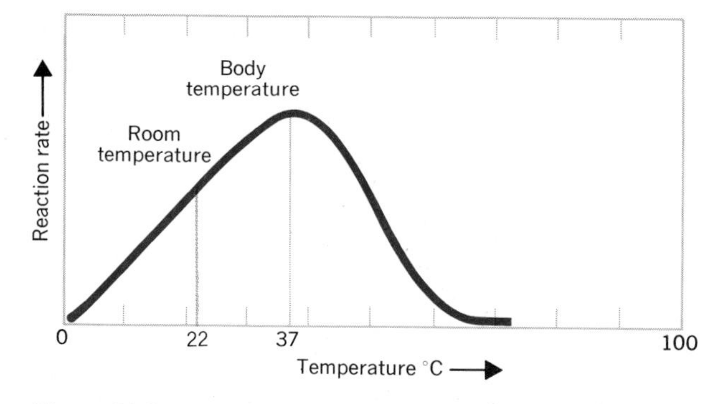

Figure 22.6 Temperature affects enzyme activity. Body enzymes will have a maximum activity at temperatures ranging from 35 to 45°C.

Some enzymes can tolerate fairly large pH changes, while others are so sensitive that even a slight change in pH will greatly decrease their activity. But extreme changes in pH will denature all enzymes, and this is why body fluids contain buffer systems to protect against such changes.

Temperature

Temperature also affects the rate at which enzyme-catalyzed reactions occur. Most body enzymes have their highest activity at temperatures from 35–45°C. Above this range, the enzyme begins to denature and the reaction rate decreases. When you are running a high fever, for example, you feel ill partly because your high body temperature slows enzyme activity. Above 80°C enzymes will become permanently denatured. Below 35°C the reaction rate slowly decreases until it essentially stops as the enzymes become inactive (Figure 22.6). Enzymes, however, are not denatured at low temperatures and will resume activity if the temperature is again raised. It is this fact that allows researchers to preserve cell cultures, human tissues for transplants, and organisms such as bacteria for further studies. The preservation of food by refrigeration is based on the fact that the enzyme-catalyzed reactions that cause spoilage are slowed or stopped at low temperatures.

Exercise 22-2

1. At physiologic pH (pH 7.4) and 37°C, it takes 1 hour or more for the following reaction to reach equilibrium.

$$CO_2 + H_2O \rightarrow H_2CO_3$$

If the enzyme carbonic anhydrase is added to the system, equilibrium is reached in less than a minute.

(a) How is this enzyme able to increase the reaction rate?
(b) If the pH is raised to a pH of 9, what effect (if any) would this have on the rate of the reaction? Why?

2. Pepsin and trypsin are both digestive enzymes that catalyze the hydrolysis of proteins. Would you expect these enzymes to hydrolyze all peptide bonds? Why or why not?

VITAMINS

22.10 What Are Vitamins?

The need for vitamins has been recognized for over 200 years. Long ago, British sailors were given the slang name "limeys" from the lime juice they drank to prevent scurvy. The symptoms of scurvy—swollen gums, painful joints, hemorrhages under the skin, loss of weight, and muscular weakness—are prevented by vitamin C, which was first isolated in 1930.

But exactly what are vitamins? **Vitamins** are small organic molecules that are essential to maintain health, but that either cannot be synthesized by the body or are synthesized in insufficient amounts. Therefore, vitamins must be present in the diet (in trace amounts) for proper cellular function, growth, and reproduction. Some vitamins or derivatives of vitamins are known to serve as coenzymes in cellular reactions, but the exact cellular function of other vitamins has yet to be determined. When vitamins are lacking in the diet, however, specific deficiency diseases result (Figure 22.7). The National Research Council of the National Academy of Sciences has developed recommended dietary allowances (abbreviated RDA) for

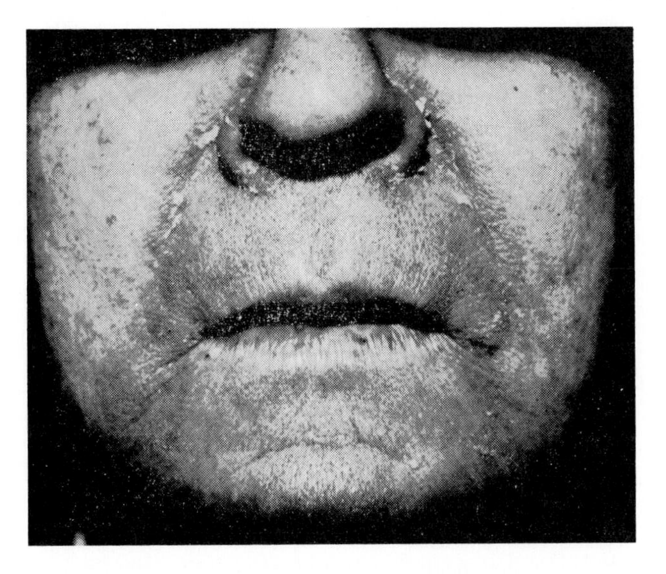

Figure 22.7 The tough, red skin around the nose and chin and the greasy scaliness of the skin are caused by a deficiency of vitamin B_6. (Courtesy R.W. Vilter et al., *J. Lab. Med.* **42**:355, 1953.)

vitamins, set to provide adequate nutrition for healthy individuals. Most healthy people who eat a balanced diet will obtain this required amount of vitamins without taking vitamin supplements (Table 22.3).

Vitamins are classified into two main groups on the basis of their solubility: they are either soluble in water or soluble in fat.

Table 22.3 Vitamins Required in Human Nutrition

Name	Recommended Dietary Allowances (RDA)[a]	Dietary Sources	Function	Symptoms of Deficiency
Water Soluble				
Vitamin C (ascorbic acid)	60 mg	Citrus fruits, leafy vegetables, tomatoes, potatoes, cabbage	Synthesis of collagen, amino acid metabolism	Scurvy, bleeding gums, loosened teeth, swollen joints
Vitamin B_1 (thiamine)	1.5 mg (1.1 mg)	Whole grains, organ meats, legumes, nuts, pork	Coenzyme in carbohydrate metabolism	Beriberi, heart failure, mental disturbances
Vitamin B_2 (riboflavin)	1.7 mg (1.3 mg)	Milk, eggs, liver, leafy vegetables; synthesized by bacteria in the gut	Coenzyme in oxidation reactions	Fissures of the skin, visual disturbances, anemia
Niacin (nicotinic acid)	19 mg (14 mg)	Yeast, liver, lean meat, fish, whole grains, eggs, peanuts	NAD, NADP; coenzymes in redox reactions	Pellagra, skin lesions, diarrhea, dementia
Vitamin B_6 (pyridoxine)	2.2 mg (2.0 mg)	Whole grains, glandular meats, pork, milk, eggs	Coenzyme for amino acid and fatty acid metabolism	Convulsions in infants, skin disorders in adults
Vitamin B_{12} (cyanocobalamin)	3 μg	Liver, brain, kidney; synthesized by bacteria in the gut	Coenzymes in nucleic acid, amino acid, and fatty acid metabolism	Pernicious anemia, retarded growth
Pantothenic acid	4–7 mg[b]	Yeast, meats, whole grains, legumes, milk, vegetables, fruits	Forms part of coenzyme A (CoA)	Neuromotor, digestive, and cardiovascular disorders

Table 22.3 Vitamins Required in Human Nutrition (Continued)

Name	Recommended Dietary Allowances (RDA)[a]	Dietary Sources	Function	Symptoms of Deficiency
Folacin[c] (folic acid)	0.4 mg	Yeast, leafy vegetables, liver, fruit, wheat germ	Coenzymes in nucleic acid and amino acid metabolism	Anemia, inhibition of cell division, digestive disorders
Biotin	0.1–0.2 mg[b]	Liver, egg yolk, legumes; synthesized by bacteria in the gut	Part of enzymes important in carbohydrate and fat metabolism	Skin disorders, anorexia, mental depression
Fat Soluble				
Vitamin A	1000 μg (800 μg) of Retinol	Green and yellow vegetables and fruits, fish oils, eggs, dairy products	Formation of visual pigments, maintenance of mucous membranes, transport of nutrients across cell membranes	Night blindness, skin lesions, eye disease (Excess: hyperirritability, skin lesions, bone decalcification, increased pressure on the brain)
Vitamin D	7.5 μg of Cholecalciferol	Fish oils, liver; provitamins in skin activated by sunlight	Regulates calcium and phosphate metabolism	Rickets (Excess: retarded mental and physical growth in children)
Vitamin E (tocopherol)	10 mg (8 mg) of α-Tocopherol	Green leafy vegetables, vegetable oils, wheat germ	Maintenance of cell membrane	Increased fragility of red blood cells
Vitamin K	70–140 μg[b]	Green leafy vegetables; synthesized by bacteria in the gut	Synthesis of prothrombin and other blood clotting factors in the liver	Failure of coagulation of blood (Excess: hemolytic anemia and liver damage)

[a] For men and women (in parentheses), 19 to 22 years, as set by the Food and Nutrition Board, National Academy of Sciences–National Research Council, 1980.

[b] The estimated safe and adequate dietary intake.

[c] Folacin is the generic term used for compounds having nutritional properties and a chemical structure similar to that of folic acid.

22.11 Water-Soluble Vitamins

Because the vitamins in this group are water-soluble, they are easily eliminated from the body in the urine. Therefore, they must be present in the diet daily. Fruits and vegetables contain many of these vitamins. Such foods should be cooked (if necessary) quickly in a very small amount of water to prevent both the loss of the vitamins in the water and the destruction of the vitamin's activity by the heat.

Vitamin C

Vitamin C, or ascorbic acid, is the most unstable of all the vitamins. It is easily destroyed by oxidation, a process that is speeded up by the presence of heat or a base.

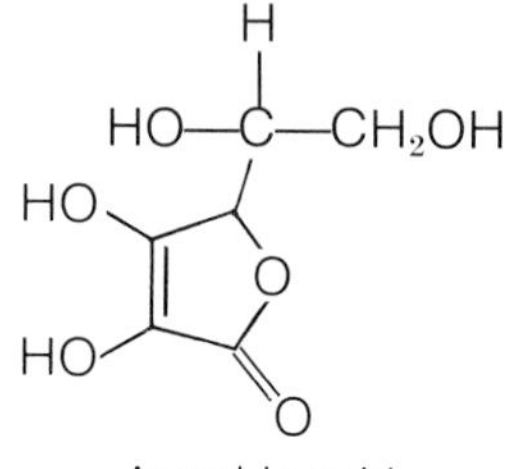

Ascorbic acid

Vitamin C has been the subject of much controversy over the last decade. Various people have recommended large doses of vitamin C to prevent colds, heart disease, and cancer. However, much more research is needed to substantiate these claims and to determine all the functions of vitamin C in the body. It is known that vitamin C is necessary for the formation of cells that produce tissues, teeth, and bone. It plays an important role in the metabolism of carbohydrates and proteins and in the formation of norepinephrine, serotonin, hemoglobin, and collagen. Collagen is the "cement" that holds tissues together, and a lack of it results in the symptoms of scurvy (Figure 22.8). Vitamin C may also aid in the absorption of iron from the intestines and its storage in the liver, as well as assisting in the excretion of adrenal hormones. At present, there are no proven toxic effects from large doses of vitamin C. However, large doses can cause disturbances in the digestive tract.

The B Vitamins

The vitamins in the B family are thiamine, riboflavin, niacin, pyridoxine, cyanocobalamin, pantothenic acid, biotin, and folacin (compounds having a structure and function similar to folic acid). They are all grouped together under the same letter because at the beginning of this century the vitamins in this group were thought to be a single vitamin. Many B vitamins act as coenzymes in reactions essential to life. For example, the body converts thiamine (vitamin B_1) to thiamine pyrophosphate, a diphosphate ester (Table 22.4). Thiamine pyrophosphate is an essential coenzyme for several reactions involved in carbohydrate and amino acid metabolism. A lack of thiamine results in beriberi, a disease that affects nerves and

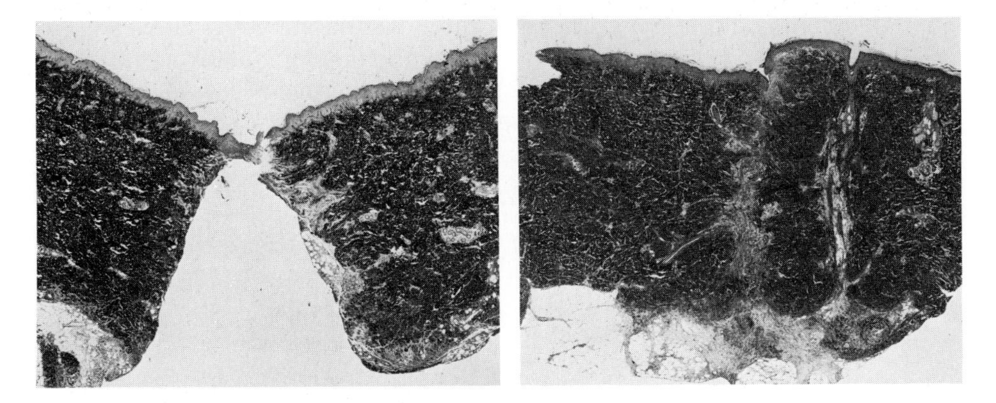

Figure 22.8 These photographs show the first scientifically controlled experiment on human scurvy. The left photograph shows a biopsy 10 days after a wound was made in the back of a subject who had eaten an ascorbic acid-free diet for 6 months. The wound shows no healing except in the outer layer of skin cells. The right photograph shows a biopsy taken after 10 days of treatment with oral doses of ascorbic acid. The wound is healing, with abundant formation of connective tissue. (Courtesy J.H. Crandon and the Upjohn Company.)

causes paralysis of the legs, enlargement of the heart, slowing of the heartbeat, and a reduction in appetite. No single food source is rich enough in vitamin B_1 to provide the complete daily requirement, so it must be obtained from many foods.

Three coenzymes formed from the B vitamins serve as hydrogen carriers in oxidation-reduction reactions. Two are derivatives of niacin (or nicotinic acid) and are called NAD^+ (nicotinamide adenine dinucleotide) and $NADP^+$ (nicotinamide adenine dinucleotide phosphate). Note that both these molecules carry a positive charge (Table 22.4). The other coenzyme, a derivative of riboflavin (vitamin B_2), is called FAD (flavin adenine dinucleotide) (Table 22.4). We can indicate how these coenzymes act as hydrogen carriers by writing the following equation.

$$NAD^+ + XH_2 \xrightarrow{\text{Enzyme}_1} NAD\text{—}H + H^+ + X$$

In this equation NAD^+ represents the positively charged NAD molecule, and XH_2 is a general way of representing some hydrogen-containing reactant molecule. The NADH produced by this reaction can then serve as a hydrogen donor in a reaction requiring hydrogen. In this way, one molecule of NAD^+ can be used over and over again in the cell.

$$NADH + H^+ + Y \xrightarrow{\text{Enzyme}_2} NAD^+ + YH_2$$

Riboflavin and niacin are found in meat and vegetables. Riboflavin is also produced by the bacteria that are always present in the intestine. Therefore, a deficiency of riboflavin in humans is rare. A deficiency in niacin results in pellagra, a disease that causes dermatitis, diarrhea, and dementia.

Pantothenic acid is a necessary starting material for the synthesis by the cell of coenzyme A, one of the most important molecules in the metabolism of all nutrients (Figure 22.9). Pantothenic acid is widely distributed in food, and there is little evidence of pantothenic acid deficiency in humans.

Table 22.4 Some Coenzymes Formed From Vitamins in the B Family

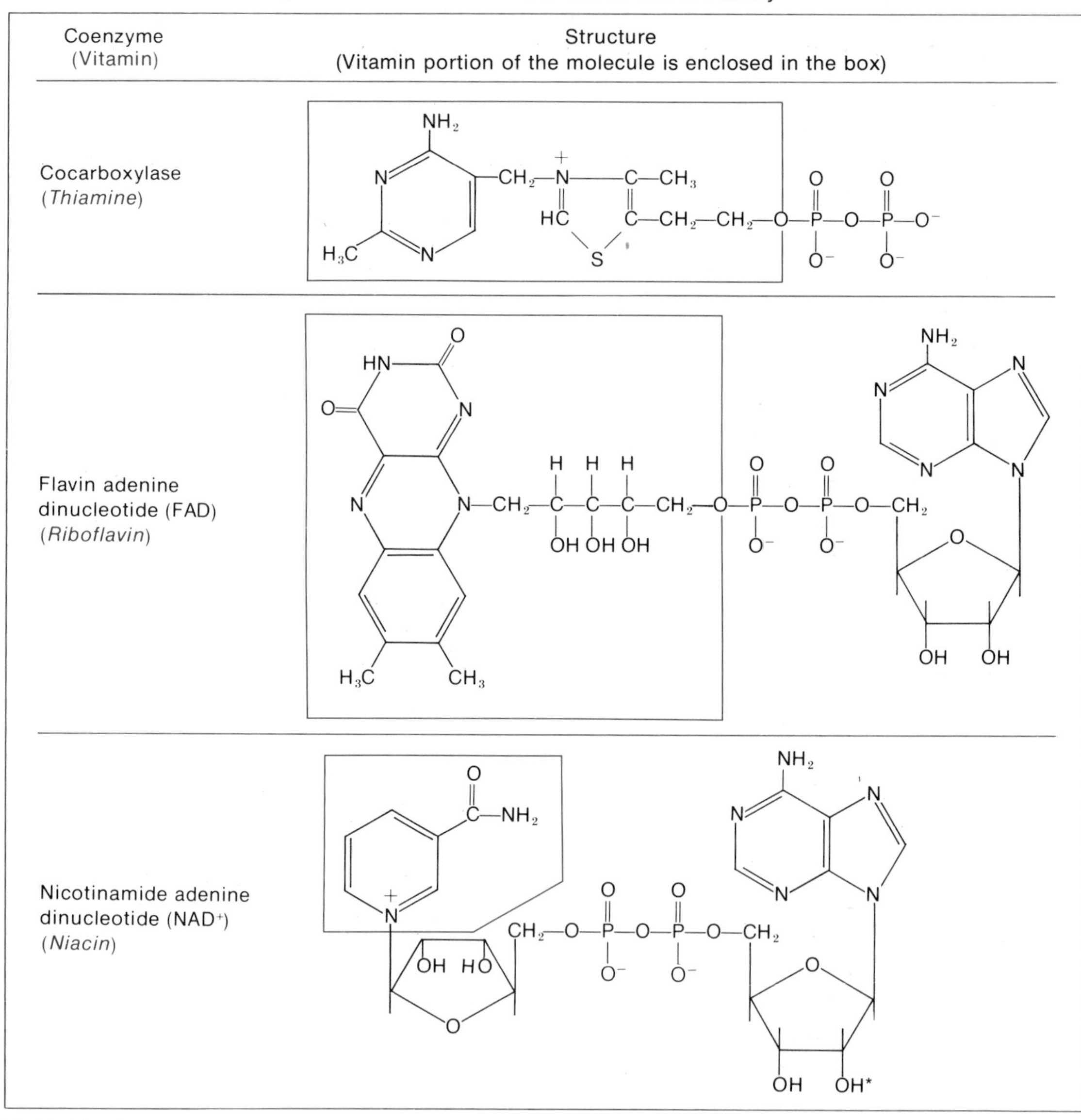

* In NADP+ this —OH is replaced by a phosphate group.

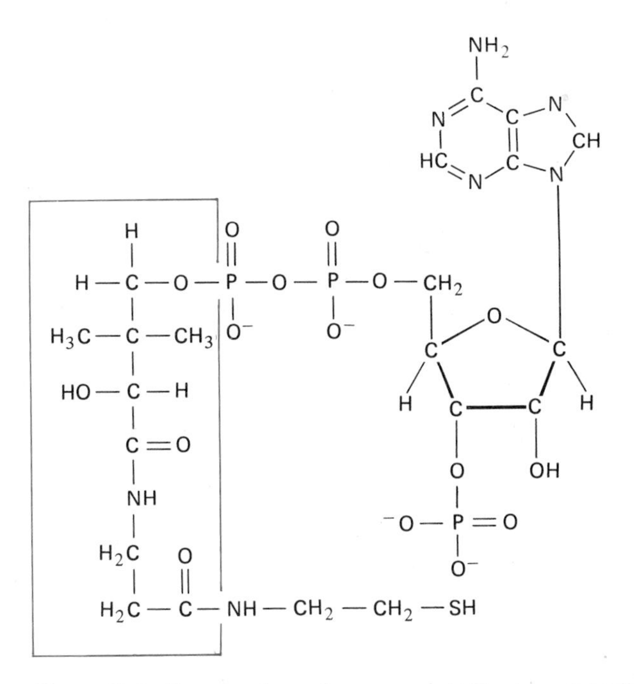

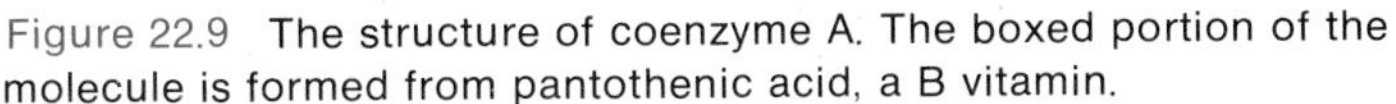

Figure 22.9 The structure of coenzyme A. The boxed portion of the molecule is formed from pantothenic acid, a B vitamin.

22.12 Fat-Soluble Vitamins

The fat-soluble vitamins (vitamins A, D, E, and K) are fairly stable, are not destroyed by heat, and are not soluble in water. Any excess that is ingested is stored in the liver for release when the body requires additional vitamins (rather than being excreted by the kidneys, as with the water-soluble vitamins). As a result, too much of a fat-soluble vitamin can be just as toxic as too little. For example, vitamin A is required for normal growth, for vision in dim light, and for maintaining healthy skin and the mucous linings of the body. It is, however, a very toxic compound. As little as 7.5 mg each day over a period of 25 days can cause such toxic effects as increases in cerebral spinal fluid pressure, headaches, irritability, patchy loss of hair, and dry skin with ulcerations. In a similar manner, a lack of vitamin D produces rickets, a disease that results from the body's inability to absorb calcium and phosphorus from the intestinal tract (see Figure 18.5). But excessive vitamin D (2.5 mg for adults and 0.1 mg for children) is toxic and will cause loss of appetite, thickening of the bones, and calcification of joints and soft tissues.

Exercise 22-3

1. Why should the dietary intake of fat-soluble vitamins be carefully controlled?
2. Egg whites contain the compound avidin, which combines with biotin and prevents it from being absorbed in the intestines.
   (a) To which class of compounds does biotin belong?
   (b) What effect might a diet of raw eggs have on a person's health? Why?

REGULATION OF ENZYME ACTIVITY

22.13 Regulatory Enzymes

Enzymes give living systems the ability to act. But if all enzymes were equally active in the cell all the time, the cell would probably "burn itself out" and die. Living systems, therefore, not only have the ability to act, but also the ability to control this action. For example, enzymes in the bloodstream catalyze the formation of blood clots when we are bleeding, but they are not active and do not form clots under normal conditions. Enzymes catalyze the contraction of muscle fibers when we walk, but are inactive when we sit or rest. The control mechanisms of the living system involve the control of enzyme concentrations and the control of enzyme activity.

We will see that most biological chemical reactions occur in a sequence of reactions that eventually produces a specific metabolic result. Because each reaction in such a sequence is catalyzed by a separate enzyme, these sequences of reactions are called **multienzyme systems.** For example, a multienzyme system is involved in the breakdown of glucose to lactic acid in muscle cells, and other multienzyme systems are involved in the synthesis of different amino acids.

In most multienzyme systems, the enzyme that catalyzes the first reaction of the series is the **regulatory** or **allosteric enzyme.** This enzyme controls the rate of the entire process. Regulatory enzymes are usually complex, high-molecular-weight molecules containing several polypeptide chains and cofactors. These enzymes usually have more than one site for the attachment of molecules—one for the active site, and one or more for regulatory molecules. A site for a regulatory molecule is called a **regulatory** or **allosteric** (meaning other space or location) **site.** The **regulatory molecule** itself can either inhibit or increase the activity of the enzyme. Such an allosteric enzyme is a flexible molecule, and the regulatory molecule causes a slight change in the shape of the enzyme. This, in turn, changes the shape of the active site, making it either more or less receptive to the substrate.

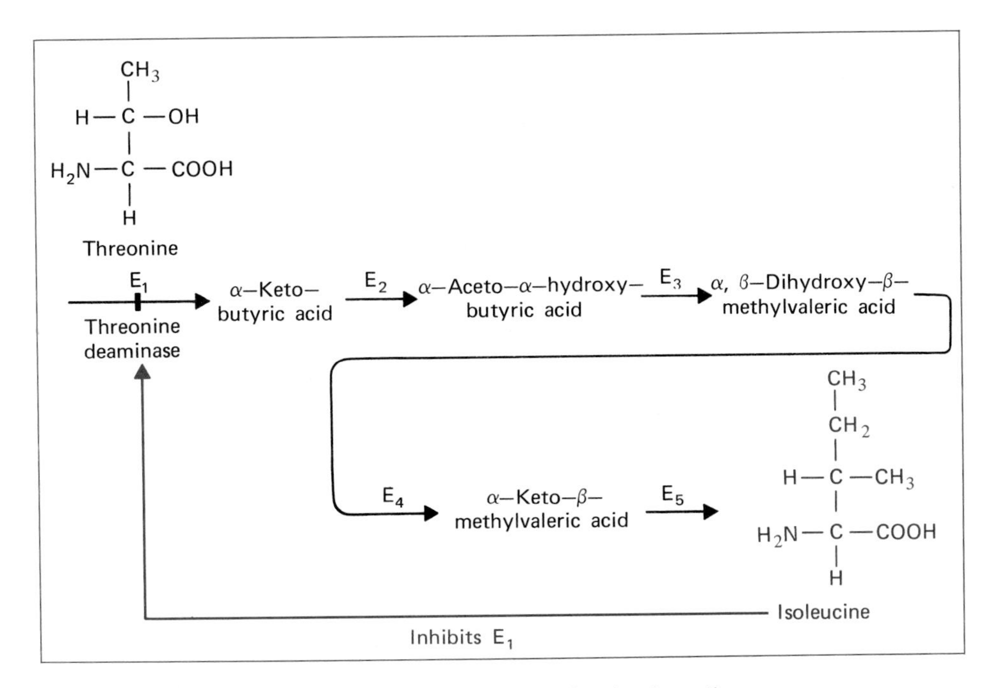

Figure 22.10 In the synthesis of the amino acid isoleucine from the amino acid threonine, the first enzyme (threonine deaminase) is the regulatory enzyme. It is inhibited by the end product (isoleucine) of the multienzyme system.

Many regulatory enzymes are inhibited by the end product of the multienzyme system. This type of regulation is called **feedback control.**

$$A \xrightarrow{E_1} B \xrightarrow{E_2} C \xrightarrow{E_3} D \xrightarrow{E_4} F \xrightarrow{E_5} G$$

In the above example, E_1 is the regulatory enzyme for the process, and it will be inhibited by high concentrations of product G (Figure 22.10).

Isocitrate dehydrogenase is a regulatory enzyme in the citric acid cycle, a series of reactions in which the compound AMP (adenosine monophosphate) is converted into the energy-rich molecule ATP (adenosine triphosphate). Specifically, this enzyme catalyzes the conversion of isocitric acid (in the form of the negative ion isocitrate) to α-ketoglutaric acid (in the form of the negative ion α-ketoglutarate).

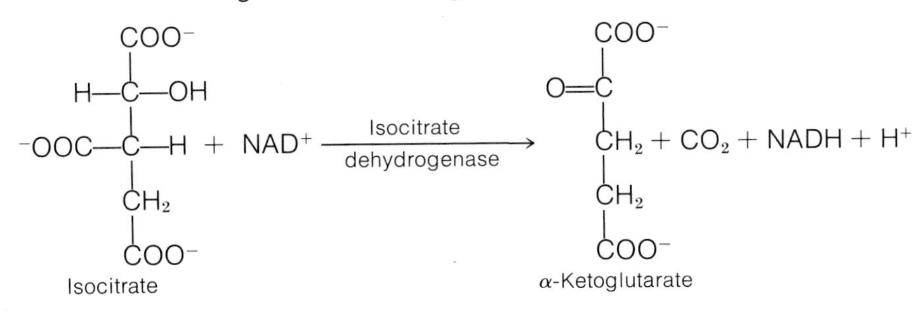

Isocitrate dehydrogenase is affected by five regulatory molecules. High concentrations of citric acid, NAD^+, or AMP will increase the activity of the enzyme, whereas high concentrations of NADH or ATP will inhibit the enzyme.

22.14 Regulatory Genes

As we will see in Chapter 24, the function of a gene is to direct the synthesis of a protein molecule. Genes, therefore, also direct the synthesis of cellular enzymes. One way in which a cell can control its metabolic activities is to switch on and off the genes that direct the synthesis of cellular enzymes. In this way the cell can control the concentration of enzymes within itself.

22.15 Regulatory Hormones

The body has control systems in addition to those within individual cells; these systems coordinate the actions between cells in multicellular systems. This higher level of communication between cells, tissues, and organs involves the nervous system and the endocrine system. The endocrine system is a group of glands that produces and secretes chemical messengers called **hormones** into the body fluids, particularly the blood (Figure 22.11 and Table 22.5). Each hormone has target organs

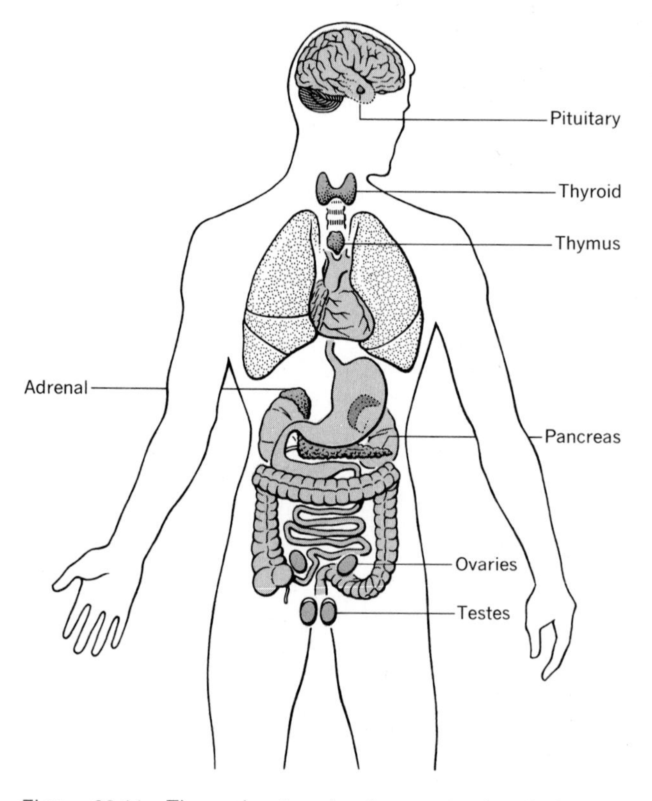

Figure 22.11 The endocrine glands secrete chemical messengers called hormones.

Table 22.5 Hormones of the Principal Endocrine Glands[a]

Gland or Tissue	Hormone	Major Function of the Hormone
Thyroid	1. Thyroxine 2. Thyrocalcitonin	1. Stimulates rate of oxidative metabolism and regulates general growth and development. 2. Lowers level of calcium in the blood.
Parathyroid	Parathormone	Regulates the levels of calcium and phosphorus in the blood.
Pancreas (Islets of Langerhans)	1. Insulin 2. Glucagon	1. Decreases blood glucose level. 2. Elevates blood glucose level.
Adrenal medulla	Epinephrine	Various "emergency" effects on blood, muscle, temperature.
Adrenal cortex	Cortisone and related hormones	Control carbohydrate, protein, mineral, salt, and water metabolism.
Anterior pituitary	1. Thyrotropic 2. Adenocorticotropic 3. Growth hormone 4. Gonadotropic (two hormones) 5. Prolactin	1. Stimulates thyroid gland functions. 2. Stimulates development and secretion of adrenal cortex. 3. Stimulates body weight and rate of growth of skeleton. 4. Stimulate gonads. 5. Stimulates lactation.
Posterior pituitary	1. Oxytocin 2. Vasopressin	1. Causes contraction of some smooth muscles. 2. Inhibits excretion of water from the body by way of urine.
Ovary (follicle)	Estrogen	Influences development of sex organs and female characteristics.
Ovary (corpus luteum)	Progesterone	Influences menstrual cycle; prepares uterus for pregnancy; maintains pregnancy.
Uterus (placenta)	Estrogen and progesterone	Function in maintenance of pregnancy.
Testis	Androgens (testosterone)	Responsible for development and maintenance of sex organs and secondary male characteristics.
Digestive system	Several gastrointestinal	Integration of digestive processes.

[a] From G. E. Nelson, G. G. Robinson, and R. A. Boolootian, *Fundamental Concepts of Biology*, Second Edition, copyright © 1970 by John Wiley and Sons, Inc., New York, page 114. Used by permission.

or cells that are influenced by its presence. The secretion of a particular hormone by an endocrine gland may be triggered by the nervous system, as in the release of adrenalin triggered by hearing a nearby explosion, or by the concentration of a specific chemical compound, as in the release of insulin triggered by an increase in blood glucose levels.

Hormones are thought to regulate cellular processes in two ways. Steroid hormones such as estrogen and testosterone will enter the cell and bind to a receptor site on a large protein molecule. This steroid-receptor group is then transferred to the nucleus of the cell, where it activates specific genes. On the other hand, peptide hormones such as insulin, human growth hormone, and hormones that regulate gastric and kidney secretions attach to specific receptor sites on the surface of the target cell membrane, triggering changes within the cell. The number of receptor sites on a target cell determines the level of response by that cell, and the number may increase or decrease depending upon the concentration of hormones in the blood.

Cyclic AMP

Cyclic AMP is a molecule that acts within the cell as a secondary messenger, or intracellular hormone. Its formation is triggered by the attachment of a hormone molecule to a target cell. The system works as follows: A hormone, the primary messenger, is released into the bloodstream. It travels to a target cell, where it attaches to a specific receptor site on the outside of the cell membrane. This attachment causes a conversion of the enzyme adenyl cyclase from an inactive form to an active form on the inside of the cell membrane. This enzyme then catalyzes the formation of cyclic AMP from ATP (Figure 22.12). The cyclic AMP spreads through the cell as a secondary messenger, instructing the cell to respond to the hormone in the particular manner characteristic of that cell (Figure 22.13). For example, the hormone thyrotropin is secreted by

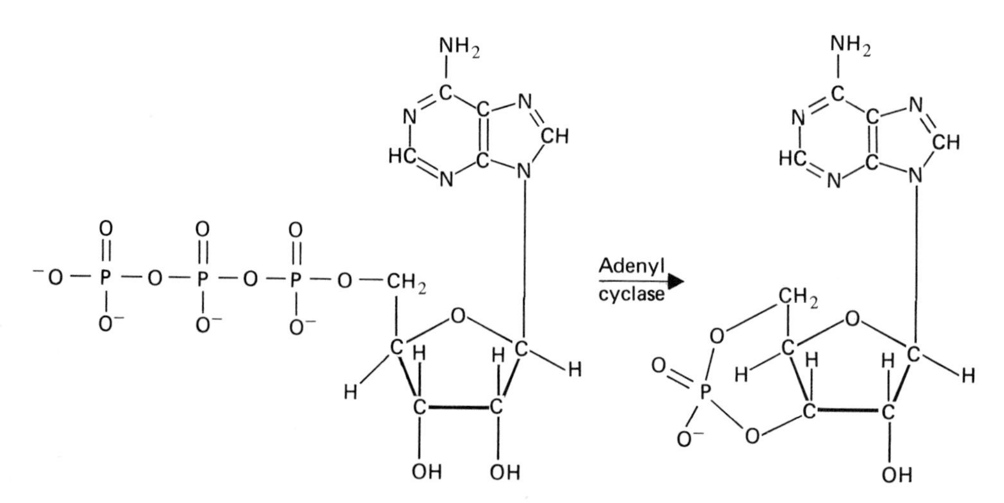

Figure 22.12 The formation of cyclic AMP from ATP.

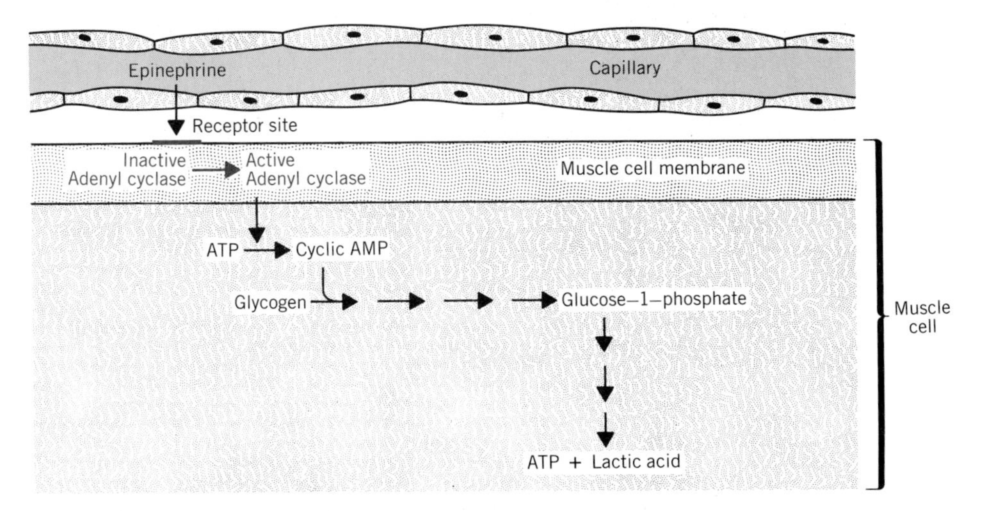

Figure 22.13 Epinephrine stimulates glycogen breakdown and the metabolism of glucose in muscle cells by attaching to a receptor site on the muscle cell membrane, which triggers the formation of cyclic AMP. The cyclic AMP sets off a series of reactions that result in production of the energy-rich molecule ATP, which supplies the energy for the contraction of the muscle cell.

the pituitary gland and travels to target cells in the thyroid gland. There it triggers the formation of cyclic AMP, to which the thyroid cells respond by secreting more thyroxine.

Prostaglandins

Prostaglandins are a class of compounds that resemble hormones in their effects, but which are chemically quite different. They are 20-carbon fatty acids that are synthesized in the cell membrane from unsaturated fatty acids such as linoleic and arachidonic acid (Figure 22.14). As with cyclic

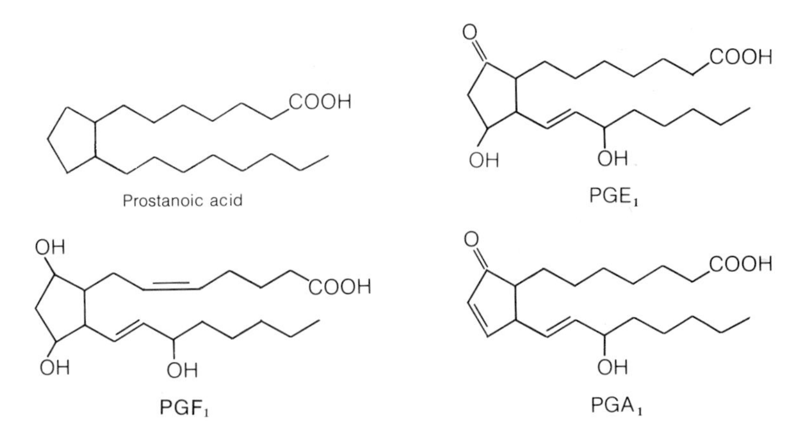

Figure 22.14 Some important prostaglandins. In general, the prostaglandins are variants of the basic structure of prostanoic acid.

AMP, prostaglandins are present in almost all cells and tissues and are released by the action of hormones, nerves, and muscles. They are among the most potent of all biological agents—as little as one billionth of a gram is biologically active—yet a very small change in their structure can completely alter their biological function. Prostaglandins can raise or lower blood pressure and can regulate gastric secretions. They seem to play a role in both the process that prevents blood platelets from clumping together in healthy blood vessels, and the process that makes them clump together (to form a blood clot) when a vessel is damaged. Fevers and inflammatory reactions are produced by prostaglandins; aspirin and steroid antiinflammatory drugs such as hydrocortisone work by inhibiting prostaglandin synthesis.

Knowledge of the functions of prostaglandins has led to effective treatment for a great number of medical conditions. Prostaglandins cause uterine contractions and thus are used to induce labor. Drugs that block prostaglandin formation are now being prescribed to treat severe menstrual cramps. Prostaglandins are being used to inhibit the secretion of stomach acid in people suffering from peptic ulcers and to treat ulcers on the hands and feet of people suffering from Raynaud's disease, diabetes, or atherosclerosis. Other prostaglandins relax the smooth muscles and are used to relieve asthma and to treat high blood pressure. Blue babies are being treated with prostaglandins to keep a fetal duct open, which increases the oxygen content of their blood until they are strong enough to undergo surgery.

Exercise 22-4

1. Construct a table showing the similarities and differences in functions, sites of synthesis, and sites of action among hormones, prostaglandins, and cyclic AMP.
2. The synthesis of cholesterol from acetyl CoA requires many enzyme-catalyzed steps. In one of the middle steps, β-hydroxymethylglutaryl CoA (HMG CoA) is converted to mevalonate by the enzyme HMG CoA reductase. Cholesterol inhibits this enzyme.

   (a) The sequence of reactions that produces cholesterol is an example of a __________ system.
   (b) The process by which an end product inhibits one of the enzymes in a sequence of reactions is called __________ control.
   (c) HMG CoA reductase is a __________ enzyme.
   (d) HMG CoA will bind to the __________ site on the enzyme molecule, and cholesterol will bind to the __________ site.

INHIBITION OF ENZYME ACTIVITY

22.16 Irreversible Inhibition

The activity of enzymes can be inhibited in several ways. Research into these processes of enzyme inhibition has produced a great deal of knowledge about the specificity of enzymes and the nature of their active sites. The action of many poisons and drugs is due to their ability to inhibit specific enzymes.

Irreversible inhibition of enzyme activity occurs when a functional group or cofactor required for the activity of the enzyme is destroyed or modified. For example, cholinesterase is an enzyme that catalyzes a reaction taking place at the juncture of nerve cells; the enzyme is necessary for normal transmission of nerve impulses. But compounds in nerve gases will combine with the —OH group on a serine molecule that is vital to the active site of the cholinesterase enzyme. In such a case the enzyme then loses its ability to catalyze the reaction, so that animals poisoned by nerve gas become paralyzed (Figure 22.15).

Some enzymes depend upon sulfhydryl groups to form tight covalent bonds with metal cofactors. Heavy metals such as mercury, lead, and silver (and even such essential metals as iron and copper in excess amounts) are toxic because they bind irreversibly with free —SH functional groups on enzymes; the sulfhydryl groups, then, are no longer available to bind with the necessary cofactor. As we saw in Chapter 21, lead inactivates

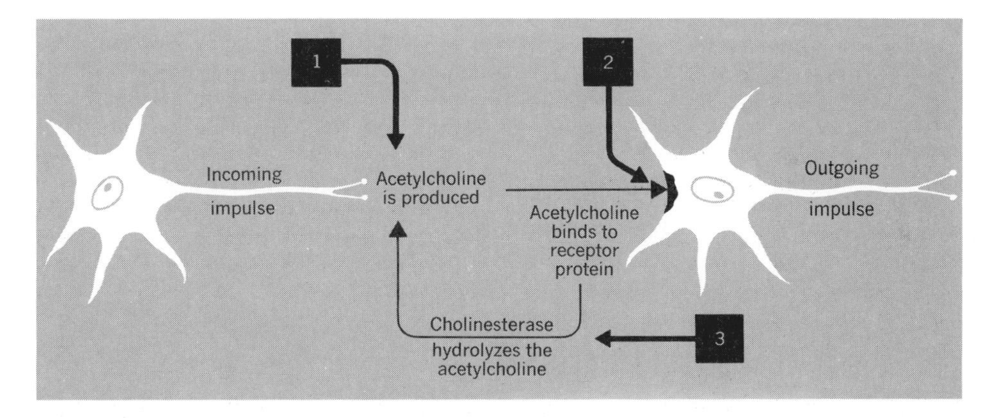

Figure 22.15 The nerve poisons. The formation and breakdown of acetylcholine must occur for normal nerve transmission. (1) Botulism toxin blocks the synthesis of acetylcholine so that nerve signals are not sent. Death is then caused by respiratory failure. (2) Nicotine, curare, atropine, morphine, codeine, cocaine, and local anesthetics such as procaine combine with the receptor protein, blocking its reaction with acetylcholine. Therefore, the second cell does not receive the impulse, and it is not sent on. (3) Anticholinesterases are poisons such as nerve gases, organophosphate insecticides, and some mushroom toxins that inhibit cholinesterase, causing the overstimulation of nerve cells by acetylcholine. This results in irregular heart rhythms, convulsions, and death.

the enzyme that catalyzes the insertion of iron into the heme molecule, thereby affecting the synthesis of hemoglobin and causing anemia. Lead also inhibits essential enzymes and proteins in the muscles and nervous system, resulting in muscle weakness and lack of coordination. One treatment for lead poisoning consists of ingesting chelating agents that bind tightly to the lead. The complex that is formed is soluble and can be excreted by the kidneys.

22.17 Reversible Inhibition

Reversible inhibition of enzyme activity occurs in two ways: noncompetitive inhibition and competitive inhibition (Figure 22.16). **Noncompetitive inhibition** occurs when the inhibitor combines reversibly with some portion of the enzyme (other than the active site) that is essential to the enzyme's function. In the muscles, for example, iodoacetic acid inhibits the conversion of glucose to lactic acid by attaching reversibly to the sulfhydryl groups found on the enzyme glyceraldehyde phosphate dehydrogenase.

$$\underset{\text{Active enzyme}}{E{-}SH} + \underset{\text{Iodoacetic acid}}{ICH_2COOH} \rightleftharpoons \underset{\text{Inactive enzyme}}{E{-}S{-}CH_2COOH} + HI$$

Competitive inhibition occurs when a compound with a structure very similar to the substrate competes with the substrate for the active site on the enzyme. When such inhibitors become bound to active sites, there are fewer enzyme molecules available to the substrate. Therefore, enzyme activity will decrease.

$$\{\underset{\text{Enzyme}}{E} + \underset{\text{Inhibitor}}{I} \rightleftharpoons EI\} \text{ competes with } \{\underset{\text{Enzyme}}{E} + \underset{\text{Substrate}}{S} \rightleftharpoons ES\}$$

For example, succinic acid and malonic acid have very similar structures. Succinate dehydrogenase catalyzes the removal of hydrogen from succinic acid, and this reaction is inhibited by malonic acid.

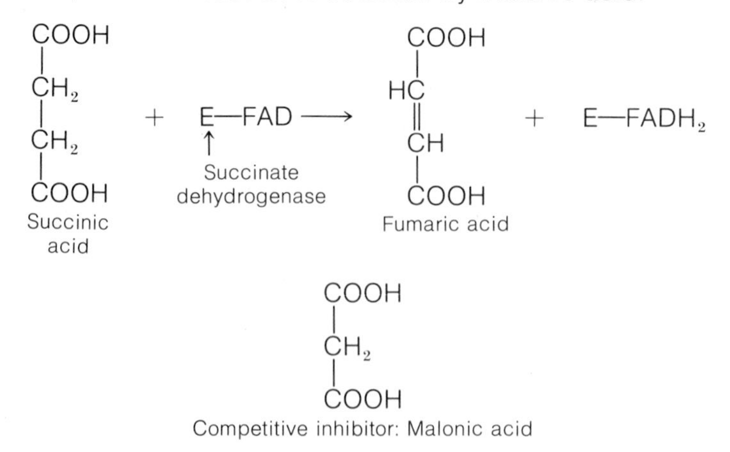

Antibiotics

Antibiotics are chemicals that are extracted from organisms such as molds, bacteria, and yeasts and that inhibit growth or destroy other

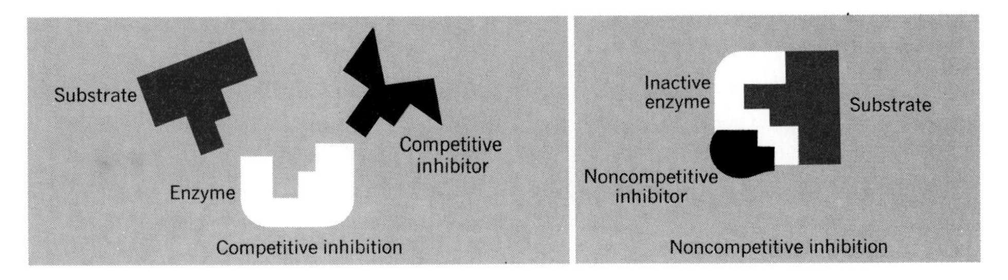

Figure 22.16 In competitive inhibition, the inhibitor and the substrate compete for the active site on the enzyme. In noncompetitive inhibition, the inhibitor combines with some other portion of the enzyme molecule that is essential for the activity of the enzyme.

microorganisms. They belong to a large class of chemicals called **antimetabolites,** which inhibit enzyme function. Antibiotics are used to treat many diseases in humans and animals, and in small amounts they are used to speed the growth of poultry and livestock.

Antibiotics can function in many ways to prevent growth or to destroy a disease-causing organism. For example, the sulfa drug sulfanilamide prevents bacterial growth by competitive inhibition. Bacteria use *p*-aminobenzoic acid in the synthesis of the vitamin folic acid.

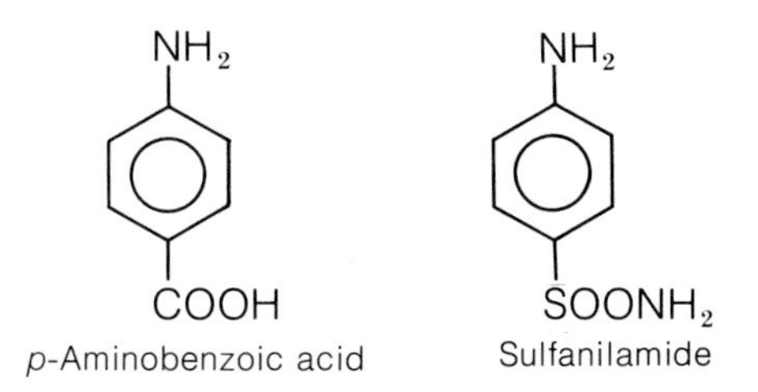

The structure of sulfanilamide closely resembles that of *p*-aminobenzoic acid, and sulfanilamide molecules compete for the active site on the bacterial enzyme. This, then, inhibits folic acid synthesis.

The most widely used antibiotics, the penicillins and cephalosporins, belong to the class of compounds called beta-lactams. These compounds work by interfering in the process used by bacteria to construct their cell walls. The structure of the beta-lactams is very similar to that of the compounds normally used by the bacteria in building the framework of the cell wall. The beta-lactam antibiotics successfully compete for the active site on an enzyme necessary to this process, critically weakening the cell wall and killing the bacteria.

Chemotherapy

As we learned at the beginning of this chapter, antimetabolites are also being used in cancer therapy. Methotrexate, an antimetabolite whose structure resembles the vitamin folic acid, inhibits the enzyme dihydrofolate reductase. By competing with folic acid for the active site on the enzyme and then binding tightly to the enzyme, methotrexate ties up a pathway for the synthesis of DNA in rapidly dividing cancer cells.

Most antitumor agents have been discovered by a chance observation or the random screening of chemicals. But Charles Heidelberger of the University of Wisconsin designed and synthesized the antimetabolite 5-fluorouracil (5-FU) after observing that malignant cells used the base uracil more efficiently than normal cells. 5-Fluorouracil is used to treat many solid tumors including tumors of the breast, colon, and ovaries. The fluorine–carbon bond in 5-FU is much stronger than the hydrogen–carbon bond in uracil and will not undergo the substitution reaction necessary for the formation of thymine (an important part of the DNA molecule to be discussed in Chapter 24).

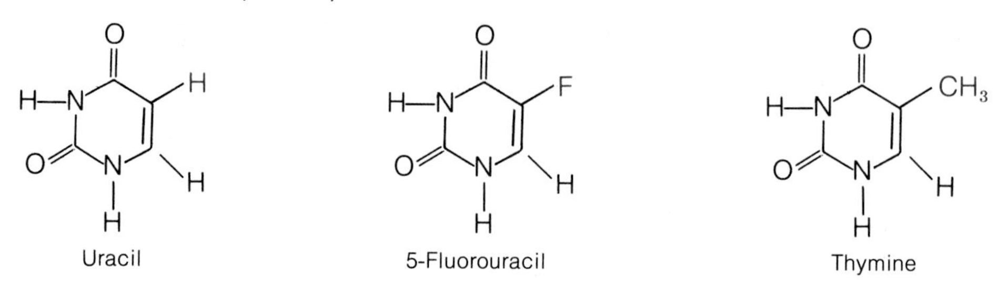

Uracil 5-Fluorouracil Thymine

Cancer cells use both 5-FU and uracil in their synthetic pathways, but the compounds formed from 5-FU inhibit the enzyme thymidylate synthetase and disrupt DNA synthesis, thus killing the cell. These 5-FU compounds are also incorporated into the structure of RNA (another key molecule to be discussed in Chapter 24) and disrupt the function of RNA in the cell.

Exercise 22-5

1. Sodium arsenite ($NaAsO_2$) has a strong attraction for sulfhydryl groups.

   (a) What effect might this attraction have on cellular enzymes?
   (b) What type of enzymes will be most affected?

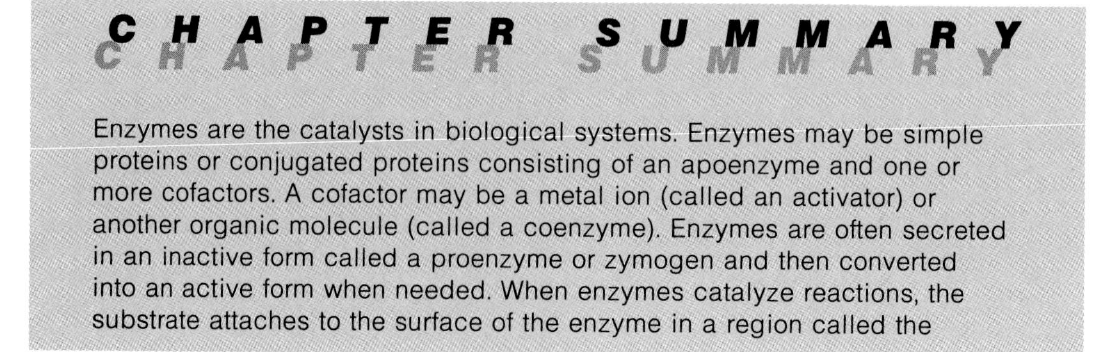

CHAPTER SUMMARY

Enzymes are the catalysts in biological systems. Enzymes may be simple proteins or conjugated proteins consisting of an apoenzyme and one or more cofactors. A cofactor may be a metal ion (called an activator) or another organic molecule (called a coenzyme). Enzymes are often secreted in an inactive form called a proenzyme or zymogen and then converted into an active form when needed. When enzymes catalyze reactions, the substrate attaches to the surface of the enzyme in a region called the active site. This activates the substrate, causing the formation of products that are then released from the surface of the enzyme. Enzymes catalyze only very specific reactions because the correct substrate fits in the active site just as only one key fits into a specific lock. The activity of an enzyme will be affected by all the factors that can alter the structure of a protein. Each enzyme has an optimal pH at which it is the most active.

Metabolic processes are the result of sequences of enzyme-catalyzed reactions. Such sequences are controlled by a regulatory or allosteric enzyme—often the enzyme that catalyzes the first step in the sequence. In addition to the active site, this enzyme contains allosteric sites to which a regulatory molecule can attach, thereby inhibiting or increasing the activity of the enzyme. Enzyme activity can be inhibited irreversibly when a functional group or cofactor required for the enzyme's activity is permanently destroyed or modified. The activity can be inhibited reversibly when another molecule competes for the active site, or when some portion of the enzyme required for its action is modified reversibly.

Enzyme activity is regulated at many levels, both inside and outside the cell. Hormones, produced by the endocrine system, are one of the body's means of intercellular control. Hormones work in one of two ways: they can attach to a special area on the membrane of a target cell (activating the formation of a substance within the cell, such as cyclic AMP or a prostaglandin), or they can themselves enter the cell and activate specific genes. Vitamins are organic molecules that are not produced by the body, but are required in small amounts for normal cellular function. They can be classified as water-soluble or fat-soluble. The lack of a vitamin in the diet will cause a specific deficiency disease. Vitamins often function as coenzymes of important cellular enzymes.

EXERCISES AND PROBLEMS

1. In general terms, describe the similarities and differences between the action of enzymes and inorganic catalysts.

2. Define each of the following terms:
   (a) metabolism
   (b) catabolic reaction
   (c) anabolic reaction
   (d) enzyme
   (e) prosthetic group
   (f) isoenzyme
   (g) multienzyme system
   (h) feedback control
   (i) antimetabolite

3. (a) Explain how the lock-and-key comparison describes the method of enzyme action.
   (b) How does the induced-fit theory change the lock-and-key theory?

4. Why doesn't pepsin, the enzyme released in the stomach to break down proteins, continue to function in the small intestine?

5. Why is boiling water an effective sterilizing agent?
6. Apples picked before they have completely ripened may be stored at low temperatures and under an atmosphere of nitrogen to keep them fresh for many months. Explain why these storage conditions slow the ripening process.
7. Each of the enzymes necessary for causing the formation of blood clots is present in the blood plasma at all times. Why aren't blood clots forming constantly in a normal blood vessel?
8. (a) What is an allosteric enzyme?
   (b) Why are allosteric enzymes necessary for the normal operation of the cell?
   (c) What effect does the regulatory molecule have on the allosteric enzyme?
9. Describe three ways that reaction rates are controlled within the human body.
10. What are the differences between vitamins and hormones? (Include both their sources and functions in your discussion.)
11. Name the vitamin whose deficiency causes each of the following diseases.
   (a) pernicious anemia
   (b) hemolytic anemia
   (c) rickets
   (d) scurvy
   (e) pellagra
12. What three coenzymes are formed from the B vitamins and serve as hydrogen carriers?
13. What important coenzyme is synthesized from pantothenic acid?
14. Prostaglandins are among the most potent biological agents. Describe some of their various physiological effects. In what way do they function like cyclic AMP?
15. What is an antimetabolite? Give two examples of compounds that act as antimetabolites.
16. Phosphoglucomutase catalyzes the conversion of glucose-1-phosphate to glucose-6-phosphate.

   $$\text{Glucose-1-phosphate} \xrightleftharpoons{\text{Phosphoglucomutase}} \text{Glucose-6-phosphate}$$

   (a) In general terms, describe the four steps by which phosphoglucomutase accomplishes this conversion.
   (b) Phosphoglucomutase has a turnover number of 1×10^3. How many molecules of glucose-1-phosphate would be converted in 1 hour by 1 molecule of phosphoglucomutase? How many moles?
17. Under normal conditions a human sperm cell can live 24 to 36 hours. However, several doctors have reported successful artificial insemination using human sperm that have been frozen up to 6

months. In one case, pregnancy resulted from the use of sperm that had been frozen at $-196.5°C$ for 6 months. Explain how it is possible for sperm cells to survive for 6 months and then resume normal activity.

18. Excess vitamin A and D is stored in the body, but excess vitamin C and B_1 is readily excreted.
    (a) What property allows vitamins A and D to be stored in the body, whereas vitamins C and B_1 are excreted?
    (b) Explain the statement that an excess of vitamin D in the diet is potentially as dangerous as a deficiency.

19. Beriberi used to be very common in the Far East where the main food of the diet was polished rice (rice with its outer coating removed). The introduction of enriched rice has mostly eliminated the disease.
    (a) What causes beriberi?
    (b) Why was beriberi occurring among this population?
    (c) With what was the rice enriched?

20. An increasing number of foods are being fortified with vitamins. Breakfast foods such as cereal, bread, margarine, and milk may all be fortified with vitamin D as well as other vitamins.
    (a) Explain why this might pose a health hazard to young children who eat a daily breakfast of cereal with milk and toast with butter.
    (b) Why should fortification of foods with vitamins be carefully controlled?

21. A 17-year-old girl was admitted to a hospital suffering from headaches, blurred vision, and ringing in her ears. These symptoms may indicate a brain tumor, but none was found to be present. The only drug she was taking was vitamin A prescribed by a dermatologist for her acne. After being questioned by the doctor, the girl admitted taking three 15-mg vitamin A pills per day instead of only one as prescribed.
    (a) What was the cause of the girl's symptoms?
    (b) How could her symptoms be alleviated?

22. The insecticide Parathion must be handled with great care because it is converted by the liver to a molecule that resembles nerve gas in its action. Explain how Parathion acts to poison human beings.

23. Public health officials estimate that more than 200,000 children become ill from lead poisoning each year, many of them from eating lead-based paint chips. What is the chemical action of lead that causes this illness?

24. Both ethanol and methanol are oxidized in the body: ethanol to acetic acid and methanol to formic acid. It is formic acid that causes the acidosis of methanol poisoning. The first step in the oxidation of both ethanol and methanol involves the enzyme alcohol dehydrogenase. One therapy for methanol poisoning is the use of a nearly intoxicating dose of ethanol. Use your knowledge of enzyme inhibition to explain why this therapy is effective.

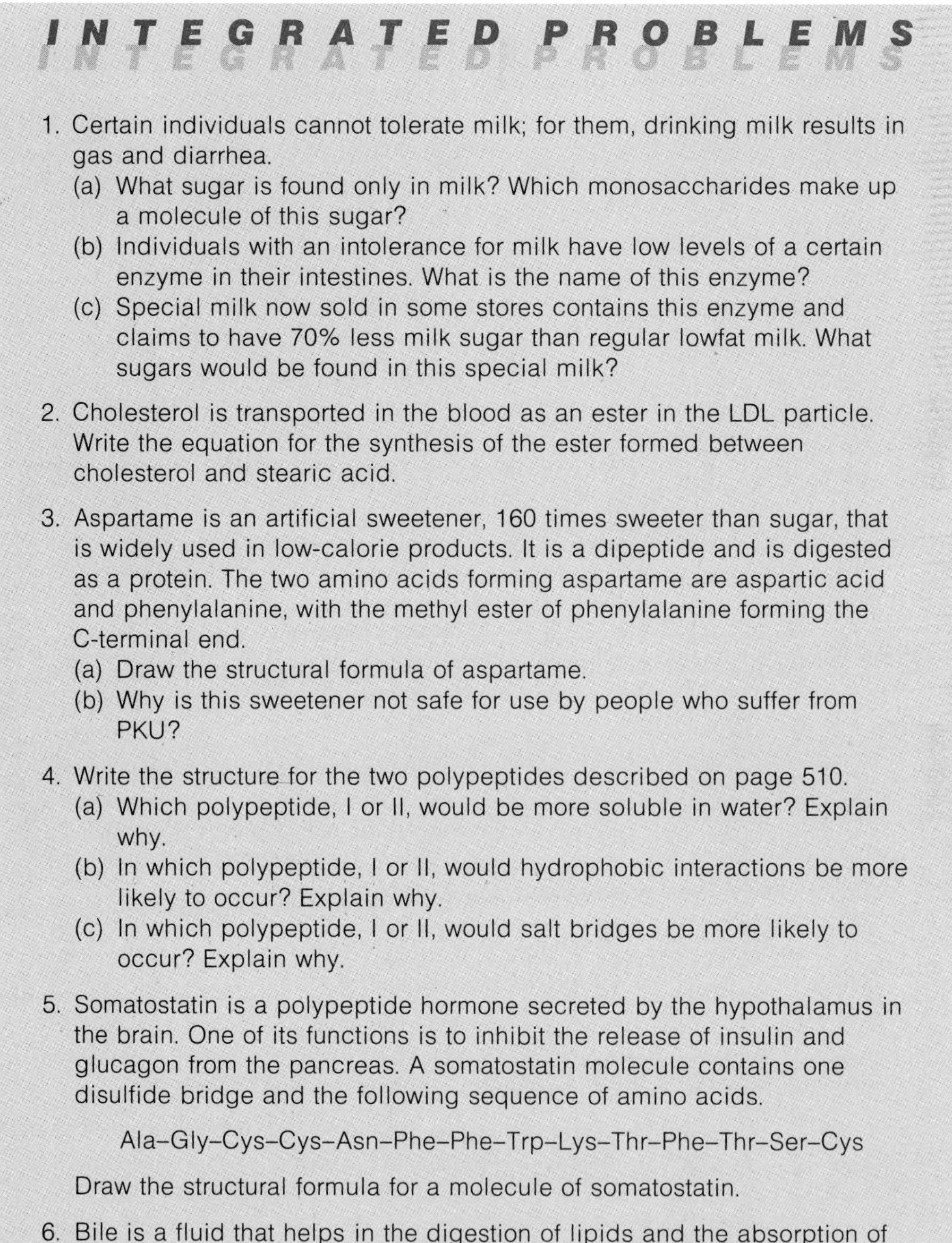

INTEGRATED PROBLEMS

1. Certain individuals cannot tolerate milk; for them, drinking milk results in gas and diarrhea.
   (a) What sugar is found only in milk? Which monosaccharides make up a molecule of this sugar?
   (b) Individuals with an intolerance for milk have low levels of a certain enzyme in their intestines. What is the name of this enzyme?
   (c) Special milk now sold in some stores contains this enzyme and claims to have 70% less milk sugar than regular lowfat milk. What sugars would be found in this special milk?

2. Cholesterol is transported in the blood as an ester in the LDL particle. Write the equation for the synthesis of the ester formed between cholesterol and stearic acid.

3. Aspartame is an artificial sweetener, 160 times sweeter than sugar, that is widely used in low-calorie products. It is a dipeptide and is digested as a protein. The two amino acids forming aspartame are aspartic acid and phenylalanine, with the methyl ester of phenylalanine forming the C-terminal end.
   (a) Draw the structural formula of aspartame.
   (b) Why is this sweetener not safe for use by people who suffer from PKU?

4. Write the structure for the two polypeptides described on page 510.
   (a) Which polypeptide, I or II, would be more soluble in water? Explain why.
   (b) In which polypeptide, I or II, would hydrophobic interactions be more likely to occur? Explain why.
   (c) In which polypeptide, I or II, would salt bridges be more likely to occur? Explain why.

5. Somatostatin is a polypeptide hormone secreted by the hypothalamus in the brain. One of its functions is to inhibit the release of insulin and glucagon from the pancreas. A somatostatin molecule contains one disulfide bridge and the following sequence of amino acids.

   Ala–Gly–Cys–Cys–Asn–Phe–Phe–Trp–Lys–Thr–Phe–Thr–Ser–Cys

   Draw the structural formula for a molecule of somatostatin.

6. Bile is a fluid that helps in the digestion of lipids and the absorption of fat-soluble nutrients. Patients whose bile ducts are blocked, preventing bile from entering the small intestine, have blood that does not clot as readily as that of healthy individuals.
   (a) What vitamin is important in the clotting process?
   (b) Why might a lack of bile in the intestines result in a failure of blood clotting?

CHAPTER TWENTY-THREE
PATHWAYS OF METABOLISM

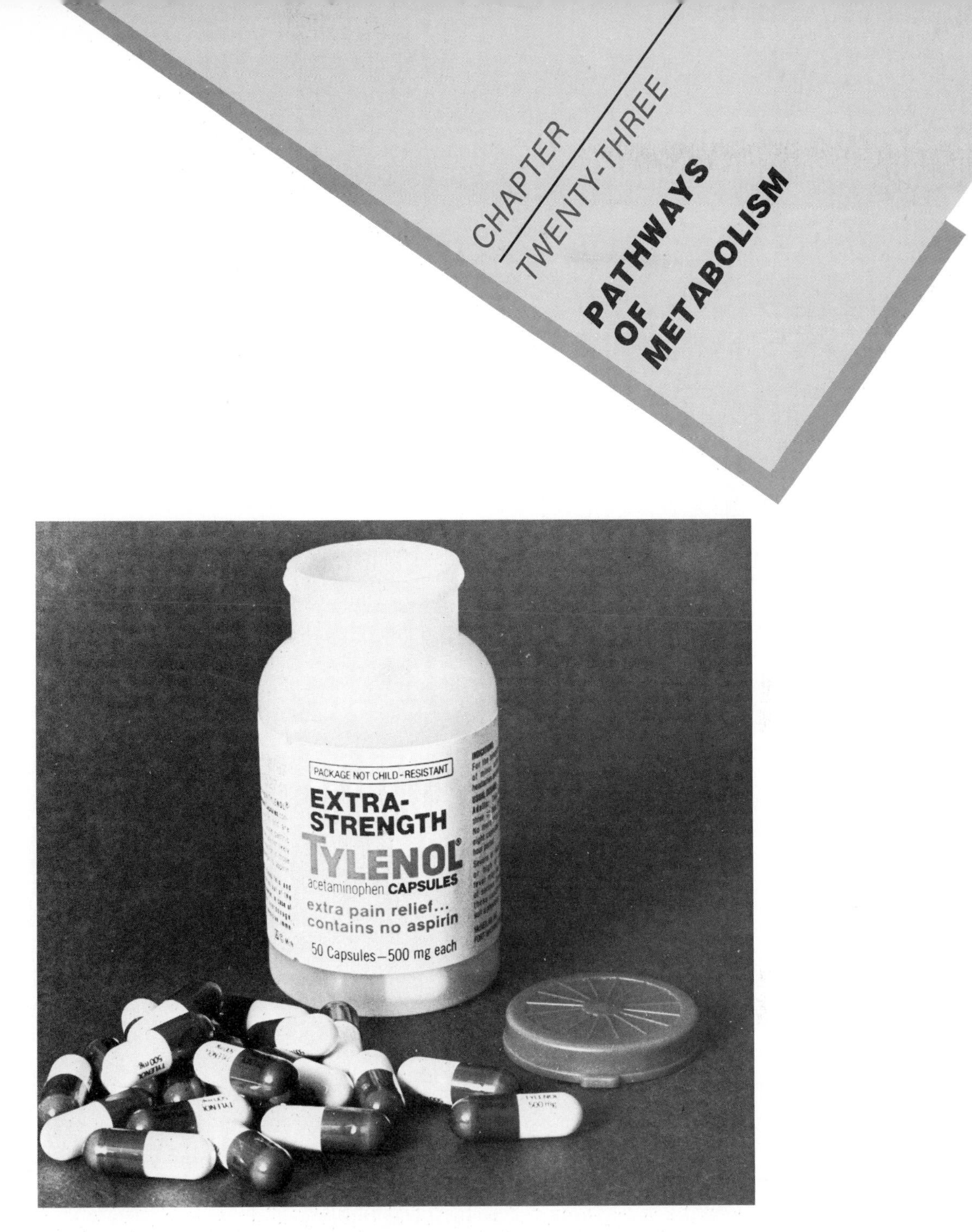

It was the morning of October 6, 1982, and Adam Janus had a nagging pain in his chest. Having no medication handy, he stopped by his neighborhood drugstore in the Chicago suburb of Arlington Heights and picked up some Extra-Strength Tylenol. About an hour later, Adam collapsed in his home. He was rushed to the hospital, where doctors tried frantically to revive him, but nothing seemed to help. He died of a sudden cardiopulmonary collapse. That evening, Adam's grieving relatives gathered at his house. Suffering from headaches brought on by the emotional strain, Adam's brother Stanley and his sister-in-law Theresa found a bottle of Tylenol in the kitchen. Shortly after taking some of these capsules, they both collapsed. Stanley died that evening, and Theresa died the next afternoon.

The connection between these deaths and the capsules of Extra-Strength Tylenol was eventually discovered by two alert off-duty firemen, but not before seven people in the Chicago area had died from sudden cardiopulmonary collapse. The capsules that they had taken were found to have contained potassium cyanide in quantities several thousand times the usual fatal dose.

Cyanide ($C{\equiv}N^-$) is one of the most potent and rapid-acting poisons known. It binds to the Fe^{3+} in the heme group of cytochrome a,a_3 in the respiratory chain, blocking the final reaction with oxygen. The respiratory chain is the major energy-producing process for all cells; when this series of reactions is blocked, cells die very quickly. Death from cyanide poisoning occurs as a result of the massive death of cells in the central nervous system. In the case of a sublethal dose of cyanide, the patient can be given various nitrites. These nitrites convert the Fe^{2+} of hemoglobin to Fe^{3+}, in the form of a compound called methemoglobin. The methemoglobin will then compete with cytochrome a,a_3 for the cyanide.

The production of energy by the cell is critical for the maintenance of life. In this chapter, we will study the major pathways in the metabolism of carbohydrates, lipids, and proteins and examine the production of energy by the cell.

LEARNING OBJECTIVES

By the time you have finished this chapter, you should be able to:

1. Describe the digestion and metabolic uses of potato starch, vegetable oil, and steak.
2. Define *high-energy bond* and give examples of compounds in which such bonds are found.
3. Write a general equation for each of the following processes:
   (a) Glycogenesis
   (b) Glycogenolysis
   (c) Glycolysis
   (d) Fermentation
   (e) β-Oxidation
   (f) Oxidative deamination
   (g) Transamination
   (h) Urea cycle
4. Describe the two stages in the oxidation of glucose, indicating where they occur in the cell and in which step the most energy is produced.
5. Describe fatty acid oxidation and the conditions that will result in ketosis.
6. List six metabolic uses of acetyl CoA.
7. List four metabolic uses of amino acids.
8. Describe how fasting affects carbohydrate, lipid, and protein metabolism.

CELLULAR ENERGETICS

23.1 Cellular Energy Requirements

The body has many energy requirements: energy is needed for the many synthesis reactions within cells, for muscle movement, and for the production of heat to maintain the body's temperature. Such energy is produced by the controlled combustion of compounds such as glucose within the cell. The cell, however, does not operate like a steam engine—converting chemical energy to heat energy and then using the heat energy to do work. Instead, the cell operates much more efficiently, using chemical energy directly to do such work as the contraction of muscles, the transmission of nerve impulses, and biosynthesis. For these tasks the cell requires **high-energy,** or **energy-rich,** compounds such as adenosine triphosphate, ATP (Figure 23.1). A molecule of ATP (as is the case with other similar molecules such as guanosine triphosphate, GTP, and uridine triphosphate, UTP) contains two oxygen-to-phosphorus bonds that are called **high-energy phosphate bonds,** often represented by a wavy line.

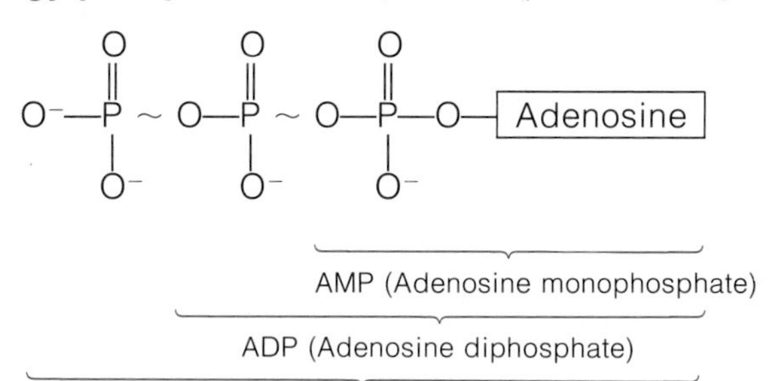

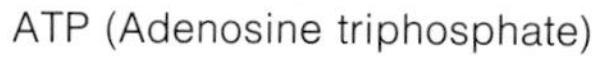

The hydrolysis of the high-energy phosphate bonds in compounds such as ATP is a highly exothermic reaction that produces about twice as much energy (7 kcal) as the hydrolysis of compounds containing low-energy phosphate bonds (2 to 5 kcal). The large amount of chemical energy stored in these high-energy phosphate bonds results from the

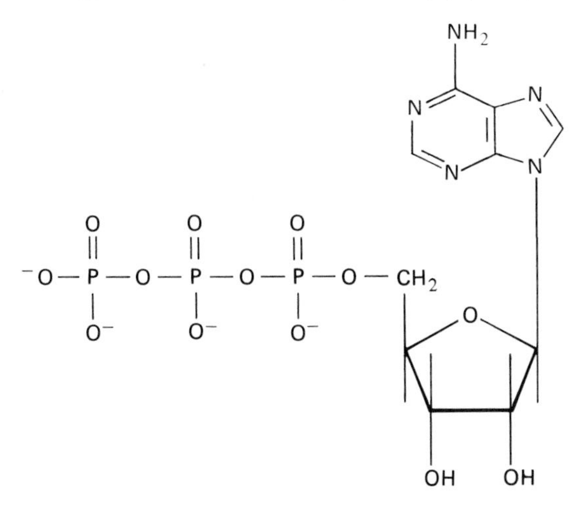

Figure 23.1 The structure of adenosine triphosphate, ATP.

specific arrangement of atoms in the molecule. When hydrolyzed, ATP forms ADP (adenosine diphosphate) and an inorganic phosphate (represented by P_i) and energy.

$$ATP + H_2O \longrightarrow ADP + P_i + 7.3 \text{ kcal}$$

(The end phosphate group on ADP can also be hydrolyzed, yielding adenosine monophosphate (AMP) and the same amount of energy as the hydrolysis of ATP.)

If the hydrolysis of ATP is coupled with an endothermic, or energy-requiring, reaction in the cell (such as the contraction of a muscle fiber or the synthesis of a molecule), the hydrolysis of ATP will supply the energy necessary for the other reaction to occur. Any excess energy provided by the hydrolysis of ATP is released as heat.

$$\text{Relaxed muscle} + ATP + H_2O \longrightarrow \text{Contracted muscle} + ADP + P_i$$

ATP also provides the energy for reactions in which the phosphate group is transferred from ATP to another molecule. For example,

$$\text{Glucose} + ATP \longrightarrow \text{Glucose-6-phosphate} + ADP$$

DIGESTION

23.2 Digestion and Absorption

Before a cell can produce ATP from the nutrients contained in our diets, these nutrients must be digested and absorbed. **Digestion** is the process by which complex foods are broken down into simple molecules. Table 23.1 summarizes the digestion of carbohydrates, lipids, and proteins.

Table 23.1 Nutrient Digestion

Site of Digestion	Source of Enzymes	Nutrients Digested ⟶ Products Formed
Mouth	Salivary glands	Starch ⟶ dextrins, maltose
Stomach	Stomach	Proteins ⟶ peptides, amino acids Emulsified fats ⟶ monoacylglycerols, fatty acids, glycerol
Small intestine	Pancreas	Starch ⟶ maltose Fats ⟶ monoacylglycerols, fatty acids, glycerol Proteins ⟶ peptides Peptides ⟶ amino acids
	Intestine	Peptides ⟶ amino acids Sucrose ⟶ glucose, fructose Maltose ⟶ glucose Lactose ⟶ glucose, galactose

Carbohydrates

The digestion of carbohydrates begins in the mouth, where teeth break large pieces of food into smaller ones. This food is then mixed with saliva, which contains an enzyme that begins the breakdown of starch to dextrins and maltose. When the food is swallowed, it passes into the stomach, where these salivary enzymes become inactivated by the low pH. The churning action of the stomach further reduces the size of the food particles. As the food then passes into the small intestine, the pancreas and cells in the intestinal wall secrete juices containing enzymes that completely hydrolyze the polysaccharides and disaccharides in the food. Glucose, fructose, and galactose—the monosaccharides formed by this process—pass directly through the intestinal wall into the bloodstream and are carried to the liver. There, the galactose is converted to glucose by liver enzymes; the fructose may either be converted to glucose or may enter into other metabolic reactions. The glucose that is absorbed or synthesized from other sugars may be used to meet immediate cellular energy needs, or it may be stored as glycogen in the liver and the muscles.

Lipids

While some digestion of already emulsified food fats such as milk or eggs may occur in the stomach, the main digestion of lipids takes place in the small intestine. There, lipids are mixed with bile salts, which emulsify the fats. This provides the maximum surface area for the action of enzymes in the pancreatic and intestinal juices, which hydrolyze the fat into glycerol, fatty acids, and monoacylglycerols and diacylglycerols.

Bile is a fluid that is continuously manufactured in the liver and stored in the gall bladder. It is released from the gall bladder by a muscle contraction that is triggered by a hormone produced when food enters the small intestine. Bile is composed of bile salts, bile pigments, phospholipids, and cholesterol. Bile salts are synthesized from cholesterol in the liver and aid in the digestion of fats and in the absorption of fatty acids, fat-soluble vitamins, and other products of fat digestion. Bile salts are very efficiently reabsorbed from the intestine and then extracted from the blood by the liver to be used again. Bile pigments, mainly bilirubin, do not play a role in fat digestion. Rather, they are the waste products of the breakdown of hemoglobin in the liver and are the substances that give color to the feces.

If the bile duct becomes blocked, or if the breakdown of hemoglobin occurs at a very fast rate, bilirubin will build up in the blood. This will make the skin appear more yellow, one of the symptoms of jaundice. In some people the mucous membranes of the gall bladder will absorb water, concentrating the bile. Under these conditions the cholesterol in bile, which is not very water-soluble, will crystallize out of solution together with bile salts and bile pigments, forming gall stones. These stones can cause infection and pain and obstruct the flow of bile, resulting in jaundice.

The glycerol, fatty acids, and monoacylglycerols and diacylglycerols that are formed from the hydrolysis of fats are absorbed by the intestinal cells. There they are reformed into triacylglycerols and packaged as tiny droplets of lipoprotein called chylomicrons, which are transported by means of the lymph system to points where they can enter the blood. After a meal rich in fat, the chylomicrons cause the blood to have a milky, opalescent appearance.

Proteins

The digestion of proteins involves yet another process. Proteins that are in their native state are not easily digested because their peptide bonds cannot be reached by digestive enzymes. Therefore, dietary proteins must first be denatured by heat (in cooking), by chewing, and by the acids in the stomach. Protein digestion itself begins in the stomach. The protein-digesting enzymes are secreted by the cells of the stomach lining as zymogens and are activated by the HCl present in stomach acid. This process prevents the enzymes from digesting the proteins making up the cells of the stomach and allows time for the secretion of mucous, which protects the cells lining both the stomach and intestines from the enzymes. Protein digestion continues in the small intestine, where enzymes in the pancreas and intestinal juices catalyze the hydrolysis of the proteins. The amino acids and some very small peptides that are produced are absorbed through the intestinal wall and enter the bloodstream.

CARBOHYDRATE METABOLISM

23.3 Glycogenesis and Glycogenolysis

We have said that carbohydrates—specifically glucose—are the major source of energy for cells. Most tissues can also use fatty acids to supply their energy needs, but the cells of the brain and nervous system can use only glucose. The brain, which contains no storehouse of glucose, depends completely upon the glucose in the blood for its energy supplies. Therefore, the blood glucose level is critical to normal brain function. The normal level of glucose in the blood is 60 to 100 mg/100 ml of blood. Any significant variation from normal, either too high or too low, can affect the brain and nervous system. But, of course, we do not supply our bodies continuously with glucose. Instead, we eat large meals and then have relatively long periods of fasting in between. To assure a continuous supply of glucose to meet the energy needs of brain cells, the body has developed complicated control mechanisms for keeping the blood sugar level fairly constant.

Right after a meal, the blood sugar level is elevated. To deal with this condition, the beta cells of the pancreas produce the hormone insulin, which aids the passage of glucose into liver, muscle, and fatty tissue cells, and encourages the storage of glucose in the liver in the form of glycogen. The process of converting glucose to glycogen is called **glycogenesis.** This biosynthetic process involves the series of reactions shown in Figure 23.2, requiring several enzymes and two high-energy molecules: ATP and UTP. As time passes and the cells of the body absorb and metabolize the glucose in the blood, the blood sugar level falls. This causes the pancreas to stop producing insulin and to start producing another hormone, glucagon. Glucagon has an effect opposite that of insulin: it stimulates the liver to hydrolyze glycogen so as to form glucose. The series of reactions by which this takes place is called **glycogenolysis,** and the glucose that it provides is released into the blood to maintain the blood sugar at normal levels (Figure 23.2).

Hyperglycemia is the condition that results when blood sugar levels are too high. Mild hyperglycemia occurs after meals. In severe hyperglycemia, the blood sugar level can rise so high that the glucose level exceeds the amount tolerated by the kidneys (called the renal threshold, 160 mg/100 ml); glucose will then be excreted in the urine (Figure 23.3). Glycosuria, sugar in the urine, can result from conditions such as diabetes mellitus, emotional stress, kidney failure, or the administration of certain drugs.

Hypoglycemia is the condition that results when blood sugar levels are below normal. Under such conditions the brain becomes starved for glucose. Mild hypoglycemia produces irritability, dizziness, lethargy, grogginess, and fainting. Severe hypoglycemia can produce convulsions, shock, and coma. Such severe hypoglycemia can be produced by the

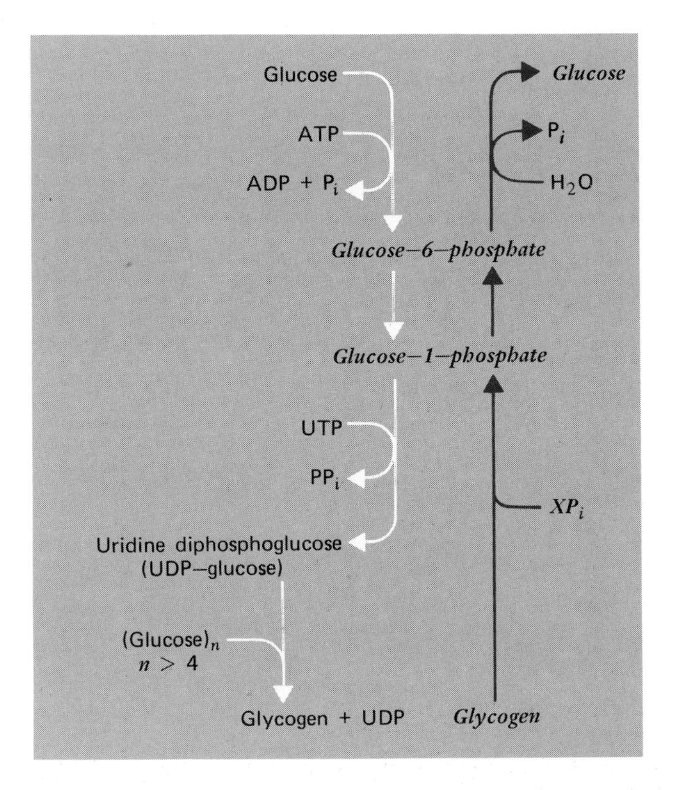

Figure 23.2 The reactions of glycogenesis and glycogenolysis (P_i = phosphate, and PP_i = pyrophosphate).

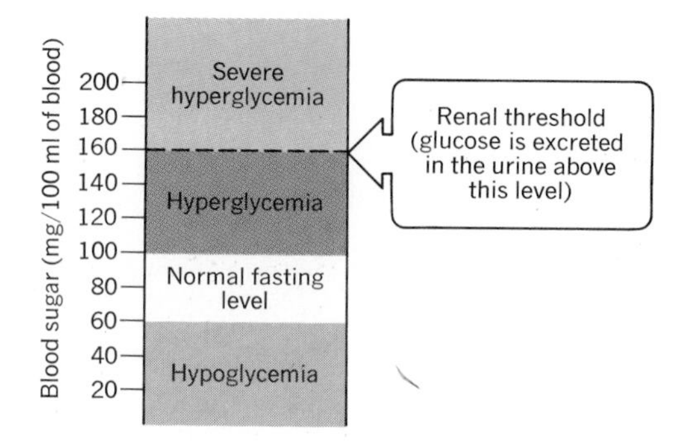

Figure 23.3 Blood sugar levels and the conditions of hypoglycemia and hyperglycemia.

overproduction or overinjection of insulin, a condition called hyperinsulinism, in which the blood sugar level drops extremely rapidly and the person goes into convulsions and coma called *insulin shock*. For most of us, our body's regulatory mechanisms keep blood sugar levels within normal ranges even when we eat a meal rich in refined carbohydrates, such as the sugars in candy and cake. It is preferable, however, to eat complex carbohydrates such as whole grain cereals and pastas, which release glucose into the blood slowly as they are digested.

23.4 Oxidation of Carbohydrates

The glucose in the blood also diffuses into tissue cells. There it can be used in biosynthetic reactions or can be broken down to yield useful cellular energy through a process called **cellular respiration.** In cellular respiration, glucose is oxidized to form carbon dioxide, water, and ATP. The cellular processes of respiration are quite efficient; about 44% of the energy released from the oxidation of glucose is trapped in the high-energy bonds of ATP molecules and can be used to do work. For each molecule of glucose that is oxidized, 36 ATPs are produced. But what, then, happens to the other 56% of the energy that is released? It is given off as heat in quantities more than enough to maintain body temperature at 37°C. Any excess heat must be eliminated from the body through cooling mechanisms such as the evaporation of sweat.

The reactions of cellular respiration are summarized in Figure 23.4. The rate of each reaction is controlled by the concentration of ATP and AMP in the cell. Cellular respiration is a very complex process, involving many steps and requiring many enzymes and coenzymes. A specific description of each step must be left to other biochemistry courses, but we can give a general overview of this process.

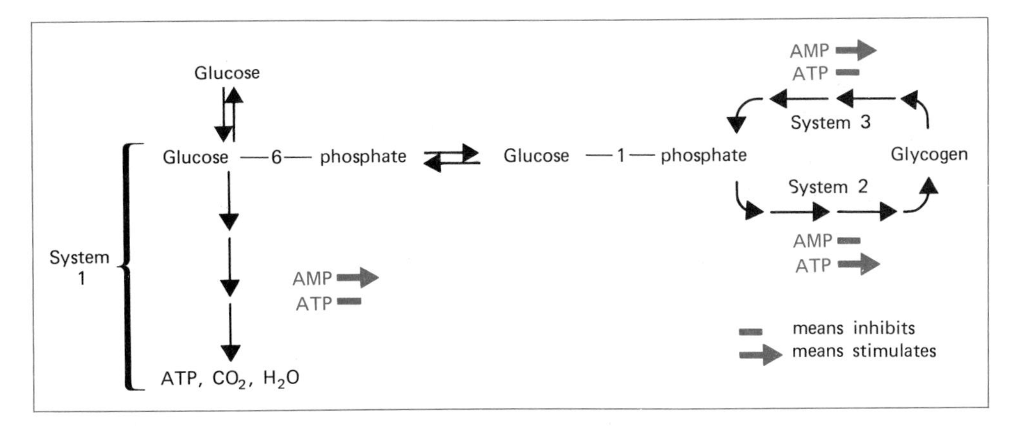

Figure 23.4 ATP and AMP have opposite regulatory effects on the three multienzyme systems controlling the level of ATP in the cell. System 1 involves the oxidation of glucose. System 2 involves glycogenesis, and system 3 glycogenolysis.

Anaerobic Stage: Glycolysis

The oxidation of glucose can be divided into two stages. The first, an anaerobic stage, requires no oxygen and occurs in the cytoplasm of the cell. This stage is called **glycolysis,** and involves the breakdown of glucose to form two molecules of lactic acid (in the form of the negative ion lactate; Figure 23.5). The net energy production from this process is two ATPs, which can be summarized by the following overall reaction.

$$\underset{\text{Glucose}}{C_6H_{12}O_6} + 2ADP + 2P_i \longrightarrow 2\ \underset{\text{Lactate}}{CH_3—\overset{\overset{\large OH}{|}}{C}H—\overset{\overset{\large O}{\|}}{C}—O^-} + 2ATP + 2H_2O + 2H^+$$

Glycolysis is the emergency energy-producing process (or pathway) for the cells. When oxygen is cut off to the tissues, ATP levels can be maintained for a short time through glycolysis. During childbirth, for example, the amount of oxygen available to the infant (and, therefore, present in the blood) is severely reduced. In order to assure an adequate supply of oxygen to the brain, the circulation of the infant's blood decreases in all tissues except for the brain. These other tissues, then, depend upon glycolysis for their energy until the circulation returns to normal after the delivery is over.

Aerobic Stage: Citric Acid Cycle and Respiratory Chain

The second part of glucose oxidation, the aerobic stage, requires oxygen. This is the stage in the oxidation of glucose in which most of the energy is produced. The reactions of this stage also serve as the final stage of oxidation for other molecules that are used by the cell for energy

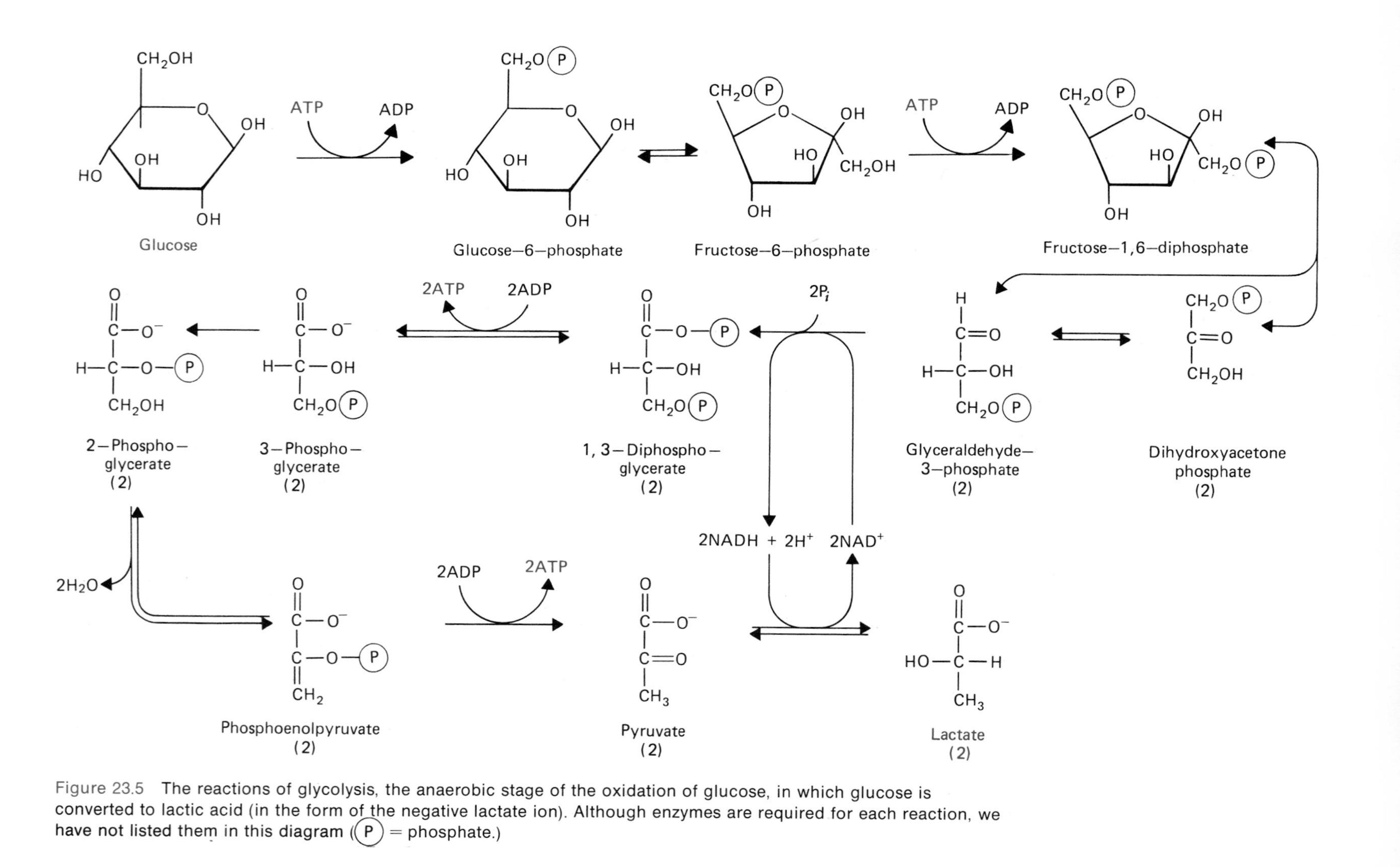

Figure 23.5 The reactions of glycolysis, the anaerobic stage of the oxidation of glucose, in which glucose is converted to lactic acid (in the form of the negative lactate ion). Although enzymes are required for each reaction, we have not listed them in this diagram ((P) = phosphate.)

production. The aerobic stage takes place in the mitochondria, often called the "power plants" of the cell (Figure 23.6). These structures are located near the parts of the cell that require energy, such as the contractile fibers of muscle cells.

This aerobic stage involves two series of reactions. The first series is called the **citric acid cycle;** it results in the final breakdown of the fuel molecule to carbon dioxide. (These reactions are also often called the **Krebs cycle**—after Hans Krebs, who suggested the basic features of this pathway in 1937—or the **tricarboxylic acid (TCA) cycle.**) This breakdown of the fuel molecule consists of a series of oxidation reactions that produce hydrogen atoms. These hydrogen atoms, in turn, are used in a second series of redox reactions, called the **respiratory chain** or **electron transport chain.** The respiratory chain requires oxygen and produces ATP and water (Figure 23.7).

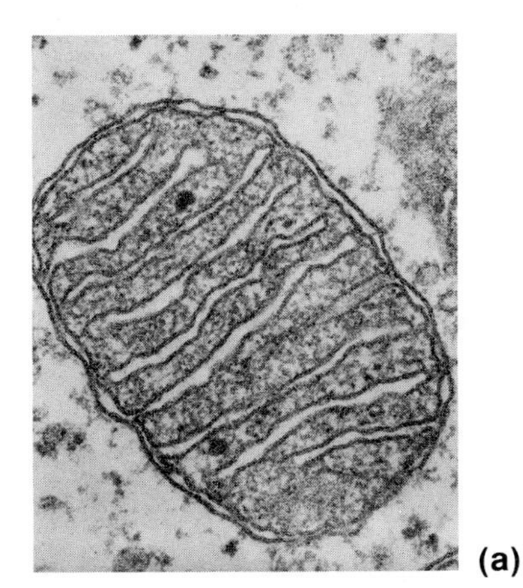

(a)

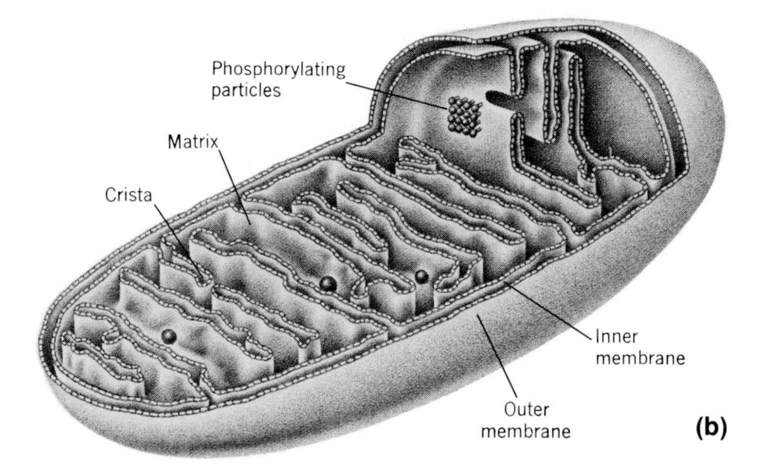

(b)

Figure 23.6 A mitochondrion. (a) An electron micrograph of a mitochondrion. (b) A model of a mitochondrion. The inner folds of the mitochondrion, called the cristae, contain the enzymes of the citric acid cycle and the respiratory chain. [(a) Courtesy Dr. George Palade.]

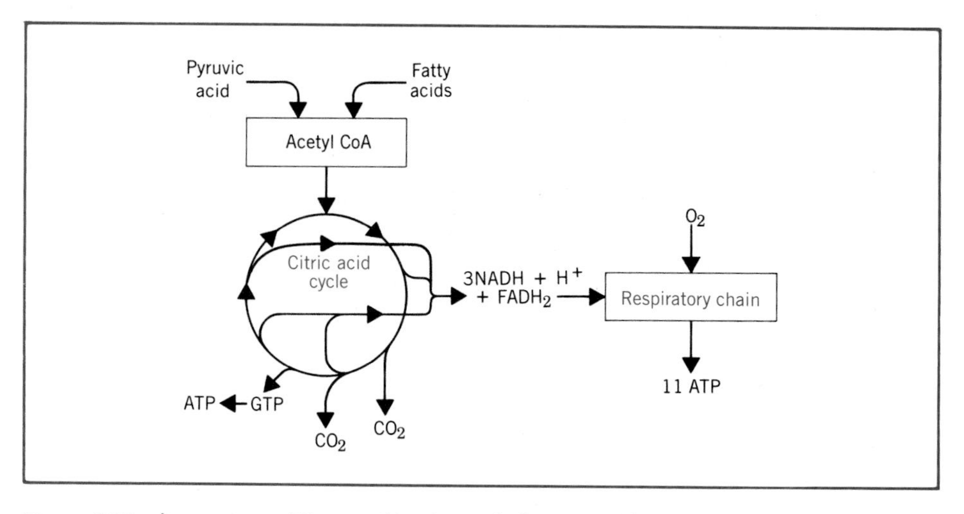

Figure 23.7 A summary of the aerobic stage of glucose oxidation.

The steps in the oxidation of glucose, the primary fuel molecule, can be summarized as follows:

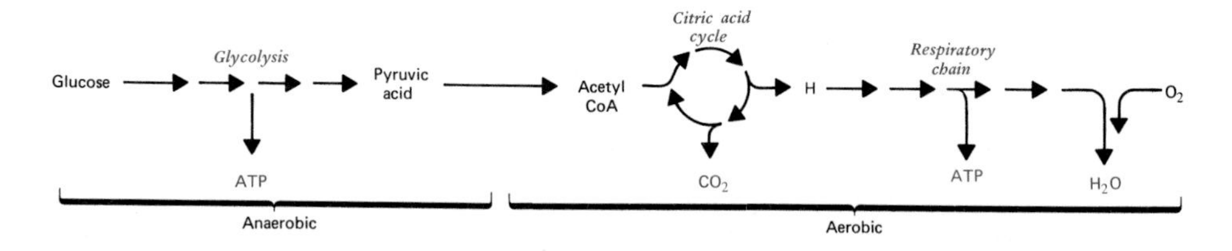

23.5 Citric Acid Cycle

If there is sufficient oxygen available, the last step of glycolysis—the conversion of pyruvic acid (in the form of the negative ion pyruvate) to lactate—does not take place (Figure 23.5). Instead, pyruvate is oxidized to acetic acid, which forms a thioester bond with coenzyme A; the resulting complex is called acetyl CoA (see Section 18.6). This reaction requires the hydrogen carrier molecule NAD^+ in addition to coenzyme A; its other products are NADH and H^+, together with one molecule of CO_2.

$$\text{Pyruvate} + NAD^+ + \text{Coenzyme A} \xrightarrow[\text{and coenzymes}]{\text{Other enzymes}} \text{Acetyl CoA} + NADH + H^+ + CO_2$$

The acetyl CoA formed in this reaction is then available to enter the citric acid cycle.

The citric acid cycle, shown in Figure 23.8, oxidizes the two-carbon acetyl group to carbon dioxide. This process produces four pairs of hydrogen atoms and one high-energy phosphate bond in the form of GTP (which transfers a phosphate to ADP, thus forming ATP). Three of the hydrogen atom pairs are captured and carried by the coenzyme NAD^+ and one by the coenzyme FAD. Cells, however, do not have an unlimited supply of NAD^+ and FAD; therefore, the citric acid cycle cannot occur

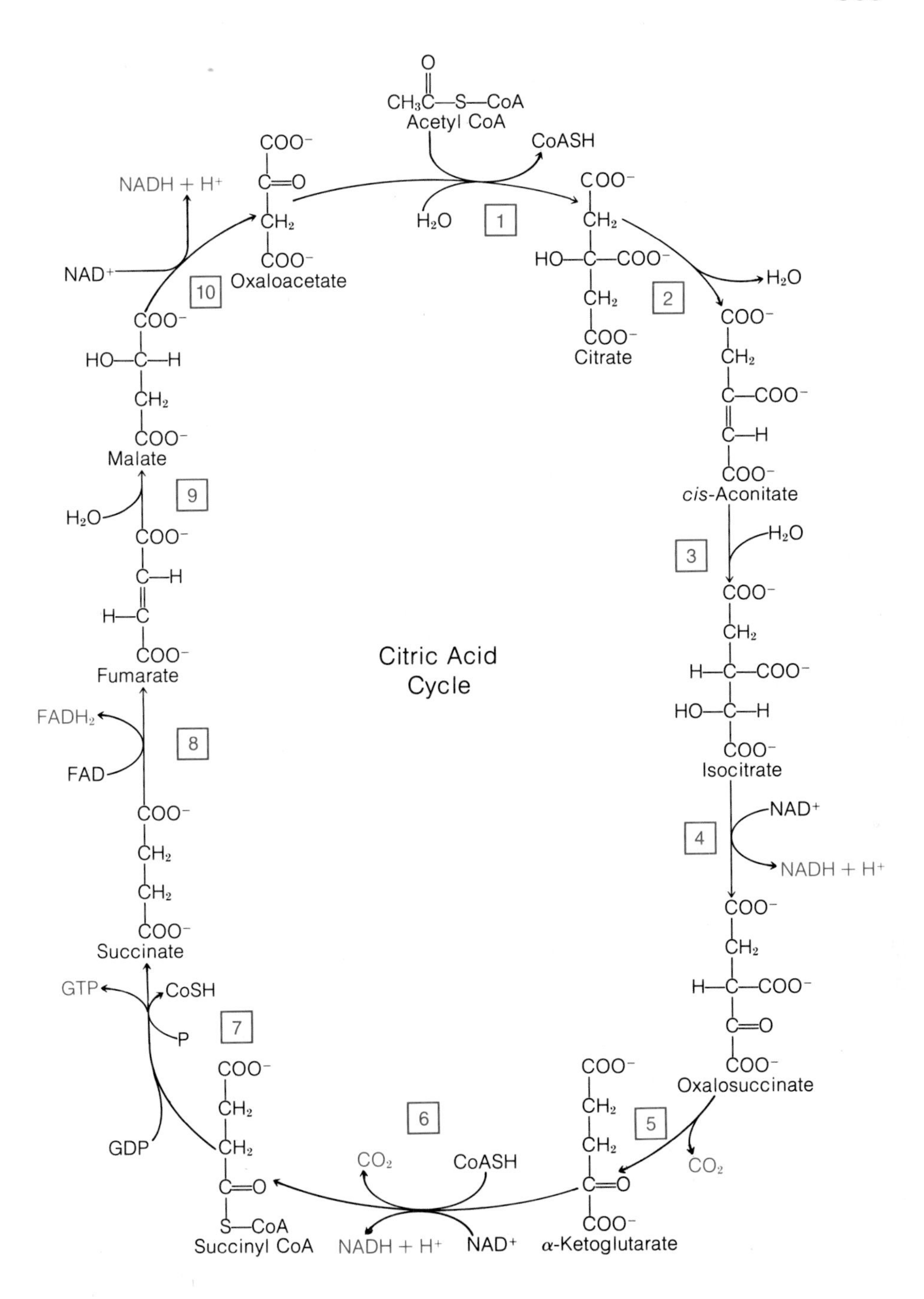

Figure 23.8 The citric acid cycle. Two carbon atoms enter the cycle in the form of acetyl CoA and are joined to oxaloacetate to form citrate (Step 1). During the series of reactions that then occur, two molecules of carbon dioxide are formed (Steps 5 and 6), and four pairs of hydrogens are produced and carried to the respiratory chain (Steps 4, 6, 8, and 10). At the end of the cycle a molecule of oxaloacetate is formed, ready for another turn of the cycle (Step 10).

independently of the respiratory chain (in which these coenzymes donate their hydrogens to other hydrogen carriers and become available again as hydrogen acceptors in the citric acid cycle.) It is important that you notice the cyclic (or circular) nature of the reactions in the citric acid cycle. In the first reaction, acetyl CoA is joined with oxaloacetate to form citrate (remember that all of the acids found in the citric acid cycle exist as negative ions at cellular pH). The last reaction again produces oxaloacetate, which is then available to combine with another acetyl CoA.

23.6 Respiratory Chain and ATP Formation

The final hydrogen acceptor, or carrier, in aerobic oxidation is the oxygen we breathe. To arrive at this point, the hydrogens carried by NAD^+ and FAD from the citric acid cycle enter a series of coupled reactions called the **respiratory chain** or **electron transport chain.** In this series of redox reactions, the hydrogens (and later just their electrons) are passed between a series of compounds until they are combined with oxygen to form water. This process produces most of the ATP molecules formed in the oxidation of glucose by the cell. Each transfer is part of an oxidation–reduction reaction in which some energy is released. As the electrons flow through this series of reactions, they can do work, just as the electrons flowing through a copper wire can do work. The kind of work done by the electrons in the cell, however, is the production of ATP from ADP and inorganic phosphate. The exact way in which this process works is not well understood, but it is known at what points in the transport chain ATP is produced. From Figure 23.9 we see that each molecule of NADH produces three ATPs and each molecule of $FADH_2$ produces two ATPs.

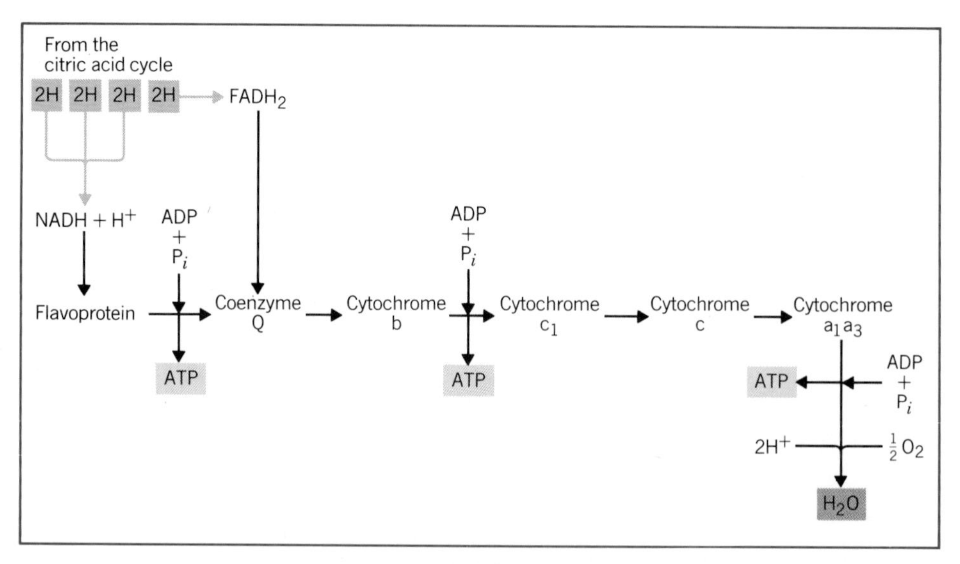

Figure 23.9 The respiratory chain. The hydrogens produced in the citric acid cycle are passed between a series of compounds until they are combined with oxygen to form water. It is this series of oxidation-reduction reactions that produces most of the ATP formed in the oxidation of glucose.

Therefore, as it passes through the citric acid cycle, one molecule of acetyl CoA will generate 12 ATPs: 1 from GTP produced in the citric acid cycle and 11 from NADH and $FADH_2$ in the respiratory chain (3 NADHs produce 9 ATPs and 1 $FADH_2$ produces 2 ATPs).

23.7 Lactic Acid Cycle

A superbly trained athlete waits at the starting blocks for the gun, sprints 400 meters at top speed, breaks the tape in record time, and then collapses on the track (Figure 23.10). Her legs feel like rubber, she painfully gasps for breath, and she feels nauseated. If the athlete is well trained, why does it require a rest of $1\frac{1}{2}$ hours to completely recover from the exhaustion of this extreme effort?

When undergoing only moderate exercise, muscle cells have enough oxygen to carry out the respiration process aerobically. But during strenuous exercise the blood cannot supply oxygen to the muscles fast enough, and the muscle cells must rely upon a backup system—the production of energy by glycolysis. The lactic acid produced by glycolysis builds up in the muscle cells to the point where it hampers muscle performance, causing muscle fatigue and exhaustion. Such a lactic acid buildup produces mild acidosis, which causes nausea, headache, lack of appetite, and impairment of oxygen transport, resulting in the difficult, painful gulping of air experienced by athletes after extreme physical efforts. Muscle cells are slow to recover from this condition. The lactic acid must be removed either by conversion to pyruvic acid in the muscle cells, or by movement from these cells to the liver. In the liver, about 25% of the lactic acid is converted to pyruvate and oxidized through the citric acid cycle; the remainder is converted to glycogen. The heavy breathing that occurs after strenuous exercise helps supply the oxygen necessary to oxidize the lactic acid (Figure 23.11). A major difference between a well-conditioned athlete and a nonathlete is that the athlete can supply her muscle cells with

Figure 23.10 The tremendous exertion required to beat a world record has taken its toll. (Thomas Hopker/Woodfin Camp.)

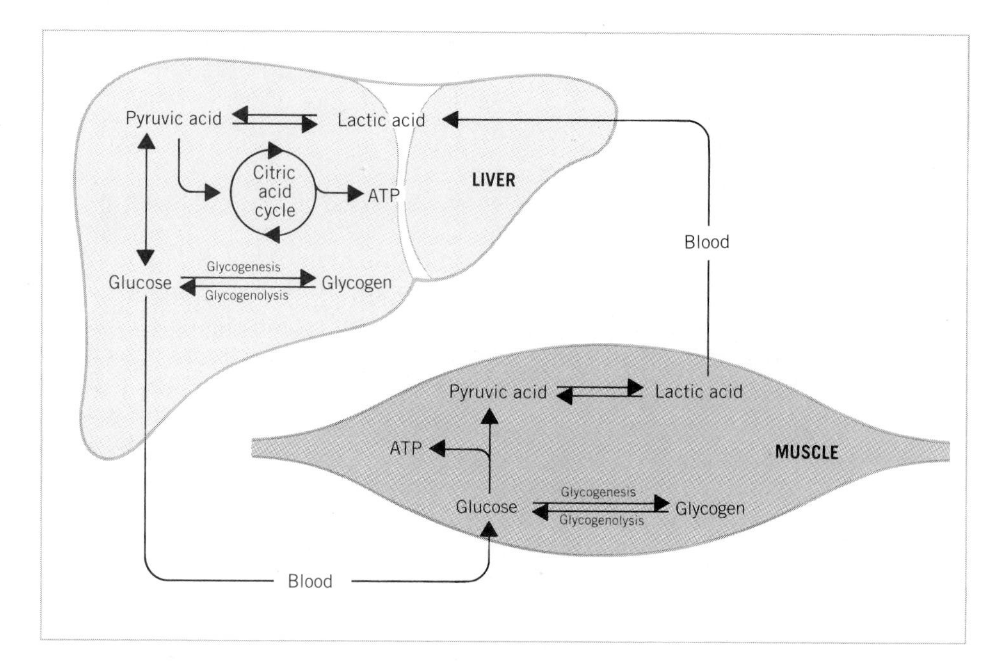

Figure 23.11 The lactic acid cycle. Lactic acid formed in strenuous exercise must be removed either by conversion to pyruvic acid in the muscle cells or by migration from these cells to the liver, where it is converted to glycogen. Some of the lactic acid (about 25%) must be oxidized in the liver to provide the energy necessary for this conversion.

the oxygen necessary to maintain aerobic respiration for a longer period than can the nonathlete. This means that her muscle cells will require glycolysis for their supply of ATP only at much higher levels of exertion.

23.8 Fermentation

Microorganisms such as yeast are able to live quite well without oxygen. Glycolysis occurs in yeast cells in much the same way as it does in animal cells. The process differs in yeast cells only in the last step, where pyruvic acid is converted to ethanol and carbon dioxide rather than to lactic acid. The name given to this form of glycolysis in yeast cells is **fermentation,** and the overall reaction is as follows:

$$\underset{\text{Glucose}}{C_6H_{12}O_6} + 2ADP + 2P_i \xrightarrow[\text{Enzymes}]{\text{Yeast}} \underset{\text{Ethanol}}{2CH_3CH_2OH} + 2CO_2 + 2ATP$$

LIPID METABOLISM

23.9 Lipogenesis

Carbohydrates, fats, and proteins that we consume in excess of our energy requirements are oxidized to acetyl CoA, which is readily converted to fatty acids. These fatty acids (along with other fatty acids from the diet) are stored as triacylglycerols in the fat cells of adipose tissue under the skin and around major organs. This synthesis of lipids is called **lipogenesis.** Storage fat serves several functions: it is an energy reserve, a support and a shock absorber for inner organs, and heat insulation for the body. Although the glycogen reserves in the liver and muscles are sufficient to last only a few hours, a 70-kg male has enough stored fat to sustain him for 70 days if there is water available.

The lipids in adipose tissue are in dynamic equilibrium with the lipids in the blood. That is, stored fatty acids are constantly being exchanged with food fatty acids. The particular composition of the storage fat differs for different organisms, but in animals it can be altered by controlling the diet. For example, experiments are being carried out to determine if cattle raised on special diets will develop higher concentrations of unsaturated fats in their muscle tissue.

23.10 Oxidation of Fatty Acids

Upon oxidation, fats produce much more energy than carbohydrates or proteins; fats yield 9 kcal/g, whereas glycogen, starch, and proteins yield about 4 kcal/g. In vertebrates, oxidation of fatty acids provides at least half the energy needed by the liver, kidneys, heart, and the skeletal muscles at rest. In hibernating animals and migrating birds, fat is the only energy source.

The process of fat oxidation starts when fats stored in the adipose tissues are hydrolyzed to fatty acids and glycerol. These compounds are then carried to energy-requiring tissues where they will be oxidized. The glycerol enters the glycolysis pathway, as we discussed in Section 23.4. The fatty acids are oxidized, mainly in the mitochondria of the liver, heart, and skeletal muscles, by means of a repeating series of reactions called the **fatty acid cycle** or **β-oxidation.** This series of reactions is shown in Figure 23.12. We can briefly describe each of the steps in this process as follows:

1. The first step involves joining the fatty acid with coenzyme A (CoA). This requires one ATP and produces AMP and a diphosphate (or pyrophosphate, denoted PP_i).
2. This step is a dehydrogenation reaction that produces a double bond between the alpha and beta carbon. The hydrogens that are removed in this reaction are attached to the hydrogen carrier molecule FAD, producing $FADH_2$.
3. In this step water is added to the double bond, forming an alcohol group on the beta carbon.
4. The alcohol group on the beta carbon is then oxidized (from which comes the name *beta-oxidation*), producing NADH and H^+.
5. In the last step, the bond between the alpha and beta carbon is broken by coenzyme A. This produces one acetyl CoA and a new fatty acid (having two fewer carbon atoms) joined with coenzyme A.

6. This fatty acid will again enter the fatty acid cycle at Step 2.

The cycle will continue to remove two-carbon units from the fatty acid until it has been completely oxidized. Each turn of the cycle produces one $FADH_2$ and one NADH that can enter the respiratory chain, and one molecule of acetyl CoA that can be used in the citric acid cycle for the production of ATP. Overall, about 50% of the energy released in the complete oxidation of a fatty acid is trapped in molecules of ATP.

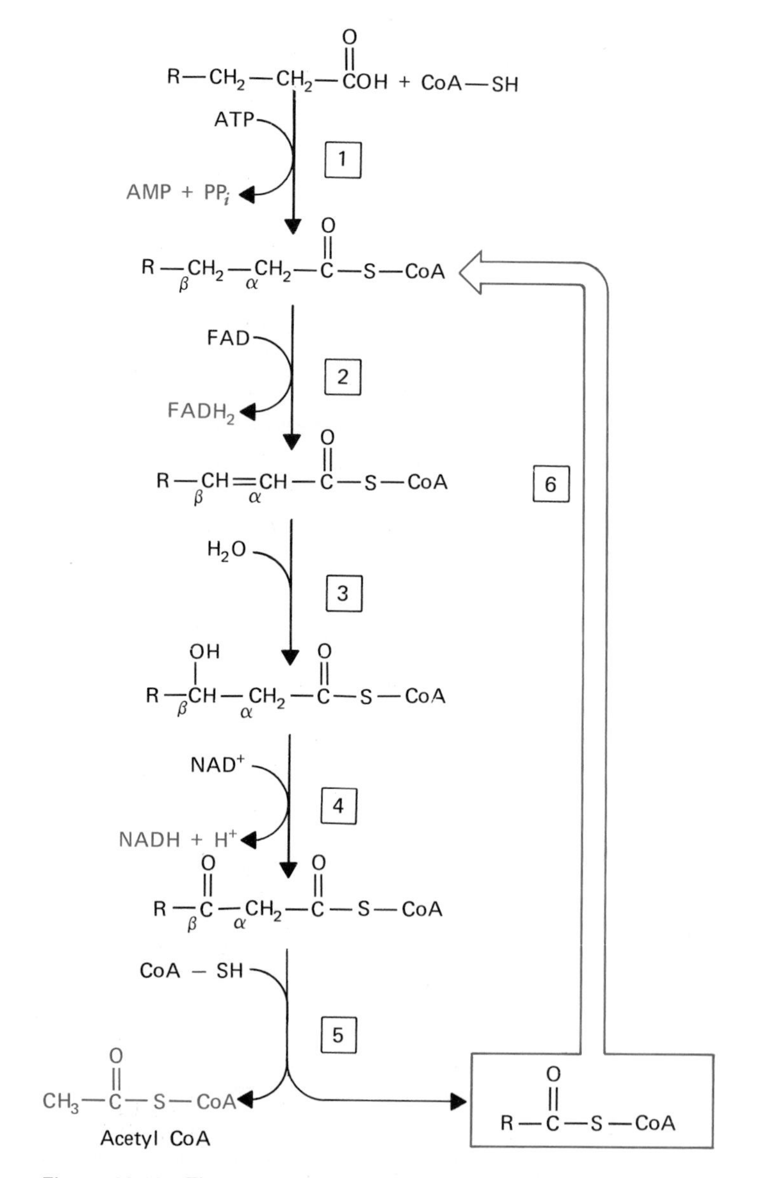

Figure 23.12 The reactions of the fatty acid cycle.

EXAMPLE 23-1

How many ATPs would be produced by the complete oxidation of one molecule of palmitic acid?

(a) The formula of palmitic acid is

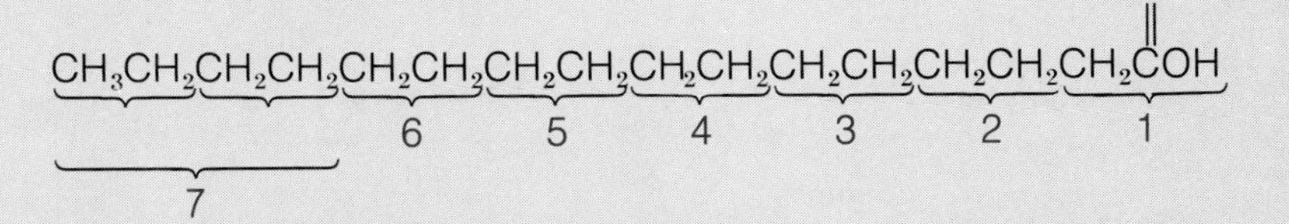

(b) Because each turn of the fatty acid cycle removes a two-carbon unit, palmitic acid will require seven turns of the cycle. (Notice that the seventh turn will produce two molecules of acetyl CoA.)

(c) Therefore, seven molecules each of $FADH_2$ and NADH that can enter the electron transport chain are formed (see Figure 23.9).

$$7\ FADH_2 \longrightarrow 14\ ATP$$

$$7\ NADH \longrightarrow 21\ ATP$$

(d) Eight molecules of acetyl CoA that can enter the citric acid cycle are produced. From Section 23.6 we have seen that each acetyl CoA produces 12 ATP. Therefore,

$$8\ \text{acetyl CoA} \longrightarrow 96\ ATP$$

(e) The total number of ATPs formed, therefore, is 14 + 21 + 96 = 131 ATP. Remember, however, that one ATP was needed at Step 1 for palmitic acid to enter the cycle, so the actual number of new ATPs available is 130.

Exercise 23-1

How many ATPs will be produced in the complete oxidation of one molecule of myristic acid?

23.11 The Metabolic Role of Acetyl CoA

Acetyl CoA plays a central role in the metabolism of nutrients in the cell. In addition to entering the citric acid cycle, acetyl CoA can be used in the biosynthesis of compounds required by the cell (such as cholesterol, steroids, and long-chain fatty acids) or, when present in excess, can be

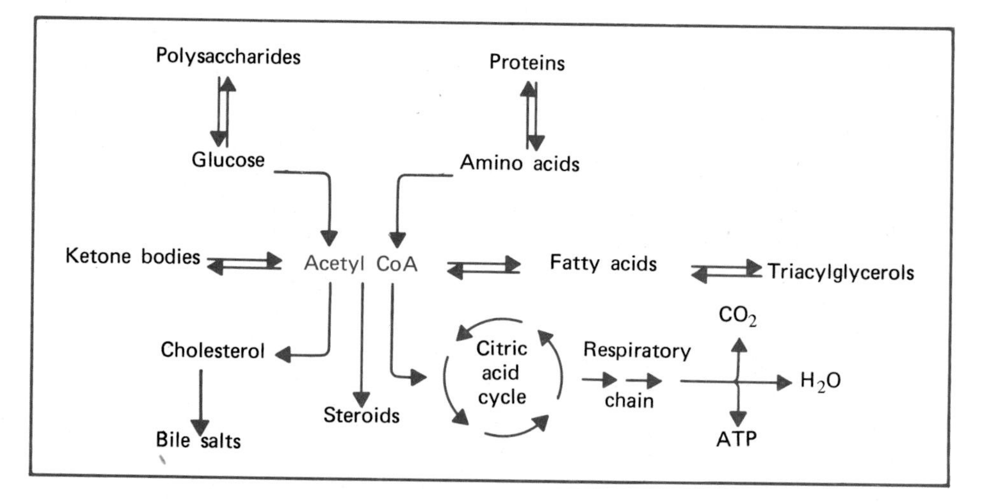

Figure 23.13 Acetyl CoA plays a central role in the anabolic and catabolic reactions of cellular metabolism.

used to synthesize ketone bodies. The key role played by acetyl CoA in metabolism is summarized in Figure 23.13.

23.12 Ketone Bodies

Excess acetyl CoA, produced by the oxidation of fatty acids, can be converted by the liver and the kidneys to acetoacetic acid (in the form of its negative ion, acetoacetate), acetone, and β-hydroxybutyric acid (in the form of its negative ion, β-hydroxybutyrate). These compounds are called **ketone bodies,** even though β-hydroxybutyrate is not a ketone (Figure 23.14). Ketone bodies normally circulate in the blood in low concentrations (about 1 mg/100 ml) and are used by skeletal and heart muscles to produce ATP.

Any disruption of normal metabolism (such as liver damage, diabetes mellitus, starvation, or extreme dieting) that causes a restriction or decrease in glucose metabolism will increase the rate of fat metabolism and,

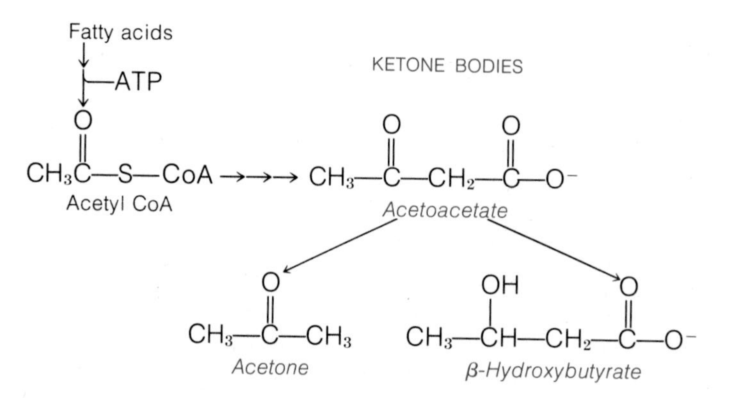

Figure 23.14 Ketone bodies are compounds formed from excess acetyl CoA.

therefore, increase the production of ketone bodies. If the level of ketone bodies in the blood exceeds the amount that can be used by the tissues, the ketone bodies accumulate in the blood causing a condition called **ketosis** and are excreted in the urine. Because two of the ketone bodies are acids, in ketosis the pH of the blood drops and acidosis occurs.

In diabetes, the glucose in the blood is not absorbed into the cells, but is instead excreted in the urine. To supply the body with energy, therefore, the untreated diabetic will metabolize large amounts of fats, resulting in a high production of ketone bodies. Acetone can be smelled on the breath of such persons, and acetoacetic acid and β-hydroxybutyric acid will be found in high concentrations in their blood and urine. These conditions result in severe acidosis, producing nausea, dehydration, depression of the central nervous system, and, in extreme cases, coma and death.

PROTEIN METABOLISM

23.13 Metabolic Uses of Amino Acids

The amino acids absorbed into the blood from the intestines have many uses in the body. They are our main dietary source of nitrogen, an element essential to life. Amino acids are utilized by cells to synthesize tissue protein used in the formation of new cells or in the repair of old cells. The largest amounts of protein are required during periods of rapid growth (such as during infancy, adolescence, pregnancy, or when a mother is nursing) and during periods of extensive repair (such as after surgery, burns, hemorrhage, or infections). Certain amino acids are also used in the synthesis of nonessential amino acids, enzymes, hormones, antibodies, and nonprotein nitrogen-containing compounds such as nucleic acids or heme groups (Figure 23.15).

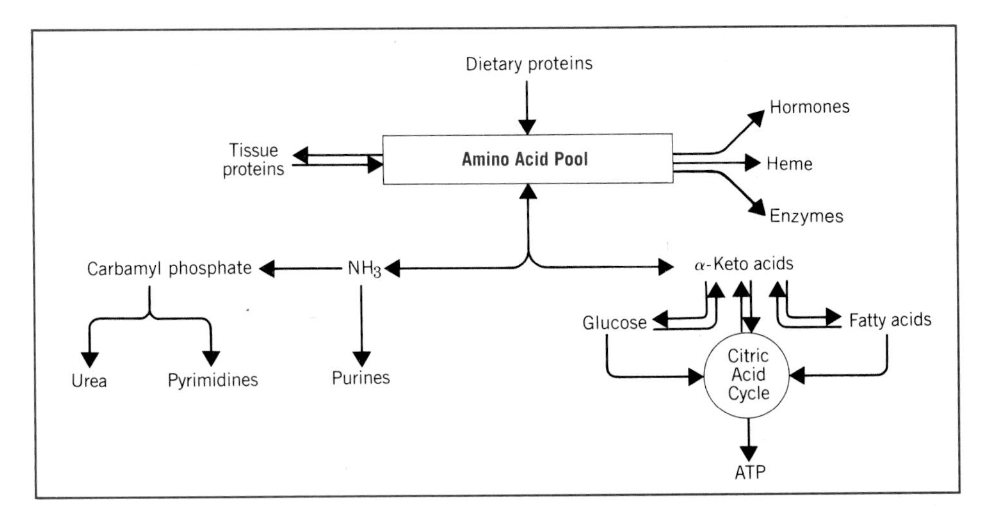

Figure 23.15 The amino acid pool in the human body is constantly changing as proteins are used for various functions in the body.

There is no storage form of amino acids in the body, as there is for carbohydrates (in the form of glycogen) and lipids (in the form of fat). Instead, the body maintains a constantly changing pool of amino acids, as tissue protein is continually being broken down and resynthesized. The turnover of amino acids in the liver and blood is relatively rapid—half will be replaced every 6 days. This turnover is much slower in muscles and supportive tissue—half the amino acids are replaced every 180 days in muscles and every 1000 days in the collagen of supportive tissues.

The amino acids not used in such a synthesis can be catabolized, or broken down, for energy. When this occurs, the amino group is removed from the amino acid through a process called oxidative deamination and will enter the urea cycle (which we will discuss shortly). The remainder of the molecule—called the carbon skeleton of the amino acid—can enter the citric acid cycle to supply needed cellular energy, or can be converted into body fat. This means that it is possible to become fat by eating too much protein, just as from eating too much lipid or carbohydrate.

Healthy adults maintain a nitrogen balance within their bodies; they excrete as much nitrogen (in the form of urea and other compounds) as they absorb from their diet. Rapidly growing young children will be in a state of positive nitrogen balance. They will not excrete as much nitrogen as they take in, because they need the amino acids to synthesize new tissues. A negative nitrogen balance, in which more nitrogen is excreted than absorbed, occurs in starvation, malnutrition, wasting diseases, and fevers, when amino acids are being used to supply the energy needs of the cell.

23.14 Amino Acid Synthesis

Transamination

The major way in which amino groups are removed from amino acids is through a process called **transamination.** Transamination involves the transfer of an amino group from an amino acid to an α-keto acid, thereby producing a new amino acid. (An α-keto acid is an acid containing a ketone functional group on the α-carbon, the carbon next to the carboxyl group.)

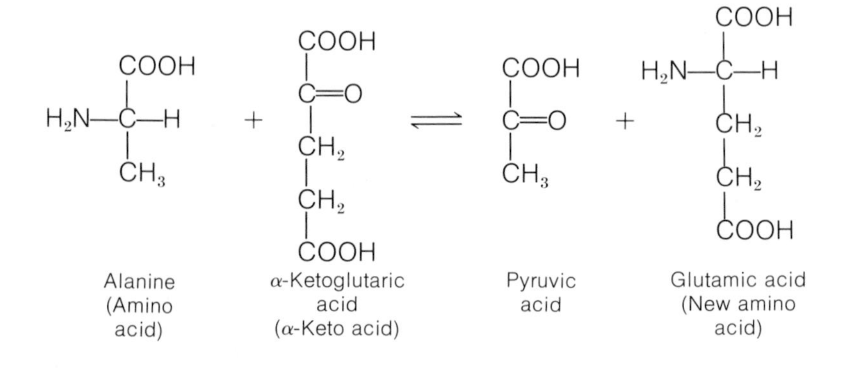

Specific Syntheses

Amino acids play a role in the synthesis of many metabolic compounds. The following are just a few examples. Tyrosine is used to produce the hormones epinephrine, norepinephrine, and thyroxine, as well as the skin pigment melanin. Tryptophan is used in the synthesis of the neurotransmitter serotonin and the coenzymes NAD^+ and $NADP^+$. Serine is converted to ethanolamine, which is found in lipids, and cysteine is used in the synthesis of bile salts.

23.15 Amino Acid Catabolism: Oxidative Deamination

Amino acids (mainly glutamic acid produced through transamination) can be converted in the liver to ammonia, carbon dioxide, water, and energy. **Oxidative deamination** is the process by which the amino group is removed, in the form of ammonia, from an amino acid.

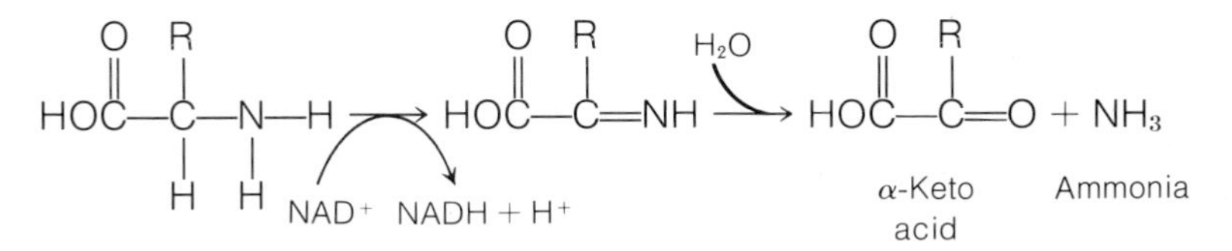

The α-keto acids that are formed in oxidative deamination can be used in several ways. They may be oxidized in the citric acid cycle to produce energy, or they may be converted to other amino acids through transamination. These α-keto acids may also be used in the synthesis of carbohydrates and fats. (See Figure 23.15)

23.16 Urea Cycle

The ammonia that is formed in oxidative deamination is toxic to cells; a concentration of 5 mg/100 ml of blood is toxic to humans. The liver disposes of this ammonia by converting it to the nontoxic urea in a cyclic reaction called the **urea cycle,** or **Krebs ornithine cycle.** Ammonia enters the cycle as carbamyl phosphate and is joined to the amino acid ornithine. In each turn of the cycle, one molecule of urea is produced and a molecule of ornithine is regenerated to begin the next turn (Figure 23.16). The net equation for the reaction is

$$2NH_3 + CO_2 \longrightarrow \underset{\text{Urea}}{H_2N\text{—}\overset{\overset{\displaystyle O}{\|}}{C}\text{—}NH_2} + H_2O$$

The urea can then be safely transported through the body for elimination by the kidneys. Any condition that impairs the elimination of urea by the kidneys can lead to uremia, a buildup of urea and other nitrogen wastes in the blood, which can be fatal. A person suffering from uremia will feel nauseated, irritable, and drowsy. His blood pressure will be elevated and he may be anemic. He may experience hallucinations and in serious cases may lapse into convulsions and coma. To reverse

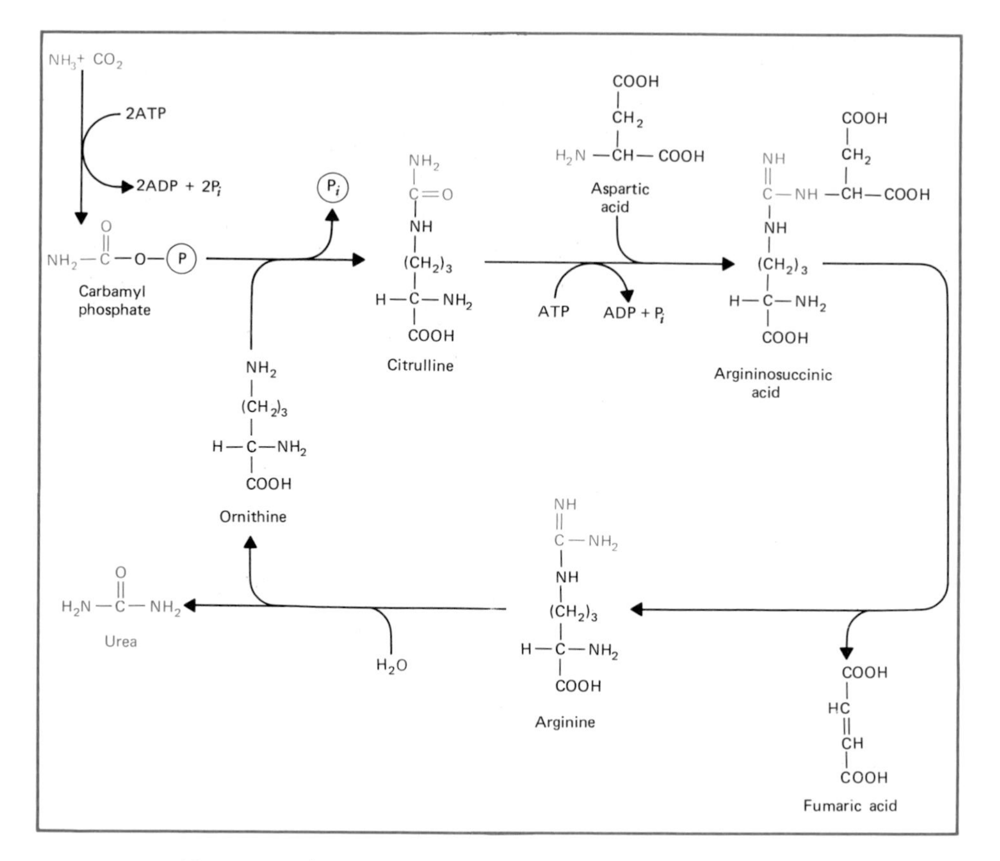

Figure 23.16 The urea cycle.

this condition, either the cause of kidney failure must be removed or the patient must undergo hemodialysis to remove the nitrogen wastes from the blood.

23.17 Pathways of Metabolism

We have discussed the metabolism of carbohydrates, lipids, and proteins separately in this chapter, but these pathways are all intricately interrelated, as shown in Figure 23.17. To illustrate the relationships among these pathways, let's consider the case of a college freshman who, when she entered college, was 5 feet 2 inches tall and weighed 167 pounds. After about a month of dormitory living, she decided she would have to lose weight. To do it as quickly as possible, she began a fasting diet of water, black coffee, and two multiple vitamin pills daily. After several days on this diet she became listless and stopped going to class; on the evening of the eighth day she collapsed while taking a shower and had to be rushed to the hospital. An examination at the hospital revealed the odor of acetone on her breath, a urinalysis that was positive for acetone but not glucose, and a urine pH of 5.5. Her blood glucose was 60 mg/100 ml. What was the cause of this sudden metabolic crisis?

Under normal conditions, the body tissues of a 70-kg person contain energy reserves as follows: 80 kcal of glucose in the blood, 280 kcal of liver glycogen, 480 kcal of muscle glycogen, 14,000 kcal of adipose fat, and 24,000 kcal of muscle protein. These body reserves become the energy source for a fasting individual. For a short while, the body can maintain a metabolic balance by using the circulating glucose. As the blood glucose reaches fasting level, the secretion of insulin ceases and that of glucagon begins; this stimulates glycogenolysis, releasing glucose from the glycogen reserves in the liver and muscles. When these glycogen reserves become depleted, the body must then find other sources of energy and blood glucose (remember, the brain requires glucose for normal function and must have 110 to 145 g of glucose a day). The body will next begin to break down muscle and tissue proteins. The amino acids produced in this tissue breakdown are deaminated in the liver, and are then used to synthesize the needed glucose, to produce ATP, and to synthesize other compounds needed for the maintenance of the citric acid cycle. The large amounts of

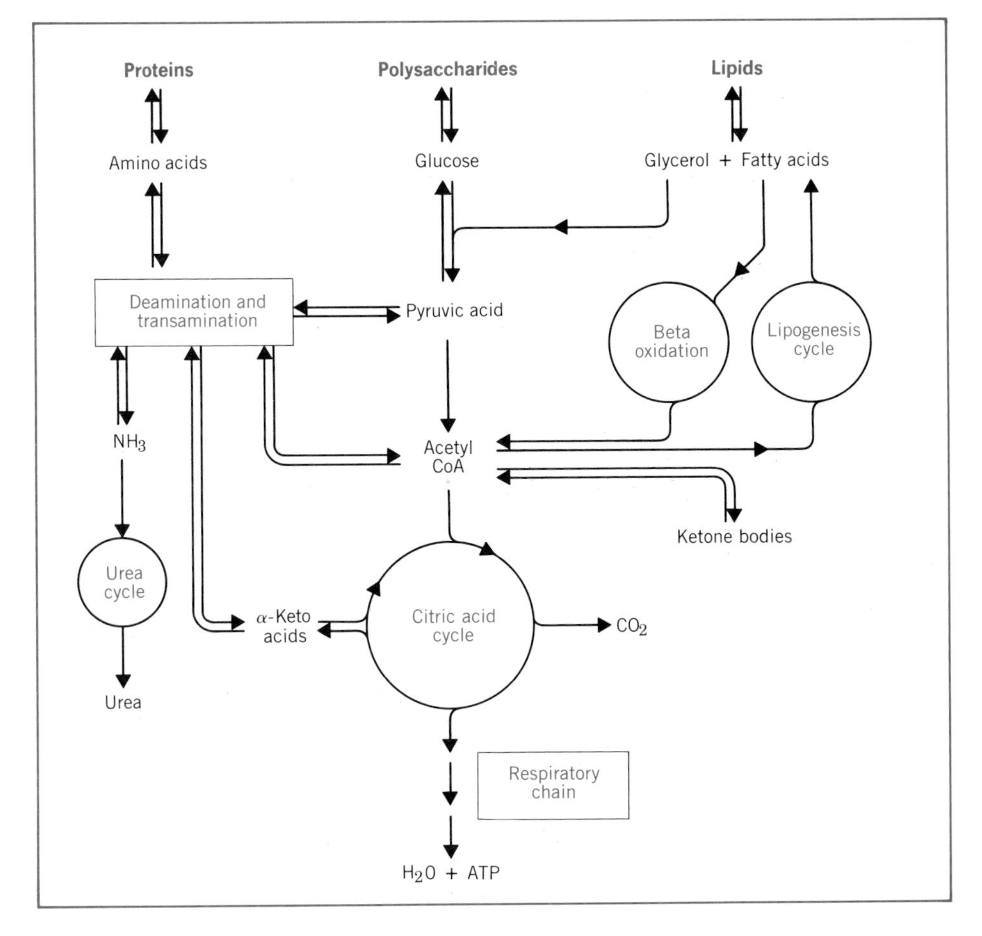

Figure 23.17 A summary of the pathways of carbohydrate, lipid, and protein metabolism.

ammonia produced by this type of deamination are excreted in the urine in the form of the ammonium ion, NH_4^+, rather than as urea. With the breakdown of all this protein and the excretion of its nitrogen as ammonia, a negative nitrogen balance develops in the body.

When insufficient glucose is available, the body uses fatty acids to provide the tissues with energy. This results in such a high level of fatty acid oxidation that the citric acid cycle is not able to handle the large amounts of acetyl CoA that are produced. The liver cells, therefore, convert this excess acetyl CoA into ketone bodies. Two ketone bodies, acetoacetic acid and β-hydroxybutyric acid, are acids; a large increase of these ketone bodies places a strain on the buffering capacity of the blood—finally exceeding it and creating acidosis. After several days of fasting, the production of ketone bodies becomes so high that they are excreted in the urine, producing further electrolyte imbalances.

You've probably gotten the idea by now that fasting is not a particularly effective means of weight reduction! The chemical imbalances that fasting produces slow down the basal metabolism rate, reducing concentration and causing physical movements to become sluggish and uncoordinated. Fasting puts a great strain on the liver (from increased fatty acid metabolism) and the kidneys (from the excretion of the waste products of amino acid and fatty acid metabolism). The acidosis and electrolyte imbalances from prolonged fasting can produce coma and death. It is important even when dieting, therefore, to eat sufficient amounts of complex carbohydrate to prevent the breakdown of tissue proteins and the formation of ketone bodies.

CHAPTER SUMMARY

CHAPTER SUMMARY

The digestion of foods begins in the mouth, where food is broken down into smaller pieces and mixed with saliva, whose enzymes begin the digestion of starch. The food then passes to the stomach, where it is further reduced in size and protein digestion begins. Pancreatic and intestinal juices (containing digestive enzymes) and bile are secreted as the food enters the small intestine. Fats are emulsified by the bile and hydrolyzed by enzymes to fatty acids, glycerol, and mono- and diacylglycerols. These compounds are absorbed, reformed, and carried by the lymph system to the bloodstream. Also in the small intestine, proteins are hydrolyzed to amino acids, and carbohydrates to monosaccharides. They are absorbed directly into the bloodstream and carried to the liver, which regulates the concentrations of these components in the blood.

The glucose that is absorbed may be converted by the liver to glycogen (through the process of glycogenesis), may be used by cells to synthesize other compounds, or may be oxidized by the cell to produce energy through the series of reactions called cellular respiration. Cellular respiration involves an anaerobic stage called glycolysis, in which glucose is broken down to lactic acid, and an aerobic stage, which involves the reactions of the citric acid cycle and the respiratory chain. This second stage produces most of the energy-rich molecules (ATP) that come from the oxidation of glucose. When there is not enough oxygen, cells use glycolysis to produce ATP. However, the lactic acid that is synthesized in glycolysis will build up in the tissues, producing mild acidosis. Microorganisms such as yeast oxidize glucose under anaerobic conditions, but the end products of this process of fermentation are ethanol and carbon dioxide, rather than lactic acid.

Fats are the main energy reserve of the body, producing 9 kilocalories per gram when oxidized. Lipids that are eaten can be stored in the fat cells of adipose tissue, or can be hydrolyzed to glycerol (which enters the glycolysis pathway) and fatty acids (which are oxidized to acetyl CoA in the fatty acid cycle). The acetyl CoA that is produced may enter the citric acid cycle or may be used in biosynthesis. If acetyl CoA is produced in excess, it is converted to ketone bodies by the liver. In high concentrations ketone bodies cause acidosis and metabolic disruption.

The amino acids absorbed from the small intestine can be used to synthesize tissue protein, nonessential amino acids, enzymes, hormones, antibodies, and nucleic acids. Amino acids are synthesized in the cell from other amino acids and α-keto acids by transamination. Amino acids are catabolized by first removing the amino group in a process called oxidative deamination. The ammonia that is produced is converted to urea through a series of reactions called the urea cycle. The α-keto acids produced then enter the citric acid cycle or other biosynthetic pathways.

The maintenance of the blood glucose level is critical to the normal functioning of tissues, especially the brain. This level is affected by diet and by the hormones insulin, glucagon, and epinephrine.

Hyperglycemia, or too much glucose in the blood, results from diabetes mellitus, emotional stress, kidney failure, and the administration of certain drugs. Hypoglycemia, or too little glucose in the blood, can occur with an overdose or overproduction of insulin or with starvation or fasting. When the blood glucose level is critically low, the body compensates by metabolizing proteins to synthesize glucose and by oxidizing fatty acids to produce energy. High levels of fatty acid oxidation lead to the formation of ketone bodies, which can cause acidosis and electrolyte imbalances.

Important Equations

Glycogenesis

$$n\text{Glucose} + n\text{UTP} + n\text{ATP} \longrightarrow \text{Glycogen} + n\text{UDP} + n\text{ADP} + 2n\text{P}_i$$

Glycogenolysis

$$\text{Glycogen} + (n\text{-}1)\text{H}_2\text{O} \longrightarrow n\text{Glucose}$$

Glycolysis

$$\underset{\text{(Glucose)}}{\text{C}_6\text{H}_{12}\text{O}_6} + 2\text{ADP} + 2\text{P}_i \longrightarrow \underset{\text{(Lactic acid)}}{2\text{C}_3\text{H}_6\text{O}_3} + 2\text{ATP} + 2\text{H}_2\text{O}$$

Fermentation

$$\underset{\text{(Glucose)}}{\text{C}_6\text{H}_{12}\text{O}_6} + 2\text{ADP} + 2\text{P}_i \longrightarrow \underset{\text{(Ethanol)}}{2\text{CH}_3\text{CH}_2\text{OH}} + 2\text{CO}_2 + 2\text{ATP}$$

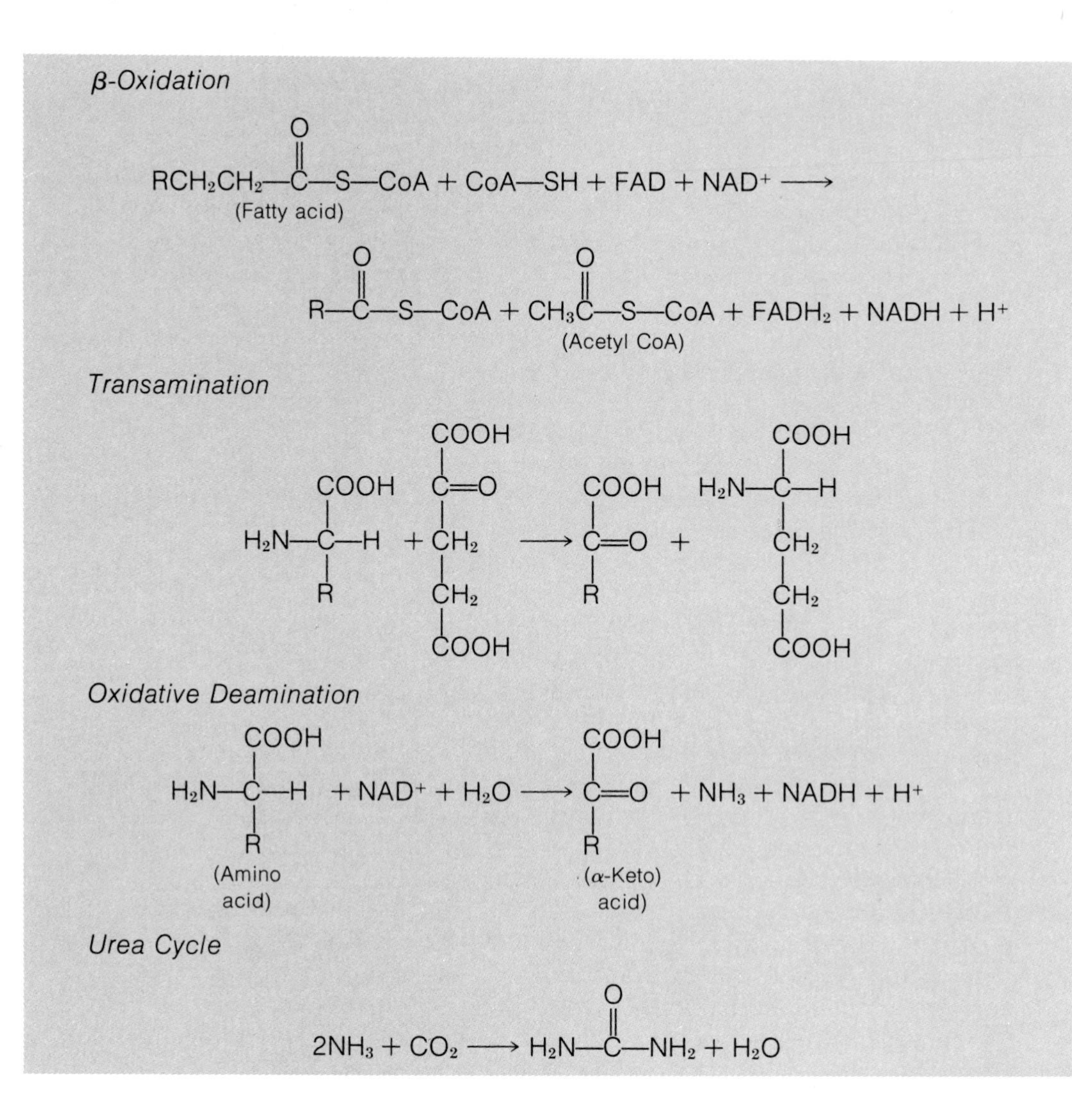

EXERCISES AND PROBLEMS

1. Describe the process by which the carbohydrates, proteins, and fats in a breakfast of bacon, eggs, and toast are digested and absorbed.

2. Why can't insulin be given to diabetics in the form of a pill?

3. (a) Draw the structure of AMP, ADP, and ATP.
   (b) What is meant by a *high-energy* phosphate bond?
   (c) Why are molecules such as ATP so important to the survival of a cell?

4. Why is it more efficient for our bodies to store excess food calories in the form of body fat than as glycogen?

5. Write a general equation for each of the following processes:

   (a) glycogenesis
   (b) glycogenolysis
   (c) glycolysis
   (d) fermentation
   (e) citric acid cycle
   (f) β-oxidation
   (g) transamination
   (h) oxidative deamination
   (i) urea cycle

6. The oxidation of glucose by a cell can be divided into two separate stages: an anaerobic and an aerobic stage.

   (a) What is the name given to the anaerobic stage?
   (b) Write the equation for the overall reaction that occurs in the anaerobic stage.
   (c) Where in the cell does each stage occur?
   (d) Which stage produces more energy?
   (e) Name the two series of reactions that occur in the aerobic stage and give a general description of the reactions that occur in each.
   (f) Write the equation for the overall reaction that occurs in the aerobic stage.

7. Red blood cells have no mitochondria. How do these cells generate the ATP necessary for their survival?

8. Describe the steps involved in the oxidation of fats by our bodies.

9. How many ATPs would be produced in the complete oxidation of one molecule of each of the following? Explain your answers.

   (a) lauric acid (b) stearic acid (c) arachidic acid

10. Write the equation for the transamination reaction between threonine and pyruvic acid.

11. Write the equation for the deamination of alanine.

12. Describe in general terms the metabolic fate of the ammonia produced in question 11.

13. Give several examples of the importance of acetyl CoA in the anabolic and catabolic reactions of metabolism.

14. Describe four possible metabolic uses of the amino acids contained in a hamburger.

15. What is ketosis? What metabolic changes cause it to occur? What are the effects of ketosis on the body?

16. A rare inherited disease called *maple syrup urine disease* results when a person lacks the ability to metabolize the α-keto acids that result from the transamination of valine, leucine, and isoleucine.

   (a) Write the structures of the α-keto acids produced in the transamination of valine, leucine, and isoleucine.
   (b) Suggest a possible reason for the name of this disease.

17. Your alarm clock doesn't go off and you wake up late for a final exam. You jump out of bed, dress quickly, run across campus, and race up three flights of stairs to your classroom. You collapse in your seat gasping for breath, and your legs feel like rubber. Describe the events that have occurred in your muscle cells during this experience. Why do you gasp for breath, and why are your muscles so weak?

18. Your brain requires a constant supply of glucose. Explain how your body maintains a fairly constant blood sugar level even though you eat (and thus obtain glucose) only a few times a day.

19. Explain why a reducing diet of 1200 calories a day should contain a minimum of 100 grams of carbohydrate.

20. A 70-kg person can survive without food for about 70 days, but not without water. What processes that are important in the metabolism of a fasting individual require water?

21. What is *insulin shock?* How does it occur? Explain why severe hyperinsulinism is more dangerous to a patient than the lack of insulin over a short period.

22. Suggest a diet for a patient suffering from hypoglycemia.

23. The glucose tolerance test is useful for diagnosing diabetes mellitus. In this test, a person who has fasted for 8 hours is given a drink containing 50 to 100 grams of glucose in solution. The patient's blood sugar level is measured over the next several hours. The data for two patients are shown below. Which patient may be suffering from diabetes mellitus? State the reason for your choice.

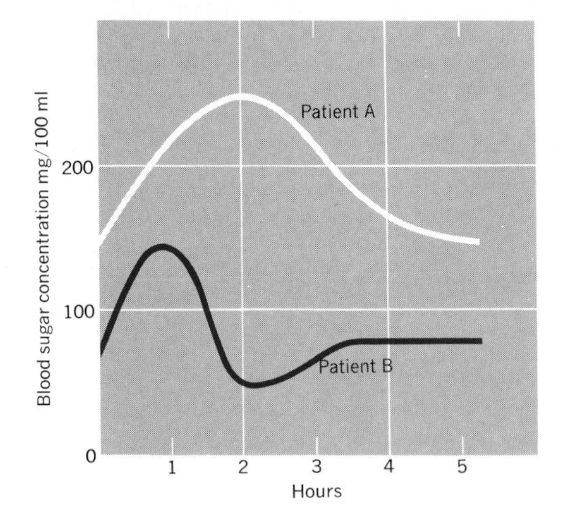

CHAPTER TWENTY-FOUR
NUCLEIC ACIDS

Mrs. Smith watched her son George walk slowly down the street to the school bus stop. She was puzzled and concerned. He'd been complaining on and off for a year about pains in his joints and abdomen. Twice he'd run a low fever and suffered from vomiting, and she had kept him home from school. He was the youngest of her four children, and the only sickly one among them.

Early that afternoon Mrs. Smith received a call from the school nurse, who had just taken George to the hospital—he had collapsed during a vigorous game of kickball, screaming about a pain in his stomach. When Mrs. Smith reached the hospital, she found George under sedation and receiving liquids intravenously. The doctor first thought that George might have appendicitis, but after talking with Mrs. Smith and learning of George's symptoms over the last year, he decided to run several blood tests. As he had suspected, the doctor found that George had sickle cell anemia and was suffering from a sickle cell crisis. The pain in George's abdomen and the swelling in his joints slowly went away, and a week later he was able to return to school.

A few weeks later George and his mother went to a neighboring town for consultation with doctors at a large clinic specializing in the diagnosis and treatment of sickle cell anemia. There they learned that sickle cell anemia is an inherited disease that affects red blood cells. They were told that long ago a mutation had occurred in the evolution of the African black population, causing an error in the gene for hemoglobin, the molecule in red blood cells that carries oxygen to the body. At present, one out of every two African blacks carries this defective gene, and one out of every 10 American blacks carries this trait. Children who inherit two defective genes from their parents will suffer from sickle cell anemia, and almost all of their hemoglobin will be defective. They will experience periodic sickle cell crises, resulting from the clogging of their capillaries by abnormal red blood cells, and causing severe pain, fever, and swelling of the joints. Such crises can be set off by an infection, cold weather, trauma, strenuous exercise, or emotional stress. Children having sickle cell anemia are anemic, often jaundiced, susceptible to disease, and will usually die in childhood.

People with only one defective gene will carry the sickle cell trait. About 40% of their hemoglobin is defective, but under normal conditions they will show no clinical signs of the disease. However, unusual stress can bring on a crisis in these individuals also.

Sickle cell anemia was first described clinically in 1910, but it was not until 1949 that Linus Pauling demonstrated the molecular basis of the disease. He showed that sickle cell hemoglobin (abbreviated HbS) has a different mobility in an electric field than does normal hemoglobin (HbA). Hemoglobin (which has a molecular weight of 64,458) is a molecule consisting of a protein part, called globin, and a nonprotein part, called heme. The globin part consists of four polypeptide chains: two alpha chains containing 141 amino acid units and two beta chains containing 146 amino acid units. Each chain is wrapped around a heme group containing an Fe^{2+} atom; this is the group that actually carries the oxygen (see Figure 18.10 and Figure 21.11). Therefore, each hemoglobin molecule can carry four molecules of oxygen.

In 1956, Ingram showed that the entire difference between HbS and HbA was in one amino acid on the beta chain. In HbS, a valine molecule is erroneously substituted for a glutamic acid molecule in position 6 on the beta chain.

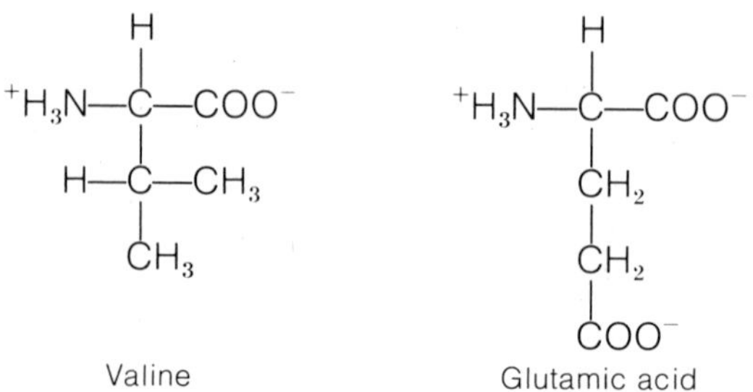

At pH 7, the glutamic acid will have a negative charge, whereas the valine will have no charge. Thus, a molecule of HbA will have two more negative charges than a molecule of HbS, accounting for the difference in mobility of the two hemoglobin molecules in an electric field.

At low oxygen concentrations, red blood cells containing HbS will take on abnormal shapes, some resembling sickles (Figure 24.1). In deoxygenated HbS, the nonpolar valine that replaced the polar glutamic acid undergoes hydrophobic bonding with a valine on a neighboring molecule. This causes the HbS molecules to polymerize into long fibers that distort the cell. The resulting abnormal cell shape usually corrects itself when the blood becomes oxygenated again, but some cells remain irreversibly sickled and are removed by the liver from the blood.

Red blood cells must be flexible, for they have to squeeze through capillaries smaller than they are. Although red blood cells containing HbS will sickle when they give up their oxygen, there is a short delay between the moment HbS releases its oxygen to the tissues and the time the cell sickles. Under normal conditions, this delay is long enough for the cell to squeeze through the capillary and enter the vein. But under stress, the HbS molecule will release its oxygen sooner than usual. This can cause the rigid, sickled cells to wedge in the capillaries and block the blood flow, depriving the tissues of further oxygen. Such blockage brings about the clinical symptoms of a sickle cell crisis.

Although people suffering from sickle cell anemia (and carriers of the trait) can be identified through blood screening tests, treatment is limited to supportive care during a crisis. Researchers are currently experimenting with drugs that inhibit the polymerization of HbS and that decrease the concentration of deoxygenated HbS by increasing the affinity of HbS for oxygen. One promising experimental treatment uses a cancer drug, 5-azacytidine, to induce the bone marrow to produce fetal hemoglobin (HbF), a hemoglobin normally produced only in fetuses. A high concentration of HbF seems to depress the synthesis of HbS.

Figure 24.1 (a) Normal red blood cells. (b) A sickled red blood cell. [(a) Courtesy Francois Morell, from *J. Cell Biol.* **48**:91–100, 1971; (b) Courtesy Springer-Verlag, from *Corpuscles* by Marcel Bessis.]

By the time you have finished this chapter, you should be able to:

1. List the three components of nucleotides.
2. Name the sugars and bases found in DNA and RNA.
3. Describe the structure of a DNA molecule and its method of replication.
4. List three types of RNA, and describe their functions.
5. Describe the genetic code and its relationship to amino acids and polypeptide chains.
6. Describe the steps that occur in protein synthesis.
7. Define *mutation,* and indicate several ways, both positive and negative, in which mutations can affect the normal functions of an organism.

DNA, DEOXYRIBONUCLEIC ACID

24.1 Molecular Basis of Heredity

Sickle cell anemia is just one of more than 2000 human diseases known to be caused by disorders in genes. Genes are specific segments of molecules called DNA. Each gene contains the information necessary to make one polypeptide chain. To understand how the information carried by genes actually directs the formation of a polypeptide chain, however, we must first become familiar with the molecular nature of the genetic material DNA.

24.2 Nucleotides

DNA (deoxyribonucleic acid) and RNA (ribonucleic acid) belong to a class of compounds called **nucleic acids.** Nucleic acids are polymers of monomer units called **nucleotides.** However, unlike the monomer units we have studied before, nucleotides can be further hydrolyzed to produce three components: a nitrogen-containing base, a five-carbon sugar, and phosphoric acid.

The Nitrogen-Containing Bases

There are two classes of nitrogen-containing bases found in nucleotides: **pyrimidines** and **purines.** The bases derived from pyrimidine are cytosine (C), thymine (T), and uracil (U). Those derived from purine are adenine (A) and guanine (G). The base uracil is found only in nucleotides of RNA, and the base thymine is found only in nucleotides of DNA.

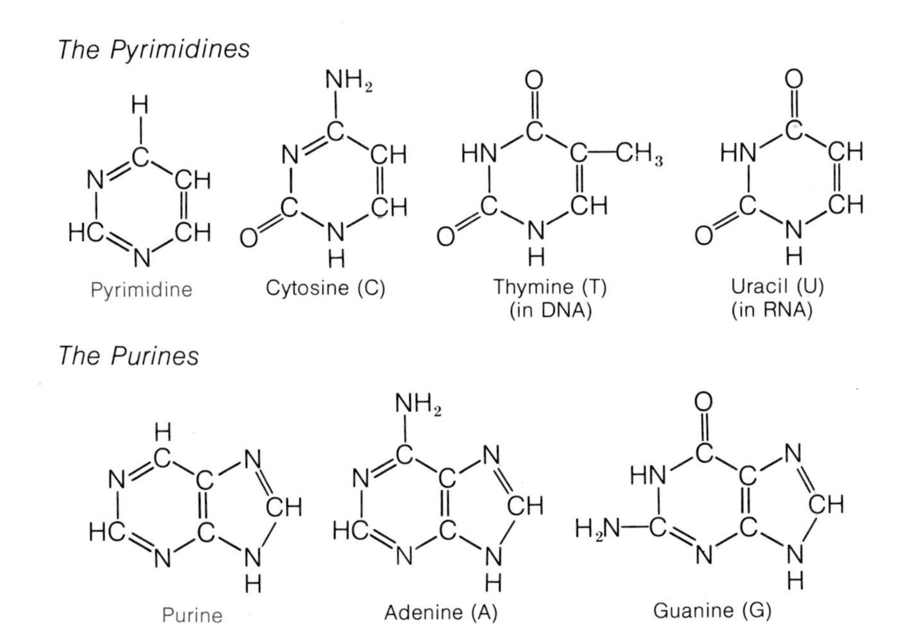

The Pentose Sugars

The second component of nucleotides is the pentose sugar. RNA contains the sugar ribose, and DNA contains a derivative of ribose, 2-deoxyribose.

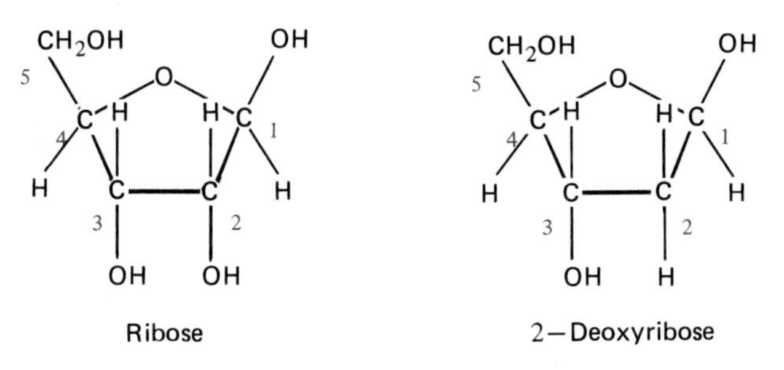

The Structure of a Nucleotide

The three components of the nucleotide are joined together in the following manner.

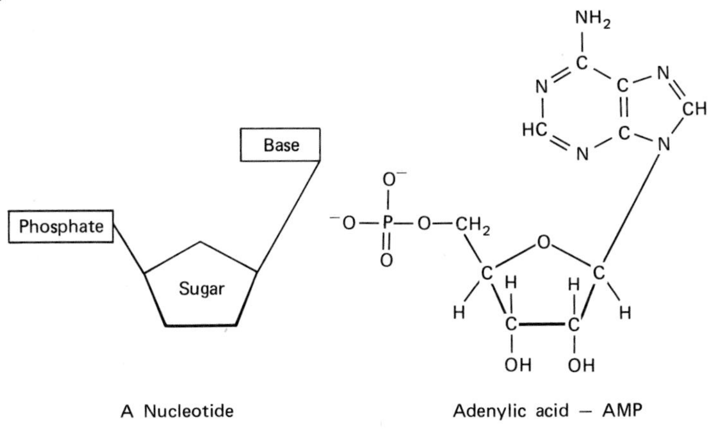

This example shows the nucleotide adenylic acid or adenosine monophosphate (AMP), which is made up of adenine, ribose, and one phosphate. Other ribonucleotides will have similar structures, but different bases. The deoxyribonucleotides will contain the sugar deoxyribose instead of ribose.

Free nucleotides are found in large numbers in the cell and perform many functions. In addition to being the basic structural unit of nucleic acids, they also participate in biosynthetic reactions, serve as coenzymes, and are important in the transport of energy from energy-releasing reactions to energy-requiring reactions. ATP (adenosine triphosphate), GTP (guanosine triphosphate), and UTP (uridine triphosphate) are all energy-carrying mononucleotides.

Exercise 24-1

1. What are the three components that make up the nucleotides of DNA? Of RNA?
2. Draw the structure of the nucleotide guanosine monophosphate, which contains the sugar ribose.
3. What nucleotide is an essential component of the coenzyme FAD?

24.3 The Structure of DNA

Each human cell contains about two meters of the nucleic acid DNA, packed into a set of 46 chromosomes having a total length of only about 200 micrometers. This reduction in length is possible because the DNA molecule wraps and folds itself around proteins (called histones) that are tightly bound to the DNA. DNA was first isolated in 1868, but the structure of the molecule was not determined until 1953, when J. D. Watson and F. H. C. Crick proposed a structure that explained the physical and chemical properties of DNA. The structure proposed by Watson and Crick consists of two helical polynucleotide chains coiled around the same axis, forming a double helix (Figure 24.2). The hydrophilic (attracted to water) sugar and phosphate components of the nucleotides are found on the outside of the helix, and the hydrophobic bases are found on the inside.

The nucleotides making up each strand of DNA are connected by ester bonds between the phosphate group and the deoxyribose sugar. This forms the "backbone" of each DNA strand, from which the bases extend. (See Figure 24.2.) The bases of one strand of DNA will pair with bases on the other strand by means of hydrogen bonding. This hydrogen bonding is very specific: Adenine's structure permits it to hydrogen bond only with thymine, and guanine will bond only with cytosine (Figure 24.3). As a result, the two strands of DNA are not identical, but rather are

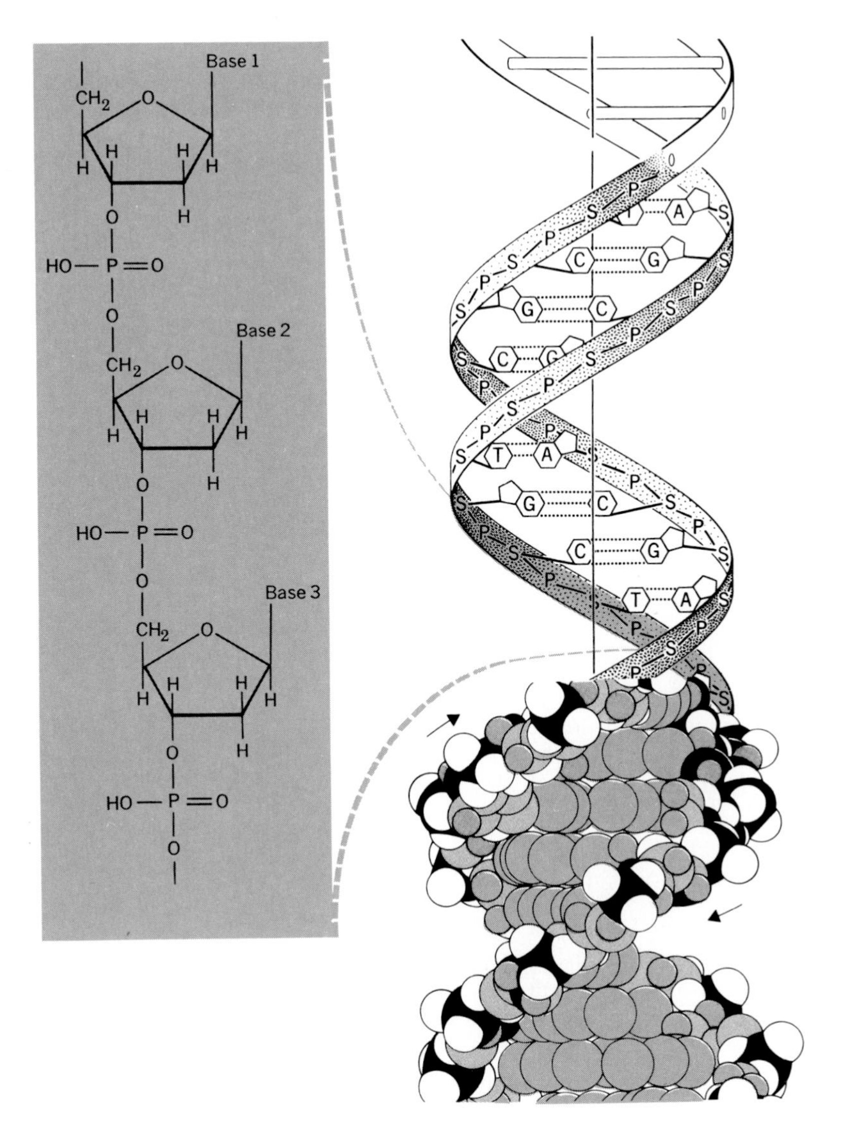

Figure 24.2 The double helix of the DNA molecule. The backbone of the DNA molecule consists of the sugar and phosphate groups of the nucleotides.

complementary—where thymine appears on one strand, adenine will appear on the other.

24.4 Replication of DNA

When a cell divides, the DNA molecules must **replicate** (that is, must make exact copies of themselves) so that each daughter cell will have DNA identical to the parent cell. The replication of DNA is catalyzed by the enzyme DNA polymerase. In this process, the two strands of the DNA helix unwind, and each strand serves as a template or pattern for the

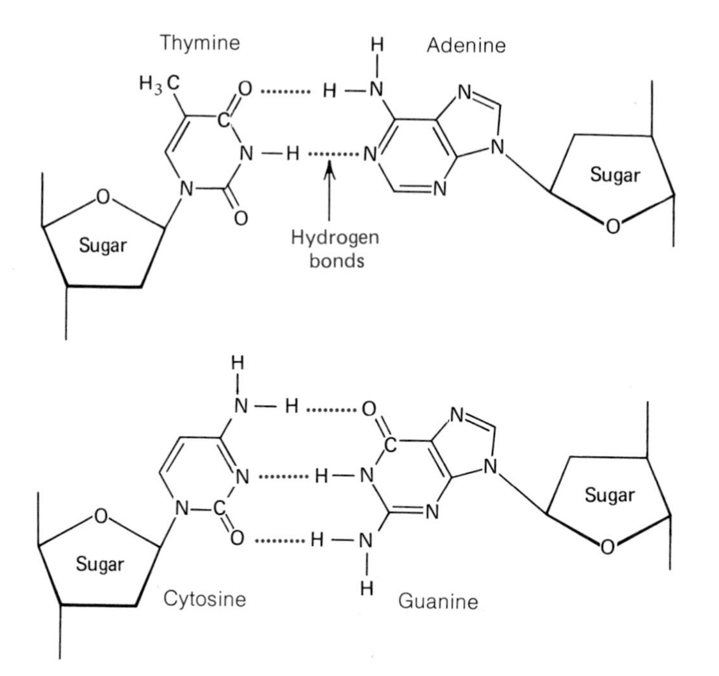

Figure 24.3 Hydrogen bonding in DNA. The bases on the nucleotides of DNA extend toward the inside of the helix, and the two strands of the helix are held together by hydrogen bonding between the bases.

synthesis of a new strand of DNA (Figure 24.4). Each of the two daughter helixes will contain one original DNA strand and one newly made strand. The genetic information for the cell is contained in the sequence of the bases A, T, C, and G in the DNA molecule. Anything that alters the order of the sequence will cause a change, or mutation, in the genes of the cell.

EXAMPLE 24-1

What is the sequence of bases on a strand of DNA that would be complementary to a strand having the following sequence of bases?

AATCGTAGGCAC

Because adenine (A) pairs only with thymine (T), and cytosine (C) only with guanine (G), we can write the sequence of bases as follows:

Original strand	AATCGTAGGCAC
Complementary strand	TTAGCATCCGTG

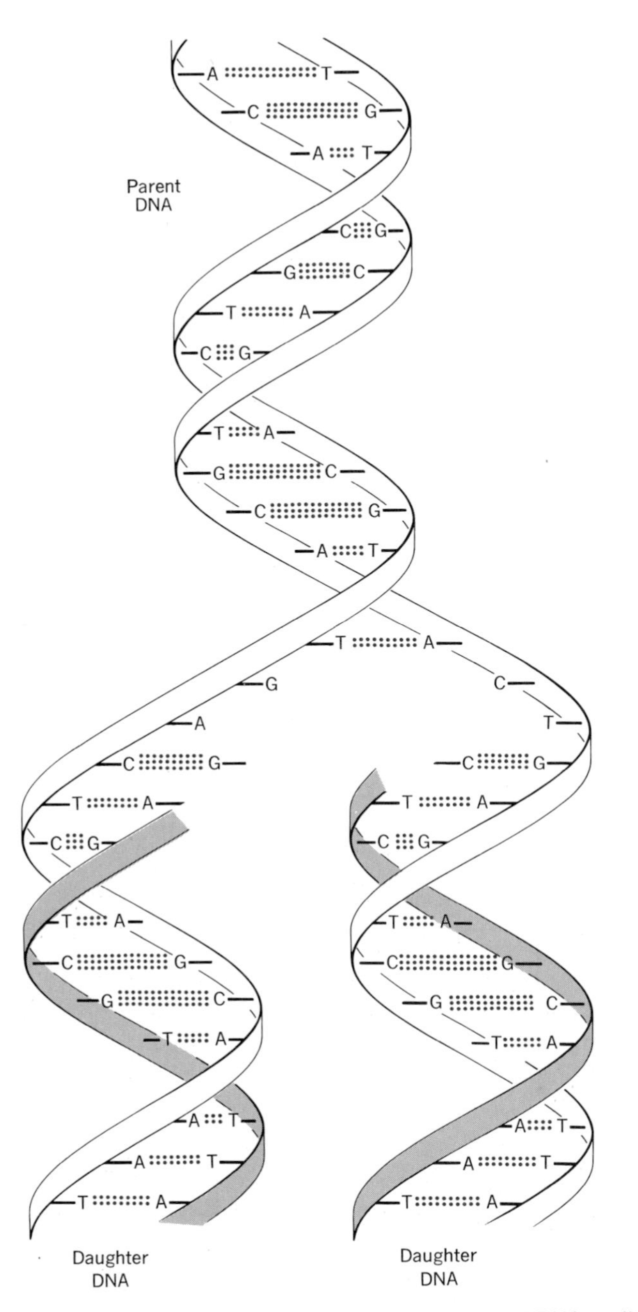

Figure 24.4 Replication of DNA. In one theory of DNA replication, the parent strand is forced apart and new strands of DNA are built on each of the old strands. Each daughter DNA will contain one old strand and one new strand.

Exercise 24-2

1. Write the sequence of bases that would be on the strand of DNA complementary to the following DNA strand:

TACTGAAAACCGATT

2. Show the sequence of bases present on the DNA found in each of the daughter cells formed when the cell containing this DNA divides.

RNA, RIBONUCLEIC ACID

Molecules of RNA make up 5 to 10% of the total weight of the cell. The nucleotides of RNA contain ribose instead of deoxyribose and the base uracil in place of thymine. Unlike DNA, RNA is not always double-stranded; it usually consists of a single strand of nucleic acid. The additional hydroxyl group on the ribose is very important in forming hydrogen bonds that stabilize the tertiary structure of the nonhelical regions of the RNA molecule. Three of the types of RNA found in the cell are **messenger RNA (*m*RNA), ribosomal RNA (*r*RNA),** and **transfer RNA (*t*RNA).**

24.5 Messenger RNA (*m*RNA)

Our first knowledge of the structure and function of *m*RNA came from *E. coli,* a common bacterium found in the intestines. In *E. coli, m*RNA is synthesized on one strand of the DNA (Figure 24.5). Therefore, it will have a sequence of bases that is complementary to the sequence on the DNA strand (remember, though, that in RNA the base uracil is substituted for the base thymine). *m*RNA is synthesized by the cell whenever it is needed. After being synthesized, *m*RNA migrates to the ribosomes, where it serves as a template or pattern for sequencing amino acids in the synthesis of proteins.

EXAMPLE 24-2

What will be the sequence of bases on the *m*RNA molecule that is synthesized on the following strand of DNA?

DNA TATCTACCTGGA

Using the same procedure as in Example 24-1, but remembering that *m*RNA contains the base uracil in place of

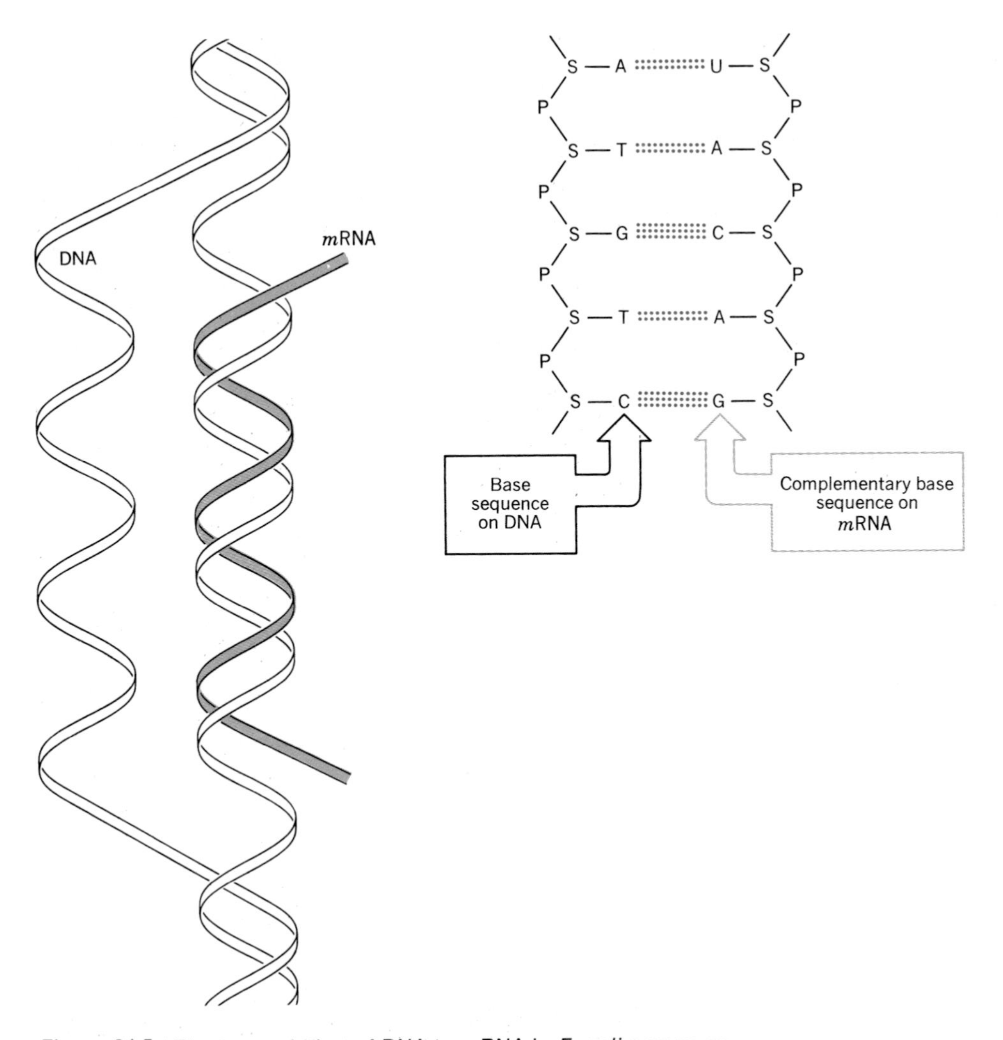

Figure 24.5 The transcription of DNA to *m*RNA in *E. coli* occurs on one strand of DNA. The sequence of bases on the *m*RNA is complementary to the sequence of the DNA. In making RNA, however, the cell uses nucleotides containing the base uracil instead of those containing thymine.

thymine, we have

DNA	T A T C T A C C T G G A
*m*RNA	A U A G A U G G A C C U

Exercise 24-3

What is the sequence of bases on the *m*RNA synthesized on the strand of DNA shown in Exercise 24-2?

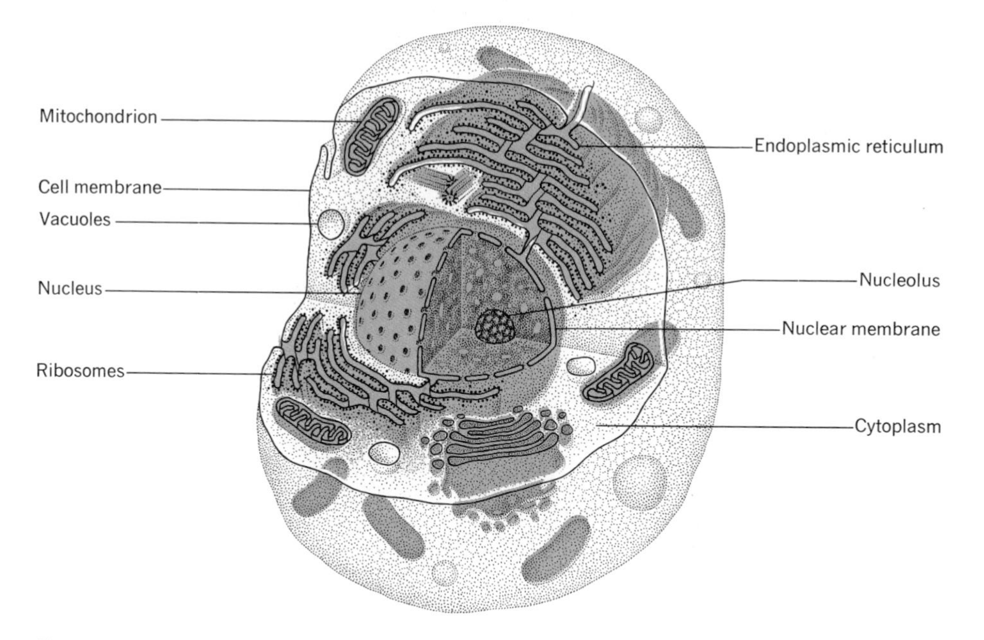

Figure 24.6 The structure of an animal cell.

24.6 Ribosomal RNA (*r*RNA)

As with all RNA, *r*RNA is synthesized in the nucleus from a DNA template. It migrates to the cytoplasm of the cell where, with proteins, it forms the ribosomes. The **ribosomes** are the sites of protein synthesis, and are often located on the endoplasmic reticulum (Figure 24.6). The ribosomes are made up of two subunits that combine with *m*RNA to form the "factory" for the production of proteins.

24.7 Transfer RNA (*t*RNA)

*t*RNA is the smallest of the RNA molecules; it is water-soluble and moves easily within the cell. *t*RNA molecules are synthesized in the nucleus of the cell, and each is specifically designed for a particular amino acid. A *t*RNA molecule becomes *charged* when a specific amino acid is joined to the terminal adenine nucleotide present on each *t*RNA polynucleotide chain (Figure 24.7). The *t*RNA molecule then carries this amino acid to the ribosomes, where the amino acid is used in protein synthesis.

PROTEIN SYNTHESIS

24.8 The Genetic Code

A **gene** is a region on a DNA molecule comprised of a sequence of bases that will translate into the specific sequence of amino acids making up a

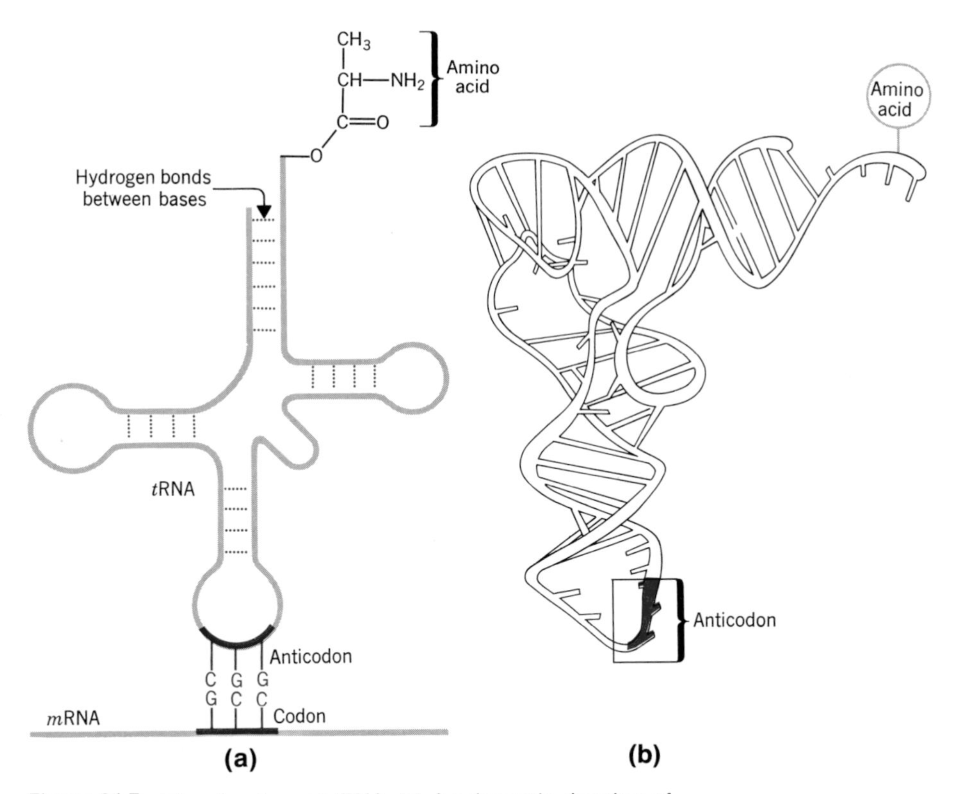

Figure 24.7 The structure of *t*RNA. (a) A schematic drawing of a *t*RNA molecule. The three bases in the anticodon region are complementary to a codon on the *m*RNA. (b) The three-dimensional structure of *t*RNA. (Redrawn from G.J. Quigley and A. Rich, *Science* **194,**796–806, November 19, 1976, Copyright © 1976 by the American Association for the Advancement of Science.)

protein. The **genetic code** is the general term used to describe the sequence of bases found in the gene. In the early 1960s the "language" of the genetic code was discovered through extensive studies using artificially synthesized *m*RNA molecules. It was discovered that a three-base sequence is necessary to code for each amino acid. This three-base sequence on *m*RNA is called a **codon.** There is more than one codon for most amino acids, as shown in Table 24.1. For example, GGU, GGC, GGA, and GGG are all codons for the amino acid glycine; UUU and UUC are codons for the amino acid phenylalanine. You will notice from the table that the first two bases in the codon for a given amino acid are usually the same. The third base can vary and is less important in determining the amino acid being specified. The genetic code on the messenger RNA is not punctuated; that is, you read the code one triplet after another, without a break from one end of the *m*RNA molecule to the other.

*m*RNA: G G U C A G U G C U C C . . .

Amino acid: Gly - Gln - Cys - Ser -. . .

Table 24.1 The Codons for the Amino Acids

UUU Phe	UCU Ser	UAU Tyr	UGU Cys
UUC Phe	UCC Ser	UAC Tyr	UGC Cys
UUA Leu	UCA Ser	UAA *Stop*	UGA *Stop*
UUG Leu	UCG Ser	UAG *Stop*	UGG Trp
CUU Leu	CCU Pro	CAU His	CGU Arg
CUC Leu	CCC Pro	CAC His	CGC Arg
CUA Leu	CCA Pro	CAA Gln	CGA Arg
CUG Leu	CCG Pro	CAG Gln	CGG Arg
AUU Ile	ACU Thr	AAU Asn	AGU Ser
AUC Ile	ACC Thr	AAC Asn	AGC Ser
AUA Ile	ACA Thr	AAA Lys	AGA Arg
AUG Met (*Start*)	ACG Thr	AAG Lys	AGG Arg
GUU Val	GCU Ala	GAU Asp	GGU Gly
GUC Val	GCC Ala	GAC Asp	GGC Gly
GUA Val	GCA Ala	GAA Glu	GGA Gly
GUG Val	GCG Ala	GAG Glu	GGG Gly

The codon AUG codes for methionine and serves as a "start" codon to signal the beginning of the amino acid sequence. Three codons (UAA, UAG, UGA) do not code for any amino acid, but serve as "stop" or terminal codons. They cause the completed protein to be released from the ribosome. Each transfer RNA molecule contains a special three-base sequence, called the **anticodon,** that is complementary to one of the codons on a messenger RNA molecule (Figure 24.7).

24.9 Introns and Exons

Bacteria, such as *E. coli,* consist of a single cell having no nucleus. Their DNA is found throughout the cell, with genes located in one continuous stretch along the DNA. Until recently, it was thought that the steps involved in the transcription of DNA to *m*RNA (that is, the passage of genetic information from DNA to *m*RNA) and then the translation of *m*RNA to protein (that is, the expression of the genetic information in the amino acid sequence of the protein)—all of which had been determined from research on *E. coli*—occurred in an identical fashion in all living cells. However, several characteristics of eukaryotic cells (cells with nuclei) bothered scientists working in this field. These cells appear to have much too much DNA—about nine times more than is needed in a human cell to code for the proteins necessary for life. They also seem to have too much RNA. The nuclei of such cells have been found to contain huge strands of RNA, a far greater amount of RNA than that which actually exits the nucleus as *m*RNA. This nuclear RNA is called ***hn*RNA** (for **heterogeneous nuclear RNA**).

The solution to this mystery was found in 1977: the genes on the DNA of eukaryotic cells are "interrupted" by segments of DNA that do not code for amino acids. These intervening sequences of DNA have been given the name **introns;** the sequences of DNA that do code for amino acids are called **exons.** The unexpectedly large amount of DNA in eukaryotic cells is explained by the finding that the exons are much shorter in length than the introns (exons are 100 to 300 bases in length, and introns are about 1000.) The gene region that codes for a protein (in all but two of the proteins studied so far) is made up of alternating regions of exons and introns. Recent research has shown that genes for different proteins contain some of the same exons. Thus, the function of the introns may be to permit the "shuffling" of exons between different genes.

The *hn*RNA found in the nucleus of eukaryotic cells is a complete transcription of the entire DNA base sequence, including all introns and exons. This long molecule is then cut and spliced by special enzymes in the nucleus of the cell in a series of reactions that produce *m*RNA (Figure 24.8). The *m*RNA produced through this process contains only the sequence of bases that actually codes for the amino acids in the protein.

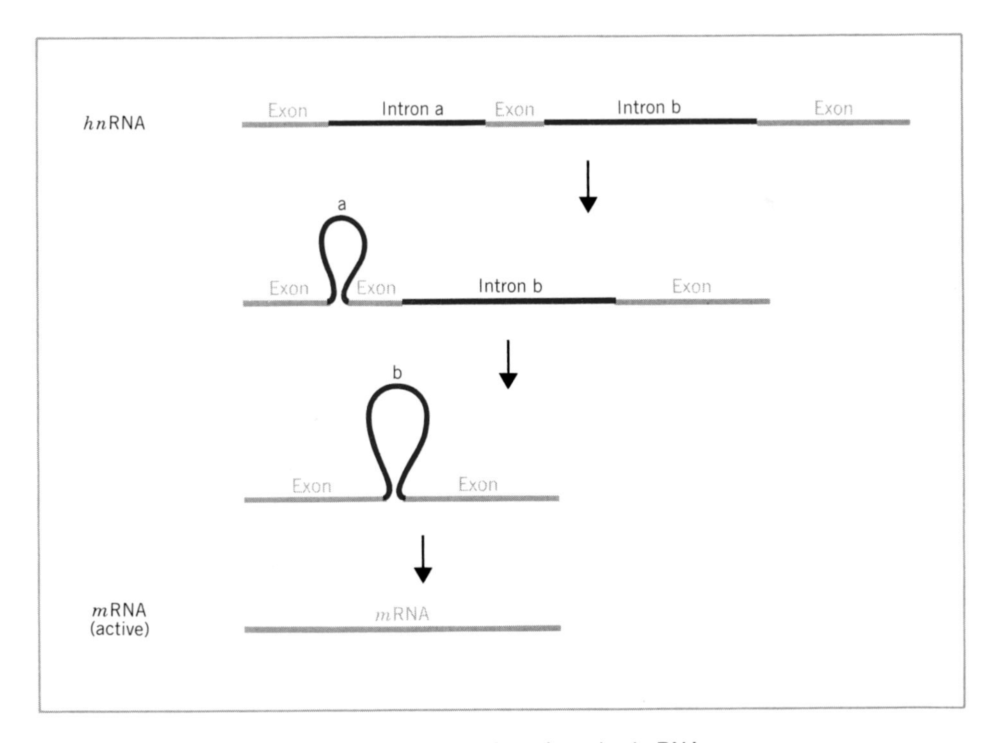

Figure 24.8 In a eukaryotic cell the introns found on the *hn*RNA are removed by special enzymes in a series of steps that produce *m*RNA.

24.10 The Steps in Protein Synthesis

Ribosomes, as we have stated, are the sites of protein synthesis. Actually, many ribosomes can be producing proteins from the same strand of *m*RNA. When viewed under a microscope, these groups of ribosomes (called polyribosomes, or **polysomes**) appear as a series of dots.

Before protein synthesis can begin, *m*RNA must be produced. Then *t*RNAs attach themselves to amino acids, catalyzed by enzymes that are specific for each amino acid and each *t*RNA. The attachment of the amino acid to the *t*RNA (called the activation or charging of the *t*RNA) is a two-step process requiring the presence of ATP.

The specific process of protein synthesis, as it has been determined in *E. coli,* occurs as follows. First, the smaller subunit of the ribosome combines with three proteins and with the energy-rich molecule GTP, an *m*RNA, and a *t*RNA carrying a special "starting" amino acid. This complex then joins with the larger ribosomal subunit. The growth or elongation of the polypeptide chain takes place within the ribosome through a series of repeated steps in which amino acids brought by *t*RNAs (whose anticodon region matches the codon on the *m*RNA) form peptide bonds with the last amino acid in the growing chain. The elongation of the chain stops when a terminal codon is reached on the *m*RNA, at which time the ribosome separates, releasing the newly formed protein and the *m*RNA. This process of protein synthesis is summarized in Figure 24.9.

Exercise 24-4

1. Define the following terms in your own words:
   (a) gene
   (b) genetic code
   (c) codon
   (d) anticodon
   (e) transcription
   (f) translation
   (g) polysome
2. Write the amino acid sequence of the polypeptide chain coded for by the *m*RNA formed in Exercise 24-3.

24.11 Recombinant DNA

A laboratory technique called **recombinant DNA** (also called molecular cloning or DNA recombination) played a key role in solving the mystery of *hn*RNA. This technique is helping answer such fundamental questions as the exact nucleotide order of genes, how these genes are arranged on chromosomes, what turns genes off and on, and what may be the function

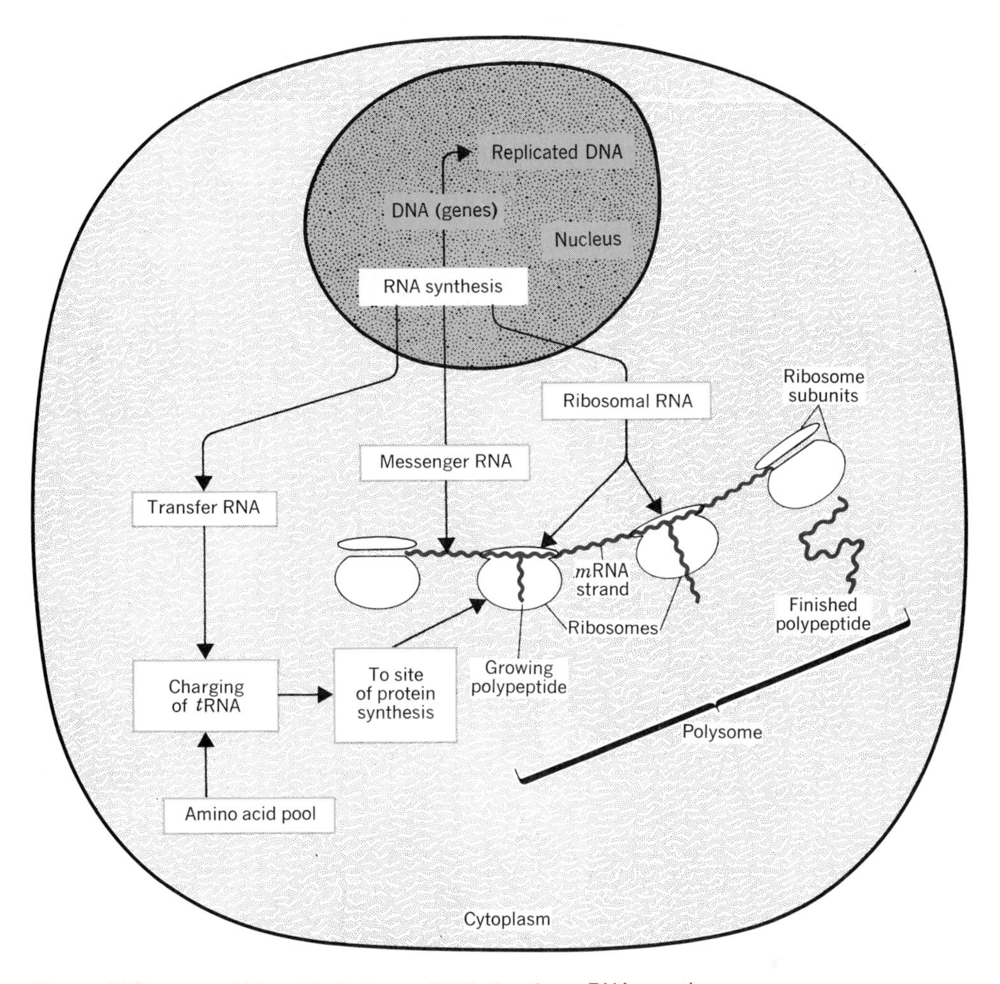

Figure 24.9 The relationship between DNA, the three RNAs, and protein synthesis. (Adapted from J.R. Holum, *Elements of General and Biological Chemistry,* copyright © 1975, John Wiley and Sons, Inc., New York. Used by permission.)

of individual genes. Recombinant DNA methods are also being used to produce human insulin, human interferon, human growth hormone, and somatostatin (a brain hormone). This technique is being developed commercially to produce vaccines, antibiotics, chemicals, and fuels and to insert desired traits (such as nitrogen fixation or resistance to frost) into plant species.

Recombinant DNA involves splicing a specific segment of DNA (such as the gene that codes for human insulin) into the plasmid (a small circular piece of DNA) of specially developed strains of *E. coli.* This new plasmid is then inserted into other *E. coli* and will be reproduced in the organism's

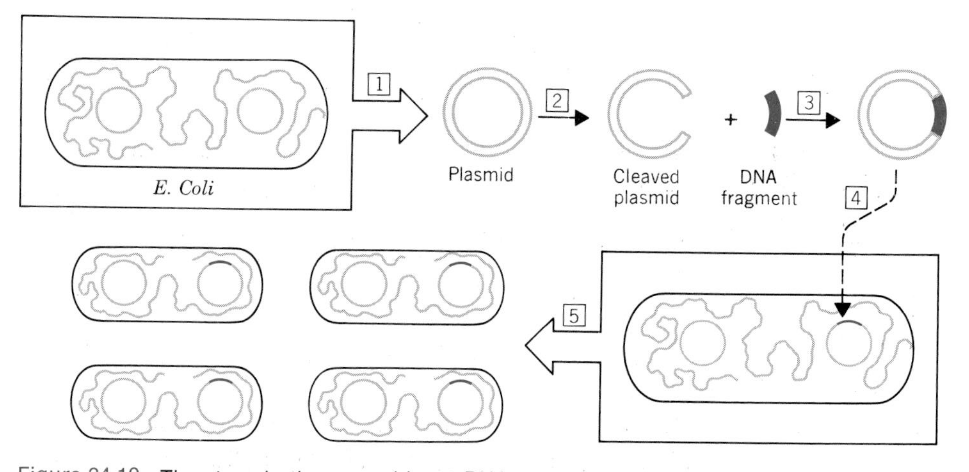

Figure 24.10 The steps in the recombinant DNA procedure. (1) The outer membranes of *E. coli* are dissolved, and small circular pieces of DNA (called plasmids) are isolated. (2) These plasmids are broken at specific positions by special enzymes that leave single-stranded "sticky" ends on the cleaved plasmids. (3) The cleaved plasmids are mixed with pieces of foreign DNA that are then attached to the plasmids by the enzyme DNA ligase. (4) The recombined plasmid with its foreign DNA is inserted into a new *E. coli* cell. (5) The segment of foreign DNA is reproduced with the rest of the native DNA in the normal reproductive cycle of the *E. coli* cells.

normal reproductive cycle (Figure 24.10). Because one *E. coli* can make billions of copies of the transplanted gene in a day, the production of large amounts of the desired DNA (or the protein it produces) is possible in a relatively short period of time. This very efficient technique, however, may have a risk: an otherwise harmless bacterium could be given some potentially dangerous properties. As a result, special strains of *E. coli* have been developed for use in recombinant DNA research, and special guidelines have been prepared for containment procedures.

24.12 Mutations

A **mutation** is a change in the sequence of bases on the DNA molecule. Some mutations may be beneficial to an organism, but most will be to varying degrees detrimental. The sequence of bases on the DNA molecule is critical and very specific. Replacing a base in the sequence, adding a base, or deleting a base can throw off the code and result in no protein or an altered or defective protein being synthesized. Such mutations can occur spontaneously or may be caused by radiation, chemical agents, or viruses.

Table 24.2 Some Diseases of Fat Metabolism That Are Hereditary[a]

Disease	Symptoms
Fabry's disease	Reddish-purple skin rash, kidney failure, and pain in legs.
Gaucher's disease	Spleen and liver enlargement, mental retardation in infantile form, and erosion of long bones and pelvis.
Generalized gangliosidosis	Mental retardation, liver enlargement, skeletal deformities, and red spot in retina in about 50% of the cases.
Niemann–Pick disease	Mental retardation, liver and spleen enlargement, and red spot in retina in about 30% of the cases.
Tay–Sachs disease	Mental retardation, red spot in retina, blindness, and muscular weakness.

[a] In these diseases, sphingolipids accumulate in tissues because the enzymes that normally catalyze the cleavage of these lipids are defective.

The majority of such changes in the DNA are repaired by special enzyme systems that are functioning constantly in the nucleus. However, any breakdown in these enzyme systems, or any change in the DNA that is not successfully repaired, can cause a mutation.

The altered proteins produced through a mutation might improve the organism's chances of survival by providing new alternative chemical pathways, or they may have no biological activity, resulting in the death of the cell. In other cases, the defective gene may cause abnormalities or disease. At the beginning of this chapter we saw that a mutation in the gene for hemoglobin occurred among the African black population, resulting in an error in one amino acid on the beta hemoglobin chain. This causes the production of the abnormal hemoglobin found in sickle cell anemia. Among the other diseases that result from mutations in genes are such maladies as cystic fibrosis, albinism, hemophilia, thalassemia, galactosemia, color blindness, and PKU (Table 24.2).

24.13 PKU – A Final Look

We began this book with the story of Billy, a child suffering from the disease phenylketonuria. PKU is a disease resulting from a mutation in the gene that codes for the liver enzyme phenylalanine hydroxylase. This defective gene is carried by 2% of the population. When a child inherits the defective gene from both parents, the child will suffer from PKU. In this disease the defective phenylalanine hydroxylase produced in the liver will not catalyze the conversion of phenylalanine to tyrosine

(Figure 24.11). This results in low tyrosine concentrations and the buildup in the body of phenylalanine and PKU metabolites—substances produced by the metabolism of the excess phenylalanine.

Melanin, the dark pigment found in hair and skin, is formed from tyrosine, and the low level of tyrosine in untreated PKU children leads to the light hair and skin observed on Billy. The hormones epinephrine, norepinephrine, and thyroxine are also synthesized from tyrosine. A high level of phenylalanine and PKU metabolites (especially phenylacetic acid) blocks energy-releasing reactions in the brain of an infant, preventing cells from obtaining the amount of energy necessary for normal functioning. This high level also seems to delay the formation of the myelin sheath, the protective coating around the nerves. In addition, large amounts of

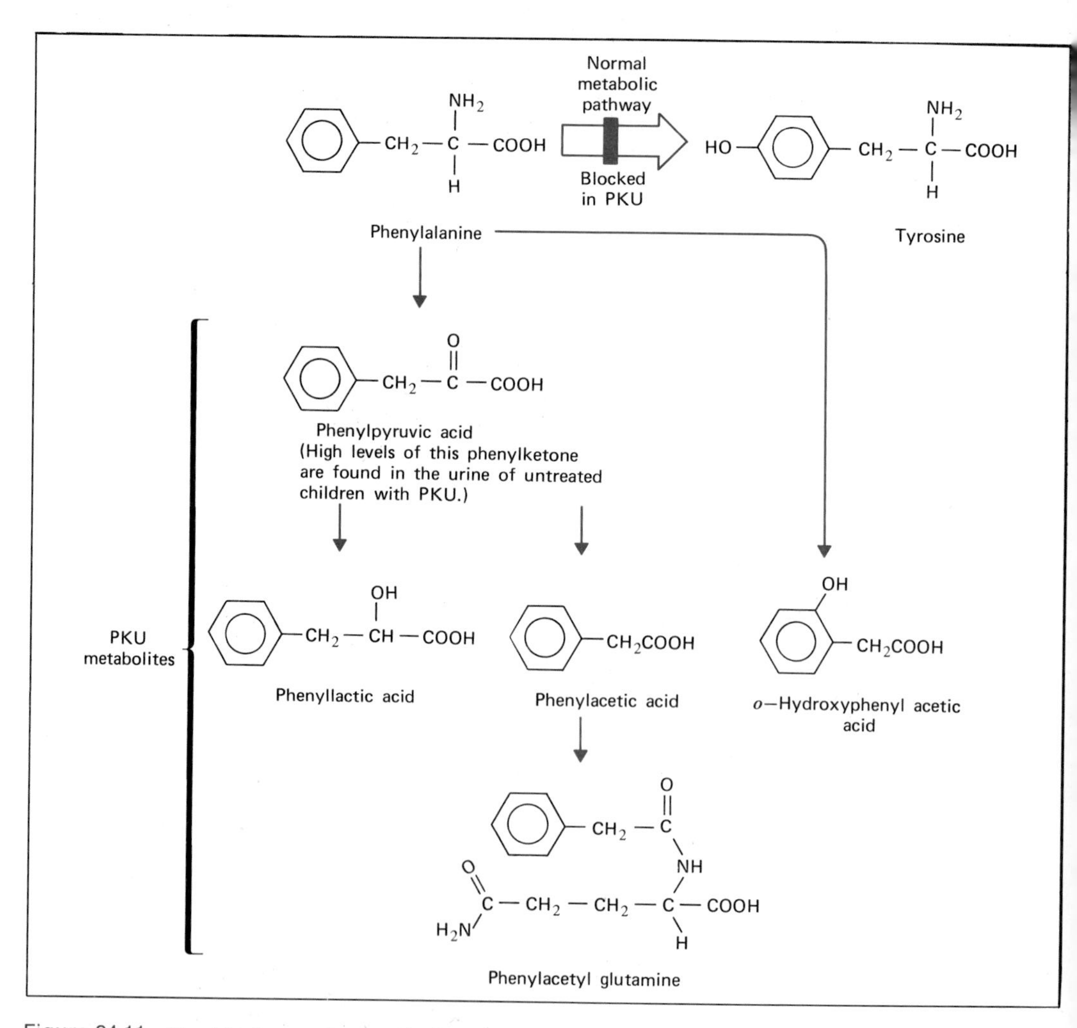

Figure 24.11 The blockage of a metabolic pathway in PKU results in the production of PKU metabolites as the body tries to metabolize phenylalanine by alternate chemical routes.

phenylalanine in the fluid of the brain prevent the normal uptake of nutrients by the brain cells, so these cells will not have the normal mix of chemicals from which to build their essential and permanent parts. The brain of an infant at birth is only 25% of its mature weight, and it grows rapidly, reaching 89% of its mature weight by 6 years of age. During these years of rapid growth, it is extremely important that the brain be surrounded by the correct chemical environment. The abnormal chemical environment of the brain of an untreated PKU child results in the formation of defective brain cells, which explains the mental retardation of such children (Figure 24.12).

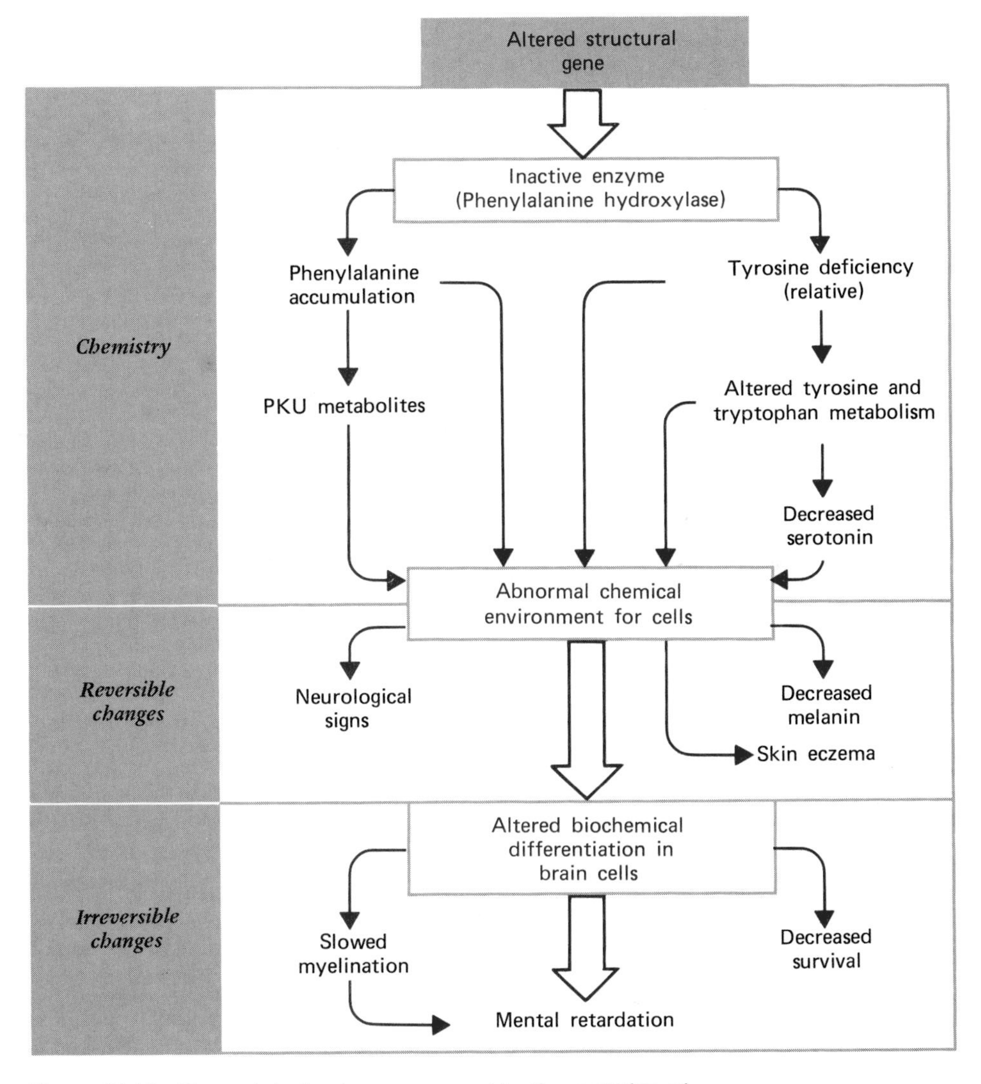

Figure 24.12 The metabolic changes caused by the mutation of PKU. (Adapted from *The Metabolic Basis of Inherited Disease* by Stanbury et al., copyright © 1972 by McGraw-Hill, Inc., New York. Used with permission of McGraw-Hill Book Company.)

CHAPTER SUMMARY

CHAPTER SUMMARY

A human tissue cell contains 46 chromosomes in its nucleus. Each chromosome is composed of molecules of DNA and proteins. The structure of the nucleic acid DNA is a double helix, consisting of two helical polynucleotide chains coiled around one another. The two chains are held together by hydrogen bonding between the bases of the nucleotides. A nucleotide consists of a sugar molecule, a nitrogen base, and a phosphate group or groups. Genetic information is carried in the sequence of bases on the DNA molecule. A gene is a sequence of such bases that codes for a specific protein molecule. DNA molecules can replicate themselves with the help of special enzymes, allowing genetic information to be passed from a parent cell to daughter cells.

RNA is the second nucleic acid found in cells. In bacteria, messenger RNA (*m*RNA) is formed on one DNA strand. In cells with nuclei, a long RNA strand (*hn*RNA) is formed on the DNA and then cut and spliced together by special enzymes to form the shorter *m*RNA. The *m*RNA then migrates to the cytoplasm where it joins with ribosomal RNA (*r*RNA) and proteins to form ribosomes, the location of protein synthesis in the cell. The function of transfer RNA (*t*RNA) is to join with specific amino acids in the cytoplasm and bring them to the ribosome. There, the anticodon region on the *t*RNA is matched with the codon on the *m*RNA, so that a protein having a specific amino acid sequence is synthesized. Figure 24.9 summarizes the steps in protein synthesis.

A mutation is any change in the order of bases on the DNA molecule. This change could result in no protein being synthesized, an insufficient amount of protein being synthesized, or the synthesis of a protein with an altered amino acid sequence. A mutation can produce a change that improves an organism's chances of survival, that results in the death of the organism, or that produces a specific genetic disease such as PKU.

EXERCISES AND PROBLEMS

EXERCISES AND PROBLEMS

1. What are the three components making up a nucleotide?
2. What is the difference between a nucleic acid and a nucleotide?
3. List three functions of nucleotides in the cell.
4. Which nitrogen bases found in DNA are pyrimidines? Which are purines?
5. (a) From the structure of deoxyribose and ribose, explain the meaning of the prefix *deoxy-*.
   (b) What is the important function of the hydroxyl group on carbon 2 of ribose in RNA molecules?

6. How are the two polynucleotide chains held together in the double helix of DNA?

7. "The two strands of DNA are not identical, but rather are complementary." Explain this statement.

8. In general terms describe the replication of a DNA molecule.

9. What are the three main structural differences between a molecule of DNA and a molecule of RNA?

10. What are three types of RNA found in the cell? Describe the function of each type.

11. How is genetic information carried on the DNA molecule?

12. Define the terms *exon* and *intron*. Explain how genetic information is transferred from DNA to *m*RNA in eukaryotic cells.

13. Write the structure of the nucleotides containing the following:

    (a) The sugar deoxyribose, the base cytosine, and one phosphate.
    (b) The sugar ribose, the base guanine, and three phosphates.

14. The following is a sequence of bases that might be found on the gene that codes for the human pituitary hormone, oxytocin:

    DNA strand: TACACAATGTAAGTTTTGACGGGGGACCCTATC

    (a) What is the sequence of bases on the complementary strand of DNA that would form an α helix with this strand?
    (b) What would be the sequence of bases found on the *m*RNA molecule synthesized on the original DNA strand?
    (c) Mark off the codons on the *m*RNA strand in (b) and list the bases found on the anticodons of the *t*RNA molecules that would join with the first three codons on the *m*RNA.
    (d) What is the sequence of amino acids in a molecule of oxytocin? (The initiating methionine is not part of the oxytocin molecule.)

15. List the sequence of bases on a molecule of DNA that would code for methionine enkephalin (discussed in Exercise 21-2).

16. Sickle cell anemia results from a mutation of a specific region of DNA that codes for the polypeptide chain of a hemoglobin molecule.

    (a) Suggest three possible causes of this mutation.
    (b) The *m*RNA for the β-chain of normal hemoglobin has the base triplet GAA or GAG in the 6th position. What is the change in this base sequence that results in the production of the β chain of sickle cell hemoglobin?
    (c) In general terms describe the steps involved in the production of one normal β-chain.
    (d) Explain the difference in mobility of normal and sickle cell hemoglobin in an electric field.
    (e) Explain how the substitution of one amino acid in the β-chain of hemoglobin can result in the symptoms of sickle cell anemia.

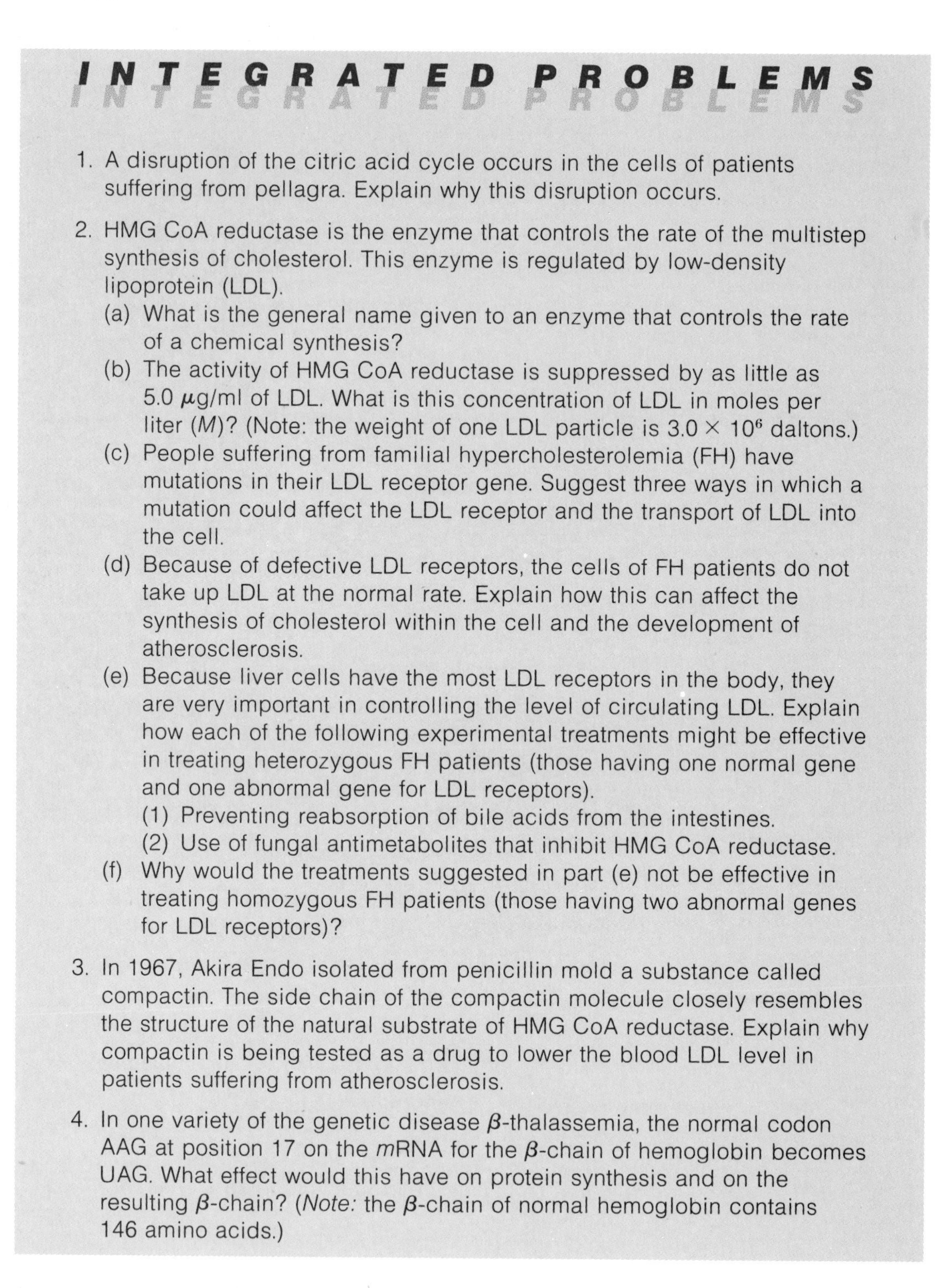

INTEGRATED PROBLEMS

1. A disruption of the citric acid cycle occurs in the cells of patients suffering from pellagra. Explain why this disruption occurs.

2. HMG CoA reductase is the enzyme that controls the rate of the multistep synthesis of cholesterol. This enzyme is regulated by low-density lipoprotein (LDL).
   (a) What is the general name given to an enzyme that controls the rate of a chemical synthesis?
   (b) The activity of HMG CoA reductase is suppressed by as little as 5.0 μg/ml of LDL. What is this concentration of LDL in moles per liter (*M*)? (Note: the weight of one LDL particle is 3.0×10^6 daltons.)
   (c) People suffering from familial hypercholesterolemia (FH) have mutations in their LDL receptor gene. Suggest three ways in which a mutation could affect the LDL receptor and the transport of LDL into the cell.
   (d) Because of defective LDL receptors, the cells of FH patients do not take up LDL at the normal rate. Explain how this can affect the synthesis of cholesterol within the cell and the development of atherosclerosis.
   (e) Because liver cells have the most LDL receptors in the body, they are very important in controlling the level of circulating LDL. Explain how each of the following experimental treatments might be effective in treating heterozygous FH patients (those having one normal gene and one abnormal gene for LDL receptors).
      (1) Preventing reabsorption of bile acids from the intestines.
      (2) Use of fungal antimetabolites that inhibit HMG CoA reductase.
   (f) Why would the treatments suggested in part (e) not be effective in treating homozygous FH patients (those having two abnormal genes for LDL receptors)?

3. In 1967, Akira Endo isolated from penicillin mold a substance called compactin. The side chain of the compactin molecule closely resembles the structure of the natural substrate of HMG CoA reductase. Explain why compactin is being tested as a drug to lower the blood LDL level in patients suffering from atherosclerosis.

4. In one variety of the genetic disease β-thalassemia, the normal codon AAG at position 17 on the *m*RNA for the β-chain of hemoglobin becomes UAG. What effect would this have on protein synthesis and on the resulting β-chain? (*Note:* the β-chain of normal hemoglobin contains 146 amino acids.)

APPENDIX ONE
NUMBERS IN EXPONENTIAL FORM

Whenever we are working with very large or very small numbers, it is convenient to write them in **exponential form** (also called scientific notation). Numbers written in exponential form are expressed as a number between one and ten, called the **coefficient,** multiplied by 10 raised to some power. The **exponent** is the power to which the number 10 is raised, and is written as a superscript next to the number 10.

$$4.75 \times 10^{3} \leftarrow \text{exponent}$$

If the exponent is a positive number, it indicates how many times the coefficient is to be multiplied by the number 10. For example,

$$10^{3} = 1 \times 10^{3} = 1 \times 10 \times 10 \times 10 = 1000$$

$$4.5 \times 10^{6} = 4.5 \times 10 \times 10 \times 10 \times 10 \times 10 \times 10 = 4{,}500{,}000$$

If the exponent is a negative number, it tells us how many times the coefficient is to be divided by the number 10. For example,

$$10^{-3} = 1 \times 10^{-3} = \frac{1}{10 \times 10 \times 10} = 0.001$$

$$4.5 \times 10^{-6} = \frac{4.5}{10 \times 10 \times 10 \times 10 \times 10 \times 10} = 0.0000045$$

The following table shows what various numbers look like in exponential form.

Number	Exponential Form	Number	Exponential Form
10	1×10^{1}	45	4.5×10^{1}
100	1×10^{2}	356	3.56×10^{2}
1000	1×10^{3}	8400	8.4×10^{3}
10,000	1×10^{4}	24,500	2.45×10^{4}
100,000	1×10^{5}	680,000	6.8×10^{5}
1,000,000	1×10^{6}	7,450,000	7.45×10^{6}
0.1	1×10^{-1}	0.5	5×10^{-1}
0.01	1×10^{-2}	0.037	3.7×10^{-2}
0.001	1×10^{-3}	0.004	4×10^{-3}
0.0001	1×10^{-4}	0.00056	5.6×10^{-4}
0.00001	1×10^{-5}	0.000082	8.2×10^{-5}
0.000001	1×10^{-6}	0.0000091	9.1×10^{-6}
0.0000001	1×10^{-7}	0.0000002	2×10^{-7}

Multiplying Numbers in Exponential Form

To multiply two numbers expressed in exponential form, you

1. *Multiply* the two coefficients.
2. Then, *add* the two exponents to determine the new power of 10 to use in the product.

For example,

(a) $(1 \times 10^{4}) \times (1 \times 10^{6}) = 1 \times 10^{(4+6)} = 1 \times 10^{10}$

(b) $(4 \times 10^{2}) \times (6 \times 10^{5}) = (4 \times 6) \times 10^{(2+5)} = 24 \times 10^{7} = 2.4 \times 10^{8}$

(c) $(2 \times 10^{4}) \times (3 \times 10^{-6}) = (2 \times 3) \times 10^{[4+(-6)]} = 6 \times 10^{-2}$

Example (b) illustrates that in exponential form we always rewrite the coefficient so that it represents a number between one and ten.

Dividing Numbers in Exponential Form

To divide two numbers expressed in exponential form, you

1. *Divide* the two coefficients.

2. Then, *subtract* the exponent in the denominator from the exponent in the numerator.

For example,

(a) $\frac{1 \times 10^6}{1 \times 10^4} = \frac{1}{1} \times 10^{(6-4)} = 1 \times 10^2$

(b) $\frac{8 \times 10^7}{2 \times 10^5} = \frac{8}{2} \times 10^{(7-5)} = 4 \times 10^2$

(c) $\frac{8 \times 10^4}{3 \times 10^{-2}} = \frac{8}{3} \times 10^{[4-(-2)]} = 2.67 \times 10^6$

(d) $\frac{4 \times 10^{-3}}{8 \times 10^2} = \frac{4}{8} \times 10^{(-3-2)} = 0.5 \times 10^{-5} = 5 \times 10^{-6}$

Exercise 1

1. Write the following numbers in exponential form:

Number	Answer
(a) 56	5.6×10^1
(b) 476.54	4.7654×10^2
(c) 0.00046	4.6×10^{-4}
(d) 75,340,000	7.534×10^7
(e) 1278	1.278×10^3
(f) 0.03	3×10^{-2}
(g) 0.6	6×10^{-1}
(h) 890,000	8.9×10^5
(i) 0.00009	9×10^{-5}
(j) 0.0000000000012	1.2×10^{-12}

2. Perform the following operations:

Problem	Answer
(a) $\frac{(3 \times 10^3)(8 \times 10^{10})}{(6 \times 10^4)(1 \times 10^6)}$	1.2×10^4
(b) $\frac{(1.5 \times 10^2)(4 \times 10^6)}{(5 \times 10^{10})(2.5 \times 10^5)}$	4.8×10^{-8}
(c) $\frac{(7.5 \times 10^{-3})(9 \times 10^6)}{(1.5 \times 10^2)(2.5 \times 10^{-8})}$	1.8×10^{10}
(d) $\frac{(2 \times 10^{-6})(4.2 \times 10^{-2})}{(1.4 \times 10^{-11})(1 \times 10^5)}$	6×10^{-2}

APPENDIX TWO

USING SIGNIFICANT FIGURES

In Section 2.8, we introduced the concept of *significant figures* or *significant digits.* Significant figures are used to indicate the precision with which a measurement is made. When solving a problem, we are often required to perform mathematical operations on experimental data. In such cases it is important to maintain the correct number of significant figures.

Determining the Number of Significant Figures in a Measurement

The following guidelines can be used to determine the number of significant figures in a given measurement.

1. The digits 1 through 9 are all significant. Therefore, for example, the number 27 has two significant figures and 3.584 has four significant figures.
2. The particular placement of the digit zero in a number will determine whether or not the zero is significant.
   (a) The zero is significant if it is located between two nonzero digits. For example, 1003 has four significant figures and 1.03 has three significant figures.
   (b) The zero is significant if it is the final digit to the *right* of the decimal point. For example, 39.0, 3.90, and 0.390 all have three significant figures.
   (c) The zero is not significant when it is used to mark the position of the decimal point in a number less than one. For example, both 0.178 and 0.00178 have three significant figures.
   (d) A zero that is used to mark the decimal place in a number greater than one is usually not significant. Specifically, if such zeros are used simply as place markers for the decimal point rather than being part of the actual measurement, they are not significant. There are several ways that we can write numbers to eliminate confusion about whether the zero is

significant. Writing the number in exponential form can show clearly whether or not the zeros are significant. For example, the number 4800 can be written as

4.8×10^3	2 significant figures
4.80×10^3	3 significant figures
4.800×10^3	4 significant figures

Exercise 1

Identify the number of significant figures in each of the following numbers:

Number	Answer
(a) 458.7	(4)
(b) 0.004	(1)
(c) 1.704	(4)
(d) 325	(3)
(e) 63.0	(3)
(f) 0.27650	(5)
(g) 3×10^3	(1)
(h) 1.0003	(5)
(i) 9.00×10^4	(3)
(j) 0.0056	(2)
(k) 45.67	(4)

Exact Numbers

Not all numbers that we use in chemical calculations come from measurements. Numbers that are given in definitions (1 meter equals 100 millimeters) or that result from counting objects (1 dozen eggs equals 12 eggs) are called **exact numbers.** When using exact numbers in calculations, we think of them as having an infinite number of significant figures. Therefore, we do not have to take exact numbers into account when determining the number of significant figures in our answer.

Electronic Calculators

A calculator can be of real help in performing mathematical computations, but its use presents a special kind of problem in calculations based on experimental data: the calculator will usually show answers with too many significant digits. The following rules will help you determine the correct number of significant figures in a calculated answer.

Rule 1: Addition and Subtraction

When numbers are added or subtracted, the number of decimal places in the answer will equal the *smallest number of decimal places* among all of the numbers added or subtracted. In this way, the answer will indicate the first decimal place in which there was uncertainty in any of the measurements.

Rule 2: Multiplication and Division

When numbers are multiplied and divided, the number of significant figures in the answer can be no more than the *smallest number of significant figures* among all of the numbers multiplied or divided.

Rule 3: Rounding Off

To obtain the correct number of significant figures in a calculated result, the nonsignificant digits in the calculator's answer are rounded off.

(a) If the nonsignificant digit is less than 5, the digit is dropped. For example, 32.233 to four significant figures is 32.23.

(b) If the nonsignificant digit is greater than 5, the digit is dropped and the last significant digit is increased by one. For example, 32.236 to four significant figures is 32.24.

(c) If the nonsignificant digit is 5 (or 5 followed by zeros), the 5 is dropped and the last significant digit is increased by one if it is odd and is not changed if it is even. For example, 32.235 to four significant figures is 32.24, but 32.225 to four significant figures is 32.22.

EXAMPLE 1

1. Add the following numbers: 25, 1.278, 127.1, and 5.45.

$$\begin{array}{r} 25 \\ 1.278 \\ 127.1 \\ 5.45 \\ \hline 158.828 \end{array} \text{ (calculator answer)}$$

↑

Uncertainty in the answer begins in the ones column (as indicated by the number 25); therefore, the answer should be rounded off to the ones place, or to three significant figures. Rounded off, the answer with the correct number of significant figures is 159.

2. Subtract 1.286 from 19.57.

$$\begin{array}{r} 19.57 \\ -\ 1.286 \\ \hline 18.284 \end{array} \text{ (calculator answer)}$$

↑

Uncertainty begins in the hundredths place, so the answer should be rounded to four significant figures: 18.28.

3. Multiply 13.6 by 0.004.

$$13.6 \times 0.004 = 0.0544 \text{ (calculator answer)}$$

The number 13.6 has three significant figures and 0.004 has one significant figure. Therefore, our answer must have one significant figure: 0.0544 rounded to one significant figure is 0.05.

4. Divide 67.0 by 563.

$$\frac{67.0}{563} = 0.1190053 \text{ (calculator answer)}$$

Both the dividend and the divisor have three significant figures. Therefore, we must round our answer to three significant figures: 0.1190053 rounded to three significant figures is 0.119.

Exercise 2

1. Round each of the following numbers to two significant figures:

Number	Answer
(a) 1.598	1.6
(b) 7.35	7.4
(c) 26.3	26
(d) 386	390 or 3.9×10^2
(e) 4.250	4.2
(f) 0.03457	0.035
(g) 0.9246	0.92
(h) 0.1486	0.15

2. Perform the following operations and round your answer to the correct number of significant figures.

Problem	Answer	Calculator answer
(a) $43.67 + 27.4 + 0.0265$	71.1	71.0965
(b) $156 + 32.7 + 4.38$	193	193.18
(c) $1.4651 - 0.53$	0.94	0.9351
(d) $256 - 139.48$	117	116.52
(e) $1.48 \times 39.1 \times 0.312$	18.1	18.054816
(f) $67.84 \div 4.6$	15	14.747826
(g) $\frac{9.50 \times 784}{1465}$	5.08	5.083959
(h) $\frac{0.036 \times 25.78}{1.4865 \times 169}$	0.0037	0.0036943

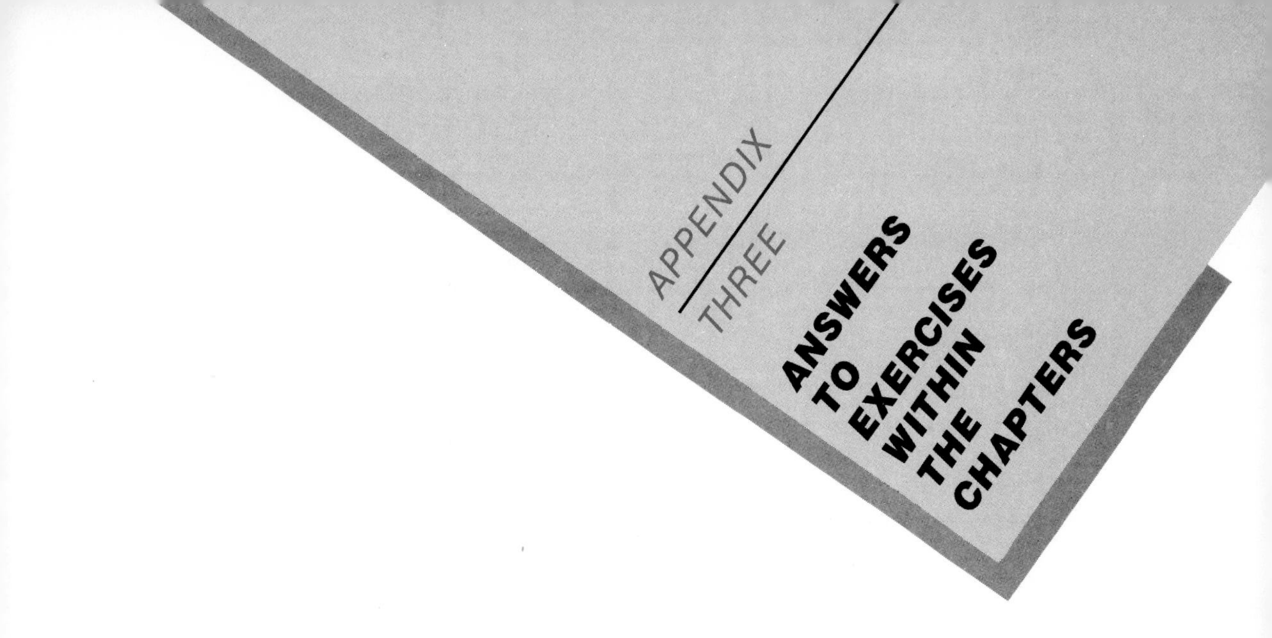

The following are the answers to all in-chapter exercises. The detailed calculations used to obtain these answers are included in the third edition of the Student Study Guide.

Chapter 2

Exercise 2-1

1. (a) 4.8 cm (b) 3200 mm (c) 30 m (d) 250 in. (e) 12 km (f) 30 yd
2. (a) 42 km (b) 4.92 min/mi (c) 3.1 min/km

Exercise 2-2

1. (a) 0.253 mg (b) 3200 g (c) 5000 mg (d) 12 oz (e) 1.5 lb (f) 680 g
2. 1.8 lb

Exercise 2-3

1. (a) 2500 ml (b) 0.345 liter (c) 25 cc (d) 1290 liters (e) 5.6 qt (f) 0.20 pt
2. Three 12-oz bottles are a better buy.

Exercise 2-4

1. (a) −65°C (b) 105°F (c) 55.6°C (d) 89.6°F (e) −131°C (f) 310 K
2. 88°F
3. 327.4°C, 600.4 K

Exercise 2-5

1. 11.3 g/cc
2. 3.0 cc
3. 373 g

Exercise 2-6

(a) 1.11 g/cc (b) 1.11 (c) The ethylene glycol is more dense than water.

Chapter 3

Exercise 3-1

1. 9200 kcal (rounding off for significant figures) 2. 16.5 kcal

Chapter 4

Exercise 4-1

0.868 atm, 660 torr

Exercise 4-2

1. $P_2 = 600$ torr 2. $V_2 = 884$ ml

Exercise 4-3

1. $T_2 = 35°C$ 2. 70 ml

Exercise 4-4

(a) $V_2 = 7.59$ liters (b) $T_2 = -68°C$ (c) $P_2 = 884$ torr

Exercise 4-5

$P_{O_2} = 731$ torr

Chapter 5

Exercise 5-1

1. (a) ^{222}Ra $p = 88$, $e^- = 88$, $n = 134$
   (b) Chromium-51 $p = 24$, $e^- = 24$, $n = 27$
   (c) $^{203}_{80}Hg$ $p = 80$, $e^- = 80$, $n = 123$
2. Atomic number = 15, mass number = 31, symbol = P

Exercise 5-2

Atomic weight = 28.1

Exercise 5-3

(a) Sodium, $1s^2\ 2s^2\ 2p^6\ 3s^1$

$1s$	$2s$	$2p_x$	$2p_y$	$2p_z$	$3s$
(↑↓)	(↑↓)	(↑↓)	(↑↓)	(↑↓)	(↑)

(b) Phosphorus, $1s^2\ 2s^2\ 2p^6\ 3s^2\ 3p^3$

$1s$	$2s$	$2p_x$	$2p_y$	$2p_z$	$3s$	$3p_x$	$3p_y$	$3p_z$
(↑↓)	(↑↓)	(↑↓)	(↑↓)	(↑↓)	(↑↓)	(↑)	(↑)	(↑)

(c) Chlorine, $1s^2\ 2s^2\ 2p^6\ 3s^2\ 3p^5$

$1s$	$2s$	$2p_x$	$2p_y$	$2p_z$	$3s$	$3p_x$	$3p_y$	$3p_z$
(↑↓)	(↑↓)	(↑↓)	(↑↓)	(↑↓)	(↑↓)	(↑↓)	(↑↓)	(↑)

Chapter 6

Exercise 6-1

1. (a) $^{226}_{88}Ra \longrightarrow ^{222}_{86}Rn + ^{4}_{2}He$

   (b) $^{28}_{13}Al \longrightarrow ^{28}_{14}Si + ^{0}_{-1}e$

   (c) $^{227}_{89}Ac \longrightarrow ^{223}_{87}Fr + ^{4}_{2}He$

2. (a) $^{45}_{20}Ca \longrightarrow ^{45}_{21}Sc + ^{0}_{-1}e$

   (b) $^{14}_{6}C \longrightarrow ^{14}_{7}N + ^{0}_{-1}e$

   (c) $^{149}_{62}Sm \longrightarrow ^{145}_{60}Nd + ^{4}_{2}He$

3. $^{137}_{55}Cs \longrightarrow ^{137}_{56}Ba + ^{0}_{-1}e + \gamma$

Exercise 6-2

After 120 hr or 5 days

Exercise 6-3

(a) $^{27}_{13}Al + ^{1}_{0}n \longrightarrow ^{24}_{11}Na + ^{4}_{2}He$

(b) $^{35}_{17}Cl + ^{1}_{0}n \longrightarrow ^{35}_{16}S + ^{1}_{1}p$

(c) $^{7}_{3}Li + ^{1}_{1}p \longrightarrow ^{7}_{4}Be + ^{1}_{0}n$

(d) $^{198}_{78}Pt + ^{1}_{0}n \longrightarrow ^{199}_{79}Au + ^{0}_{-1}e$

Chapter 8

Exercise 8-1

1. (a) Rb· (b) ·Ṡi· (c) :Ï:

2. (a) K· + ·Ï: ⟶ K^+ :Ï:$^-$

   (b) :Ï· + ·Mg· + ·Ï: ⟶ Mg^{2+} + 2[:Ï:]$^-$

Exercise 8-2

(a) KI (b) MgI_2

Exercise 8-3

(a) H:C̈l: H—Cl

(b) H:N̈:H H—N—H
 H |
        H

(c) S̈::C::S̈ S═C═S

Exercise 8-4

(a) Ca: 2^+
Cl: 1^-

(b) H: 1^+
O: 2^-
S: 6^+

(c) O: 2^-
Cl: 1^+

(d) H: 1^+
O: 2^-
C: 4^+

(e) Mn: 4^+
O: 2^-

(f) P: 5^+
O: 2^-

Exercise 8-5

(a) Magnesium iodide
(b) Iron(II) oxide or ferrous oxide
(c) Sulfur dioxide
(d) Hydrogen sulfide
(e) Sodium hydrogen sulfate or sodium bisulfate
(f) Potassium dichromate

Exercise 8-6

(a) MgO
(b) $NiCl_2$
(c) K_2S
(d) SO_3
(e) SiF_4
(f) N_2O_5
(g) $SnCl_4$
(h) $Ca(HSO_4)_2$
(i) $Ca_3(PO_4)_2$

Chapter 9

Exercise 9-1

(a) $4P + 5O_2 \longrightarrow P_4O_{10}$
(b) $2NOCl \longrightarrow 2NO + Cl_2$
(c) $CH_4 + 2O_2 \longrightarrow CO_2 + 2H_2O$
(d) $Ca(OH)_2 + 2HCl \longrightarrow CaCl_2 + 2H_2O$
(e) $2Mg + O_2 \longrightarrow 2MgO$
(f) $2PbS + 3O_2 \longrightarrow 2PbO + 2SO_2$
(g) $Na_2CO_3 + Mg(NO_3)_2 \longrightarrow MgCO_3 + 2NaNO_3$

Exercise 9-2

1. (1) Mg = element being oxidized, Br = element being reduced
   (2) Br = oxidizing agent, Mg = reducing agent
   (3) $Mg + Br_2 \longrightarrow MgBr_2$

2. (1) Cl = element being oxidized, Mn = element being reduced
   (2) Mn = oxidizing agent, Cl = reducing agent
   (3) $2KCl + MnO_2 + 2H_2SO_4 \longrightarrow K_2SO_4 + MnSO_4 + Cl_2 + 2H_2O$

3. (1) Cu = element being oxidized, N = element being reduced
   (2) N = oxidizing agent, Cu = reducing agent
   (3) $3Cu + 8HNO_3 \longrightarrow 3Cu(NO_3)_2 + 2NO + 4H_2O$

Exercise 9-3

1. (a) 9.9 g (b) 130 g (c) 3.2 g (d) 0.010 g
2. (a) 0.300 mol Ag (b) 35 mol Si (c) 0.001 mol Ne (d) 20 mol U

Exercise 9-4

1. (a) $I_2 = 253.8$ g
   (b) $HF = 20$ g
   (c) $PbS = 239.2$ g
   (d) $KClO_3 = 122.6$ g
   (e) $Al_2(SO_4)_3 = 342$ g
   (f) $C_4H_{10} = 58$ g
2. (a) $HF = 20$ g (b) $KClO_3 = 122.6$ g (c) $Al_2(SO_4)_3 = 342$ g
3. 1 mol $I_2 = 253.8$ g

Exercise 9-5

1. (a) 127 g (b) 670 g (c) 4.3 g (d) 1370 g
2. (a) 25 mol (b) 0.30 mol (c) 0.0107 mol
3. 1.5×10^{22} molecules

Exercise 9-6

1. 1 molecule of methane will react with 2 molecules of oxygen gas to produce 1 molecule of carbon dioxide plus 2 molecules of water.
2. 1 mole of methane will react with 2 moles of oxygen gas to produce 1 mole of carbon dioxide plus 2 moles of water.
3. 16 grams of methane will react with 64 grams of oxygen gas to produce 44 grams of carbon dioxide plus 36 grams of water.

Exercise 9-7

1. 177 g MgO
2. 0.15 g HCl

Chapter 10

Exercise 10-1

1. (a) Endothermic (b) Absorbed
   (c) $C_{(s)} + H_2O_{(g)} + 31.4 \text{ kcal} \longrightarrow CO_{(g)} + H_{2(g)}$
2.

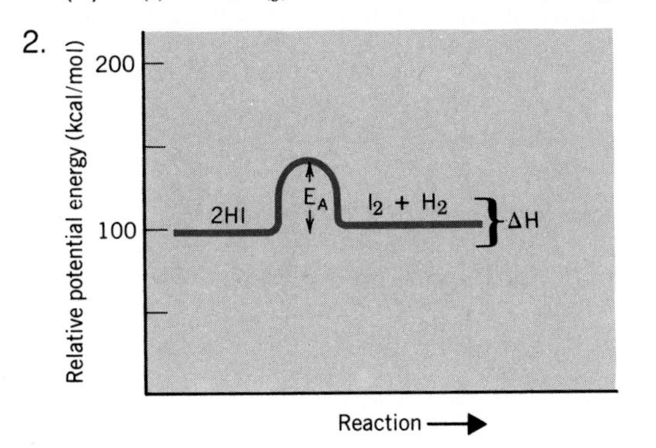

Exercise 10-2

(a) The concentration of O_2 will decrease.

(b) The concentration of O_2 will decrease.

(c) The concentration of O_2 will decrease.

(d) The concentration of O_2 will increase.

(e) There will be no change in the equilibrium concentration of O_2.

Chapter 11

Exercise 11-1

1. (a) Yes (b) No (c) Yes (d) Yes
2. (a) $Mg^{2+}_{(aq)} + 2OH^-_{(aq)} \longrightarrow Mg(OH)_{2(s)}$
   (b) $2Cl^-_{(aq)} + Pb^{2+}_{(aq)} \longrightarrow PbCl_{2(s)}$
   (c) $Ba^{2+}_{(aq)} + SO^{2-}_{4(aq)} \longrightarrow BaSO_{4(s)}$
3. $Ca^{2+}_{(aq)} + CO^{2-}_{3(aq)} \longrightarrow CaCO_{3(s)}$

Chapter 12

Exercise 12-1

1. (a) Dissolve 5.30 g Na_2CO_3 in enough water to bring up the volume to 250 ml.
   (b) Dissolve 110 g H_3PO_4 in enough water to bring up the volume to 1.5 liters.
   (c) Dissolve 14.2 g $KMnO_4$ in enough water to bring up the volume to 150 ml.
2. 625 ml

Exercise 12-2

1. (a) Start with 16.1 g NaCl and add enough water to make 350 g of solution.
   (b) Start with 600 ml of ethylene glycol and add enough water to make 2.0 liters of solution.
   (c) Start with 0.121 g K_2CO_3 and add enough water to make 55 ml of solution.
   (d) Start with 20 mg glucose and add enough water to make 25 ml of solution.
2. 80 ml blood
3. 33.0 g glucose

Exercise 12-3

CCl_4; no; 1.6 ppb

Chlordane; yes; 5.2 ppb

Lead; no; 0.044 ppm

Mercury; no; 0.84 ppb

Exercise 12-4

1. (a) 39.1 g
   (b) 35.5 g
   (c) 48 g
   (d) 48 g
2. 6.72 g/liter
3. 330 mg%

Exercise 12-5

1. (a) Dissolve 266 g NaCl in a small amount of water and then bring up the volume to 1.65 liters.
   (b) Start with 1.10 liters of the 4.12 *M* NaCl and add enough water to make 1.65 liters of solution.
2. Take 1 volume of 2.75 *M* NaCl and add 5 equal volumes of water; 0.458 *M* NaCl

Chapter 13

Exercise 13-1

1. HI: hydroiodic acid
   HIO: hypoiodous acid
   HIO_2: iodous acid
   HIO_3: iodic acid
   HIO_4: periodic acid
2. (a) $NaHSO_3$ (b) IO_4^-

Exercise 13-2

(a) $KOH_{(aq)} + HNO_{3(aq)} \longrightarrow H_2O_{(l)} + KNO_{3(aq)}$

(b) $OH^-_{(aq)} + H^+_{(aq)} \longrightarrow H_2O_{(l)}$

Exercise 13-3

(a) $[H^+] = 1 \times 10^{-11}$

(b) $[H^+] = 2 \times 10^{-9}$

(c) $[H^+] = 5 \times 10^{-12}$

(d) $[H^+] = 2.5 \times 10^{-11}$

(e) $[H^+] = 1 \times 10^{-13}$

(f) $[H^+] = 6.3 \times 10^{-8}$

Exercise 13-4

1. (a) Basic $[H^+] = 1 \times 10^{-11}$ $[OH^-] = 1 \times 10^{-3}$
   (b) Acid $[H^+] = 1 \times 10^{-2}$ $[OH^-] = 1 \times 10^{-12}$
   (c) Acid $[H^+] = 1 \times 10^{-5}$ $[OH^-] = 1 \times 10^{-9}$
   (d) Basic $[H^+] = 1 \times 10^{-9}$ $[OH^-] = 1 \times 10^{-5}$
2. pH 7

Exercise 13-5

0.027 *M* acid

Exercise 13-6

1. (a) 63 g (b) 37.0 g
2. Dissolve 0.800 g of HCl in a small amount of water and add enough water to bring up the volume to 200 ml.
3. Dissolve 0.925 g of $Ca(OH)_2$ in a small amount of water and add enough water to bring up the volume to 500 ml.

Exercise 13-7

$N = 0.050$

Chapter 14

Exercise 14-1

1. (a) $CH_3C(CH_3)_2CH_2CH_3$ (b) $CH_3(CH_2)_4CH_3$

Exercise 14-2

1. (a) Heptane (b) 2-Methyl-4-ethylhexane

Exercise 14-3

1. $CH_3CH_2CH_2CH_2CH_2CH_2CH_2CH_3$

2. $CH_3—CH(CH_3)—CH_2—CH_2—CH_2—CH_2—CH_3$

3. 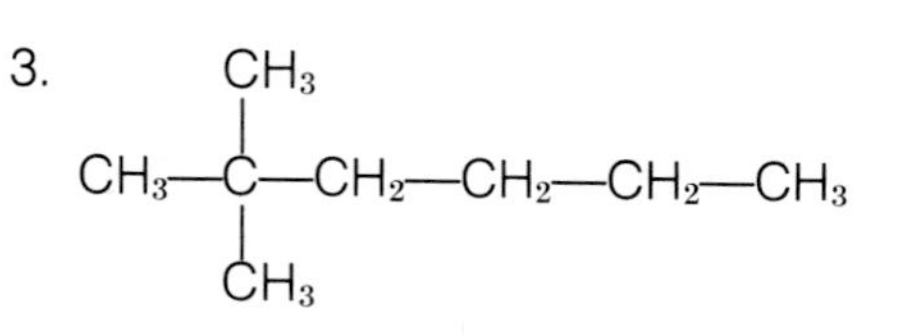

4. and 6. 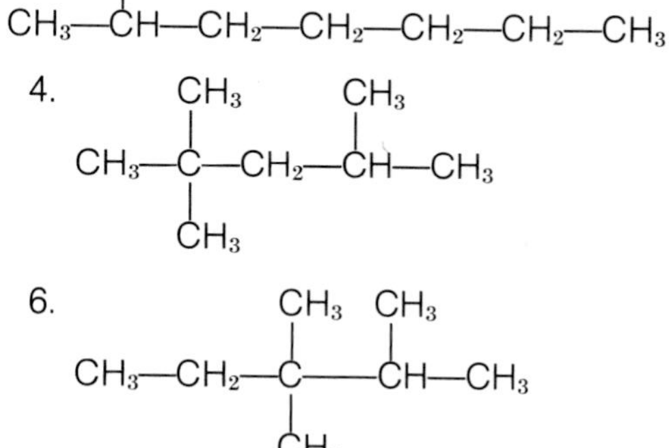

5. 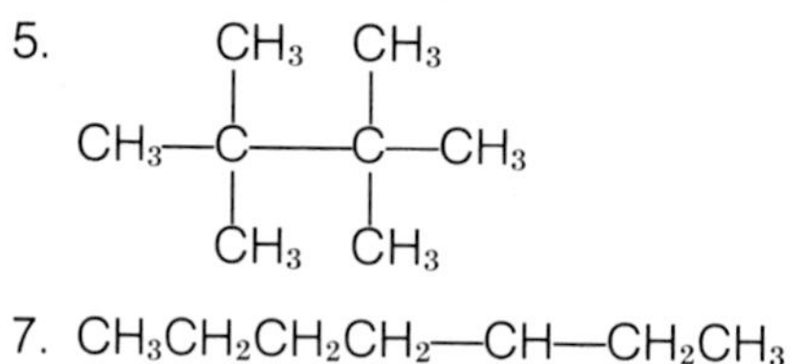

7. $CH_3CH_2CH_2CH_2—CH(CH_3)—CH_2CH_3$

8. 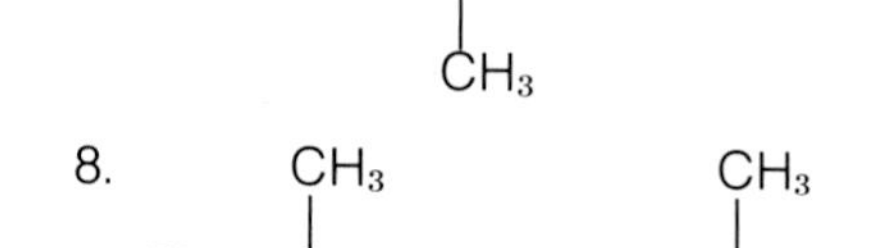

9. $CH_3—CH(CH_3)—CH(CH_3)—CH(CH_3)—CH_3$

Exercise 14-4

1. $2C_6H_{14} + 19O_2 \longrightarrow 12CO_2 + 14H_2O$ + energy

2. $C_2H_6 + Cl_2 \longrightarrow C_2H_5Cl + HCl$

Chapter 15

Exercise 15-1

1. (a) 2-Hexene
   (b) 2-Methyl-1-butene
   (c) 2-Methyl-1,3-pentadiene
   (d) 2-Methyl-4-ethyl-2-heptene

2. (a) $CH_3—CH(CH_3)—CH{=}CH—CH_2CH_3$
   (b) $CH_3—CH(Cl)—CH{=}CHCH_3$
   (c) 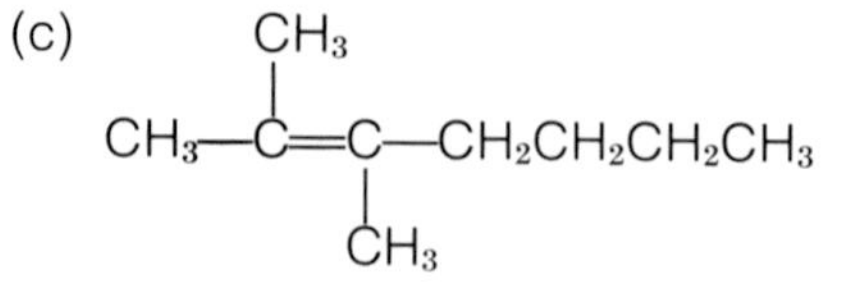
   (d) $CH_2{=}CH—CH{=}CH_2$

3. $CH_2{=}CHCH_2CH_2CH_3$ 1-pentene

   $CH_3CH{=}CHCH_2CH_3$ 2-pentene

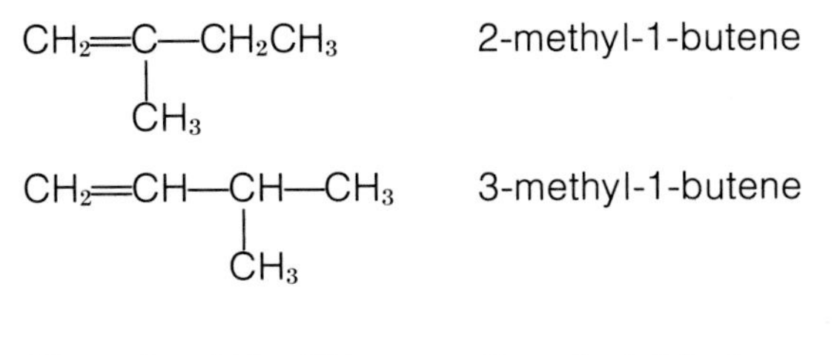

2-methyl-1-butene

3-methyl-1-butene

2-methyl-2-butene

Exercise 15-2

CH_3 CH_2CH_3 CH_3 H

C=C C=C

H H H CH_2CH_3

Exercise 15-3

1. (a) 1-Butyne
   (b) 3-Methyl-1-butyne
   (c) 6,6-Dimethyl-4-ethyl-2-heptyne

2. (a) $CH{\equiv}C{-}CH_2CH_2CH_3$

(b)

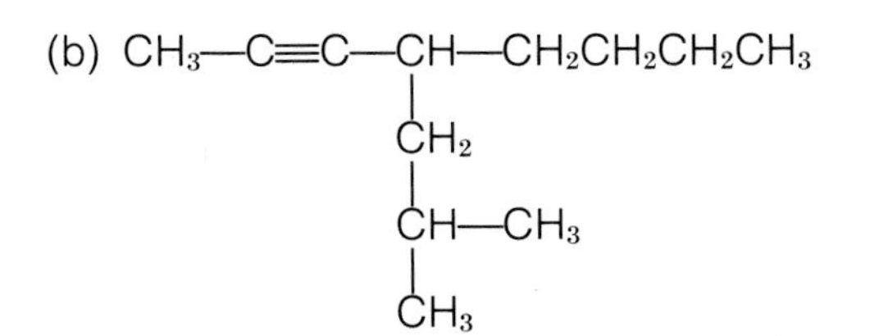

(c)

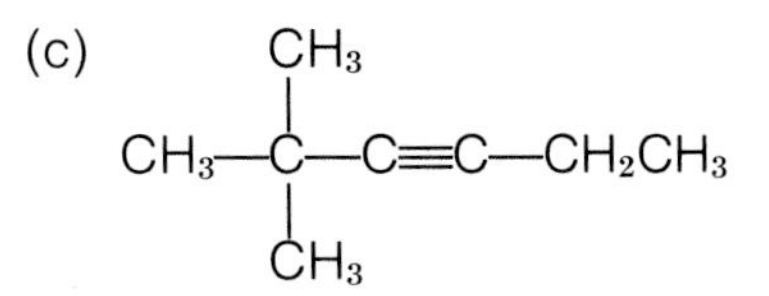

Exercise 15-4

1. (a) *m*-Dichlorobenzene (c) Aniline
   (b) 2,4,6-Tribromophenol (d) Phenylethyne

2. (a)

(c)

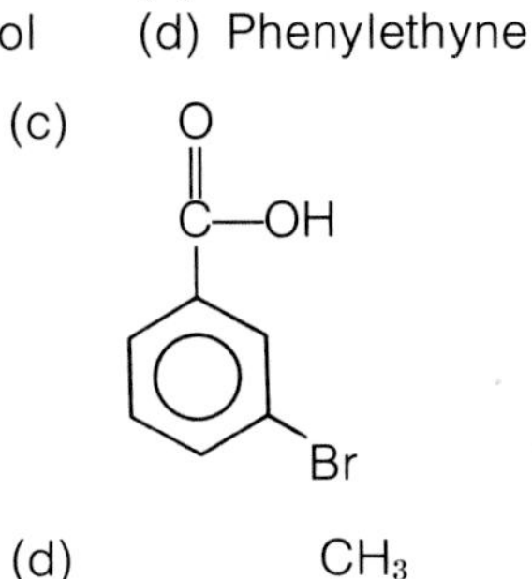

(b) (d)

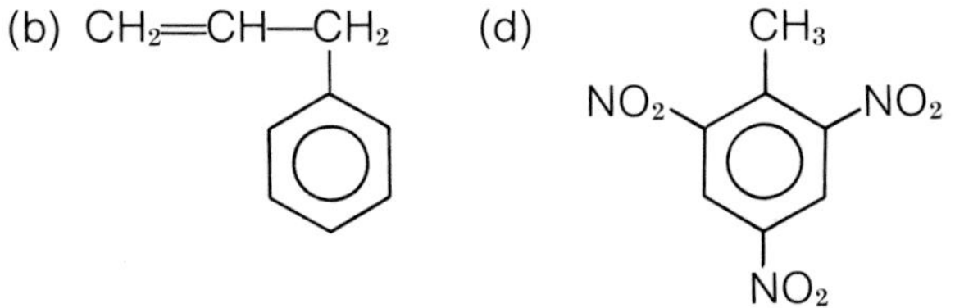

Chapter 16

Exercise 16-1

1. (a) 2-Methyl-2-hexanol, tertiary
   (b) 2,2-Dimethyl-1-butanol, primary
   (c) 3-Pentanol, secondary

2. (a)

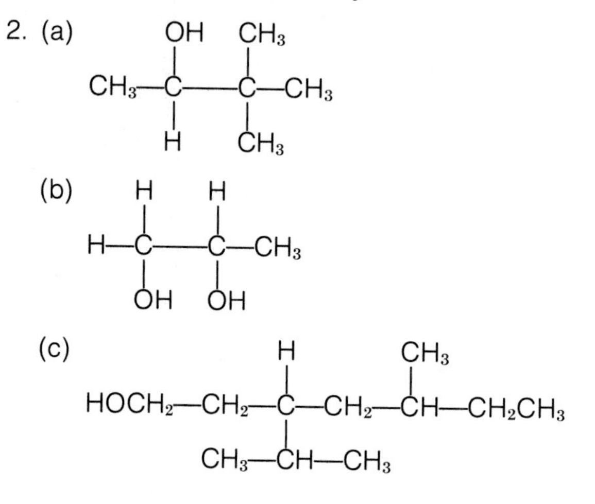

Exercise 16-2

(a) An ether
(b) An unsaturated hydrocarbon
(c) An alcohol
(d) A ketone
(e) An aldehyde

Exercise 16-3

(a) Methyl butyl ether
(b) Ethyl isopropyl ether
(c) Cyclohexyl methyl ether

Exercise 16-4

1. (a) 2-Methylpropanal (b) 1-Phenyl-2-butanone

2. (a)

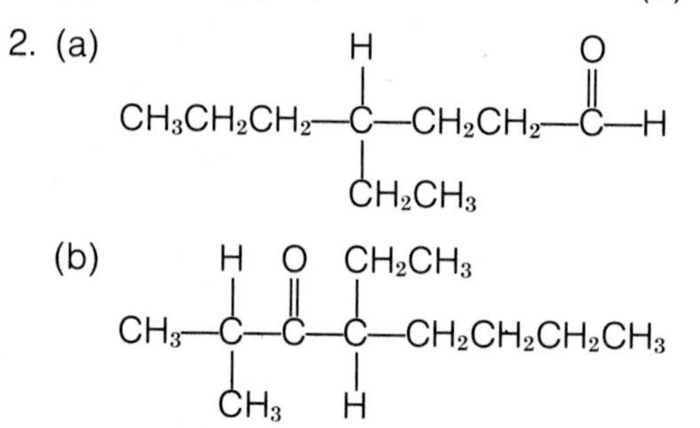

Exercise 16-5

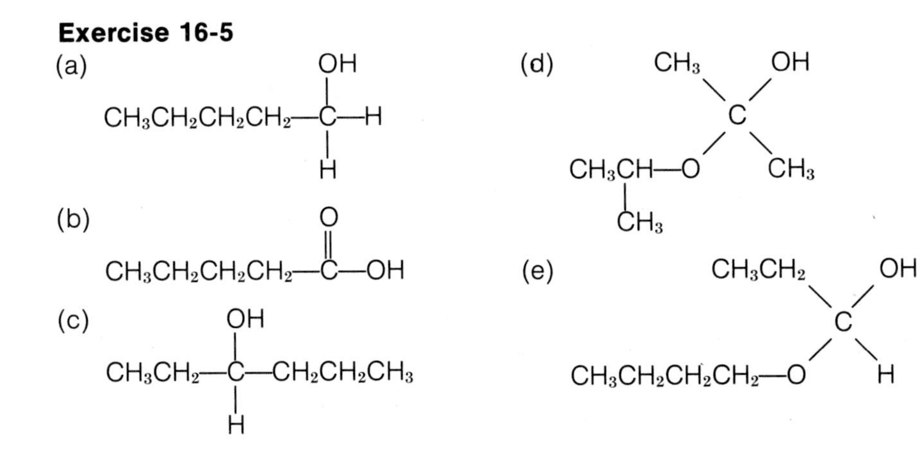

Exercise 16-6

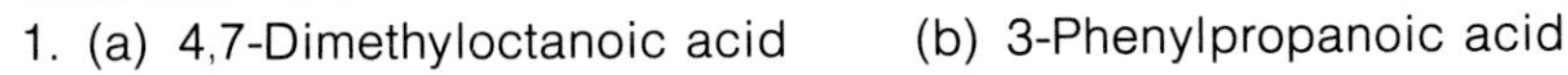

1. (a) 4,7-Dimethyloctanoic acid (b) 3-Phenylpropanoic acid

2.

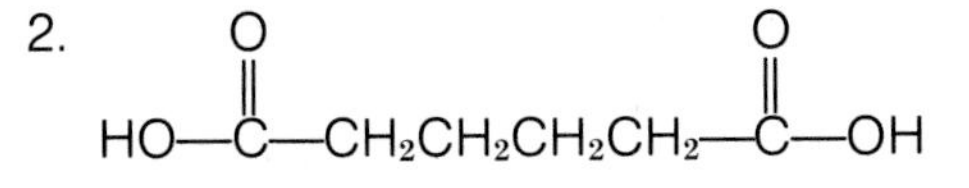

Exercise 16-7

(a)

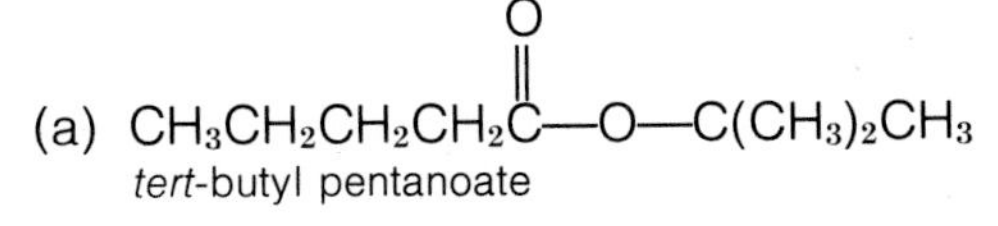

tert-butyl pentanoate

(b)

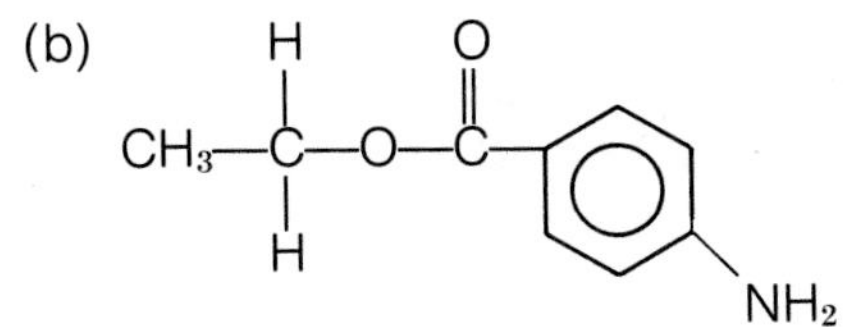

Ethyl *p*-aminobenzoate

Chapter 17

Exercise 17-1

1. (a) Octylamine
   (b) Triethylamine
   (c) *N*-Ethylbenzylamine
   (d) 2-Aminoethanol
   (e) 4-(*N*-Phenylamino)-2-butanol

2. (a)

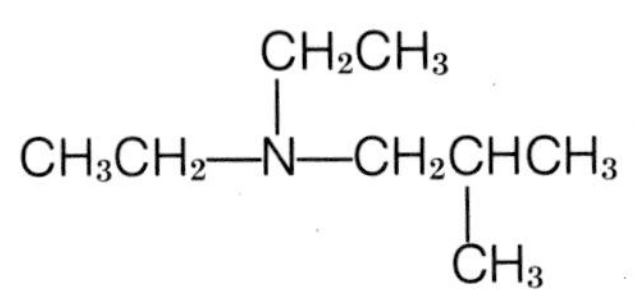

(b)

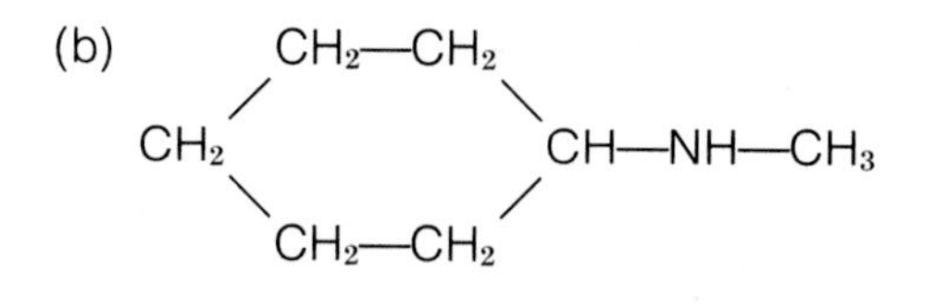

(c) $H_2NCH_2CH_2CH_2CH_2CH_2CH_2NH_2$

(d)

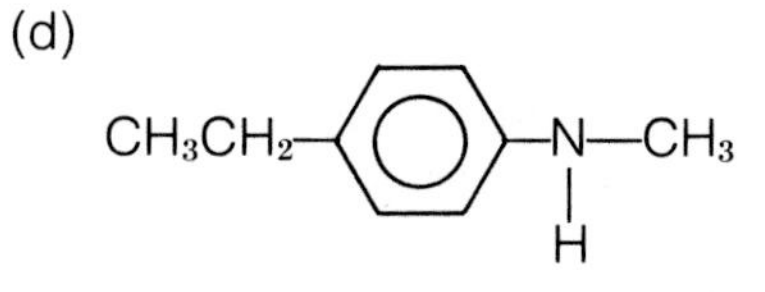

Exercise 17-2

1. (a) Acetamide
   (b) *N*-Phenylpropanamide
   (c) *N*-Ethylhexanamide
   (d) *N*-Methyl-*N*-isopropylpentanamide

2. (a)

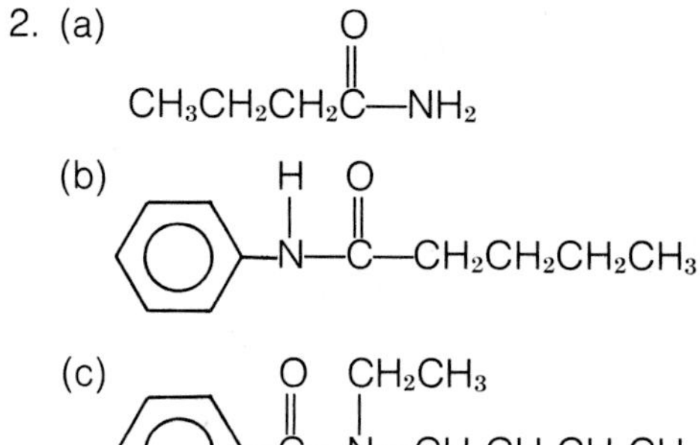

(d)

$CH_3CH_2C(=O)—N(CH_3)—CH_3$

Chapter 19

Exercise 19-1

(a) Ribose—aldopentose
(b) Fructose—ketohexose
(c) Threose—aldotetrose
(d) Galactose—aldohexose
(e) Mannose—aldohexose
(f) Ribulose—ketopentose

Exercise 19-2

Disaccharide	Source	Monosaccharide Subunits	Reducing Sugar?	Type of Linkage
Sucrose	Fruit, honey, vegetables	Glucose, fructose	No	α,1:2
Lactose	Milk	Glucose, galactose	Yes	β,1:4
Maltose	Starch, glycogen	Glucose	Yes	α,1:4

Exercise 19-3

(a) Dextran
(b) Glycogen
(c) Dextrin
(d) Starch
(e) Cellulose
(f) Glucose

Chapter 20

Exercise 20-1

1. (a) Simple lipids yield only fatty acids and an alcohol upon hydrolysis, whereas compound lipids yield fatty acids, alcohol, and some other compounds upon hydrolysis.
   (b) Essential fatty acids are essential for good health but are either not produced or not produced in an adequate supply by our bodies. Linoleic, linolenic, and arachidonic acids are essential fatty acids.
   (c) Saturated fatty acids only have carbon-to-carbon single bonds and are waxy solids at room temperatures. Unsaturated fatty acids have one or more carbon-to-carbon double bonds and are liquids at room temperatures.
   (d) Waxes are esters of long-chain fatty acids and long-chain monohydric alcohols.
   (e) Triacylglycerols, or triglycerides, are simple fats in the form of esters of glycerol and three fatty acids.

2. Since the fatty acids involved are all different, this is a mixed triacylglycerol.

$$\begin{array}{l} CH_2OC(=O)(CH_2)_{14}CH_3 \\ | \\ CHOC(=O)(CH_2)_{16}CH_3 \\ | \\ CH_2OC(=O)(CH_2)_7CH{=}CH(CH_2)_7CH_3 \end{array}$$

Exercise 20-2

1. Low iodine number. This compound is likely to be a solid at room temperature.

2.

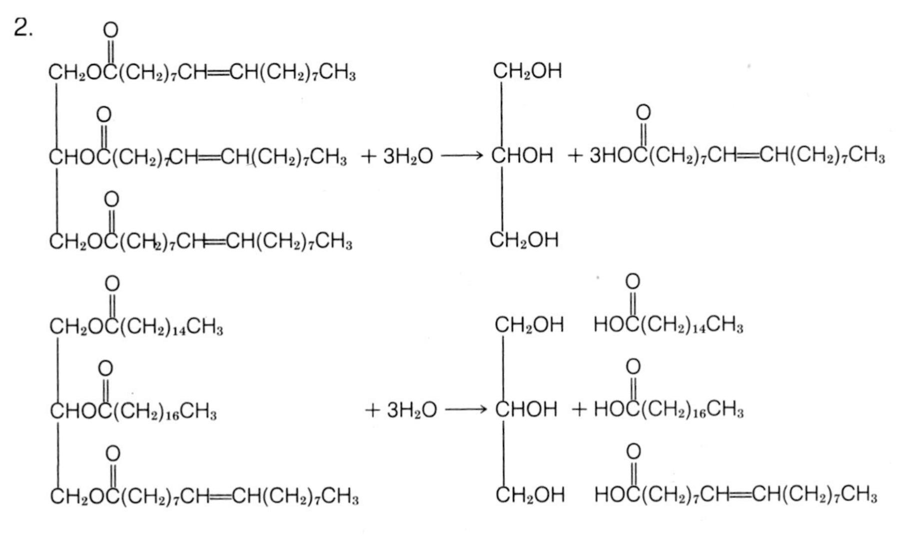

3.

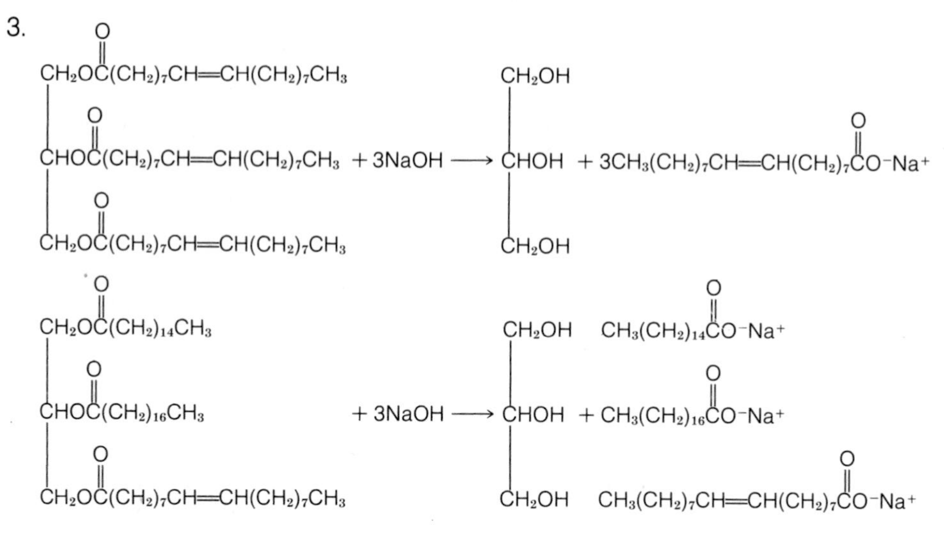

Exercise 20-3

1. (a) Glycerol, two fatty acids, a phosphate group, and a nitrogen-containing group
   (b) Glycerol, two fatty acids, a phosphate group, and choline
   (c) Glycerol, two fatty acids, a phosphate group, and ethanolamine
   (d) Sphingosine, one fatty acid, a phosphate group, and choline
   (e) Glycerol or sphingosine, one or two fatty acids, and sugar
   (f) Sphingosine, one fatty acid, and sugar

2. PC has a polar head that is soluble in the vinegar and nonpolar tails that are soluble in the salad oil, thus binding the oil and vinegar.

Exercise 20-4

1. Cholesterol, cortisone, vitamin D, testosterone, or progesterone
2. Phosphatidylcholine, phosphatidylethanolamine, sphingolipids, and cholesterol are the major components of cellular membranes. Proteins are also found in the membrane structure.

Chapter 21

Exercise 21-1

1. (a) Simple proteins produce only amino acids upon hydrolysis.
   (b) Conjugated proteins produce not only amino acids upon hydrolysis but also other organic and inorganic substances.
   (c) Prosthetic groups are the organic or inorganic substances that, along with amino acids, make up conjugated proteins.
   (d) Globular proteins are soluble in water, fragile, and have an active function.
   (e) Fibrous proteins are insoluble in water, are physically tough, and have a structural or protective function.
   (f) Amino acids, the building blocks of proteins, are carboxylic acids that have an amino group on the alpha carbon.
   (g) Molecules, such as amino acids, that contain both an acid group and a base group and, therefore, can act as either acids or bases in water are called amphoteric.

(h) Zwitterions are dipolar ions.
(i) The isoelectric point for an amino acid or a protein is the specific pH at which the molecule is electrically neutral.
(j) Electrophoresis is a laboratory method of separating proteins from one another at different pH levels on the basis of their charge.

2. (a) Amino acids can act as buffers because they act as either acids or bases in aqueous solution.

(b), (c)

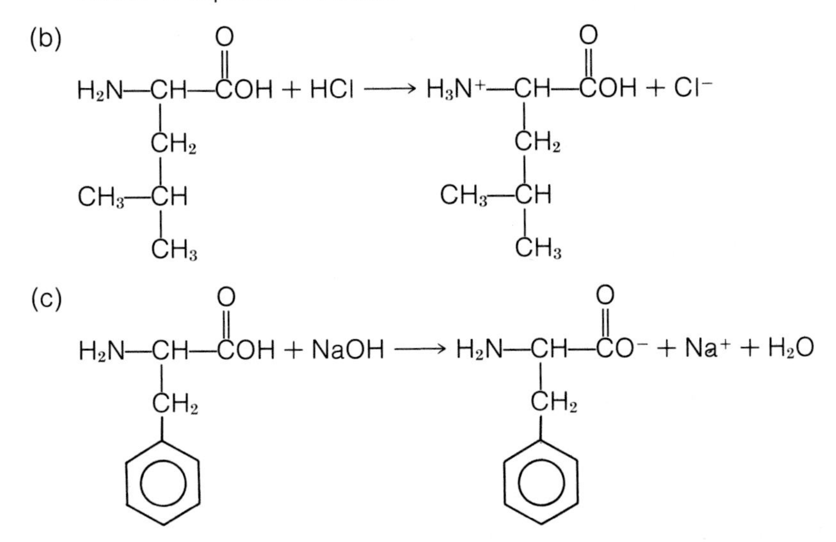

3. (a) The pI of phenylalanine is 5.5.
(b) The pH of 7 is more basic than pH 5.5; therefore, the phenylalanine will have a net negative charge and move toward the positive electrode.

Exercise 21-2

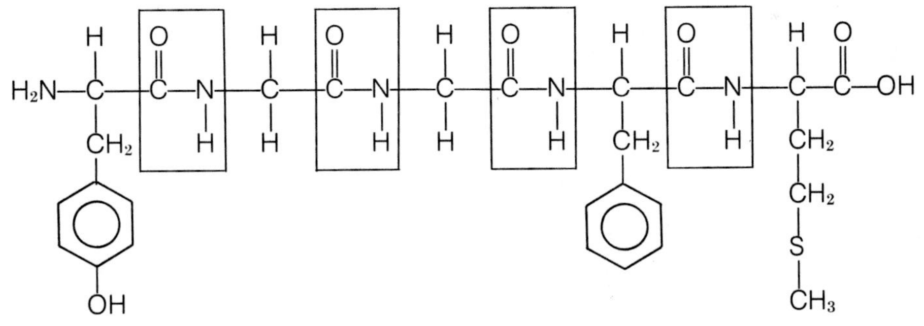

Exercise 21-3

1. (a) The primary structure of a protein is the sequence of amino acids in the protein that are held together by covalent bonds called peptide bonds.
(b) The secondary structure of a protein results from hydrogen bonding and gives the protein its specific geometric arrangement.
(c) The three-dimensional or tertiary structure of proteins is due to the interaction between the R-group side chains of the amino acids. Interactions include hydrogen bonding, disulfide linkages, salt bridges, and hydrophobic interactions.

(d) In the quaternary structure of proteins, separate polypeptide chains and prosthetic groups are held together by hydrogen bonding, hydrophobic interactions, and salt bridges.

2. The native state of a protein is the three-dimensional structure of the protein that is energetically the most stable at normal pH and temperature. Denaturation occurs when changes in the surroundings of the protein disrupt the native state.

3. (a) A strong acid will disrupt the hydrogen bonding and salt bridges of a protein.
   (b) Sunlight will disrupt the hydrogen bonding and salt bridges of a protein.
   (c) Lead ions may disrupt a protein's salt bridges by forming salt bridges between the lead ions and the protein.

Chapter 22

Exercise 22-1

1. (a) Apoenzyme is the term for the protein part of the enzyme molecule.
   (b) A coenzyme is a complex organic molecule other than a protein that is contained in and is necessary for the activity of a conjugated enzyme.
   (c) An activator is a metal ion that is contained in and is necessary for the activity of a conjugated enzyme.
   (d) Cofactors are the additional chemical groups appearing in and necessary for the activity of conjugated enzymes.
   (e) A proenzyme is the inactive form of an enzyme.
   (f) The substrate is the chemical substance or substances upon which the enzyme acts.
   (g) The active site is the specific area of the enzyme to which the substrate attaches during the reaction.
   (h) The turnover number of an enzyme is the number of substrate molecules transformed per minute by one molecule of the enzyme under optimal conditions of temperature and pH.
   (i) Hydrolases are enzymes that catalyze hydrolysis reactions.
   (j) Oxido-reductases are enzymes that catalyze oxidation–reduction reactions.
   (k) Transferases are enzymes that catalyze reactions involved in the transfer of functional groups.
   (l) Lyases are enzymes that catalyze the addition to double bonds.
   (m) Isomerases are enzymes that catalyze the interconversion of isomers.
   (n) Ligases are enzymes that, in conjunction with ATP, catalyze the formation of new bonds.

2. 1×10^{-18} mole/hr

Exercise 22-2

1. (a) Carbonic anhydrase acts as a catalyst, lowering the activation energy of the reaction.
   (b) The change in pH would slow the reaction because increasing the

pH would cause some denaturation of the enzyme, interfering with its ability to catalyze the reaction.

2. Because of the specificity necessary between enzyme and substrate, pepsin and trypsin are not able to hydrolyze all peptide bonds in proteins.

Exercise 22-3

1. Fat-soluble vitamins are not excreted from the body but are stored in the liver and can be toxic at high concentrations.
2. (a) Biotin is a vitamin.
   (b) A vitamin deficiency could result from a diet high in raw eggs because biotin would not be absorbed in the intestines.

Exercise 22-4

1.

Function	Site of Synthesis	Site of Action
Hormones Coordinate actions between cells, tissues, and organs	Endocrine glands	Circulate in the blood to target cells or organs
Prostaglandins Large variety of diverse functions	Cell membrane	Within the cell
Cyclic AMP Regulates cell's response to hormones	Cell membrane	Within the cell

2. (a) Multienzyme
   (b) Feedback
   (c) Regulatory
   (d) Active, regulatory

Exercise 22-5

1. (a) Sodium arsenite could bind with the sulfhydryl groups of an enzyme and thereby inhibit the enzyme.
   (b) Enzymes for which the sulfhydryl group is important to the activity of the enzyme.

Chapter 23

Exercise 23-1

One molecule will produce 113 ATP

Chapter 24

Exercise 24-1

1. DNA is made of 2-deoxyribose, phosphoric acid, and one of the nitrogen-containing bases: cytosine, thymine, adenine, or guanine. RNA

is made of ribose, phosphoric acid, and one of the nitrogen-containing bases: cytosine, uracil, adenine, or guanine.

2.

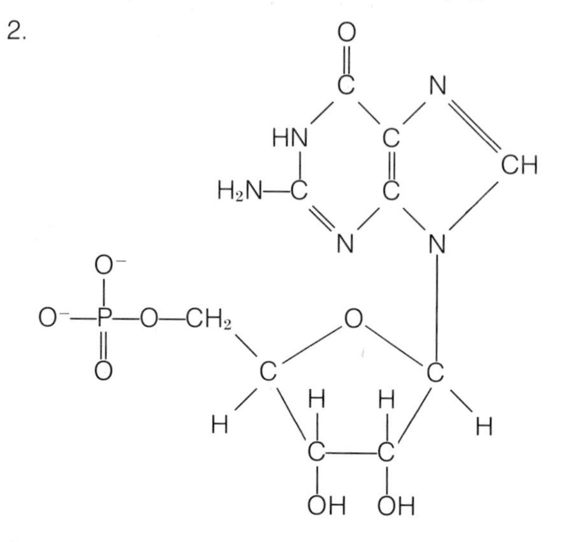

3. Adenosine diphosphate

Exercise 24-2

1. ATGACTTTTGGCTAA
2. ATGACTTTTGGCTAA
   TACTGAAAACCGATT

Exercise 24-3

AUGACUUUUGGCUAA

Exercise 24-4

1. (a) A gene is a sequence of bases on a DNA molecule that will translate into the specific sequence of amino acids making up a protein.
   (b) The genetic code is the general term used to describe the sequence of bases found in the gene.
   (c) The three-base sequence on *m*RNA that is necessary to code for an amino acid is called a codon.
   (d) An anticodon is the three-base sequence on a *t*RNA molecule that is complementary to the codon on a *m*RNA molecule.
   (e) Transcription is the passage of genetic information from DNA to *m*RNA.
   (f) Translation is the formation of a protein from the genetic information in the *m*RNA.
   (g) Polysomes are groups of ribosomes producing proteins from the same strand of *m*RNA.
2. Methionine–Threonine–Phenylalanine–Glycine

GLOSSARY

Absolute zero, 0 K or −273.15°C: The temperature at which all motion within a substance stops.

Acetal linkage, the linkage formed by a condensation reaction between a hemiacetal and an alcohol.

Accuracy, the closeness of a measured value to the true value.

Acid (Brønsted-Lowry definition), a substance that donates hydrogen ions (protons); a proton donor.

Acidosis, a condition that occurs when the blood pH falls below 7.3.

Activated complex, an unstable combination of particles that is an intermediate state between reactants and products in a chemical reaction.

Activation energy, the minimum amount of energy with which two particles must collide in order for a chemical reaction to occur.

Activator, a metal ion cofactor for an enzyme.

Active site, the area on an enzyme molecule to which the substrate attaches.

Addition reaction, in organic chemistry, a reaction in which a reagent reacts with a double or triple bond, allowing other substances to be added to the molecule.

Adenosine diphosphate (ADP), a high-energy diphosphate ester that is produced by the hydrolysis of ATP in the cell.

Adenosine monophosphate (AMP), a low-energy monophosphate ester produced by the hydrolysis of ATP and ADP. This nucleotide is also used in the formation of the nucleic acids, DNA and RNA.

Adenosine triphosphate (ATP), a very high-energy triphosphate ester that provides the energy required for the reactions of metabolism.

Adequate protein, a protein that contains all the essential amino acids required by humans.

Adipose tissue, tissue that contains fat storage cells.

ADP (see adenosine diphosphate).

Aerobic, requiring oxygen.

Alcohol, an organic compound containing a hydroxyl group (—OH) bonded to a carbon atom having four single bonds.

Aldehyde, an organic compound containing a terminal carbonyl group

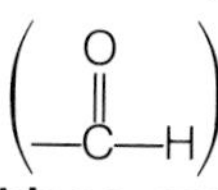

Aldose, general term for a monosaccharide containing an aldehyde group.

Alkali metals, the elements in group I on the periodic table.

Alkaline, basic (not acidic); pH > 7.

Alkaline earth metals, the elements in group II on the periodic table.

Alkaloids, a large class of complex nitrogen-containing compounds, many of which are produced by plants as part of their defense system and which often have strong physiological effects on humans.

Alkalosis, a condition that occurs when the blood pH rises above 7.5.

Alkane, any hydrocarbon containing only single bonds.

Alkene, any hydrocarbon containing one or more carbon-to-carbon double bonds.

Alkyl group, a substituted group that is derived from an alkane.

Alkyne, any hydrocarbon that contains one or more carbon-to-carbon triple bonds.

Allosteric enzyme, a regulatory enzyme that controls the rate of a series of reactions.

Allosteric site, the site on the allosteric enzyme (other than the active site) to which the regulatory molecule attaches to inhibit or increase the activity of the enzyme.

Alpha helix (α helix), a secondary configuration of proteins in which the polypeptide chain is held in a twisted coil by hydrogen bonds.

Alpha radiation, ionizing radiation consisting of streams of high-energy helium nuclei (symbol: ^4_2He).

Amide, an organic compound that contains the amide group

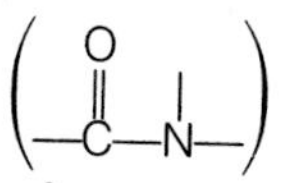

Amine, an organic compound containing a nitrogen attached to one, two, or three carbons.

Amino acid, a monomer unit of a protein, containing both an amino group and a carboxylic acid group.

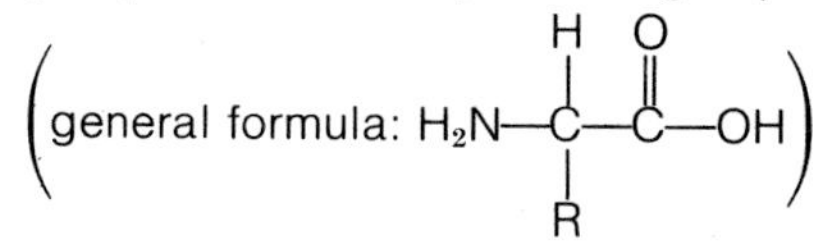

Amino group, the group of atoms —NH_2.

Amorphous solid, a noncrystalline solid whose particles are randomly arranged.

AMP (see adenosine monophosphate).

Amphoteric, having both acidic and basic properties.

Amylopectin, the highly branched polymer of α-glucose found in starch.

Amylose, the linear polymer of α-glucose found in starch.

Anabolic reactions, reactions in which the cell uses energy to produce molecules needed for growth and repair of the cell.

Anaerobic, not requiring oxygen.

Analgesic, a drug that reduces pain without causing the loss of consciousness.

Anemia, a group of diseases that result in a low number of red blood cells. Examples are hemolytic anemia, iron deficiency anemia, and pernicious anemia.

Aneurysm, a ballooning of an artery caused by atherosclerosis.

Angina pectoris, a sharp, intense pain in the chest caused by reduced blood flow to the heart.

Anion, a negatively charged ion.

Antibiotic, a drug that is extracted from microorganisms and that acts as an antimetabolite.

Anticodon, the three-base sequence on *t*RNA that is complementary to the codon on *m*RNA.

Antimetabolite, any compound that inhibits enzyme activity.

Antioxidant, a substance added to food to prevent the oxidation of unsaturated fats or oils present in the food.

Antipyretic, a drug that reduces or prevents fever.

Apoenzyme, the protein portion of an enzyme molecule.

Aqueous solution, a solution containing water as the solvent.

Aromatic hydrocarbons, the class of hydrocarbons containing benzene and its derivatives.

Arteriosclerosis, the disease of the arteries commonly known as hardening of the arteries.

Atherosclerosis, the most common form of arteriosclerosis, in which the inner layer of the arterial wall becomes thickened by lipid deposits.

Atmosphere (atm), the unit of pressure equal to the force per unit area that will support a column of mercury 760 mm high.

Atom, the smallest unit of an element having the properties of that element.

Atomic mass unit (amu), an arbitrary unit of measure established to allow comparison of the relative masses of the elements.

Atomic number, for each element, the number of protons in the nucleus of any atom of that element.

Atomic weight, the weighted average of the masses of the naturally occurring isotopes of an element, expressed in atomic mass units.

ATP (see adenosine triphosphate).

Avogadro's number, the number of particles in one mole: equals 6.02×10^{23}.

Background radiation, ionizing radiation that comes from natural sources.

Basal metabolism rate (BMR), the minimum amount of energy required daily by the body to maintain the basic processes of life.

Base (Brønsted-Lowry definition), a substance that accepts hydrogen ions (protons): a proton acceptor.

Becquerel (Bq), the SI unit that describes the activity of a radioactive source (1 Bq = 1 disintegration/sec).

Bends, a painful disorder caused by the formation of nitrogen bubbles in the tissues of deep-sea divers who are brought to the ocean's surface too quickly.

Benedict's test, the test widely used for the detection of a reducing sugar in the urine.

Benign, not malignant.

Beriberi, deficiency disease resulting from a lack of thiamine, vitamin B_1.

Beta configuration (β configuration) (see pleated sheet).

Beta oxidation (see fatty acid cycle).

Beta radiation, ionizing radiation consisting of streams of high-energy electrons (symbol: ${}_{-1}^{0}e$).

Bile, a fluid (containing bile salts, bile pigments, and cholesterol) that is produced in the liver, stored in the gall bladder, and released in the small intestine.

Binary compound, a compound containing only two elements.

Blood sugar, term often used for glucose.

Blood sugar level, the concentration of glucose in the blood (usually expressed in mg per 100 ml of blood).

Boiling point, the temperature at which a substance boils (i.e., changes state from liquid to gas). The normal boiling point of a substance is the temperature at which a substance boils when atmospheric pressure equals 760 mm Hg.

Bond axis, the imaginary line connecting two nuclei that are held together by a chemical bond.

Boyle's law, this law states that the volume of a gas is inversely proportional to the pressure when the temperature remains constant.

Brachytherapy, a procedure in which "seeds" containing a radionuclide are inserted by means of a needle into the area of tissue that requires treatment.

Breeder reactor, a type of nuclear reactor in which more nuclear fuel is produced than is used up in the process of producing energy.

Brownian movement, the erratic, random movement of particles in a colloid.

Buffer, any substance that, when added to a solution, protects against sudden changes in the pH of that solution.

Burner reactor, a nuclear reactor in which the nuclear fuel is used up as energy is produced.

Calcitonin, a hormone produced by the thyroid gland which lowers the level of calcium in the blood.

Calorie (cal), the amount of energy necessary to raise the temperature of one gram of water exactly one degree Celsius.

Calorimeter, the instrument used to determine the caloric content of a substance.

Carbohydrate, any of a class of compounds containing polyhydric aldehydes, polyhydric ketones, or their polymers, whose function is energy storage or structural support in living organisms.

Carbonyl group, the group containing a carbon double-bonded to an oxygen

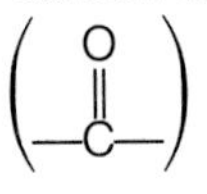

Carboxylic acid, an organic acid containing the carboxyl group

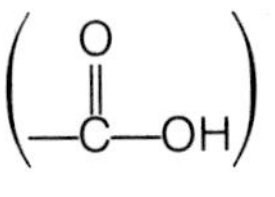

Carcinogen, any chemical that can cause cancer in animals.

Catabolic reaction, a cellular reaction in which large molecules are broken down to produce smaller molecules and cellular energy.

Catalyst, a substance that increases the rate of a chemical reaction without being consumed in the reaction.

Cation, a positively charged ion.

Cellular respiration, the series of reactions by which glucose is oxidized to form CO_2, H_2O, and ATP.

Cellulose, a linear polymer of β-glucose that is the main structural molecule in plants.

Celsius (°C), the temperature scale in the metric system with 100 degrees between the freezing point of water (set at 0°C) and the boiling point of water (100°C).

Chain reaction, a self-sustaining process in which the product of one reaction causes one or more additional reactions to occur.

Charles' law, this law states that the volume of a gas is inversely proportional to the temperature (in degrees Kelvin) if the pressure remains constant.

Chemical change, a transformation that involves a change in the basic chemical composition of the reactants.

Chemical energy, potential energy stored in substances, which results from the spatial arrangement and forces between atoms in the substance.

Chemical formula, a shorthand way of representing the composition of a substance by using the chemical symbols of the elements and subscripts to indicate the number of atoms of each element in a reacting unit of the substance.

Chemical symbol, a one- or two-letter abbreviation representing one atom of an element.

Chemotherapy, the treatment of cancer and other diseases using drugs such as antimetabolites.

Chiral carbon, a carbon that has four different groups attached to it and that, therefore, is asymmetric.

Chlorophyll, the green pigment found in plant cells that is involved in the light reactions of photosynthesis.

Chloroplast, a site of the light reactions of photosynthesis in a plant cell.

Cholesterol, a sterol that is produced by all cells and is found in cell membranes. It is used to synthesize bile salts and some hormones.

Cirrhosis, a chronic disease of the liver caused by nutritional deficiency, poisons, or previous infections.

***Cis* isomer,** the geometric isomer with the specified atoms or groups of atoms on the same side of the double bond or the ring.

Citric acid cycle, a cyclic series of reactions (occurring in the mitochondria of cells) that convert acetyl CoA into two molecules of CO_2, one ATP, and eight hydrogens that enter into the respiratory chain.

Codon, the three-base sequence on *m*RNA that codes for a specific amino acid.

Coenzyme, a cofactor that is a complex organic molecule other than a protein.

Cofactor, a metal ion or organic molecule required for an enzyme to function properly.

Colligative properties, those properties of a solution that depend only on the number of solute particles in the solution.

Colloid, a homogeneous mixture containing relatively large particles (from 1 to 1000 nm) that do not settle out.

Competitive inhibition, inhibition of an enzyme that occurs when another molecule competes with the substrate for the active site.

Complex carbohydrate, a polysaccharide.

Compound, a substance formed in a chemical change and composed of two or more elements that combine in a definite proportion by weight.

Compound lipid, a saponifiable lipid

that when hydrolyzed yields fatty acids, an alcohol, and some other compound.

Concentration, a numerical measure of the relative amount of solute in a solution.

Condensation, the conversion of a substance in the gaseous state to the liquid state.

Condensation reaction, in organic chemistry, a reaction in which a water molecule is removed from two reactant molecules, thereby forming one product molecule.

Conjugate acid, the substance formed when a base accepts a hydrogen ion.

Conjugate base, the substance formed when an acid donates a hydrogen ion.

Conjugated double bonds, an arrangement of double and single bonds alternating between carbon atoms in a hydrocarbon molecule.

Conjugated protein, a protein that produces amino acids and other organic or inorganic groups upon hydrolysis.

Conversion factor, a ratio or fraction that is formed from an equality and is equal to the number "one."

Coordinate covalent bond, a covalent bond in which one atom donates both of the electrons shared in the bond.

Covalent bond, the type of bond formed when electrons are shared between two atomic nuclei.

Covalent compound, a compound consisting of single, electrically neutral units called molecules.

Crenation, the shrinking of red blood cells when they are placed in a hypertonic solution.

Critical mass, the minimum amount of a fissionable isotope that must be present for a nuclear chain reaction to occur.

Crystal lattice, the repeating, three-dimensional arrangement of particles in a crystalline solid.

Crystalline solid, a solid whose particles are arranged in a regular repeating pattern.

Crystalloid, a substance containing particles that are small in size (less than 1 millimicron) and that form a true solution when placed in water.

Curie, the unit of measure that describes the activity of a radioactive source (1 curie $= 3.7 \times 10^{10}$ disintegrations/sec).

Cyclic AMP, a chemical messenger within cells, whose formation is triggered when a hormone attaches to a receptor site on the cell membrane.

Cyclic compound, a compound containing a ring structure.

Dalton's law, this law states that the total pressure of a mixture of gases is equal to the sum of the partial pressures of each gas in the mixture.

Decay series, a series of radioactive decays or disintegrations by which an unstable nucleus becomes a stable nucleus.

Deficiency disease, a disease resulting from too little of an essential nutrient in the diet.

Dehydration, (*a*) a medical condition resulting from excessive water loss. (*b*) In organic chemistry, a reaction involving the removal of a molecule of water from another molecule.

Dehydrogenation, in organic chemistry, a reaction in which two hydrogen atoms are removed from a molecule.

Denaturation, a reversible or irreversible disruption of the normal arrangement of atoms (the native state) of a protein.

Density, the mass of one unit of volume of a substance (commonly expressed in grams per cubic centimeter).

Deoxyribonucleic acid (DNA), the nucleic acid that is a polymer of deoxyribonucleotides, and whose base sequence carries genetic information.

Dextran, a glucose polymer that is produced by bacteria.

Dextrin, a short-chain polymer of glucose produced by the partial hydrolysis of starch.

Diabetes insipidus, a disease in which the body no longer produces the hormone vasopressin that controls the amount of water reabsorbed by the kidneys, resulting in the excretion of large quantities of urine.

Diabetes mellitus, a disease resulting from low levels or a total lack of the hormone insulin; commonly known as "diabetes."

Dialysis, the movement of ions and small molecules (but not colloidal particles) through a membrane.

Diatomic, containing two atoms.

Differentially permeable membrane, a membrane that allows water, but not solute particles, to pass through.

Diffusion, the spontaneous mixing of particles from regions of higher to regions of lower concentration.

Digestion, the process by which food is broken down into simple molecules that can be absorbed through the lining of the intestinal tract.

Dipeptide, a molecule that can be hydrolyzed to form two amino acids.

Dipolar ion, an ion, such as an amino acid, that has a positive and a negative region.

Disaccharide, a compound composed of two monosaccharides.

Disintegration series (see decay series).

Distillation, a procedure that separates the compounds in a mixture on the basis of their boiling points.

Disulfide, any organic compound containing the group (—S—S—).

DNA (see deoxyribonucleic acid).

Double bond, a covalent bond in which two pairs of electrons (four electrons) are shared by two atomic nuclei.

Double helix, the structure of a DNA molecule—two helical polynucleotide chains coiled around the same axis.

Edema, a swelling of the tissues caused by an increase in the amount of water in the extracellular fluid.

Effusion, the movement of a gas through a tiny hole to a region of lower concentration.

Electrolyte, any substance that, in water solution, conducts electricity.

Electromagnetic spectrum, the entire range of radiant energy, of which visible light is only a small part.

Electron, a subatomic particle that is found in certain regions (called orbitals) around the nucleus. It has a mass of 1/1837 amu and a negative (−1) charge.

Electron affinity, the amount of energy released when an electron is added to a neutral atom.

Electron configuration, the most stable arrangement of electrons in probability regions (orbitals) around the nucleus of an atom.

Electronegativity, a measure of the ability of an atom to attract toward itself the electrons that it shares in a covalent bond.

Electron transport chain (see respiratory chain).

Element, a pure substance that cannot be broken down by ordinary chemical processes.

Emphysema, a disease in which the lung tissue is so badly damaged that adequate levels of oxygen cannot be maintained in the blood, resulting in very labored breathing.

Endocrine gland, any of a specialized group of glands in the body that produce hormones and regulate their secretion into the blood.

Endothermic reaction, a reaction in which energy in the form of heat is required to keep the reaction occurring.

Energy, the capacity to do work.

Energy level, an energy region around the nucleus occupied by electrons.

Entropy, a measure of the randomness or disorder in a system.

Enzyme, a protein molecule that functions as a biological catalyst.

Epinephrine, an adrenal hormone more commonly known as adrenalin.

Equilibrium, a dynamic state in which the rate of the forward reaction equals the rate of the reverse reaction.

Equivalence point, the pH at which all

the hydrogen (or hydroxide) ions in a solution have been neutralized.

Equivalent (Eq), (*a*) the amount of a substance containing one mole of charge (either + or −). (*b*) The amount of acid that will donate one mole of hydrogen ions. (*c*) The amount of base that will neutralize one mole of hydrogen ions.

Erythrocyte, a red blood cell.

Essential, a term referring to ten amino acids and three fatty acids that must be present in the diet for normal growth and development.

Ester, any organic compound formed in a condensation reaction between an alcohol and an organic acid, and containing the functional group

$$\left(-\overset{\overset{\displaystyle O}{\|}}{C}-O-\overset{|}{\underset{|}{C}}-\right)$$

Esterification, the condensation reaction between an alcohol and a carboxylic acid, producing an ester.

Ether, any organic compound containing an oxygen bonded to two carbons

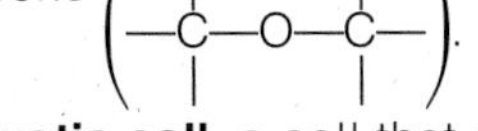

Eukaryotic cell, a cell that contains a nucleus.

Evaporation, the conversion of a substance in the liquid state to the gaseous state.

Exact number, a number that is given in a definition or that results from counting objects and, therefore, has an infinite number of significant digits.

Excited atom, an atom having one or more electrons in an energy level higher than normal.

Exons, sequences of bases on the DNA of eukaryotic cells that code for amino acids.

Exothermic reaction, a reaction in which energy in the form of heat is produced.

Exponential form, the expression of any number as a number between 1 and 10 times a power of 10.

Extracellular fluids, any fluids found in the body tissues, but not contained inside the cells.

Fahrenheit (°F), temperature scale in the English system with 180 degrees between the freezing point of water (set at 32°F) and the boiling point of water (212°F).

Familial hypercholesterolemia (FH), a genetic disease that causes blood cholesterol levels to be two to ten times higher than normal.

Family (see group).

Fat, a triacylglyerol that is a solid at room temperature and that contains mainly saturated fatty acids.

Fatty acid, an organic compound containing one carboxylic acid group and usually having a long carbon chain.

Fatty acid cycle, a repeating series of reactions in which a fatty acid is oxidized in two-carbon units.

Feedback control, regulation of multienzyme systems wherein the end product of the system inhibits the regulatory enzyme.

Fermentation, the process by which yeast cells convert glucose to ethanol, carbon dioxide, and energy.

Fibrosis, the abnormal formation of fibrous tissue.

First law of thermodynamics, this law states that energy can neither be created nor destroyed, but only changed in form.

Fissile isotope, an isotope that is easily fissioned.

Fission, the process by which a large unstable nucleus, when bombarded by neutrons, breaks apart to form two smaller nuclei, several neutrons, and a tremendous amount of energy.

Fluorescence, a type of luminescence in which a substance stops giving off light as soon as the external source of energy is removed.

Fluorosis, an enlargement of bones and abnormal bone growth caused by a high concentration of fluoride ions in the diet.

Force, a push or pull on an object that causes the object to start moving or to change its speed or direction once it is moving.

Forensic scientist, a scientist who uses

medical and scientific information and technology to help solve criminal cases.

Formula (see chemical formula).

Formula weight, the sum of the atomic weights of all the atoms appearing in the chemical formula of a substance.

Fractional distillation, a process by which compounds in a mixture are separated by means of their boiling points.

Free radical, a highly reactive uncharged particle.

Functional group, a group of atoms that gives characteristic chemical properties to all molecules containing that group.

Fusion, the process by which several small nuclei combine to form a larger, more stable nucleus and tremendous amounts of energy.

Gamma radiation, a naturally occurring high-energy electromagnetic radiation, similar to X rays, with high penetrating power.

Gene, a sequence of bases on a DNA molecule that codes for a specific protein.

Genetic code, the sequence of bases on a DNA molecule.

Genetic disease, a disease that results from a defective gene that is inherited from one or both parents.

Geometric isomers, compounds that have the same molecular formula, but that have a different geometric arrangement of atoms around a double bond or around a ring within their structure.

Glucose tolerance test, a series of tests for blood sugar level taken after the ingestion of a high dose of glucose; often used to diagnose diabetes mellitus.

Glycogen, the highly branched polymer of α-glucose that is the glucose storage molecule in animals.

Glycogenesis, the series of reactions by which glucose molecules are joined together to form glycogen.

Glycogenolysis, the series of reactions by which glycogen molecules are hydrolyzed or broken down to form glucose molecules.

Glycolipid, a compound lipid that contains fatty acids, an alcohol, and a sugar group that is either galactose or glucose.

Glycolysis, the series of reactions by which glucose is converted to two molecules of lactic acid and two molecules of ATP; occurs in the cellular cytoplasm.

Glycosidic linkage, an acetal linkage formed between two monosaccharides.

Glycosuria, a medical condition in which sugar molecules are found in the urine.

Goiter, a swelling of the thyroid gland caused by a deficiency of iodine in the diet.

Graham's law, this law states that lighter gases will effuse more rapidly than heavier gases.

Gram, the metric unit of mass; 1 gram = 0.035 oz.

Gram-equivalent weight, the weight of a substance, in grams, that contains one equivalent.

Gray (Gy), a SI unit that describes the amount of energy absorbed by irradiated tissue.

Ground state, the term applied to an atom having all of its electrons in the lowest possible energy levels.

Group, a vertical column of elements on the periodic table.

Half-life, the length of time required for one-half of the atoms of a radionuclide in a given sample to undergo radioactive decay.

Halogen, any of the elements in group VII on the periodic table.

Heat of fusion, the amount of energy (in calories) required to change one gram of a substance (at the melting point) from a solid to a liquid.

Heat of reaction (ΔH), the amount of heat energy (in kcal/mole) absorbed or released in a chemical reaction.

Heat of vaporization, the amount of

energy (in calories) required to change one gram of a substance (at the boiling point) from a liquid to a gas.

Hemiacetal, a compound formed in the reaction between an aldehyde and an alcohol.

Hemiketal, a compound formed in the reaction between a ketone and an alcohol.

Hemodialysis, the process of removing wastes from the blood by dialysis.

Hemoglobin, the oxygen-carrying molecule in red blood cells. It consists of four polypeptide chains and four nonprotein heme groups, each of which can carry one molecule of oxygen.

Hemolysis, the rupturing of red blood cells that results when the cells are placed in a hypotonic solution (or from other causes).

Henry's law, this law states that the higher the pressure, the greater the solubility of a gas in a liquid.

Heterocyclic compound, a compound having a ring structure that contains two or more different types of atoms making up the ring.

Heterogeneous, nonuniform.

Heterogeneous nuclear RNA (*hn*RNA), RNA that is synthesized on the DNA of eukaryotic cells and that contains many more bases than are found on *m*RNA.

High-energy phosphate bond, the phosphorus-to-oxygen bond, found in molecules such as ATP and UTP, which, upon hydrolysis, releases large amounts of energy.

Homogeneous, uniform throughout.

Hormone, any of the chemical messengers produced by the endocrine glands.

Hybrid orbitals, orbitals formed by a combination of *s, p,* or *d* atomic orbitals.

Hydrated, surrounded by water molecules.

Hydration reaction, in organic chemistry, the addition of a water molecule to an unsaturated bond.

Hydrocarbon, an organic compound containing only atoms of carbon and hydrogen.

Hydrogenation, in organic chemistry, the addition of a hydrogen molecule to an unsaturated bond.

Hydrogen bond, a weak force of attraction between a partially positive hydrogen and a partially negative atom such as oxygen, fluorine, or nitrogen on another molecule or on another region of the same molecule.

Hydrogen ion, a proton. This term is also often used in acid-base chemistry to mean the hydronium ion (H_3O^+).

Hydrohalogenation, the addition of a hydrohalogen (such as HCl or HBr) to an unsaturated bond.

Hydrolysis, the addition of a water molecule to a reactant, thereby breaking the reactant into two product molecules.

Hydrometer, an instrument used to measure the specific gravity of a liquid.

Hydronium ion, H_3O^+, the ion formed when a hydrogen ion (a proton) joins to a water molecule.

Hydrophilic, water-attracting; term given to substances or groups of atoms that are generally very polar or ionic.

Hydrophobic, water-repelling; term given to substances or groups of atoms that are nonpolar.

Hyper, prefix used to indicate "higher than normal." For example, hypertension, hypercholesterolemia, hypertonic, etc.

Hyperglycemia, a condition resulting from higher than normal blood glucose levels.

Hyperthyroid, a condition in which the thyroid gland produces higher than normal amounts of the hormone thyroxine.

Hypertension, high blood pressure.

Hypertonic solution, a solution with a higher solute concentration than the standard solution.

Hypo, prefix used to indicate "lower than normal." For example, hypoglycemia, hypotonic, hypothyroid, etc.

Hypoglycemia, a condition resulting from lower than normal blood glucose levels.

Hypothermia, a lowering of the body's interior temperature.

Hypotonic solution, a solution with a lower solute concentration than the standard solution.

Imine, an organic compound containing a carbon doubly bonded to a nitrogen

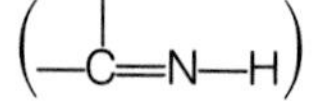

Indicator, a chemical dye that changes color at a specific hydrogen ion concentration.

Induced-fit theory, theory of enzyme action stating that the active site of some enzymes is induced by the substrate to fit the shape of the substrate molecule.

Inhibition, the prevention of, or interference with, the action of an enzyme, thus lowering its activity. May be reversible or irreversible.

Inner transition elements, the metals having atomic numbers 58 to 71 and 90 to 103, found in the two long rows at the bottom of the periodic table.

Insulin, the hormone, produced by beta cells in the pancreas, that controls blood glucose levels by increasing the absorption of glucose from the blood and the rate of glycogenesis.

Insulin shock, convulsion and coma resulting from an overproduction or overinjection of insulin, which causes the blood glucose level to decrease very fast.

Intracellular fluid, fluid found within cells.

Intravenous, administered by means of a vein.

Introns, intervening sequences of bases, found on the DNA of eukaryotic cells, that do not code for amino acids.

Inverse square law, this law states that the intensity of radiation on a given surface area decreases by the square of the distance from the source.

Iodine number, indicates the amount of unsaturation in a compound: the higher the iodine number, the more unsaturated the compound.

Ion, a positively or negatively charged particle.

Ionic bond, the attraction between ions formed when one or more electrons are transferred from one atom to another.

Ionic compound, a compound consisting of an orderly arrangement of oppositely charged ions, which are combined in a ratio such that the compound is electrically neutral.

Ionization energy, the amount of energy that must be added to an atom to remove one electron from its outermost energy level.

Ionizing radiation, radiation, such as alpha, beta, or gamma radiation, that can produce unstable and highly reactive ions in living tissue.

Ion product constant of water (K_w), $K_w = [H^+][OH^-] = 1 \times 10^{-14}$.

Isoelectric point, the pH at which an amino acid or protein is electrically neutral and will not migrate in an electric field.

Isoenzymes, enzymes that perform the same function in living organisms, but that have different sequences of amino acids in certain portions of their polypeptide chains.

Isomers, compounds having the same molecular formula, but different structures.

Isotonic solution, a solution with a solute concentration equal to the standard solution.

Isotopes, atoms of the same element that differ in the number of neutrons in their nuclei.

Jaundice, a condition caused by a high level of bilirubin in the blood, which results from a blockage of the bile duct or malfunction of the liver.

Joule (J), a unit of energy in the SI system (4.184 J = 1 cal).

Kelvin (K), the temperature scale in the SI system, with 100 degrees between the freezing point of water (273.15 K) and the boiling point of water (373.15 K).

Keratosis, skin condition characterized by thick warty growths.

Ketone, an organic compound containing a carbonyl group bonded to two other carbons

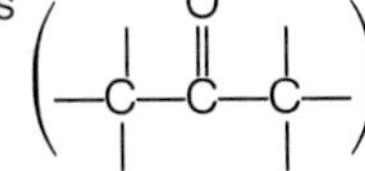

Ketone bodies, compounds produced from the metabolism of excess acetyl CoA (which, in turn, is produced when large amounts of fats are oxidized to supply cellular energy).

Ketose, the general term for a monosaccharide containing a ketone group.

Ketosis, a condition caused by higher than normal levels of ketone bodies in the blood.

Kilocalorie (Kcal), the amount of heat energy necessary to raise the temperature of 1000 grams of water one degree Celsius.

Kinetic energy, energy of motion.

Kinetic-molecular theory, the theory that explains the behavior of a gas in terms of the motion of its particles.

Kwashiorkor, a disease that results from a lack of essential amino acids in the diet.

Law of conservation of energy (see first law of thermodynamics).

Law of conservation of mass, this law states that in a chemical reaction the mass of the products equals the mass of the reactants.

Law of definite proportions, this law states that a compound is composed of specific elements in a definite proportion by weight.

$LD_{50/30}$, abbreviation for the dose of radiation sufficient to kill 50 percent of the exposed population within 30 days.

Le Chatelier's principle, this principle states that a system at equilibrium will resist changes in its temperature, pressure, or concentration of reactants or products.

LET (linear energy transfer), the amount of energy transferred to a tissue, per unit of path length traveled by ionizing radiation.

Lipid, any of a large class of nonpolar, organic compounds that have oily or waxy properties.

Lipogenesis, the synthesis of lipids in cells.

Liter (litre), a unit of volume equal to 1000 cubic centimeters; 1 liter = 1.06 quarts.

Lock-and-key theory, the theory of enzyme action stating that only a specific substrate will fit the active site of an enzyme, just as only a specific key will fit in a lock.

Low-density lipoprotein (LDL), the primary cholesterol transport molecule in the blood.

Luminescence, the release of energy as visible light by an excited atom.

Macromineral, one of seven elements (K, Mg, Na, Ca, P, S, Cl) that are required in small amounts for normal cell growth and development.

Malignant, term describing cells that are growing and dividing in an uncontrolled fashion.

Mass, a measure of the resistance of an object to a change in speed or direction.

Mass number, the sum of the number of protons and neutrons in the nucleus of an atom.

Matter, anything that has mass and occupies space.

Melanin, a brown skin pigment produced by the body to protect against the effects of ultraviolet radiation.

Melting point, the temperature at which a solid breaks down to form a liquid.

Messenger RNA (*m*RNA), the RNA that carries the genetic code for a specific protein. *m*RNA is synthesized in the nucleus and then migrates to the cytoplasm where it attaches to ribosomes and serves as the template for protein synthesis.

Metabolism, all of the enzyme-catalyzed reactions in the body.

Metal, an element that is shiny, dense, and easily worked, that has a high melting point, and that conducts electricity.

Metalloid, an element that acts like a metal in some ways and like a nonmetal in other ways. These elements are found between the metals and the nonmetals on the periodic table.

Metastable, in an energy state higher than normal.

Metastasis, the spread of cancer cells from a tumor to other parts of the body.

Meter (metre), unit of length in the metric system; 1 meter = 1.09 yards.

Metric system, a system of measure based on the decimal system.

Milliequivalent (mEq), unit of measure used to express the concentration of ions in the blood; 1000 mEq = 1 Eq.

Millimeters of mercury (mm Hg), unit of pressure equal to 1/760 atmosphere.

Mitochondria, structures in the cell where the citric acid cycle and the respiratory chain occur.

Mixture, two or more substances combined in any proportion.

Molarity (*M*), unit of solution concentration defined as the number of moles of solute per liter of solution.

Mole (mol), the amount of a substance that has the same number of particles as there are atoms in 12 grams of carbon-12 (6.02×10^{23} atoms).

Molecule, an electrically neutral unit formed when two or more atoms are joined together by covalent bonds.

Monomer, a single unit that joins with many other identical units to form a polymer.

Monosaccharide, a carbohydrate that cannot be broken into smaller units by hydrolysis.

Multienzyme system, a sequence of enzyme catalyzed reactions that produces a specific metabolic result.

Mutagen, any chemical or physical agent that is capable of producing a mutation.

Mutant, containing altered DNA.

Mutation, a change in the sequence of the four bases that are the informational code on the DNA molecule.

Myelin sheath, the protective coating surrounding nerves.

Myocardial infarction, a heart attack.

Native state (native configuration), the shape of a protein that is energetically the most stable.

Net-ionic equation, a chemical equation showing only the ions that take part in the reaction.

Neutral, term applied to a solution that has neither acidic nor basic properties, or to a particle that has no net electrical charge.

Neutralization, the process by which an acidic or basic solution is converted to a neutral solution.

Neutron, a subatomic nuclear particle with a mass of 1 amu and no charge.

Nitrile, an organic compound containing a carbon triply bonded to a nitrogen and singly bonded to another carbon

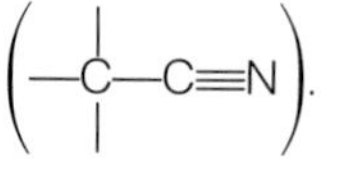

Noble gas, any of the elements in group 0 (helium, neon, argon, krypton, xenon, or radon), all of which have great chemical stability.

Nonelectrolyte, a substance that does not conduct electricity when placed in solution.

Nonmetal, an element that is brittle, has low density and a low melting point, and does not conduct electricity.

Nonpolar, the term applied to covalent bonds and covalent molecules when the centers of positive and negative charge coincide.

Nonsaponifiable lipid, any lipid that cannot be hydrolyzed by an aqueous solution of base.

Normality (*N*), the number of equivalents of acid or base per liter of solution.

Nuclear transmutation, a reaction in which a high-speed nuclear particle collides with a nucleus to produce a different nucleus.

Nucleic acid, a polymer of nucleotides: either DNA (deoxyribonucleic acid) or RNA (ribonucleic acid).

Nucleotide, the monomer unit of nucleic acids, whose structure contains a five-carbon sugar, a nitrogen-containing base, and a phosphate group.

Nucleus, (*a*) the dense center of an atom, containing protons and neutrons. (*b*) A cellular organelle surrounded by a nuclear membrane and containing chromosomes and nucleoli.

Octet rule, the tendency of elements in groups I to VII to form bonds that result in eight valence electrons in the outer energy level of each atom (except in the first energy level, where the tendency is toward two electrons).

Oil, a triacylglycerol, extracted from vegetable seeds or fruits, that is a liquid at room temperature and that contains mainly unsaturated fatty acids.

Optical isomers, compounds that are mirror images and that differ only in the way they interact with polarized light.

Orbital, a region around the nucleus of an atom in which there is a high probability of finding one or two electrons. The four kinds of orbitals are called *s*-, *p*-, *d*-, and *f*-orbitals.

Organic chemistry, the study of carbon compounds.

Osmol, 1 mole of any combination of particles.

Osmolarity, a unit of concentration that describes the total number of particles in solution; expressed in osmol/liter.

Osmosis, the movement of water molecules through a differentially permeable membrane from a region of lower solute concentration to a region of higher solute concentration.

Osmotic pressure, the amount of pressure that would have to be applied to a solution to prevent osmosis if the solution were separated from pure water by a differentially permeable membrane.

Osteoporosis, a progressive thinning and breakdown of bone tissue that happens primarily to women after menopause.

Oxidation, the loss of one or more electrons by an atom, ion, or molecule. In organic chemistry, the loss of hydrogen or the gain of oxygen by an organic molecule or ion.

Oxidation number, the charge an atom would have if all the electrons in a chemical bond were transferred to the more electronegative atom.

Oxidative deamination, the removal of an amino group from an amino acid, producing an α-keto acid and ammonia.

Oxidizing agent, a substance that causes the oxidation of a reactant molecule.

Pascal (Pa), the SI unit of pressure (133.3 Pa = 1 torr).

Partial pressure, the pressure exerted by a specified gas in a mixture of gases.

Parts per billion (ppb), unit of concentration: the number of micrograms of solute per liter of solution.

Parts per million (ppm), unit of concentration: the number of milligrams of solute per liter of solution.

Pellagra, a deficiency disease resulting from a lack of the vitamin niacin.

Peptide bond, an amide linkage formed by a condensation reaction between two amino acids.

Percent concentration

Weight/weight (w/w) percent: the number of grams of solute per 100 grams of solution.

Volume/volume (v/v) percent: the volume of solute per 100 volumes of solvent.

Weight/volume (w/v) percent: the number of grams of solute per 100 milliliters of solution.

Milligram percent (mg%): the number of milligrams of solute per 100 milliliters of solution.

Period, a horizontal row on the periodic table.

Periodicity, the repeating nature of chemical properties of the elements when they are arranged in order of atomic number.

Periodic law, this law states that many properties of the elements repeat periodically as the atomic number of the elements increases.

Periodic table, an arrangement of the elements in order of increasing atomic number that illustrates chemical similarities between groups of elements.

pH, a measure of the hydrogen ion concentration of an aqueous solution; $[H^+] = 1 \times 10^{-pH}$

Phenylalanine, an amino acid that accumulates in the body of a child with PKU.

Phenylketonuria, PKU, an inherited disease in which an enzyme responsible for the conversion of phenylalanine to tyrosine is defective.

Phosphoglyceride, a phospholipid containing the alcohol glycerol.

Phospholipid, a compound lipid whose structure contains an alcohol, fatty acids, and a phosphate group.

Phosphorescence, a type of luminescence in which the substance continues to give off light for a short period of time after the external source of energy is removed.

Photosynthesis, the process by which green plants use sunlight as the source of energy to produce glucose and oxygen from water and carbon dioxide.

Physical change, a transformation during which a substance changes form, but keeps its chemical identity.

Pi bond (π bond), a bond formed by the sideways overlap (above and below the bond axis) of two *p* orbitals.

Plaque, (*a*) in dentistry, the sticky substance produced by bacteria in the mouth that adheres to the teeth. (*b*) In cardiology, a deposit of smooth muscle cells, fats, and scar tissue on the interior of an arterial wall.

Pleated sheet, a secondary structure of a protein, in which polypeptide chains lie next to one another, held together by hydrogen bonds and having the R-groups extending above and below the sheet.

Polar, term applied to covalent bonds and covalent molecules when the center of positive charge and the center of negative charge do not coincide, thus forming an electric dipole.

Polyatomic ion, an electrically charged group of covalently bonded atoms that stays together as a unit in most chemical reactions.

Polycythemia, the excessive formation of red blood cells; called *polycythemia vera* when the increase is caused by a tumor.

Polyester, a polymer of ester molecules, used to make fibers for fabrics.

Polyhydric, containing more than one hydroxyl group.

Polymer, a very large molecule made up of repeating units called monomers.

Polymerization, a chemical reaction in which single molecules called monomers react with each other to form large molecules called polymers.

Polypeptide, a polymer composed of amino acids connected by peptide bonds.

Polyprotic acid, an acid that can donate more than one hydrogen ion.

Polysaccharide, a polymer of three or more monosaccharide molecules.

Polysome, a group of ribosomes all synthesizing protein on the same molecule of *m*RNA.

Polyunsaturated, the term describing a triacylglycerol whose molecules have two or more double bonds.

Potential energy, energy of position.

Precipitate, a solid that forms in a solution as a result of a chemical reaction.

Precision, the degree to which measurements are reproducible.

Pressure, a force exerted per unit of area.

Primary structure, the sequence of amino acids (connected by peptide bonds) in the polypeptide chain of a protein.

Product, a substance that results from a chemical reaction.

Proenzyme, the inactive form of an enzyme.

Prostaglandin, any of a class of 20-carbon fatty acids that are derived from prostanoic acid and that have a wide variety of potent physiological effects.

Prosthetic group, a cofactor that is tightly bound to an apoenzyme.

Protein, a polymer of amino acids.

Proton, a subatomic nuclear particle having a mass of 1 amu and a positive (+1) charge.

Ptomaine, any amine produced in the natural decay of living organisms.

Pulmonary, having to do with the lungs.

Purines, heterocyclic amines whose derivatives (adenine and guanine) are essential parts of DNA and RNA molecules.

Pyrimidines, heterocyclic amines whose derivatives (cytosine, thymine, and uracil) are essential parts of DNA and RNA molecules.

Quaternary structure, the overall structure of a protein that contains more than one polypeptide chain.

Rad, unit of radiation dosage used to describe the amount of energy absorbed by the irradiated tissue.

Radioactive decay, the process by which an unstable nucleus gives off nuclear particles and/or gamma radiation to become more stable.

Radioactive tracer, a chemical that contains radioactive atoms, but that has the same chemical nature and behavior as naturally occurring compounds; used to follow metabolic pathways.

Radioactivity, the giving off, or emission, of radiation from certain isotopes.

Radionuclide, a radioactive isotope.

Radiopharmaceutical therapy, the administration of radionuclides in a chemical form designed to be concentrated in certain regions of the body or in cancerous tissue.

Rancid, a term applied to foods containing fats and oils that have undergone hydrolysis or oxidation, forming substances that give the food a bad smell or taste.

Reactant, a starting substance in a chemical reaction.

Recombinant DNA, a technique that produces large amounts of the product of a gene that is inserted into a specially developed strain of *E. coli.*

Recommended dietary allowance (RDA), the amount of a nutrient needed in the diet to meet the daily requirements of a healthy individual (as established by the Food and Nutrition Board of the National Academy of Sciences).

Redox reaction, abbreviation for *reduction–oxidation* reaction, a reaction in which electrons are transferred from one reactant to another.

Reducing agent, a substance that causes the reduction of a reactant molecule.

Reducing sugar, any carbohydrate that can act as a reducing agent and produce a positive Benedict's test.

Reduction, the gaining of electrons by a reagent. In organic chemistry, reduction occurs when an organic molecule or ion gains hydrogen atoms or loses oxygen atoms.

Regulatory enzyme (see allosteric enzyme).

Regulatory site (see allosteric site).

Rem, the unit of absorbed dose of radiation that will produce the same biological effect as 1 rad of therapeutic X rays.

Renal threshold, the concentration of glucose in the blood above which glucose begins to appear in the urine.

Replicate, to make an exact copy; in cell division the DNA molecules replicate, producing two daughter cells with identical DNA.

Representative element, any element in groups I–VII and group 0 on the periodic table.

Respiratory chain, the series of oxidation–reduction reactions that is linked to the citric acid cycle and that produces the majority of ATP in the oxidation of molecules in the cell.

Retinopathy, a form of blindness caused by a chemical imbalance that results from excess glucose in the blood.

Ribonucleic acid (RNA), a group of nucleic acids, polymers of ribonucleotides, synthesized on the DNA strand and having different cellular functions.

Ribosomal RNA (*r*RNA), the RNA that, with proteins, forms granules called ribosomes in the cytoplasm.

Ribosomes, granules in the cytoplasm made up of two subunits formed by *r*RNA and protein that combine with *m*RNA to synthesize polypeptides.

Rickets, a disease that results from too little vitamin D in the diet and that causes bones to soften and bend out of shape.

RNA (see ribonucleic acid).

Saline solution, (*a*) a solution that contains salt. (*b*) A solution that is isotonic to blood plasma (normal or physiological saline).

Salt bridge, a force of attraction that occurs between the charged R-groups on the polypeptide chains of a protein; similar to the attraction between ions in an ionic crystal.

Saponifiable lipid, a lipid that can be hydrolyzed in a basic aqueous solution.

Saponification, the hydrolysis of an ester in an aqueous solution of strong base.

Saturated compound, any hydrocarbon or its derivative that contains only carbon-to-carbon single bonds.

Saturated solution, a solution that contains as many solute particles as can dissolve in the solvent at that temperature.

Scientific method, the study of nature through observation, development of a hypothesis, and testing of the hypothesis by further experimentation.

Scurvy, a deficiency disease caused by a lack of vitamin C (ascorbic acid).

Secondary structure, the shape of a protein molecule (or several protein molecules) that results from bonding forces other than the peptide bond—for example, hydrogen bonding.

Second law of thermodynamics, this law states that the entropy or disorder of the universe is increasing.

SI units, abbreviation for the International System of Units, a system of weights and measures that is the successor to the metric system.

Sickle cell anemia, anemia caused by an inherited defect in the hemoglobin molecule.

Sievert (Sv), the SI unit of absorbed dose (1 Sv = 100 rem).

Sigma bond (σ bond), a bond formed by the overlap of orbitals along the bond axis.

Significant figures (significant digits), the number of digits in a measured value that are known, plus one digit that is uncertain.

Simple lipid, a saponifiable lipid that, when hydrolyzed, yields fatty acids and an alcohol.

Simple protein, a protein that yields only amino acids upon hydrolysis.

Single bond, a chemical bond in which one pair of electrons (two electrons) are shared by two atomic nuclei.

Soap, a salt of a fatty acid, produced by the saponification of triacylglycerols.

Solubility, the amount of a solute that will dissolve in a fixed volume or weight of a solvent.

Solute, the substance being dissolved in a solution.

Solution, a homogeneous mixture of two or more substances.

Solvent, the substance in a solution in which the solute is being dissolved.

Specific gravity, a comparison of the mass of a liquid with the mass of the same volume of pure water.

Specific heat, the amount of heat energy required to raise the temperature of one gram of a substance from 15 to 16°C.

Sphingolipid, a phospholipid containing the alcohol sphingosine.

Standard atmospheric pressure, the pressure that will support a column of mercury 760 mm high at a temperature of 0°C.

Starch, a mixture of amylose and amylopectin; the energy storage molecule in plants.

Steroid, any of a large class of nonsaponifiable lipids, all of which contain a complicated four-ring framework.

STP, standard temperature and pressure: 0°C and 1 atm.

Structural formula, a diagram that shows the arrangement of atoms in the molecule.

Structural isomers, isomers that have the same molecular formula, but that differ in the sequence of atoms in their molecules.

Sublimation, the process of changing from the solid state directly to the gaseous state.

Substitution reaction, a chemical reaction in which an atom or group of atoms is substituted for some atom on a reactant molecule.

Substrate, a reactant in an enzyme-catalyzed reaction which attaches to the surface of the enzyme.

Sulfhydryl group, the functional group —S—H.

Supersaturated, the term applied to a solution that has more solute particles dissolved in it than it will hold at equilibrium at that temperature.

Surface tension, the resistance of the particles on the surface of a liquid to the expansion of that liquid.

Surfactant (surface active agent), a substance that acts to reduce the surface tension of water.

Suspension, a heterogeneous mixture having particles that, in time, will settle out.

Symbol, a shorthand way of representing one atom of an element. For example, C for carbon.

Tartar, material that forms on teeth when calcium compounds are deposited in unremoved plaque.

Teletherapy, the use of high-intensity radiation, such as gamma rays or X rays, to destroy cancerous tissue.

Temperature, a measure of the average kinetic energy of the particles of a substance.

Tertiary structure, the shape of a protein caused by the folding of the secondary structure, resulting from disulfide bridges, salt bridges, and hydrophobic interactions between the R-groups.

Thioester group, the functional group

—C(=O)—S—C—

Thioether group (organic sulfide), the functional group

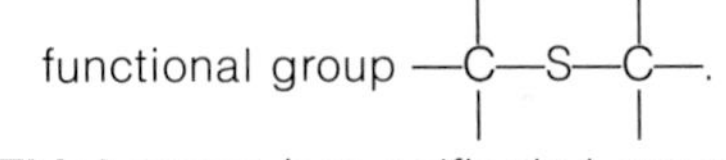

Thiol group (see sulfhydryl group).

Thyroid, the endocrine gland, located in the neck, that produces the hormones thyroxine and calcitonin.

Thyroxine, an iodine-containing hormone produced by the thyroid.

Titration, a laboratory procedure for measuring the unknown concentration of an acidic or basic solution.

Torr, a unit of pressure equal to 1 mm Hg.

Trace element, an element required in minute amounts for normal cell growth and development.

Transamination, the transfer of an amino group from an amino acid to an α-keto acid, thus producing a new amino acid.

Transcription, the transfer of genetic information from DNA to *m*RNA; the synthesis of *m*RNA on a segment of DNA.

Transfer RNA (*t*RNA), RNAs that each carry a specific amino acid to the ribosomes and place the amino acid, by pairing the bases in the anticodon region of *t*RNA with the codon of *m*RNA, in the proper sequence for the formation of the polypeptide chain.

***Trans* isomer,** the geometric isomer with the specified atoms or groups of atoms on opposite sides of the double bond or the ring.

Transition element, the metals located between groups II and III in periods 4, 5, and 6 of the periodic table.

Translation, the expression of genetic information in the amino acid sequence of a protein; the synthesis of protein molecules on one *m*RNA molecule.

Triacylglycerol, an ester of glycerol and three fatty acids; general term for fats and oils.

Triglyceride (see triacylglycerol).

Triple bond, a chemical bond in which three pairs of electrons (six electrons) are shared by two atomic nuclei.

Tyndall effect, the scattering of light by the particles in a colloid.

Turnover number, the number of substrate molecules transformed per minute by one molecule of enzyme under optimal conditions.

Tyrosine, an amino acid that is lacking in an untreated PKU child.

Unit factor (see conversion factor).

Unsaturated compound, any hydrocarbon or its derivative that contains one or more double or triple bonds.

Unsaturated solution, a solution in which more solute can be dissolved at that temperature.

Urea cycle (Krebs' ornithine cycle), the cyclic series of reactions in the liver by which urea is produced from ammonia and carbon dioxide.

Uremia, a condition, resulting from a damaged kidney, in which urea builds up in the blood.

Valence electron, an electron in the outermost energy level of an atom.

Vapor pressure, the pressure of the gas above a liquid in a container.

Vasopressin, an antidiuretic hormone; that is, a substance that controls the release of water into the urine by the kidneys.

Viscosity, a measure of how easily a liquid flows.

Vitamin, an organic nutrient that the body cannot synthesize, but which is necessary for normal body function.

Vitamin deficiency disease, a disease that results only from the lack of a specific vitamin in the diet.

Wavelength (λ), the distance between crests in the wavelike fluctuations of electromagnetic radiation.

Wax, an ester of a long-carbon-chain fatty acid and a long-carbon-chain alcohol.

Weight, a measure of the attraction of gravity on an object.

X Ray, high-energy ionizing radiation, similar to gamma rays, produced in X-ray tubes.

Zwitterion, a dipolar ion; one that has a positively charged area and a negatively charged area. For example, an amino acid.

Zymogen (see proenzyme).

PHOTO CREDITS

Chapter 1 Opener	Stefan Bloomfield.
Chapter 2 Opener	The Eugene Register-Guard.
Chapter 3 Opener	Alan Dorow.
Chapter 4 Opener	Stefan Bloomfield.
Chapter 5 Opener	Courtesy of Argonne National Laboratory.
Chapter 6 Opener	Novosti/Gamma-Liaison.
Chapter 7 Opener	Frank Siteman/Stock, Boston.
Chapter 8 Opener	UPI.
Chapter 9 Opener	Hazel Hankin/Stock, Boston.
Chapter 10 Opener	AP/Wide World Photos.
Chapter 11 Opener	Sarah Putnam/The Picture Cube.
Chapter 12 Opener	Bernard Lawrence. Reprinted with permission of McNiel Pharmaceutical.
Chapter 13 Opener	Alice Kandell/Photo Researchers.
Chapter 14 Opener	Photo by Holger W. Jannash. Reprinted with permission of the American Institute of Biological Sciences.
Chapter 15 Opener	Barbara Alper/Stock, Boston.
Chapter 16 Opener	Ira Kirschenbaum/Stock, Boston.
Chapter 17 Opener	Michael Gordon/Picture Group, Inc.
Chapter 18 Opener	Culver Pictures.
Chapter 19 Opener	Earl Presnell/Picaro Gallerie.
Chapter 20 Opener	Stefan Bloomfield.
Chapter 21 Opener	Stan Goldblatt/Photo Researchers.
Chapter 22 Opener	Stefan Bloomfield.
Chapter 23 Opener	Diego Goldberg/Sygma.
Chapter 24 Opener	Courtesy of Springer-Verlag, Publishers. From *Corpuscles* by Marcel Bessis.

INDEX

**The page numbers in italics indicate pages with tables containing the entry.*

64
68
72
88
76
92
90
81

COMMON CONVERSION FACTORS

	Unit Factors	
Length		
1 inch (in) = 2.54 centimeters (cm)	$\left(\frac{2.54 \text{ cm}}{1 \text{ in}}\right)$	$\left(\frac{1 \text{ in}}{2.54 \text{ cm}}\right)$
1 yard (yd) = 0.914 meter (m)	$\left(\frac{0.914 \text{ m}}{1 \text{ yd}}\right)$	$\left(\frac{1 \text{ yd}}{0.914 \text{ m}}\right)$
1 mile (mi) = 1.61 kilometers (km)	$\left(\frac{1.61 \text{ km}}{1 \text{ mi}}\right)$	$\left(\frac{1 \text{ mi}}{1.61 \text{ km}}\right)$
Mass		
1 ounce (oz) = 28.4 grams (g)	$\left(\frac{28.4 \text{ g}}{1 \text{ oz}}\right)$	$\left(\frac{1 \text{ oz}}{28.4 \text{ g}}\right)$
1 pound (lb) = 454 grams (g)	$\left(\frac{454 \text{ g}}{1 \text{ lb}}\right)$	$\left(\frac{1 \text{ lb}}{454 \text{ g}}\right)$
1 pound (lb) = 0.454 kilogram (kg)	$\left(\frac{0.454 \text{ kg}}{1 \text{ lb}}\right)$	$\left(\frac{1 \text{ lb}}{0.454 \text{ kg}}\right)$
Volume		
1 pint = 0.473 liter	$\left(\frac{0.473 \text{ liter}}{1 \text{ pint}}\right)$	$\left(\frac{1 \text{ pint}}{0.473 \text{ liter}}\right)$
1 quart (qt) = 0.946 liter	$\left(\frac{0.946 \text{ liter}}{1 \text{ qt}}\right)$	$\left(\frac{1 \text{ qt}}{0.946 \text{ liter}}\right)$
1 gallon (gal) = 3.78 liters	$\left(\frac{3.78 \text{ liters}}{1 \text{ gal}}\right)$	$\left(\frac{1 \text{ gal}}{3.78 \text{ liters}}\right)$
Pressure		
1 atmosphere (atm) = 760 mm Hg	$\left(\frac{760 \text{ mm Hg}}{1 \text{ atm}}\right)$	$\left(\frac{1 \text{ atm}}{760 \text{ mm Hg}}\right)$